Entrepreneurship with Microorganisms

Developments in Applied Microbiology and Biotechnology

Entrepreneurship with Microorganisms

Edited by

Amritesh C. Shukla
Department of Botany, University of Lucknow, Lucknow, India

ELSEVIER

ACADEMIC PRESS
An imprint of Elsevier

Academic Press is an imprint of Elsevier
125 London Wall, London EC2Y 5AS, United Kingdom
525 B Street, Suite 1650, San Diego, CA 92101, United States
50 Hampshire Street, 5th Floor, Cambridge, MA 02139, United States
The Boulevard, Langford Lane, Kidlington, Oxford OX5 1GB, United Kingdom

ISBN: 978-0-443-19049-0

For information on all Academic Press publications
visit our website at https://www.elsevier.com/books-and-journals

Publisher: Stacy Masucci
Acquisitions Editor: Kattie Washington
Editorial Project Manager: Billie Jean Fernandez
Production Project Manager: Swapna Srinivasan
Cover Designer: Greg Harris

Typeset by STRAIVE, India

Contents

Contributors

Raed O. AbuOdeh
Department of Medical Laboratory Sciences, College of Health Sciences, University of Sharjah, Sharjah, United Arab Emirates

Sunita Aggarwal
Department of Microbiology, Institute of Home Economics, University of Delhi, New Delhi, India

Kena P. Anshuman
Department of Microbiology, Sir P.P. Institute of Science, MK Bhavnagar University, Bhavnagar, Gujarat, India

C. Aswini Sivadas
National Institute of Technology, Calicut, Kerala, India

Zeenat Ayoub
Lab of Molecular Biology, Department of Botany, Dr. HarisinghGour Vishwavidyalaya, A Central University, Sagar, MP, India

Sabyasachi Banerjee
Department of Pharmaceutical Chemistry, Gupta College of Technological Sciences, Asansol, West Bengal, India

Santanu Banerjee
Department of Pharmacology, Gupta College of Technological Sciences, Asansol, West Bengal, India

Subhasis Banerjee
Department of Pharmaceutical Chemistry, Gupta College of Technological Sciences, Asansol, West Bengal, India

Sankhadip Bose
School of Pharmacy, The Neotia University, Sarisa, West Bengal, India

Nagendra Kumar Chandrawanshi
School of Studies In Biotechnology, Pt. Ravishankar Shukla University, Raipur, Chhattisgarh, India

Naorem Bidyaleima Chanu
Department of Vegetable Science, College of Horticulture and Forestry, CAU, Pasighat, Arunachal Pradesh, India

Ngathem Taibangnganbi Chanu
Department of Basic Sciences and Humanities, College of Horticulture and Forestry, CAU, Pasighat, Arunachal Pradesh, India

Avik Das
Department of Pharmacology, Gupta College of Technological Sciences, Asansol, West Bengal, India

Madhuchanda Das
Department of Life Science and Bioinformatics, Assam University, Silchar, India

Subrata Das
Department of Life Science and Bioinformatics, Assam University, Silchar; Department of Botany and Biotechnology, Karimganj College, Karimganj, India

Sulabha B. Deokar
Department of Biotechnology, Nowrosjee Wadia College, Pune, Maharashtra, India

Joystu Dutta
Department of Environmental Science, Sant Gahira Guru Vishwavidyalaya, Sarguja Ambikapur, Chhattisgarh, India

Alka Ekka
Department of Biotechnology, Guru Ghasidas Vishwavidyalaya, Bilaspurr, Chhattisgarh, India

Jannathul Firdous
Faculty of Medicine, Royal College of Medicine Perak, Universiti Kuala Lumpur, Ipoh, Perak, Malaysia

Jentu Giba
Department of Biotechnology, APSCS & T CoE for Bioresources and Sustainable Development, Kimin, Arunachal Pradesh, India

Suman Kumar Halder
Department of Microbiology, Vidyasagar University, Midnapore, West Bengal, India

J. Basilio Heredia
Food and Development Research Center A.C. Culiacán, Sinaloa, Mexico

C. Honey Chandran
Synthite Industries Pvt. Ltd., Kadayiruppu, Kerala, India

Luis Alfonso Jiménez-Ortega
Food and Development Research Center A.C. Culiacán, Sinaloa, Mexico

Gayathri A. Kanu
Department of Medical Laboratory Sciences, College of Health Sciences, University of Sharjah, Sharjah, United Arab Emirates

K. Kaur
Department of Microbiology, Punjab Agricultural University, Ludhiana, Punjab, India

Maninder Jeet Kaur
Department of Microbiology, Panjab University, Chandigarh, India

Ashish Kumar
Department of Biotechnology, Guru Ghasidas Vishwavidyalaya, Bilaspurr, Chhattisgarh, India

C. Kumar
Amity Institute of Organic Agriculture, Amity University Uttar Pradesh, Noida, Uttar Pradesh, India

Krishna Kumar
Division of Crop Protection, ICAR-Indian Institute of Pulses Research, Kanpur, India

Arti Kumari
Department of Microbiology, Institute of Home Economics, University of Delhi, New Delhi, India

Khaling Lallemmoi
Department of Floriculture, College of Horticulture and Forestry, CAU, Pasighat, Arunachal Pradesh, India

Akshita Maheshwari
Microbial Technology Division, CSIR-National Botanical Research Institute, Lucknow, India

Abhinav Mehta
R. C. Patel Institute of Pharmaceutical Education and Research, Shirpur, MH, India

Monika Mishra
Division of Crop Protection, ICAR-Indian Institute of Pulses Research, Kanpur, India

Raj K. Mishra
Division of Crop Protection, ICAR-Indian Institute of Pulses Research, Kanpur, India

Ahmed A. Mohamed
Department of Chemistry, University of Sharjah, Sharjah, United Arab Emirates

Keshab Chandra Mondal
Department of Microbiology, Vidyasagar University, Midnapore, West Bengal, India

Subhadeep Mondal
Centre for Life Sciences, Vidyasagar University, Midnapore, West Bengal, India

Deepa Nath
Department of Botany, Gurucharan College, Silchar, India

Rajat Nath
Department of Life Science and Bioinformatics, Assam University, Silchar, India

Rina Ningthoujam
Department of Vegetable Science, College of Horticulture and Forestry, CAU, Pasighat, Arunachal Pradesh, India

Alma Karen Orozco-Ochoa
Food and Development Research Center A.C. Culiacán, Sinaloa, Mexico

Akhilesh Kumar Pandey
Mycological Research Laboratory, Department of Biological Science, Rani Durgawati University, Jabalpur, Madhya Pradesh, India

Rakesh Pandey
Experimental Botany and Nutraceutical Lab, Department of Botany, DDU Gorakhpur University, Gorakhpur, Uttar Pradesh, India

Sonika Pandey
Division of Crop Protection, ICAR-Indian Institute of Pulses Research, Kanpur, India

Vaibhav Sharan Pandey
Experimental Botany and Nutraceutical Lab, Department of Botany, DDU Gorakhpur University; Plant Resource Centre, Department of Botany, SVM Mahila Mahavidyalaya, Arya Nagar, Gorakhpur, Uttar Pradesh, India

Vashist Narayan Pandey
Experimental Botany and Nutraceutical Lab, Department of Botany, DDU Gorakhpur University, Gorakhpur, Uttar Pradesh, India

Gagan Kumar Panigrahi
School of Applied Sciences, Centurion University of Technology and Management, Khurda, Odisha, India

Girish R. Pathade
Department of Microbiology, Haribhai V. Desai College, Pune, Maharashtra, India

Jayanta Kumar Patra
Research Institute of Biotechnology & Medical Converged Science, Dongguk University, Republic of Korea

Deepak K. Rahi
Department of Microbiology, Panjab University, Chandigarh, India

Sonu Rahi
Department of Botany, Govt. P.G. College, A.P.S. University, Rewa, India

A.K. Rana
Department of Microbiology, Punjab Agricultural University, Ludhiana, Punjab, India

Utkarsh Singh Rathore
Division of Crop Protection, ICAR-Indian Institute of Pulses Research, Kanpur, India

Gokul Shankar Sabesan
Faculty of Medicine, Manipal University College Malaysia, Melaka, Malaysia

Annie Jeyachristy Sam
Faculty of Medicine, Royal College of Medicine Perak, Universiti Kuala Lumpur, Ipoh, Perak, Malaysia

Kunja Bihari Satapathy
School of Applied Sciences, Centurion University of Technology and Management, Khurda, Odisha, India

S. Saxena
Department of Biotechnology, Babasaheb Bhimrao Ambedkar University, Lucknow, Uttar Pradesh, India

Santosh Kumar Sethi
School of Biotechnology, Gangadhar Meher University, Sambalpur, Odisha, India

M. Sharma
Department of Biotechnology, Babasaheb Bhimrao Ambedkar University, Lucknow, Uttar Pradesh, India

Neha Sharma
Clinical Research Division, Department of Biosciences, School of Basic and Applied Sciences, Greater Noida, Uttar Pradesh, India

Amritesh C. Shukla
Department of Botany, University of Lucknow, Lucknow, Uttar Pradesh, India

Ajay Kumar Singh
Mycological Research Laboratory, Department of Biological Science, Rani Durgawati University, Jabalpur, Madhya Pradesh, India

Mayanglambam Chandrakumar Singh
Department of Basic Sciences and Humanities, College of Horticulture and Forestry, CAU, Pasighat, Arunachal Pradesh, India

C.S. Siva Prasath
Synthite Industries Pvt. Ltd., Kadayiruppu, Kerala, India

Sonal Srivastava
Microbial Technology Division, CSIR-National Botanical Research Institute, Lucknow, India

Suchi Srivastava
Microbial Technology Division, CSIR-National Botanical Research Institute, Lucknow, India

T.V. Suchithra
National Institute of Technology, Calicut, Kerala, India

Anupam Das Talukdar
Department of Life Science and Bioinformatics, Assam University, Silchar, India

Sumpam Tangjang
Department of Botany, Rajiv Gandhi University, Doimukh, Arunachal Pradesh, India

Udita Tiwari
Department of Biochemistry, School of Life Sciences, Dr. Bhimrao Ambedkar University, Agra, Uttar Pradesh, India

Devinder Toor
Amity Institute of Virology and Immunology, Amity University Uttar Pradesh, Noida, India

Octavio Valdez-Baro
Food and Development Research Center A.C. Culiacán, Sinaloa, Mexico

J. Verma
Department of Biotechnology, Babasaheb Bhimrao Ambedkar University, Lucknow, Uttar Pradesh, India

P. Vyas
Department of Microbiology, Punjab Agricultural University, Ludhiana, Punjab, India

Tonlong Wangpan
Department of Botany, Rajiv Gandhi University, Doimukh, Arunachal Pradesh, India

CHAPTER 1

Microorganisms as biofactories of powerful agents against plant diseases

Luis Alfonso Jiménez-Ortega, Alma Karen Orozco-Ochoa, Octavio Valdez-Baro, and J. Basilio Heredia

Food and Development Research Center A.C. Culiacán, Sinaloa, Mexico

1.1 Introduction

Entrepreneurship through microorganisms is a profitable and efficient option for multiple uses in agriculture. Agriculture produces large volumes of biomass and by-products each harvest season, and large amounts of synthetic agrochemicals are used, which are not always sustainable, biodegradable, and safe for the population. That is why new alternatives are urgently needed, since being one of the most profitable and relevant industries for society, it must also be sustainable, circular, and capable of promoting the use of new green technologies such as biotechnology.

The challenge for scientists and farmers is to develop a sustainable bio-economy model for agriculture, which can be approached from the sustainable use of land, water, and cultural practices, which address recovery, crop rotation, generation of new products based on by-products (biomass), and the use of agrochemicals produced from green methodologies and of a natural origin such as microorganisms (Fenibo et al., 2021; Sillanpää and Ncibi, 2017b).

Agriculture has had to face the destructive activities of numerous pests like fungi, weeds, and insects, which seriously affect feed production as global crop yield is reduced by 20%–40% annually due to plant pests and diseases (FAO, 2012). Various pests in many parts of the world are attacking crops and forests. In recent decades, extensive use of chemical pesticides has caused environmental pollution of soil and aqueous ecosystems and resulted in pesticide resistance in many insect species (Choe et al., 2022). The control of pests and weeds in agriculture and insects that cause human diseases is traditionally managed with chemical insecticides. However, using these products has given rise to many problems, such as increased resistance of pathogens and pests to various chemical compounds, environmental pollution, and its effects on human health, such as cancer and several immune system disorders. As a result of the threat that the chemicals caused by their direct action and residual impact on human health and the environment, consumers increasingly demand pesticide-free foods (Hernández-Rosas et al., 2020).

Entrepreneurship with Microorganisms. https://doi.org/10.1016/B978-0-443-19049-0.00011-6

Biological pesticides or biopesticides represent a range of bio-based substances acting against invertebrate pests with different mechanisms of action. Based on a technical definition provided by the United States Environmental Protection Agency (EPA) (EPA, 2022), they can be classified into three main classes:

(i) Naturally occurring biochemicals that act through nontoxic mechanisms.
(ii) Microbial entomopathogens.
(iii) Plant-incorporated protectants derived from genetically engineered plants.

Currently, only a few bacterial genera, *Bacillus*, *Pseudomonas*, *Agrobacterium*, and *Streptomyces*, are used as biopesticides (Thakur et al., 2020). Looking into highly diverse microbial environments, such as organic carbon-rich soil and food spoilage, may enhance new biopesticide discoveries that could potentially improve food production (Booth et al., 2022). This chapter underscores the utility of biocontrol agents composed of microorganisms, including bacteria, fungi, and viruses, that control pest (Fig. 1.1).

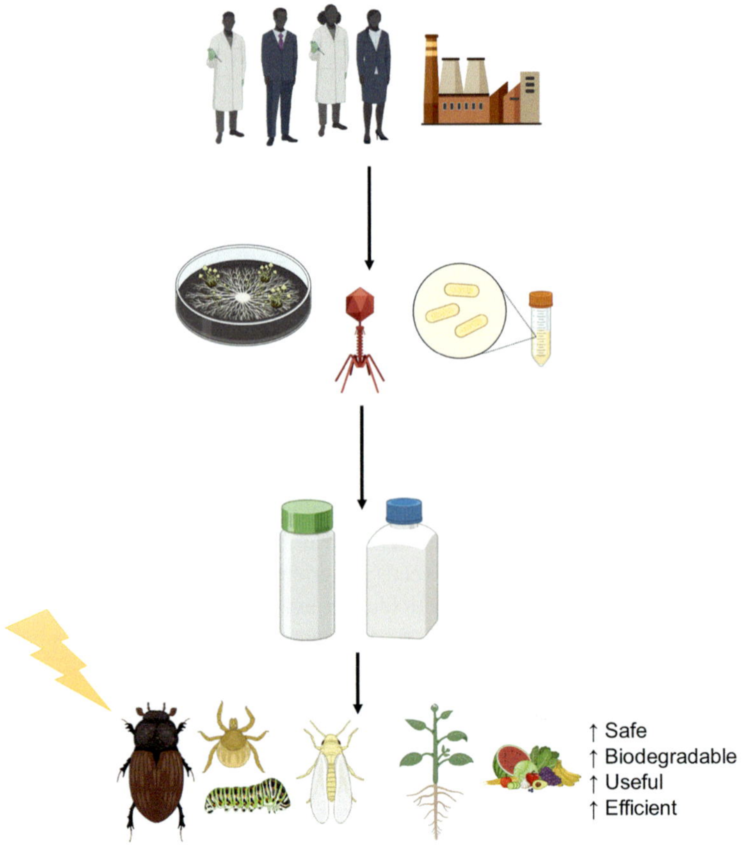

FIG. 1.1

Entrepreneurship through microorganisms with bioactive potential for agriculture (Created in BioRender).

1.2 Global burden problem due to agricultural pests: Food security

Currently, the demand for quality food is increasing due to increases in population levels and the development of countries that were previously marginalized and today can guarantee greater access to food for their population. According to the most up-to-date statistical data from the Food and Agriculture Organization of the United Nations (FAO), the most produced crops (in million tons (Mt)) that are necessary to satisfy this demand are sugarcane (18690.72 Mt), corn (1171.13 Mt), wheat (762.74 Mt), rice (756.74 Mt), potato (359.07 Mt), soybean (353.46 Mt), cassava (302.66 Mt), beetroot (302.66 Mt), tomato (186.82 Mt), and barley (157.03 Mt) (FAO, 2022a).

Due to these crops' importance, their production increases each year. However, these turn out to be insufficient to satisfy 100% of the food requirements of the entire world, so one of the challenges of the FAO for the year 2050 is to have greater food security, with the prediction of lowering the rates of malnutrition from 16.3% (823 million people) to 4.8% (370 million people), were to achieve these objectives it is necessary to apply effective strategies such as reduction of poverty, safety nets, rural development programs and avoiding the loss of crops caused by the attack of phytopathogens (FAO, 2022b).

The attack of phytopathogens is one of the biggest problems that affect crops, and that prevents guaranteeing food security, concentrating the most losses in areas with food deficits, rapidly growing populations, and places where climatic conditions favor the emergence of pathogens (Savary et al., 2019).

Among the phytopathogenic agents that represent the greatest threat to the losses of crops, we find fungi, oomycetes, bacteria, nematodes, and viruses. These agents cause damage to crops in various plant organs such as leaves, stems, flowers, roots, and fruits, causing diseases such as leaf wilting, chlorosis, necrosis, blight, rot (fruits, vascular bundle, and roots), canker, galls, mosaics, dwarfism, leaf curling, among others.

Within the groups or genera that produce more diseases are fungi (rust, smut, powdery mildew, *Fusarium* spp., *Alternaria* spp., *Colletotrichum* spp., *Botrytis* spp., *Verticillium* spp., *Sclerotinia* spp., *Mycosphaerella* spp., *Macrophomina* spp., *Rhizoctonia* spp., among others), oomycetes (downy mildew, *Phytophthora* spp., *Pythium* spp., among others), bacteria (*Pseudomonas* spp., *Clavibacter* spp., *Ralstonia* spp., *Xhantomonas* spp., *Erwinia* spp., *Agrobacterium* spp., among others), nematodes (*Meloidogyne* spp., *Pratylenchus* spp., *Globodera* spp., *Heterodera* spp., *Ditylenchus* spp., *Rotylenchulus* spp., among others), virus (family *Geminiviridae*), among others (do Rêgo Barros et al., 2022; Oliver, 2012; Roumagnac et al., 2022; Setubal et al., 2005; Sanjeev Sharma et al., 2021).

1.3 Pesticides in modern agriculture: Challenges, problems, and alternatives

There are different management types for the control of the phytopathogenic agents that cause losses and damage to crops. However, chemical pesticides are the most used since they are very effective in eradicating pests' phytopathogens, in addition to being corrective, thus managing to reduce or avoid crop losses, which makes them essential to maintaining the required production levels. However, the bad practices that have been carried out since its first use, added to its residuality and high toxicities, have caused numerous damages to the environment, such as contamination to a large extent of soils, water, and air; in addition to the contamination of the food that is consumed, since some agricultural products postharvest pesticides are applied to maintain a much longer fresh life, which the constant consumption of pesticides has caused the appearance of various diseases, especially damage to the endocrine system and cancer (Sharma et al., 2019).

Various international organizations have been in charge of regulating the use of pesticides to control their negative effects, which vary between territories. For example, in northwestern Africa, the regulatory body for pesticides is the Sahelian Pesticide Committee (CSP) (CSP, 2018). In the European Union, these are regulated by various regulations approved by the European Parliament, such as Directive 2009/128/EC, Regulation (CE) No. 1107/2009, and Regulation (EC) No. 396/2005. In the United States, pesticides are regulated by the EPA (EPA, 2020). In Australia, the regulation is governed by the Australian Academy of Technological Sciences and Engineering (ATSE) (ATSE, 2002). In Latin American and Asian countries, the regulation is based on what is described by the FAO (FAO, 2013). Other countries have their internal rules subject to what their government dictates as adequate levels of pesticide use.

Due to all these damages caused by the use of pesticides, in recent years, alternatives have been proposed for the management of phytopathogenic agents, resulting in a new line of trade for agricultural products called agriculture of organic products. These alternatives are based on the reduction or elimination of the use of chemical pesticides, where the other options used are biocontrol agents that can parasitize phytopathogenic agents, compete for space and nutrients or produce secondary metabolites that eradicate them (Balaes and Tanase, 2016; Kong, 2018; Marin et al., 2019), application of viruses that infect only phytopathogenic agents (Holtappels et al., 2021); the use of botanical extracts of plants, which have phytochemicals capable of having a pesticidal effect (Sousa et al., 2021); and the use of genetic/metabolic engineering for the production of chemical compounds with a pesticidal effect, which is produced by genetically modified microorganisms to focus their metabolism on the biosynthesis of certain target molecules with potential use in agriculture for the management of phytopathogens (Raman et al., 2022; Sousa et al., 2021).

1.4 Biopesticide

As defined by the EPA, biopesticides are certain types of pesticides derived from such natural materials as animals, plants, bacteria, and certain minerals. In commercial terms, biopesticides include microorganisms that control pests (microbial pesticides), naturally occurring substances that control pests (biochemical pesticides), and pesticidal substances produced by plants containing added genetic material (plant-incorporated protectants). Biopesticides are used in agriculture for insect control, disease control, weed control, nematode control, and plant physiology and productivity (Fenibo et al., 2022; Kachhawa, 2017; Seiber et al., 2014). In addition, biopesticides, which are pest management agents based on living microorganisms or natural products, offer great promise in controlling yield loss without compromising product quality (Sharma et al., 2019).

According to Usta, (2013), the advantages of using biopesticides in place of other chemical ones are based on these factors:

- Ecological benefit; inherently less harmful and less environmental load.
- Target specificity; designed to affect only one specific pest or, in some cases, a few target organisms.

- Environmental beneficence; is often effective in very small quantities and usually decomposes quickly, resulting in lower exposures and largely avoiding pollution problems.
- Suitability; when used as a component of integrated pest management programs, biopesticides can contribute greatly.

1.5 Microorganisms as biofactories of biopesticides

Microbial pesticides are also known as Biological Control Agents (BCA). In this category, the active ingredient is a microorganism that either occurs naturally or is genetically engineered. The pesticide action may be from the organism itself or the substance it produces (Kachhawa, 2017). The pesticide action may be from the organism itself or the substance it produces. They offer the advantages of higher selectivity and less or no toxicity compared to conventional chemical pesticides. They come from naturally occurring or genetically altered bacteria, fungi, algae, viruses, or protozoans. Microorganisms, e.g., a bacterium, fungus, virus, or protozoan as the active ingredient, can control many kinds of pests. However, each separate active ingredient is relatively specific for its target pest. For example, some fungi control certain weeds and other fungi that kill particular insects (Usta, 2013). One bacterial species, *Bacillus thuringiensis* (Bt), has been widely used in pest control. Still, the large-scale production of this biopesticide is expensive because of the high cost of the medium and production methods (Choe et al., 2022).

The most commonly used microbial biopesticides are pathogenic organisms for the pest of interest. These include biofungicides (*Trichoderma, Pseudomonas, Bacillus*), bioherbicides (*Phytophthora*), and bioinsecticides (Bt) (Gupta and Dikshit, 2010).

Knowledge of the mechanisms of biocontrol in a microbial pesticide is a key factor in achieving an efficient reduction of the pathogen in their host (Kumar et al., 2021). Several strains cover a single mechanism; others use a combination of them (Bonaterra et al., 2012). Some bacteria and fungi can induce defense responses in plants by producing either elicitors (e.g., cell wall components) or messenger molecules (e.g., salicylic acid) (Spadaro and Gullino, 2004). Some BCAs can inhibit plant pathogens by degradation of chemical signal messengers necessary for quorum sensing (e.g., acyl homoserine lactones) used to start the infection process by the pathogen (Molina et al., 2003).

Some of the crops established in orchards are ideal for biopesticides, whereas the use of baculoviruses stands out mainly. In greenhouses and nurseries, it can be used to combat insects such as aphids, whiteflies, mites, and thrips, among others. It is biodegradable, safe to use, and nontoxic qualities allow workers to handle the products in this environment; however, it is always recommended to read the safety labels carefully. The use of biopesticides on a large scale has been documented in the cultivation of corn and cotton; mainly, Bt has been used to control Lepidoptera. Other potential uses are in forestry and vegetable production (Arthurs and Dara, 2019).

1.5.1 Bacteria-based biocontrol

Bacteria were among the first microbial control agents to be developed and continue to dominate the microbial pesticide market due to the popularity of Bt products for controlling caterpillars. Bacteria are typically produced in large fermenters and contain a mixture of spores, crystal proteins, and inert carriers (Sanahuja et al., 2011).

Generally, they are used as insecticides, although they can be used to control the growth of plant pathogenic bacteria and fungi. As an insecticide, they are generally specific to individual species of moths and butterflies or species of beetles, flies, and mosquitoes. To be effective, they must come into contact with the target pest and may be required to be ingested. In insects, bacteria disrupt the digestive system by producing endotoxins that are often specific to the insect pest. When used to control pathogenic bacteria or fungi, the bacterial biopesticide colonizes the plant and crowds out the pathogenic species (O'Brien et al., 2009).

Various bacterial species and subspecies, especially *Bacillus* and *Pseudomonas* (Pandya and Saraf, 2015), have been established as microbial pesticides which control insect pests and plant diseases (Table 1.1). The most salient among these are insecticides based on several subspecies of Bt. These include *B. thuringiensis* sp. kurstaki and *B. thuringiensis* sp. Aizawa, which are highly toxic to Lepidopteran larval species, and *B. thuringiensis israelensis*, with activity against mosquito larvae, black fly (simuliid), and fungus gnats (Pathak et al., 2017).

For instance, cotton cultivation is one of the most prominent economically speaking due to its importance in the textile industry. In addition to being a crop that uses more than 60% of the globally marketed agrochemicals due to its high susceptibility to pests. Numerous control methods based on microorganisms have been proposed, observing promising results that may eventually replace those currently sold. The case of Bt, which has been followed to be capable of inhibiting the activity of *Bemisia tabaci* larvae, has also been observed to inhibit *Earias biplaga* up to 88% and is selective against some pathogens such as *Alabama argillacca* and *Aphis gossypii*, *Helicoverpa armigera,* and *Heliothis* spp. to mention a few (Malinga and Laing, 2022).

The modes and mechanisms of action by which some types of bacteria, such as *Agrobacterium radiobacter, Bacillus popilliae, Bacillus sphaericus, Bacillus subtilis, B. thuringiensis, Pseudomonas syringae,* are through antagonistic action and stomach poisoning in insects and larvae, however, they have also presented fungicidal and bactericidal activity (Malinga and Laing, 2022).

Some metabolites derived from microorganisms such as *Actinomycetales bacterium* and *Saccharopolyspora spinosa* have shown bioactive properties against insects, mainly caterpillars and thrips, which attack ornamental, fruit, and vegetable crops mostly. Derivatives of *Proteobacteria* are

Table 1.1 A broad description of some common bacterial species as microbial pesticides (Kachhawa, 2017; Sillanpää and Ncibi, 2017a).

Bacillus variety	Target pest
B. popilliae	Larvae of beetles
B. sphaericus	Mosquito larvae
A. radiobacter	*A. tumefaciens* (Crown galls)
B. subtilis	*Rhizoctonia, Fusarium, Alternaria, Pythium*
P. syringae	Molds and rots
Bt subsp. *aizawai*	Moth larvae
Bt subsp. *israelensis*	Mosquito and blackflies
Bt subsp. *kurstaki*	Lepidopteran larvae, leaf beetles
Bt subsp. *tenebrionis*	Colorado potato beetle. Coleopteran beetles
Bt subsp. *galleriae*	Lepidopteran larvae, *Helicoverpa armigera, Manduca sexta*
B. moritai	Diptera

capable of exerting broad-spectrum insecticidal activities, mostly *Burkholderia rinojensis* strain A396 and *Chromobacterium subtsugae* strain PRAA4-1 have shown insecticidal, nematicidal, and acaricidal activities. *Pasteuria* spp. exerts potent nematicidal activities, a couple of products have been successful in the turf grass market from Nortica 5WG and Clariva (Arthurs and Dara, 2019).

Other types of *Bacillus,* such as the *amyloliquefaciens*, *licheniformis*, and *pumilus* species, have shown antifungal activity, as have their metabolites. Some of the diseases susceptible to these compounds are present in the root, leaves, aerial parts, and postharvest, which commonly affect avocado, tomato, cucurbit, and strawberry crops, which are of economic and nutritional relevance. Products have also been developed from bioactive strains against pathogens that cause damping off and soft rots. In the United States, it can be found as Kodiak and Ballad Plus (Mishra et al., 2015).

1.5.2 Fungi-based biocontrol

Entomopathogenic fungi play a vital role as a biological control agent of insect populations (Table 1.2) (Sharma and Malik, 2012). The infection process normally starts with the germination of conidia or spores that have come into contact with the host cuticle. Due to a combined enzymatic and mechanical

Table 1.2 Selection of commercial products based on fungi (Ruiu, 2018; Sillanpää and Ncibi, 2017a).

Active substances	Commercial names	Main targets
Beauveria bassiana	Bio-Power, Biorin/Kargar, Botanigard, Daman, Naturalis, Nagestra, Beauvitech-WP, Bb-Protec, Racer, Mycotrol	Wide range of insects and mites
Beauveria brongniartii	Bas-Eco	*Helicoverpa armigera*, Berry borer, Root grubs
Alternaria destruens	Smolder G (Sylvan Bioproducts)	Parasitic plants of the genus *Cuscuta*
Trichoderma viride	Ecosom TV (Agri Life)	Rot diseases
Gliocladium virens	Soil Guard12G (Certis)	Soil pathogens (damping off and root rot)
Streptomyces griseoviridis	Mycostop (Verdera Oy)	Wilt, seed and stem rot
Hirsutella thompsonii	No-Mite	Spider mites
Isaria fumosorosea	Nofly	Whitefly
Metarhizium anisopliae	Biomet/Ankush, Bio-Magic, Devastra, Kalichakra, Novacrid, Met52/BIO1020 granular, Pacer, Bio-Blast (EcoScience), Placer MA (AgriLife)	Beetles and caterpillar pests; grasshoppers, termites, locusts
Metarhizium brunneum	Attracap	*Agriotes* spp.
Paecilomyces lilacinus	Bio-Nematon, MeloCon, Mytech-WP, Paecilo	Plant pathogenic nematodes
Paecilomyces fumosoroseus	Bioact WG, No-Fly-WP, Paecilomite	Insects, Mites, Nematodes, Thrips
Verticillium lecanii	Bio-Catch, Mealikil, Bioline/Verti-Star	Mealy bugs and sucking insects
Lecanicillium lecanii	Lecatech-WP, Varunastra	Aphids, leafminers, mealybugs, scale insects, thrips, whiteflies
Myrothecium verrucaria	DiTera (Valent Biosciences)	Nematodes

action, the fungus penetrates the host body, and the mycelium develops internally, often producing different types of conidia or spores colonizing the host. During vegetative growth, the fungus may produce and release a variety of metabolites, favoring its growth or acting as virulence factors or toxins. New conidia or spores will be produced outside the infected host to ensure spread in the environment. Before this stage, the host affected by the fungus's biochemical and mechanical action normally dies. The infection is triggered by the first conidium or spore germination, which normally requires specific environmental conditions (i.e., temperature and relative humidity). Main fungal entomopathogens include species in the following phyla: *Chytridiomycota*, *Zygomycota*, *Oomycota*, *Ascomycota*, and *Deuteromycota* (Ruiu, 2018; Vega et al., 2012). Entomopathogenic fungi have great potential as control agents, as they constitute a group with over 750 species and, when dispersed in the environment, provoke fungal infections in insect populations (Kachhawa, 2017).

Different products may refer to different microbial strains. The representative trade names are those shown on the relevant company websites to which reference should be made for details. For example, the *Beauveria bassiana* fungus has been widely used as a pest control agent. It is even considered a broad-spectrum agent against insects since it can penetrate the exoskeleton or cuticle, causing irreversible damage. For the cultivation of cotton, it has been observed that it is capable of mitigating and eradicating pests caused by *Tetranychus urticae, H. armigera, B. tabacci*, cotton bollworm, and *A. gossyppi* to mention a few (Malinga and Laing, 2022).

Another fungus with the greatest industrial use is *Metarhizium rileyi*, which is highly effective against Lepidopteran insects. However, this fungus is capable of penetrating the body of the hosts through the cuticle or by its ingestion at the time the larva is fed, causing damage to the internal tissues, and leading to its death. Insects of the order Coleoptera may also be susceptible to *B. tabaci* and *H. armigera* (Malinga and Laing, 2022).

1.5.3 Viruses-based biocontrol

Viruses have been isolated from more than 1000 species of insects from at least 13 different insects order (Table 1.3) (Srivastava and Dhaliwal, 2010). Among the insect viruses found in nature, those belonging to the baculovirus family (*Baculoviridae*) were considered to develop the most commercial

Table 1.3 A broad description of some common viruses species as microbial pesticides (Kachhawa, 2017; Sillanpää and Ncibi, 2017a).

Nature of virus	Commercial names	Host
Nuclear polyhedrosis virus (NPV)	AfMNVP (Certis), Multigen (Embrapa), Polygen (Agrogen), Biotrol (Certis), Elcar (Novartis), NPVSf (Certis)	*Helicoverpa armigera, Spodoptera litura, Amsacta albistriga, Spilosoma obliqua, Pericallia ricini, Pseudaletia separata, Spodoptera mauritia, Corcyra cephalonica, Plusia chalcites, Antheraea mylitta, Dasychira mendosa, Plusia peponis*
Granulosis virus	Capex (Andermatt), Cyd-X (Certis)	*Cnaphalocrocis medinalis, Pericallia ricini, Achaea janata, Phthorimaea operculella* and *Chilo infuscatellus*
Cytoplasmic polyhedrosis virus		*Helicoverpa armigera*

viral biopesticides (Kachhawa, 2017). Species in the family *Baculoviridae* represents DNA viruses establishing pathogenic relationships with invertebrates and showing potential in biological control (Clem and Passarelli, 2013; Haase et al., 2015). The virus infectivity is associated with the production of crystalline occlusion bodies containing infectious particles within the host cell. Based on the morphology of these occlusion bodies, Baculoviruses are divided into two main groups: the nucleopolyhedroviruses (NPVs), in which these bodies are polyhedron-shaped and develop in cell nuclei, and the granuloviruses (GVs), in which these bodies are granular-shaped (Ruiu, 2018). Baculoviruses are infectious per Os (by mouth) and exhibit efficient horizontal transmission. When an insect consumes occlusion bodies (OBs), the alkaline environment of the midgut triggers the dissolution of polyhedra (OBs) and the release of virions into the midgut lumen (Adams and McClintock, 1991).

Also, baculoviruses are seen as attractive BCA against insect crop pests for many reasons. First, they have a long and detailed history of research, so basic knowledge of their taxonomy, biology, and pathogenicity is available. Additionally, they have an established safety profile and environmental acceptability. They are highly efficient pathogens of some of the world's most important crop pests, such as *Heliothis*/*Helicoverpa* species, *Spodoptera* spp., and *Plutella xylostella*. Finally, their use as biological pesticides is feasible because commercially viable mass production systems are well-advanced for many baculoviruses (Wilson et al., 2020).

The first nucleopolyhedrovirus-based product on the market was in China in 1993, turning out to be an excellent product to combat pests in the field. This product is usually efficient against *Helicoverpa armigera*, and *Helicoverpa zea*, in citrus and cotton crops (Malinga and Laing, 2022).

1.5.4 Bacteriophage-based biocontrol

Bacteriophages or phages are the natural enemies of bacteria and can be successfully used as therapeutic or prophylactic agents against bacteria-related plant diseases (Frampton et al., 2012). New phages have been isolated and characterized for different pathovars of different bacteria. For example, (Kazantseva et al., 2022) have characterized *Pseudomonas* phage Pf-10, a component of the biopesticide "Multiphage" used for bacterial diseases of crops caused by *P. syringae*. According to their results, the phage Pf-10 is the first phage from the "Multiphage" that has been sequenced and subjected to a thorough analysis, including TEM analysis, stability test, genome analyses (the genome organization and the genome packaging strategy), phylogenetic analysis and determination of phage infection parameters such as the rate of adsorption, the life cycle and size of phage progeny on the *P. syringae* BIM B-268. The results of the study contribute to the accumulation of data on phage biology that can help in the formulation of the most efficient phage combinations in phage-based preparations to combat plant diseases caused by *Pseudomonas*.

1.5.5 Others

Some nematodes have been documented as good insect controllers. The most widely used and marketed are *Heterorhabditis bacteriophora*, *Heterorhabditis megidis*, *Phasmarhabditis hermaphrodite,* and *Steinerma carpocapsae*. Susceptible pests are Lepidopteran larvae, Japanese beetles, black vine weevils, soil insects, slugs, weevils, cutworms, and termites. The elucidated modes of action are that these may be entomopathogens and slug-eating nematodes. Products available on the market are Heteromask and Hortscan (BioLogic), Terranem and larvae (Koppert), Nemaslug (Becker), Underwood and Bio-Safe-N (Certis) (Sillanpää and Ncibi, 2017a).

Some protozoa of *Mattesia* and *Nosema* species have been active against *Cryptolestes ferrugineus* and *Cryptolestes pusillus* larvae. Several studies have demonstrated the in vitro bioactivity of protozoa against larvae; what follows is to evaluate them on an in vivo and pilot scale (Wakefield, 2018).

1.6 Microorganism pest control agent's industrial development

According to (Bonaterra et al., 2022), developing bacterial biocontrol agents requires several steps (Fig. 1.2). It includes:

(i) The isolation and selection of strains using screening methods able to analyze a high number of microorganisms.
(ii) The characterization of the bacterial biocontrol agents, including the identification, the determination of phenotypic and genotypic traits, and the mechanisms of action, biocontrol efficacy in pilot tests and improvement.
(iii) Mass production and an appropriate formulation allow increasing biocontrol activity and ensure its stability.

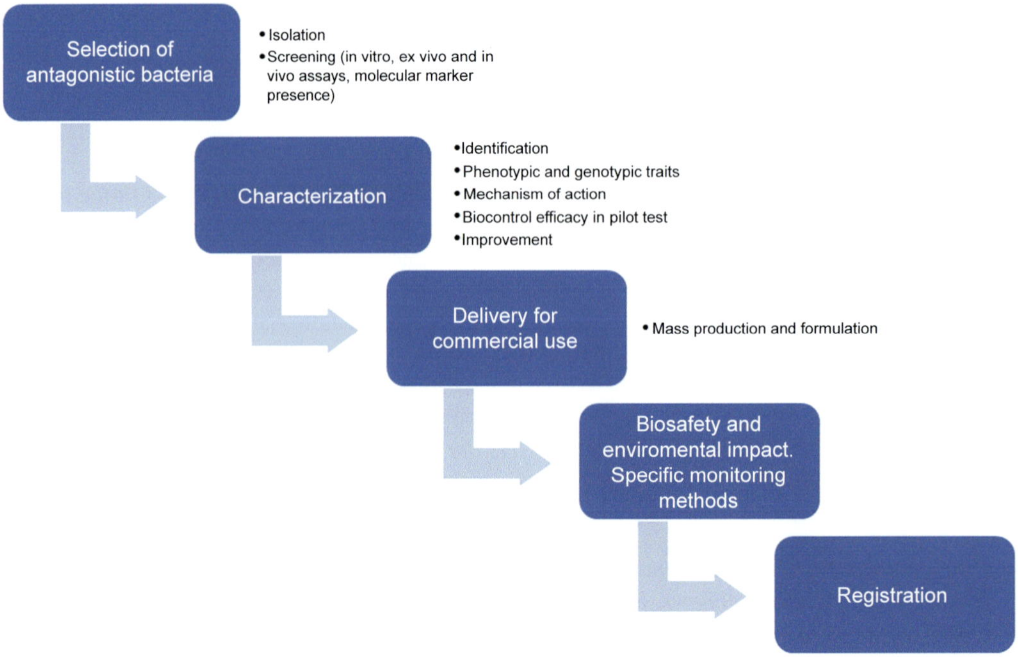

FIG. 1.2

Flowchart of actions for bacterial biocontrol agents' development.

Finally, developing a monitoring system to detect and quantify the bacterial biocontrol agents in the environment and to make more extensive toxicology tests or environmental impact studies register for use is required.

The industrial development of biocontrol agents is an area of tremendous opportunity for entrepreneurs and researchers. Several successful companies can be cited, such as the Development of Certis with its products Delfin, Javelin, Condor, and Crymax, based on Bt (*kurstaki*). Valent Biosciences, with its products Dipel, Biobit, Florbac, and XenTari also based on Bt (*kurstaki)* and *aizawai*. Andermatt PHP with its products BotaniGard, Mycotrol, Nomu-Protect, and Bolldex, based on *B. bassiana*, *M. rileyi,* and NPV, respectively. And other big companies like BASF with its broadband product made from *B. bassiana* (Malinga and Laing, 2022).

The industrial development of biopesticides is not easy because it represents a challenge if the companies do not have the financial and human resources to support it correctly. The greatest requirements are energy and water since the facilities must ensure good manufacturing and safety practices, which implies rigorous temperature controllers (Singh, 2017). In addition, a challenge that must be considered when developing a virus-based biocontrol agent is its sensitivity to UV radiation; therefore, the applications must be in the evening or at night or add a protector that prevents their degradation.

1.7 Regulations on the use of biological pest control agents

The regulation of biopesticides must be fast and efficient, especially their registration with the corresponding regulatory entities. For example, the FAO, in the guide for the registration of microbial, botanical, and semichemical agents for pest control (FAO, 2017), mentions the qualities and characteristics that a product of this type must meet, where it stands out that it must have low toxicity, formulated in inert materials and be used in doses similar to other products already on the market, in addition to being consistent with the chemistry green.

Before your registration, both parties must maintain good communication, submitting the appropriate and duly completed preregistration forms. Among other things, all the information and tests must be supported under rigorous scientific methodologies. Specifically, for bioproducts generated by microorganisms, the label must clarify the amount of the active ingredient, in this case as colony-forming units per kilogram or liter of the product (UFC/kg/L). Or in terms of biopotency. Entrepreneurs who wish to patent a product are encouraged to follow the guide mentioned above to obtain a successful registration process.

Specifically, in the United States, the registration of a new biopesticide must be submitted to the EPA's biopesticides and pollution prevention division, which through various studies based on the secondary effects on humans, animals, and the environment, will fulfill the regulatory process. Although the process is usually fast and efficient, the review periods can be at least a year or a year and a half to be approved, being much less than a process of regulation and registration of synthetic pesticides (3 years) (Arthurs and Dara, 2019).

Biopesticide regulatory agencies facilitate the registration and manufacturing procedures for products with microorganisms. Among the most relevant are China http://www.gov.cn/zhengce/content/2017-04/01/content_5182681.htm, India https://legislative.gov.in/sites/default/files/A1968-46.pdf, USA https://www.epa.gov/, Europe https://eurlex.europa.eu/legalcontent/EN/ALL/?uri=celex%3A31991L0414, South Korea https://www.rda.go.kr/foreign/ten/ and Australia https://apvma.gov.au/ (Mishra et al., 2015).

1.8 Trends and trade worldwide in biological pest control agents

Undoubtedly, the United States and other North American regions led the biopesticides market; it was estimated that during 2015 it presented a value of more than 539 million dollars, with projections for 2022 of 1.67 billion. The United States is one of the largest consumers of biopesticides, encompassing the different modes of action from bioinsecticides, acaricides, fungicides, herbicides, bactericides, and nematicides, among others. Its market value was estimated at USD 1.1 billion in 2016; however, it is expected to continue growing up to 17% CAGR (compound annual growth rate), which would exceed the CAGR for synthetic pesticides (Arthurs and Dara, 2019). The markets for these products can expand exponentially, not only in North America, which demands almost half of those produced globally, if not the European and Asian ones, which are strong consumers of important crops for nutrition (Mishra et al., 2015).

Records date back to 1971 on the development of commercial biopesticide formulations, highlighting some companies in the market such as Bayer Biologics, BASF, Dow, Monsanto BioAG, Valent, Certis, Phyllom BioProducts, Becker Microbial Products, Bayer, Valent Biosciences, Marrone Bio Innovations, St Gabriel Organics, Syngenta, Lam International, Troy BioSciences, BioSafe, Terragena, BASF, Andermatt, Novozymes, Futureco, M&R Durango, bioTEPP, Arysta, Agrivir, Sylvar, and US Forest Service (Arthurs and Dara, 2019). In addition, more than 2 million dollars were invested in acquiring new patents and biopesticides between 2012 and 2015.

The most commercialized vehicles for these agents are usually powdered, emulsifiable concentrates or suspensions, wettable granules, and soluble in water. These products are typically used for ornamental crops, followed by fruit trees, berries, cucurbits, stone fruits, nuts, leafy vegetables, cereals, herbs and spices, citrus fruits, and legumes (Arthurs and Dara, 2019). The products that lead the biopesticide market are those made from bacteria (74%), fungi (10%), viruses (5%), and others (3%) (Mishra et al., 2015).

1.9 Conclusions and future outlook

The use of pest control agents in the field is vital since the most important thing is safeguarding the food security of the entire world. However, it is also very important to consider with what type of product this objective is met. Biopesticides made from microorganisms have become a feasible and profitable alternative. In addition to having low toxicity, they are safe and do not represent an environmental hazard. However, their use is limited due to a lack of dissemination, knowledge, and marketing.

Future projections mention that by 2050 these products may surpass synthetics in use and commercialization. Therefore, scientific dissemination is necessary, and other researchers are encouraged to investigate their modes and mechanisms of action. As mentioned earlier, there is a great variety of biocontrol agents, such as bacteria, fungi, viruses, phages, and nematodes, each with specific characteristics.

Like all products, these agents have their disadvantages, such as sensitivity to UV radiation and specificity, which, although advantageous for the control of specific pathogens, could see their broad-spectrum activities limited, having to use more products to control all pathogens. In addition, slow action in the articular of baculoviruses, temperature, and adverse weather conditions can degrade their effects, such as humidity and rain. That is why intelligent encapsulation methods must be developed that can release the compounds for a long time and simultaneously serve as protection.

This type of product's industrial and technological development must be efficient and supported by health, agriculture, and government organizations. Its environmental and health benefits and goodness should be considered to replace synthetic agrochemicals that afflict public health in the future.

References

Adams, J.R., McClintock, J.T., 1991. Baculoviridae. Nuclear polyhedrosis viruses. In: Adams, J.R., Bonami, J.R. (Eds.), Atlas of Invertebrate Viruses, first ed. CRC Press, Boca Raton, p. 140.

Arthurs, S., Dara, S.K., 2019. Microbial biopesticides for invertebrate pests and their markets in the United States. J. Invertebr. Pathol. 165, 13–21.

ATSE, 2002. Pesticide Use in Australia. Retrieved from https://www.atse.org.au/wp-content/uploads/2019/01/pesticide-use-australia.pdf.

Balaes, T., Tanase, C., 2016. Basidiomycetes as potential biocontrol agents against nematodes. Rom. Biotechnol. Lett. 21 (1), 11185–11193.

Bonaterra, A., Badosa, E., Cabrefiga, J., Francés, J., Montesinos, E., 2012. Prospects and limitations of microbial pesticides for control of bacterial and fungal pome fruit tree diseases. Trees 26 (1), 215–226.

Bonaterra, A., Badosa, E., Daranas, N., Francés, J., Roselló, G., Montesinos, E., 2022. Bacteria as biological control agents of plant diseases. Microorganisms 10 (9).

Booth, J., Schenk, P.M., Mirzaee, H., 2022. Microbial biopesticides against bacterial, fungal and oomycete pathogens of tomato, cabbage and chickpea. Appl. Microbiol. 2 (1), 288–301.

Choe, S.G., Maeng, H.R., Pak, S.J., Song Nam, U., 2022. Production of *Bacillus thuringiensis* biopesticide using penicillin fermentation waste matter and application in agriculture. J. Nat. Pest. Res. 2, 100012.

Clem, R.J., Passarelli, A.L., 2013. Baculoviruses: sophisticated pathogens of insects. PLoS Pathog. 9 (11), e1003729.

CSP, 2018. Pesticides Finder. Retrieved from http://csp.dev4u.it/search.cfm.

do Rêgo Barros, F.M., Pedrinho, A., Mendes, L.W., Freitas, C.C.G., Andreote, F.D., 2022. Interactions between soil bacterial diversity and plant-parasitic nematodes in soybean plants. Appl. Environ. Microbiol. 88 (17), e00963-00922.

EPA, 2020. Federal Pesticide Regulation. Retrieved from http://npic.orst.edu/reg/regfed.html.

EPA, 2022. Biopesticides. Retrieved from https://www.epa.gov/pesticides/biopesticides#:~:text=Protection%20Act%20(FQPA)-,What%20are%20Biopesticides%3F,%2Dincorporated%20protectants)%20or%20PIPs.

FAO, 2012. Global Pact Against Plant Pests Marks 60 Years in Action. Retrieved from https://www.fao.org/news/story/en/item/131114/icode/.

FAO, 2013. FAO Statistical Yearbook 2013: World Food and Agriculture. Retrieved from https://www.fao.org/3/i3107e/i3107e.pdf.

FAO, 2017. Guidelines for the Registration of Microbial, Botanical and Semiochemical Pest Control Agents for Plant Protection and Public Health Uses. Retrieved 14 December from https://www.fao.org/documents/card/en/c/4e84d2c6-df73-430a-82ef-0aed35856e0e/.

FAO, 2022a. Crops and Livestock Products. Retrieved from https://www.fao.org/faostat/en/#data/QCL.

FAO, 2022b. World Agriculture in the Perspective of the Year 2050. Retrieved from https://www.fao.org/fileadmin/templates/wsfs/docs/Issues_papers/Issues_papers_SP/La_agricultura_mundial.pdf.

Fenibo, E.O., Ijoma, G.N., Matambo, T., 2021. Biopesticides in Sustainable Agriculture: A Critical Sustainable Development Driver Governed by Green Chemistry Principles. 5.

Fenibo, E.O., Ijoma, G.N., Nurmahomed, W., Matambo, T., 2022. The potential and green chemistry attributes of biopesticides for sustainable agriculture. Sustainability 14 (21).

Frampton, R.A., Pitman, A.R., Fineran, P.C., 2012. Advances in bacteriophage-mediated control of plant pathogens. Int. J. Microbiol. 2012, 326452.

Gupta, S., Dikshit, A.K., 2010. Biopesticides: an ecofriendly approach for pest control. J. Biopestic. 3 (1), 186–188.
Haase, S., Sciocco-Cap, A., Romanowski, V., 2015. Baculovirus insecticides in Latin America: historical overview, current status and future perspectives. Viruses 7 (5), 2230–2267.
Hernández-Rosas, F., Figueroa-Rodríguez, K.A., García-Pacheco, L.A., Velasco-Velasco, J., Sangerman-Jarquín, D.M., 2020. Microorganisms and biological pest control: an analysis based on a bibliometric review. Agronomy 10 (11).
Holtappels, D., Fortuna, K., Lavigne, R., Wagemans, J., 2021. The future of phage biocontrol in integrated plant protection for sustainable crop production. Curr. Opin. Biotechnol. 68, 60–71.
Kachhawa, D., 2017. Microorganisms as a biopesticides. J Entomol Zool Stud 5 (3), 468–473.
Kazantseva, O.A., Buzikov, R.M., Pilipchuk, T.A., Valentovich, L.N., Kazantsev, A.N., Kalamiyets, E.I., Shadrin, A.M., 2022. The bacteriophage Pf-10—A component of the biopesticide "multiphage" used to control agricultural crop diseases caused by *Pseudomonas syringae*. Viruses 14 (1).
Kong, Q., 2018. Marine microorganisms as biocontrol agents against fungal phytopathogens and mycotoxins. Biocontrol Sci. Tech. 28 (1), 77–93.
Kumar, J., Ramlal, A., Mallick, D., Mishra, V., 2021. An overview of some biopesticides and their importance in plant protection for commercial acceptance. Plan. Theory 10 (6).
Malinga, L.N, Laing, M.D, 2022. Role of microbial biopesticides as an alternative to insecticides in integrated pest management of cotton pests. In: Rebolledo, R.R.E. (Ed.), Insecticides. IntechOpen, pp. 1–23.
Marin, V.R., Ferrarezi, J.H., Vieira, G., Sass, D.C., 2019. Recent advances in the biocontrol of *Xanthomonas* spp. World J. Microbiol. Biotechnol. 35 (5).
Mishra, J., Tewari, S., Singh, S., Arora, N.K., 2015. Biopesticides: where we stand? In: Arora, N.K. (Ed.), Plant Microbes Symbiosis: Applied Facets. Springer India, New Delhi, pp. 37–75.
Molina, L., Constantinescu, F., Michel, L., Reimmann, C., Duffy, B., Défago, G., 2003. Degradation of pathogen quorum-sensing molecules by soil bacteria: a preventive and curative biological control mechanism. FEMS Microbiol. Ecol. 45 (1), 71–81.
O'Brien, K.P., Franjevic, S., Jones, J., 2009. Green Chemistry and Sustainable Agriculture: The Role of Biopesticides. Retrieved from https://silo.tips/download/green-chemistry-and-sustainable-agriculture-the-role-of-biopesticides.
Oliver, R., 2012. Genomic tillage and the harvest of fungal phytopathogens. New Phytol. 196 (4), 1015–1023.
Pandya, U., Saraf, M., 2015. Antifungal compounds from pseudomonads and the study of their molecular features for disease suppression against soil borne pathogens. In: Arora, N.K. (Ed.), Plant Microbes Symbiosis: Applied Facets. Springer India, New Delhi, pp. 179–192.
Pathak, D.V., Yadav, R., Kumar, M., 2017. Microbial pesticides: development, prospects and popularization in India. In: Singh, D.P., Singh, H.B., Prabha, R. (Eds.), Plant-Microbe Interactions in Agro-Ecological Perspectives: Microbial Interactions and Agro-Ecological Impacts. 2. Springer Singapore, Singapore, pp. 455–471.
Raman, J., Kim, J.S., Choi, K.R., Eun, H., Yang, D., Ko, Y.J., Kim, S.J., 2022. Application of lactic acid bacteria (LAB) in sustainable agriculture: advantages and limitations. Int. J. Mol. Sci. 23 (14).
Roumagnac, P., Lett, J.-M., Fiallo-Olivé, E., Navas-Castillo, J., Zerbini, F.M., Martin, D.P., Varsani, A., 2022. Establishment of five new genera in the family *Geminiviridae*: *Citlodavirus*, *Maldovirus*, *Mulcrilevirus*, *Opunvirus*, and *Topilevirus*. Arch. Virol. 167 (2), 695–710.
Ruiu, L., 2018. Microbial biopesticides in agroecosystems. Agronomy 8 (11).
Sanahuja, G., Banakar, R., Twyman, R.M., Capell, T., Christou, P., 2011. Bacillus thuringiensis: a century of research, development and commercial applications. Plant Biotechnol. J. 9 (3), 283–300.
Savary, S., Willocquet, L., Pethybridge, S.J., Esker, P., McRoberts, N., Nelson, A., 2019. The global burden of pathogens and pests on major food crops. Nat. Ecol. Evol. 3 (3), 430–439.
Seiber, J.N., Coats, J., Duke, S.O., Gross, A.D., 2014. Biopesticides: state of the art and future opportunities. J. Agric. Food Chem. 62 (48), 11613–11619.

Setubal, J.C., Moreira, L.M., da Silva, A.C.R., 2005. Bacterial phytopathogens and genome science. Curr. Opin. Microbiol. 8 (5), 595–600.

Sharma, S., Malik, P., 2012. Biopesticides: types and applications. Int. J. Adv. Pharm. Biol. Chem. 1 (4).

Sharma, A., Kumar, V., Shahzad, B., Tanveer, M., Sidhu, G.P.S., Handa, N., Thukral, A.K., 2019. Worldwide pesticide usage and its impacts on ecosystem. SN Appl. Sci. 1 (11), 1446.

Sharma, S., Sundaresha, S., Bhardwaj, V., 2021. Biotechnological approaches in management of oomycetes diseases. 3 Biotech 11 (6), 274.

Sillanpää, M., Ncibi, C., 2017a. Biochemicals. In: Sillanpää, M., Ncibi, C. (Eds.), A Sustainable Bioeconomy: The Green Industrial Revolution. Springer International Publishing, Cham, pp. 141–183.

Sillanpää, M., Ncibi, C., 2017b. Bioeconomy: the path to sustainability. In: Sillanpää, M., Ncibi, C. (Eds.), A Sustainable Bioeconomy: The Green Industrial Revolution. Springer International Publishing, Cham, pp. 29–53.

Singh, B.K., 2017. Creating new business, economic growth and regional prosperity through microbiome-based products in the agriculture industry. Microb. Biotechnol. 10 (2), 224–227.

Sousa, R., Cunha, A.C., Fernandes-Ferreira, M., 2021. The potential of Apiaceae species as sources of singular phytochemicals and plant-based pesticides. Phytochemistry 187.

Spadaro, D., Gullino, M.L., 2004. State of the art and future prospects of the biological control of postharvest fruit diseases. Int. J. Food Microbiol. 91 (2), 185–194.

Srivastava, K.P., Dhaliwal, G.S., 2010. A Textbook of Applied Entomology. Vol. 1 Kalyani Publishers.

Thakur, N., Kaur, S., Tomar, P., Thakur, S., Yadav, A.N., 2020. In: Rastegari, A.A., Yadav, A.N., Yadav, N. (Eds.), Microbial biopesticides: current status and advancement for sustainable agriculture and environment. Elsevier, New and Future Developments in Microbial Biotechnology and Bioengineering, pp. 243–282.

Usta, C., 2013. Microorganisms in biological pest control—A review (bacterial toxin application and effect of environmental factors). In: Marina, S.-O. (Ed.), Current Progress in Biological Research. IntechOpen, pp. 287–317.

Vega, F.E., Meyling, N.V., Luangsa-ard, J.J., Blackwell, M., 2012. Fungal entomopathogens. In: Vega, F.E., Kaya, H.K. (Eds.), Insect Pathology. Academic Press, San Diego, pp. 171–220.

Wakefield, M.E., 2018. Microbial biopesticides. In: Athanassiou, C.G., Arthur, F.H. (Eds.), Recent Advances in Stored Product Protection. Springer Berlin Heidelberg, Berlin, Heidelberg, pp. 143–168

Wilson, K., Grzywacz, D., Curcic, I., Scoates, F., Harper, K., Rice, A., Dillon, A., 2020. A novel formulation technology for baculoviruses protects biopesticide from degradation by ultraviolet radiation. Sci. Rep. 10 (1), 13301.

CHAPTER

Fungi: A microbial community with promising potential of bioremediation

2

Naorem Bidyaleima Chanu[a], Mayanglambam Chandrakumar Singh[b], Rina Ningthoujam[a], Khaling Lallemmoi[c], and Ngathem Taibangnganbi Chanu[b]

[a]*Department of Vegetable Science, College of Horticulture and Forestry, CAU, Pasighat, Arunachal Pradesh, India,* [b]*Department of Basic Sciences and Humanities, College of Horticulture and Forestry, CAU, Pasighat, Arunachal Pradesh, India,* [c]*Department of Floriculture, College of Horticulture and Forestry, CAU, Pasighat, Arunachal Pradesh, India*

2.1 Introduction

Planetary mechanization, urbanization, and industrialization have led to contaminate and accumulate the environment with pollutants of different types creating serious environmental threats worldwide. A series of remediation strategies including chemical/thermal treatments were used to rectify the scenario but various issues like expensiveness, lower efficiency, creation of secondary pollutants, etc. have developed out of the methods (Arthur et al., 2005). Phytoremediation, a method of remediate pollutants done by plants, has gained acceptance from many researchers because of its cost effectiveness and environmental friendliness (Cunningham et al., 1995; Singh et al., 2003). Moreover, use of plants as remediator benefits the environment as plants can readily accumulate wide spectrum of harmful metals, metalloids, and organic pollutants. Meagher (2000) reported that plant tissues can transform toxic contaminants into less poisonous or nontoxic products through their metabolic processes. Several forms of phytoremediations like phytovolatilization, phytostabilization, rhizofiltration, etc. are being used.

Gunarathne et al. (2019) mentioned that about 400 species of hyperaccumulator plants are used in phytoremediation. These plants can assimilate high doses of pollutants by their root systems. These plants can accumulate 50–100 times more concentration of metals than nonhyperaccumulator plants (Chaney et al., 1997). Watanabe (1997) proposed that an ideal hyperaccumulator plant must have extensive root system, grow fast and be able to produce high biomass. Moreover, they must be resistant to pests and diseases. Plant characteristics like fast growth, elevated biomass, hardiness, and resistance to pollutants are the most exploited traits for phytoremediation remediation but most plants used for phytoremediation lack one or other such properties. Thus, there comes the need of biotechnological approach to enhance the performance of plants for phytoremediation through genetic manipulation and plant-transforming technologies. These technologies led to the production of transgenic plants for enhancing phytoremediation. Key et al. (2008) defined transgenic plants as genetically transform plants through recombinant DNA technology to manifest a desired gene that is not of the plant. Generally,

Entrepreneurship with Microorganisms. https://doi.org/10.1016/B978-0-443-19049-0.00001-3

genes of interest are identified from hyperaccumulators or any other suitable sources and then transferred to those plants with fast growth rate, extensive root system, resistance to pests/diseases/climatic conditions, etc. The transformed plants are generated through tissue culture.

Van Aken (2008) highlighted that genetically modified plants to be used for phytoremediation were designed for enhancing heavy metal tolerance. The chapter focuses on representing an overview of enhanced phytoremediation of pollutants by the use of transgenic plants.

2.2 Categories of pollutants

There are different categories of pollutants polluting the atmosphere. Inorganic and organic pollutants are the major classes of pollutants. Strategies of phytoremediation of different classes of pollutants vary as per the nature of the pollutants.

2.2.1 Inorganic pollutants

Inorganic pollutants include heavy metals and nonmetallic pollutants like mercury, lead, cadmium, arsenic, etc. Several other radionuclides like uranium, phosphorous, cesium, etc. and plant fertilizers like nitrates and phosphates also come under the class of inorganic pollutants. Many metals perform a vital role in the regular growth and development of plants but at higher doses, metals turn harmful leading to the generation of free radicals and reactive oxygen species (ROS) that can damage the plant cells (Jones et al., 1991; Cho and Park, 2000). The inorganic pollutants occur mainly as charged ions. Some ions can cause disturbance in the structure and functioning of proteins in the plants and moreover higher doses of inorganic pollutants may replace other essential nutrients in plants. There certain methods available for the phytoremediation of inorganic pollutants are phytostabilization (immobilization), phytoextraction or rhizofiltration, and phytovolatilization. Biotechnological approaches have been able to successfully change the capability of plants on more tolerance and accumulation of inorganic pollutants (Pilon-Smits, 2005; Doty, 2008).

2.2.2 Organic pollutants

The organic pollutants viz petroleum hydrocarbons, halogenated hydrocarbon, chlorinated solvents, etc. Organic pollutants are mostly man made and not commonly expected to be present in organisms (Xerobiotic). Generally, organic pollutants pollute the environment due to accidental release, industrial activities, agricultural practices, military activities, etc. Ensley and Raskin (1999) revealed that polluted sites are maximally contaminated with organic as well as inorganic pollutants. Most of the organic pollutants are toxic to the plants and therefore strategies for phytoremediation need the necessary biodegradation of the organic pollutants. A certain group of organic pollutants like chemical pesticides can cause serious abnormalcies in humans and can reduce the quality of environment. Phytoremediation of pesticides by transgenic plants can be enhanced through bioremediation of pesticides using microorganisms capable of degrading such pollutants (Hussein et al., 2007). Direct dumping of used solvents leads to the contamination of the underground water. TCE (trichloroethylene) is one of the most abundant organic pollutants in the nature (Dhankher et al., 2012). Microorganism assisted phytoremediation or genetic engineering can serve as the solution to improve phytoremediation of organic pollutants like TCE.

2.3 Types of fungi in bioremediation

The utilization of life forms to degrade hazardous environmental pollutants to less harmful or non-harmful forms with minimal inputs constitutes bioremediation. Bacteria as well as fungi are important bioremediators, while the former is known for its partial or inefficient decomposition compared to the later one. Therefore, fungi are more often exploited by microbes in bioremediation than the bacterial groups. Many attempts are made practically by mycologists to identify noble fungal agents for organic pollutants in the past decades. The pace of bioremediation has advanced with the discovery of potential myco-remedial fungi like *Phanerochaete chrysosporium* (White rot fungi) which has proven to give satisfying practical results (Singh, 2006). Myco-remediation is a novel technology where fungal metabolism is exploited for removing potentially hazardous pollutants or minimizing xenobiotics. Fungi are well-known fast decomposers; mushrooms are basidiomycetes group of fungi with immense potential to decompose organic and inorganic pollutants to environment safer outputs. Fungi functions by producing mycelial enzymes bearing the ability to digest several forms of pollutants. For instance, mushroom fungi release several extracellular enzymes like pectinase, cellulase, ligninase, peroxidases, and oxidases (Kulshrestha et al., 2014). While on the other hand some groups of fungi are designed to accumulate toxic heavy metals from the polluted sites reducing environmental pollution (Jagtap et al., 2003). The extend of mycoremediation ranges from contaminated water, soil, oil spill, industrial effluents, and many more (Bennet et al., 2001). Some mushroom fungi species are reported to degrade plastic polymers in the recent past (Kulshrestha et al., 2014).

The global headache to fight climate change can be worsening if the sole focus is made for pollution management. An integrative approach involving physical, biochemical, mechanical, and biological means must be adopted to have an efficient pollution control with little or no environmental impacts. Such approaches are made more possible to execute by expanding researches on fungus involved in bioremediation of several pollutant sources. The immense capability of several fungi in remediation of plastic or insoluble pollutants to a simpler soluble form upon acted by the applied fungus pays a landmark achievement in operational bioremediation till date. An excellent number of fungi are known and have been reported to be an effective tool for bioremediation. Some potential fungi species utilized in bioremediation are listed in Table 2.1 with details on their specific compound source for remediation.

2.3.1 White rot fungi (WRF) in bioremediation

Fungi are microbes with gigantic family groups, among the group WRF are true fungi belonging to the basidiomycetes subdivision, Hymenomycetes class, and Holobasidiomycetidae subclass (Hawksworth et al., 1995). WRF got its name from the white appearance of the product after the decomposition of woods after removing tough lignin and leaving simpler cellulose and hemicellulose. Various kinds of extracellular enzymes catalyze lignin degradation, naming some are laccases (Leontievsky et al., 1997), peroxidases (Camarero et al., 1999), and oxides releasing peroxides (Guillén et al., 1992). They are especially known for their vast decomposing ability of lignin and lignin associated substances, some of the popular WRF are conks, puff balls, a variety of mushrooms, and some other crust-like fungi. These groups of fungi produce massive hyphae growth covering the substrates thus emitting several enzymes and digesting insoluble pollutant substrate to soluble or nearly soluble fewer toxic outputs. Upon nitrogen starvation of these fungi, they tend to feed on lignin under the secondary metabolism pathway. Types of lignin decomposition, variety of substrate fed, functioning power, and

Table 2.1 List of some potential mycoremediators.

Sl. no.	Fungi species	Compound for mycoremediation	References
1	*Doratomyces* spp., *Phoma eupyrena*, *Thermoascus crustaceus*, *Aspergillus niger*	Polychlorinated biphenyls	Mouhamadou et al. (2013), Marco-Urrea et al. (2015)
2	White rot fungi	Polychlorinated dibenzofurans	Wu et al. (2013)
3	*Aspergillus*, *Curvularia*, *Acremonium*, *Pythyme* *Aspergillus flavus*	Heavy Metals	Akhtar et al. (2013), Kurniati et al. (2014)
4	*Aspergillus* spp.	Chlorpyriphos	Silambarasan and Abraham (2013)
5	*Gongronella* sp. and *R. stolonifera*	Metalaxyl and Folpet	Martins et al. (2013)
6	*A. niger*, *Rhizopus* sp., *Candida* sp., *Penicillium* sp., *Mucor* sp.	Crude oil	Damisa et al. (2013)
7	*Pleurotus ostreatus* *Trametes versicolor*	Diphenyl ether	Rosales et al. (2013)
8	*Armillaria* sp.	Anthracene	Hadibarata et al. (2013b)
9	*Chrysosporium keratinophilum*, *Gliocladium roseum*, *Fusarium solani*, *A. restrictus*, *Penicillium*, and *Stemphylium*	Caffeine	Nayak et al. (2013)
10	White rot fungi *Pleurotus eryngii*	Naphthalene	Hadibarata et al. (2013a)

enzymes involvement may vary from species to species of WRF in mycoremediation (Hatakka, 2001). These fungal groups are gaining quite a popularity in bioremediation, the sole reason behind this is the capacity to resist the toxicity of the hazardous organic source pollutants without malfunctioning of its own fungal system (Aust et al., 2003). Target pollutants of WRF constitute contaminants of complex admixtures, absorbed forms, higher molecular weight, inorganic pollutants as well as organic pollutants (Azadpour et al., 1997).

2.3.1.1 WRF degradable compounds

2.3.1.1.1 Complex lignin

Lignin peroxidase and glyoxal oxidase are the main extracellular enzyme produced by the WRF for lignin degradation. Glyoxal oxidase metabolizes glyoxal as well as methyl glyoxal and reduces O_2 to H_2O_2, this process activates lignin degrading enzyme "Lignin peroxidase." Nonphenolic aromatic nuclei contents in lignin get oxidized by the enzyme releasing electrons which catalyzes many reactions to degrade lignin nonenzymatically. This reaction results in breaking of polymers giving out aromatic and aliphatic outputs which are mineralized by the fungal hyphae. Phenolic compound in lignin gets degraded by Mn peroxidase in the presence of peroxides; however, the actual reactions are yet to be documented.

The most important and widely exploited lignin degraders are the basidiomycetes group of fungi, concerning more on WRF, some mushrooms, and some ascomycetes. Among all WRF have been the most exploited and intensively studied lignin degraders in bioremediation (Kirk et al., 1992).

2.3.1.1.2 Polycyclic aromatic hydrocarbons (PAHs)

PAHs are complex formed by fusing four or more rings of benzene. These complexes persist long in soil and environment causing various toxicity and are difficult to degrade. PAH deposits are originated from vegetation decomposition, transportation efflux, industrial emission, wood burning, oil deposits, and power generation from fossil fuels by heating.

Numerous studies have been conducted and found that effective remediation of PAH can be done by utilizing WRF, naming some of them are *P. chrysosporium*, *Chrysosporium lignorum*, *Phanerochaete sordida*, *Pleurotus ostreatus*, *Bjerkandera* sp., *Phanerochaete laevis*, *Dichomitus squalens*, *Irpex lacteus*, and *Coriolus versicolor*. *Pleurotus* species have also been reported to use successfully in PAH bioremediation (Baldrian et al., 2000). *Pleurotus tuber-regium* was reported to be used effectively in soil amelioration of polluted soil with crude oil polluted and after the treatment, the soil supports seeds germination and seedling growth of cowpea (Isikhuemhen et al., 2003).

2.3.1.1.3 Synthetic textile dyes

Synthetic dyes are not easily degraded in the environment after releasing from textile industries and moreover decolorization remains a challenge in wastewater treatment plans (Michaels and Lewis, 1985). Recent research on WRF for successful decolorization of textile dyes has made possible by the use of fungi such as *Bjerkandera* sp., *Ceriporia metamorphosa*, *Daedalea flavida*, *Daedalea confragosa*, *Lentinus tigrinus*, *Mycoacia nothofagi*, *P. chrysosporium*, *P. sordida*, *Phellinus pseudopunctatus*, *Phlebia* spp., *Piptoporus betulinus*, *Pleurotus eryngii*, *P. ostreatus*, *Pleurotus sajor-caju*, *Polyporus ciliatus*, *Polyporus sanguineus*, *Pycnoporus sanguineus*, *Stereum hirsutum*, *Stereum rugosum*, *Trametes (Coriolus) versicolor*, *I. lacteus*, *Geotrichum candidum*, and *D. squalens*. *P. chrysosporium* recorded 23.1%–48.1% mineralization rates for an extensive range of azo dyes subsequently at 12 days of incubation (Spadaro et al., 1992).

2.3.1.1.4 TNT (2,4,6-trinitrotoluene)

Selected white rot basidiomycetes fungi have been reported to mineralize TNT, for instance: *P. chrysosporium* (Hodgson et al., 2000). Degradation of TNT results information of DNTs (dinitrotoluenes) such as 2-amino-4,6-dinitrotoluene, 2,4-diamino-6-nitrotoluene, 2,6-diamino-4-nitrotoluene, and 4-amino-2,6-dinitrotoluene. These DNTs are nondegradable and are more persistent than TNTs (Bumpus, 1989). Considerably, only white rot basidiomycetes fungi have shown the potential to degrade and mineralize finally to CO_2 (Hodgson et al., 2000).

2.3.1.1.5 Pesticides

Indiscriminate use of pesticides in agricultural managements willingly or unknowingly has affected the environment severely. However, the insecticide residues are mostly nonpersistent and hence can be remediated by several means. *P. chrysosporium* has been reported to mineralize radiolabeled ^{14}C pesticides like chlorpyrifos, fonofos, and terbufos up to 12.2%–27.5% during 18 days treatment (Bumpus, 1989).

2.3.1.1.6 Herbicides

Herbicides such as atrazine (2-chloro-4-ethylamino-6-isopropylamino-1,3,4-triazine) are recalcitrant in nature after its deposition in the environment. However, WRF like *Pleurotus pulmonarius* and *P. chrysosporium* (Mougin et al., 1994) have been reported to convert atrazine to less recalcitrant product forms of hydroxylated and N-dealkylated metabolites.

2.3.1.1.7 Other compounds

WRF can further degrade several other hazardous pollutants from the environment and mineralize it to an ecosystem friendly product. Nylon polymers are known to be transformed to biodegradable forms by various enzymes dissolution by WRF. Chlorophenols are complexes that are persistent in the environment, yet there are several reports on the transformation of such complexes to nontoxic forms by certain WRF.

2.3.2 Mycorrhizal fungi in bioremediation

Mycorrhizae are symbiotic association of tree roots and soil inhabiting hyphae of fungi. Their relationship is mutualistic in nature, where bidirectional nutrients exchange took place for resuming growth and development. Nitrogen, phosphorus, and other nutrients are captured by the fungal partner and supplied to the plant system and in return they receive photosynthates from the plants. Mycorrhizal associations are widely classified on the basis of the fungal species involved, resulting root structures and some other parameters. Some of them are as ECM (ectomicorrhiza), ERM (ericoid mycorrhiza), AM (arbuscular mycorrhizas), ectendo mycorrhizas, ARM (arbutoid mycorrhizas), and orchid associated mycorrhizas.

2.3.2.1 Heavy metal tolerance mechanism in arbuscular mycorrhizal fungi

The amelioration process of heavy metals or toxic pollutants by mycorrhizal fungi constitutes mycorrhizal-bioremediation (Hildebrandt et al., 2007). The role of mycorrhiza is very vast in bioremediation; it not only transforms toxic pollutants to environment acceptable forms but it also prepares plants to tolerate toxicity without impairing the normal functioning. For instance, poplar tree saplings with mycorrhizal association are less vulnerable to heavy metal toxins as the fungal hyphae restrict the toxins only to the rhizosphere zone preventing from translocation in the plant system.

Naturally AM fungi are known to be habitat in heavy metal pollutes environment and are recovered from such sites to test their tolerance capacity to heavy metals, yet their distinct phenomena behind such interactions are not properly documented. Some mycorrhizal fungi are reported to absorb and translocate micronutrients as well as toxic elements directly in the plant system; such as Glomalean hyphae are reported to uptake and translocate Cu, Zn, and Cd to host roots (Li et al., 1991; Bukert and Robson, 1994). Some reports have also suggested that heavy metals get immobilized in the extra radical fungal hyphae preventing the plant roots from toxic effects of heavy metals (Kaldorf et al., 1999). Agents such as chitin from the fungal cell wall activate the metal binding ability of the fungal hyphae enhancing the adsorption of the heavy metals on the extra radicular hyphal growth and escaping the harmful effects on the plant roots (Zhou, 1999).

2.3.2.2 Mycorrhizal fungi in remediation of toxic pollutants

2.3.2.2.1 Lignin degradation

Ericoid and ectomycorrhizal fungi are recently reported to mineralize lignin (Paul and Clark, 1989). Vesicular-arbuscular mycorrhiza (VAM) has been reported to give a meager contribution in lignin degradation studies as it cannot be pure cultured in media. Some of the ectomycorrhizal fungi are capable of degrading lignin, hemicellulose, and lignocellulose in vitro conditions. This group of fungi when supplied with sole carbon source pollutants degrades lignin more efficiently to simpler absorbable forms (Trojanowski et al., 1984).

2.3.2.2.2 Polychlorinated biphenyl (PCB) degradation

Twenty-one numbers of mycorrhizal fungi have been reported to metabolize PCBs successfully. For instance, ectomycorrhizal fungi such as *Radiigera atrogleba* and *Hysterangium gardneri* showed degradation capacity up to 80%. They converted complex PCBs to simpler dichlorophenol. While ericoid mycorrhizal fungi, namely, *Hymenoscyphus ericae* and *Oidiodendron griseum* didn't gave the satisfactory metabolizing ability of PCBs as that of ectomycorrhizal fungi (Donnelly and Fletcher, 1995).

2.3.2.2.3 Heavy metals mineralization

The presence of mycorrhizal association in heavy metal contamination, oil spills, and mining site imparts less toxicity to the plants rather than in its absence. These sites are generally saturated with Zn, Mn, Cu, and Cd heavy metals. Plants grown in such areas with mycorrhizal association are often protected from toxic effects imparted by the heavy metals present. Such fungi are believed to be selected overtime for their tolerance to heavy metal toxicity and are now being exploited for bioremediation research purposes. *H. ericae*, an ericoid mycorrhizal fungus reported to metabolize phytotoxic heavy metal compounds in in vitro conditions. Mycorrhizal fungus imparts host protection against hazardous heavy metals toxins and allows plants to resume normal growth and development process (Sylvia et al., 1987).

2.3.2.2.4 Herbicide degradation

Many studies have been conducted on herbicidal bioremediation using mycorrhizal fungi. Several fungi are reported to possess the ability to degrade several chlorinated aromatic herbicides such as 2,4-D and atrazine namely (Donnelly et al., 1993). Ectomycorrhizal fungus like *Rhizopogon vinicolor* and an ericoid mycorrhizal fungus such as *H. ericae* were shown to degrade 2,4-D herbicide effectively. While ericoid groups of mycorrhizal fungi like *H. ericae* and *O. griseum* proved successful degradation of atrazine weedicide. Also, some ectomycorrhizal fungi like *Rhizopogon vulgaris* and *Gautieria crispa* were also shown the ability to metabolize atrazine effectively (Donnelly et al., 1993).

2.3.3 Mushrooms

2.3.3.1 *Phanerochaete chrysosporium*

P. chrysosporium is ideal for bioremediation. It has been well-known to degrade lignin macro molecules, many natures of organo pollutants, and toxic xenobiotics, namely, PAHs, polychlorinated biphenyls and dioxins, chlorophenols, chlorolignins, nitrocranditics, synthetic dyes, and different pesticides. Polyphenol oxidases and lignin peroxidases are involved in the degradation process. It has been observed to affect the bioleaching of organic dyes (Nigam et al., 1995; Barr and Aust, 1994). It produced the first extracellular enzyme (ligninase) discovered to depolymerize lignin, and lignin substructured compounds in vitro (Aitken and Irvine, 1989).

2.3.3.2 *Phanerochaete flavido-alba*

Phanerochaete flavido-alba has the potential to decolorize olive oil mill wastewater (OMW), a main waste product of olive oil extraction for subsequent use in bioremediation assays. Nitrogen-limited *P. flavido-alba* cultures containing 40 mg/L Mn(II) which made it the most efficient at decolorizing OMW. The decolorization process also decreases OMW phenolic content by 90%. Apart from concentrated extracellular fluids, mycelium binding forms part of the decolorization process.

2.3.3.3 *Trametes versicolor*

T. versicolor produced three ligninolytic enzymes that have efficient degradation capacity on lignin, PAH, polychlorinated biphenyl mixture, and a number of synthetic dyes (Tanaka et al., 1999; Novotny et al., 2004). It has also reported to delignify and bleach pulp which may make it a potential eco-friendly technology for pulp-paper industry and it also acts as a biocatalyst for decolorization *T. versicolor* can also be used for wastewater treatment in dye industry (Gamelas et al., 2005; Selvam et al., 2002; Amaral et al., 2004).

2.3.3.4 Pleurotus ostreatus *(Jacq. Fr.) P. Kumm*

Recent studies have shown that *P. ostreatus* has the potential to degrade a variety of PAHs (Sack and Gunther, 1993). It degrades PAH in nonsterile soil both in the presence and in the absence of cadmium and mercury, catalyzed humification of anthracene, benzo(*a*)pyrene, and flora in two PAH—contaminated soils from a manufactured gas facility and an abandoned electric cooping plant (Bojan et al., 1999).

2.3.3.5 Pleurotus tuber-regium *(Fries) Singer*

The white rot fungus, *P. tuber-regium* is a fungus having the capacity to ameliorate crude oil polluted soil. The contaminated soil shows good plant growth after treating the soil with fungi.

2.3.3.6 Lentinus squarrosulus *(Mont.) Singer*

Lentinus squarrosulus mineralized soil contaminated with various concentrations of crude oil resulting in increased nutrient contents except for potassium in treated soil. The rapid mycelia growth and enhanced enzyme production by *L. squarrosulus* have biotechnological applications for wood and pulp, textile, and tanning, as well as in the bioremediation of oil spills (Isikhuemhen et al., 2010; Adenipekun and Fasidi, 2005; Adenipekun and Isikhuemhen, 2008).

2.3.3.7 *Pleurotus pulmonarius*

P. pulmonarius perform well in the management of cement and battery polluted soils. An increase in the carbon content, organic matter, phosphorus, and potassium and a decrease in the percentage of nitrogen, calcium, and pH have been observed after 10 weeks of incubation. Significant decrease in the copper, manganese, nickel, and polyaromatic hydrocarbons (PAH) contents of the soils while the lead concentration remains the same (Adenipekun and Lawal, 2011).

2.4 Biodegradation of different pollutants by Fungi

The contamination of our surrounding rises gradually due to the development of many industrial and manufacturing works. Their excretion and residues polluted the soils, water sources, and air with dangerous and harmful substances. The contemporary system of agriculture, the chemistry industry, and the production of different forms of energy are also the main causes of the liberation of organic compounds in our biosphere. The use of fuel for combustion in diverse industries and automobiles is also one of the reasons for increasing atmospheric carbon dioxide, which leads to global warming (Tortella et al., 2005). Likewise, the changes in biogeochemical cycles are directly related with the input of inorganic fertilizers, pesticides, insecticides, and other chemicals in agriculture (Ogunseitan, 2003). In past years, different steps have been taken up to decrease the release the toxic compounds in the environment but still, the wiping of environmental pollutants is a current issue and burden to society. According to Barr and Aust (1994), the implementation of chemical and physical methods to regain the environment would be expensive and result in environmental stress even after the pollutants are removed. Consequently, a proper effective method and practice are required for both present and future generations.

A unique process known as bioremediation can be considered as a solution to cope up the problems. It is a process of transforming toxic organic contaminants to fewer toxic substances with the use of microorganisms (Arun et al., 2008). The biodegradation of organic compounds is enhancing with the incorporation of living organisms. The organism in the environment and genetically engineered microorganisms plays an important role in biodegradation of organic compounds. Bacteria and fungi are the type of microorganisms generally used for bioremediation. While, in the recent years back the concern for exploitation of fungi in bioremediation is evident as it provides efficiency toward enzymatic and metabolic potentiality. The metabolism of fungus provides an easy preparative method for the generation of metabolites in large quantity. Moreover, the fungal hyphae have the capacity to penetrate into the contaminated soil extending up to pollutants (Husaini et al., 2008). In addition to this, it is reported that some saprotrophic soil fungi have the ability to metabolize pollutants when exposed to contamination (Pinto et al., 2012). Fungi are microscopic organisms and most of them can be cultured for identification. There are about 63,500 identified fungal species but 13,500 species are associated with algae as lichens. Fungi have the ability to survive in diverse conditions, starting from simple to complex soil and freshwater to marine (Deshmukh et al., 2016). Additionally, they have the capacity to thrive very well in soils of different climatic conditions and can also be reproduced through the dispersal of spores in the air (Anastasi et al., 2013). They even have the tendency to live in effluent treatment plants (ETPs) by treating different wastewaters (Badia-Fabregat et al., 2015).

Classification of fungi for biodegradation can be done based on their enzymatic machinery. They are commonly classified as WRF, brown rot fungi, and soft rot fungi. WRF have the potential to decompose lignin whereas brown rot fungi decompose only cellulose. However, soft rot fungi can decompose cellulose on damp woods surface. WRF are the chief agents of degrading endocrine disrupting chemicals (EDCs) and TrOCs like pharmaceuticals and body care products. The storing of these products for long causes acute and chronic toxicity to aquatic organisms and also to human health. *P. chrysosporium*, *P. ostreatus*, *T. versicolor* and *Bjerkandera* sp. BOL13 have the ability to degrade phenolic compounds like nonylphenols (NP) and bisphenol A (BPA). The main reason for the success of biodegrading by WRF is due to secretion of different ligninolytic enzymes like laccases and peroxidases (dos Santos Bazanella et al., 2016). Ligninolytic enzymes help in promoting the activity of microorganisms and play a very important role in transforming toxic compounds from contaminated waters (Rodríguez-Rodríguez et al., 2013). The remediation of cresolate contaminated soil can also be achieved by using WRF with the process of bioaugmenting two strains *T. versicolor* and *L. tigrinus* (Lladó et al., 2013). The vast potentiality of white rot fungus also includes the degradation of polychlorophenols (PCBs) using *Phlebia brevispora*. Polychlorophenols (PCBs) are an important group of phenols, which are incorporated in fungicides, herbicides, insecticides, and in the synthesis of other pesticides. Further too, the combustion of fossil fuels like petroleum, coal tar, and shale oil induces PAHs that are tenacious in nature and human health. A filamentous fungus and nonligninolytic soil fungus *Cunninghamella elegans* is capable of PAHs mineralization and transformation. Other diverse groups of fungi like *Cyclothyrium* sp., *Penicillium simplicissimum*, and *Psilocybe* sp. also have the capacity to degrade PAHs compounds such as pyrene, anthracene, phenanthrene, and benzo[*a*]pyrene (Da Silva et al., 2003). The microbial activity of *Polyporus* sp. S133 has the bioremediation potential of degrading contaminated soil with crude oil at a rate of 93% (Kristanti et al., 2011). Besides 2,4,6-trinitrotoluene (TNT) and RDX (Hexahydro-1,3,5-trinitro-1,3,5-triazine) are present in explosive components. When an explosion occurs its remaining act as pollutants and causes harmful to surroundings. However, the degradation of these compounds can be successfully degraded by the use of fungi like *Aspergillus*, *Coniothyrium*, *Trichoderma* sp., *Paecilomyces*, and *Penicillium*. Different metals also cause toxicity to our soil and it has become a serious problem for agricultural cultivation. The residues of heavy metal (Zn, Cd, Pb, Fe, Ni, Ag, Th, Ra, and U) and nondegradable chemical contaminate the soil and are a major dangerous threat to the environment due to their long-term availability and diffusion into underground water resources. The presence of these toxic compounds in the soil can be reduced by the process of fungal bioaccumulation/biosorption. Most of the common filamentous fungi can absorb heavy metals and possess metal binding potential. It leads to an economical and sustainable method for the removal and recovery of heavy metals. *P. chrysosporium*, *Ganoderma carnosum*, and *Aspergillus parasiticus* are some of the common fungi species popularly known for their efficiency in metal remediation. Other than this, marine fungi also inherent the capacity in bioremediation of hydrocarbons and heavy metals (Damare et al., 2012). There are some exceptional benefits of marine fungi than terrestrial fungi due to their great adaptability to high saline concentration and pH value. Moreover, it is reported that marine fungi have the ability to tolerate heavy metals at high concentration (Gazem and Nazareth, 2013). The removal of heavy toxic metals from contaminated area can also be dealt with the filamentous fungi. Their capacity of high tolerance and remediation toward heavy metals like Cd, Cu, and Ni (up to 1500 mg/L) results significance for bioremediation from contaminated soil and wastewater. Members of *Aspergillus* group including *A. flavus*, *A. niger*, and *A. foetidus* have been reported for their capacity to reduce heavy metals (Bennett et al., 2013). *A. foetidus* was found to be tolerant to high concentrations of lead (Pb) up to 200 mg/L (Chakraborty et al., 2013) and also able to recover 98% mercury in presence of 10 mg/L mercury in the medium (Kurniati et al., 2014). The heavy metal-contaminated soils for cultivation can be overcome by the attachment of AM fungi in the roots of plants. It helps in enhancing the immobilization of heavy metals and makes the plants to grow in metal-contaminated soils (Garg and Bhandari, 2014; Yang et al., 2015). The accumulation of tones of municipal solid waste is also a part of polluting our atmosphere by emitting a foul smell and increasing pathogens in the soil (Soobhany et al., 2015). However, the use of these wastes in a beneficial form can reduce pollution, and the use of fungi for the process of composting and biomethanation by anaerobic digestion is a desirable solution. The treatment of MSW converts the complex polymeric substances to simple compounds, which are precursors for VFA and biogas production. For this process, many fungal groups can be used and a strain of *A. niger* can be utilized for solid-state fermentation of cellulolytic, hemicellulolytic, pectinolytic, and amylolytic enzymes (Janveja et al., 2013). The decomposition of willow and rice straw can be done by application of a fungal consortium consisting of two fungi *Armilleria gemina* and *Pholiota adiposa* (Dhiman et al., 2015). Hence, fungi can be used as a possible microorganism on biodegradation process to manage various levels of pollutants remediating the adulterated environment (Fig. 2.1).

2.4.1 Aromatic hydrocarbons

Aromatic hydrocarbons are also known as arenes. It has distinctive aromas and has one or more benzene rings in the molecule. The derivatives of the aromatic hydrocarbon groups such as toluene ($C_6H_5CH_3$) and the isomeric dimethyl benzene ($CH_3C_6H_4CH_3$) are used as solvents as well as in the synthesis of

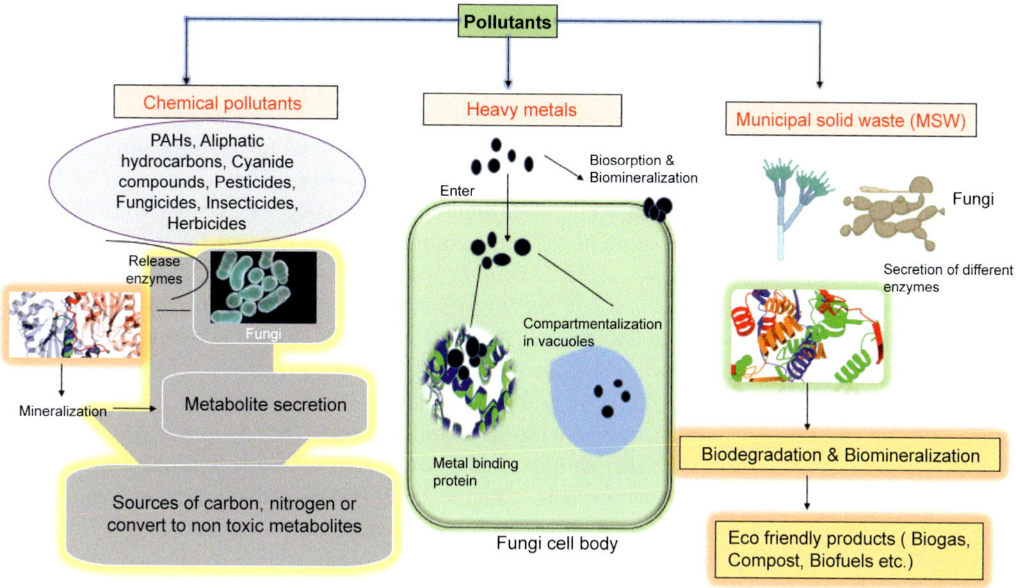

FIG. 2.1

Biodegradation processes of different pollutants by fungi.

drugs, dyes, and plastics (Speight, 2016). Naphthalene is formed when two condensed rings of aromatic hydrocarbon are fused. It is derived from coal tar and used in the synthesis of other compounds. Naphthalene is a crystalline solid with a powerful odor and is used for disinfectants. Benzo(*a*)pyrene another derivative of aromatic hydrocarbon is produced in small amounts by the combustion of organic substances and found to cause carcinogenic. Not only this, but also other derivatives of aromatic hydrocarbons like benzene, toluene, styrene, and xylene are also found harmful to nature. So, the degradation performance of aromatic hydrocarbons and their derivatives by fungi was studied. *P. chrysosporium* KFRI 20742, *T. versicolor* KFRI 20251, and *Daldinia concentrica* KFRI 40-1 are used for biodegradation of monomeric styrene and produce benzoic acid, cyclohexadiene-1,4-dione, butanol, 2-phenyl ethanol, and succinic acid as end products (Lee et al., 2006). Another problem for dirtying the environment is oil pollution. Biodegradation of diesel pollution from aqueous solution can be done by using the fungus *Cladosporium* (Lee et al., 2007), whereas contamination of soil with crude oil can be recovered by exploitation of *P. tuber-regium* (Ogbo and Okhuoya, 2008). Furthermore, *Fusarium solani* SZFWT02 has great ability in biodegradation of contaminated areas with fuel oil (Yoshioka and Komuro, 2006).

PAHs are well distributed in our surroundings due to incomplete combustion of organic materials such as oil, petroleum gas, wood, municipal, and urban waste (Juhasz and Naidu, 2000). They have resistant nature to degradation and accumulation leads to health hazards. PAHs are formed by fusing three or more benzene rings (Di Toro et al., 2000) and are extremely stable due to their strong negative resonance energy. The contact of PAHs causes acute symptoms such as irritation in the eyes, vomiting, and nausea. Exposure to high concentration may damage in kidney and liver, burn in skin, weakens immune reactions, and embryotoxicity during pregnancy. As well as it is reported for causing genotoxic, carcinogenic, mutagenic, and teratogenic effects (Rengarajan et al., 2015). Thus, the remediation of PAHs has become a problematic issue and attract great attention to many researchers. Bioremediation of PAHs by fungi is a method to control the problems. Some fungi which come under the genus *Aspergillus*, *Penicillium*, *Paecilomyces*, *Coriolus*, *Pycnoporus*, *Pleurotus*, *Fomitopsis*, *and Daedalea* have positive result degrading PHAs in both soil and aquatic environments (Sanyal et al., 2006). Other WRF belonging to genera *Phanerochaete*, *Irpex*, *Polyporus*, *Stereum*, *Lentinus*, *Bjerkandera*, *Irpex*, *Pleurotus*, and *Phlebia* have found the ability to remediate contaminated soils (Valentin et al., 2006). *C. versicolor*, *Trichoderma* sp., *A. niger*, and *Fusarium* sp. are some of the most common effective fungi for the bioremediation of pyrene (Arun et al., 2008). However, *Absidia cylindrospora*, *Absidia fusca*, *C. elegans*, *Aspergillus terreus*, *Cladosporium herbarum*, *Penicillium chrysogenum*, *Rhodotorula glutinis*, and *Saccharomyces cerevisiae* have the quality to change anthracene (AC) to 1,4-dihydroxyanthraquinone (Guiraud et al., 2008), 1,2,3,4-tetrahydronaphthalene (THN) gives out 3,4-dihydro-4-hydroxy-1(2*H*)-naphthalenone, 3,4-dihydro-1(2*H*)-naphthalenone, 1,2,3,4-tetrahydro-1-naphthalenol, and 1,2,3,4-tetrahydro-1,2-naphthalenediol as byproducts when treated with the marine fungus *Hypoxylon oceanicum* (Li et al., 2005). Phenol and its other form are also an important derivative of aromatic hydrocarbons and cause contamination when uncover in the environment. Chemical and fuel-producing industries are the main sources of phenol. In agriculture, the residues of pesticides, insecticides, fertilizers, and herbicides develop toxins in the exposed area. These harmful phenols can be biodegrading or bio transform into the less toxic compound. It is found that *Fusarium* sp. HJ01 was able to survive in phenol using carbon as a resource and catechol was produced as a biotransformation product. *T. versicolor* was also able of biodegrading phenol compounds. *P. chrysosporium*, *P. ostreatus*, *T. versicolor*, and *Bjerkandera* sp. BOL13 achieved successful biodegradation of nonylphenol (Soares et al., 2005). Moreover, a strain of *A. niger* "PSH" has the properties to break down tannins into gallic acid and ellagic acid (Ventura et al., 2008). *Myrioconium* sp. strain UHH 1-13-18-4 and *Clavariopsis aquatica* are used for the degradation of two micropollutants from aquatic galaxolide (HHCB) and tonalide (AHTN) (Martin et al., 2007). Besides, it is reported that *Polyporus brumalis* was applied to degrade dibutyl phthalate (DBP) (Lee et al., 2007).

2.4.2 Aliphatic hydrocarbons

Aliphatic hydrocarbons are hydrocarbons that have chains of carbon atoms. Accordingly, there are three types of aliphatic hydrocarbons. They are alkanes, alkenes, and alkynes. Alkanes are simple aliphatic hydrocarbons with only one covalent bond. Alkenes are hydrocarbons that have a minimum double bond of carbon while alkynes are hydrocarbons that have a C–C triple bond. Furthermore, there is another aliphatic hydrocarbon containing a ring of C atoms. These hydrocarbons are known as cycloalkanes or cycloalkenes or cycloalkanes. Alkanes are also called saturated hydrocarbons because it contains the maximum number of H atoms. Many aliphatic hydrocarbons are produced from fuels and their derivative causes considerable hazards to biological receptors. It produced toxic and carcinogenic metabolites. Presently, "bioremediation" is a process to mitigate the problem. Bioremediation process is cost-effective and eco-friendly when compared to other methods. In the activity process of aliphatic hydrocarbon biodegradation, many indigenous microorganisms are utilized to transform/mineralize hydrocarbon contaminants. Biodegradation of hydrocarbons by fungal possesses various packs of enzymes. They utilize different hydrocarbons as sole carbon/energy sources. Fungi are likely to have potential in degrading *n*-alkanes including tridecane, tetradecane, pentadecane, hexadecane, heptadecane, octadecane (C13-C18), and crude Omani oil. *A. niger*, *Aspergillus ochraceus*, *Trichoderma asperellum* strain TUB F-1067 (SA4), *T. asperellum* strain Tr48 (SA5), *T. asperellum* strain TUB F-756 (SA6), *Penicillium* (P1), and *Aspergillus* species (P9) are some of the fungi that play a great role in bioremediation of aliphatic hydrocarbons (Elshafie et al., 2007). According to many reports, P1 strain shows considerable potential in breaking down aliphatic hydrocarbon compounds of already utilize motor oil (Husaini et al., 2008). "*n*-Eicosane" is considered as most common aliphatic hydrocarbon pollutants, and *Trichoderma* sp. S019 is the desirable type of strain, which can degrade *n*-eicosane, releasing nonadecanoic acid, *n*-octadecane, hexadecanoicacid, oleic acid and stearic acid as reaction products (Hadibarata et al., 2007). Additionally, *Gliocladium roseum*, *Penicillium brevicompactum*, *Penicillium funiculosum*, *Phialophora fastigiata*, and *Verticillium lecanii* are used for degenerating Imidazolium compounds (ICs) and quaternary ammonium compounds (QACs) (Zabielska-Matejuk and Czaczyk, 2006). Due to modern society, pollution level has risen up day by day and different water bodies has become contaminated. The percentage of available drinking water has reduced and an important source of water, i.e., groundwater is not fit for use in some sites due to the emission of pollutants on the ground. Carbon tetrachloride (CT), trichloroethylene (TCE), and perchloroethylene (PCE) are the most common pollutants in groundwater. Carbon tetrachloride is a clear, colorless, and volatile hydrocarbon. It is added as a solvent for fats and oils, used as a refrigerant, and as a dry cleaning agent. Exposure and inhalation of its vapors cause defects in the activity of the central nervous system and cause deterioration of the liver and kidneys. Trichloroethylene (TCE) and perchloroethylene (PCE) are two chlorinated hydrocarbons. They are widely used in industry as degreasers, dry cleaning agents, paint removers, solvents for chemical extraction, and components of adhesives and lubricants. Most significant exposure to TCE and PCE occurs in the workplace. The inhalation of tetrachloroethylene cause irritation of the upper respiratory tract and eyes, kidney dysfunction, and neurologically. *T. versicolor* was used for an experiment to degrade PCE and TCE under aerobic condition and resulted in the formation of trichloroacetic acid (TCA) from PCE while 2,2,2-trichloroethanol and CO_2 are released as main byproducts from TCE degradation. Furthermore, experiment on using fungi *Ganoderma lucidum* and *I. lacteus* to degrade perchloroethylene (PCE) and trichloroethylene (TCE) in pure culture were also reported (Marco-Urrea et al., 2007). Other than these fungi *Bjerkandera adusta*, WRF were found to have the capacity on degrading hexachlorocyclohexane (HCH) isomers discharging 1-(3-chloro-4-methoxyphenyl) ethanone and (2,4-dichloro-3-methoxy)-1-benzenecarbonyl chloride compounds. In addition, *Botrytis cinerea*, *Curvularia lunata* AS3.3589, and *Absidia coerulea* CICC40302 were also reported for utilizing in biotransformation products (Daoubi et al., 2006).

2.5 Advanced technologies used in fungal bioremediation

Significant progress has been achieved in molecular biology related to fungi, related to the extraction of genetic material (RNA and DNA), gene cloning, and genetic engineering of fungi. The development of biotechnology for using WRF for environmental pollution control has been implemented to treat various refractory wastes and to remediate contamination (Gao et al., 2010).

Advancement in biotechnology, enzymology, and molecular biology is the reason for the development of fungi improved through genetic engineering and enzymes for mycoremediation. Most of which is to add desired metabolic pathways for enzyme production by manipulating the strain using molecular toolboxes and genomic sequences. Fungi have a great advantage over other organisms for use in bioremediation as they are easier to genetically engineered, transport, and scale up (Obire et al., 2008).

Fungal genes can be cloned to meet the objectives of mycoremediation. Fungal mutants that oversecrete specific enzymes can be produced, and various processes using such mutants may be designed and scaled up in the treatment of wastes and wastewaters. Fungal protoplasts can be exploited to enhance processes related to mycoremediation. Thirty fungal species are screened for genes that encode lignin peroxidase and a dendrogram illustration of sequence relationship among 32 fungal peroxidases has been presented (Singh, 2006). A solid foundation for work in the disciplines of agriculture, industry, medicine, and remediation has been laid by genomic sequencing. In a paper for fungal comparative genomics, the fungal genome initiative (FGI) steering committee identified a coherent set of 44 fungi as immediate targets for sequencing (Birren et al., 2003).

Another advanced method is the application of nanotechnology for fungal bioremediation, fungi like *Aspergillus*, *Fusarium*, *Penicillium*, and *Verticillium* are used for metal nanoparticles. The production of metals like silver, gold, gold-silver alloy, platinum, selenium, tellurium, palladium, silica, titanium, quantum dots, zirconium, usnic acid, magnetite, uraninite, and cadmium telluride nanoparticles through the process of mycosynthesis has been reported. It has been observed that fungal enzymes interact with metal ions and reduce to form metal nanoparticles (Mukherjee et al., 2002; Gholami-Shabani et al., 2016).

There are growing needs for remediation of the different increasing pollutants daily for which bioremediation is one of the methods that is popular today. Bioremediation using fungi is also very effective in the field. The advancement in the field is of utmost essential today. Hence, further research on the potential and incorporation of fungi in bioremediation is required. The prime factors affecting the microbial bioremediation are air (oxygen) availability, moisture content, nutrient levels, matrix pH, and ambient temperature. Nutrient availability, oxygenation, and the presence of other inhibitory contaminants can play an important role in determining the suitability of bioremediation, but these are more specific to the individual application. Bioremediation tends to rely on the natural abilities of indigenous soil organisms and so treatment can occur between 0°C and 50°C, for the greatest efficiency but the ideal range for the bioremediation process is around 20–30°C, as this tends to optimize enzyme activity and pH of 6.5–7.5 is the optimum ranges which depend on the individual species involved. Sands and gravels are the most suitable soil types for the bioremediation of soil pollutants while compare to heavy clays and those with a high organic content, like peaty soil.

2.6 Factors affecting bioremediation

2.6.1 Nutrient availability

Inorganic nutrients such as nitrogen and phosphorus are necessary for microbial activity and cell growth for the proper functioning of the biodegradation process. Treating petroleum-contaminated soil with nitrogen can increase cell growth rate and help to maintain microbial populations at high activity levels and thus increase the rate of hydrocarbon degradation. However, excessive amounts of nitrogen in soil may also cause microbial inhibition.

2.6.2 Moisture content

Moisture is needed for cell growth and function of all the soil microorganisms. Soil affects the circulation of water and soluble nutrients into and out of microorganism cells. However excess moisture reduces the amount of available oxygen for aerobic respiration.

2.6.3 Temperature

Temperature is also one of the important factors, which influence the rate of biodegradation by controlling the rate of enzymatic reactions within microorganisms. An enzymatic reaction in the cell approximately doubles for each 10°C rise in temperature. Temperature also has a direct effect on the log growth rate of the microorganisms, in addition to the degradation rate of the hydrocarbons, depending upon their specific characteristics. Temperature affects bioremediation by changing the properties and viscosity in the case of oil, causing the oil to thicken which changes the toxicity and solubility of the oil, depending upon its chemical content (Amnan, 2010). The surface area of the oil is also a significant factor in the success of bioremediation. Soil bacteria degrade petroleum hydrocarbons, which have an optimum temperature ranging from 2°C to 45°C. Thermophilic bacteria which are normally found in hot springs and compost loads exist indigenously in cool soil environments and can be activated to degrade hydrocarbons with an increase in temperature to 60°C. The best range of pH is around 6.5–7.5 to ensure good microbial growth and timely biodegradation.

2.6.4 Microbial population

Microorganisms (fungi and some bacteria) are the main biotic factor for proper biodegradation of all types of contaminants.

2.6.5 Oxygen

Oxygen is required for aerobic *biodegradation* (about 2% oxygen in the gas phase or 0.4 mg/L in the soil or water).

2.6.6 Water

Water is also one of the important facts for the proper degradation of soil pollutants (pollutants 50%-70% of the water holding capacity of the soil).

2.6.7 Energy sources

One of the primary sources for the activity of bacteria and fungi is the availability of reduced organic materials to serve as energy sources. Whether a contaminant is used as an energy source for an aerobic heterotrophic organism which can be the function of the average oxidation state of the carbon in the material. The outcome of every degradation process depends on microbial (biomass concentration, population diversity, enzyme activities), substrate (physio-chemical characteristics, molecular structure, and concentration), availability of electron acceptors, and carbon and energy sources.

2.6.8 Bioavailability

The rate of conversion of contaminants during bioremediation depends on the rate of contaminant uptake and metabolism and the rate of transfer to the cell. The bioavailability of a contaminant is controlled by a number of physio-chemical processes such as sorption and desorption, diffusion, and dissolution). A reduced bioavailability of contaminants in soil is caused by the slow mass transfer to the degrading microbes. Contaminants become unavailable when the rate of mass transfer is 0. The decrease of the bioavailability in the course of time is often referred to as aging or weathering. These bioavailability problems can be overcome by the use of food-grade surfactants (Boopathy and Manning, 1999), which increase the availability of contaminants for microbial degradation.

2.6.9 Bioactivity

The bioactivity is defined as the operating state of microbiological processes. Improving bioactivity implies that system conditions are adjusted to optimize biodegradation (Blackburn and Hafker, 1993). In nature, the ability of organisms to transfer contaminants to both simpler and more complex molecules is very diverse. Favorable or unfavorable biochemical conversions are evaluated in terms of whether individual or groups of parent compounds are removed, whether increased toxicity is a result of the bioremediation process, and sometimes whether the elements in the parent compound are converted to measurable metabolites. These biochemical activities can be controlled in an in situ operation when one can control and optimize the conditions to achieve a desirable result.

2.7 Conclusion

Different groups of microbial flora are exploited for bioremediation studies. Among those groups, fungi are one of the potential remediators owing to their effective and fast decomposition of different sources of pollutants. A variety of fungal, bacterial, and other soil microbe species have been studied for reclaiming polluted environments, while their combined or sole interaction with environmental factors in in situ conditions needs a thorough study on how it actually works in the bioremediation process. However, a multidisciplinary approach involving scientific researchers, governments, industrialists, and climate activists needs to go hand in hand to offer a more sustainable yet highly efficient bioremediation strategy.

2.8 Future prospects

Bioremediation is one of the most effective tools for managing polluted soil and water system in the environment as it is target specific and nonharmful in nature. Effective destruction of organic and inorganic components can be achieved in no time. Agricultural waste management including pesticides, fungicides, and herbicides is made meticulously effective through mycoremediation. Also, many bacteria apart from fungi have also been reported to give positive impacts on bioremediation. However, a combination effect of fungi and bacteria, more species of fungi, bacteria, and other microbes need to be explored in the near future to broaden the impact of microbial remediation. Moreover, successful laboratory experiments need to be practically executed in contaminated environments.

References

Adenipekun, C.O., Fasidi, I.O., 2005. Bioremediation of oil polluted soil by *Lentinus subnudus*, a Nigerian white rot fungus. Afr. J. Biotechnol. 4 (8), 796–798.

Adenipekun, C.O., Isikhuemhen, O.S., 2008. Bioremediation of engine oil polluted soil by the tropical white-rot fungus, *Lentinus squarrosulus* Mont. (Singer). Pak. J. Biol. Sci. 11 (12), 1634–1637.

Adenipekun, C.O., Lawal, Y., 2011. Mycoremediation of crude oil and palm kernel contaminated soils by *Pleurotus pulmonarius* Fries (Quelet). Nat. Sci. 9 (9), 125–131.

Aitken, M.B., Irvine, R.L., 1989. Stability testing of ligninase and Mn peroxidase from *Phanerochaete chrysosporium*. Biotechnol. Bioeng. 34, 1251–1260.

Akhtar, S., Mahmood-ul-Hassan, M., Ahmad, R., Suthor, V., Yasin, M., 2013. Metal tolerance potential of filamentous fungi isolated from soils irrigated with untreated municipal effluent. Soil Environ. 32, 55–62.

Amaral, P.F.F., Fernandes, L.A.F.D., Tavares, A.P.M., Xavier, A.M.R., Cammarota, M.C., Coutinho, J.A.P., Coelho, M.A.Z., 2004. Decolorization of dyes from textile wastewater by *Trametes versicolor*. Environ. Technol. 25 (11), 1313–1320.

Amnan, A., 2010. What Is Bioremediation—Various Factors Involved in the Process. Biotech Articles.

Anastasi, A., Tigini, V., Varese, G.C., 2013. The bioremediation potential of different ecophysiological groups of fungi. In: Fungi as Bioremediators. Springer, Berlin, Heidelberg, pp. 29–49.

Arthur, E.L., Rice, P.J., Rice, P.J., Anderson, T.A., Baladi, S.M., Henderson, K.L.D., Coats, J.R., 2005. Phytoremediation-an overview. Crit. Rev. Plant Sci. 24 (2), 109–122.

Arun, A., Raja, P.P., Arthi, R., Ananthi, M., Kumar, K.S., Eyini, M., 2008. Polycyclic aromatic hydrocarbons (PAHs) biodegradation by basidiomycetes fungi, *Pseudomonas* isolate, and their cocultures: comparative in vivo and in silico approach. Appl. Biochem. Biotechnol. 151 (2), 132–142.

Aust, S.D., Swaner, P.R., Stahl, J.D., 2003. Detoxification and metabolism of chemicals by white-rot fungi. In: Zhu, J.J.P.C., Aust, S.D., Lemley Gan, A.T. (Eds.), Pesticide Decontamination and Detoxification. Oxford University Press, Washington, DC, pp. 3–14.

Azadpour, A., Powell, P.D., Matthews, J., 1997. Use of lignin degrading fungi in bioremediation. Remediation 997, 25–49.

Badia-Fabregat, M., Lucas, D., Gros, M., Rodríguez-Mozaz, S., Barceló, D., Caminal, G., Vicent, T., 2015. Identification of some factors affecting pharmaceutical active compounds (PhACs) removal in real wastewater. Case study of fungal treatment of reverse osmosis concentrate. J. Hazard. Mater. 283, 663–671.

Baldrian, P.C., Der Viesche, C., Gabriel, S., Nerud, F., Zadrazil, F., 2000. Influence of cadmium and mercury on activities of ligninolytic enzymes and degradation of polycyclic aromatic hydrocarbons by *Pleurotus ostreatus* in soil. Appl. Environ. Microbiol. 66, 2471–2478.

Barr, B.P., Aust, D., 1994. Mechanisms of white-rot fungi use to degrade pollutant. Environ. Sci. Technol. 28, 78–87.

Bennet, J.W., Connick, W.J., Daigle, D., Wunch, K., 2001. Formulation of fungi for In Situ Bioremediation. In: Gadd, G.M. (Ed.), Fungi in Bioremediation. British Mycological Society Symposium Series, pp. 97–108.

Bennett, R.M., Cordero, P.R.F., Bautista, G.S., Dedeles, G.R., 2013. Reduction of hexavalent chromium using fungi and bacteria isolated from contaminated soil and water samples. Chem. Ecol. 29 (4), 320–328.

Birren, B., Fink, G., Lander, E., 2003. Fungal Genome Initiative: A White Paper for Fungal Comparative Genomics. Center for Genome Research, Cambridge, MA.

Blackburn, J.W., Hafker, W.R., 1993. The impact of biochemistry, bioavailability, and bioactivity on the selection of bioremediation technologies. Trends Biotechnol. 11, 328–333.

Bojan, B.W., Lamar, R.T., Burjus, W.D., Tien, M., 1999. Extent of humification of anthrecene, fluoranthene adbenzo (a) pyrene by *Pleurotus ostreatus* during growth in PAH-contaminated soils. Lett. Appl. Microbial 28, 250–254.

Boopathy, R., Manning, J., 1999. Surfactant-enhanced bioremediation of soil contaminated with 2,4,6-trinitrotoluene in soil slurry reactors. Water Environ. Res. 71, 119–124.

Bukert, B., Robson, A., 1994. Zinc uptake in subterranean clover (*Trifolium subterraneum* L.) by three vesicular arbuscular mycorrhizal fungi in a root free sandy soil. Soil Biol. Biochem. 26, 1117–1124.

Bumpus, J.A., 1989. Biodegradation of polycyclic aromatic hydrocarbons by *Phanerochaete chrysosporium*. Appl. Environ. Microbiol. 55, 154–158.

Camarero, S., Sarkar, S., Ruiz-Dueñas, F.J., Martínez, M.J., Martinez, A.T., 1999. Description of a versatile peroxidase involved in natural degradation of lignin that has both Mn-peroxidase and lignin-peroxidase substrate binding sites. J. Biol. Chem. 274, 10324–10330.

Chakraborty, S., Mukherjee, A., Das, T.K., 2013. Biochemical characterization of a lead-tolerant strain of *Aspergillus foetidus*: an implication of bioremediation of lead from liquid media. Int. Biodeterior. Biodegradation 84, 134–142.

Chaney, R.L., Malik, M., Li, Y.M., Brown, S.L., Brewer, E.P., Angle, J.S., Baker, A.J.M., 1997. Phytoremediation of soil metals. Curr. Opin. Biotechnol. 8 (3), 279–284.

Cho, U.H., Park, J.O., 2000. Mercury-induced oxidative stress in tomato seedlings. Plant Sci. 156, 1–9.

Cunningham, S.D., Berti, W.R., Huang, J.W., 1995. Phytoremediation of contaminated soils. Trends Biotechnol. 13 (9), 393–397.

Da Silva, M., Cerniglia, C.E., Pothuluri, J.V., Canhos, V.P., Esposito, E., 2003. Screening filamentous fungi isolated from estuarine sediments for the ability to oxidize polycyclic aromatic hydrocarbons. World J. Microbiol. Biotechnol. 19 (4), 399–405.

Damare, S., Singh, P., Raghukumar, S., 2012. Biotechnology of marine fungi. Prog. Mol. Subcell. Biol. 53, 277–297.

Damisa, D., Oyegoke, T.S., Ijah, U.J.J., Adabara, N.U., Bala, J.D., Abdulsalam, R., 2013. Biodegradation of petroleum by fungi isolated from unpolluted tropical soil. Int. J. Appl. Biol. Pharm. Technol. 4, 136–140.

Daoubi, M., Durán-Patrón, R., Hernández-Galán, R., Benharref, A., Hanson, J.R., Collado, I.G., 2006. The role of botrydienediol in the biodegradation of the sesquiterpenoid phytotoxin botrydial by Botrytis cinerea. Tetrahedron 62 (35), 8256–8261.

Deshmukh, R., Khardenavis, A.A., Purohit, H.J., 2016. Diverse metabolic capacities of fungi for bioremediation. Indian J. Microbiol. 56 (3), 247–264.

Dhankher, O.P., Pilon-Set, E.A.H., Meagher, R.B., Doty, S., 2012. Biotechnological approaches for phytoremediation. In: Plant Biotechnology and Agriculture. Academic Press, Elsevier Inc, pp. 309–328.

Dhiman, S.S., Haw, J.R., Kalyani, D., Kalia, V.C., Kang, Y.C., Lee, J.K., 2015. Simultaneous pretreatment and saccharification: green technology for enhanced sugar yields from biomass using a fungal consortium. Bioreosur. Technol. 179, 50–57.

Di Toro, D.M., McGrath, J.A., Hansen, D.J., 2000. Technical basis for narcotic chemicals and polycyclic aromatic hydrocarbon criteria. I. Water and tissue. Environ. Toxicol. Chem. Int. J. 19 (8), 1951–1970.

Donnelly, P.K., Fletcher, J.S., 1995. PCB metabolism by ectomycorrhizal fungi. Bull. Environ. Contam. Toxicol. 54 (4), 507–513.

Donnelly, P.K., Entry, J.A., Crawford, D.L., 1993. Degradation of atrazine and 2, 4-dichlorophenoxyacetic acid by mycorrhizal fungi at three nitrogen concentrations in vitro. Appl. Environ. Microbiol. 59 (8), 2642–2647.

dos Santos Bazanella, G.C., Araújo, C.A.V., Castoldi, R., Maria, G., Maciel, F.D.I., Marques, C.G., Peralta, R.M., 2016. Ligninolytic enzymes from white-rot fungi and application in the removal of synthetic dyes. In: Fungal Enzymes. CRC Press, Boca Raton, FL, p. 258.

Doty, S.L., 2008. Tansley Review: enhancing phytoremediation through the use of transgenics and endophytes. New Phytol. 179, 318–333.

Elshafie, A., AlKindi, A.Y., Al-Busaidi, S., Bakheit, C., Albahry, S.N., 2007. Biodegradation of crude oil and n-alkanes by fungi isolated from Oman. Mar. Pollut. Bull. 54 (11), 1692–1696.

Ensley, B.D., Raskin, I., 1999. Rationale for use of phytoremediation. In: Phytoremediation of Toxic Metals: Using Plants to Clean Up the Environment. Wiley, New York.

Gamelas, J.A.F., Tavares, A.P.M., Evtuguin, D.V., Xavier, M.R.B., 2005. Oxygen bleaching of Kraft pulp with polyoxometaltes and laccase applying a novel multi-stage process. J. Mol. Catal. B Enzym. 33 (36), 57–64.

Gao, D., Du, L., Yang, J., Wu, W.M., Liang, H., 2010. A critical review of the application of white rot fungus to environmental pollution control. Crit. Rev. Biotechnol. 30, 70–77.

Garg, N., Bhandari, P., 2014. Cadmium toxicity in crop plants and its alleviation by arbuscular mycorrhizal (AM) fungi: an overview. Plant Biosyst. 148, 609–621.

Gazem, M.A., Nazareth, S., 2013. Sorption of lead and copper from an aqueous phase system by marine-derived *Aspergillus* species. Ann. Microbiol. 63 (2), 503–511.

Gholami-Shabani, M., Shams-Ghahfarokhi, M., Gholami-Shabani, Z., Razaghi-abyaneh, M., 2016. Microbial enzymes: current features and potential applications in nanobiotechnology. In: Prasad, R. (Ed.), Advances and Applications Through Fungal Nanobiotechnology. Springer, Heidelberg.

Guillén, F., Martinez, A.T., Martinez, M.J., 1992. Substrate specificity and properties of the arylalcohol oxidase from the ligninolytic fungus *Pleurotus eryngii*. Eur. J. Biochem. 209, 603–611.

Guiraud, P., Bonnet, J.L., Boumendjel, A., Kadri-Dakir, M., Dusser, M., Bohatier, J., Steiman, R., 2008. Involvement of *Tetrahymena pyriformis* and selected fungi in the elimination of anthracene, and toxicity assessment of the biotransformation products. Ecotoxicol. Environ. Saf. 69 (2), 296–305.

Gunarathne, V., Mayakaduwa, S., Ashiq, A., Weerakoon, S.R., Biwas, J.K., Vithanage, M., 2019. Transgenic plants: benefits, application and potential risks in phytoremediation. In: Transgenic Plant Technology for Remediation of Toxic Metals and Metalloids. Academic Press, pp. 89–102.

Hadibarata, T., Tachibana, S., Itoh, K., 2007. Biodegradation of n-eicosane by fungi screened from nature. Pak. J. Biol. Sci. 10 (11), 1804–1810.

Hadibarata, T., Teh, Z.C., Zubir, M.M., Khudhair, A.B., Yusoff, A.R., Salim, M.R., Hidayat, T., 2013a. Identification of naphthalene metabolism by white-rot fungus *Pleurotus eryngii*. Bioprocess Biosyst. Eng. 24, 728–732.

Hadibarata, T., Zubir, M.M., Rubiyabto, T.Z., Chuang, T.Z., Yusoff, A.R., Fulazzaky, M.A., Seng, B., Nugroho, A.E., 2013b. Degradation and transformation of anthracene by white-rot fungus *Armillaria* sp. F022. Folia Microbiol. 58, 385–391.

Hatakka, A., 2001. Biodegradation of lignin. In: Hofrichter, M., Steinbüchel, A. (Eds.), Biopolymers. vol. 1. Wiley-VCH, Weinheim.

Hawksworth, D.L., Kirk, P.M., Sutton, B.C., Pegler, D.N., 1995. Ainsworth DN and Bisby's Dictionary of the fungi. CAB, Oxon.

Hildebrandt, U., Regvar, M., Bothe, H., 2007. Arbuscular mycorrhiza and heavy metal tolerance. Phytochemistry 68 (1), 139–146.

Hodgson, J., Rho, D., Guiot, S.R., Ampleman, G., Thiboutot, S., Hawari, J., 2000. Tween 80 enhanced TNT mineralization by *Phanerochaete chrysosporium*. Can. J. Microbiol. 46, 110–118.

Husaini, A., Roslan, H.A., Hii, K.S.Y., Ang, C.H., 2008. Biodegradation of aliphatic hydrocarbon by indigenous fungi isolated from used motor oil contaminated sites. World J. Microbiol. Biotechnol. 24 (12), 2789–2797.

Hussein, S., Ruiz, O.N., Terry, N., Daniell, H., 2007. Phytoremediation of mercury and organomercurials in chloroplast transgenic plants: enhanced root uptake, translocation to shoots and volatilization. Environ. Sci. Technol. 41, 8439–8446.

Isikhuemhen, O., Anoliefo, G., Oghale, O., 2003. Bioremediation of crude oil polluted soil by the white rot fungus *Pleurotus tuber regium* (Fr.) Sing. Environ. Sci. Pollut. Res. 10, 108–112.

Isikhuemhen, O.S., Adenipekun, C.O., Ohimain, E.I., 2010. Preliminary studies on mating and improved selection in the tropical culinary medicinal mushroom *Lentinus squarrosulus* Mont. (Agaricomycetideae). Int. J. Med. Mushrooms 12 (2), 177–183.

Jagtap, V.S., Sonawane, V.R., Pahuja, D.N., Rajan, M.G., Rajashekharrao, B., Samuel, A.M., 2003. An effective and better strategy for reducing body burden of radio strontium. J. Radiol. Prot. 23 (3), 317–326.

Janveja, C., Rana, S.S., Soni, S.K., 2013. Kitchen waste residues as potential renewable biomass resources for the production of multiple fungal carbohydrases and second generation bioethanol. J. Technol. Innov. Renew. Energy 2 (2), 186–200.

Jones, P., Kortenkamp, A., Brien, P.O., Wang, G., Yang, G., 1991. Evidence for the generation of hydroxyl radicals from a chromium (V) intermediate isolated from the reaction of chromate with glutathione. Arch. Biochem. Biophys. 286, 652–655.

Juhasz, A.L., Naidu, R., 2000. Bioremediation of high molecular weight polycyclic aromatic hydrocarbons: a review of the microbial degradation of benzo [a] pyrene. Int. Biodeterior. Biodegradation 45 (1–2), 57–88.

Kaldorf, M., Kuhun, A.J., Schroder, W.H., Hilderbrandt, U., Bothe, H., 1999. Selective element deposits in maize colonized by a heavy metal tolerance conferring arbuscular mycorrhizal fungus. J. Plant Physiol. 154, 718–728.

Key, S., Ma, J.K.-C., Drake, P.M.W., 2008. Genetically modified plants and human health. J. R. Soc. Med. 101 (6), 290–298.

Kirk, T.K., Lamar, R.T., Glaser, J.A., 1992. The potential of white-rot fungi in bioremediation. In: Mongkolsuk, S., et al. (Ed.), Biotechnology and Environmental Science: Molecular Approaches. Plenum Press, New York, pp. 131–138.

Kristanti, R.A., Hadibarata, T., Toyama, T., Tanaka, Y., Mori, K., 2011. Bioremediation of crude oil by white rot fungi *Polyporus* sp. S133. J. Microbiol. Biotechnol. 21 (9), 995–1000.

Kulshrestha, S., Mathur, N., Bhatnagar, P., 2014. Mushroom as a product and their role in mycoremediation. AMB Express 4 (29), 1–7. Springer.

Kurniati, E., Arfarita, N., Imai, T., Higuchi, T., Kanno, A., Yamamoto, K., Sekine, M., 2014. Potential bioremediation of mercury-contaminated substrate using filamentous fungi isolated from forest soil. J. Environ. Sci. 26, 1223–1231.

Lee, J.W., Lee, S.M., Hong, E.J., Jeung, E.B., Kang, H.Y., Kim, M.K., Choi, I.G., 2006. Estrogenic reduction of styrene monomer degraded by *Phanerochaete chrysosporium* KFRI 20742. J. Microbiol. 44 (2), 177–184.

Lee, S.M., Lee, J.W., Koo, B.W., Kim, M.K., Choi, D.H., Choi, I.G., 2007. Dibutyl phthalate biodegradation by the white rot fungus, *Polyporus brumalis*. Biotechnol. Bioeng. 97 (6), 1516–1522.

Leontievsky, A.A., Vares, T., Lankinen, P., Shergill, J.K., Pozdnyakova, N.N., Myasoedova, N.M., Kalkkinen, N., Golovleva, L.A., Cammack, R., Thurston, C.F., Hatakka, A., 1997. Blue and yellow laccases of ligninolytic fungi. FEMS Microbiol. Lett. 156, 9–14.

Li, X.L., Marschner, H., George, E., 1991. Acquisition of phosphorus and copper by VA-mycorrhizal hyphae and root to shoot transport in white clover. Plant Soil 136, 49–57.

Li, H., Lan, W., Lin, Y., 2005. Biotransformation of 1,2,3,4-tetrahydronaphthalene by marine fungus *Hypoxylon oceanicum*. Fenxi Ceshi Xuebao 24 (4), 45–47.

Lladó, S., Covino, S., Solanas, A.M., Viñas, M., Petruccioli, M., D'annibale, A., 2013. Comparative assessment of bioremediation approaches to highly recalcitrant PAH degradation in a real industrial polluted soil. J. Hazard. Mater. 248, 407–414.

Marco-Urrea, E., Caminal, G., Gabarrell, X., Vicent, T., Reddy, C.A., 2007. Aerobic degradation/mineralization of trichloroethylene and perchloroethylene by white-rot fungi. In: Proceedings of the International In Situ and On-Site Bioremediation Symposium, United States, 2007, pp. H44/1–H44/6.

Marco-Urrea, E., García-Romera, I., Aranda, E., 2015. Potential of non-ligninolytic fungi in bioremediation of chlorinated and polycyclic aromatic hydrocarbons. New Biotechnol. 2 (6), 620–628.

Martin, C., Moeder, M., Daniel, X., Krauss, G., Schlosser, D., 2007. Biotransformation of the polycyclic musks HHCB and AHTN and metabolite formation by fungi occurring in freshwater environments. Environ. Sci. Technol. 41 (15), 5395–5402.

Martins, M.R., Pereira, P., Lima, N., Cruz-Morais, J., 2013. Degradation of Metalaxyl and Folpet by filamentous fungi isolated from Portuguese (Alentejo) vineyard soils. Arch. Environ. Contam. Toxicol. 65, 67–77.

Meagher, R.B., 2000. Phytoremediation of toxic elemental and organic pollutants. Curr. Opin. Plant Biol. 3 (2), 153–162.

Michaels, G.B., Lewis, D.L., 1985. Sorption and toxicity of azo and triphenylmethane dyes to aquatic microbial populations. Environ. Toxicol. Chem. 4, 45–50.

Mougin, C., Laugero, C., Asther, M., Dubroca, J., Frasse, P., 1994. Biotransformation of the herbicide atrazine by the white-rot fungus *Phanerochaete chrysosporium*. Appl. Environ. Microbiol. 60, 705–708.

Mouhamadou, B., Faure, M., Sage, L., Marçais, J., Souard, F., Geremia, R.A., 2013. Potential of autochthonous fungal strains isolated from contaminated soils for degradation of polychlorinated biphenyls. Fungal Biol. 117, 268–274.

Mukherjee, P., Senapati, S., Mandal, D., Ahmad, A., Khan, M.I., Kumar, R., Sastry, M., 2002. Extracellular synthesis of gold nanoparticles by the fungus *Fusarium oxysporum*. ChemBiolChem 3, 461–463.

Nayak, V., Pai, P.V., Pai, A., Pai, S., Sushma, Y.D., Rao, C.V., 2013. A comparative study of caffeine degradation by four different fungi. Bioremediat. J. 17, 79–85.

Nigam, P., Banat, I.M., McMullan, G., Dalel, S., Marchant, R., 1995. Microbial degradation of textile effluent containing Azo, Diazo and reactive dyes by aerobic and anaerobic bacterial and fungal cultures. In: 36th Annual Conference AMI, Hisar, pp. 37–38.

Novotny, C., Svobodova, K., Erbanova, P., Cajthaml, T., Kasinath, A., Lange, E., Sasek, V., 2004. Ligninolytic fungi in bioremediation: extracellular enzyme production and degradation rate. Soil Biol. Biochem. 36 (10), 1545–1551.

Obire, O.E., Anyanwu, C., Okigbo, R.N., 2008. Saprophytic and crude oil-degrading fungi from cow dung and poultry droppings as bioremediating agents. Int. J. Agric. Technol. 4, 81–89.

Ogbo, E.M., Okhuoya, J.A., 2008. Biodegradation of aliphatic, aromatic, resinic and asphaltic fractions of crude oil contaminated soils by *Pleurotus tuber-regium* Fr. Singer—a white rot fungus. Afr. J. Biotechnol. 7 (23), 4291–4297.

Ogunseitan, O.A., 2003. Biotechnology and industrial ecology: new challenges for a changing global environment. Afr. J. Biotechnol. 2 (12), 596–601.

Paul, E.A., Clark, F.E., 1989. Soil Microbiology and Biochemistry. Academic Press, Inc., San Diego, CA, p. 273.

Pilon-Smits, E., 2005. Phytoremediation. Annu. Rev. Plant Biol. 56, 15–39.

Pinto, A.P., Serrano, C., Pires, T., Mestrinho, E., Dias, L., Teixeira, D.M., Caldeira, A.T., 2012. Degradation of terbuthylazine, difenoconazole and pendimethalin pesticides by selected fungi cultures. Sci. Total Environ. 435, 402–410.

Rengarajan, T., Rajendran, P., Nandakumar, N., Lokeshkumar, B., Rajendran, P., Nishigaki, I., 2015. Exposure to polycyclic aromatic hydrocarbons with special focus on cancer. Asian Pac. J. Trop. Biomed. 5 (3), 182–189.

Rodríguez-Rodríguez, C.E., Castro-Gutiérrez, V., Chin-Pampillo, J.S., Ruiz-Hidalgo, K., 2013. On-farm biopurification systems: role of white-rot fungi in depuration of pesticide-containing wastewaters. FEMS Microbiol. Lett. 345, 1–12.

Rosales, E., Pazos, M., Ángeles, S.M., 2013. Feasibility of solid-state fermentation using spent fungi-substrate in the biodegradation of PAHs. Clean Soil Air Water 41, 610–615.

Sack, U., Gunther, T., 1993. Metabolism of PAH by fungi and correction with extracellular enzymatic activities. J. Basic Microbiol. 33, 269–277.

Sanyal, P., Samaddar, P., Paul, A.K., 2006. Degradation of poly (3-hydroxybutyrate) and poly (3-hydroxybutyrate-co-3-hydroxyvalerate) by some soil *Aspergillus* spp. J. Polym. Environ. 14 (3), 257–263.

Selvam, K., Swaminathan, K., Song, M.H., Chae, K.S., 2002. Biological treatment of a pulp and paper industry effluent by *Fomes lividus* and *Trametes versicolor*. World J. Microbiol. Biotechnol. 18 (6), 523–526.

Silambarasan, S., Abraham, J., 2013. Ecofriendly method for bioremediation of chlorpyrifos from agricultural soil by novel fungus *Aspergillus terreus* JAS1. Water Air Soil Pollut. 224, 1369.

Singh, H., 2006. Mycoremediation: Fungal Bioremediation. Wiley, New York.

Singh, O.V., Labana, S., Pandey, G., Budhiraja, R., Jain, R.K., 2003. Phytoremediation: an overview of metallic ion decontamination from soil. Appl. Microbiol. Biotechnol. 161 (5–6), 405–412.

Soares, A., Jonasson, K., Terrazas, E., Guieysse, B., Mattiasson, B., 2005. The ability of white-rot fungi to degrade the endocrine-disrupting compound nonylphenol. Appl. Microbiol. Biotechnol. 66 (6), 719–725.

Soobhany, N., Mohee, R., Garg, V.K., 2015. Comparative assessment of heavy metals content during the composting and vermicomposting of municipal solid waste employing *Eudrilus eugeniae*. Waste Manag. 39, 130–145.

Spadaro, J.T., Gold, M.H., Renganathan, V., 1992. Degradation of azo dyes by the lignin-degrading fungus *Phanerochaete chrysosporium*. Appl. Environ. Microbiol. 58, 2397–2401.

Speight, J.G., 2016. Environmental Organic Chemistry for Engineers. Butterworth-Heinemann.

Sylvia, D.M., Hung, L.L., Graham, J.H., 1987. Mycorrhizae in the next decade: practical applications and research priorities: proceedings of the 7th North American Conference on Mycorrhizae, May 3-8, 1987, Gainesville, FL, USA. In: 7th North American Conference on Mycorrhizae, Gainesville, Fla (USA). Institute of Food and Agricultural Sciences, University of Florida.

Tanaka, H., Itakura, S., Enoki, A., 1999. Hydroxyl radical generation by an extracellular low-molecular-weight substance and phenol oxidase activities during wood degradation by the white-rot basidiomycetes *Trametes versicolor*. J. Biotechnol. 75 (1), 57–70.

Tortella, G.R., Diez, M.C., Durán, N., 2005. Fungal diversity and use in decomposition of environmental pollutants. Crit. Rev. Microbiol. 31 (4), 197–212.
Trojanowski, J., Haider, K., Hüttermann, A., 1984. Decomposition of 14 C-labelled lignin, holocellulose and lignocellulose by mycorrhizal fungi. Arch. Microbiol. 139 (2), 202–206.
Valentin, L., Feijoo, G., Moreira, M.T., Lema, J.M., 2006. Biodegradation of polycyclic aromatic hydrocarbons in forest and salt marsh soils by white-rot fungi. Int. Biodeterior. Biodegradation 58 (1), 15–21.
Van Aken, B., 2008. Transgenic plants for phytoremediation: helping nature to clean up environmental pollution. Trends Biotechnol. 26 (5), 225–227.
Ventura, J., Belmares, R., Aguilera-Carbo, A., Gutiérrez-Sanchez, G., Rodríguez-Herrera, R., Aguilar, C.N., 2008. Fungal biodegradation of tannins from creosote bush (*Larrea tridentata*) and tar bush (*Fluorensia cernua*) for gallic and ellagic acid production. Food Technol. Biotechnol. 46 (2), 213–217.
Watanabe, M.E., 1997. Phytoremediation on the brink of commercialization. Environ. Sci. Technol. 31 (4), 182A–186A.
Wu, J., Zhao, Y., Liu, L., Fan, B., Li, M., 2013. Remediation of soil contaminated with decarbrominated diphenyl ether using white-rot fungi. J. Environ. Eng. Landsc. Manag. 21, 171–179.
Yang, Y., Liang, Y., Ghosh, A., Song, Y., Chen, H., Tang, M., 2015. Assessment of arbuscular mycorrhizal fungi status and heavy metal accumulation characteristics of tree species in a lead–zinc mine area: potential applications for phytoremediation. Environ. Sci. Pollut. Res. 22 (17), 13179–13193.
Yoshioka, T., Komuro, M., 2006. The biodegradation of the oil using fungi isolated from polluted oils. Kaijo Hoan Daigakko Kenkyu Hokoku, Rikogaku-kei 50 (1–2), 1–9.
Zabielska-Matejuk, J., Czaczyk, K., 2006. Biodegradation of new quaternary ammonium compounds in treated wood by mould fungi. Wood Sci. Technol. 40 (6), 461.
Zhou, J.L., 1999. Zn biosorption by *Rhizopus arrhizus* and other fungi. Appl. Microbiol. Biotechnol. 51, 686–693.

Further reading

Alloway, J.B., Ayres, D.C., 1993. Chemical Principles of Environmental Pollution. Chapman and Hall, London.
Del Val, C., Barea, J.M., Azcon-Aguilar, C., 1999. Assessing the tolerance to heavy metals of arbuscular mycorrhizal fungi isolated from sewage sludge-contaminated soils. Appl. Soil Ecol. 11, 261–269.
Joner, E.J., Leyval, C., 1997. Uptake of 109Cd by roots and hyphae of a *Glomus mosseae/Trifolium subterraneum* mycorrhiza from soil amended with high and low concentrations of cadmium. New Phytol. 135, 353–360.
Lux, H.B., Cumming, J.R., 2001. Mycorrhizae confer aluminum resistance to tulip-poplar seedlings. Can. J. For. Res. 31 (4), 694–702.
Masaphy, S., Levanon, D., Vaya, J., Henis, Y., 1993. Isolation and characterization of a novel atrazine metabolite produced by the fungus *Pleurotus pulmonarius*, 2-chloro-4-ethylamino-6-(1-hydroxyisopropyl) amino-1,3,5-triazine. Appl. Environ. Microbiol. 59, 4342–4346.
Prabu, P.C., Udayasoorian, C., 2005. Biodecolorization of phenolic paper mill effluent by Ligninolytic Fungus *Trametes versicolor*. J. Biol. Sci. 5 (5), 558–561.
Quintero, J.C., Lu-Chau, T.A., Moreira, M.T., Feijoo, G., Lema, J.M., 2007. Bioremediation of HCH present in soil by the white-rot fungus *Bjerkandera adusta* in a slurry batch bioreactor. Int. Biodeterior. Biodegradation 60 (4), 319–326.
Sasek, V., Cajthaml, T., 2005. Mycoremediation. Current state and perspectives. Int. J. Med. Mushrooms 7 (3), 360–361.
Van Aken, B., 2009. Transgenic plants for enhanced phytoremediation of toxic explosives. Curr. Opin. Biotechnol. 20 (2), 231–236.

CHAPTER

Role of fungi in biotechnology

3

Sunita Aggarwal and Arti Kumari
Department of Microbiology, Institute of Home Economics, University of Delhi, New Delhi, India

3.1 Introduction

Fungi are diverse, eukaryotic organisms, present primarily in terrestrial environments. About 90,000 fungal species are known and it is expected that the number of fungal species existing is 10 times more than the known ones (Willey et al., 2014). These have different shapes and sizes and are single celled (yeasts) or multicellular (mold) organisms. The fungal species are distributed widely, present in soil, mud, fresh water, plants, and animals to even extreme environments including polar regions to deep sea. Nutritionally, most of the fungi are saprophytic and are important decomposers. By secreting extracellular enzymes, these degrade dead organic matter in the environment to release inorganic and small organic molecules thus playing a significant role in the recycling of elements and nutrients. Many ascomycetes and basidiomycetes can degrade even many chemically stable and recalcitrant compounds. The capability of many chytrides to degrade keratin enables the degradation of crustacean exoskeletons. Approximately 5000 species are pathogenic and may be responsible for causing serious infections to plants, humans, and animals (Willey et al., 2014). Examples include *Aspergillus fumigatus* associated with asthma and sinusitis, mucormycetes with mucormycosis and *Clavicep purpurea* with ergotism (St. Anthony's fire). The toxins (aflatoxins) released by *Aspergillus flavus* can cause liver cancer. Few fungi are opportunists, for example, *Candida* sp. causing infection in immunocompromised individuals. Many fungi form ecologically beneficial associations in the form of lichens and mycorrhizae.

Fungi also have a value as model research organisms and being exploited by microbiologists, biochemists, genetics, and physiologists to unravel various fundamental biological processes. *Saccharomyces cerevisiae* is a model organism used to understand the cell cycle during mitosis and the process of development of cancer when control is lost on cell cycle. *Aspergillus nidulans* is used to reveal the questions related to development biology.

The fungi belong to six major groups: Chytridomycota, Zygomycota, Glomeromycota, Ascomycota, Basidiomycota, and Microsporida. During their life cycle, fungi produce two types of metabolites. While primary metabolites, for example, enzymes, fats, alcohols, and organic acids are necessary for the vegetative growth of fungi, secondary metabolites are produced during the stationary phase and are associated with differentiation and sporulation. Their production is by a restricted number of organisms and contains antibiotics, statins, alkaloids, cyclosporine, and many others. Indeed, the fungi have contributed enormously to human life. The fungi are used in various industrial fermentation processes to produce a number of valuable products of economic significance having numerous biotechnological

Entrepreneurship with Microorganisms. https://doi.org/10.1016/B978-0-443-19049-0.00009-8

and therapeutic applications. These include food products (beverages, cheese), enzymes (amylases, cellulases, proteases), organic acids (citric acid, gallic acid, malic acid), drugs (ergometrine, cortisone), antibiotics (penicillin, griseofulvin), pharmaceutical compounds, immunosuppressive drugs (cyclosporine), pigments, heterologous (Recombinant) proteins, vaccines (hepatitis B), vitamins (riboflavin, ascorbic acid), amino acids (tryptophan, phenylalanine), biofabrication, biofuels, polysaccharides, lipids, glycolipids, plant growth regulators, and so on (Mukherjee et al., 2018). The list is still expanding. Moreover, fungi find their applications in bioremediation and biotransformations. Recombinant DNA technology and molecular manipulations have led to increased product yield and the use of fungi as microbial cell factories. Now, mycotechnology occupies a significant place in the economies of many countries. In this chapter, we will deliberate the industrial application of different fungi and their biomolecules in human welfare.

3.2 Some important fungi and their role in the following topics under this chapter

3.2.1 Organic acid

Organic acids are compounds that are acidic in nature. They usually contain mild acidic groups like carboxylic acid which is linked with other compounds to become more advanced (Magnuson and Lasure, 2004). They could be readily recovered from natural resources like plants, animals, and microbes. However, their large-scale production is limited to microbial origin owing to their ease of multiplication, genetic alteration, and low economic expenditure. Microbes like bacteria, yeast, and fungi produce a variety of organic acids which includes simple unsubstituted acid like acetic acid, complex acids like isovaleric acid, etc. (Kubicek, 2001). Among all microbes, filamentous fungi are good producers of organic acids as they prefer to grow in an acidic environment ranging from pH 3 to 5 (Ruijter et al., 2011). The extent of organic acid accumulation in fungi depends upon the incomplete substrate oxidation (Mattey, 1992). Some well-known organic acid producers are *Aspergillus niger*, *Yarrowia lipolytica*, *Saccharomyces. cerevisiae,* and *Rhizopus oryzae* (Ruijter et al., 2011). These fungi can produce various acids in two different ways; one is from the intermediate of the metabolic pathway and the second is from the direct oxidation of glucose (Mattey, 1992). This article includes the short description of commonly produced organic acid from the above two aforesaid pathways.

3.2.1.1 Organic acid from metabolic pathways

3.2.1.1.1 Citric acid

Citric acid is the natural ingredient of most of the citrus fruits and for many years these fruits have been used as a basic source for procuring this. Moreover, lately microorganisms such as bacteria, yeast, and fungi were exploited to isolate citric acid. Of them fungi were able to produce industrially acceptable amounts, this includes for example a few species of *Aspergillus*, *Penicillium*, *Mucor,* and *Candida*. However, the mutant of *A. niger*, *C. lipolytica,* and *Penicillium* strain produce citric acid at the commercial level (Plassard and Fransson, 2009). In addition, these mutant strains are able to suppress the other byproducts like gluconic acid, oxalic acid, and isocitric acid. Hence, it is mandatory to use that strain which meets the requirement of the manufacturer. In addition to this careful media selection and optimization of fermentation conditions need to be standardized to get the maximum benefit from these strains. In most of the industries, *A. niger* is used as the principal fungus for the production of citric acid as it can utilize a cheap carbon source for its growth and is easy to manipulate which makes the process more economic. The yield from these strains may reach even up to 70% of the carbon source used (Sauer et al., 2008).

Citric acid accumulation by fungi takes place in a controlled fermentation process. Presently, citric acid production can be carried out by the submerged-aerated technique as well as surface processes. It is a primary metabolite formed in idiophase in the tricarboxylic acid cycle where glucose is the main carbon source. During the growth phase, citric acid producers have the ability to utilize 80% of the glucose by glycolysis, and 20% of it is used in the pentose phosphate pathway (Dörsam et al., 2017). Regulation of these pathways will also be dependent on the kind of nutrient media used in the fermentation processes. For carbon source, a variety of carbohydrate sources such as potato starch, sugarcane syrup, sugar cane molasses, beet molasses, glucose syrup from saccharified starch, and other starch hydrolysate can be used (Plassard and Fransson, 2009). In case fermentation media contains starch as a carbon source, amylase needs to be added to convert it into glucose and if any hydrolysate, molasses, or syrup is used a cation exchanger needs to be used to remove the cation impurities from the media. In addition to this, trace elements, such as iron, magnesium, manganese, copper, molybdenum, zinc, are needed for achieving the improved yield of citric acid. Besides the composition of nutrient media, pH also plays an important role in increasing the citric acid yield and it may vary between pH 3.0 and 5.0. During idiophase, the pH must be below 3.0 to suppress oxalic acid and gluconic acid formation. Contrarily, to this in the beginning of trophophase pH should be near to 5.0 (Sauer et al., 2008). Another advantage of this varying pH is to decrease the probability of contamination. The recovery of citric acid from the fermentation broth is again a challenging process due to the sideways production of oxalic acid. However, this can be overcome by reducing the pH during the fermentation processes as well as using calcium oxalate for precipitation during downstream processing. Later on, calcium salts of citrate can be purified by rotating filters or centrifuge.

Owing to its flavor, crystalline nature and acidity, citric acid can be used in various sectors like pharma, food, and chemical industries. Due to its crystal forming nature it is extensively used as seizing agent in industries and as an anticoagulant for preserving the blood.

3.2 1.1.2 Kojic acid

Kojic acid is a pyrone having the chemical name 5-hydroxy-2-hydroxymethyl-4-pyrone. It lacks carboxyl group; therefore, its hydroxyl group attributes to its acidic properties. It is a metabolic product of genus *Aspergillus* which is commonly known as koji in Japan. This acid was first isolated in 1907 as a bi-product of fermented malting rice. *Aspergillus flavus* and *A. oryzae* are the main producers of kojic acid which yields ranging from 70% to 90% depending upon the fermentation conditions. For decades it has been produced by direct fermentation process, however, submerged fermentation could also be an alternative approach for its production. Other media components such as glucose, ammonium nitrate, sucrose, and xylose favor the kojic acid production (Ola et al., 2019). Kojic acid production can be increased by increasing the amount of glucose in the media. Moreover, organic nitrogen is preferred over the inorganic one. The C/N ratio must be around 100 for maximum yield of kojic acid. Contrary to this, iron reduces the yield of kojic acid by interacting with it to form a red color component. Besides these other factors such as aeration, pH, and temperature may also cause fluctuations in the kojic acid yield (El-Kady et al., 2014).

Due to its acidic nature, it acquires the property of antibacterial, antifungal and hampers the viral multiplication via unknown mechanisms. It also inhibits the tyrosinase enzyme which plays an important role in melanin synthesis. In addition, it also prevents oxidative browning of cut fruits and vegetables. Therefore, it is widely used in cosmetics and food industries.

3.2.1.2 Organic acid from direct oxidation

3.2.1.2.1 Gluconic acid

Gluconics acid is produced by the dehydrogenation of glucose by glucose- oxidase which can be carried out by many microorganisms. It is a two-step process where δ-D gluconolactone formed in the first step can be hydrolyzed spontaneously or enzymatically to produce gluconic acid. This reaction can be carried out chemically, during fermentation or separately by purified enzymes. The first industrial manufacturing of gluconic acid was done by fermentation using *Penicillium luteum-purpurogenum* in 1928 (Ramachandran et al., 2006). Nowadays submerged fermentation processes are used for its production by utilizing *A. niger*. It involves fed-batch cultivation with stepwise addition of glucose where pH and temperature are kept at 6.0–6.5 and 34 °C, respectively (Godjevargova et al., 2004). Its production is directly related to the pH, oxygen, and the glucose oxidase activity. It was noticed that at pH greater than 4.0 glucose oxidase activity was greatly enhanced, while it decreases below pH 2.0. Oxygen is one of the substrates of glucose oxidase during bioconversion of glucose. Other organisms also produce gluconic acids but do not have the potential to reach up to the industrial level such as *Penicillium*, *Endomycopsis*, and *Gonatobotrys*. However, production of gluconic acid using immobilized enzymes seems to me more attractive due to its ease of purification (Ramachandran et al., 2006). Yet, this approach is not common in industries because of low yield.

The physicochemical properties of gluconic acid have made it an industrially valuable compound. It is water soluble and can form complexes with di- and trivalent metal ions showing very corrosivity and toxicity. Because of these properties it is used for cleaning metal surfaces. It is also used as sequestering agent in many detergents and useful as food additives and in medicines to act as counter ions for metal deficiency therapies.

3.2.1.2.2 Acetic acid

Production of acetic acid from microorganisms is as old as the production of wine. Until the middle ages, the vinegar was produced for drinking on special occasions by Romans and Greeks. In ancient times the production was carried out in flat open vats. This was a slow process and took time to get fermented by the microorganism naturally. During the nineteenth century an improved process was developed where the use of trickling generators was introduced (Sujatha et al., 2002). Later in the middle of the century submerged processes were introduced and from then both the types of fermentation processes have been used worldwide. However, the submerged process gets advanced with time by utilizing the potential of acetic acid bacteria such as *Gluconobacter* and *Acetobacter*. They both oxidize ethanol to acetic acid while *Acetobacter* over oxidizes the ethanol to CO_2 and H_2O. Therefore, acetic acid production is an incomplete oxidation process (Sujatha et al., 2002). In the first oxidation step ethanol gets converted into acetaldehyde by alcohol dehydrogenase enzyme. In the second step acetaldehyde dehydrogenase converts acetaldehyde to acetaldehyde hydrate and finally to acetic acid. In both the steps NADP gets converted into $NADPH_2$, so 1 M of ethanol produces 1 M of acetic acid along with 6 ATP molecules. Production of acetic acid by *Gluconobacter* and *Acetobacter* sps is so high that no other microorganism replaces them at an industrial scale. The recovery of acetic acid produced by the fermentation process can be initiated by filtration followed by the use of $K_4 (Fe (CN)_6)$ to decolorize the final product (Budak et al., 2014).

Acetic acid readily dissolves in water to form H^+ ions and could be a good solvent for many industrial processes (Budak et al., 2014). Owing to its chemical nature it found application in cosmetic industries specially perfumes, textiles, food industries in soft drinks and pesticides, etc.

3.2.2 Vitamins

Vitamins are the organic compounds required in trace amounts for the normal growth of the human body. Various microorganisms can be exploited for the production of these valuable components such as folic acid, thiamine, pyridoxal phosphate, biotin, β-carotene, ergosterol and riboflavin, etc. Several vitamins can be produced by normal metabolism of the microorganism through the fermentation process. However, some of the vitamins can be produced by the biotransformation reactions such as ascorbic acid and tocopherol. Here we have focused the vitamins produced by fungi such as vitamin B12, riboflavin, and β-carotene.

Vitamin B_{12} is also known as cyanocobalamin having corrin rings. It is exclusively synthesized by some microorganisms such as bacteria and mushroom (fungi) but not by animals and plants (Watanabe and Bito, 2018). The first fermentation for commercial production of B12 was carried out by using *Streptomyces griseus* where it formed as a byproduct of streptomycin production. Among fungi *Craterellus cornucopioides* (black trumpet), *Cantharellus cibarius* (golden chanterelle), *Agaricus bisporus* (button mushroom), and *Lentinula edodes* (shiitek mushroom) are good source of vitamin B12. Interestingly, B12 was also found in their composts.

Riboflavin is another vitamin of group B. It is very important for the reproductive growth of animals and humans. It is a byproduct of acetonebutanol fermentation. Various fungi are also known to produce riboflavin such as *Candida flareri*, *C. guilliermondia*, *Eremothecium ashbyii*, and *Ashbya gossypii*. However, its commercial production is done with ascomycetes fungi *Eremothecium ashbyii*, and *Ashbya gossypii*, by direct fermentation (Watanabe and Bito, 2018). In addition, these fungi are plant pathogens therefore it is mandatory to sterilize the fermentation broth before employing it for commercial use. Their nutrient media majorly consist of glucose, corn oil, and other organic nutrients, where corn oil stimulates the production of riboflavin. In these fungi the riboflavin fermentation takes place in three different phases. The first phase is a rapid growth phase where glucose consumption is high due to which pH falls down as the phase proceeds small amounts of riboflavin also formed along with pyruvic acid. By the end of this phase glucose is consumed and growth ceases (Strzelczyk and Leniarska, 1985). In the beginning of the second phase sporulation takes place and pyruvate concentration decreases due to which pH becomes alkaline which results in the synthesis of cell bound riboflavin in the form of flavin adenine dinucleotide and flavin mononucleotide. In the final phase autolysis occurs that releases the riboflavin in the medium.

Vitamin A also known as β-carotene is naturally found in most of the agricultural products. Although β-carotene can also be synthesized by microorganism, yield is not compatible to make it commercial (Sun et al., 2019). It is mainly produced by Phycomycetes especially the Choanephoraceae family such as *Choanephora cucurbitarum*, *Phycomyces blakesleeanus*, and *Blakeslea trispora*. Researchers have found that the yield of β-carotene is high when both plus and minus mating types of these fungi grow together in the same fermentation media. The fermentation medium must contain β-ionone, fats, oils, waxes, and fatty acids. Out of them, β-ionone is an immediate precursor of β-carotene but it is not directly incorporated into the molecule. It (β-ionone) acts as "Steering" factor which stimulates all the enzymes which participate in the synthesis of β-carotene. The inoculums of both plus and minus strand must be grown separately and 5% of it must be added either simultaneously or in step wise manner in the same fermentation medium.

3.2.3 Recombinant proteins and vaccines

There are a number of proteins and other biomolecules that are produced naturally by different living organisms and find their applications in industries related to food, chemical, detergents, paper, pharmaceutical textile, and many others. However, mass production of these biomolecules from natural sources may be a challenge because of low production, high cost, and poor purity. The heterologous (recombinant) proteins are the ones that are produced by using host systems in which the product is not made naturally. Recombinant protein production in the microbes such as bacteria, yeast, and fungi is the immediate product of recombinant DNA technology. Till date many bacterial (e.g., *Escherichia coli*), yeast (*S. cerevisiae, Pichia pastoris*), and fungal hosts (*Aspergillus, Penicillium, Neurospora*) have been exploited for the production of recombinant protein. One of the major advantages of these systems is the over production of desired protein along with known genome sequence and easy genetic manipulation. This paves the way for the generation of many therapeutic proteins such as growth hormones, vaccines, and antimicrobial agents. Human insulin was the first pharmaceutical product that started marketing in 1982 by using recombinant DNA technology. The market for heterologous proteins has surged thereafter and by 2022 is expected to reach $2850.5 million (Vieira Gomes et al., 2018), the largest share being held by pharmaceutical and biotechnology companies.

Though *E. coli* was used initially for producing heterologous proteins like insulin and growth hormones, lack of posttranslational modifications limited its use as a suitable host. These recombinant proteins are best expressed in the eukaryotic systems such as yeast and fungi as they have machinery for posttranslational modifications like correct folding and glycosylation typical of eukaryotes. Therefore, their demand to produce high titer protein in the biopharmaceutical industries is increasing.

Being unicellular organisms with limited nutritional requirements just like prokaryotes, yeasts are preferential hosts for easier and cheaper production of heterologous proteins and are used for commercial production of ethanol, vitamins, vaccines, organic acids, and single cell proteins. Among yeasts, *S. cerevisiae* was the commonly used host system since 1980s for expression of the majority of heterologous proteins including hepatitis vaccines, Insulin, Glucagon, GM-CSF (granulocyte monocyte colony stimulating factor), HPV (human papiloma virus) vaccines, and HGH (human growth hormone). Keasling's company, Amyris Biotechnologies developed a genetically engineered *S. cerevisiae* strain for producing artemisinin, a semisynthetic antimalarial drug.

The problem of hyper-glycosylation, partial degradation of product, and requirement of sophisticated fermentation equipment in the use of *S. cerevisiae,* however, led to the search of other expression host systems. Also, ethanol toxicity limits the cellular density and thus the amount of heterologous proteins. Lately, other nonconventional yeasts including *Pichia pastoris, Hansenula polymormpha*, and *Kluyveromyces lactis* emerged as advantageous hosts for foreign protein expression (Vieira Gomes et al., 2018).

P. pastoris is one of the popularly used yeast strains for the expression of more than 100 heterologous proteins. It is a methylotroph and can be grown easily on simple media at a very high cellular biomass. Besides providing a strong, inducible promoter system, it is possible to regulate expression of foreign protein efficiently and precisely. Moreover, these secrete the required protein in bulk in the correct folding and with desired posttranslational modifications and glycosylation just like humans. *P. pastoris* expression kits are available commercially from companies such as Mobitec ATUM, Research Corporation Technologies, and Thermo Fischer Scientific. Using *P. pastoris* as host, broad-spectrum of heterologous proteins are produced such as enzymes (cellulases, hemicellulases, and lipases), animal feed additives (recombinant phytase), antimicrobial protein (Nanobody ALX00171), phospholipase C, and so on (Veeresh and Chuan, 2017). These recombinant proteins are used in the treatment of diabetes (Insugen-human insulin), rheumatoid arthritis (Nanobody ALX-0061-anti-IL6 receptor single domain antibody fragment), hereditary angioedema (Kalbitor-kallikrein inhibitor protein), hepatitis C and cancer treatment (Shanferon-interferon-alpha 2b) and blood volume expansion.

The yeast *K. lactis* is recognized as generally regarded as safe (GRAS) by FDA and has been used in the food industry and production of enzyme lactase for decades. Just like *P. pastoris*, it has been used to produce a diverse range of fully active heterologous proteins originated from bacteria, viruses, fungi, and higher organisms; the most common being commercially produced bovine milk coagulation enzyme chymosin. It provides a potential expression system because of its advantageous characteristics like-known genome sequence, use of episomal expression vectors, carry genetic manipulation easily, can be grown efficiently on cheap substrates, and unlike *P. pastroris* does not require explosion proof fermenters. Some of the recombinant proteins produced by *K. lactis* include laccase, inulinase, xylanases, phaseolamin, macrophage colony stimulating factor, and serum albumin (Van Ooyen et al., 2006).

Most of the recombinant proteins made by using yeast system contain O-mannosylation. To overcome these kinds of problems, recombinant proteins were expressed in fungi *A. niger* and *Trichoderma reesei* which shows promising results. The antibody produced in *A. niger* showed the same pharmacokinetic behavior as shown by the mammalian cell-derived antibody. Moreover, filamentous fungi have high secretory capabilities and unlike the yeast produce proteins with N- and O-linked glycans having small oligomannose and high mannose N-linked structures. They also have different kinds of glycosidases and glycosyltransferases for the wide range of glycosylation patterns along with the addition of single glucose, galactose, sulfate, and phosphate groups to their proteins. Also, there is a possibility to express recombinant secretory proteins that are synthesized by the ribosome attached to the endoplasmic reticulum and expelled out of the membrane by signal recognition particles. Comparatively, filamentous fungi are easy to manipulate and modify at genetic level to change the glycosylation pathways same as that of humans, but the use may be restricted because of difficulty in cultivation and high production cost (Ntana et al., 2020).

Number of other yeasts and filamentous fungi are also exploited for the expression of recombinant proteins. Examples include *Hansenula polymorpha* (HBV vaccine), *Y. lipolytica* (pancrelipase), *T. reesei* (feruloyl esterases, human α-galactosidase A, laccases, lipases), *Neurospora crassa* (Human antibody fragment), *Penicillium chrysogenum* (Cysteine rich antifungal proteins), *A. niger* (Glyoxal Oxidases), *Penicillium griseoroseum* (Phytases), *Schizosaccharomyces pombe* (Human transferrin), *A. oryzae* (β-1,4-Endoglucanases), and *Penicillium canescens* (Laccases) (Van Ooyen et al., 2006; Ntana et al., 2020).

The illegitimate and inappropriate use of antibiotics since the discovery of the first antibiotic in the 1940s has led to the development of antibiotic-resistant microbial strains. Also, the development of new antibiotics is a slow process. In such a scenario, vaccines seem to be the important prophylactic measure against infectious diseases. Most of the conventional licensed vaccines, except peptide derived ones, are inactivated or attenuated microorganisms, which require specific storage conditions. Though these vaccines are able to fight infectious diseases, there are certain concerns like the possibility to cause allergic reaction or reversion to virulent strain, need for adjuvant and multiple boosters, maintenance of cold chain during distribution, and limitation in using live vaccines in immune-compromised individuals (Kumar and Kumar, 2019).

These drawbacks necessitate the need for an alternative strategy for the preparation of novel vaccine development against infectious diseases and cancer. Yeasts are found to be suitable hosts for heterologous protein expression and have been used successfully for the expression of a number of antigens from viruses, fungi, nematodes, bacteria, and cancer. Hepatitis B was the first recombinant vaccine produced commercially using *S. cerevisiae*. Features like inherent adjuvant properties because of polysaccharide beta-1,3-D-glucan and mannan in cell wall as natural immunostimulators, ability to activate cell mediated immunity and interleukin production, nontoxic and expression of foreign antigen on cell surface make them a right candidate for oral vaccines. Hepatitis A and B and HPV vaccines have been prepared through heterologous expression in *S. cerevisiae*.

Use of fungi as cell factories for the production of different recombinant proteins and vaccines of value through cost-effective processes is a subject of major research in the coming years.

3.2.4 Foods

Filamentous fungi and yeasts have been employed in food production since prebiblical times. Edible mushrooms like *A. bisporus* have found a place in the human diet and are used as medicine for centuries. Being approximately 90% water and 10% protein and carbohydrate, these are the rich source of vitamins and minerals and are low in fat. Their culinary value in human food is because of the presence of more than 150 different volatile aromatic compounds including terpenoids, sulphur compounds, benzaldehyde, cyclooctenol, and benzenol that give characteristic flavor and aroma. The market of mushrooms has bloomed in recent years and millions of tons of mushrooms are raised globally using agricultural wastes like saw dust, cereal bran, wheat, or rice straw composted with animal dung and wood logs. The most popular ones being *A. bisporus* and *L. edodes*, commonly called white button mushroom and white rot fungus (Shiitake), respectively. Besides being nutritious, mushrooms also provide other health benefits. There are claims of their utility as antitumor agents, in reducing serum cholesterol, and as stimulators of the immune response. These are used in traditional Chinese medicines for a long time. A specific compound Lentinan (β-glucan) in Shiitake is found to have medicinal properties. Medicinal uses of mushrooms like *Polyporus umbellatus* and *Hericium erinaceus* have been suggested for gastric ulcers, cirrhosis, and hepatitis B infections (Money, 2015).

The use of yeasts in brewing and baking has been practiced since ancient times. Yeasts, being facultative anaerobe, convert sugar into carbon dioxide and alcohol in the absence or scarcity of oxygen or in the presence of high sugar concentration. The practice has been used since ages in the baking and brewing sector catering a huge global market. Various species of *Saccharomyces* like *S. cerevisiae, S pastorianus*, and *S. bayanus* are associated with the production of the majority of alcoholic drinks, the dominant being *S. cerevisiae* (Money, 2015). A great variety of alcoholic beverages can be made on the basis of the sources of sugar used as a raw material, namely, wine and cider from fruit juices, rum from sugar cane, beer from starchy crop or barley malt, brandy from grape juice, whiskey from barley malt, palm wine from wine sap, gin from wheat, corn, rye, and so on. Inclusion of skin of purple or black grapes results in red wine while white grapes or juice extracted without skin produces white wine during fermentation. Fermentation results in the accumulation of alcohol up to 10%–12% concentration and the number of other components during yeast growth that contribute to characteristic aroma and flavor. Some alcoholic beverages rely on storage after fermentation in wooden casks or other containers for varying time period.

S. cerevisiae (Baker's yeast) in the form of freeze-dried granules, compressed cake, or concentrated liquid is employed for the fermentation of dough in bakeries. The amylase present in dough hydrolyzes the starch into simple sugars, which is then converted to carbon dioxide and alcohol by yeast. Carbon dioxide raises the dough and gluten gives the sponginess and elasticity to bread.

Proteolytic and lipolytic activities of yeasts and fungi are also involved in the flavoring and ripening of a variety of commercial cheeses. The enzyme chymosin involved in the coagulation of milk in commercial cheese production is obtained from recombinant *E. coli*, *A. niger* var. *awarmori*, and *Kluyveromyces lactis*. A range of ripened and unripened cheese with different textures and characteristics can be produced using different species of fungi and yeasts, viz. *Penicillium camemberti* in surface ripened soft cheeses—brie and camembert cheese; *P. roqueforti* in Roquefort and Danish blue cheese, *Debaromyces hansenii* in Limburger and *Geotrichum candidum* in soft cheeses (Money, 2015).

Microbial succession in cacao fermentation involves yeasts *Hanseniaspora guilliermondii* and *H. opuntiae* followed by *S. cerevisiae*, *Pichia kudriavzevii*, and other yeast strains.

Food and Drug Administration (FDA) has authorized the use of yeast extracts as natural flavoring and seasoning agent in food industry. These are used in a number of food products like sauces, soups, sea products, etc. These also find their application as feed supplements and in baby food formulations as a rich nutritional source of vitamins, minerals, amino acids, and peptides.

Fungal fermentation (pure fungal culture/combination of fungi and yeasts) is also responsible for the production of a variety of traditional fermented foods across the world (Table 3.1) (Jay et al., 2006). These foods, besides increased shelf life, also have improved digestibility, more nutritious, increased vitamins and minerals content, antioxidant properties, and better flavor and appearance. Yeasts increase Vitamin B content in fermented products, for example, Vitamin B12 content increased from 0.15 μg/g in soybean to 5 μg/g in tempeh. Yogurt decreases cholesterol level in blood and has an anticancer effect. Moreover, antinutritional compounds like lectins, phytates, cyanogenic components, stachyose present in raw material are removed during fermentation.

Fungi are also the source of food additives like flavoring agents, PUFAs (poly unsaturated fatty acids), colorants, sugar substitutes, etc. (Copetti, 2019). Thaumatin and xylitol are approved and recognized as safe food additive by FDA (food and drug administration) and are utilized as low-calorie sweeteners in food products like soft drinks, chewing gum, medicated products, etc. Thaumatin is normally produced by *A. niger* var. *awamori* and *P. roqueforti* whereas *Pichia stipites* produces xylitol. Recombinant *S. cerevisiae* is capable of producing xylitol in higher concentration (Copetti, 2019).

The research is being continued to exploit the microorganisms, especially fungi, in a better way in the food sector for taking care of the growing demand for food and to improvise the human nutrition and animal feeding.

3.2.5 Enzymes

Enzymes are proteinaceous in nature and produced by all living beings from unicellular to multicellular organisms. They are biological catalysts to perform metabolic processes leading to the growth and survival of the organism. They are complete biomolecules and continue to function even if they are separated from the source without disturbing their structure and chemical nature. Keeping this in mind enzymes from microorganisms such as bacteria, fungi, and yeast have been widely isolated and commercialized to perform the industrially important reaction in vitro. In addition, microbial enzymes could be adaptive or constitutive in nature, that is, their concentration can be manipulated depending upon the requirement of the cell. Hence, microbes are easy to manipulate genetically or metabolically by media engineering to get over production of desired enzymes. Due to these reasons commercial microbial enzyme production is focused only on fungi, yeast, and bacteria. Among them, most of the fungi grown at acidic pH that further strengthened their candidature for commercialization. The nature

Table 3.1 Traditional fermented food with their sources and associated microorganisms.

S. No.	Product	Raw material	Country	Microorganisms involved
1.	Tempeh	Soybean	Java, Indonesia New Guinea	*Rhizopus oligosporus* *Rhizopus oryzae*
2.	Furu/sufu	Soybean	China, Taiwan	*Actinomucor elegans, Mucor disperses*
3.	Soy Sauce (Shoyu)	Soybeans Wheat	Japan China	Koji—*Aspergillus oryzae* *A. sojae* Maromi—Yeast and lactic acid bacteria (*Zygosaccharomyces rouxii, Lactobacillus delbrueckii*)
4.	Bongkrek	Coconut presscake	Indonesia	*Rhizopus oligosporus*
5.	Miso	Soybeans, Rice, Barley	Japan China	*A. oryzae, Z. rouxii*
6.	Natto	Whole soybeans	Japan	*Bacillus natto*
7.	Dosa and Idli	Rice, Dehulled black gram	India	Yeast, *Leuconostoc mesenteroides*
8.	Laochao	Waxy variety of rice	China, Indonesia	*Amylomyces rouxii* *Rhizopus chinensis, R. oryzae*
9.	Pickles	Cucumbers	Worldwide	*P. cerevisiae*, *L. plantarum*
10.	Mahewu	Maize	S. Africa	*L. delbrueckii*
11.	Red koji/yeast rice (*angkak*)	Rice	China	*Monoascus purpureus*
12.	Chinese liquor/ Baijiu	Sorghum, Rice, Wheat, Corn, Millet	China	*Bacillus, Lactobacillus, Saccharomyces, Candida, Mucor, Absidia, Penicillium, Aspergillus*
13.	Black olives	Olives	Mediterranean region	Yeasts—*Saccharomyces, Hansenula, Candida, Torulopsis, Debaryomyces, Pichia, Kluyveromyces, Cryptococcus*
14.	Kenkey	Corn	Ghana, Nigeria	*Aspergillus* spp., *Penicillium* spp., Lactobacilli, Yeasts
15.	Sake	Rice	Japan	*A. oryzae (koji preparation)* *Saccharomyces sake, S. cerevisiae*
16.	Coffee beans	Coffee cherries	Brazil, Congo, Hawaii, India	*Erwinia dissolvens*, *Saccharomyces* spp.
17.	Cider	Apples, others	Worldwide	*Saccharomyces* spp.
18.	Vodka	Potatoes	Russia, Scandinavia	*Yeast*
19.	Scotch whiskey	Barley	Scotland	*S. cerevisiae*
20.	*Kombucha*	Sweetened black tea	China Russia, Germany	Bacteria and Yeasts *Acetobacter xylinum, Candida, Brettanomyces, Saccharomyces, Pichia, Zygosaccharomyces*
21.	Lebanon bologna—semidry sausage	Beef	Lebanon, Pennsylvania	*Pediococcus cerevisiae* *P. acidilactici*
22.	*Kefir/Koumiss*	Milk	Central Asian and Eastern European regions	*Lactococcus lactis, Lactobacillus acidophilus, Candida kefir, Kluveromyces marxianus, S. cerevisiae*
23.	*Ontjam*	Peanut press cake	Indonesia	*Neurospora intermedia*

of enzymes may vary from one species to another with respect to their pH and temperature optima (Aunstrup, 1979).

The gross market of microbial enzymes in the world should reach $7.0 billion by 2023 with an annual growth rate of 4.9% for the period 2018–2023. Here we will discuss only those enzymes which are currently being produced commercially from the fungi which include amylase, protease, cellulose, xylanase, penicillin acylase, asparaginase, lipase, etc. (Aunstrup, 1979).

3.2.5.1 Amylase

Amylases are the important enzymes to perform starch-saccharification. They hydrolyze starch into dextrins, then to maltose, and finally to glucose. On the basis of their mode of action, these amylases can be categorized into three different types: α-amylase, β-amylase, and glucoamylase (Gopinath et al., 2017).

Enzyme α-amylases also known as 1,4-α-glucan-glucanohydrolases which hydrolyzes α-1,4-glycosidic bonds present in the interior of the starch chain. They are produced by many bacteria and fungi. Few examples of fungi that can produce amylase are *Aspergillus*, *Candida*, *Rhizopus*, *Mucor*, and *Penicillium* (Gopinath et al., 2017). All these organisms produce α-amylase constitutively.

Enzyme β-amylases are also known as α-1,4-glucan-maltohydrolases and widely expressed in plants. However, microbial sources are also there such as *Bacillus cereus*, *B. polymixa*, *B. megaterium* from bacteria and *Rhizopus japanicus* from fungi (Pandey et al., 2000).

Glucoamylases are also known as α-1,4-glucan-glucohydrolases but unlike α-amylases, they split glucose from nonreducing ends during starch hydrolysis. They are more prominent in fungi such as *Rhizopus formosaensis*, *R. niveus*, *A. oryzae*, and *A. niger* (Gopinath et al., 2017; Pandey et al., 2000). Nowadays industrial fermentation of amylases from fungi has been carried out by submerged fermentation techniques due to their increased demand in the industries. Most of the industrial demand of fungal amylase is fulfilled by *A. oryzae*. However, for *A. oryzae* stationary fermentation using wheat bran is equally productive (Pandey et al., 2000).

In industries, amylases are widely used as sizing agents in detergents as well as in the brewing industries. They are highly demanding in fructose syrup production which is formed by hydrolysis of starch by the amylase. Glucose is not as sweet as fructose, therefore, one additional enzyme glucose isomerase is also required in the process along with amylase.

3.2.5.2 Protease

These are the second most important enzymes in industries after amylase. These are the enzymes that can cleave the amide bonds. Proteases are categorized on the basis of their pH optima into alkaline, acidic and neutral proteases (Macchione et al., 2008).

Alkaline proteases are stable in the alkaline pH and widely used in detergent industries where the pH is in the range of pH 9–11. They are stable at high temperature and cannot be affected by the presence of any chelating agent such as EDTA (ethylene diamine tetraacetic acid) since they consist of serine at their active site. They are widely produced by both bacteria and fungi (Novelli et al., 2016). Fugal producers include *A. niger*, *A. flavus*, *A. sojae,* and *A. oryzae*. Fed-batch process is used for the production of these enzymes because it is mandatory to lower the concentration of nitrogenous materials such as amino acids and ammonium ions as they suppress the protease production.

Neutral proteases have stability in a very narrow range; therefore, sodium and calcium salts have been added to maximize their stability. Since, they are not very stable so they are not popular at the industrial scale (Xu et al., 2019).

Acidic proteases have maximum stability in the pH 2–4. One such example is rennin like proteases which are widely used in cheese production. They are a very good alternative in the production of soy sauce and in breakdown of gluten in baking industries. They are also used in meat tenderization, silk industries, and textile industries. Various fungi produce acidic proteases, namely, *Mucor delemar*, *A. oryzae*, and *A. flavus* (Yue et al., 2019). Submerged fermentation could be a good alternative for their production at a commercial scale.

3.2.5.3 Cellulase

Cellulases are glycoside hydrolase, which cleaves the glycosidic bonds present between two carbohydrates or noncarbohydrate components. Most of the fungi produced three different types of cellulases: *endo*-β-1,4-glucanase (EG), cellobiohydrolase (CBH), and β-glucosidase (Okal et al., 2020).

Endo-β-1,4-glucanases breakdown the -β-1,4-glycosidic bonds present in the cellulose to generate the cellodextrins, cellotetraose and cellobiose as a major product. These enzymes first attach to the substrate to arrange themselves in a proper orientation, for the reaction to occur (Kyu et al., 2020).

Cellobiohydrolase (CBH) cleaves the reducing and nonreducing ends of the cellulose to liberate cellobiose. Fungal cellobiohydrolase shows *exo-exo* synergism where CBH I act on the reducing end, while CBH II on the nonreducing end of cellulose. This synergism enhances the cellulose degradation by several times. Among fungi, *Trichoderma* is the rich producer of cellobiohydrolase which in total produce 50%–60% of CBH I and 10%–15% of CBH II (Momeni et al., 2015).

Enzyme β-glucosidase cleaves cellobiose to glucose and makes the important component of cellulose degradation system of fungi. Fungal β-glucosidase cannot cleave insoluble cellulose therefore it needs it in solubilized form such as cellobiose. Interestingly, these enzymes are inhibited by higher concentration of glucose which limits the fungal cellulose system to commercialize. However, *Aspergillus* sp. produces such β-glucosidase which can tolerate moderate amounts of glucose therefore they are often used in the industrial processes.

Other fungal cellulose producers include *Trichoderma reesei*, *T. atroviride*, *T. harzianum*, *Aspergillus niger*, *A. fumigatus*, *A. flavus*, *A. ochraceu*, *A. nidulans*, *Penicillium glaucum*, *P. candidum*, and *P. expansum*, etc. Among them *Aspergillus* sps produces all three types of cellulases and they have been extensively studied and exploited at industrial scale. Cellulases are the third largest industrial enzymes commercialize globally. They contribute approximately 20% of the total demand of the enzyme market. They are widely used in biofuel, paper, pulp, food, textile, and detergent industries. The use of cellulase in the biofuel industries has increased its demand, but the contradiction in the production cost of the enzyme has limited its further uses.

3.2.5.4 Xylanase

Xylan is the most abundant polysaccharide present in the wood which can be degraded by enzyme xylanase (Bajaj and Mahajan, 2019). Traditionally, xylanase best works in combination with the cellulases to degrade the lignocellulosic biomass especially from agricultural waste. They are commercially used for the production of oligosaccharides from xylan, which further can be used as sweeteners in food industries. Along with pectinase they are also used in the clarification of fruit juices, modifying baking products, and enhancing the palatability of animal feed.

Xylanases are produced by a wide variety of fungi but the commercial production is restricted to *Trichoderma* and *Aspergillus* sps. The maximum production of these enzymes was achieved by submerged fermentation using economic carbon sources by adding inducer substrates such as xylan or cellulose. Alternatively, one can use less expensive substrates such as lignocellulosic biomass to reduce the production cost (Bakri et al., 2003).

3.2.5.5 Lipase

These are triacylglycerol hydrolases which catalyze the esterification reaction between fatty acid and glycerol under micro-aqueous conditions and the transesterification reaction under aqueous conditions. In addition, most of the yeast and fungal lipases have unique properties of enatio-selectivity and region-selectivity, which make them industrially valuable enzymes especially in pharma sectors. They find applications in various other sectors also such as detergents, cosmetics, textile, food, leather, and agriculture. Owing to their unique properties, their demand in the industries is increasing day by day. The market value of lipases in 2018 was $425 million and expected to increase by $590 in 2023.

They can be procured from bacteria, fungi, and yeast. Most of the fungi such as *Penicillium*, *Geotrichum*, *Yarrowia*, *Candida*, *Mucor*, *Aspergillus*, and *Rhizopus* sps produce extracellular lipases (Sethi et al., 2016). In almost all cases, lipase production can be induced by adding fats and oils in the production medium. However, when lipase acts on these fats and oils, they produce glycerol as a byproduct that represses the enzyme production. Mostly, these lipases are produced as isozymes with varying substrate specificity, pH, and temperature optima.

3.2.6 Pigment

Color has been an integral part of life since human civilization. The use of dyes and pigments began thousands of years ago to provide an artistic value. A Neolithic site in France gave evidence of using kermes and word (blue) for dyeing yarn threads. Plant-derived indigo dyed fabric more than 4000 years old was found in Egyptian tomb. Dyeing garments with madder dye was a routine practice during the Indus valley era around 2500 BC (Venil et al., 2020). Cochineal dye from cactus insects found its use in food items. All these dyes were natural as developed from sources such as plants, trees, insects, and mollusks without any chemical processing. The first natural dye from oak bark was patented in 1775 in America and later on dye cudbear from lichens was patented (Yusof et al., 2017).

With the increase in demand due to rapid industrialization and the discovery of the first synthetic dye (Mauve) in 1856 by a British scientist, William H. Perkin, a shift occurs from natural to synthetic dyes. The natural dyes are gradually replaced by synthetic dyes as being more cost-effective, readily available, and easy to produce and apply, and the number of these dyes and pigments used industrially at present exceeds 10,000. However, their production and usage have detrimental effects on the environment as these usually require the use of strong acids, alkali, and metals which are toxic and lead to water and environmental pollution because of their low biodegradability. Besides affecting aquatic life, they pose a threat to human health as many triggers allergy and are mutagenic and/or carcinogenic and immunosuppressive. So, the solution lies in natural dyes and pigments (Venil et al., 2020). Limitations in synthetic dye usage raised interest among consumers, researchers, and industrialists in natural dyes which are less toxic, less polluting, and nonpoisonous. Majority of these natural colors, for example, turmeric, kachnar, dhao, indigo, henna, beet root, catechu, saffron, etc. are derived from animals, insects, and different plant parts like roots, leaves, bark, and woods. Due to several disadvantages in plants and animal-based colorants, for example, having low yield, nonstandardized, seasonal variation, high cost, and large space requirements, microbial sources like bacteria, fungi, and algae are potential alternative dye resources. The fungi are emerging as novel producers of natural dyes with remarkable stability, which finds their uses commercially in industries including food, cosmetic, textile, leather, paper, and pharmaceutical. These are fast growing and capable of providing vivid colors with no seasonal and geographical variations. The process is manageable and scalable and can be grown in fermenters even by using cheap renewable waste. Moreover, most pigments are water soluble, making the extraction process easy without the use of any solvent.

The fungi produce a wide range of chemically different pigment classes including carotenoids, polyketides, melanins, phenazines, flavins, and monascin that are receiving attention worldwide (Table 1.2) (Lagashetti et al., 2019). Research is being done globally to exploit filamentous fungi for colorant production to meet the increasing demand of colorants. Ascolor Biotech, Czech Republic produced the first commercial colorant Arpink red from *Penicillium oxalicum* (Yusof et al., 2017).

Fungal carotenoids have been approved by European Union to be used as food colorants. Number of mucorales fungi produces carotenoids that are suitably used in a variety of foods. Since 800 BC, *Monoascus purpureus* was used to produce red mold rice (ang-kak) in China. The red carotenoid pigment (astaxanthin) from yeast *Phaffia rhodozyma* imparts an orange red color to farm salmonid white flesh on adding yeast to their diet. Orange red to violet red edible pigments (e.g., monascorubramine, monacolins, ergosterol) produced by *Monoascus* species during fermentation implying different culture media can be used as replacement of commonly used food dyes fd & c red no. 2 and red no. 4. *Monoascus ruber* and *M. purpureus* were reported to be the potential pigment producers; the only concern is the occasional presence of a mycotoxin citrinin in fermented products. Its use in food is approved in countries like China and Japan (Yusof et al., 2017).

The fungi thus can be a good source for the commercial production of food colorants and dyes. β carotene is produced industrially by the fungus *B. trispora* during the sexual mating process. Interestingly, besides food, *Monoacus* pigments also found their application in other sectors like cosmetics, dyeing, printing, solar cells, textiles, etc. In addition to carotenoids, filamentous fungi also produce many other pigments including polyketides, quinones, dihydroxy naphthalene melanin, flavin, etc. in abundance. Number of researchers worldwide reported natural pigments from fungi. These pigment-producing fungi can be isolated from different ecosystems such as marine, mangrove, and terrestrial (Kalra et al., 2020).

The textile industry is another major consumer of dyes. Fungal pigments can be used as textile dyes directly or may produce dye intermediates. Pigments isolated from various fungal species, namely, *Aspergillus, Trichoderma, Drechslera, Penicillium, Rhizopus, Sclerotinia, Fusarium, Alternaria, Curvularia, Monoascus,* developed different hues of colors in different fabrics (Table 3.2). These show excellent color fastness to washing, rubbing and perspiration, high dye uptake and stability to heat, light, and pH, and generally nontoxic. Natural dyes (yellow and red) synthesized by *Penicillium murcianum* and *Talaromyces australis* (wood inhabiting fungi) are reported to be promising industrial candidates for wool dyeing (Hernández et al., 2019). Moreover, many pigments can absorb UV light and provide protection to the skin. The antimicrobial properties of these pigments suggest their futuristic use in wound dressings and bandages. Nonetheless, the need is to standardize the process for maximum color production, perform more toxicological studies, and maximize the industrial dyeing procedure for their use as a substitute of conventional dyes.

In addition to their use as a colorant, fungal pigments display enormous valuable biological benefits such as antioxidants (*Carotenoids, Azaphilones*), antidiabetic, antiinflammatory (*Naphthoquinones, Azaphilones*), anticancererous (*Polyketides*), antimicrobial (*Carotenoids, Polyketides*), antiviral activity (*Hydroxyanthraquinone*), antifungal (*Naphthoquinones*), antileishmanial activity (*Hydroxyanthraquinone*),

Table 3.2 Fungal species producing different color pigments.

S. No.	Pigment class	Pigments	Color	Fungi
1.	*Carotenoids*	β-carotene Lycopene Xanthophylls	Yellow Orange Red	*Phycomyces blakesleeanus, Blakeslea trispora, Penicillium oxalicum, P. murcianum, Rhodosporidium Aspergillus giganteus, Sclerotium rolfsii, Sclerotinia sclerotiorum, Sporidiobolus pararoseus, Talaromyces australis*
2.	*Anthraquinones*	Alterporriol K, L, M Bostrycin Tetrahydrobostrycin Cynodontin Helminthosporin	Yellow Orange Red, Reddish-brown, Brown Violet	*Alternaria* sp., *Aspergillus* sp., *Eurotium* sp., *Emericella* sp., *Fusarium* sp., *Penicillium* sp., e.g., *P. oxalicum, Microsporum* sp., *Trichoderma virens, Thermomyces* sp., *Curvularia lunata, Dreschelera* sp.
3.	*Hydroxyanthraquinone*	Altersolanol Macrosporin Dactylariol Tetrahydroaltersolanol B Catenarin Rubrocristin Aspergiolide B Physcion Emodin Tritisporin Pachybasin	Yellow Orange Red Red Bronze Maroon Brownish-red	*Alternaria* sp., e.g., *A. solani, A. porri, Aspergillus* sp., e.g., *A. glaucus, A. nidulans, A. versicolor* *Curvularia lunata, Dreschlera* sp., e.g., *D. dictyoides, D. rostrate, Eurotium* sp., e.g., *E. amstelodami, E. rubrum, Trichoderma* sp., *Haloresellinia, Verticicladiella* sp.
4.	*Naphthoquinones*	Trypethelonamide A 7- methoxy trypethelone Viopurpurin Rubrosulfin Aurofusarin Bostrycoidin 8-*O*-methyl fusarubin	Yellow Red Violet Purple	*Trichoderma* sp., *Chlorocibori* sp., *Arthrographis cuboidea, Trypethelium eluteriae, Aspergillus sulphureus, Fusarium* sp., e.g., *F. culmorum, F. oxysporum*
5.	*Azaphilones*	Monascorubrin Ankaflavine Monascin Rubropunctatin Monascorubramin Rubropunctamine	Yellow Orange Pink red Reddish-brown Purple-red	*Penicillium* sp., e.g., *P. aculeatum, P. purpurogenum, Monascus* sp., e.g., *M. purpureus, M. rubropunctatus, Talaromyces* sp., *Chaetomium, Aspergillus clavatus, Isaria farinose Epicoccum nigrum, Fusarium verticillioides, Emericella nidulans*
6.	Dihydroisocoumarin	Xanthomegnin	Orange	*Aspergillus ochraceus, Penicillium* sp., *P. cyclopium, P. viridicatum*
7.	Chromene	Citromycetin Citromycin	Yellow	*Penicillium bilaii*
8.	Isoquinoline	Panaefluorolines A, B, C	Yellowish-green	*Amygdalaria panaeola*
9.	Quinone	Viomellein Variecolorquinone A	Yellow Reddish-brown	*Aspergillus* sp., e.g., *A. ochraceus, A. variecolor*
10.	1,8-Dihydroxynaphthalene	Melanin	Dark-brown	*Aspergillus fumigatus*

herbicidal and insecticidal (*Anthraquinones, Naphthoquinones*), cytotoxic activities (*Hydroxyanthraquinone*), membrane stabilizers (*Carotenoids*), immuno-suppressors (*Polyketides*), controlling cholesterol and obesity, and precursors to vitamins (*Carotenoids*) (Hernández et al., 2019). This marks their promising avenues in the health and pharmaceutical industries.

3.2.7 Biofabrication

Biofabrication is a novel emerging, fast-growing technical field to produce environmentally responsible biomaterial with low carbon footprints using different organisms. The field is still in the experimental stage and relies on genetic engineering advancements. In the last 5 years, substantial commercial and academic interest has been generated in the use of fungal mycelium for the production of various environmentally sustainable useful biological products.

The use of fungal biomass for bio fabrication can be traced back to 1957 when mycelium-derived paper sheets having writing and printing characteristics comparable to paper sheets made of wood pulp were developed and patented by the Institute of paper chemistry. Addition of small amount of cellulose to sheets fabricated from mycelia was noted to provide the unusual flexibility to these sheets (Conkey et al., 1957). Further, the possibility of growing fungal mycelium on paper mill spent liquor and then their use in paper making provided a resource recovery process (Johnson and Carlson, 1978). The process of sheet making was coupled with the disposal of high BOD (biological oxygen demand) containing paper mill effluent where fungal mycelia containing chitin –β glucan was grown inexpensively on the spent liquor.

Because of the structural similarity of chitin (fungal cell wall component) and cellulose, industrial application of fungal pulp was also suggested in the textile and pharmaceutical industries (biomedical fields). The best-known example of biofabrication is the preparation of tissues for treating damaged joints and organs. Limitations of conventional dressings like irritation, ineffectiveness in curing chronic wounds, and poor biocompatibility led to the search for alternative new low cost and mass producible wound management technologies. The dressing material made from fungal chitin glucan filaments has found its application in wound management and exhibits cell binding and pro-proliferate activity. Sacchachitin membranes derived from the fruiting body of *Ganoderma taugae* accelerate wound healing by increased proliferation and normal differentiation of keratinocytes (Su et al., 2005). Because of their beneficial or desired properties of being nontoxic, biodegradability, antimicrobial activity, and hemostatic activity, chitin and its derivative chitosan are gaining attention in a number of biomedical applications. Recently, the production of chitosan from fungi is gaining importance.

The concept was also expanded to other industries as well for the preparation of products like food packaging material, foam-like construction material, textile fibers, leather-like substances, for example, Styrofoam for insulation and soundproofing, sportswear, luxury products, and so on (Jones et al., 2020). The packaging material is being created from mushroom mycelium by growing it on agricultural waste. It is environmentally sustainable (biodegradable), besides being light, strong, and durable and being used by many big companies.

Leather is another bio-product fabricated from collagen grown by yeast and can be custom designed into various textures, color, and thickness. Leather is a commonly used commodity for furniture, clothing, shoes, and accessories like valet, belts, etc. The leather industry captures a huge market and produces leather naturally from animal skin after physical and chemical treatment or synthetically from polyvinyl chloride or polyurethane. However, production and processing of leather is not environment friendly. Besides ethical and social concerns, livestock farming or rearing results in the loss of animal habitats due to deforestation, green gas emission, and generation of animal waste (Jones et al., 2021). Furthermore chemicals like chromium salt used during leather tanning generate significant quantities of hazardous waste sludge as well as unsafe for workers' health. Although no livestock farming or tanning is required for synthetic leather and also it is more environment friendly, its production relies on the use of fossil fuels and other hazardous substances like diphenyl isocyanate (Baur et al., 1994). Moreover, these are nonbiodegradable. Genetically engineered yeast and fungi are getting attention as an alternative cost-effective and environmentally responsible source for leather-like material. Primarily consisting of chitin, it is derived from fungal biomass grown on low-cost agricultural waste, tree mulch, saw dust, and black strap molasses by submerged or solid-state fermentation followed by physical and chemical treatments processing to improve its strength, elasticity, and decay resistance. Many biotechnology companies have already launched their promotional products like handbags, shoes, key chains, pouches, wallets, etc. made from fungal mycelium-based leather-like material in the market. To name a few include MycoTech, Bolt Threads Inc., Mogu S.r.l., Myco Works Inc., and Ecovative Design LLC (Jones et al., 2021). The products are eco-responsible, as no/minimal energy input is required as well as the carbon is not emitted but rather stored. The process is involved in up-cycling of byproducts and moreover, the product is completely biodegradable. The main challenge faced in the development of technology is getting a homogenous mycelium.

Creating microbe clothing/fabric from fungi is another sustainable, eco friendly, nonpolluting futuristic approach with a low carbon footprint, and different brands like muskin, mylo, and myela are in use for making accessories like bags, shoes, belts, even dresses, etc.

One example is the MycoTex fabric created from mushroom mycelium. The fabric is found to be comfortable, skin friendly, biodegradable, and nontoxic and can be used without sewing. Moreover, the process used has environmental benefits as it requires no chemicals and very little water for production. However, lesser acceptance by consumers and the production process used currently being time consuming and labor intensive are some of the challenges associated with fungus-grown clothes.

3.2.8 Biofuels

Biofuel is an alternative to fossil fuels made from the microbial digestion of biomass such as agricultural waste, vegetable waste, and lignocellulosic biomass. It has the ability to replace conventional fuel in the transport industry (Fatma et al., 2018; Bhattarai et al., 2011).

Biofuel could be of two types: gaseous biofuel which includes biogas and biohydrogen and liquid biofuel consist of biodiesel, bioethanol, and biobutanol (Bhattarai et al., 2011).

Mainly, biogas is formed by the anaerobic digestion of lignocellulosic biomass by anaerobic microorganisms such as methanogenic bacteria and anaerobic fungi. Biohydrogen, on the other hand, is a relatively new type of gaseous biofuel. Carbohydrates present in the raw material get converted into hydrogen and organic acid by heterotrophic fermentation (Bhattarai et al., 2011).

Liquid biofuel is produced by the fermentation of sugars. The main enzymes involved in the conversion of sugars to ethanol are invertase and zymase which were isolated from yeast and fungi. For this forest leftover is the best substrate for fermentation as it is rich in lignocellulose. However, pretreatment is required for these substrates which makes the process expensive and limits its uses at the industrial scale (Robak and Balcerek, 2018).

Another liquid biofuel that is biodiesel can be obtained by the trans-esterification reaction of vegetable oil in the presence of alcohol. The biocatalyst used here could be lipase procured from fungi.

As a fuel Biobutanol seems to be better among all liquid biofuels because it is less volatile and hygroscopic. *Clostridia* sp. is known for its bulk production of biobutanol under anaerobic conditions. Clostridia convert the sugars first into organic acid and then to other solvents in the ratio 3:6:1 for acetone:butanol:ethanol, respectively (Robak and Balcerek, 2018).

In spite of all these researches, biofuel could not replace conventional fossil fuel in the current scenario. The main reason behind this is the cost of production. To overcome this, we need to use the raw resources which are available at a cheaper rate. At present the majority of biofuel production is based on agro-industrial waste or agricultural waste. However, the forest biomass remains untouched and the use of microbes for fermenting these waste products needs to be further explored for lowering the cost and making it available for the large-scale production.

3.2.9 Antibiotics

Antibiotics are secondary metabolites produced by microorganisms that at a low concentration either inhibit growth (−static) or kill (−cidal) other microorganisms. There are a number of antibiotics produced by fungi.

Penicillin, the first antibiotic produced by *Penicillium notatum* and discovered by Sir Alexander Fleming in 1928, marks the beginning of the "antibiotic era." It belongs to the β lactam group of antibiotics that constitutes a significant portion of the antibiotic market. The penicillins contain a four membered β lactam ring (4-atom cyclic amide) fused to five atom thiazolidine ring with different side chain substitutions and include both classical (penicillin G and V) and semisynthetic (ampicillin, amoxicillin, oxacillin, methicillin) antibiotics. The semisynthetic penicillins can be easily prepared by replacing the side chains of natural ones. Penicillins may be polar or nonpolar in nature; nonpolars are synthesized by filamentous fungi only (Horgan and Murphy, 2018).

The antibiotic penicillin works during bacterial growth by binding to penicillin-binding proteins (PBP-transpeptidases) associated with the bacterial cell wall. This inhibits the transpeptidation reaction, and thus inhibits the cross-linking of the growing peptidoglycan layer of the cell wall and causes cellular lysis. Penicillins are used for treating throat infections, meningitis, syphilis, pneumonia, UTI (urinary tract infection), and other infections. Allergic reactions with penicillin may occur in some individuals.

Both Gram positives and Gram negatives are sensitive to penicillins, but many bacteria develop resistance to them by producing a group of enzymes called penicillinases including β lactamases and acylases. While β lactamases inactivate antibiotics by attacking β lactam rings, acylases on the other hand bring inactivation by cleaving the acylamino side chain of penicillins. To solve the problem, semisynthetic penicillins were developed, few of these, for example, methicillins were found to be insensitive to penicillinase enzymes.

Another β lactam antibiotic, consisting of a six-membered dihydrothiazine ring associated with a β lactam ring is cephalosporin, produced by the fungus *Acremonium chrysogenum*. The production pathway and action mechanism of cephalosporin are analogous to penicillin. Like penicillin, cephalosporin works by associating with PBPs to cause cell lysis and has broad-spectrum activity against a number of diseases. The susceptibility of cephalosporin to β lactamase enzymes led to the development of different generations of cephalosporins, e.g., cefamandole, cefixine, cefapine. Ceftobiprole, an approved cephalosporin, actively works against methicillin-resistant *S. aureus* (Horgan and Murphy, 2018).

Fusidic acid is a narrow spectrum antibiotic produced by *Fusidium coccineum* that is effective against penicillin and cephalosporin resistant Gram negative bacteria. It is used for the treatment of common skin infections and works by inhibiting bacterial protein synthesis by interfering with the activity of elongation factor G.

Penicillium griseofulin produces antifungal antibiotic griseofulvin, which inhibits fungal growth by inhibiting mitosis. Echinocandins (e.g., Caspofungin, Micafungin) are broad-spectrum antifungal compounds that inhibit the activity of 1,3-β-D-glucan synthase enzymes associated with fungal cell wall synthesis and eventually lead to osmotic lysis and fungal cell death (Horgan and Murphy, 2018).

In addition to antibiotics, a number of other metabolites synthesized by filamentous fungi and yeasts find their extensive applications as antitumor, antimalarial, antifungal, immunosuppressant, enzyme inhibitors, and hypocholesterolemic drugs (Table 3.3). These bioactive agents play a dramatic role in improving human health as well as alleviating infectious and noninfectious diseases (Adrio and Demain, 2003; Pradeep et al., 2019).

Taxol and camptothecin are the commonly used anticancer drugs produced by endophytic fungi. Taxol (paclitaxel) was approved in 1992 and employed for breast cancer and Kaposi's sarcoma treatment. It inhibits the multiplication of cancerous cells by preventing cell division through binding and polymerizing microtubules. Many endophytic fungi such as *Colletotrichum, Cladosporium, Taxomyces, Tubercularia*, and *Fusarium* are the producers of texol. Camptothecin synthesized by *Entrophospora infrequens* and *Trichoderma atroviridi* is clinically used for lung, uterine, and ovarian cancers. The

Table 3.3 Antibiotics and other drugs from fungi.

S. No.	Antibiotic/bioactive drugs	Microbial source	Activity
1.	Penicillin	*Penicillium chrysogenum* *P. nalgiovense, P. rubens*	Antibiotic
2.	Cephalosporin	*Acremonium chrysogenum*	Antibiotic
3.	Fusidic acid	*Fusidium coccineum*	Antibiotic
4.	Cyclosporin A	*Tolypocladium nivenum*	Immunosuppressant Antifungal peptide
5.	Fingolimod (FTY720)	*Isaria sinclairii*	Immunosuppressant Multiple sclerosis
6.	Mycophenolate mofetil Novartis Myfortic	*Penicillium stoloniferum*	Immunosuppressant
7.	Lovastatin Simvastatin	*Aspergillus terreus*	Hypolipidemic drugs Hypocholesterolemic agents
8.	Pravastatin	*Penicillium citrinum* and *Streptomyces carbophilus*	Hypocholesterolemic agents
9.	Griseofulvin	*Penicillium griseofulvum*	Antifungal
10.	Cerulenin	*Cephalosporium caerulens*	Antifungal
11.	Parnafungin	*Fusarium lavarum*	Antifungal
12.	Caspofungin	*Glarea lozoyensis*	Antifungal
13.	Micafungin	*Coleophama empedri*	Antifungal activity
14.	Pleuromutilins	*Pleurotus mutilus*	Inhibit Protein synthesis

target being cellular Type 1 DNA topoisomerases. Irinotecan and topotecan, the water-soluble derivatives of camptothecin, are also used as anticancer drugs against cervical and colorectal cancer. An antiangiogenesis compound, Fumagillin which stops tumor development by stopping blood vessel formation is produced by *A. fumigatus*. Endophytic fungi seem to be an effective source for new chemotherapeutic agents (Demain and Martens, 2017).

Presence of high levels of cholesterol and fat, if untreated, increases the risk of atherosclerosis, heart diseases, and strokes in many people. Statins, a group of secondary fungal metabolites, can be widely used as hypocholesterolemic drugs that lower the plasma cholesterol level by 20%–40% by inhibiting HMG-CoA (hydroxymethyl glutaryl-coenzyme A reductase), the key enzyme involved in the synthesis of cholesterol in the liver. Clinically used statins may be natural (Lovastatin, Compactin), semisynthetic (Simvastatin, Pravastatin), and synthetic (Atorvastatin, Fluvastatin) (Manzoni and Rollini, 2002). Lovastatin produced by *Aspergillus terreus* and *M. ruber* was the first commercialized statin approved by FDA in 1987. It is used to produce Simvastatin (Zocor) which besides lowering the bad cholesterol and improving the good one is reported to have other beneficial activities like antioxidants, antiinflammatory, and antiviral activity. Pravastatin (Pravacol) is obtained by a two-step process: synthesis of Compactin (Mevastatin) by *Penicillium citrinum* followed by biotransformation involving *Streptomyces carbophilus*. Liquid or solid-state fermentation can be employed for producing these compounds.

The ergot alkaloid, Ergometrine and methylergometrine, synthesized semisynthetically from *Claviceps paspali* is therapeutically used for treating postpartum hemorrhage. Immunosuppressive drugs made the organ transplantation easy and successful. Cyclosporine A produced from *Tolypocladium nivinum* was approved in 1983 as an immunosuppressant. It inhibits specific T cell activation by interfering with interleukin production and is employed in the transplantation of heart, liver, and kidney. Cyclosporine is also reported to have antiviral activity and found its use in psoriasis and eczema. Mycophenolate mofetil (ester derivative of mycophenolic acid) is another immunosuppressive agent approved for kidney (in 1995) and heart (in 1998) transplantation. It can be used along with cyclosporine A during transplantation. FDA also approved the use of Novartis Myfortic acid (mycophenolic acid delayed release tablets) and Fingolimod (FTY720) as immunosuppressive drugs in 2004 and 2010, respectively. Therapeutic potential of fungi is immense and provides a multibillion-dollar market worldwide (Mostafa et al., 2020).

3.2.10 Polysaccharides

Polysaccharides are complex carbohydrate molecules formed by the bonding of many same or different sugar molecules. They are water soluble and could be linear and branched. Fungal polysaccharides are widely used in the food and pharma sectors.

Most of the Basidiomycetes are known for their anticancer and antitumor properties which are contributed by the polysaccharides produced by them. For example, *Tremella fuciformis*, *Ganoderma lucidum*, *Griflola frondosa*, *H. erinaceus*, *L. edodes*, *Inonotus obliquus*, and *Coriolus versicolor*. The antitumor activity was first checked on S-180 tumor cells (Sarcoma) by Byerrum et al. (1957). Since then, many components, especially polysaccharides, were isolated from fungi using various purification methods.

In the past few decades, the research on novel fungal polysaccharides has been accelerated to intensify the usage and marketing of these components. Most of the research articles claim that the antitumor and anti-cancerous properties of the fungal polysaccharides are due to their immunomodulation property. Fungal polysaccharides can be classified on the basis of structure into linear and branched, sugar composition into homo and heteropolysaccharides, and type of bonds into α-(1 → 3), β-(1 → 3), β-(1 → 6), etc. (Xiao et al., 2020).

Linear polysaccharides have linear, uniform, and flat structure that exist in ribbon like structure like cellulose a chain of β-D-glucopyranosyl units, and chitin polymer of 2-acetamido-2-deoxy-β-D-glucopyranosyl units (Xiao et al., 2020).

Branched polysaccharides may consist of extensive branch points in the structure. One such example is gum arabica produced by *Agaricus blazi*, *G. frondosa*, *G. lucidum*, and *L. edodes*.

Homopolysaccharide is a polymer of the same subunit; an example may include the molecules like glucagon, starch, cellulose, etc. Most of the antitumor polysaccharide consists of β-glucans having glucose in the backbone.

Heteropolysaccharides are formed by different kinds of sugar subunits joined by glycosidic bonds. For example, Pullulan secreted by *Aureobasidium pullulans*. Xanthan secreted by *Xanthomonas campastris*, scleroglucan by *Sclerotium glucanicum*, etc. (Xiao et al., 2020).

Pullulan consists of maltotrios unit each joined by α-1,6 glycosidic bond where these maltotrios formed by joining of three glucose units by α-1,4 glycosidic bonds. Pullulan strengthened the cell by providing resistance against the heat and predation. They are also used commercially for oral hygiene products like Listerine, etc.

Xanthan gum consists of D-glucose having β-(1 → 4)-linkage and at each third position glucose is bound with a trisaccharide side chain having two mannose residues and one glucuronic acid. Due to this structure, they have immense application in the food industries, and commercial production began in late 1964 after the authorization by US FDA (Kumar et al., 2018). It is produced commercially by submerged batch fermentation having starch, sucrose, and glucose in the medium (Kumar et al., 2018). The maintenance of pH is important during the fermentation process because at pH below 5.0 the production of xanthan gum ceases. For making food-grade gum, the fermentation process must be stopped as soon as the carbohydrate is exhausted in the medium. Then it is precipitated by isopropyl alcohol and dried for commercialization. The total market is expected to increase by $972.0 million in 2022 from $737.0 million in 2017 with an annual growth rate of 5.7%.

3.2.11 Lipids and glycolipids

Lipids and glycolipids are major constituents of the fungal membrane, where they can help in transferring the intra-extracellular signals. Various types of fungal lipids are sphingolipids, fatty acids, oxylipins, glycolipids, phospholipids, sterols, oils, and fats.

Fatty acids are carboxylic acids having an aliphatic tail, serving as an essential source of energy (Morita et al., 2013). Lipids, such as triacylglycerols, consist of three fatty acids attached to three carbon glycerol molecules. They could be saturated without double bonds and polyunsaturated with many double bonds, also known as polyunsaturated fatty acids (PUFA). PUFAs play an important role in human physiology such as it helps in the synthesis of prostaglandins, strengthening the immune system, and maintaining the blood pressure. Fungal sources of PUFA include *Mucor phytium* and *Mortierella* (Morita et al., 2013). Till date three PUFAs are available commercially using microorganisms which includes γ-linolenic acid (GLNA) by *Mucor circinelloides*, docosahexaenoic acid (DHA), and arachidonic acid (ARA) by *Mortierella* spp. (Tauk-Tornisielo et al., 2009). Out of them GLNA is widely used in the treatment of eczema, depression, hyperactivity, and peroxisomal disorders.

Among all lipids triacylglycerols, a storage lipid is the major component that may be utilized for the growth and development. The second major lipid component is sterols and squalene which provide a liquefying effect to the fungal membrane. They may also help in the sexual reproduction of the fungi by synthesizing steroid hormones.

Phospholipids are another important structural component which helps in signaling. Other lipid molecules are also there to help fungi in the signaling, while oils and fats are used for the commercial purpose (Patton-Vogt and de Kroon, 2020). Since most of the oil and fats demand is fulfilled by the plant and animal sources, we need to have some alternative that can be cultivated in lesser time duration on a large scale and manipulated as per the requirement such as microbial sources. Therefore, in the past few years, microbial lipids have attracted the industrialist as alternative renewable carbon sources for practical applications.

3.2.12 Bioremediation

The process of removal of pollutants from the environment by using a variety of organisms is called bioremediation. A wide range of pollutants is released into the soil and aquatic environment by industries such as paper and pulp, textiles, agricultural, cosmetics, pharmaceuticals, and others. These pollutants include dyes, pesticides, herbicides, insecticides, fungicides, textile auxiliaries, chemicals, metals, salt, surfactant, etc.

Textile industry is one of the most important sources of environmental pollution. It is the fast-growing industry across the globe with a more than 4.0% growth rate in the market annually due to increased consumption and demand of textile products (Sivaram et al., 2019). The key activities during the preparation of textile fibers/yarns to knitting/weaving to final products involve processes like cleaning, carding, sizing, decolorizing, texturing, mercerizing, dyeing, printing, finishing, etc. (Sivakumar et al., 2014). Many of these are wet processes having a demand of a large volume of water, thus in turn generating a huge amount of wastewater effluent loaded with pollutants. The load and composition of pollutants vary with the type of processes and the chemicals used; the primary one being residual synthetic dyes, chemicals, chromium ions (Cr^{6+}), surfactants, and raw material impurities. Moreover, ~75% of the salts used (70–90 g/L) during the dyeing process enters into wastewater that was difficult to remove by traditional methods (Bisschops and Spanjers, 2003).

Rapid industrialization and urbanization in the nineteenth century led to the voluminous use of chemicals including dyes in different industries, viz., food, cosmetics, agriculture, paper, and textile. Commercially, over 1 lakh (100,000) dyes are produced globally with more than 7×10^5 metric tons annual production (Campos et al., 2001). A large amount of these dyes is released in the industrial effluents as spent dye, many of which are reported to be toxic, mutagenic, and carcinogenic. Their presence in water poses a health hazard to humans as well as a threat to aquatic life.

The increase in demand for agriculture products because of population explosion, control of pests and insects to protect vegetation, and to improve the quality of human life results in the large-scale use of pesticides including herbicides, insecticides, and fungicides. The need is to have pesticides that are nontoxic and readily degradable. The natural organic compounds though readily biodegradable by natural microflora and do not persist in the environment for long, are not able to meet the required demand. The first synthetic pesticide used in 1939 was DDT (1,1,1-trichloro-2, 2-bis (*p*-chlorophenyl) ethane) which led to the start of the modern era of synthetic pesticides. Thereafter, other compounds such as pentachlorophenol, polychlorinated biphenyls (PCBs), chlorophenoxyalkanoates, triazines, chlorophenols, polychlorinated dibenzofurans (PCDFs), organophosphorus compounds, and phenols have been in use. These compounds are released in soil and water columns and load the environment with large amounts of chemicals and recalcitrant compounds (Maloney, 2001).

These inorganic and organic pollutants including dyes, pesticides, surfactants, etc. are of great concern for environment safety and living organisms. Most of these are toxic and have very low/no biodegradability and remain in the environment for long times (recalcitrant). Moreover, many of these may enter into food chains and accumulate in organisms at various trophic levels (bioaccumulation) and pose threat to human health, for example, surfactants used during raw material preparation and finishing process in textiles accumulate in fat tissues of aquatic animals. These reduce photosynthesis and lowers dissolved oxygen concentration in water bodies and thus can destroy aquatic life. High salt concentration in wastewater effluent causes salinization of soil and underground and surface freshwater, thereby affecting agricultural activity and drinking water resources. It also alters the aquatic ecosystems equilibrium. In addition, the scarcity of water resources emphasizes the recycling and reuse of wastewater in industries.

Various physico-chemical and biological treatment processes such as chemical- and electro-coagulation, oxidation, chemical precipitation, membrane separation, flotation, and reverse osmosis are commonly employed. However, the problem of their high operational costs, low efficiency, time consuming, and huge sludge production and disposal difficulties associated with existing physicochemical methods necessitates the urgent need for simple, effective, and low-cost techniques, such as bioremediation to decontaminate liquid wastes. Ubiquitous presence of microbes like bacteria and fungi makes them a prospective alternative for pollution control (Sivakumar et al., 2014). Many valuable characteristics of these microorganisms, like easy and fast growth, wide range of metabolic capabilities, and ease of genetic manipulations make their use for bioremediation as an inexpensive, permanent, and ecologically safe way for removal of these organic pollutants from soils and water columns.

The fungi are the organisms of choice for bioremediation because of their robust metabolic capabilities (Gomaa et al., 2012). White rot fungi are capable of metabolizing a number of structurally diverse compounds. *Phanerochaete chrysosporium* (wood-rotting fungus) is the most widely studied for xenobiotic biodegradation. The soil fungi such as *P. chrysosporium*, *Trametes versicolor*, *Bjerkandera adjusta*, and *Pleurotus* sp. through secretion of extracellular lignin-degrading enzymes efficiently degrade environmentally persistent compounds-lignin in natural lignocellulosic substrates. Moreover, these lignolytic enzymes (e.g., laccases and peroxidases) have been reported to degrade many xenobiotic compounds (Wesenberg et al., 2003).

There are a number of advantages of using fungi species in bioremediation. The first is their cell wall, which can bind effectively to diverse chemicals and metals. Moreover, it is easy to separate fungal mat from growing media. This allows the potential use of fungal mat as biosorbent—the biological material used for removal of pollutants from solution—that allows the treatment of large volumes of effluent cost effectively (Kaushik and Malik, 2009). The biomass of the fungus *Cunninghamella elegans* was found to be an excellent biosorbent in decreasing the content of dyes, salts, surfactants, heavy metals, and other molecules in wastewater (Tigini et al., 2011). It seems to be a promising candidate that can be used alone or along with other physicochemical treatment techniques like activated carbon, ozonation, etc. for pollutant removal.

In addition, fungus can be grown easily on cheap nutrients, even on various industrial byproducts, and can adapt readily to various extreme environmental conditions, yielding high biomass. Further, these possess a variety of biochemical reactions including oxidation, reduction, methylation, dehydrogenation, dehalogenation, aromatic ring cleavage, hydrolysis, etc. that help in the modification or removal of hazardous chemicals (Deshmukh et al., 2016). Also, the extra- and intracellular enzymes secreted by fungi are involved in the decontamination of contaminated areas. Fungi are used extensively and new ones are being investigated for their use in bioremediation because of their high tolerance to large concentrations of pollutants. Examples include enzymes laccase used for the decolorization of spent dyes and removal of pollutants; and protease enzymes for decomposing hides, flesh, and animal residue in the effluent. Other enzymes of importance are peroxidases and cytochrome 450 for detoxification and biodegradation. Extensive research is being carried out for the use of fungal enzymes in different biotechnological applications. The degradation of Malachite green, Nigrosin, and basic fuchsin dyes by *A. niger* and *P. chrysosporium* concluded their viable application in waste and wastewater treatment (Rani et al., 2014). Similarly, involvement of *A. niger* and *A. flavus* in environmental decontamination indicated their potential use in mycoremediation (Sivakumar et al., 2014).

Many other fungi, viz., *Aspergillus tamari, A. solani, A. ochraceus, Penicillium purpurogenum, Fusarium oxysporum, Trichoderma lignorum, Tramates pubescens, Pleurotus ostreatus, T. versicolor, P. chrysosporium,* and even some basidiomycetous fungi found to be involved in pollutants detoxification and decolorization of dyes, making feasibility of their potential application in various industrial processes (Rani et al., 2014; Singh et al., 2015). Additionally, many species demonstrate both enzyme-mediated degradation and biosorption ability for efficient pollutant decontamination (Park et al., 2007). The need is to screen more fungal species for their bioremediation capabilities and production of extracellular enzymes and also to optimize the working conditions to maximize effluent treatment economically.

3.3 Methods to enhance the fungal properties

The development of improved strains for industrial application is a prerequisite to build up economical processes in biotechnology. Mainly strain improvement is done by mutagenesis, RDT, and recombination (Weber et al., 2012).

In the usual course of events, RDT (recombinant DNA technology) has immense potential for strain improvement as it has given the basic process of cloning in microbial populations. These cloning processes now have been standardized for many microbes such as bacteria, yeast, and filamentous fungi. With the advent of advanced sequencing facilities, DNA manipulation becomes easy which helps in the infusion of desired character in the industrially important microbes. In addition, metabolic engineering also contributes to improving the yield and quality of final fermentation products formed by various fungi. There are many examples in which strain improvement has increased the yield of desired product such as the production of penicillin by improved *P. chrysogenum* strain (Weber et al., 2012). Aforesaid strain was formed by the introduction of an additional *penDE* gene into the genome of high-penicillin-yielding strains which results in a considerable rise in penicillin production (Weber et al., 2012). In another example, an endoglucanase gene from *T. reesei* was introduced in *S. cerevisiae*, by using recombinant DNA techniques which allowed brewer's yeast to hydrolyze β-glucans (Chidananda et al., 2008). Similarly, with the help of RDT a starch utilizing strain was developed which lowers the acidity during the fermentation process. Another example is, the addition of the amyloglucosidase gene in *A. niger* helps in the breakdown of dextrins during the beer production (Chidananda et al., 2008).

Other than genetic engineering the use of mutagen also helps in the development of improved strains. Even though it is a random process having ambiguity to identify the site of action in the target genome, it is still used to develop improved strains. The major advantage of this process is to stop the functioning of undesirable proteins. However, the screening process is time consuming, labor intensive and needs lots of resources. One such example is *Aspergillus niger* CFTRI 1105 strain which was mutated by the application of UV radiation and nitrous acid for enhancing the production of asperenone. Asperenone is an inhibitor of the enzyme lipoxygenase (Chidananda et al., 2008). Same mutagens were used to improve the *A. terreus* strain for the production of lovastatin which is used to lower the bad cholesterol in the blood (Parekh et al., 2000).

In some of the fungi parasexual recombination can be used to improve the strain. Since sexual reproduction is limited in these microorganisms so genetic recombination is usually done by parasexual mode. However, parasexual mechanisms generally result in low frequencies and can be performed in limited organisms. Additionally, it requires a genetic marker for the selection of desirable mutants. Because of this reason, this technique is not very popular for developing the improved strain for industrial purposes. However, the protoplast fusion technique is a breakthrough for getting high recombination frequencies by parasexual recombination. So, some expectations are there to get interspecies hybrids for improving the strains. One strain was developed in *Aspergillus awamori* to improve the improved chymosin production (Ward et al., 1990).

3.4 Future perspectives in fungal biotechnology

In the last few decades, there has been immense progress in the field of fungal biotechnology. It has progressed on five major fronts which include progress at the industrial level, screening of new strains having valuable properties, advancement in genetic engineering, development in genomics tools, and development in the fermentation processes. However, there is still a space to carry forward these progresses for developing better technology. For example, many antibiotics isolated from fungi have been discovered at lab scale that need to be expanded further to the industries, and for achieving this we need to ease out the technology transfer processes. In addition, with the emergence of drug-resistant microorganisms, there is always a need for new antibiotics, therefore, an omic approach that is proteomics, genomics, and metabolomics needs to be utilized in this field.

Nowaday's sustainable production is the need of the day so the potential of these GRAS (generally recognized as safe) fungi needs to be exploited for reusing the biodegradable waste. Most of the oleaginous fungi have a natural ability to accumulate lipids which can be used for the production of biofuels to replace the conventional fossil fuels.

If we look at the past development of the fungal biotechnology, it shows that out of all known fungi only 5% of the fungal species have been used for making valuable products. Our nature still consists of numerous unknown fungal species that need to be identified by using the approach of metagenomics. There is always a space to develop new methods for culturing those fungal species which are hard to grow in the normal culture conditions. As we advance in this field more functional new proteins will be revealed by utilizing bioinformatics tools or functional biology. These approaches will help us to identify new targets for screening the novel fungal products. In addition, there is always a need for enzymes with improved properties for any biological reaction whether in vivo or in vitro which can be identified by directed evolution. Primary and secondary metabolites can be improved by genome shuffling of whole cells along with the metabolic engineering.

It is now obvious that fungal biotechnology has many possibilities for improvement and many directions to proceed in the future. Working in all the aspects at a time is not possible, so critical thinking, proper management, and evaluation of market value of the product need to be there for improving the use of fungi in the industries.

References

Adrio, J.L., Demain, A.L., 2003. Fungal biotechnology. Int. Microbiol. 6, 191–199.

Aunstrup, K., 1979. Production, isolation, and economics of extra cellular enzymes. In: Wingard, L., Katchalski-katzir, G. (Eds.), Applied Biochemistry and Bioengineering. Academic Press, New York, pp. 27–69.

Bajaj, P., Mahajan, R., 2019. Cellulase and xylanase synergism in industrial biotechnology. Appl. Microbiol. Biotechnol. 103 (22), 8711–8724.

Bakri, Y., Jacques, P., Thonart, P., 2003. Xylanase production by *Penicillium canescens* 10-10c in solid-state fermentation. Appl. Biochem. Biotechnol. 108, 737–748.

Baur, X., Marek, W., Ammon, J., Czuppon, A.B., Marczynski, B., Raulf-Heimsoth, M., Roemmelt, H., Fruhmann, G., 1994. Respiratory and other hazards of isocyanates. Int. Arch. Occup. Environ. Health 66 (3), 141–152.

Bhattarai, K., Stalick, W.M., McKay, S., Geme, G., Bhattarai, N., 2011. Biofuel: an alternative to fossil fuel for alleviating world energy and economic crises. J. Environ. Sci. Health A Tox. Hazard. Subst. Environ. Eng. 46 (12), 1424–1442.

Bisschops, I., Spanjers, H., 2003. Literature review on textile wastewater characterisation. Environ. Technol. 24, 1399–1411.

Budak, N.H., Aykin, E., Seydim, A.C., Greene, A.K., Guzel-Seydim, Z.B., 2014. Functional properties of vinegar. J. Food Sci. 79 (5), 757–764.

Byerrum, R.U., Clarke, D.A., Lucas, E.H., Ringler, R.L., Stevens, J.A., Stock, C.C., 1957. Tumor inhibitors in *Boletus edulis* and other *Holobasidiomycetes*. Antibiot. Chemother. 7 (1), 1–4.

Campos, R., Kandelbauer, A., Robra, K.H., Cavaco-Paulo, A., Gübitz, G.M., 2001. Indigo degradation with purified laccases from *Trametes hirsuta* and *Sclerotium rolfsii*. J. Biotechnol. 89 (3), 131–139.

Chidananda, C., Kumar, C.M., Sattur, A.P., 2008. Strain improvement of *Aspergillus niger* for the enhanced production of asperenone. Indian J. Microbiol. 48 (2), 274–278.

Conkey, W.H., Van Horn, W.M., Shema, B.F., Shockley, W.H., 1957. Sheets Comprising Filaments of Fungi. US patent 2,811,442.

Copetti, M.V., 2019. Fungi as industrial producers of food ingredients. Curr. Opin. Food Sci. 25, 52–56.

Demain, A.L., Martens, E., 2017. Production of valuable compounds by molds and yeasts. J. Antibiot. (Tokyo) 70 (4), 347–360.

Deshmukh, R., Khardenavis, A.A., Purohit, H.J., 2016. Diverse metabolic capacities of fungi for bioremediation. Indian J. Microbiol. 56 (3), 247–264.

Dörsam, S., Fesseler, J., Gorte, O., Hahn, T., Zibek, S., Syldatk, C., Ochsenreither, K., 2017. Sustainable carbon sources for microbial organic acid production with filamentous fungi. Biotechnol. Biofuels 10, 242.

El-Kady, I.A., Abdel Zohri, N.A., Hamed, S.R., 2014. Kojic acid production from agro-industrial by-products using fungi. Biotechnol. Res. Int. 2014, 1–11.

Fatma, S., Hameed, A., Noman, M., Ahmed, T., Shahid, M., Tariq, M., Sohail, I., Tabassum, R., 2018. Lignocellulosic biomass: a sustainable bioenergy source for the future. Protein Pept. Lett. 25 (2), 148–163.

Godjevargova, T., Dayal, R., Turmanova, S., 2004. Gluconic acid production in bioreactor with immobilized glucose oxidase plus catalase on polymer membrane adjacent to anion-exchange membrane. Macromol. Biosci., 950–956.

Gomaa, O.M., Fathey, R., Kareem, H.A.E., 2012. Biotechnological applications of fungi in textile waste water bioremediation. In: Non-Conventional Textile Waste Water Treatment. Nova Publishers, New York, pp. 79–96.

Gopinath, S.C., Anbu, P., Arshad, M.K., Lakshmipriya, T., Voon, C.H., Hashim, U., Chinni, S.V., 2017. Biotechnological processes in microbial amylase production. Biomed. Res. Int., 1272193.

Hernández, V.A., Galleguillos, F., Müller, R.T.A., 2019. Fungal dyes for textile applications: testing of industrial conditions for wool fabrics dyeing. J. Text. Inst. 110, 61–66.

Horgan, K.A., Murphy, R.A., 2018. Pharmaceutical and chemical commodities from fungi. In: Kavanagh, K. (Ed.), Fungi Biology and Applications. John Wiley & Sons, Inc, pp. 169–201.

Jay, J.M., Loessner, M.J., Golden, D.A., 2006. Non-dairy fermented foods and products. In: Modern Food Microbiology. CBS Publishers and Distributors, India, pp. 175–195.

Johnson, M.A., Carlson, J.A., 1978. Mycelial paper: a potential resource recovery process. Biotechnol. Bioeng. 20, 1063–1084.

Jones, M., Kujundzic, M., John, S., Bismarck, A., 2020. Crab vs. mushroom: a review of crustacean and fungal chitin in wound treatment. Mar. Drugs 18 (1), 64.

Jones, M., Gandia, A., John, S., et al., 2021. Leather-like material biofabrication using fungi. Nat. Sustain. 4, 9–16.

Kalra, R., Conlan, X.A., Goel, M., 2020. Fungi as a potential source of pigments: harnessing filamentous fungi. Front. Chem. 8, 358–369.

Kaushik, P., Malik, A., 2009. Fungal dye decolourization: recent advances and future potential. Environ. Int. 35, 127–141.

Kubicek, C.P., 2001. Organic acids. In: Ratledge, C., Kristiansen, B. (Eds.), Basic Biotechnology. Cambridge University Press, Cambridge, pp. 305–324.

Kumar, R., Kumar, P., 2019. Yeast-based vaccines: new perspective in vaccine development and application. FEMS Yeast Res. 19, 1–7.

Kumar, A., Rao, K.M., Han, S.S., 2018. Application of xanthan gum as polysaccharide in tissue engineering: a review. Carbohydr. Polym. 15, 128–144.

Kyu, M.T., Nishio, S., Noda, K., Dar, B., Aye, S.S., Matsuda, T., 2020. Predominant secretion of cellobiohydrolases and endo-β-1,4-glucanases in nutrient-limited medium by *Aspergillus* spp. isolated from subtropical field. J. Biochem. 168 (3), 243–256.

Lagashetti, A.C., Dufossé, L., Singh, S.K., Singh, P.N., 2019. Fungal pigments and their prospects in different industries. Microorganisms 7 (12), 604.

Macchione, M.M., Merheb, C.W., Gomes, E., da Silva, R., 2008. Protease production by different thermophilic fungi. Appl. Biochem. Biotechnol. 146 (3), 223–230.

Magnuson, J.K., Lasure, L.L., 2004. Organic acid production by filamentous fungi. In: Tkacz, J.S., Lange, L. (Eds.), Advances in Fungal Biotechnology for Industry Agriculture and Medicine. Kluwer, Acadernicl Plenurn Publishers, pp. 110–115.

Maloney, S.E., 2001. Pesticide degradation. In: Gadd, G.M. (Ed.), Fungi in bioremediation. Cambridge University Press, pp. 188–223.

Manzoni, M., Rollini, M., 2002. Biosynthesis and biotechnological production of statins by filamentous fungi and application of these cholesterol-lowering drugs. Appl. Microbiol. Biotechnol. 58 (5), 555–564.

Mattey, M., 1992. The production of organic acids. Crit. Rev. Biotechnol. 12, 87–132.

Momeni, M.H., Ubhayasekera, W., Sandgren, M., Ståhlberg, J., Hansson, H., 2015. Structural insights into the inhibition of cellobiohydrolase Cel7A by xylo-oligosaccharides. FEBS J. 282 (11), 2167–2177.

Money, N.P., 2015. The fungi and biotechnology. In: Watkinson, S.C., Boddy, L., Money, N.P. (Eds.), Fungi and Biotechnology. Academic Press, pp. 401–424.

Morita, T., Fukuoka, T., Imura, T., Kitamoto, D., 2013. Production of annosylerythritol lipids and their application in cosmetics. Appl. Microbiol. Biotechnol. 97 (11), 691–700.

Mostafa, A., Kandeil, A.M., Elshaier, Y., Kutkat, O., Moatasim, Y., Rashad, A.A., Shehata, M., Gomaa, M.R., Mahrous, N., Mahmoud, S.H., GabAllah, M., Abbas, H., Taweel, A.E., Kayed, A.E., Kamel, M.N., Sayes, M.E., Mahmoud, D.B., El-Shesheny, R., Kayali, G., Ali, M.A., 2020. FDA-approved drugs with potent in vitro antiviral activity against severe acute respiratory syndrome coronavirus 2. Pharmaceuticals (Basel) 13 (12), 443.

Mukherjee, D., Singh, S., Kumar, M., Kumar, V., Datta, S., Dhanjal, D.S., 2018. Fungal biotechnology: role and aspects. In: Gehlot, P., Singh, J. (Eds.), Fungi and their Role in Sustainable Development: Current Perspectives. Springer Nature, Singapore, pp. 37–46.

Novelli, P.K., Barros, M.M., Fleuri, L.F., 2016. Novel inexpensive fungi proteases: production by solid state fermentation and characterization. Food Chem. 198, 119–124.

Ntana, F., Mortensen, U.H., Sarazin, C., Figge, R., 2020. *Aspergillus*: a powerful protein production platform. Catalysts 10 (9), 1064.

Okal, E.J., Aslam, M.M., Karanja, J.K., Nyimbo, W.J., 2020. Mini review: advances in understanding regulation of cellulase enzyme in white-rot basidiomycetes. Microb. Pathog. 147, 104410.

Ola, A.R.B., Metboki, G., Lay, C.S., Sugi, Y., Rozari, P.D., Darmakusuma, D., Hakim, E.H., 2019. Single production of kojic acid by *Aspergillus flavus* and the revision of Flufuran. Molecules 24, 1–6.

Pandey, A., Nigam, P., Soccol, C.R., Soccol, V.T., Singh, D., Mohan, R., 2000. Advances in microbial amylases. Biotechnol. Appl. Biochem. 2, 135–152.

Parekh, S., Vinci, V.A., Strobel, R.J., 2000. Improvement of microbial strains and fermentation processes. Appl. Microbiol. Biotechnol. 54 (3), 287–301.

Park, C., Lee, M., Lee, B., Kim, S.W., Chase, H.A., Lee, J., Kim, S., 2007. Biodegradation and biosorption for decolorization of synthetic dyes by *Funalia trogii*. Biochem. Eng. J. 36, 59–65.

Patton-Vogt, J., de Kroon, A.I.P.M., 2020. Phospholipid turnover and acyl chain remodeling in the yeast ER. Biochim. Biophys. Acta Mol. Cell Biol. Lipids 1865 (1), 158462.

Plassard, C., Fransson, P., 2009. Regulation of low-molecular weight organic acid production in fungi. Fungal Boil. Rev. 23, 30–39.

Pradeep, P., Manju, V., Ahsan, M.F., 2019. Antiviral potency of mushroom constituents. In: Agrawal, D., Dhanasekaran, M. (Eds.), Medicinal Mushrooms. Springer, Singapore, pp. 275–297.

Ramachandran, S., Fontanille, P., Pandey, A., Larroche, C., 2006. Gluconic acid: properties, applications and microbial production. Food Technol. Biotechnol. 44 (2), 185–195.

Rani, B., Kumar, V., Singh, J., Bisht, S., Teotia, P., Sharma, S., Kela, R., 2014. Bioremediation of dyes by fungi isolated from contaminated dye effluent sites for bio-usability. Braz. J. Microbiol. 45 (3), 1055–1063.

Robak, K., Balcerek, M., 2018. Review of second generation bioethanol production from residual biomass. Food Technol. Biotechnol. 56 (2), 174–187.

Ruijter, G.J.G., Kubicek, C.P., Visser, J., 2011. Production of organic acids by fungi. In: Esser, K., Bennet, J.W. (Eds.), The Mycota, Vol. X Industrial Application. Springer-Verlag Berlin Heidelberg, pp. 213–230.

Sauer, M., Porro, D., Mattanovich, D., Branduardi, P., 2008. Microbial production of organic acids: expanding the market. Trends Biotechnol. 26 (2), 100–108.

Sethi, B.K., Nanda, P.K., Sahoo, S., 2016. Characterization of biotechnologically relevant extracellular lipase produced by *Aspergillus terreus* NCFT 4269.10. Braz. J. Microbiol. 47 (1), 143–149.

Singh, R.L., Singh, P.K., Singh, R.P., 2015. Enzymatic decolorization and degradation of azo dyes—a review. Int. Biodeterior. Biodegradation 104, 21–31.

Sivakumar, D., Gayathri, G., Nishanthi, R., Vijayabharathi, V., Das, S., Kavitha, R., 2014. Role of fungi species in colour removal from textile industry wastewater. Int. J. ChemTech Res. 6, 4366–4372.

Sivaram, N.M., Gopal, P.M., Barik, D., 2019. Toxic waste from textile industries. In: Barik, D. (Ed.), Energy from Toxic Organic Waste for Heat and Power Generation. Woodhead Publishing Series in Energy, Woodhead Publishing, pp. 43–54.

Strzelczyk, E., Leniarska, U., 1985. Production of B-group vitamins by mycorrhizal fungi and actinomycetes isolated from the root zone of pine (*Pinus sylvestris*). Plant Soil 86, 387–394.

Su, C.H., Liu, S.H., Yu, S.Y., Hsieh, Y.L., Ho, H.O., Hu, C.H., Sheu, M.T., 2005. Development of fungal mycelia as a skin substitute: characterization of keratinocyte proliferation and matrix metalloproteinase expression during improvement in the wound-healing process. J. Biomed. Mater. Res. A 72, 220–227.

Sujatha, E., Girisham, S., Reddy, S.M., 2002. Production of indole acetic acid and free amino acids by three thermophilic fungi. Hindustan Antibiot. Bull. 44 (4), 37–41.

Sun, L., Kwak, S., Jin, Y.S., 2019. Vitamin A production by engineered *Saccharomyces cerevisiae* from xylose *via* two-phase *in situ* extraction. ACS Synth. Biol. 8 (9). 20.

Tauk-Tornisielo, S.M., Arasato, L.S., de Almeida, A.F., Govone, J.S., Malagutti, E.N., 2009. Lipid formation and γ-linolenic acid production by *Mucor circinelloides* and *Rhizopus* sp., grown on vegetable oil. Braz. J. Microbiol. 40 (2), 342–345.

Tigini, V., Prigione, V., Donelli, I., Anastasi, A., Freddi, G., Giansanti, P., Mangiavillano, A., Varese, G.C., 2011. *Cunninghamella elegans* biomass optimisation for textile wastewater biosorption treatment: an analytical and ecotoxicological approach. Appl. Microbiol. Biotechnol. 90 (1), 343–352.

Van Ooyen, A.J., Dekker, P., Huang, M., Olsthoorn, M.M., Jacobs, D.I., Colussi, P.A., Taron, C.H., 2006. Heterologous protein production in the yeast *Kluyveromyces lactis*. FEMS Yeast Res. 6 (3), 381–392.

Veeresh, J., Chuan, W.J., 2017. Heterologous protein expression in *Pichia pastoris*: latest research progress and applications. ChemBioChem 18, 7–21.

Venil, C.K., Velmurugan, P., Dufossé, L., Devi, P.R., Ravi, A.V., 2020. Fungal pigments: potential coloring compounds for wide ranging applications in textile dyeing. J. Fungi 6 (2), 45–68.

Vieira Gomes, A.M., Souza Carmo, T., Silva Carvalho, L., Mendonça Bahia, F., Parachin, N.S., 2018. Comparison of yeasts as hosts for recombinant protein production. Microorganisms 6 (2), 29–38.

Ward, M., Wilson, L.J., Kodama, K.H., Rey, M.W., Berka, R.M., 1990. Improved production of chymosin in *Aspergillus* by expression as a glucoamylase-chymosin fusion. Biotechnology 8 (5), 435–440.

Watanabe, F., Bito, T., 2018. Vitamin B_{12} sources and microbial interaction. Exp. Biol. Med. 243 (2), 148–158.

Weber, S.S., Polli, F., Boer, R., Bovenberg, R.A., Driessen, A.J., 2012. Increased penicillin production in *Penicillium chrysogenum* production strains via balanced overexpression of isopenicillin N acyltransferase. Appl. Environ. Microbiol. 78 (19), 7107–7113.

Wesenberg, D., Kyriakides, I., Agathos, S.N., 2003. White-rot fungi and their enzymes for the treatment of industrial dye effluents. Biotechnol. Adv. 22 (2), 161–187.

Willey, J.M., Sandman, K., Wood, D., 2014. The fungi. In: Prescott's Microbiology. McGraw Hill Higher Education, USA, pp. 588–603.

Xiao, Z., Zhou, W., Zhang, Y., 2020. Fungal polysaccharides. Adv. Pharmacol. 87, 277–299.

Xu, L., Chen, B., Geng, X., Feng, C., Meng, J., Chang, M., 2019. A protease resistant α galactosidase characterized by relatively acid pH tolerance from the shitake mushroom *Lentinula edodes*. Int. J. Biol. Macromol. 128, 324–330.

Yue, X., Chen, P., Zhu, Y., Zeng, Y., Liu, H., Liu, H., Wang, M., Sun, Y., 2019. Heterologous expression and characterization of *Aspergillus oryzae* acidic protease in *Pichia pastoris*. Sheng Wu Gong Cheng Xue Bao 35 (3), 415–424.

Yusof, M., Shabbir, M., Mohammad, F., 2017. Natural colorants: historical, processing and sustainable prospects. Nat. Prod. Bioprospect. 7, 123–145.

CHAPTER

Use of fungi in pharmaceuticals and production of antibiotics

4

Zeenat Ayoub[a] and Abhinav Mehta[b]

[a]Lab of Molecular Biology, Department of Botany, Dr. HarisinghGour Vishwavidyalaya, A Central University, Sagar, MP, India, [b]R. C. Patel Institute of Pharmaceutical Education and Research, Shirpur, MH, India

4.1 Introduction

The worldwide biome diversity results in a huge affluence in plants, animals, and microorganisms including lower fungi (yeasts and molds) and higher fungi (mushrooms). Fungi are eukaryotic heterotrophic organisms, seen more or less in all kinds of habitats on Earth. Most of the fungal species can acclimatize to extreme environments of oxygen, temperature, salinity, pH, organic/inorganic complexes, metal, water, and wastewater. They obtain the carbon and other nutrients from other organisms like that of animals. Some fungi are biotrophs, that is, they get their nutrients from the live organisms (plants or animals). Some are necrotrophs and some live as saprotrophs, that is, they find their nutrition by killing of host cells and from dead and decaying organisms, respectively. They were classified in the plant kingdom in earlier times. As far as the fungal mode of nutrition intake, their distinctive physical and structural characteristics, and lack of chlorophyll—a photosynthetic pigment, they were placed as an isolated kingdom like plants and animals (Mohmand et al. 2011). Studies revealed that presently less than 5% of fungal diversity is recognized (Muddassir et al. 2020). They are primordial living things with immense significance as (i) they are necessary for the existence of several organisms associated to them with the mutual organization; (ii) they play an important ecological role in the nutrient recycling and overall carbon cycle; (iii) act as main terrestrial ecosystem decomposers; (iv) fungi show huge prospective for agriculture, biotechnology, and biological production; (v) most of the fungi act as pathogens against plants, animals, and even humans; and (vi) for molecular-based biologists, fungi act as powerful genetic system models (Chambergo and Valencia, 2016).

To facilitate fungal survival, they themselves developed various approaches for their protection; the production of a number of secondary metabolites is among one of the most important approaches. Because of these fungal and fungal-provoked components in plants amplify the capacity to resist the attack of diseases, parasites, and predators in plants. Same as plants, fungal strains will probably be present in the under sampled areas, but our familiarity about fungi is not so complete as new fungal strains will be revealed in compactly populated regions like Netherland and United Kingdom as well as in the well-examined areas (De Kesel and Haelewaters 2019). Undeniably, about fifty newly identified fungal species were included to the British mycota in 2019. Several species among them are not perfectly named and are reported in introductory findings but nine species were latest to science. New fungal strains can be originated more or less everywhere from the Antarctica rocks to the sheep dung,

Entrepreneurship with Microorganisms. https://doi.org/10.1016/B978-0-443-19049-0.00017-7

Table 4.1 Top 10 countries where new fungal species described with numbers of new species published in 2019.

S. No.	Country	No. of new fungal species
1	China	377
2	Thailand	129
3	United States	105
4	Australia	96
5	Brazil	85
6	Spain	75
7	Italy	63
8	South Africa	62
9	India	61
10	Germany	47

sand dunes, leaf cutting ants nests, and air of basements (Cheek et al. 2020). Asia is a leading region in new fungal species discoveries. In 2019, the huge mainstream of new fungal species described are from China (377) and Thailand (129) as shown in Table 4.1, but records about the fungi from Central Asia are inadequate and other regions including Malaysia, Indonesia, and Myanmar. The top countries in the new fungal descriptions in 2019 are mostly equal as in 2017, the top three countries were China, Thailand, and Australia (Niskanen et al., 2018).

Numerous fungal species of miscellaneous morphological characteristics having abundant bioactive components make some of the fungal species as our foes and some as our friends. Most of these bioactive components became mycotoxins and possess detrimental consequences. Mainly fungi are well known to be the food spoilers as they cause harm to the food grains as well as cooked food items. They are also known as plant pathogens as they cause various severe diseases to plants including late blight of Potato and rice blast caused by *Phytophthora infestans* and *Magnaportheoryzae*, respectively. Nevertheless, fungi are valuable to humankind as they are the manufacturers of alkaloids such as ergot alkaloids, enzymes like lipase, cellulose, etc., food colorant, aroma and antibiotics such as cyclosporine, cephalosporin, penicillin, etc. as well. The medicinal value of mushrooms has supported the fact that they contain a variety of bioactive compounds such as ergothioneines, glutathiones, phenolics, terpenoids, lectins, sacharides, indole compounds, etc. that possessed a variety of biological properties such as antimicrobial, antioxidative, antiinflammatory, anticancer, etc. (Muszyńska et al. 2018). They help in biological nematode control. They also act as the nutritious foods for a mankind and possess numerous health benefits being proteins, vitamin D, potassium, selenium, niacin, and riboflavin sources. They also played a great role in the treatment or even in the prevention of hypertension, stroke, cancers, Alzheimer, and Parkinson diseases. All the beneficial as well as the harmful effects of fungi (Fig. 4.1) are recognized because of the abundant metabolites present in them. In view of the production of pharmaceuticals, enzymes and food pigments, fungi and fungal products are supposed to be the upcoming microbial cell factories and nowadays most of the industries are based on them. Due to the escalating requirements of the fungal metabolites the significant production can be attained via modern biotechnology equipments and proper use of fermentation techniques (Goyal et al. 2016).

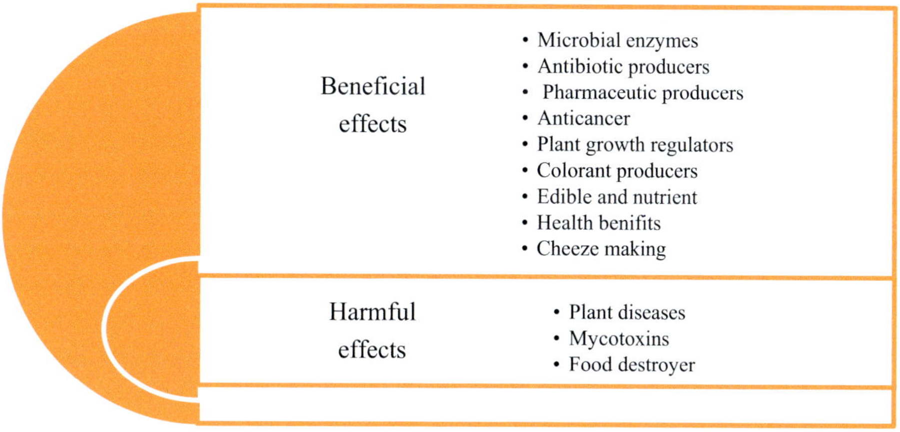

FIG. 4.1

Beneficial and harmful effects of fungi.

4.2 Production of secondary metabolites

Microbial fermentation is one of the most common technology generally customized in industries to obtain secondary metabolites and are utilized in making appropriate goods. This method is done in submerged environments in crops growing in liquid and this process is greatly easy and is effortless to scaling up the formation of goods and to influence the control factors. So as for the production of bioactive compounds under agro-industrial, microbial culture cultivation involves the solid-state condition fermentation technology. This kind of technology involves the solid support with less moisture content. This method is suitable for fungal species as they grow in low water content and is cost effective. In addition to primary metabolites, this method is also useful for the production of secondary metabolites (Kour et al. 2019; Yadav et al. 2019).

4.3 Biosynthesis of fungal metabolites

Fungi make a large number of secondary metabolites organizing from antibiotics to mycotoxins and the biosynthetic pathways involved in their synthesis are also diverse. Their synthesis involves three common metabolic pathways such as mevalonic acid pathway—which involve the synthesis of steroids, terpenoids, etc., the shikimic acid pathway—which involve the synthesis of alkaloids, aromatic amino acids, and the acetate pathway—which involve the synthesis of fatty acids, polyketides, etc. The pathways are generally said as after the involvement of enzymes or their intermediates and are usually used to categorize secondary metabolites. The enzymes which take part in these pathways are terpene cyclizes, geranylgeranyl diphosphate synthases, dimethylallyl tryptophan synthetases, polyketide synthases, nonribosomal peptide synthetases, etc. in order to obtain different fungal products, these enzymes use amino acids, acetyl-coA, mevalonate as building blocks. The real vital building block in steroids, terpenes, and gibberellins is dimethylallyl diphosphate, synthesized by Mevalonic acid by connecting three acetyl co-A molecules. Most of the hybrid metabolites are obtained by the cyclic acts of these enzymes, building blocks, and cofactors (Lee et al. 2016; Goyal et al. 2016). The number of secondary metabolites is escalating with each passing time. A great variety of secondary metabolites are generated in biological synthetic pathway by slight alterations having several uses in food, agriculture, and pharmaceuticals. The production of secondary metabolites is dependent on primary metabolite precursors like acetyl-CoA and amino acids. In one of the report, beauvericin was obtained by the modification of *Aspergillus niger* expression system (Richter et al. 2014). Another study involves the biosynthetic production of artemisinin by *Escherichia coli*, *Saccharomyces cerevisiae*, and *Artemisia annua* engineering, collected into two operons and then converted into a host strain of *E. coli* to attain proficient compound production (Ro et al. 2006).

4.4 Pharmaceuticals applications of fungi

In the medicinal and biotechnological fields, a new period has arrived with therapeutic applications, by the penicillin innovation from the fungus *P. notatum* and this innovation inspired many researchers to find new antibiotics. Then in the mid of 1950s, the microorganism screening had started in most of the pharmaceutical companies (Dreyfuss and Chapela 1994). Fungi from all the screened microorganisms proved to be the supreme sources for bioactive substances and were employed as a shield against other pathogenic fungi and bacteria (Deshmukh et al. 2017). Various secondary metabolites identified from fungi are having a variety of activities including antibiotic, antiinflammatory and anticancer, etc. and all the fungal metabolites have to be chosen as per the respective disease. The fungal alkaloids secreted from *Penicillium, Aspergillus, Chromocleista, and Pestalotiopsis* (Ma et al. 2016) are showing antiparasitic, antimicrobial, and antiHIV properties. The phenolics show a broad physiological continuum against allergies, oxidative stress, inflammation, etc. A great concern in pharmaceutical applications is regarded to resistance toward antibiotics, which ultimately makes the conventional therapeutic ineffective. The compounds derived from microorganisms are employed to amplify the vulnerability of drugs in mediators responsible for causing diseases. Fungi including mushrooms are showing various biological activities because of the presence of secondary metabolites, for example, caspofungin, taxol, penicillin, cyclosporine, lovastatin is having antifungal, anticancer, antibacterial, immunosuppressant, and cholesterol lowering properties, respectively (Macheleidt et al. 2016). Bioactive compound Griseofulvin, isolated from *Penicillin griseofulvum*, is used in the healing of nail and skin infections (Shu 2007). Mycophenolic acid, an antibiotic, isolated from the fungus *Penicillium brevicompactum*, has an antiproliferative potential toward a number of mice and rat tumors by acting as an effectual uncompetitive, reversible inhibitor of a rate-limiting enzyme, that is, inosine monophosphate dehydrogenase (IMPDH) which is responsible for de novo synthesis of guaosine nucleotides (Clutterbuck et al. 1932). Furthermore flavonoids synthesized from the fungus *Alternaria alternate* have a big role against phytopathogens in defense of plants and are involved in various metabolic activities of plants such as reproduction, growth, and cell signaling pathways. In New Zealand and Australia, natural products derived from fungi play an important role in the field of medicine, and 45 drugs derived from them reached to market level. Atorvastatin among all being the most extensively sold pharmaceutical drug (Beekman and Barrow 2014). Isocoumarins are the most important fungal bioactive compounds reported to possess numerous biological activities such as cytotoxic, antimicrobial, immunomodulatory, etc. Three hundred and seven derivatives of isocaumarins mostly isolated from *Penicillium* and *Asperigillus* genera were reported from 2000 to 2019 (Noor et al. 2020).

Moreover, since ancient times, mushrooms are believed as the sources of many medicines and are also utilized as folk medicines throughout the world. However, it is evident from the earliest literature that most of the mushrooms also possessed various biological activities including anticancer, antimicrobial, antiinflammatory, antidiabetic, hepatoprotective, etc. The mushroom extracts are used as chemotherapeutic tools widely in Japan, China, and Korea. The medicinal mushroom complex of polysaccharides-protein increases innate immunity resulting in anticancer efficacy in humans as well as in animals (Mohmand et al. 2011). Extensive research conducted throughout the world, substantiating the medical efficacy of mushrooms and the identification of their bioactive components such as proteins, glycoproteins, terpenoids, and polysaccharides. The most important therapeutic polysaccharides such as lentinan, schizophyllan, Grifron-D, and polysaccharopeptide (PSP) isolated from *Lentinula edodes*, *Schizophyllum commune*, *Grifola frondosa*, and *Trametes versicolor*, respectively, undergone widespread clinical trials against cancer (Thakur and Singh 2013). Some of the recent examples of bioactive compounds isolated from fungi having biological properties are presented in Table 4.2.

4.4.1 Fungi as sources of antibiotics

Fungi are extensively used in fermentation productions of organic acids, ethanol, enzymes, and antibiotics. Fungi such as *Cenococcum* sp., *Penicillium chrysogenum*, and *P. notatum* are used in antibiotic production (Yadav et al. 2019). In 1941, Selman Waksman first time used the term antibiotic for a small particle made by a microbe that provokes the development of other microbes. Alexander Fleming during the cleaning of the *Staphylococcus* loaded plates on September 3, 1928; has seen a clean part of contamination grown around *Penicillium notatum* (mold). He concluded that the mold inhibited the growth of bacteria on the plate and the mold has some substance responsible for bacterial killing. However, Ernst Chain and Howard Florey after 10 years isolated that substance from the mold and named as penicillin while working at Oxford University (Goyal et al. 2016). Alexander Fleming's penicillin innovation marked the basic point of the age of antibiotic chemotherapy and this innovation motivated other investigators to isolate novel chemical entities from fungi. Various antibiotics were revealed over the next 50 years, among them, comparatively few classes of compounds were obtained from fungi (Karwehl and Stadler 2017). Then the cephalosporins of the same class of penicillins specifically the beta-glucan antibiotics, as well as the pleuromutilin and fusidic acid were included (Hyde et al. 2019). Cephalosporins are all-purpose antibiotics, which are generally utilized by the patients having allergies to penicillin and also used against penicillin resistant infectivity. Cephalosporin C is reported to possess a broader range of antibacterial potential as compared to penicillins. Fusidic acid, a fusidane-type triterpenoid, was isolated from *Fusidium coccineum* by Godtfredsen et al. (1962) by testing the antibiotic activity of some fungal extracts. The antibiotic screening revealed efficient activity toward *Bacillus subtilis*, *Streptococcus pneumoniae*, *Mycobacterium tuberculosi*, *Corynebacterium diphtheria*, *Staphylococcus aureus*, *Clostridium tetani*, and *Neisseria meningitides* (Beekman and Barrow 2014). The same metabolite isolated from a fungus *Acremonium fusidioides* has been taken orally as an antibiotic and reported to be effective against some Gram positive bacterial infections.

In the 1950s the detection and therapeutic usage of antibiotics modernized the cure plus suffering and improved the life period. Antibiotics are valuable in the therapeutic of protozoal, bacterial, and fungal infectivity and several physiological disorders including cholesterol lowering (Berdy 2012). Because of the growing antibiotic utilization, their resistance started expanding and resulted in the search of novel antibiotics. The quantity of antibiotics mostly their analogs escalated in the exponential form.

Table 4.2 Some recent examples of bioactive compounds and their bioactivities.

Metabolites	Sources	Bioactivities	References
1,2-Dehydro-terredehydroaustin	*Aspergillus terreus*	Antiinflammatory	Liu et al. (2018a,b)
Asperisocoumarin C	*Aspergillus* sp.	Antioxidant	Chen et al. (2016)
Citreoisocoumarinol	*Peyronellaea glomerata*	Antioxidant	Zhao et al. (2016)
Mucorisocoumarin A and B and Peyroisocoumarin A, B and C			
Desmethyldichlorodiaportintone	*Ascomycota* sp.	Antiinflammatory	Chen et al., (2018)
Dichlorodiaportintone	*Ascomycota* sp.	Antibacterial	Chen et al., (2018)
Diaportinol	*Peyronellaea glomerata*	Antioxidant	Zhao et al. (2016)
Desmethyldichlorodiaportin	*Ascomycota* sp.	Antibacterial	Chen et al., (2018)
Pestalactone C	*Pestalotiopsis* sp.	Antifungal	Song et al. (2017)
Penicolinate A	*Bionectria* sp.	Anticancer	Kamdem et al. (2018)
Cyclohexene derivative; Thiazole	*Colletotrichum gloeosporioides*	Anticancer	Liu et al. (2018a,b)
2′-Deoxyribolactone, hexylitaconic acid, ergosterol	*Curvularia* sp.	Antimicrobial, antioxidant	Kaaniche et al. (2019)
Steroids, quinones, terpenoids, peptides, Xanthones	*Endolichenic* sp.	Anticancer, antiviral, antibacterial, antifungal, and antiAlzheimer's disease	Kellogg and Raja (2017)
β-1,3-Glucanolytic	*Fusarium solani*	Antifungal	Awad et al. (2018)
Polyene, diketopiperazine	*Penicillium crustosum*	Cytotoxic	Liu et al. (2019)
β-1,6-Glucanase	*Rhizoctonia solani*	Antifungal	Awad et al. (2018)
α-1,3-Glucanases	*Sclerotium rolfsi*	Antifungal	Awad et al. (2018)
Limonene and guaiol	*Trichoderma viride*	Antimicrobial	Awad et al. (2018)
Brefeldin A	*Penicillium* sp.	Antiviral	Raekiansyah et al. (2017)
Engyodontochones A–F	*Engyodontium album*	Antimicrobial	Wu et al. (2016)
2,6-Dimethyl-3-*O*-methyl-4-(2-methylbutyryl) phloroglucinol	*Aspergillus flocculosus*	Antineuroinflammatory	Choi et al. (2021)
Pretrichodermamide A	*Trichoderma harzianum*	Antimicrobial	Harwoko et al. (2021)
Epicorazine A	*Epicoccum nigrum*	Anticancer	Harwoko et al. (2021)
Vinblastine	*Nigrospora sphaerica*	Anticancer	Ayob et al. (2017)

8-(aminomethyl)-7-hydroxy-1-(1-hydroxy-4-(hydroxylmethoxy)-2,3-dimethylbutyl)-2 methyldodecahydro phenanthren-9(1H)-one; 1-((E)-2-ethylhex-1-en-1- yl)2-((E)-2-ethylidenehexyl) cyclohexane-1,2-dicarboxylate and 5-amino-2-(6- (2-hydroxyethyl)-3-oxononyl) cyclohex-2-enone	*Streptomyces coelicoflavus*	Antimicrobial Antioxidant Antiinflammatory	Rao et al. (2017)
Periconone B	*Periconia* sp.	Antiviral	Liu et al. (2017)
Stachbotrysin A and G	*Stachybotrys chartatum*	Antiviral	Zhao et al. (2017)
1,2-Dehydro-terredehydroaustin	*Aspergillus terreus*	Antiinflammatory	Liu et al. (2018a,b)
Aculeatusquinone C; 3-methyl-orsellinic acid aculeatus-quinone A	*Penicillium sclerotiorum*	Antiinflammatory	Zhao et al. (2020)
13β-Hydroxy conidiogenone C.	*Penicillium* sp.	Antiinflammatory	Li et al., (2020)
Minutaside A	*Tagetes minuta*	Antidiabetic, antioxidant	Ibrahim et al. (2015)
Erinacine C	*Hericium erinaceus*	Antiinflammatory	Wang et al. (2019a,b)
Antcin A	*Antrodia cinnamomea*	Anticancer	Kumar et al. (2019)
Erinacine A	*Hericium erinaceus*	Anticancer	Wang et al. (2019a,b)
Lentinan	*Lentinus edodes*	Anticancer	Deng et al. (2018)
Ganodermanondiol	*Ganoderma lucidum*	Melanogenesis inhibition	Kim et al. (2016)
Ganoderic acid D	*Ganoderma lucidum*	Antiaging, antioxidant	Xu et al. (2020)
2-Chloro-1,3-dimethoxy-5-methyl benzene	*Hericium* sp.	Antifungal	Song et al. (2020a,b)
Grifolin; neogrifolin and confluentin	*Albatrellus flettii*	Anticancer	Yaqoob et al. (2020)
HEP3 (single-band protein)	*Hericium erinaceus*	Immunomodulatory	Diling et al. (2017)
Ergosterol peroxide	*Grifola frondosa*	Antidiabetic	Wu et al. (2020)
Spiromentins C and B	*Tapinella atrotomentosa*	Antioxidant	Béni et al. (2018)
5-Hydroxy-hex-2-en-4-olide; Osmundalactone and Spiromentin C	*Tapinella atrotomentosa*	Antimicrobial	Béni et al. (2018)

In order to eradicate resistance to antibiotics, numerous semisynthetic antibiotics were developed by pharmaceutical industries, which lead to the new area of discovery of antibiotics (Bérdy 2005). In fact, there is much requirement for the novel antibiotic development having less toxicity, more effectiveness and showing lesser resistance risks while using against lethal infections of microbes (Wiese and Imhoff, 2019).

Fungi are the main sources of antibiotics taking part in human health and improving sufferings, as Streptomyces being prime manufacturer of large number of secondary metabolites and antibiotics (~80% of total known). In addition to antibiotics, ergot alkaloids, lipase, and statin cholesterol synthesis inhibitor are imperative pharmaceutical commodities of fungal sources. It is multifaceted to figure out the quantity of secondary metabolites obtained from fungi. Most of the fungal products can potentially be upcoming remedies and can be taken as medicines. Some of the examples of bioactive substances with antibiotic properties are the bioactive compound polyketide monocerin isolated from an endophytic fungus—*Exserohilum rostratum* of an Amazon plant species *Bauhinia guianensis*—which has an adversary activity toward some bacterial species such as *S. aureus*, *Salmonella typhimurium*, *E. coli*, *Bacillus subtilis*, and *Pseudomonas aeruginosa* (Pinheiro et al. 2017).

Fungi and fungal products from marine habitat are also rich sources of a number of antibiotics, for example, the marine fungal species *Xylaria psidii* KT30 from *Kappaphycus alvarezii*—a seaweed proved to exhibit potential antibacterial efficacy against Gram positive bacteria such as *S. aureus and B. subtilis* (Indarmawan et al. 2016). Six new antibiotic polyketides of basic configuration of an anthraquinone-xanthone, that is, engyodontochones A-F, obtained from *Engyodontium album* LF069, isolated from a Mediterranean Sea sponge known as *Cacospinga scalaris*. These bioactive compounds revealed effects against bacterial species including *S. aureus* and *Staphylococcus epidermidis* having IC_{50} values ranging from 0.2 μM to 6.8 μM. Furthermore, engyodontochones B among all demonstrated 10 times stronger inhibitory activity against *S. aureus* which is more resistant to methicillin than chloramphenicol (Wu et al. 2016). The antibiotic activity of a polyketide known as Lindgomycin which was isolated from a fungal strain KF970 of the Lindgomycetaceae family, obtained from the Arctic seawater diminished the growth of some Gram positive bacterial species including *S. aureus*, *S. epidermidis*, and *Propionibacterium acnes*, plus the yeast *Candida albicans* having IC_{50} values from 2.7 μM to 5.7 μM. Talaromycesone A and B, the two dimeric forms of oxaphenalenone isolated from fungus *Talaromyces* sp. derived from a Mediterranean Sea sponge of sponge *Axinella verrucosa* were also effective toward *S. epidermidis* and methicillin-resistant *S. aureus* presenting the IC_{50} values near to 18.4 μM and 4.6 μM and, respectively (Wu et al. 2015). Asperamide A is a sphingolipid derived from *Aspergillus niger* obtained from *Colpomenia sinuosa*—a marine brown alga showed antibiotic activity against *C. albicans* (Abad et al. 2011). Moreover, several experimental prospects in the whole world had started the investigation to isolate the novel antimicrobial agents from mushrooms. For instance, the molecules like physcion, austrocortilutein, emodin, 6-methylxanthopurpurin-3-*O*-methyl ether, erythroglaucin, and torosachrysone were isolated from *Cortinarius* sp. exhibited antibacterial activity toward *S. aureus* (Beattie et al. 2010). Another species *Flammulina velutipes* contained the bioactive compounds enokipodins A–D, which were effective against *S. aureus* and *B. subtilis* (Ishikawa et al. 2001). Lots of mushrooms are capable to reveal antifungal effects also, for example, 2-chloro-1,3-dimethoxy-5-methyl benzene, a derivative of chlorinated orcinol isolated from the mycelia of *Hericium* sp. is effective against *Candida neoformans* and *C. albicans* (Song et al. 2020a,b).

Viruses on the other hand in continuing cause to severe epidemics on well-being and mortality-debilitating disorders throughout the world, particularly in the places where antiviral chemotherapies and vaccines are not available, or if they are available but enough. Also, the present state of diseases caused by viruses is extensively limiting in effectiveness of drugs through the drug resistant strains emergence. Therefore, the imperative necessity is required to recognize and build up drugs inspired by natural entities that can assist to control infections caused by viruses. A superfluity of very effective and active substances of fungal sources was isolated and then examined for antiviral activity, but not any among them achieved the market level. This entrance observes on natural substances showing strong activity on some viruses that are pathogenic to humans such as respiratory syncytial virus, hepatitis virus, enterovirus-71, herpes virus, influenza virus, and human immunodeficiency virus (Hyde et al. 2019). *Pleurotus* mushroom also showed antiviral properties either directly or indirectly as a result of immune stimulatory action. An antiviral agent named Ubiquitin obtained from the fruiting body of oyster mushroom. $-glucans and their derivatives isolated from *Pleurotus tuberregium* were forceful against type-1 and type-2 simplex viruses (Patel et al. 2012).

4.4.2 As anticancer agents

Cancer, a multifactorial disorder is the principal cause of mortality and morbidity around the world and in 2018, it has been estimated that 9.6 million deaths occurred because of cancer. It is regarded to be related to control the cell division as well as cell proliferation with the loss of growth factors (GBD 2015). Various remedies for cancer are available, which are administered as per the growing circumstances of such disease. Among various treatments chemotherapy and radiotherapy became conventional for the treatment of cancer. Though various negative aspects are familiar in chemotherapeutic drugs including their high resulting failure, severe side effects, and less effectiveness (Hyde et al. 2019). In order to find the suitable cure, natural products illustrated a broad range of properties against cancer via antiproliferative, pro-apoptotic, angiogenic, and antimigratory effects through a variety of pathways (Evidente et al. 2014). Fungi are the imperative sources of secondary metabolites, though mainly anticancer drug agents are obtained only from bacteria and plants. The secondary metabolites of fungal origin have yet to partake with an approved medication against cancer, cytoskyrins, and indole alkaloids isolated from *Curvularia lunata* and *Penicillium aurantiogriseum*, respectively, are shown to possess anticancer activity (Chadha et al. 2014; Abo-Kadoum et al. 2013). The dimeric diketopiperazines isolated from the *Leptosphaeria* sp., collected from the marine alga *Sargassum tortile* is a secondary metabolite of the leptosin family which is effective anticancer agent (Bugni and Ireland 2004). Gallic acid isolated from the *Fusarium* sp. exhibits anticancer potential (Pan et al. 2017). It has also been reported that a fungal species *Chaetomium* obtained from the leaves of plant species *Sapiumellipticum* produces chaetocochin A–C, which was used against cancer (Akone et al. 2016). Furthermore in another study, deoxypodophyllotoxin isolated from *Aspergillus fumigates* has been shown to possess a strong anticancer potential (Khaled et al. 2013). Another anticancerous bioactive compound (+)-epiepoxydon, isolated from the extracts of *Apiospora montagnei* obtained from the red alga known as *Polysiphonia violacea* against human cancer cell lines (Klemke et al. 2004). Tetrapeptide from the *Fusarium* sp. of green alga *Codium fragile* revealed to show anticancer activity (Raghukumar 2008). Three bioactive compounds such as cis-9-Octadecenoic acid, Heptadecanoic acid, 16methyl-, methyl ester, and 9,12-Octadecadienoic acid were isolated from the two strains of *Hypocrea* sp. The effect of these compounds was evaluated on a skin cancer protein namely 4,5-Diarylisoxazole Hsp90 Chaperone on the basis of docking scores. Among all compounds, Heptadecanoic acid, 16 methyl, methyl ester showed more inhibition of a skin cancer protein with −11.4592 Kcal/moL docking score than that of the Dyclonine—a known inhibitor of such protein having −10.088 Kcal/moL docking score (Kandasamy et al. 2012).

Additionally, several mushrooms possess promising anticancer properties. The hetero beta-glucans and their protein complexes, terpenoids, dietary fibers, and lectins are the anticancer polysaccharides of medicinal mushrooms. Most of the anticancer substances have been obtained from the culture media, mycelia, and fruiting bodies of therapeutic mushrooms such as *Flammulina velutipes*, *Inonotus obliquus*, *Schizophyllum commune*, *Lentinus edodes*, *Trametes versicolor*, and *Ganoderma lucidum* (Wasser and Weis 1999). It has been reported that the ethanol extract of the fruiting body of mushroom *Pleurotus ferulae* possessed noteworthy anticancer activity toward lung carcinoma (Choi et al. 2004). The fruiting body of *Pleurotus ostreatus* contained higher amounts of flavonoids which are responsible for the cytotoxic activity of this mushroom toward human leukemia (HL-60) cell line in vitro (Maiti et al. 2011). The in vivo anticancer activity of beta-glucan obtained from *Ganoderma lucidium* caused primary lung cancer metastasis inhibition in C57BL/6 mice (Chen et al. 2014). One of the medicinal mushrooms, *Hericium erinaceus*, has been broadly examined in animal models in vivo (Li et al. 2014). The ethanol extract of this species is reported to inhibit colon, liver, and gastric cancers in xenograft tumor of mouse model. HEP3 protein isolated from this species also demonstrated the capability to reduce the colon cancer cell growth of xenograft tumors in mouse (Diling et al. 2017).

Some of the promising compounds obtained from fungi that are in the clinical as well as in preclinical developmental stage include Irofulven, a semisynthetic derivative of illudin S, obtained from *Omphalotus illudens* interferes with cell division and also with the complexes of DNA replication during DNA formation. The promising consequences of this bioactive compound in I and II phase of clinical trials against different types of cancers such as cancers of the lungs, pancreas, breast, ovarian, sarcoma, blood, colon, brain, and central nervous system (Devi et al. 2020; Topka et al. 2018). A recent study by Sandargo et al. 2019 reported that illudin conjugate in comparison to irufulven show greater activities in vitro and is under preclinical development in present times. Leptosins F and C, other leading anticancer compounds derived from *Leptoshaeria* sp., these compounds showed the anticancer potential when examined in embryos of mouse.

4.4.3 As antioxidant agent

Antioxidants are commonly known as inhibitors of oxidation and play a very dynamic role in fighting reactive oxygen species (ROS). The detection of effective antioxidants which can fight or avoid any disease naturally is a continuing practice. As part of the ongoing innovative trials for novel and safe antioxidants from natural resources, fungal species, and their secondary metabolites were considered to be the prospective sources of antioxidants. The secondary metabolites including coumarin, isopestacin, borneol, salidroside, pestacin, *p*-tyrosol, 2,14-dihydroxy-7-drimen-12,11-olide, lapachol, 2,3,6,8-tetrahydroxy-1-methylxanthone, rutin, phloroglucinol, 5-(hydroxymethyl)-2-furanocarboxylic acid, and corynesidones A and B, isolated from fungi can be a probable cause of novel natural antioxidants that possess various biological activities including anticancer. As far as the antioxidants are concerned they are regarded as the most prominent chemo-defensive agents against a variety of cancers and the fungal antioxidant metabolites are of immense importance in diminishing the disease risks caused by oxidative stress (Huang et al. 2007; Gupta et al. 2020). The importance of the bioactive compounds with antioxidant activity sets on the reality that they are extremely valuable toward ROS damage and other free radicals derived from oxygen, which take part in a number of pathologies (Bhagobaty and Joshi 2012).

Most of the biological activities such as antiartherosclerotic, antiinflammatory, antiviral, and antimutagenic of antioxidants propose promising remedies toward avoidance and cure of diseases related to ROS (Pimentel et al. 2011). The fungal species, like *Aspergillus candidus*, *Penicillium roquefortii*, *Emericella falconensis* and, *Mortierella* sp., are said to produce bioactive substances with antioxidant property. Two novel derivatives of *Acremonium* sp. were found to show very potent antioxidant activity (Abdel-Lateff., 2004). A novel sesquiterpenoid isolated from marine *Halorosellinia oceanica* endophytic fungus which itself was isolated from mangrove of Mai Po located in Hong Kong presented a special method of biological oxidation devoid of interrupting the stimulating alicyclics skeleton through bio reformation of 1, 2, 3, 4-tetrahydronaphthalene into four oxidative products (Pan et al. 2008). The bioactive compounds such as erythroglaucin, chaetopyranin, and isotetrahydroauroglaucin obtained from the endophyte *Chaetomium globosum*, secluded from marine red alga *Polysiphonia urceolata* were shown to possess clement radical scavenging activity (Wang et al. 2006). The polysaccharides of *Penicillium* sp. were revealed to show noteworthy antioxidant ability toward hydroxyl and superoxide scavenging radicals (Mayer et al. 2013). A marine fungus *Curvularia tuberculata* showed excellent antioxidant potential in hydroxyl radical and reducing power scavenging essays with 11.69% and 62.15% inhibition, respectively (Venkatchalam et al. 2011). Parasitenone, a bioactive compound isolated from *Aspergillus parasiticus*—a marine fungus, which itself has been obtained from red alga, namely, *Carpopeltis cornea* has shown to possess free radical scavenging property (Abdel-Lateff 2004).

Researchers have also focused on the antioxidant activity of mushrooms. One study reported that the ethanolic extract and syringic acid isolated from *Elaphomyces granulates* fruiting body proved the powerful antioxidant activity against myelomonocytic HL-60 cells having IC_{50} values 41 μg/mL and 0.7 μg/mL for extract and syringic acid, respectively. The antioxidant activity of syringic acid was effective and was nearly equal to the activity of vitamin C a known antioxidant having the IC_{50} value 0.5 μg/mL (Stanikunaite et al. 2009). Another study evaluated for the antioxidant activity of 11 species of mushrooms. Among all *Scleroderma laeve* showed effective activity having IC_{50} value less than 20 μg/mL. Some species such as *Rhizopogon couchii*, *R. pedicellus*, *Geopora clausa*, and *Leucogaster rubescens* revealed moderate activity and their IC_{50} values ranges in between 20 and 50 μg/mL. Other species, such as *Melanogaster tuberiformis*, *Gautieria monticola*, *Elaphomyces muricatus*, *E. granulatus*, *Rhizopogon subaustralis*, and *R. nigrescens*, showed weak antioxidant activity with IC_{50} values of more than 50 μg/mL (Stanikunaite et al. 2007).

4.4.4 As antidiabetic agents

Diabetes also known as diabetes mellitus is a chronic disorder having high glucose concentration in the blood because of imbalance of insulin in the body or simply we can say that diabetic persons cannot produce or efficiently utilize insulin in the body. Commonly there are two types of diabetes which are type-1 and type-2. Type-1, that is, insulin dependent diabetes, the patients having type-1 diabetes are not able to produce insulin as in the pancreas, the insulin secreting beta cells cannot function properly (Meier et al. 2005). 5–10% of the total cases of type-1 are reported throughout the world and it affects mostly the adolescents and children. Type-2 diabetes, that is, noninsulin dependent diabetes, the patients having type-2 diabetes are not able to produce enough insulin or cannot efficiently metabolize it. This kind of disorder accounts 90–95% and is mostly associated with elderly people (Hameed et al. 2015). 7% of World's adult population is affected by this disorder. In a report of 2017, nearly 73 million cases of diabetes were traced in India, biggest number, that is, 114 million diabetic patients were recorded in China and 30 million in the United States (https://www.statista.com/statistics/281082/countries-withhighest-number-ofdiabetics). Various negative outcomes including blindness, cancer, kidney failure, cardiovascular disorders, and even fatality can develop if this disorder is not treated on time (Huang et al. 2018). Most fungal species like *Cyclocybe cylindracea*, *C. aegerita*, *Agaricus bisporus*, and *Tremella fuciformis* of Basidiomycota are utilized for type-2 diabetes treatment as these species have lesser amounts of carbohydrates in diet which are digestible and that characteristic feature is responsible to avoid higher glucose levels in patients (Poucheret et al. 2006).

Various bioactive metabolites secluded from therapeutically lower as well as from higher fungi (whole mushrooms, fruiting bodies, mycelia) possess antihyperglycemic activity in diabetes treatment either in the form of extracts or in the form of isolated compounds. The bioactive metabolite Aquastatin-A isolated from the *Cosmospora* sp. of endophytic fungi exhibited antidiabetic activity toward type-2 diabetes (Bugni Ireland, 2004). This compound inhibits the protein tyrosine phosphatases which are responsible for the modulation of cellular processes dependent on tyrosine phosphorylation. *Pseudomassaria* sp., another endophytic fungus contains insulin mimetic which has the ability to diminish glucose level in blood and can be considered as the novel treatment for diabetes (Strobel and Daisy 2003). β-Glucan isolated from dried fruiting bodies of *Agaricus subrufescens* was reported to possess the antidiabetic activity (Kim et al. 2005). Polysaccharide obtained from the mycelia of *Ophiocordyceps sinensis* has been reported to possess the antidiabetic activity in Alloxan and Streptozotocin induced diabetic mice where glucose level decreased and insulin levels increased in the blood (Li et al. 2006). The *Antrodia cinnamomea* fruiting bodies can be used to produce drugs having antidiabetic properties (Huang et al. 2018). The *Inocutis levis* extracts have therapeutic ability against diabetes as they escalate insulin sensitivity as well as resistance and also increase glucose uptake, thus facilitate to manage glucose levels in blood (Ehsanifardet al. 2017). The extract of fruiting body of *Agaricus bisporus* has been reported to possess antidiabetic activity by reducing the blood glucose levels in Streptozotocin-injected diabetic Sprague Dawley rats (Jeong et al. 2010). Some medicinal antidiabetic mushroom products such as capsules of *Ophiocordyceps sinensis*, Tremella, SX-Fraction, and Reishi-Max capsules are sold for diabetes as remedial products which are stated to reduce the fasting blood glucose concentrations in type-2 diabetes (De Silva et al. 2012). Tremella obtained from *Tremella fuciformis* is utilized in Chinese medicine, primarily used for declining cholesterol and blood glucose levels (Li et al. 2004). Another mushroom product, that is, SX-Fraction is regarded as a foremost substitute for increasing insulin sensitivity (Preuss et al. 2007). Future examinations are required to illuminate the continuing effects of taking products of curative mushroom with other drugs.

4.4.5 As antiinflammatory agents

Inflammation is a process of living organisms, when become deregulated can cause various types of disorders such as inflammatory bowel disease, psoriasis, rheumatoid arthritis, cancer, cardiovascular, and alzeihmers diseases. Various therapeutic aspects are available to treat the inflammation related diseases like antihistamines and antiinflammatory drugs of steroidal and nonsteroidal nature. In spite of some prominent success, there is still much more requirement to treat inflammatory disorders. Modern healing advances to cure the inflammatory disorders are dependent on cyclooxygenase enzyme that converts the arachidonic acid into prostaglandins. Although the marvelous progresses of synthetic drugs and our dependence on modern drugs, 80% of the global populace depend on traditional medicine and cannot afford western drugs. Fungi and their metabolites are important sources of natural products with antiinflammatory activity (Deshmukh et al. 2012). The bioactive compounds such as diaporthein B, diaporthein C, penidepsidone A, penisclerotiorin A, 3-methylorsellinic acid, aculeatusquinone

A, aculeatusquinone C, 3-methylorsellinic acid, and [3,8-dihydroxy1,4,6,9-tetramethyl-dibenzo and dioxepin-11-one were isolated from the fungus *Penicillium sclerotiorum* and their structures were characterized by X-ray diffraction, spectroscopic methods and quantum chemical calculations. All the compounds were evaluated on the nitric oxide production in lipopolysaccharide induces microgial cells for their antiinflammatory activity. Among all the isolated compounds, aculeatusquinone C, aculeatusquinone A, and 3-methylorsellinic acid revealed the powerful antiinflammatory activity than indomethacin—a known antiinflammatory (positive control) drug (Zhao et al. 2020). Another recently studied work reports three new bioactive components from a fungus *Penicillium* sp. TJ403-2 derived from sea sediment. The compounds were 13β Hydroxy conidiogenone C; 12β-Hydroxy conidiogenone C and 12β-Hydroxy conidiogenone D. Their structures were elucidated by one- and two-dimensional NMR analyses, HRESIMS, and X-ray crystallography experiment. All the compounds were examined for antiinflammatory effects against LPS-induced NO production. All the compounds were effective against inflammation, but 13β Hydroxy conidiogenone C exhibited outstanding inhibitory influence having IC_{50} value as 2.19 ± 0.25 μmol/L, then indomethacin a known antiinflammatory (positive control) drug with IC_{50} values 8.76 ± 0.92 μmol/L. The immunofluorescence and western blot tests revealed that this compound NF-κB-activated pathway and emphasized that this compound will be the promising initial point for new antiinflammatory agent development (Li et al. 2020). Mushrooms are also good sources of antiinflammatory agents, for example, ergosterol is a precursor of vitamin D, found in large quantities in edible mushrooms including *Agaricus bisporus* fruiting bodies and is known to exhibit antiinflammatory activity. Pleuran, a secondary metabolite isolated from the oyster mushroom fruiting bodies showed antiinflammatory activity (Muszyńska et al. 2018). Two small molecules such as syringic acid and syringaldehyde as well as the ethanolic extract of *Elaphomyces granulatus* showed antiinflammatory activity by COX-2 (an inflammation responsible enzyme) inhibition in Raw264.7 cells. The syringic acid and syringaldehyde caused more COX-2 inhibition having IC_{50} values 0.4 and 3.5 μg/mL respectively. The NS-398, a known inhibitor of COX-2 (positive control) showed the IC_{50} value as 0.2 μg/mL. Its ethanolic extract showed 68% COX-2 inhibition at 50 μg/mL (Stanikunaite et al. 2009)

4.5 Conclusions

Fungi (including lower and higher fungi) have been studied and reported to be used for their nutritional and medicinal values for centuries. Biosphere is rich in different fungal species which are the reservoirs of numerous bioactive compounds. These bioactive compounds play important roles in human history as they have immense importance in food, agriculture, etc. Interesting biological activities including antibiotic, anticancer, antioxidant, antidiabetic, antiinflammatory, etc. are also associated with the fungal bioactive compounds. As large number of bioactive compounds and their derivatives are obtainable in market for various agricultural and medical applications, but profound examinations in this field are essential to meet the rising necessities for novel and improved drugs and drug commodities. It is eminent that a little portion of the diversity of fungi has been explored for their compound prevalence throughout the world. No doubt they are available in all parts of the Earth and there is more possibility to find out new unexplored species and their novel metabolites. As for as the applications of fungal metabolites are concerned, further widespread collections, their culturing techniques, and chemical analysis of fungal species for the presence of secondary metabolites are required, which may lead to accomplish the influential need toward new global therapies and also used in different industries with no detrimental consequences. Furthermore, it is essential to know the functions of a respective fungal metabolite and its molecular mechanism which may assist in a better perception toward the control of a disease.

References

Abad, M.J., Bedoya, L.M., Bermejo, P., 2011. Marine compounds and their antimicrobial activities. In: Méndez-Vilas, A. (Ed.), Science Against Microbial Pathogens: Communicating Current Research and Technological Advances. vol. 51. FORMATEX, Badajoz, pp. 1293–1306.

Abdel-Lateff, A.A.-A.M., 2004. Secondary metabolites of marine-derived fungi: natural product chemistry and biological activity. Dissertation.

Abo-Kadoum, M.A., Abo-Dahab, N.F., Awad, M.F., Abdel-Hadi, A.M., 2013. Marine-derived fungus, *Penicillium aurantiogriseum* AUMC 9757: a producer of bioactive secondary metabolites. J. Basic Appl. Mycol. 4 (1), 77–83.

Akone, S.H., Mandi, A., Kurtan, T., Hartmann, R., Lin, W., Daletos, G., et al., 2016. Inducing secondary metabolite production by the endophytic fungus Chaetomium sp. through fungal–bacterial co-culture and epigenetic modification. Tetrahedron 72, 6340–6347.

Awad, N.E., Kassem, H.A., Hamed, M.A., El-Feky, A.M., Elnaggar, M.A., Mahmoud, K., et al., 2018. Isolation and characterization of the bioactive metabolites from the soil derived fungus *Trichoderma viride*. Mycology 9, 70–80.

Ayob, F.W., Simarani, K., Zainal Abidin, N., Mohamad, J., 2017. First report on a novel *Nigrosporasphaerica* isolated from *Catharanthus roseus* plant with anticarcinogenic properties. Microbbiotechnol 10 (4), 926–932.

Beattie, K.D., Rouf, R., Gander, L., May, T.W., Ratkowsky, D., Donner, C.D., Gill, M., Tiralongo, E., 2010. Antibacterial metabolites from Australian macrofungi from the genus *Cortinarius*. Phytochemistry 71, 948–955.

Beekman, A.M., Barrow, R.A., 2014. Fungal metabolites as pharmaceuticals. Aust. J. Chem. 67 (6), 827–843.

Béni, Z., Dékány, M., Kovács, B., Csupor-Löffler, B., Zomborszki, Z.P., Kerekes, E., Szekeres, A., Urbán, E., Hohmann, J., Ványolós, A., 2018. Bioactivity-guided isolation of antimicrobial and antioxidant metabolites from the mushroom *Tapinellaatrotomentosa*. Molecules 23 (5), 1082.

Bérdy, J., 2005. Bioactive microbial metabolites. J. Antibiot. (Tokyo) 58, 1–26.

Berdy, J., 2012. Thoughts and facts about antibiotics: where we are now and where we are heading. J. Antibiot. 65, 385–395.

Bhagobaty, R.K., Joshi, S.R., 2012. Antimicrobial and antioxidant activity of endophytic fungi isolated from ethnomedicinal plants of the "Sacred forests" of Meghalaya, India. Med. Mycol. J. 19 (1), 5–11.

Bugni, S., Ireland, C., 2004. Marine-derived fungi, a chemically and biologically diverse group of microorganisms. R. Soc. Chem. Adv. 21, 143–163.

Chadha, N., Mishra, M., Prasad, R., Varma, A., 2014. Root endophytic fungi: research update. J. Biol. Life. Sci. 5, 135–158.

Chambergo, F.S., Valencia, E.Y., 2016. Fungal biodiversity to biotechnology. Appl. Microbiol. Biotechnol. 100 (6), 2567–2577.

Cheek, M., Nic Lughadha, E., Kirk, P., Lindon, H., Carretero, J., Looney, B., Douglas, B., Haelewaters, D., Gaya, E., Llewellyn, T., Ainsworth, A.M., 2020. New scientific discoveries: plants and fungi. Plants, People, Planet 2 (5), 371–388.

Chen, S.N., Chang, C.S., Hung, M.H., Chen, S., Wang, W., Tai, C.J., Lu, C.L., 2014. The effect of mushroom beta glucans from solidculture of *Ganoderma lucidum* on inhibition of the primary tumor metastasis. Evid. Based Complement Alternat Med. 2014, 1–7.

Chen, S., Cai, R., Hong, K., She, Z., 2016. New furoisocoumarins and isocoumarins from the mangrove endophytic fungus *aspergillus* sp. 085242. Beilstein J. Org. Chem. 12 (1), 2077–2085.

Chen, Y., Liu, Z., Liu, H., Pan, Y., Li, J., Liu, L., She, Z., 2018. Dichloroisocoumarins with potential anti-inflammatory activity from the mangrove endophytic fungus *Ascomycota* sp. CYSK-4. Mar. Drugs 16 (2), 54.

Choi, D.B., Cha, W.S., Kang, S.H., Lee, B.R., 2004. Effect of *Pleurotusferulae* Extracts on viability of human lung cancer and cervical Cancer cell lines Biotechnol. Bioprocess Eng. 9, 356–361.

Choi, B.K., Cho, D.Y., Choi, D.K., Trinh, P.T., Shin, H.J., 2021. Two new phomaligols from the marine-derived fungus *aspergillus flocculosus* and their anti-neuroinflammatory activity in BV-2 microglial cells. Mar. Drugs 19 (2), 65.

Clutterbuck, P.W., Oxford, A.E., Raistrick, H., Smith, G., 1932. Studies in the biochemistry of micro-organisms, the metabolic products of the *Penicillium brevicompactum* series. Biochem. J. 26 (5), 1441.

De Kesel, A., Haelewaters, D., 2019. Laboulbeniales (Fungi, Ascomycota) of cholevine beetles (*Coleoptera, Leiodidae*) in Belgium and the Netherlands. Sterbeeckia 35, 60–66.

De Silva, D.D., Rapior, S., Hyde, K.D., Bahkali, A.H., 2012. Medicinal mushrooms in prevention and control of diabetes mellitus. Fungal Divers. 56, 1–29.

Deng, S., Zhang, G., Kuai, J., Fan, P., Wang, X., Zhou, P., Yang, D., Zheng, X., Liu, X., Wu, Q., Huang, Y., 2018. Lentinan inhibits tumor angiogenesis via interferon γ and in a T cell independent manner. J. Exp. Clin. Cancer Res. 37 (1), 1–2.

Deshmukh, S.K., Verekar, S.A., Periyasamy, G., Ganguli, B.N., 2012. Fungi: a potential source of anti-inflammatory compounds. Microorganism in Sustain Agricul and Biotechnol, 613–645.

Deshmukh, S.K., Prakash, V., Ranjan, N., 2017. Recent advances in the discovery of bioactive metabolites from Pestalotiopsis. Phytochem. Rev. 16, 883–920.

Devi, R., Kaur, T., Guleria, G., Rana, K.L., Kour, D., Yadav, N., Yadav, A.N., Saxena, A.K., 2020. Fungal secondary metabolites and their biotechnological applications for human health. In: New and Future Developments in Microbial Biotechnology and Bioengineer. Elsevier, pp. 147–161.

Diling, C., Chaoqun, Z., Jian, Y., Jian, L., Jiyan, S., Yishen, X., Guoxiao, L., 2017. Immunomodulatory activities of a fungal protein extracted from Hericiumerinaceus through regulating the gut microbiota. Front. Immunol. 8, 666.

Dreyfuss, M., Chapela, I.H., 1994. Potential of fungi in the discovery of novel, low-molecular weight pharmaceuticals. In: Gullo, V.P. (Ed.), Discovery of Novel Natural Products with Therapeutic Potential Newnes, pp. 49–80. Boston, MA.

Ehsanifard, Z., Mir-Mohammadrezaei, F., Safarzadeh, A., Ghobad-Nejhad, M., 2017. Aqueous extract of Inocutislevis improves insulin resistance and glucose tolerance in high sucrose-fed Wistar rats. J. Herbmed. Pharmacol. 6, 160–164.

Evidente, A., Kornienko, A., Cimmino, A., Andolfi, A., Lefranc, F., Mathieu, V., et al., 2014. Fungal metabolites with anticancer activity. Nat. Prod. Rep. 31, 617–627.

GBD, 2015. Mortality and Causes of Death Collaborators Global, regional, and national life expectancy, all-cause mortality, and cause-specific mortality for 249 causes of death, 1980–2015: a systematic analysis for the Global Burden of Disease Study 2015. Lancet 388 (10053), 1459–1544.

Godtfredsen, W.O., Jahnsen, S., Lorck, H., 1962. Fusidic acid: a new antibiotic. Nature 193, 987.

Goyal, S., Ramawat, K.G., Mérillon, J.M., 2016. Different shades of fungal metabolites: an overview. In: Fungal Metabolites. Springer, pp. 1–29.

Gupta, S., Chaturvedi, P., Kulkarni, M.G., Van Staden, J., 2020. A critical review on exploiting the pharmaceutical potential of plant endophytic fungi. Biotechnol. Adv. 39, 107462.

Hameed, I., Masoodi, S.R., Mir, S.A., Nabi, M., et al., 2015. Type 2 diabetes mellitus: from a metabolic disorder to an inflammatory condition. World J. Diabetes 6, 598–612.

Harwoko, H., Daletos, G., Stuhldreier, F., Lee, J., Wesselborg, S., Feldbrügge, M., Müller, W.E., Kalscheuer, R., Ancheeva, E., Proksch, P., 2021. Dithiodiketopiperazine derivatives from endophytic fungi *Trichoderma harzianum* and *Epicoccum nigrum*. Nat. Prod. Res. 35 (2), 257–265.

Huang, W.Y., Cai, Y.Z., Xing, J., Corke, H., Sun, M., 2007. A potential antioxidant resource: endophytic fungi from medicinal plants. Econ. Bot. 61, 14–30.

Huang, H., Wang, S.L., Nguyen, V., Kuo, Y.H., 2018. Isolation and identification of potent antidiabetic compounds from Antrodiacinnamomea—an edible Taiwanese mushroom. Molecules 23, 1–12.

Hyde, K.D., Xu, J., Rapior, S., Jeewon, R., Lumyong, S., Niego, A.G., Abeywickrama, P.D., Aluthmuhandiram, J.V., Brahamanage, R.S., Brooks, S., Chaiyasen, A., 2019. The amazing potential of fungi: 50 ways we can exploit fungi industrially. Fungal Divers. 97 (1), 1–36.

Ibrahim, S.R., Mohamed, G.A., Abdel-Latif, M.M., El-Messery, S.M., Shehata, I.A., 2015. Minutaside A, new-amylase inhibitor flavonol glucoside from *Tagetes minuta*: antidiabetic, antioxidant, and molecular modeling studies. Starch-Stärke 67, 976–984.

Indarmawan, T., Mustopa, A.Z., Budiarto, B.R., Tarman, K., 2016. Antibacterial activity of extracellular protease isolated from an algicolous fungus *Xylariapsidii* KT30 against gram-positive bacteria. Hayati J. Biosci. 23 (2), 73–78.

Ishikawa, N.K., Fukushi, Y., Yamaji, K., Tahara, S., Takahashi, K., 2001. Antimicrobial cuparene-type sesquiterpenes, enokipodins C and D, from a mycelial culture of flammulinavelutipes. J. Nat. Prod. 64, 932–934.

Jeong, S.C., Jeong, Y.T., Yang, B.K., Islam, R., et al., 2010. White button mushroom (Agaricus bisporus) lowers blood glucose and cholesterol levels in diabetic and hypercholesterolemic rats. Nutr. Res. 30, 49–56.

Kaaniche, F., Hamed, A., Abdel-Razek, A.S., Wibberg, D., Abdissa, N., El Euch, I.Z., 2019. Bioactive secondary metabolites from new endophytic fungus *Curvularia* sp. isolated from *Rauwolfia macrophylla*. PloS One 14, e0217627.

Kamdem, R.S., Wang, H., Wafo, P., Ebrahim, W., Özkaya, F.C., Makhloufi, G., Janiak, C., Sureechatchaiyan, P., Kassack, M.U., Lin, W., Liu, Z., 2018. Induction of new metabolites from the endophytic fungus Bionectria sp. through bacterial co-culture. Fitoterapia 124, 132–136.

Kandasamy, S., Sahu, S.K., Kandasamy, K., 2012. In silico studies on fungal metabolite against skin Cancer protein (4,5-Diarylisoxazole HSP90 chaperone). Int. Sch. Rres. Network. 626214.

Karwehl, S., Stadler, M., 2017. Exploitation of fungal biodiversity for discovery of novel antibiotics the antibiotic crisis-facts, challenges, technologies & future perspective. Curr. Top. Microbiol. Immunol. 398, 303–338.

Kellogg, J.J., Raja, H.A., 2017. Endolichenic fungi: a new source of rich bioactive secondary metabolites on the horizon. Phytochem. Rev. 16, 271–293.

Khaled, M., Jiang, Z.Z., Zhang, L.Y., 2013. Deoxypodophyllotoxin: a promising therapeutic agent from herbal medicine. J. Ethnopharmacol. 149 (1), 24–34.

Kim, Y.W., Kim, K.H., Choi, H.J., Lee, D.S., 2005. Anti-diabetic activity of β-glucans and their enzymatically hydrolyzed oligosaccharides from *Agaricus blazei*. Biotechnol. Lett. 27, 483–487.

Kim, J.W., Kim, H.I., Kim, J.H., Kwon, O., Son, E.S., Lee, C.S., Park, Y.J., 2016. Effects of ganodermanondiol, a new melanogenesis inhibitor from the medicinal mushroom *Ganoderma lucidum*. Int. J. Mol. Sci. 17 (11), 1798.

Klemke, C., Kehraus, S., Wright, A.D., Konig, G.M., 2004. New secondary metabolites from the marine endophytic fungus *Apiosporamontagnei*. J. Nat. Prod. 67 (6), 1058–1063.

Kour, D., Rana, K.L., Kumar, A., Rastegari, A.A., Yadav, N., Yadav, A.N., et al., Gupta, V.K., Singh, B.N., 2019. Extremophiles for hydrolytic enzymes productions: Biodiversity and potential biotechnological applications. In: Molina, G., Gathergood, N. (Eds.), Bioprocessing for Biomolecules Production. Wiley, Hoboken NJ, pp. 321–372.

Kumar, K.S., Vani, M.G., Hsieh, H.W., Lin, C.C., Wang, S.Y., 2019. Antcin-a modulates epithelial-to-mesenchymal transition and inhibits migratory and invasive potentials of human breast cancer cells via p53-mediated miR-200c activation. Planta Med. 85 (09/10), 755–765.

Lee, S.Y., Kim, M., Kim, S.H., et al., 2016. Transcriptomic analysis of the white rot fungus *Polyporusbrumalis* provides insight into sesquiterpene biosynthesis. Microbiol. Res. 182, 141–149.

Li, W.L., Zheng, H.C., Bukuru, J., De Kimpe, N., 2004. Natural medicines used in the traditional Chinese medical system for therapy of diabetes mellitus. J. Ethnopharmacol. 92, 1–21.

Li, S.P., Zhang, G.H., Zeng, Q., Huang, Z.G., Wang, Y.T., Dong, T.T.X., Tsim, K.W.K., 2006. Hypoglycemic activity of polysaccharide, with antioxidation, isolated from cultured *Cordyceps* mycelia. Phytomed 13, 428–433.

Li, G., Yu, K., Li, F., Xu, K., Li, J., He, S., Cao, S., Tan, G., 2014. Anticancer potential of *Hericiumerinaceus* extracts against human gastrointestinal cancers. J. Ethnopharmacol. 153, 521–530.

Li, F., Sun, W., Zhang, S., Gao, W., Lin, S., Yang, B., Chai, C., Li, H., Wang, J., Hu, Z., Zhang, Y., 2020. New cyclopiane diterpenes with anti-inflammatory activity from the sea sediment-derived fungus *Penicillium* sp. TJ403–2. Chin. Chem. Lett. 31 (1), 197–201.

Liu, J., Zhan, D., Zhang, M., Chen, R., et al., 2017. Periconones B-E, new meroterpenoids from endophytic fungus Periconia sp. Chin. Chem. Lett. 28, 248–252.

Liu, Z., Liu, H., Chen, Y., She, Z., 2018a. A new anti-inflammatory meroterpenoid from the fungus *Aspergillus terreus* H010. Nat. Prod. Res. 32 (22), 2652–2656.

Liu, H.X., Tan, H.B., Chen, Y.C., Li, S.N., Li, H.H., Zhang, W.M., 2018b. Secondary metabolites from the *Colletotrichum gloeosporioides* A12, an endophytic fungus derived from *Aquilaria sinensis*. Nat. Prod. Res. 32, 2360–2365.

Liu, C.C., Zhang, Z.Z., Feng, Y.Y., Gu, Q.Q., Li, D.H., Zhu, T.J., 2019. Secondary metabolites from Antarctic marine-derived fungus *Penicillium crustosum* HDN153086. Nat. Prod. Res. 33, 414–419.

Ma, Y.M., Liang, X.A., Kong, Y., Jia, B., 2016. Structural diversity and biological activities of indole diketopiperazine alkaloids from fungi. J. Agric. Food Chem. 64, 6659–6671.

Macheleidt, J., Mattern, D.J., Fischer, J., Netzker, T., Weber, J., Schroeckh, V., et al., 2016. Regulation and role of fungal secondary metabolites. Annu. Rev. Genet. 50, 371–392.

Maiti, S., Mallick, S.K., Bhutia, S.K., Behera, B., Mandal, M., Maiti, T.K., 2011. Antitumor effect of culinary-medicinal oyster mushroom, *Pleurotusostreatus* (Jacq.: Fr.) P. Kumm., derived protein fraction on tumor-bearing mice models. Int. J. Med. Mushrooms 13, 427–440.

Mayer, A., Rodríguez, A., Taglialatela-Scafati, O., Fusetani, N., 2013. Marine pharmacology in 2009–2011: marine compounds with antibacterial, antidiabetic, antifungal, anti-inflammatory, antiprotozoal, antituberculosis, and antiviral activities; affecting the immune and nervous systems, and other miscellaneous mechanisms of action. Mar. Drugs 11 (7), 2510–2573.

Meier, J.J., Bhushan, A., Butler, A.E., Rizza, R.A., Butler, P.C., 2005. Sustained beta cell apoptosis in patients with long-standing type 1 diabetes: indirect evidence for islet regeneration? Diabetologia 48, 2221–2228.

Mohmand, A.Q., Kousar, M.W., Zafar, H., Bukhari, K.T., Khan, M.Z., 2011. Medical importance of Fungi with special emphasis on mushrooms. Isr. Med. J. 31.

Muddassir, M., Ahmed, M., Butt, F., Basirat, U., 2020. Fungi-an amalgam of toxins and antibiotics: a Mini-review. Pak. J. Surg. Med. 1 (1), 52–55.

Muszyńska, B., Grzywacz-Kisielewska, A., Kała, K., Gdula-Argasińska, J., 2018. Anti-inflammatory properties of edible mushrooms: a review. Food Chem. 15 (243), 373–381.

Niskanen, T., Douglas, B., Kirk, P., Crous, P., Lücking, R., Matheny, P.B., Cai, L., Hyde, K., Cheek, M., Willis, K.J., 2018. New discoveries: species of fungi described in 2017. State World's Fungi, 18–23.

Noor, A.O., Almasri, D.M., Bagalagel, A.A., Abdallah, H.M., Mohamed, S.G., Mohamed, G.A., Ibrahim, S.R., 2020. Naturally occurring isocoumarins derivatives from endophytic fungi: sources, isolation, structural characterization, biosynthesis, and biological activities. Molecules 25 (2), 395.

Pan, J.H., Jones, E.B.G., She, Z.G., Pang, J., Lin, Y., 2008. Review of bioactive compounds from fungi in the South China Sea. Bot. Mar. 51, 179–190.

Pan, F., Su, T.J., Cai, S.M., Wu, W., 2017. Fungal endophyte-derived *Fritillaria unibracteata* var. wabuensis: diversity, antioxidant capacities *in-vitro* and relations to phenolic, flavonoid or saponin compounds. Sci. Rep. 7 (42008).

Patel, Y., Naraian, R., Singh, V.K., 2012. Medicinal properties of *Pleurotus* species (oyster mushroom): a review. World J. Fungal Plant Biol. 3 (1), 1–2.

Pimentel, M.R., Molina, G., Dionísio, A.P., Roberto, M., Junior, M., Pastore, G.M., 2011. The use of endophytes to obtain bioactive compounds and their application in biotransformation process. Biotechnol. Res. Int.

Pinheiro, E.A., Pina, J.R., Feitosa, A.O., Carvalho, J.M., Borges, F.C., Marinho, P.S., et al., 2017. Bioprospecting of antimicrobial activity of extracts of endophytic fungi from *Bauhinia guianensis*. Rev. Argent. Microbiol. 49, 3–6.

Poucheret, P., Fons, F., Rapior, S., 2006. Biological and pharmacological activity of higher fungi: 20-year retrospective analysis. Cryptogam. Mycol. 27, 311–333.

Preuss, H.G., Echard, B., Bagchi, D., Perricone, N.V., Zhuang, C., 2007. Enhanced insulin-hypoglycemic activity in rats consuming a specific glycoprotein extracted from maitake mushroom. Mol. Cell. Biochem. 306, 105–113.

Raekiansyah, M., Mori, M., Nonaka, K., Agoh, M., Shiomi, K., Matsumoto, A., Morita, K., 2017. Identification of novel antiviral of fungus-derived brefeldin A against dengue viruses. Trop. Med. Health 45 (1), 1–7.

Raghukumar, C., 2008. Marine fungal biotechnology: an ecological perspective. Fungal Divers. 31, 19–35.

Rao, K.V., Mani, P., Satyanarayana, B., Rao, T.R., 2017. Purification and structural elucidation of three bioactive compounds isolated from *Streptomyces coelicoflavus* BC 01 and their biological activity. Biotech 7 (1), 24.

Richter, L., Wanka, F., Boecker, S., Storm, D., Kurt, T., Vural, O., et al., 2014. Engineering of *Aspergillus niger* for the production of secondary metabolites. Fungal Biol. Biotechnol. 1, 4.

Ro, D.K., Paradise, E.M., Ouellet, M., Fisher, K.J., Newman, K.L., Ndungu, J.M., 2006. Production of the antimalarial drug precursor artemisinic acid in engineered yeast. Nature 440, 940.

Sandargo, B., Chepkirui, C., Cheng, T., Chaverra-Muñoz, L., Thongbai, B., Stadler, M., Hüttel, S., 2019. Biological and chemical diversity go hand in hand: Basidomycota as source of new pharmaceuticals and agrochemicals. Biotechnol. Adv.

Shu, C.H., 2007. Fungal fermentation for medicinal products. In: Bioprocessing for value-added products from renewable resources. Elsevier, pp. 447–463.

Song, R.Y., Wang, X.B., Yin, G.P., Liu, R.H., Kong, L.Y., Yang, M.H., 2017. Isocoumarin derivatives from the endophytic fungus, *Pestalotiopsis* sp. Fitoterapia 122, 115–118.

Song, X., Gaascht, F., Schmidt-Dannert, C., Salomon, C.E., 2020a. Discovery of antifungal and biofilm preventative compounds from mycelial cultures of a unique north American *Hericium* sp. fungus. Molecules 25 (963).

Song, X., Gaascht, F., Schmidt-Dannert, C., Salomon, C.E., 2020b. Discovery of antifungal and biofilm preventative compounds from mycelial cultures of a unique north American *Hericium*sp. Fungus. Molecules 25 (4), 963.

Stanikunaite, R., Trappe, J.M., Khan, S.I., Ross, S.A., 2007. Evaluation of therapeutic activity of hypogeous ascomycetes and basidiomycetes from North America. Int. J. Med. Mushrooms 9, 7–14.

Stanikunaite, R., Khan, S.I., Trappe, J.M., Ross, S.A., 2009. Cyclooxygenase-2 inhibitory and antioxidant compounds from the truffle *Elaphomycesgranulatus*. Phytother. Res. 23, 575–578.

Strobel, G., Daisy, B., 2003. Bioprospecting for microbial endophytes and their natural products. Microbiol. Mol. Biol. Rev. 4 (67), 491–502.

Thakur, M.P., Singh, H.K., 2013. Mushrooms, their bioactive compounds and medicinal uses: a review. Med. Plants 5 (1), 1–20.

Topka, S., Khalil, S., Stanchina, E., Vijai, J., Offit, K., 2018. Preclinical evaluation of enhanced irofulven antitumor activity in an ERCC3 mutant background by *in-vitro* and *in-vivo* tumor models. AACR 78, 3258.

Venkatchalam, G., Venkatchalam, A., Suryanarayanan, T.S., Doble, M., 2011. Isolation and characterization of new antioxidant and antibacterial compounds from algicolous marine fungus *Curvulariatuberculata*. In: International Conference on Bioscience, Biochemistry and Bioinformatics. vol. 5, pp. 302–304.

Wang, S., Li, X., Teuscher, F., Li, D., Diesel, A., Ebel, R., Wang, B., 2006. Chaetopyranin, a benzaldehyde derivative, and other related metabolites from *Chaetomium globosum*, an endophytic fungus derived from the marine red alga Polysiphonia urceolata. J. Nat. Prod. 69 (11), 1622–1625.

Wang, L.Y., Huang, C.S., Chen, Y.H., Chen, C.C., Chen, C.C., Chuang, C.H., 2019a. Anti-inflammatory effect of erinacine C on NO production through down-regulation of NF-κB and activation of Nrf2-mediated HO-1 in BV2 microglial cells treated with LPS. Molecules 24 (18), 3317.

Wang, X.Y., Zhang, D.D., Yin, J.Y., Nie, S.P., Xie, M.Y., 2019b. Recent developments in *Hericiumerinaceus* polysaccharides: extraction, purification, structural characteristics and biological activities. Crit. Rev. Food Sci. Nutr. 59, 96–115.

Wasser, S.P., Weis, A.L., 1999. Medicinal properties of substances occurring in higher basidiomycetes mushrooms: current perspectives. Int. J. Med. Mushrooms 1 (1).

Wiese, J., Imhoff, J.F., 2019. Marine bacteria and fungi as promising source for new antibiotics. Drug Develop. Res. 80 (1), 24–27.

Wu, B., Wiese, J., Labes, A., Kramer, A., Schmaljohann, R., Imhoff, J.F., 2015. Lindgomycin, an unusual antibiotic polyketide from a marine fungus of the Lindgomycetaceae. Mar. Drugs 13, 4617–4632.

Wu, B., Wiese, J., Wenzel-Storjohann, A., Malien, S., Schmaljohann, R., Imhoff, J.F., 2016. Engyodontochones, new antibiotics from the marine fungus *Engyodontium album* strain LF069. Chemistry 22, 7452–7462.

Wu, S.J., Tung, Y.J., Ng, L.T., 2020. Anti-diabetic effects of *Grifolafrondosa* bioactive compound and its related molecular signaling pathways in palmitate-induced C2C12 cells. J. Ethnopharmacol. 260, 112962.

Xu, Y., Yuan, H., Luo, Y., Zhao, Y.J., Xiao, J.H., 2020. Ganoderic Acid D Protects Human Amniotic Mesenchymal Stem Cells against Oxidative Stress-Induced Senescence through the PERK/NRF2 Signaling Pathway. Oxid Med CellLongev.

Yadav, A.N., Kour, D., Rana, K.L., Yadav, N., Singh, B., Chauhan, V.S., et al., 2019. Metabolic engineering to synthetic biology of secondary metabolites production. In: Gupta, V.K., Pandey, A. (Eds.), New and Future Developments in Microbial Biotechnology and Bioengineering. Elsevier, Amsterdam, pp. 279–320.

Yaqoob, A., Li, W.M., Liu, V., Wang, C., Mackedenski, S., Tackaberry, L.E., Massicotte, H.B., Egger, K.N., Reimer, K., Lee, C.H., 2020. Grifolin, neogrifolin and confluentin from the terricolous polypore *Albatrellusflettii* suppress KRAS expression in human colon cancer cells. PloS One 15 (5), e0231948.

Zhao, Y., Liu, D., Proksch, P., Yu, S., Lin, W., 2016. Isocoumarin derivatives from the sponge-associated fungus *Peyronellaea glomerata* with antioxidant activities. Chem. Biodivers. 13, 1186–1193.

Zhao, J., Feng, J., Tan, Z., Liu, J., et al., 2017. Stachybotrysins A-G, Phenylspirodrimane derivatives from the fungus *Stachybotryschartarum*. J. Nat. Prod. 80, 1819–1826.

Zhao, M., Ruan, Q., Pan, W., Tang, Y., Zhao, Z., Cui, H., 2020. New polyketides and diterpenoid derivatives from the fungus *Penicillium sclerotiorum* GZU-XW03-2 and their anti-inflammatory activity. Fitoterapia 1 (143), 104561.

CHAPTER

Fungal metabolites and their importance in pharmaceutical industry

5

Subrata Das[a,b], Madhuchanda Das[a], Rajat Nath[a], Deepa Nath[c], Jayanta Kumar Patra[d], and Anupam Das Talukdar[a]

[a]*Department of Life Science and Bioinformatics, Assam University, Silchar, India,* [b]*Department of Botany and Biotechnology, Karimganj College, Karimganj, India,* [c]*Department of Botany, Gurucharan College, Silchar, India,* [d]*Research Institute of Biotechnology & Medical Converged Science, Dongguk University, Republic of Korea*

5.1 Introduction

Fungi (singular, fungus) are eukaryotic, heterophytes, achlorophyllous means devoid of green coloring pigment called chlorophyll. Fungi are placed as separate kingdoms like plant kingdom and animal kingdom. Yeast, mold, and mushrooms are different forms of fungi (Mohmand et al., 2011). Fungi are cosmopolitan, that is, have a universal distribution, and are found in almost all types of extreme environments and habitats (Solomon et al., 2019). Taxonomists have been described around 75,000 species of fungi by till now, yet, some estimations suggest that a total of 1.5 million species may be there (Webster and Weber, 2007). Examples of several subdivisions of the kingdom fungi are—Basidiomycota, the division includes toadstools and mushrooms; Ascomycota, in this group the maximum number of species are included such as cup fungi and flask fungi; Zygomycota mostly microscopic species; Oomycota is mostly water molds and some vital pathogens; Deuteromycota (fungi imperfecti) includes molds (*Alternaria, Aspergillus, Penicillium*); Microsporidiomycota is spore forming unicellular parasites (Goyal et al., 2016).

The natural products obtained from microorganisms have many therapeutic uses and these are produced via primary and secondary metabolism. Among them, 50–60% are produced by plants and ~10% have been obtained from microbes with biological activity (Demain and Martens, 2017). Fungi produce different types of secondary metabolites, such metabolites are enough to strive with other organisms or cohabit with other existing species in nature (Demain, 2014). Secondary metabolites are the product of the intermediary metabolism of primary metabolites. Secondary metabolites are usually produced as low molecular weight bioactive compounds. Harold Raistrick in 1922 studied and characterized more than 200 secondary metabolites of mold but after the discovery of Penicillin, the prevalent attention was focused on the isolation of metabolites from fungi. Search for biologically active compounds from secondary metabolites of fungi has continued, and thousands of compounds that obstruct the growth of parasites, protozoa, viruses, and bacteria have been revealed (Keller et al., 2005). From antibiotics to mycotoxins diverse range of metabolic compounds are produced by fungi but their biosynthetic paths are also diverse, three mostly found pathways are (1) the shikimic acid pathway (alkaloid, aromatic amino acid), (2) the pathway of mevalonic acid (steroids, terpenoids,

Entrepreneurship with Microorganisms. https://doi.org/10.1016/B978-0-443-19049-0.00021-9

etc.), (3) the acetate pathway (fatty acids, polyketides, etc.). Some enzymes which are linked with these pathways are polyketide syntheses (PKSs), nonribosomal peptide synthetase (NRPSs), terpene cyclases (TCs), geranyl-geranyl diphosphate synthases (GGPPs), dimethylallyl tryptophan synthetases (DMATs), etc. (Goyal et al., 2016). The pharmaceutical importance of fungal secondary metabolites has been recognized for hundreds of years and each year millions of patients are treated with fungal-derived medicines (Langdon and Pearce, 2017). Some molds and fungi are the source of Penicillin, Griseofulvin, Lovastatin, Streptomycin, and other medicines. The antibiotic Penicillin is derived from a fungus called *Penicillium*. Many other fungi also produce antibiotic substances which are now used worldwide. *Claviceps purpurea* is a mold that causes disease on Rye crop known as Ergot which produces Ergotamine and LSD (Lysergic acid diethylamide). Ergotamine is used in childbirth. Anticancer drug Paclitaxel is obtained from *Nodulisporium salviforme* and *Taxomyces andreanae*. *Penicillium griseofulvum* produces Griseofulvin, it has also anticancer properties. Some fungi which parasite caterpillars have been traditionally used in China as a tonic for hundreds of years. Medicinal mushrooms have anticancer, cardiovascular, antibacterial, antiviral, antiinflammatory, antiparasitic, hepatoprotective, and antidiabetic activities (Mohmand et al., 2011).

5.2 History of fungal medicine

Bioactivity of secondary metabolites of fungi has been known for the last twelfth centuries but now based on research led to the development of some of the most important drugs of twenty-first century. In China around 800 AD use of fungi as medicinal recorded was as red yeast rice by a yeast *Monascus purpurea*. In the later twentieth-century chemical analysis of red yeast rice exhibited that they comprise a diversity of organic compounds associated with Compactin and Statins, known as mevastatin, which is the first statin accepted by the United states FDA use for blood cholesterol-lowering agent. According to Chinese researcher the red yeast rice had diverse bioactivities including cardiovascular problems. The second historic fungal product is the Psilocybin found in the Mushrooms *Psilocybe mexicana*. It was useful in religious formalities in Central America by Aztec Indians and other religions around the world for terminal illness (Langdon and Pearce, 2017). A bacteriologist, Alexander Fleming was at St.Mary's Hospital, and one day he was speaking to a coworker by observing an agar plate containing invading fungus on bacteria culture, where bacterial growth did not occur. By isolating the mold from the agar plate, it was identified that belonging to the genus *Penicillium* and named its active agent as Penicillin. Fleming published his findings in 1929, he found that Penicillin possessed antibacterial activity on Staphylococci and other such pathogens that cause diseases, like Gonorrhea, Diphtheria, Pneumonia, Meningitis, and Scarlet fever (Gaynes, 2017). In 1940, two fellow researchers Howard Florey and Ernst Chain were able to produce Penicillin in large quantities for use in World War II (Bennett and Chung, 2001).

The second member of the beta-lactam antibiotic family is Cephalosporin and it was discovered from a fungus *Cephalosporium* in 1948 by Italian Scientist Giuseppe Brotzu. The natural Cephalosporin was never used clinically but analogs were prepared with superior activity. There are multiple generations of cephalosporins with improved activity.

Fusidic acid is the third clinically important antibiotic produced by fungi *Fusidium coccineum*. It was developed in 1960 for the curing of Gram positive bacterial infection and, by the Danish drug company Leo Pharma.

Lentinan, a polysaccharide obtained from Shiitake mushroom, native to East Asia. Lentinan is effective in enhancing the immune system. It has other properties like lowers cholesterol, kills microbes and viruses.

From *Claviceps purpurea,* Ergotamine was first isolated by Arthur stall at Sandoz (now Novartis) in 1918. The use of Ergotamine began in the sixteenth century at that time it was used to accelerate parturition.

Lysergic acid was discovered from *Claviceps purpurea* in 1938 but it is not active itself, and Lysergic acid diethylamide (LSD) is a highly potent CNS active compound producing hallucinations.

Cyclosporin A is a powerful immunosuppressant used in organ transplants to overcome organ rejection by the new host. Discovered from *Tolypocladium inflatum* by Sandoz research workers.

Compactin and Mevacor obtained from *Penicillium compactum* and *Aspergillus terreus* are the first statins discovered, these compounds inhibit cholesterol biosynthesis. Another statin named Pravastatin formed by stereoselective microbial hydroxylation of Compactin was discovered by Sankyo Pharma in 1970.

Pravastatin was formed by the stereoselective microbial hydroxylation of Compactin and discovered by Sankyo Pharma in the 1970s. It is marketed in the US, as Pravachol and reached 1.3 billion dollars in the US in 2005.

Cancidas-R was discovered from the Fungus *Glarea lozoyensis* and developed by Merck in the 1990s is a new class of antifungal agents that inhibit glycan biosynthesis.

Fingolimod is the most recent fungal metabolite and marketed by Novartis. It was approved in 2011 for the treatment of multiple sclerosis. Fingolimod is a synthetic derivative of myriocin a powerful immunosuppressant and a metabolite from *Isaria sinclairii* and *Cordyceps sinclairii* which colonizes insects. The culture broth of this fungus was used in traditional Chinese medicine as an "eternal youth" elixir (Langdon and Pearce, 2017).

5.3 Fungi in producing natural compounds and secondary metabolites

Secondary metabolites have a broad range of biological activities, and many important pharmaceuticals have been discovered through the study of fungal chemistry. Fungal-derived lead compounds have been found to be effective in many disorders, such as against cancer, malaria, bacterial infection, neurological, cardiovascular diseases, and autoimmune disorders. Some best-known examples are the beta-lactam antibiotics, penicillins, and cephalosporins (Bills and Gloer, 2017). After the discovery of Penicillin (Penicillin F) in 1929 and Penicillin G (Fig. 5.1) in 1940, research work has been made widespread in fungal secondary metabolites for the discovery of thousands of compounds that have different pharmaceutical values (Table 5.1) (Hansson, 2013). Secondary metabolites besides antibiotic, statins, inhibitor of cholesterol synthesis, lipase, and Ergot alkaloids are some vital pharmaceutical goods also derived from fungi.

5.3.1 Mycotoxins

Fungal mycotoxins are considered to be as good secondary metabolite, but their functions still not clear and some are described as they play a role in eliminating other microorganisms from fungal colony. Mycotoxins are very toxic to health and even a single mycotoxin can produce various adverse effects

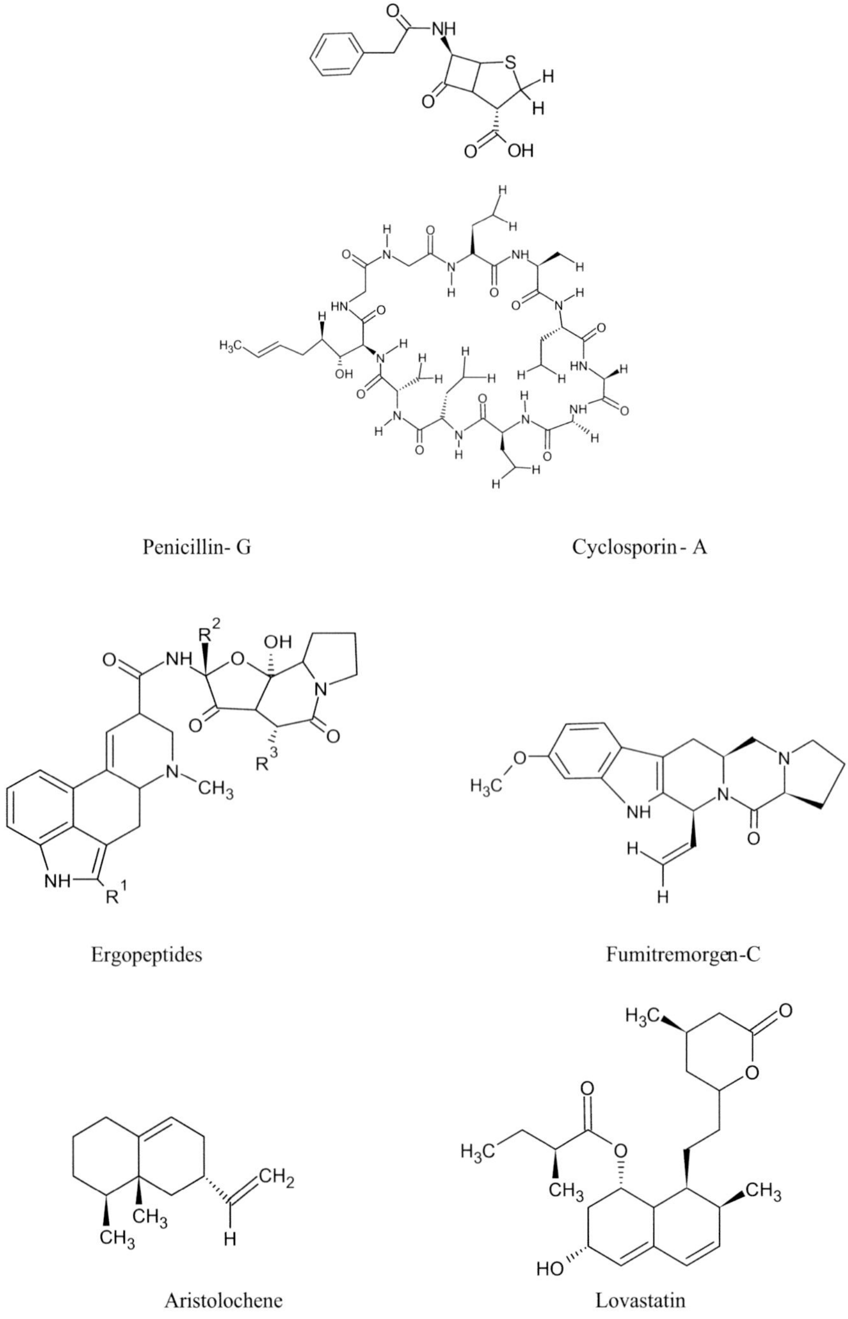

FIG. 5.1

Structure of fungal secondary metabolites.

(Continued)

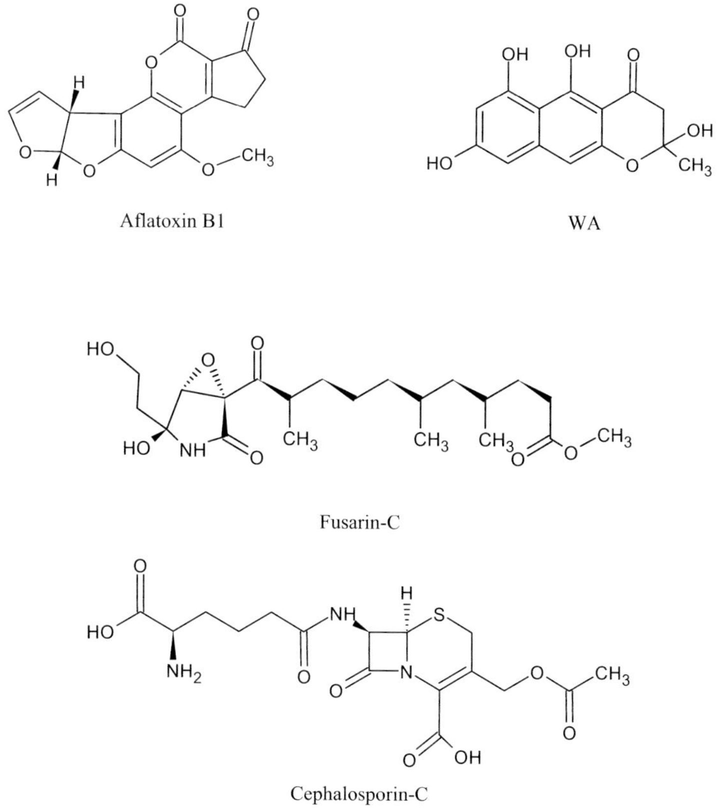

FIG. 5.1, CONT'D

such as carcinogenic, vasoactive, and some case central nervous system damage (Brase et al., 2009). However, they have been useful in treating medical conditions of hypertonia, angina pectoris, in agalactorrhea inhibition of protein release, serotonin-associated disturbances, decrease in blood loss following childbirth, and inhibition of nidation in premature pregnancy. And also, by inhibition of serotonin, noradrenalin, adrenalin, and smooth muscles contraction of the uterus and some possesses antibiotic effect (Demain and Martens, 2017). Earliest report of Chinese writings in~1100 BC, it has been noted that have agonist action in adrenergic, dopaminergic, and serotogenic receptors. Nowadays different commercial derivatives of naturally occurring Ergot alkaloids which are designated by FDA for the curing of various diseases. Ergonovine, an alkaloid from *Claviceps purpurea* is an agon of tryptaminergic sense organ in smooth muscles, and due to this characteristic, it is used for uterine stimulus in managing hemorrhage and postnatal uterine atony. Dihydroergotamine is peptide alkaloids, derivative of Ergotamine, used for miscarriages and inhibition of vascular headaches like histaminic cephalea, cluster headaches, migraines, and migraine variants. Bromocriptine is a semisynthetic derivative of peptide alkaloid marketed toward the curing of Parkinsonism juxtaposed with L-dopa treatment in patients suffering a failing signal to the medicine, it is also used in the therapy of hyperprolactinemia.

Table 5.1 Some examples of fungal metabolites and their activities (Goyal et al., 2016).

Fungal Secondary metabolites	Source	Bioactivity
6-Deoxyfusarubin and Ascomycone-B	CCF 4378 of *Biatriospora* sp.	Cytotoxicity (Stodůlková et al., 2015).
Beauvericin	*Fusarium* sp.	Trypanocidal effect (Campos et al., 2015).
Apicidin-F	*F. fujikuroi*	Antimalarial (Von Bargen et al., 2013).
Citrinin	*Sponge associated Penicillium* sp.	Antibacterial (Subramani et al., 2013).
Cladosin-C	*Cladosporium sphaerospermum* 2005-01-E3	Antiviral action; influenza A H1N1 virus (Wu et al., 2014).
1-(2,6-dihydroxyphenyl)-pentan-1-one	*Cryptosporiopsis* sp.	Antibacterial (Zilla et al., 2013)
Dihydronaphthalenone2	*Nodulisporium* sp.	Antimicrobial activity (Prabpai et al., 2015)
Ganoleucoins-C and A	*Ganoderma leucocontextum*	Repressive effect against HMG-CoA reductase (Wang et al., 2015)
4-Hydroxymellein	*Phoma* sp.	Inhibitory activity against cells of P388 murine leukemia (Santiago et al., 2014)
Herqueidiketal	*Penicillium* sp.	Involve in action against *Staphylococcus aureus* (Julianti et al., 2013)
Hispidin	*Phaeolus schweinitzii*	Antioxidant effect (Han et al., 2013)
Isosclerone	*Aspergillus fumigatus*	Antiproliferative against MCF-7 human breast carcinoma cells (Li et al., 2014a)
Nodulisporiviridin-G	*Nodulisporium* sp.	Amyloid beta-42 aggregation inhibitory activities (Zhao et al., 2015)
Neoechinulin-A	*Eurotium* sp.	Antiinflammatory activity (Kim et al., 2013)
Polyporusterone-B	*Polyporus umbellatus*	Antitumor effect (Zhao et al., 2010)
Phenylpyropenes-F and E	*P. concentricum*	Cytotoxic in MGC-803 cells (Ding et al., 2015)
Pinazaphilones-B and (+−)-penifupyrone	*Penicillium* sp.	Alpha-glucosides Inhibitory (Liu et al., 2015)
Verrucosidin	*Penicillium* sp.	Antimycobacterial effect (Bu et al., 2016)

The trichothecenes are chemically related mycotoxins used as antiviral, especially for herpes simplex virus, antibiotic, antimalarial, antileukemic, and immunotoxin. Zearalenone is mycoestrogen or nonsteroidal estrogen. It has been known for the treatment of postmenopausal syndromes in woman. Zearalenol and Zearalenone (Fig. 5.2) are proprietary as an oral prophylactic agent. Citrinin shows antibiotic action to Gram positive bacteria and in 1952, it was designated as the "antibiotic of the future." Citrinin was tested against Leishmania and displayed inhibition of growth (Brase et al., 2009).

5.4 Major groups of fungi producing different classes of antibiotics

The term antibiotic means an organic element, formed by a microorganism, which prevents the growth of some other microorganisms. Sir Alexander Fleming was the first to establish the role of fungi in manufacturing antibiotic substances in 1929 (Solomon et al., 2019).

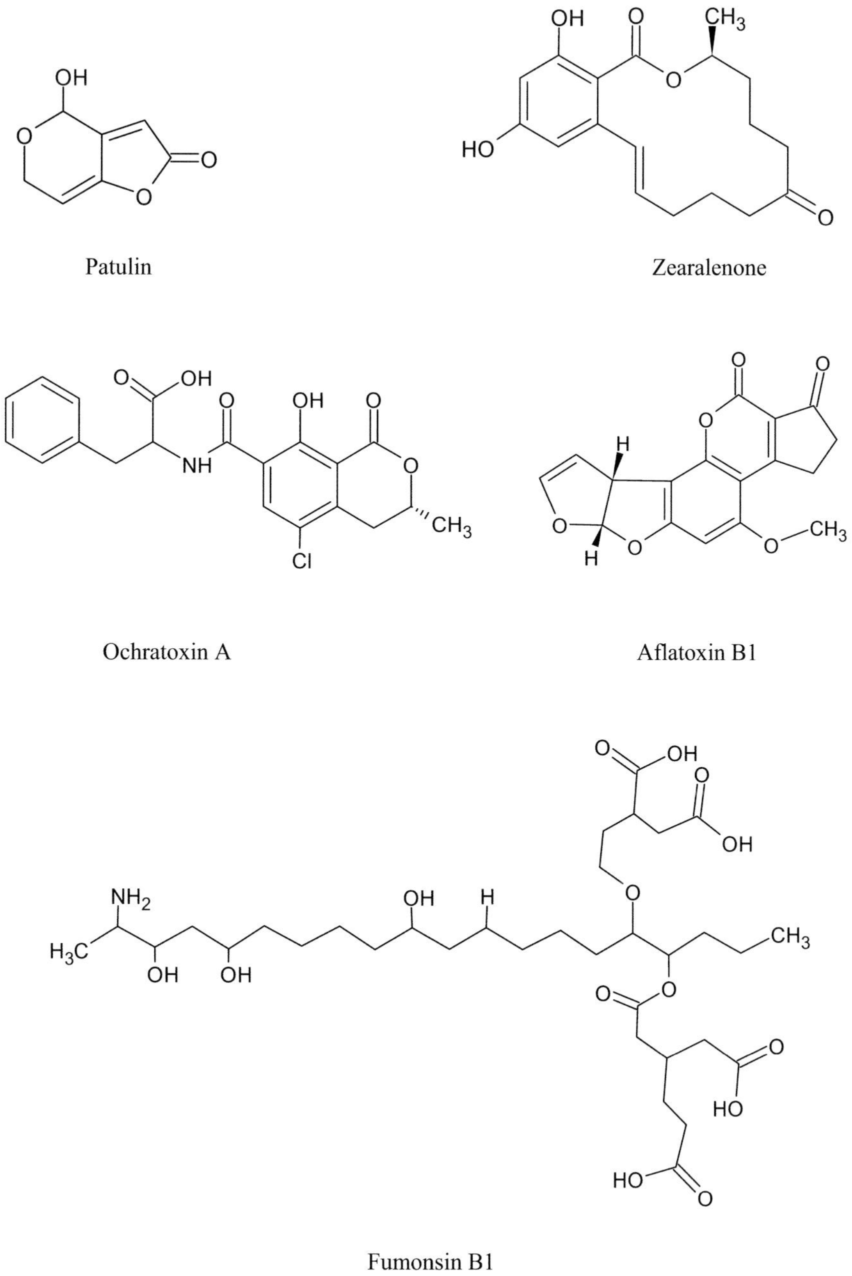

FIG. 5.2

Structures of mycotoxins.

Out of 12,000 antibiotics identified in 1995, 22% are formed by filamentous fungi. The most important class of antibiotic is beta-lactam. They comprise major portion of antibiotics in the market. Examples are Penicillin and Cephalosporins (Demain, 2014). *P. chrysogenum* and *P. notatum* are the two improved strains of Penicillium from which Penicillin is obtained on a commercial scale worldwide including India. *Streptomyces griseus* produces Streptomycin and it can destroy Gram negative bacteria. A number of antibiotics have been isolated from cultures of Aspergillus, but their effectiveness has not been as Penicillin. Griseofulvin is extracted from mycelium of *Penicillium griseofulvum,* which has antifungal characteristics. It affects the hyphae and interferes with wall formation. It is therefore effective against fungal skin diseases like athlete's foot disease and ringworms (Solomon et al., 2019).

5.4.1 Some antibiotics and their mode of action

Antimicrobial power of different classes of antibiotic is focused on some exceptional features of bacterial cell structure and metabolic processes. The mechanism of action of antibiotics are:

- Cell wall synthesis inhibition.
- Breaking down of cell membrane assembly and function function.
- Breakdown of nucleic acids structure and function.
- Protein synthesis inhibition.
- Obstruction of metabolic pathways (Etebu and Arikekpar, 2016).

5.4.1.1 Penicillin

Fungi have been a rich source of therapeutic agents and that offers the discovery of Penicillin from *Penicillium notatum* (Abdel-Razek et al., 2020). Penicillin, the first broad spectrum antibiotic altered the exercise of medication and transformed the path of pharmaceutical research and that saved countless lives in the World War II (Keller et al., 2005). Penicillin is consisting of family of antibiotics that includes Penicillin G (Fig. 5.3), Penicillin F, and Penicillin X as well as Amoxicillin, Ampicillin, Ticarcillin, and Nafcillin. Penicillin was first discovered from the green mold Penicillium but nowadays formed through synthetic means.

Penicillin is classified as Penicillinase resistant Penicillin (Cloxacillin, Methicillin), natural Penicillin (Penicillin G, Penicillin V), extended spectrum Penicillin (Carboxypenicillin), beta-Lactamase combinations (Augmentin). Used in the treatment of Meningococcal infections, Streptococcal infections, Syphilis, and prophylaxis for Scarlet fever (Sharma et al., 2013).

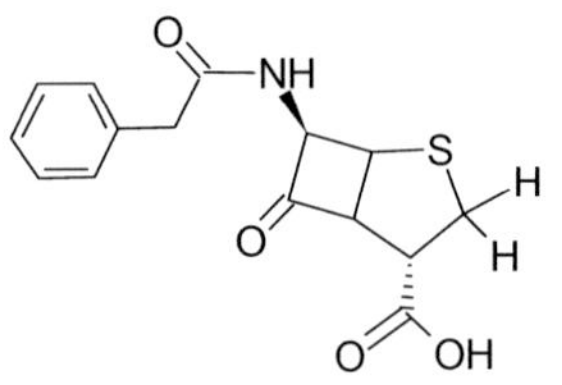

FIG. 5.3

Structure of Penicillin G.

5.4.1.2 Cephalosporin

The antibiotic Cephalosporin (Fig. 5.4) was first obtained from the mold, they inhibit the formation of the bacterial cell wall. The antibiotic Cephalosporin comprises Cefoxitin, Cefazolin, Cefuroxime, Cefotaximine, and moxalactam, which are effective in Gram positive bacteria but the newer generation of Cephalosporin antibiotics are also active against Gram negative bacteria.

Members of this cluster are analogous to Penicillin in function and structure. Cephalosporins consist of a 7-aminocephalosporanic acid nucleus and adjacent chains comprising 3,6-dihydro-2H-1,3 thiazane rings and a diversity of side chains that allow them to attach to diverse Penicillin-binding proteins (PBPs).

5.4.1.3 Tetracycline

Tetracycline (Fig. 5.5) was extracted from the fungus Actinomycetes, the tetracycline antibiotics comprise Doxycycline, Minocycline, and tetracycline. The antibiotic tetracycline is broad spectrum medicines that prevent the growth of some Gram positive bacteria, Chlamydiae Rickettsiae, and Gram negative bacteria (Darken et al., 1960).

5.4.1.4 Griseofulvin

This antibiotic produced from *Penicillium griseofulvum* is a FDA approved drug of choice against *Tinea capitis*. It acts as an inhibitor of microtubule assembly. It interacts with microtubules to affect the formation of the mitotic spindle and ultimately inhibits mitosis in dermatophytes. Griseofulvin (Fig. 5.6) serves as a fungistatic against Trichophyton, Microsporum, and *Epidermophyton* sp. (Aly et al., 2011).

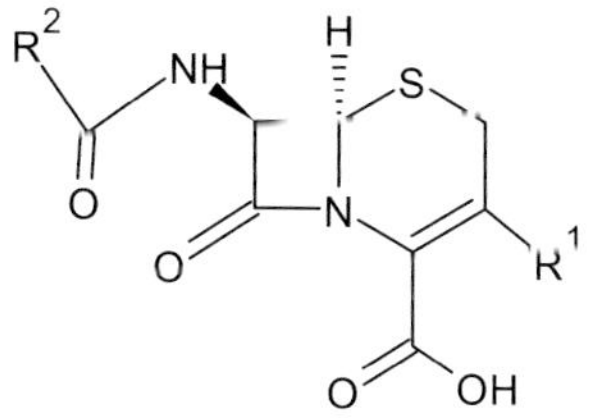

FIG. 5.4

Structure of Cephalosporin.

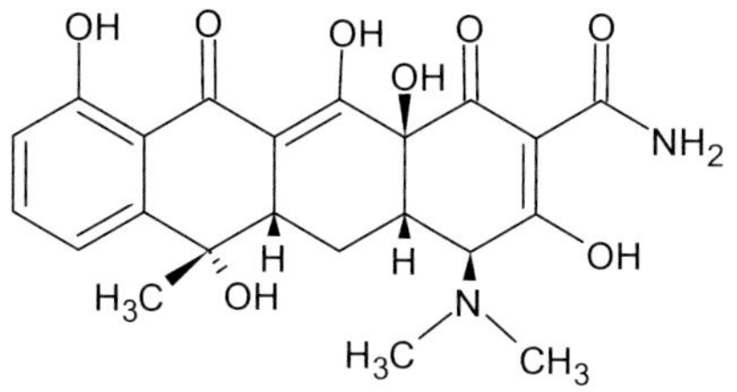

FIG. 5.5

Structure of Tetracycline.

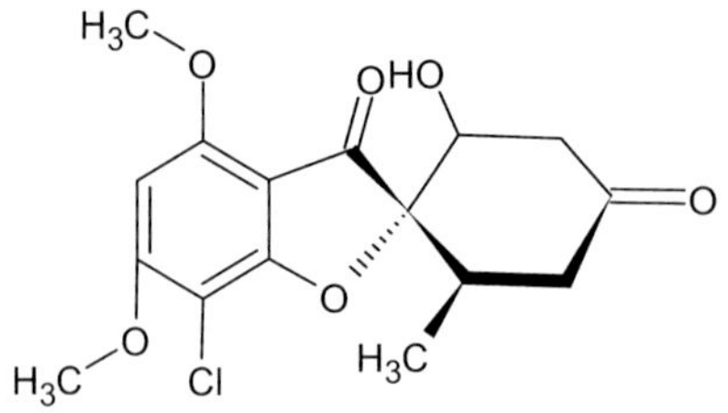

FIG. 5.6

Structure of Griseofulvin.

5.4.1.5 Marine fungi producing antibiotic

Since 1980, the number of antibiotics obtained from marine fungi is quickly growing. Approximately 350 antimicrobials marketed worldwide from marine fungi and they include natural products. The wide uses of semisynthetic and synthetic antibiotics give rise to the growth of resistant microbes. Worryingly, many pharmaceutical companies concentrated or entirely terminated R&D programs on antibiotics. So the number of researches has been going on to isolate the natural compounds from the marine fungi. Fig. 5.7 shows the structure of some compounds originated from marine fungi and Table 5.2 showing some compounds from marine fungi and their antibiotic activity.

5.5 Fungi as antimicrobial

Various mushrooms produce antifungal and antibacterial compounds to guard them from pathogenic bacteria, insects and protozoa. Distillation of these bioactive molecules have great medicinal value. Fungal components have antimicrobial action in contrast to Gram negative and Gram positive bacteria including yeast, food-grown pathogenic bacterial strains, and mycelial fungi (Table 5.3), for example, dermatophytes.

5.5.1 Antibacterial activity

Ascomycota and Basidiomycota show antibacterial action toward a series of bacteria. A number of polysaccharides from fungus show antibacterial effects, like lentinan from the mycelia of *Tinea versicolor*, and mycelial extracts like cortinelin from various strains of *Lentinula edodes* have exposed the capability to prevent the growing of Gram negative and Gram positive bacteria, for example, *Escherichia coli, Salmonella typhimurium, Listeria monocytogenes, Staphylococcus aureus,* and *Bacillus cereus* (Smith, 2014). Endophytic fungi are capable of producing bioactive molecules; these may be a vital source of producing new antibiotics. There are seven types of fungi isolated from mangrove plants, of these, *Rhizophora apiculata* and *Bruguiera gymnorrhiza* (L.) Lamk. showed antibacterial activity against *S. typhi*, *Penicillium* sp. (Rossiana et al., 2016). Sesquiterpenoid hydroquinone like ganomycins formed by *Ganoderma pfeifferi* Bres. prevent the growing of *S. aureus* (methicillin-resistant) and other

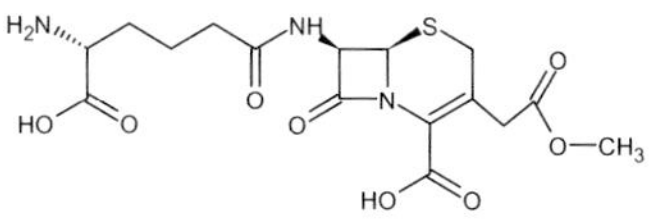

Cephalosporin C

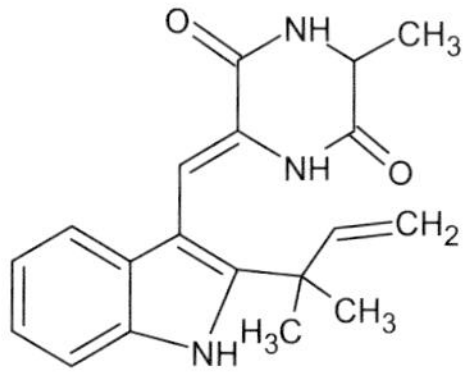

Chrysogenazine

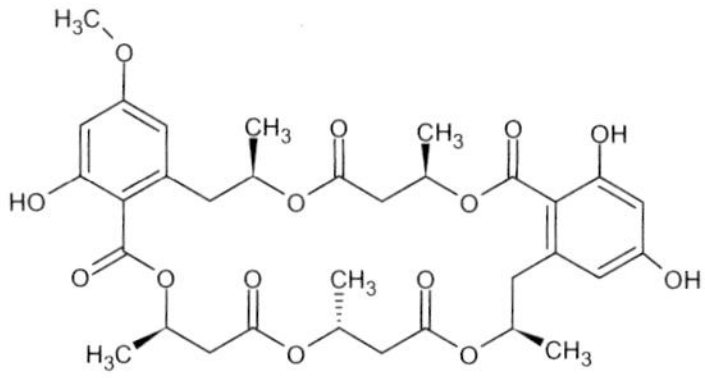

Calcaride A

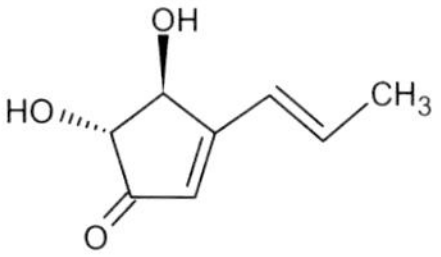

(+)-Terrein

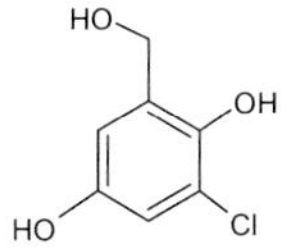

3-Chloro-2,5-dihydroxy benzyl alcohol

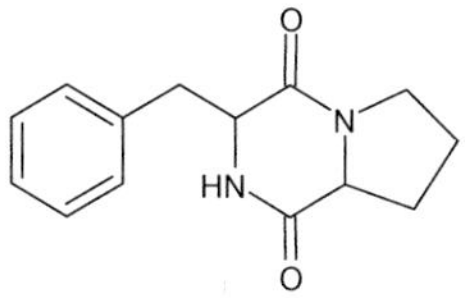

Cyclo-(Pro-Phe)

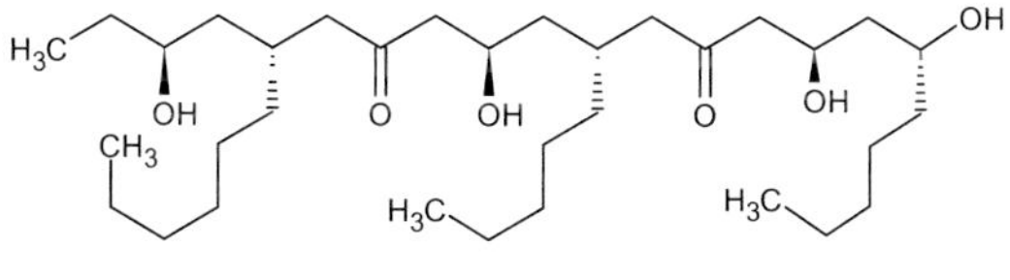

Exophilin A

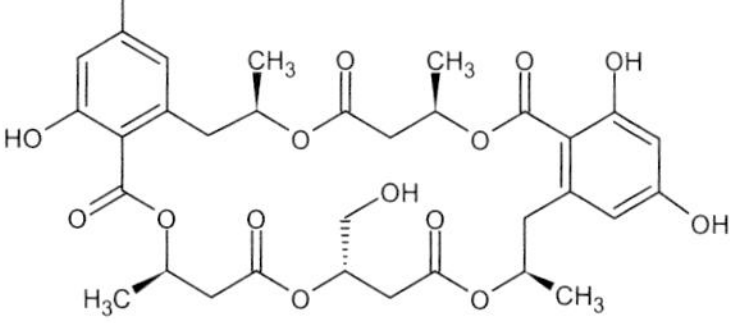

15G265alpha

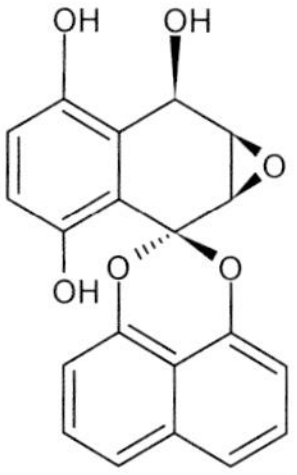

Ascochytatin

FIG. 5.7

Structure of some of the compounds from marine fungi.

Table 5.2 Some compounds from marine fungi and their antibiotic activity (Silber et al., 2016).

Compound	Producer	Antibiotic activity against
15G265 alpha, beta, gamma macrocycloic polylactones, and lipodepsipeptide	*Hypoxylon oceanicum* LL-15G256, mangrove	*Staphylococcus epidermidis, Xanthomonas campestris, Propionibacterium acnes* (Schlingmann et al., 1998).
Ascochytatin, Spirodioxynaphthalene	*Ascochyta* sp. NGB4,	Bacterial two-component regulatory system (Kanoh et al., 2008)
Bis(2-ethylhexyl) phthalate	*Cladosporium* sp.	*Rhodovulum* sp., *M. luteus, Loktanella hongkongensis, Ruegeria* sp. (Xiong et al., 2009)
Tetramic acid, Ascosetin	*Halichondria panacea,* Lindgomycetaceae	*S. epidermis, S. aureus, P. acnes, Xanthomonas campestris* (Wu et al., 2015)
Benzene derivative, 3-Chloro-2,5-dihydroxy benzyl alcohol	Marine biofilm, *Ampelomyces* sp.	*Vibrio* sp., *Micrococcus* sp., *S. aureus, Pseudoalteromonas* sp., *Staphylococcus haemolyticus* (Qian et al., 2012)
Chrysogenazine, diketopiperazine	*Penicillium chrysogenum, Porteresia coarctata*	*Vibrio cholera* (Naik et al., 2005)
Cephalosporin, beta-lactam	*Cephalosporium chrysogenum,* sea water	Broad spectrum (Elander, 2003)
Corollosporin and derivatives, Phthalide derivatives	*Corollospora maritima*	*Candida maltose, E coli, Pseudomonas aeruginosa, Bacillus subtilis* (Mikolasch et al., 2008)
Enniatines, Cyclodepsipeptides	*Halosarpheia* sp., mangrove	*Enterococcus faecium, E coli, Shigella dysentriae, Listeria monocytogenes, Salmonella enteric, Yersinia enterocolitica* (Zobel et al., 2016)
Exophilin A,3,5-dihydroxydecanoic polyester	Exophiala pisciphila, *Mycale adhaerens* (sponge)	*E. faecium, S. aureus, Enterococcus faecalis* (Doshida et al., 1996)
Lindgomycin, tetramic acid	Lindgomycetaceae, Halichondria panacea (sponge from Baltic sea)	*S. faecium, P. acnes, X. campestris, Sciara tritici* (Wu et al., 2015)

Table 5.3 Antimicrobial actions of some filamentous fungi.

Fungal species	Effective against
Ganoderma sp.	Gram positive bacteria: *Bacillus subtilis, B. cereus* and *Staphylococcus aureus*. Gram negative bacteria: *Pseudomonas aeruginosa* and *E. coli* (Suay et al., 2000)
Lentinula edodes	*Streptococcus aureus, Streptococcus pyogenes, Candida albicans, Bacillus megaterium, Actinomyces* sp., *Streptococcus, Lactobacillus* sp., *Porphyromonas* sp. and *Prevotella* sp. of bacteria (Hatvani, 2001)
Coprinus sp.	Multidrug resistant Gram positive bacteria (Johansson et al., 2001)
Pleurotus ostreatus	*P. aeruginosa, Aspergillus niger, Vibrio cholera, Fusarium oxysporum,* and *Salmonella typhi* (Gerasimenya et al., 2002)
Pleurotus esyngii	*S. aureus, Enterococcus faecium,* and *B. subtilis* (Suay et al., 2000; Wang and Ng, 2004)
Pleurotus sajor-caju	*M. arachidicola, F. oxysporun, S. aureus*, and *P. aeruginosa* (Ngai and Ng, 2004)
Monascus sp.	*Aspergillus, Penicillium, Mucor, E. coli, Fusarium, Botrytis*, and *Alternaria* (Ferdes et al., 2009)

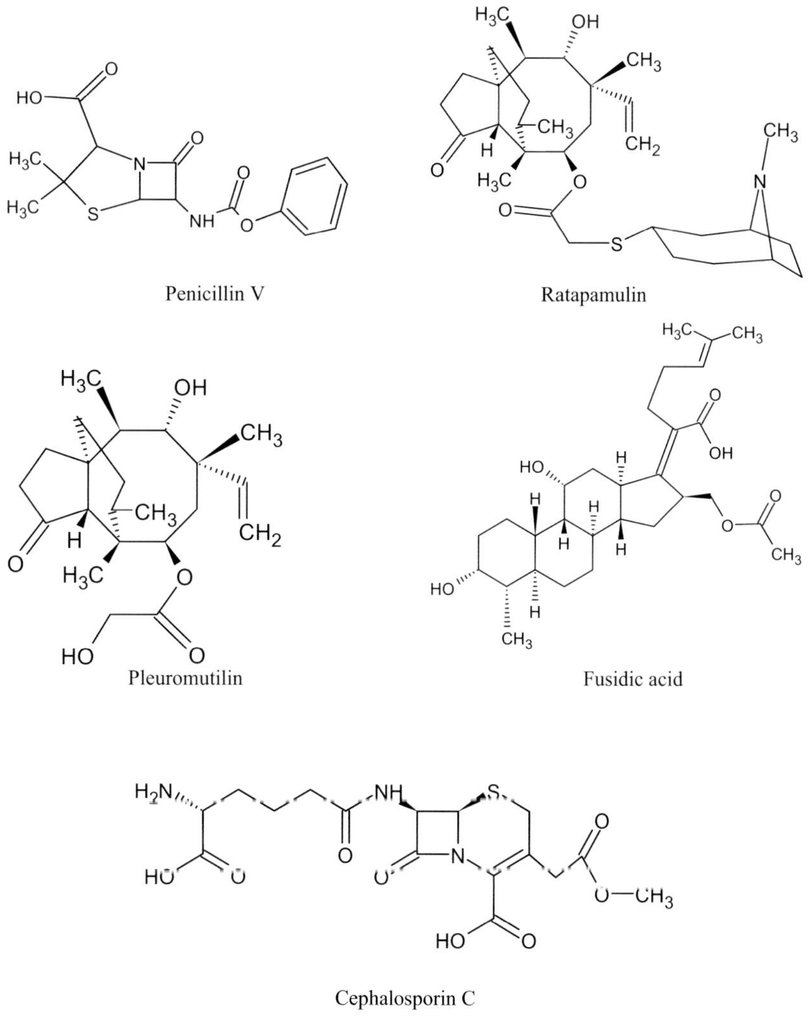

FIG. 5.8

Chemical structure of Fungal antibacterial drugs (Hyde et al., 2019).

bacteria. Total extracts of *G. pfeifferi* can prevent the growth of microorganism (*Pityrosporum ovale, Staphylococcus epidermidis*) causing skin problem (Lindequist et al., 2005). Fig. 5.8 showing chemical structure of some fungal antibacterial medicine.

5.5.2 Antifungal activity

Antifungal compounds (Fig. 5.9, Table 5.4) have been mainly found in Actinomycetes and Basidiomycetes groups. Lentin extracted from mycelia of *Lentinula edodes* of Basidiomycetes has effective antifungal action. This protein inhibits the growing of mycelium in various pathogenic fungi, for example, *Physalospora piricola*, *Mycosphaerella arachidicola*, and *Botrytis cinerea*. Lentin has preventive action on HIV-1 reverse transcriptase and formation of leukemia cells.

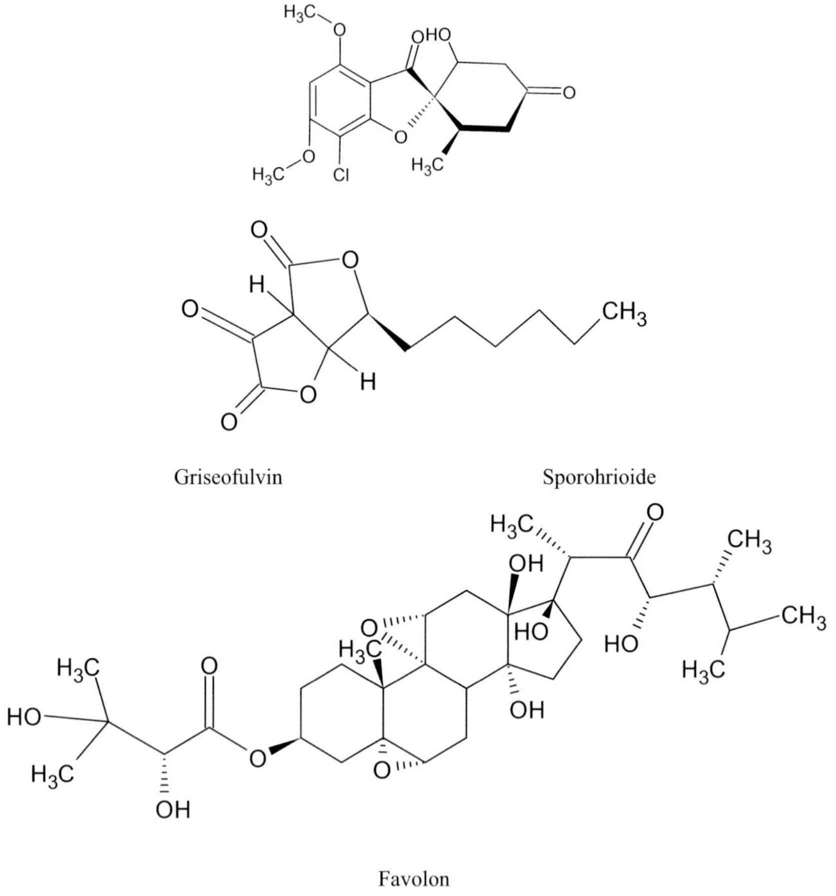

FIG. 5.9

Chemical structure of fungal-derived antifungal drug (Hyde et al., 2019).

Ascomycetes has also possessed antifungal protein. These proteins are extremely promising because they do not exert any toxic and inflammatory effect on mammalian cells and synergistically can interact with other medicines.

5.5.3 Antiviral activity

Generally, antibiotics are employed to treat bacterial infections but in viral infections very efficacious molecules are essential to fight against the virus. Antiviral drugs (Table 5.5) act by the prevention of function of viral enzymes and formation of nucleic acid. Basidiomycetes produces several polysaccharides, which are efficient inhibitors against the growth of viral infections. Fig. 5.10 is showing antiviral compounds of fungal origin.

Table 5.4 Antifungal compounds from endophytic fungi (Deshmukh et al., 2018).

Fungus	Isolated compounds	Biological activity
Pestalotiopsis fici	FicipyroneA	Compound active against *G. zeae*, Ketoconazole (Liu et al., 2013)
Phomopsis sp.	(14beta,22E)-9,14-dihydroxyergosta-4,7,22-triene-3,6-dione), (5alpha,6beta,15beta,22E)-6-ethoxy-5,15-dihydroxyergosta-7,22-dien-3-one, calvasterols A, ganodermaside D	Compound effective against *H. compactum*, *C. albicans* and *A. niger* (Chapla et al., 2014)
Rhizopycnis vagum Nitaf 22	Rhizopycnin D and TMC-264	Inhibit the spore propagation of *M. oryzae* (Lai et al., 2016)
Colletotrichum sp.	Colletonoic acid	Compound is active against *M. violaceum* (Hussain et al., 2014b)
Coniothyrium sp.	1,6-Dihydroxy-3-methyl-9,10-anthraquinone(phomarin), 1,7-dihydroxy3-methyl-9,10-anthraquinone, Coniothyrinones A-D and 1-hydroxy-3-hydroxymethyl-9,10-anthraquinone	Active against *B. cinerea* and *M. violaceum* (Sun et al., 2013)
Xylaria feejeensis	Xyolide	Active against *Pythium ultimum* (Baraban et al., 2013)
Chaetomium globosum	Chaetoglobosin A, Chaetomugilin A, Chaetomugilin D, Chaetoglobosin B, Chaetoglobosin E,F and Penochalasin G	Active against *P. herbarum* (Li et al., 2016)
Chaetomium cupreum	Ergosta-5,7,22-trien-3-beta-ol	Effective against *B. cinerea* and *S. sclerotiorum* (Wang et al., 2013)
Plectophomella sp.	(−)-Mycorrhizin A	Molecule is effective against *E. repens* and *Ugandatrichia violacea* (Hussain et al., 2014a)

The bioactive molecules from Basidiomycetes have been tested in Herpes simplex virus (HSV) and Human immunodeficiency Virus (HIV). Proteoglycan PSK (polysaccharide-k) and PSP (polysaccharide-peptide) obtained from *T. versicolor* demonstrate antiviral action against cytomegalovirus and HIV in vitro. Neuraminidase enzymes are treated as drug targets for the inhibition of influenza infection (Smith, 2014). Triterpenoids of *Ganoderma lucidum* are ganodermanontriol, ganoderiol F, and ganoderic acid B, which are antiviral toward HIV type1. Lucidadiol, ganodermadiol, and applanoxidic acid G obtained from *G. pfeifferi* have antiviral efficacy against the influenza virus. Ganodermadiol is effective to HSV type1. Lignin (water soluble) obtained from *Inonotus obliquus* known as Chaga inhibit HIV protease. AntiHIV activities were reported from mycelia of L. *edodes* (Lindequist et al., 2005).

5.6 Fungi as hepatoprotective

Mushrooms have accumulation of metabolites, that is, phenolic compounds, polysaccharides, steroids, and terpenes. Many of them have potential bioactivities and specific pharmacological properties

Table 5.5 Antiviral activities of fungal metabolites (Moghadamtousi et al., 2015) (Fig. 5.16).

Virus	Antiviral compound	Isolated from	Chemical class
EV71	Grisephenone A	*Stachybotrys* sp.	Xanthone (Qin et al., 2015)
	3,6,8-Trihydroxy-1-methylxanthone	*Stachybotrys* sp.	Peptide (Qin et al., 2015)
HSV	Arisugacin A	*Aspergillus terreus* SCSGAF0162	Lactone (Nong et al., 2014)
	Isobutyrolactone II	*Aspergillus terreus* SCSGAF0162	Lactone (Nong et al., 2014)
	Balticolid	*Ascomycetous strain*222	Macrolide (Nong et al., 2014)
	Halovirs A-E	*Scytalidium* sp.	Peptide (Rowley et al., 2003)
HIV	Integric acid	*Xylaria* sp.	Acylated Eremophilane Sesquiterpenoid (Rowley et al., 2004)
	Phomasetin	*Phoma* sp.	Tetramic acid (Singh et al., 1999)
	Oxoglyantrypine	*Cladosporium* sp.	Indole alkaloid (Peng et al., 2013)
	Norquinadoline-A	*Cladosporium* sp.	Indole alkaloid (Peng et al., 2013).
	Deoxynortryptoquivaline	*Cladosporium* sp.	Alkaloid (Peng et al., 2013)
	Deoxytryptoquivaline	*Cladosporium* sp.	Alkaloid (Peng et al., 2013)
	Tryptoquivaline	*Cladosporium* sp.	Alkaloid (Peng et al., 2013)
	Quinadoline-B	*Cladosporium* sp.	Alkaloid (Peng et al., 2013).

(Table 5.6). Water extracts of *Lentinula edodes, Volvariella volvacea, Auricularia auricular, Flammulina velutipes, Grifola frondosa,* and *Tremella fuciformis* were tested for hepatoprotective effectiveness for paracetamol induced liver injury in rats as a model. Paracetamol caused hepatic toxicity and raised concentration of serum transaminases (ALT and AST). 100 mg/kg weight of all the extracts exert hepatoprotective effect by reducing the quantity of the AST and ALT, while water extract of *T. fuciformis* mycelia showed hepatoprotective effect exclusively at a greater dosage of 300 mg/kg body weight (Soares et al., 2013).

5.6.1 Some molecules related with the Hepatoprotective activity of Fungi

Anthraquinol isolated from *Actinomycetospora cinnamomea* ethanolic mycelia extracts (Fig. 5.11).

Antrodia cinnamomea has hepatoprotective activity against several drugs, toxins, and liver diseases caused by alcohol. Compound potentially prevented the ethanolic induced ALT, AST, ROS, MDA, NO formation, and GSH reduction in human hepatic carcinoma cells (HepG2 cells) (Kumar et al., 2011).

Terpenoids, Peptides, and Polysaccharides from *Ganoderma lucidum* (Fig. 5.12).

Nearly, 400 chemical compounds have been extracted from *G. lucidum* and some of these are triterpenoids, polysaccharides, ergosterol, nucleosides, proteins/peptides, and fatty acids. From these, mainly, triterpenoids and polysaccharides have bioactive elements responsible for protection toward toxin-induced hepatic damage (Gao et al., 2003; Zhou et al., 2002).

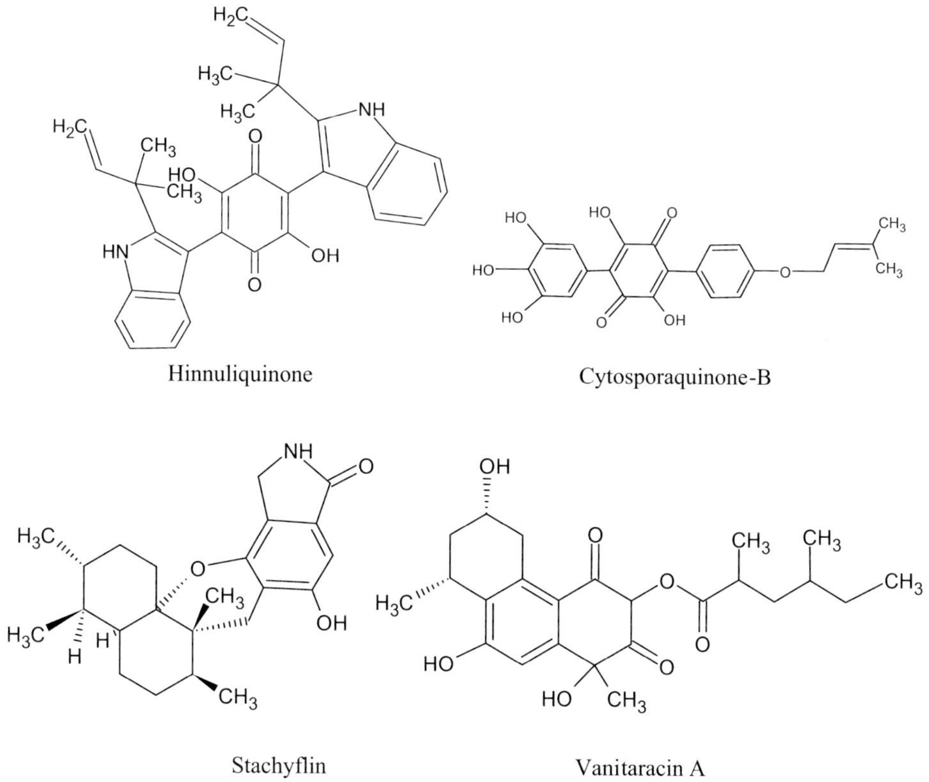

FIG. 5.10

Antiviral compounds of fungal origin.

5.6.1.1 Polysaccharide from mycelium of Pleurotus ostreatus

From *Pleurotus ostreatus,* polysaccharides of nonstarch and insoluble forms were assessed as pretreatment to inhibit CCl4 tempted hepatic injury in rats (Nada et al., 2010).

5.6.1.2 Lectin obtained from Pleurotus florida

Lectin obtained from *Pleurotus florida* is effective in relapsing the arsenic-induced hepatic injury in rats. Treatment with lectin showed significant alternations in the antioxidant enzymes level, SOD2 gene expression profile, and oxidative stress intermediates on arsenic exposure but this condition is reinstated by co-administration of Lectin (Rana et al., 2012).

5.7 Fungi as antidiabatic

Diabetes mellitus is a type of chronic metabolic disorder. Many Basidiomycota fungi, for example, *Cyclocybe aegerita, Agaricus bisporus, Tremella fucifomis,* and *C. cylindracea* are exploited as drug for diabetes type 2. Extracts of *Inocutis levis* elevate insulin sensitivity, insulin resistance, and uptake

Table 5.6 Hepatoprotective activity of fungal species.

Mushroom	Drug that induces hepatic damage	Main contributions
Ganoderma tsugae	CCl4	Isolates show antifibrotic activity, reduce the quantity of AST and ALT and decrease the prothrombin time (Wu et al., 2004)
Ganoderma lucidum	CCl4 Benzo-pyrene Ethanol	Extracts have antoxident, radical scavenging effect, which contributes to hepatoprotection (Shieh et al., 2001)
Coprinus comatus	Alloxane CCl4	The extract showed antioxidant potential (Popović et al., 2010)
Astraeus hygrometricus	CCl4	Decrease in the quantity of classical markers of liver damage (Biswas et al., 2011)
Phellinus rimosus	CCl4	Free radical scavenging activity (Ajith and Janardhanan, 2002)
Calocybe indica	CCl4	The ethanoic extract rejuvenates the antioxidant status of liver (Chatterjee et al., 2011).
Antrodia camphorata	Ethanol	Water infusion decreases the levels of ALT, ALP, AST, and bilirubin (Hsiao et al., 2003)
Pleurotus florida	Thioacetamide paracetamol	Hepatic defenses are assessed by managing the plasma levels of ALT, AST, bilirubin, ALP, protein, and cholesterol (Minnady et al., 2010)
Lentinula edodes	Galactosamine dimethyl-nitrosamine	Mushroom's ethanol and hot water extracts decrease the quantity of classical markers of liver damage (Akamatsu et al., 2004)
Pleurotus cornucopiae	CCl4	Major constituent in the extract is ergosterol, mannitol, D-beta-(1—3) glucans, linoleic acid, phenolic compounds, peptide, and other carbohydrates (El Bohi et al., 2009)

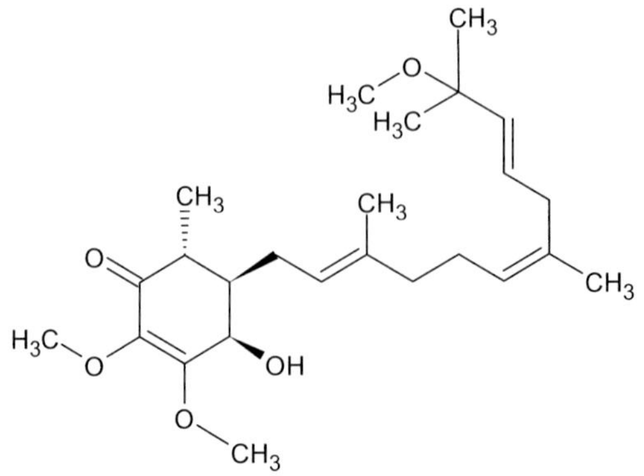

FIG. 5.11

Structure of Anthraquinol.

of glucose in tissues and support to regulate glucose level of blood. Fruiting bodies of *Antrodia cinnamomea* have antidiabetes properties. Extracts of *Grifola frondosa* are effective for hyperglycemia and hyperinsulinema. Capsules made of *Ophiocordyceps sinensis*, Reishi-Max capsules, SX-fraction, and Tremella are a few examples of antidiabetic agents formed from mushrooms. SX-Fraction is used for enhancing insulin sensitivity (Table 5.7) (Hyde et al., 2019).

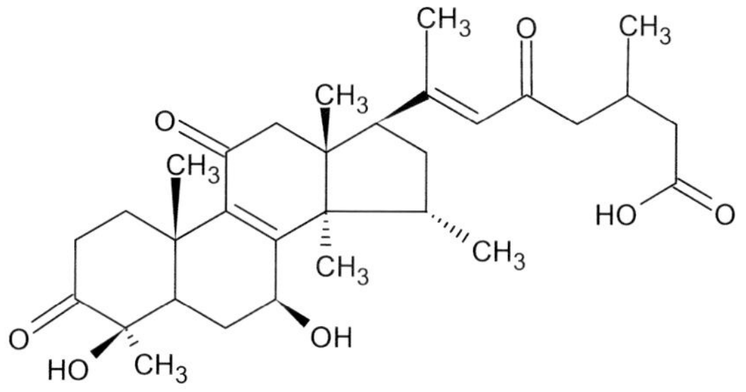

FIG. 5.12

Structure of Ganoderenic acid.

Table 5.7 Antidiabetic activity of medicinal mushroom.

Mushroom species	Common name	Bioactive compounds	Observed effects
Agaricus campestris	Field mushroom	Aqueous extract fruiting body	Reduced blood glucose level (Gray and Flatt, 1998)
A. bisporus	White button mushroom	Dry fruiting body extract	Decreased blood sugar level (Jeong et al., 2010)
Iomus obliquus	Chaga mushroom	Ethyl acetate extract from dehydrated Fruiting bodies	Decreased blood glucose level (Lu et al., 2010)
Amphaces subrufescens	Almond mushroom	Beta-glucans from dehydrated fruiting bodies	Reduced blood glucose level (Kim, 2005)
Ophiocordyceps sinensis	Caterpillar fungus	Polysaccharide of Cordyceps	Decreased blood sugar level/increased level of insulin in the blood (Li et al., 2003)
Hericium erinaceus	Caterpillar fungus	Methanol extract of dehydrated fruiting body	Reduced blood glucose level (Wang et al., 2005)
T. mesenteric	Brain mushroom (Yellow)	Extract of fruiting body	Elevated secretion of insulin and metabolism of glucose (Lo et al., 2006)

Flavanone glycosides from *Citrus grandis* were transformed to aglycones by hesperidinase and naringinase and to hydroxylated states by *Aspergillus saitoi*, which have higher antioxidant and antidiabetic activity. Thus, it is exploited for the growth of therapeutic nutrients to regulate the level of blood sugar of diabatic patients by preventing α-glucosidase and α-amylase in the intestinal tract (Godavari and Amutha, 2016).

5.8 Fungi as anticancer

After cardiovascular disease cancer holds the second place in terms of mortality. There are several treatments for cancer, radiation therapy, chemotherapy, immunotherapy, and surgery but only some of them precisely target tumor cells with more toxicity. Fungi are an important source of anticancer natural product discovery. Table 5.8 is showing fungi producing anticancer compounds.

Table 5.8 Fungal-derived anticancer drugs (Gohar et al., 2020).

Metabolites	Fungi	Mode of action
Taxol	*Taxomyces andreanae*	Induce G2/M phase arrest of the cell cycle (Strobel et al., 1997)
Terrein	*A. terreus*	Inducer of apoptotic cycle in breast carcinoma (Liao et al., 2012)
BrefeldinA	*P. brefeldianum*	Inducer of apoptosis in leukemia (Gnägi et al., 2019)
Penicimutanolone, penicimutanin A, penicimutanin B and Penicimutatin	*Penicillium purpurogenum*	Cytotoxicity against tumor cell lines (Fang et al., 2014)
Chondrosterin	*Chondrostereum* spp.	Cytotoxicity against tumor cell lines (Li et al., 2014b)
Asperlin	*A. nidulans*	Induce G2/M cell cycle arrest (He et al., 2011)

Irofulven is a semisynthetic derivative of illudin S, which is isolated from *Omphalotus illudin* is a natural toxin. It inhibits DNA replication complexes and DNA synthesis in cell division and the anomalous cells in S-phase turn to death of cells through apoptosis. It gives promising result against cancers of brain, central nervous system, colon, blood, breast, prostate, sarcoma, pancreas, lungs, and ovary. Aphidicolin is a tetracyclic diterpene having anticancer potential obtained from *Akanthomyces muscarius* or *Cephalosporium aphidicola* and *Nigrospora sphaerica*. Other anticancer compounds include Leptosins C and F (Fig. 5.13) extracted from *Leptoshaeria sp.* (Hyde et al., 2019).

5.9 Fungi as neuroprotection

Fungi possess a potent neuroprotective activity through various compounds from different fungal species. Medicinal mushroom consists of neuroactive compound which improves nerve functioning. Number of medicinal mushrooms, for example, *Antrodia camphorata*, *Ganoderma sp.*, *Hericum erinacus*, *Lignosus rhinocerotis,* and *Pleurotus giganteus* useful in improving the peripheral nervous system. *Hericium erinaceus* yield two exceptional terpenoid classes, viz., erinacines and hericenones (Fig. 5.14) from the maturing body and mycelia of fungus. That can instigate the formation of nerve growing factor through the TrkA/Erk1/2 pathway (Thongbai et al., 2015). Sclerotium extract of medicinal mushroom of Malaysia, *Lignosus rhinocerotis* contains neuroprotective compound, that initiates neurite outgrowth (Eik et al., 2012). *G. neo-japonicum* stimulates neuronal outgrowth (Seow et al., 2013). *P. giganteus* comprises a high quantity of uridine, which shows neuronal outgrowth stimulatory activity (Phan and Sabaratnam, 2012). *Laxitextum incrustatum* and *Dentipellis fragilis* of the family Hericicaceae comprise molecules of erinacine type; these cyanide diterpenoids enhance nerve growth factors (Gong et al., 2020). Fingolimod compound produced by the insect linked ascomycetes *Isaria sinclairii* is an effective immunosuppressant, it was accepted by US FDA in 2010 and used as a novel treatment in multiple sclerosis this compound is also used for the treatment of cancer and organ transplants (Hyde et al., 2019).

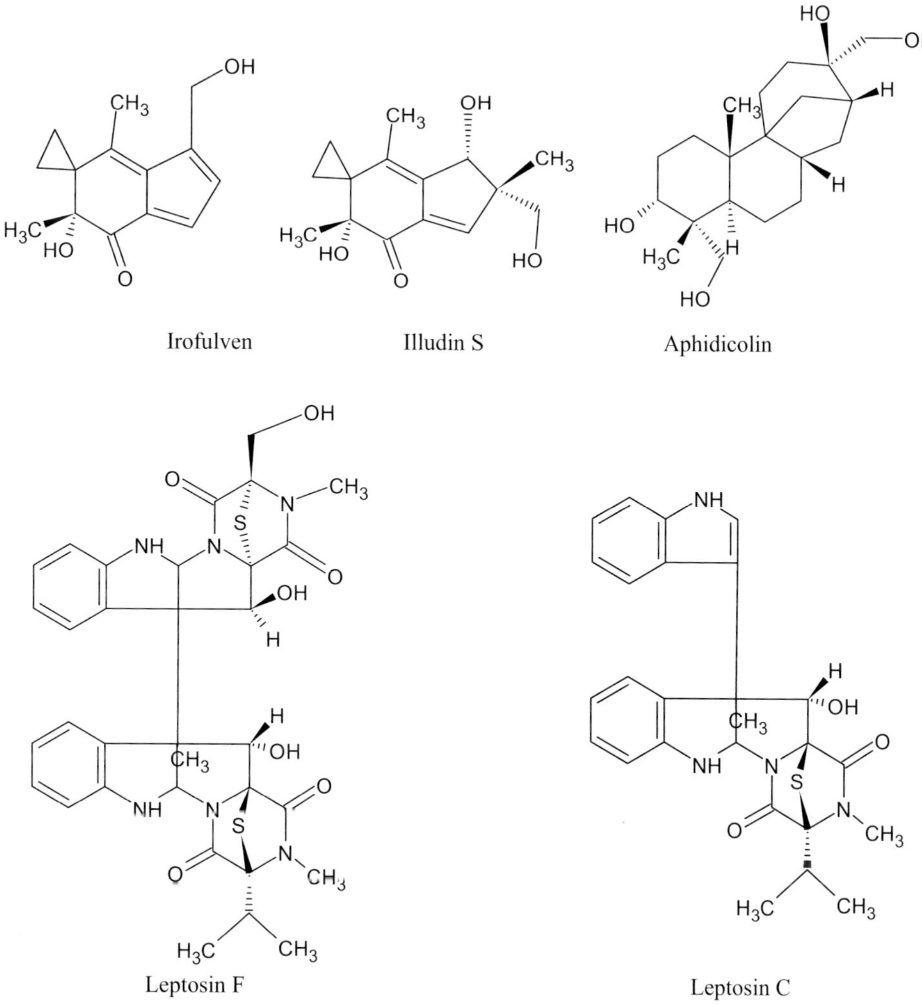

FIG. 5.13

Structure of anticancer compounds isolated from fungi.

5.10 Fungi as anticardiovascular drugs

Cardiovascular diseases comprise diseases of heart, blood vessels, and vascular diseases of the brain. Elevated levels of plasma cholesterol cause atherosclerosis, the hardening, and clogging of arteries. Cholesterol is an important compound both structurally and functionally. Cholesterol level remains within a certain limit but if this level increases, then this led to hypercholesterolemia which can cause serious effects in the body (Hyde et al., 2019). Statins are the largest selling class of drugs. Different species of the genus *Monascus* and *Aspergillus* synthesize natural statins (Gohar et al., 2020). Some filamentous fungi also synthesize natural statins such as *Penicillin*, *Eupenicillin*, *Doratomyces*,

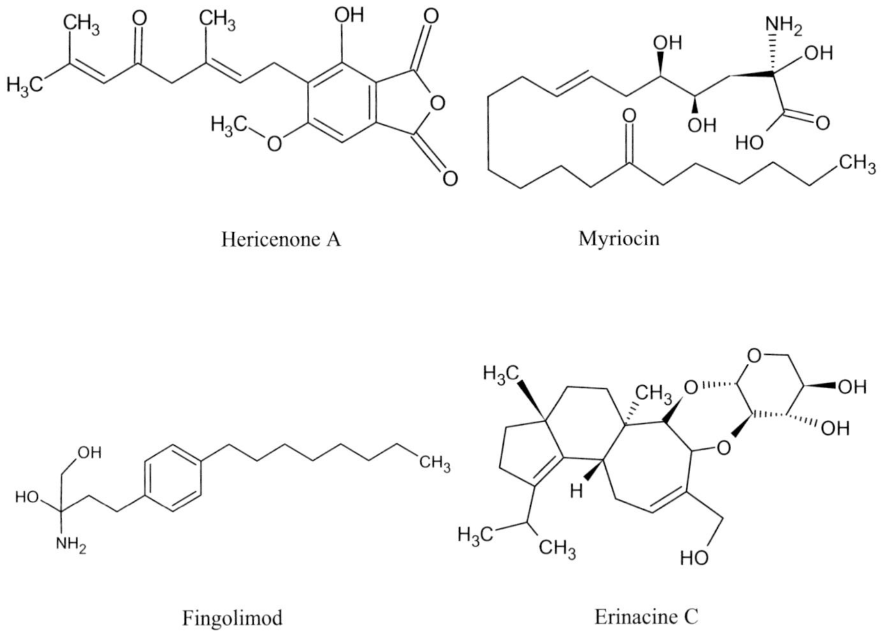

FIG. 5.14

Chemical structure of neurotrophic compounds.

Hypomyces, Gymnoascus, Trichoderm, and *Pleurotus*. From all statins, mevastatin was the first discovery of its kind and names ML-236B. Lovastatin (Fig. 5.15) is an anticholesterolemic drug that was made available in the pharmaceutical market with the approval of the United state FDA. It was synthesized from the genus *Monascus* (Gohar et al., 2020). Table 5.9 depicts cholesterol-lowering drugs from fungi.

5.11 Fungi as immunosuppressive drugs

Immune system of human body is efficient in distinguishing between foreign and native antigens and response only against the former. Immune response suppression either by radiation or drugs so that to avoid the refusal of transplants to regulate autoimmune ailments is immunosuppression. From fungal metabolites (Fig. 5.16) immunosuppressant drugs have been discovered (Demain, 2014).

5.11.1 Mycophenolic acid (MPA)

Mycophenolic acid is produced by the fermentation using the fungus *Penicillium brevicompactum* (Patel et al., 2017), it has dual function and exhibits antimicrobial and immunosuppressive properties. Mycophenolic acid an antiproliferative drug prevents the synthesis of inosine 5-monophosphate dehydrogenase enzyme. Mycophenolic acid inhibits the lymphocyte proliferation, prevents the precursor

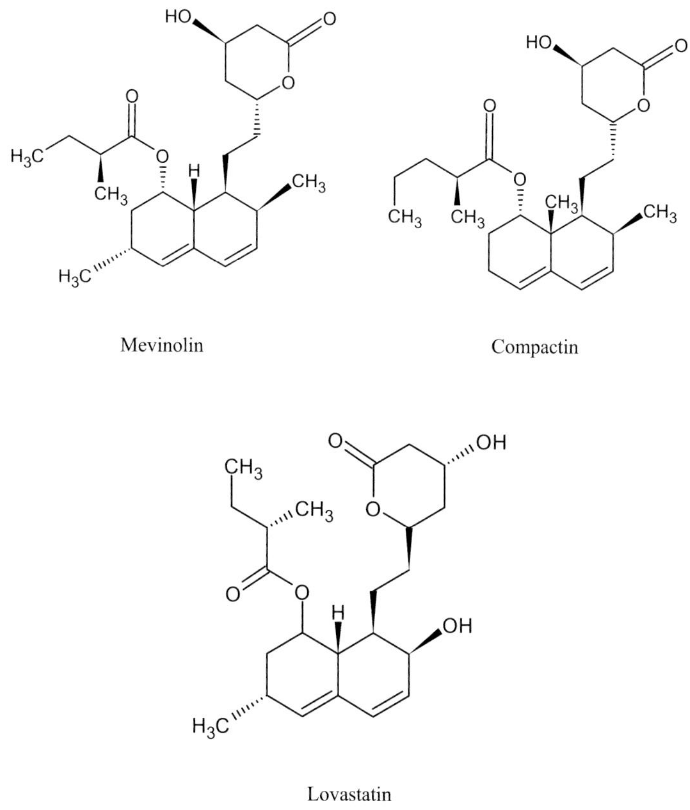

FIG. 5.15

Structure of cholesterol-lowering drugs derived from fungi.

Table 5.9 Fungi producing cholesterol-lowering drugs.

Fungal products	Mode of action	Fungi
Lovastatin	Prevent hydroxymethylglutaryl-coenzyme (HMG-CoA) reductase	*Aspergillus terreus* (Tobert, 1987)
Compactin		*Penicillium citrinum* (Boruta and Bizukojc, 2017)
Lovastatin	Inhibit 3-hydroxy-3-methylglutaryl-coenzymeA (HMG-CoA) reductase	*Aspergillus terreus* (Boruta and Bizukojc, 2017)
Itaconic acid	Conversion of cis-aconitate to itaconate by an enzymatically catalyzed decarboxylation	*Aspergillus terreus* (Boruta and Bizukojc, 2017)
Lovastatin	HMG-CoA reductase inhibitor	*Monascus purpureus* (Seenivasan et al., 2018)
Statins	Inhibit HMG-CoA reductase	*Aspergillus terreus* (Subhan et al., 2016)

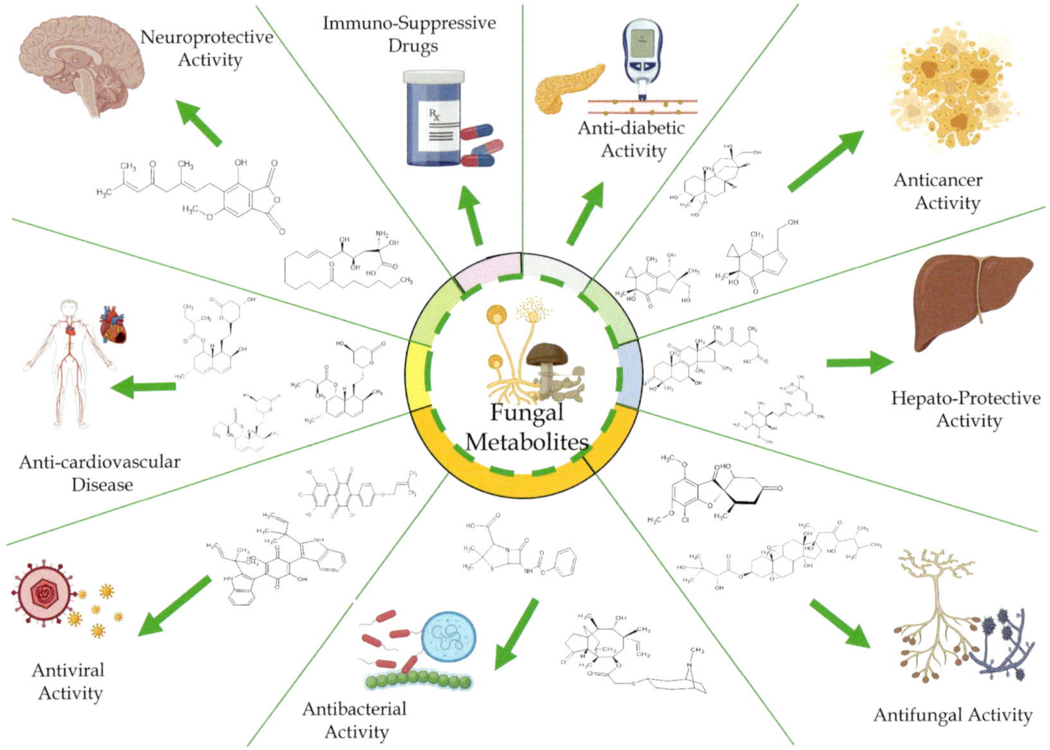

FIG. 5.16

Fungal Metabolites and their Bioactivity.

for the deoxyribonucleic acid synthesis, and blocks immunoreactions (Heischmann et al., 2017). It has also antiviral properties, effective against multiple viruses, for example, West Nile, Chikungunya virus, Yellow fever, and HCV (Fang et al., 2017).

5.11.2 Gliotoxin

Gliotoxin is produced from fungus *Aspergillus fumigates* and various mold species it is a mycotoxin and shows immune-suppressive action (Coméra et al., 2007). It has been proved that Gliotoxin prevents the thiol requiring enzymes and disrupts the NADPH oxidase activity to inhibit the respiratory bursts in the neutrophils (Tsunawaki et al., 2004).

5.11.3 Cyclosporin A

Cyclosporin A was first produced from soil fungus *Tolypocladium inflatum*. Cyclosporin A is considered an immunosuppressive drug, which has transformed the organ transplantation to a lifesaving procedure. Cyclosporin A antagonizes the activity of calcineurin, a calcium dependent serine-threonine phosphate that dephosphorylates the transcription factor NF-AT activated T-cell nuclear factor to stimulate IL2 expression, so that the dephosphorylation of NFAT is inhibited and proliferation of IL-2 dependent T-cells is suppressed. Cyclosporin A further interferes with the p38 and JNK signaling cascade (Gohar et al., 2020).

5.12 Present and future scope of the study

Fungi have been a great model organism in the biological science since long back and in recent times it is also playing a great role in the development of natural therapeutics. In modern pharmaceutical industry, fungal-derived secondary metabolites are the leading primary source of lead molecule. Based on the necessity of natural therapeutics, number of new chemical and pharmaceuticals industry are setting up in the developed and developing countries. Several governmental projects have been sanctioned by different countries in their territory for the research and development of therapeutics based on fungal metabolites. In India too, DBT govt. of India, DST govt. of India, CSIR, ICMR, and ICAR are several governmental wings, which sanctions number of projects in the R&D segments for the development of therapeutics from fungal products. Nowadays mushroom cultivation and isolation of chemical compound therein is one of the hot topics in the scientific research. Mushroom cultivation is getting a vibrational acceptance in the remote areas too of many cities in India. Newer entrepreneurs are developing in the field of mushroom cultivation and selling it to the market for human consumption may be fresh or dry condition.

5.13 Entrepreneurship opportunity from fungi

Fungi, especially mushroom cultivation is a very important field of entrepreneur development. The health-aware society is much concerned to include mushroom in their diet. Mushroom is very rich in protein; therefore, the vegetarians choose to eat mushrooms alternative to meat and chicken. People today love to eat dishes of mushroom with various recipes. And also, mushroom is one of the most important medicine supplements in the common health-conscious society of modern generation. As mushroom is an exceptional foundation of carbohydrates, proteins, fibers, vitamins, folic acid, and minerals and also consists of good quantity of iron for anemic patients. It is known as one of the good mediators of the conversion of agro-wastes into superior proteins. The mushroom can also be consumed by the heart and diabetic patients as it contains low calorific value. These good properties have transformed the mushroom commercial scheme for cultivation. The mushroom grows naturally and if required can be grown in a room-controlled temperature. The commercialization of mushroom cultivation in India has started recently mainly in the hill states as it requires colder temperature. From the mushroom a number of secondary metabolites are getting isolated day by day for the development of natural therapeutics for different ailments. So, there will be a good practice of development of entrepreneurship in this field.

5.14 Conclusions

Only small portion of the assessed fungal diversity has been explored for bioactive lead molecule. Fungi possess the capability to acclimatize to almost all niches of the environment and the probability of discovering new species and novel bioactive metabolites from them is very high. Consequently, more widespread collections of fungal species and culturing methods need to be developed. This will help to achieve the requirement for new therapies worldwide. Fungi produce antibacterial, antifungal, antiviral, immunosuppressive, antidiabetic, and hepatoprotective agents from various species including marine fungi, endophytic fungi, soil fungi, environmental fungi, and use these as pharmaceutical purposes (Aly et al., 2011). After the discovery of beta-lactam antibiotic Penicillin, now fungi have become a source of modern medicine including immune-pharmacology and organ transplantation. So, this advancement and development of medicine from secondary metabolites from fungi play role in semisynthetic modification of natural products to improve their activity. Therefore, it is concluded that fungi are potential source of biopharmaceutical and secondary metabolites (Gohar et al., 2020). And also, it can be concluded that based on mushroom cultivation and fungal metabolite isolation newer entrepreneur development in the society is possible.

References

Abdel-Razek, A.S., El-Naggar, M.E., Allam, A., Morsy, O.M., Othman, S.I., 2020. Microbial natural products in drug discovery. PRO 8 (4), 470.

Ajith, T., Janardhanan, K., 2002. Antioxidant and antihepatotoxic activities of Phellinus rimosus (Berk) Pilat. J. Ethnopharmacol. 81 (3), 387–391.

Akamatsu, S., Watanabe, A., Tamesada, M., Nakamura, R., Hayashi, S., Kodama, D., Kawase, M., Yagi, K., 2004. Hepatoprotective effect of extracts from Lentinus edodes mycelia on dimethylnitrosamine-induced liver injury. Biol. Pharm. Bull. 27 (12), 1957–1960.

Aly, A.H., Debbab, A., Proksch, P., 2011. Fifty years of drug discovery from fungi. Fungal Divers. 50 (1), 3–19.

Baraban, E.G., Morin, J.B., Phillips, G.M., Phillips, A.J., Strobel, S.A., Handelsman, J., 2013. Xyolide, a bioactive nonenolide from an Amazonian endophytic fungus, Xylaria feejeensis. Tetrahedron Lett. 54 (31), 4058–4060.

Bennett, J.W., Chung, K.-T., 2001. Alexander Fleming and the Discovery of Penicillin. Advances in Applied Microbiology. vol. 49 Academic Press, pp. 163–184. https://doi.org/10.1016/S0065-2164(01)49013-7.

Bills, G.F., Gloer, J.B., 2017. Biologically active secondary metabolites from the fungi. In: Heitman, J., Howlett, B.J., Crous, P.W., Stukenbrock, E.H., James, T.Y., Gow, N.A.R. (Eds.), The Fungal Kingdom. ASM Press, pp. 1087–1119.

Biswas, G., Sarkar, S., Acharya, K., 2011. Hepatoprotective activity of the ethanolic extract of Astraeus hygrometricus (Pers.) Morg. Dig J Nanomater. Bios 6 (2), 637–641.

Boruta, T., Bizukojc, M., 2017. Production of lovastatin and itaconic acid by aspergillus terreus: a comparative perspective. World J. Microbiol. Biotechnol. 33 (2), 34.

Brase, S., Encinas, A., Keck, J., Nising, C.F., 2009. Chemistry and biology of mycotoxins and related fungal metabolites. Chem. Rev. 109 (9), 3903–3990.

Bu, Y.-Y., Yamazaki, H., Takahashi, O., Kirikoshi, R., Ukai, K., Namikoshi, M., 2016. Penicyrones A and B, an epimeric pair of α-pyrone-type polyketides produced by the marine-derived Penicillium sp. J. Antibiot. 69 (1), 57–61.

Campos, F.F., Sales Junior, P.A., Romanha, A.J., Araújo, M.S., Siqueira, E.P., Resende, J.M., Alves, T., Martins-Filho, O.A., Santos, V.L.d., Rosa, C.A., 2015. Bioactive endophytic fungi isolated from Caesalpinia echinata Lam.(Brazilwood) and identification of beauvericin as a trypanocidal metabolite from Fusarium sp. Mem. Inst. Oswaldo Cruz 110 (1), 65–74.

Chapla, V.M., Zeraik, M.L., Ximenes, V.F., Zanardi, L.M., Lopes, M.N., Cavalheiro, A.J., Silva, D.H.S., Young, M.C.M., Fonseca, L.M.d., Bolzani, V.S., 2014. Bioactive secondary metabolites from Phomopsis sp., an endophytic fungus from Senna spectabilis. Molecules 19 (5), 6597–6608.

Chatterjee, S., Dey, A., Dutta, R., Dey, S., Acharya, K., 2011. Hepatoprotective effect of the ethanolic extract of Calocybe indica on mice with CCl4 hepatic intoxication. Int. J. Pharm. Tech. Res. 3 (4), 2162–2168.

Coméra, C., André, K., Laffitte, J., Collet, X., Galtier, P., Maridonneau-Parini, I., 2007. Gliotoxin from Aspergillus fumigatus affects phagocytosis and the organization of the actin cytoskeleton by distinct signalling pathways in human neutrophils. Microbes Infect. 9 (1), 47–54.

Darken, M.A., Berenson, H., Shirk, R.J., Sjolander, N.O., 1960. Production of tetracycline by Streptomyces aureofaciens in synthetic media. Appl. Microbiol. 8 (1), 46.

Demain, A.L., 2014. Valuable Secondary Metabolites From Fungi, biosynthesis and Molecular Genetics of Fungal Secondary Metabolites. Springer, pp. 1–15.

Demain, A.L., Martens, E., 2017. Production of valuable compounds by molds and yeasts. J. Antibiot. 70 (4), 347–360.

Deshmukh, S.K., Gupta, M.K., Prakash, V., Saxena, S., 2018. Endophytic fungi: a source of potential antifungal compounds. J. Fungi. 4 (3), 77.

Ding, Z., Zhang, L., Fu, J., Che, Q., Li, D., Gu, Q., Zhu, T., 2015. Phenylpyropenes E and F: new meroterpenes from the marine-derived fungus Penicillium concentricum ZLQ-69. J. Antibiot. 68 (12), 748–751.

Doshida, J., Hasegawa, H., Onuki, H., Shimidzu, N., 1996. Exophilin A, a new antibiotic from a marine microorganism Exophiala pisciphila. J. Antibiot. 49 (11), 1105–1109.

Eik, L.-F., Naidu, M., David, P., Wong, K.-H., Tan, Y.-S., Sabaratnam, V., 2012. *Lignosus rhinocerus* (Cooke) Ryvarden: a medicinal mushroom that stimulates neurite outgrowth in PC-12 cells. In: Evidence-Based Complementary and Alternative Medicine. vol. 2012., 320308.

El Bohi, K.M., Hashimoto, Y., Muzandu, K., Ikenaka, Y., Ibrahim, Z.S., Kazusaka, A., Fujita, S., Ishizuka, M., 2009. Protective effect of Pleurotus cornucopiae mushroom extract on carbon tetrachloride-induced hepatotoxicity. Jpn. J. Vet. Res. 57 (2), 109–118.

Elander, R., 2003. Industrial production of β-lactam antibiotics. Appl. Microbiol. Biotechnol. 61 (5), 385–392.

Etebu, E., Arikekpar, I., 2016. Antibiotics: classification and mechanisms of action with emphasis on molecular perspectives. Int. J. Appl. Microbiol. Biotechnol. Res 4 (2016), 90–101.

Fang, S.-M., Wu, C.-J., Li, C.-W., Cui, C.-B., 2014. A practical strategy to discover new antitumor compounds by activating silent metabolite production in fungi by diethyl sulphate mutagenesis. Mar. Drugs 12 (4), 1788–1814.

Fang, S., Su, J., Liang, B., Li, X., Li, Y., Jiang, J., Huang, J., Zhou, B., Ning, C., Li, J., 2017. Suppression of autophagy by mycophenolic acid contributes to inhibition of HCV replication in human hepatoma cells. Sci Rep. 7 (1), 1–12.

Ferdes, M., Ungureanu, C., Radu, N., Chirvase, A.A., 2009. Antimicrobial effect of Monascus purpureus red rice against some bacterial and fungal strains. N. Biotechnol. 25, 1.

Gao, Y., Huang, M., Lin, Z.-B., Zhou, S., 2003. Hepatoprotective activity and the mechanisms of action of Ganoderma lucidum (Curt.: Fr.) P. Karst.(Ling Zhi, Reishi mushroom)(Aphyllophoromycetideae). Int. J. Med. Mushrooms 5 (2).

Gaynes, R., 2017. The discovery of penicillin—new insights after more than 75 years of clinical use. Emerg. Infect. Dis. 23 (5), 849.

Gerasimenya, V.P., Efremenkova, O.V., Kamzolkina, O.V., Bogush, T.A., Tolstych, I.V., Zenkova, V.A., 2002. Antimicrobial and antitoxical action of edible and medicinal mushroom Pleurotus ostreatus (Jacq.: Fr.) Kumm. extracts. Int. J. Med. Mushrooms 4 (2).

Gnägi, L., Martz, S.V., Meyer, D., Schärer, R.M., Renaud, P., 2019. A Short Synthesis of (+)-Brefeldin C through Enantioselective Radical Hydroalkynylation. Chem. A Eur. J. 25 (50), 11646–11649.

Godavari, A., Amutha, K., 2016. Anti diabetic activity of Bacteria and Fungi: A review. Int. J. Front. Sci. Technol. 4 (2).

Gohar, U.F., Mukhtar, H., Mushtaq, A., Farooq, A., Saleem, F., Hussain, M.A., Ghani, M.U., 2020. Fungi: A potential source of biopharmaceuticals. Agrobiol. Records 2, 49–62.

Gong, W., Wang, Y., Xie, C., Zhou, Y., Zhu, Z., Peng, Y., 2020. Whole genome sequence of an edible and medicinal mushroom, Hericium erinaceus (Basidiomycota, Fungi). Genomics 112 (3), 2393–2399.

Goyal, S., Ramawat, K.G., Mérillon, J.-M., 2016. Different shades of fungal metabolites: an overview. Fungal metabol. 1, 29.

Gray, A., Flatt, P., 1998. Insulin-releasing and insulin-like activity of Agaricus campestris (mushroom). J. Endocrinol. 157 (2), 259–266.

Han, J.-J., Bao, L., He, L.-W., Zhang, X.-Q., Yang, X.-L., Li, S.-J., Yao, Y.-J., Liu, H.-W., 2013. Phaeolschidins A–E, five hispidin derivatives with antioxidant activity from the fruiting body of Phaeolus schweinitzii collected in the Tibetan Plateau. J. Nat. Prod. 76 (8), 1448–1453.

Hansson, D., 2013. Structure and biosynthesis of fungal secondary metabolites. vol. 2013 Department of Chemistry, Swedish University of Agricultural Sciences.

Hatvani, N., 2001. Antibacterial effect of the culture fluid of Lentinus edodes mycelium grown in submerged liquid culture. Int. J. Antimicrob. Agents 17 (1), 71–74.

He, L., Nan, M.-H., Oh, H.C., Kim, Y.H., Jang, J.H., Erikson, R.L., Ahn, J.S., Kim, B.Y., 2011. Asperlin induces G2/M arrest through ROS generation and ATM pathway in human cervical carcinoma cells. Biochem. Biophys. Res. Commun. 409 (3), 489–493.

Heischmann, S., Dzieciatkowska, M., Hansen, K., Leibfritz, D., Christians, U., 2017. The immunosuppressant mycophenolic acid alters nucleotide and lipid metabolism in an intestinal cell model. Sci. Rep. 7 (1), 1–11.

Hsiao, G., Shen, M.-Y., Lin, K.-H., Lan, M.-H., Wu, L.-Y., Chou, D.-S., Lin, C.-H., Su, C.-H., Sheu, J.-R., 2003. Antioxidative and hepatoprotective effects of Antrodia camphorata extract. J. Agric. Food Chem. 51 (11), 3302–3308.

Hussain, H., Kliche-Spory, C., Al-Harrasi, A., Al-Rawahi, A., Abbas, G., Green, I.R., Schulz, B., Krohn, K., Shah, A., 2014a. Antimicrobial constituents from three endophytic fungi. Asian Pac. J. Trop. Med. 7, S224–S227.

Hussain, H., Root, N., Jabeen, F., Al-Harrasi, A., Al-Rawahi, A., Ahmad, M., Hassan, Z., Abbas, G., Mabood, F., Shah, A., 2014b. Seimatoric acid and colletonoic acid: two new compounds from the endophytic fungi, Seimatosporium sp. and Colletotrichum sp. Chin. Chem. Lett. 25 (12), 1577–1579.

Hyde, K.D., Xu, J., Rapior, S., Jeewon, R., Lumyong, S., Niego, A.G.T., Abeywickrama, P.D., Aluthmuhandiram, J.V., Brahamanage, R.S., Brooks, S., 2019. The amazing potential of fungi: 50 ways we can exploit fungi industrially. Fungal Divers. 97 (1), 1–136.

Jeong, S.C., Jeong, Y.T., Yang, B.K., Islam, R., Koyyalamudi, S.R., Pang, G., Cho, K.Y., Song, C.H., 2010. White button mushroom (Agaricus bisporus) lowers blood glucose and cholesterol levels in diabetic and hypercholesterolemic rats. Nutr. Res. 30 (1), 49–56.

Johansson, M., Sterner, O., Labischinski, H., Anke, T., 2001. Coprinol, a new antibiotic cuparane from a Coprinus species. Z. Naturforsch. C 56 (1–2), 31–34.

Julianti, E., Lee, J.-H., Liao, L., Park, W., Park, S., Oh, D.-C., Oh, K.-B., Shin, J., 2013. New polyaromatic metabolites from a marine-derived fungus Penicillium sp. Org. Lett. 15 (6), 1286–1289.

Kanoh, K., Okada, A., Adachi, K., Imagawa, H., Nishizawa, M., Matsuda, S., Shizuri, Y., Utsumi, R., 2008. Ascochytatin, a novel bioactive spirodioxynaphthalene metabolite produced by the marine-derived fungus, Ascochyta sp. NGB4. J. Antibiot. 61 (3), 142–148.

Keller, N.P., Turner, G., Bennett, J.W., 2005. Fungal secondary metabolism—from biochemistry to genomics. Nat. Rev. Microbiol. 3 (12), 937–947.

Kim, A.Y., 2005. Application of biotechnology to the production of natural flavor and fragrance chemicals. ACS Publications.

Kim, K.-S., Cui, X., Lee, D.-S., Sohn, J.H., Yim, J.H., Kim, Y.-C., Oh, H., 2013. Anti-inflammatory effect of neoechinulin a from the marine fungus Eurotium sp. SF-5989 through the suppression of NF-κB and p38 MAPK pathways in lipopolysaccharide-stimulated RAW264. 7 macrophages. Molecules 18 (11), 13245–13259.

Kumar, K.S., Chu, F.-H., Hsieh, H.-W., Liao, J.-W., Li, W.-H., Lin, J.C.-C., Shaw, J.-F., Wang, S.-Y., 2011. Antroquinonol from ethanolic extract of mycelium of Antrodia cinnamomea protects hepatic cells from ethanol-induced oxidative stress through Nrf-2 activation. J. Ethnopharmacol. 136 (1), 168–177.

Lai, D., Wang, A., Cao, Y., Zhou, K., Mao, Z., Dong, X., Tian, J., Xu, D., Dai, J., Peng, Y., 2016. Bioactive dibenzo-α-pyrone derivatives from the endophytic fungus Rhizopycnis vagum Nitaf22. J. Nat. Prod. 79 (8), 2022–2031.

Langdon, S., Pearce, C.J., 2017. The microbial pharmacy: FDA approved medicines from Fungi. Mycosynthetix 18, 1–4.

Li, S.P., Zhao, K.J., Ji, Z.N., Song, Z.H., Dong, T.T., Lo, C.K., Cheung, J.K., Zhu, S.Q., Tsim, K.W., 2003. A polysaccharide isolated from Cordyceps sinensis, a traditional Chinese medicine, protects PC12 cells against hydrogen peroxide-induced injury. Life Sci. 73 (19), 2503–2513.

Qian, P.-Y., Li, X., Kwong, F.N., Yang, L.H., Dobretsov, S.V., 2012. Use of Marine Fungus Originated Compounds as Antifouling Agents. Google Patents.

Li, Y.-X., Himaya, S., Dewapriya, P., Kim, H.J., Kim, S.-K., 2014a. Anti-proliferative effects of isosclerone isolated from marine fungus Aspergillus fumigatus in MCF-7 human breast cancer cells. Process Biochem. 49 (12), 2292–2298.

Li, H.-J., Jiang, W.-H., Liang, W.-L., Huang, J.-X., Mo, Y.-F., Ding, Y.-Q., Lam, C.-K., Qian, X.-J., Zhu, X.-F., Lan, W.-J., 2014b. Induced marine fungus Chondrostereum sp. as a means of producing new sesquiterpenoids chondrosterins I and J by using glycerol as the carbon source. Mar. Drugs 12 (1), 167–175.

Li, W., Yang, X., Yang, Y., Duang, R., Chen, G., Li, X., Li, Q., Qin, S., Li, S., Zhao, L., 2016. Anti-phytopathogen, multi-target acetylcholinesterase inhibitory and antioxidant activities of metabolites from endophytic Chaetomium globosum. Nat. Prod. Res. 30 (22), 2616–2619.

Liao, W.-Y., Shen, C.-N., Lin, L.-H., Yang, Y.-L., Han, H.-Y., Chen, J.-W., Kuo, S.-C., Wu, S.-H., Liaw, C.-C., 2012. Asperjinone, a nor-neolignan, and terrein, a suppressor of ABCG2-expressing breast cancer cells, from thermophilic aspergillus terreus. J. Nat. Prod. 75 (4), 630–635.

Lindequist, U., Niedermeyer, T.H., Jülich, W.-D., 2005. The pharmacological potential of mushrooms. Evid. Based Complement. Alternat. Med. 2 (3), 285–299.

Liu, S., Liu, X., Guo, L., Che, Y., Liu, L., 2013. 2H-Pyran-2-one and 2H-Furan-2-one Derivatives from the Plant Endophytic Fungus Pestalotiopsis fici. Chem. Biodivers. 10 (11), 2007–2013.

Liu, Y., Yang, Q., Xia, G., Huang, H., Li, H., Ma, L., Lu, Y., He, L., Xia, X., She, Z., 2015. Polyketides with α-glucosidase inhibitory activity from a mangrove endophytic fungus, Penicillium sp. HN29-3B1. J. Nat. Prod. 78 (8), 1816–1822.

Lo, H.-C., Tsai, F.-A., Wasser, S.P., Yang, J.-G., Huang, B. M., 2006. Effects of ingested fruiting bodies, submerged culture biomass, and acidic polysaccharide glucuronoxylomannan of Tremella mesenterica Retz.: Fr. On glycemic responses in normal and diabetic rats. Life Sci. 78 (17), 1957–1966.

Lu, X., Chen, H., Dong, P., Fu, L., Zhang, X., 2010. Phytochemical characteristics and hypoglycaemic activity of fraction from mushroom Inonotus obliquus. J. Sci. Food Agric. 90 (2), 276–280.

Mikolasch, A., Hessel, S., Salazar, M.G., Neumann, H., Manda, K., Gördes, D., Schmidt, E., Thurow, K., Hammer, E., Lindequist, U., 2008. Synthesis of new N-analogous corollosporine derivatives with antibacterial activity by laccase-catalyzed amination. Chem. Pharm. Bull. 56 (6), 781–786.

Minnady, M., Paulraj-Dominic, S., Thomas, S., Subramanian, S., 2010. Therapeutic role of edible mushroom Pleurotus florida on Thioacetamide induced hepatotoxicity in rats. Int. J. Curr. Res. 5, 041–046.

Moghadamtousi, S.Z., Nikzad, S., Kadir, H.A., Abubakar, S., Zandi, K., 2015. Potential antiviral agents from marine fungi: an overview. Mar. Drugs 13 (7), 4520–4538.

Mohmand, A.Q.K., Kousar, M.W., Zafar, H., Bukhari, K.T., Khan, M.Z., 2011. Medical importance of Fungi with special emphasis on mushrooms. Isr. Med. J. 31.

Nada, S.A., Omara, E.A., Abdel-Salam, O.M., Zahran, H.G., 2010. Mushroom insoluble polysaccharides prevent carbon tetrachloride-induced hepatotoxicity in rat. Food Chem. Toxicol. 48 (11), 3184–3188.

Naik, C., Devi, P., Rodrigues, E., 2005. Chrysogenazine Obtained from Fungus Penicillium Chrysogenum Having Antibacterial Activity. Google Patents.

Ngai, P.H., Ng, T., 2004. A ribonuclease with antimicrobial, antimitogenic and antiproliferative activities from the edible mushroom Pleurotus sajor-caju. Peptides 25 (1), 11–17.

Nong, X.-H., Wang, Y.-F., Zhang, X.-Y., Zhou, M.-P., Xu, X.-Y., Qi, S.-H., 2014. Territrem and butyrolactone derivatives from a marine-derived fungus Aspergillus terreus. Mar. Drugs 12 (12), 6113–6124.

Patel, G., Patil, M.D., Soni, S., Chisti, Y., Banerjee, U.C., 2017. Production of mycophenolic acid by Penicillium brevicompactum using solid state fermentation. Appl. Biochem. Biotechnol. 182 (1), 97–109.

Peng, J., Lin, T., Wang, W., Xin, Z., Zhu, T., Gu, Q., Li, D., 2013. Antiviral alkaloids produced by the mangrove-derived fungus Cladosporium sp. PJX-41. J. Nat. Prod. 76 (6), 1133–1140.

Phan, C.-W., Sabaratnam, V., 2012. Potential uses of spent mushroom substrate and its associated lignocellulosic enzymes. Appl. Microbiol. Biotechnol. 96 (4), 863–873.

Popović, M., Vukmirović, S., Stilinović, N., Čapo, I., Jakovljević, V., 2010. Anti-oxidative activity of an aqueous suspension of commercial preparation of the mushroom Coprinus comatus. Molecules 15 (7), 4564–4571.

Prabpai, S., Wiyakrutta, S., Sriubolmas, N., Kongsaeree, P., 2015. Antimycobacterial dihydronaphthalenone from the endophytic fungus Nodulisporium sp. of Antidesma ghaesembilla. Phytochem. Lett. 13, 375–378.

Qin, C., Lin, X., Lu, X., Wan, J., Zhou, X., Liao, S., Tu, Z., Xu, S., Liu, Y., 2015. Sesquiterpenoids and xanthones derivatives produced by sponge-derived fungus Stachybotry sp. HH1 ZSDS1F1-2. J. Antibiot. 68 (2), 121–125.

Rana, T., Bera, A.K., Das, S., Bhattacharya, D., Pan, D., Bandyopadhyay, S., Mondal, D.K., Samanta, S., Bandyopadhyay, S., Das, S.K., 2012. Pleurotus florida lectin normalizes duration dependent hepatic oxidative stress responses caused by arsenic in rat. Exp. Toxicol. Pathol. 64 (7–8), 665–671.

Rossiana, N., Miranti, M., Rahmawati, R., 2016. Antibacterial Activities of Endophytic fungi from Mangrove Plants *Rhizophora apiculata* L. and *Bruguiera gymnorrhiza* (L.) Lamk. On *Salmonella typhi*, AIP Conference Proceedings. AIP Publishing LLC. p. 020040.

Rowley, D.C., Kelly, S., Kauffman, C.A., Jensen, P.R., Fenical, W., 2003. Halovirs A-E, new antiviral agents from a marine-derived fungus of the genus Scytalidium. Bioorg. Med. Chem. 11 (19), 4263–4274.

Rowley, D.C., Kelly, S., Jensen, P., Fenical, W., 2004. Synthesis and structure–activity relationships of the halovirs, antiviral natural products from a marine-derived fungus. Bioorg. Med. Chem. 12 (18), 4929–4936.

Santiago, C., Sun, L., Munro, M.H.G., Santhanam, J., 2014. Polyketide and benzopyran compounds of an endophytic fungus isolated from C innamomum mollissimum: biological activity and structure. Asian Pac. J. Trop. Biomed. 4 (8), 627–632.

Schlingmann, G., Milne, L., Williams, D.R., Carter, G.T., 1998. Cell wall active antifungal compounds produced by the marine fungus Hypoxylon oceanicum LL-15G256 II. Isolation and structure determination. J. Antibiot. 51 (3), 303–316.

Seenivasan, A., Venkatesan, S., Tapobrata, P., 2018. Cellular localization and production of lovastatin from Monascus purpureus. Indian J. Pharm. Sci. 80 (1), 85–98.

Seow, S.L.-S., Naidu, M., David, P., Wong, K.-H., Sabaratnam, V., 2013. Potentiation of neuritogenic activity of medicinal mushrooms in rat pheochromocytoma cells. BMC Complement. Altern. Med. 13 (1), 1–10.

Sharma, S., Singh, L., Singh, S., 2013. Comparative study between penicillin and ampicillin. Sch. J. Appl. Med. Sci. 1 (4), 291–294.

Shieh, Y.-H., Liu, C.-F., Huang, Y.-K., Yang, J.-Y., Wu, I.-L., Lin, C.-H., Lin, S.-C., 2001. Evaluation of the hepatic and renal-protective effects of Ganoderma lucidum in mice. The American journal of. Chinas Med. 29 (03n04), 501–507.

Silber, J., Kramer, A., Labes, A., Tasdemir, D., 2016. From discovery to production: biotechnology of marine fungi for the production of new antibiotics. Mar. Drugs 14 (7), 137.

Singh, S.B., Zink, D., Polishook, J., Valentino, D., Shafiee, A., Silverman, K., Felock, P., Teran, A., Vilella, D., Hazuda, D.J., 1999. Structure and absolute stereochemistry of HIV-1 integrase inhibitor integric acid. A novel eremophilane sesquiterpenoid produced by a Xylaria sp. Tetrahedron Lett. 40 (50), 8775–8779.

Smith, H.A., 2014. Production of Antimicrobials and Antioxidants from Filamentous fungi. National University of Ireland Maynooth.

Soares, A.A., Sá-Nakanishi, D., Babeto, A., Bracht, A., Da Costa, S.M.G., Koehnlein, E.A., De Souza, C.G.M., Peralta, R.M., 2013. Hepatoprotective effects of mushrooms. Molecules 18 (7), 7609–7630.

Solomon, L., Tomii, V.P., Dick, A.-A., 2019. Importance of fungi in the petroleum, agro-allied, agriculture and pharmaceutical industries. NY Sci. J 12, 8–15.

Stodůlková, E., Man, P., Kuzma, M., Černý, J., Císařová, I., Kubátová, A., Chudíčková, M., Kolařík, M., Flieger, M., 2015. A highly diverse spectrum of naphthoquinone derivatives produced by the endophytic fungus Biatriospora sp. CCF 4378. Folia Microbiol. 60 (3), 259–267.

Strobel, G.A., Torczynski, R., Bollon, A., 1997. Acremonium sp.—a leucinostatin A producing endophyte of European yew (Taxus baccata). Plant Sci. 128 (1), 97–108.

Suay, I., Arenal, F., Asensio, F.J., Basilio, A., Cabello, M.A., Díez, M.T., García, J.B., Del Val, A.G., Gorrochategui, J., Hernández, P., 2000. Screening of basidiomycetes for antimicrobial activities. Antonie Van Leeuwenhoek 78 (2), 129–140.

Subhan, M., Faryal, R., Macreadie, I., 2016. Exploitation of aspergillus terreus for the production of natural statins. J. Fungi. 2 (2), 13.

Subramani, R., Kumar, R., Prasad, P., Aalbersberg, W., 2013. Cytotoxic and antibacterial substances against multidrug resistant pathogens from marine sponge symbiont: Citrinin, a secondary metabolite of Penicillium sp. Asian Pac. J. Trop. Biomed. 3 (4), 291–296.

Sun, P., Huo, J., Kurtán, T., Mándi, A., Antus, S., Tang, H., Draeger, S., Schulz, B., Hussain, H., Krohn, K., 2013. Structural and stereochemical studies of hydroxyanthraquinone derivatives from the endophytic fungus Coniothyrium sp. Chirality 25 (2), 141–148.

Thongbai, B., Rapior, S., Hyde, K.D., Wittstein, K., Stadler, M., 2015. Hericium erinaceus, an amazing medicinal mushroom. Mycol. Prog. 14 (10), 1–23.

Tobert, J., 1987. New developments in lipid-lowering therapy: the role of inhibitors of hydroxymethylglutaryl-coenzyme A reductase. Circulation 76 (3), 534–538.

Tsunawaki, S., Yoshida, L.S., Nishida, S., Kobayashi, T., Shimoyama, T., 2004. Fungal metabolite gliotoxin inhibits assembly of the human respiratory burst NADPH oxidase. Infect. Immun. 72 (6), 3373–3382.

Von Bargen, K.W., Niehaus, E.-M., Bergander, K., Brun, R., Tudzynski, B., Humpf, H.-U., 2013. Structure elucidation and antimalarial activity of apicidin F: an apicidin-like compound produced by fusarium fujikuroi. J. Nat. Prod. 76 (11), 2136–2140.

Wang, H., Ng, T., 2004. Eryngin, a novel antifungal peptide from fruiting bodies of the edible mushroom Pleurotus eryngii. Peptides 25 (1), 1–5.

Wang, J.C., Hu, S.H., Wang, J.T., Chen, K.S., Chia, Y.C., 2005. Hypoglycemic effect of extract of Hericium erinaceus. J. Sci. Food Agric. 85 (4), 641–646.

Wang, J., Zhang, Y., Ding, D., Yu, S., Wang, L., 2013. A study on the secondary metabolites of endophytic fungus Chaetomium cupreum ZJWCF079 in Macleaya cordata. Health Res 33, 94–96.

Wang, K., Bao, L., Xiong, W., Ma, K., Han, J., Wang, W., Yin, W., Liu, H., 2015. Lanostane triterpenes from the Tibetan medicinal mushroom Ganoderma leucocontextum and their inhibitory effects on HMG-CoA reductase and α-glucosidase. J. Nat. Prod. 78 (8), 1977–1989.

Webster, J., Weber, R., 2007. Introduction to Fungi. Cambridge University Press.

Wu, Y.-W., Chen, K.-D., Lin, W.-C., 2004. Effect of Ganoderma tsugae on chronically carbon tetrachloride-intoxicated rats. Am. J. Chin. Med. 32 (06), 841–850.

Wu, G., Sun, X., Yu, G., Wang, W., Zhu, T., Gu, Q., Li, D., 2014. Cladosins A–E, hybrid polyketides from a deep-sea-derived fungus, Cladosporium sphaerospermum. J. Nat. Prod. 77 (2), 270–275.

Wu, B., Wiese, J., Labes, A., Kramer, A., Schmaljohann, R., Imhoff, J.F., 2015. Lindgomycin, an unusual antibiotic polyketide from a marine fungus of the Lindgomycetaceae. Mar. Drugs 13 (8), 4617–4632.

Xiong, H., Qi, S., Xu, Y., Miao, L., Qian, P.-Y., 2009. Antibiotic and antifouling compound production by the marine-derived fungus Cladosporium sp. F14. J. Hydro Environ. Res. 2 (4), 264–270.

Zhao, Y.-Y., Chao, X., Zhang, Y., Lin, R.-C., Sun, W.-J., 2010. Cytotoxic steroids from Polyporus umbellatus. Planta Med. 76 (15), 1755–1758.

Zhao, Q., Chen, G.-D., Feng, X.-L., Yu, Y., He, R.-R., Li, X.-X., Huang, Y., Zhou, W.-X., Guo, L.-D., Zheng, Y.-Z., 2015. Nodulisporiviridins A–H, Bioactive Viridins from Nodulisporium sp. J. Nat. Prod. 78 (6), 1221–1230.

Zhou, C., Jia, W., Yang, Y., Bai, Y., 2002. Experimental studies on prevention of several kinds of fungi polysaccharides against alcohol-induced hepatic injury. Edible Fungi 24, 36–37.

Zilla, M.K., Qadri, M., Pathania, A.S., Strobel, G.A., Nalli, Y., Kumar, S., Guru, S.K., Bhushan, S., Singh, S.K., Vishwakarma, R.A., 2013. Bioactive metabolites from an endophytic Cryptosporiopsis sp. inhabiting Clidemia hirta. Phytochemistry 95, 291–297.

Zobel, S., Boecker, S., Kulke, D., Heimbach, D., Meyer, V., Süssmuth, R.D., 2016. Reprogramming the biosynthesis of cyclodepsipeptide synthetases to obtain new enniatins and beauvericins. ChemBioChem 17 (4), 283–287.

CHAPTER

Fungal enzymes in textile industry: An emerging avenue to entrepreneurship

6

Deepak K. Rahi[a], Sonu Rahi[b], and Maninder Jeet Kaur[a]
[a]Department of Microbiology, Panjab University, Chandigarh, India, [b]Department of Botany, Govt. P.G. College, A.P.S. University, Rewa, India

6.1 Introduction

The Indian economy since 1990 has risen pointedly. Today, the economy of India is the third biggest on the planet as estimated by purchasing power parity (PPP) with a GDP (Gross Domestic Product) of US $1.0 trillion after estimating in USD swapping scale terms. India is the second fastest developing significant economy on the planet after China, at a rate of 5% in FY20 and for FY21, the rate is anticipated to rise to 6%–6.5%. The current slowdown in growth rate is the result of slow consumption rate, decrease in fixed investment, weak trade activities, and an extreme manufacturing atmosphere (www.indiainfoline.com). The main thrusts behind this remarkable achievement are information technology (IT) and biotechnology (BT) (Mani, 2006). IT and BT establish roughly 5% of the nation's GDP, showing 23% and 40% development, separately, during the year 1999–2000. BT covers an expansive scope of specialized areas including biopharmaceutical, bioindustrial, bioservices, bioagriculture, and bioinformatics. Out of these, the bioindustrial area, which includes the harnessing of microorganisms for the creation of significant worth added bioactive ingredients (industrial enzymes, organic acids, mass chemicals, single-cell proteins, etc.), has assumed a prevalent part in the overall advancement of biotechnology after biopharmaceuticals. Both biopharmaceutical and bioindustrial enterprises produced nearly Rs. 35,700 and 4250 million incomes during the year 2004–2005, respectively (http://www.biospectrumindia.com). Around the world, Europe has extensive resources in the field of modern biotechnology. In Europe, 70% of the world enzyme industries exist with an high level of information in the field of food innovation and fine chemistry (Buchholz and Poulson, 2000). Due to always increasing work, energy, and crude material expenses, more mass assembling organizations are, nonetheless, moving to the Far East. In this time of expanded industrial development of events, the impact of modern biotechnology in the improvement of any country's economy is essential.

A new Frost and Sullivan study portrays that different agreement research models like joint research and collective research have given the BT area an effective driving force around the world. India has the twelfth best BT industry on the planet as estimated by the number of organizations. The fundamental main thrust for the commercialization of BT in India is the Department of Biotechnology

Entrepreneurship with Microorganisms. https://doi.org/10.1016/B978-0-443-19049-0.00013-X

(DBT) of the Ministry of Science and Technology. DBT played a fundamental part in the commercialization of R&D activities for biotechnology, human asset advancement, and bioinformatics programs (Mani, 2006). Enzymes have been the focal point of attention for scientists/industrialists the world over because of their wide scope of employing in the processes used in various industries. Despite the fact that enzymes have been isolated, purified, and studied from microbial, animal, and plant sources, microorganisms are the most well-known sources of enzymes because of their wide biochemical variety and possibility of large level production by using cheap sugar sources (Kirk et al., 2002; Chand and Mishra, 2003).

The enzyme market is expected to reach $14.7 billion constantly by 2025, on the back of a Compound Annual Growth Rate (CAGR) of 6.7%, in terms of value. A sharp improvement in the enzyme market is projected because of significant breakthroughs in enzyme work, introduction of genetically engineered compounds, and progressions in green science and enzyme engineering (Global Enzymes Market, 2023; www.deloitte.com; www.ficci.com).

6.2 Major industrial enzymes and their applications

Enzymes are catalysts, natural compounds produced by living microorganisms that speed up the transformation rate of the substrate to the finished product by bringing down the activation energy barrier (Bailey and Ollis, 1997). Although all enzymes are initially synthesized in the cell, some are produced through the cell wall and function in the cell's environment. Thus, we have two kinds of enzymes based on the location of activity: intracellular or endozymes (working in the cell), and extracellular or exoenzymes (working outside the cell) (Buchholz and Poulson, 2000). Diastase was the first commercialized enzyme for the production of dextrins in bread kitchens, brew, and wine from natural products in France in 1830 (Payen and Persoz, 1833). In 1874, in Denmark, Christian Hansen began the first company (Christian Hansen's Laboratory) for the sale of standardized enzymes, rennet for cheddar making (Buchholz and Poulson, 2000). In USA, J. Takamine produced bacterial amylases in the 1890s at Miles Laboratories. In 1913, Otto Rohm's patent plans about using enzymes extracted from animal pancreas to be used in washing which helps for cleaning of clothes, and the item was sold under the brand name Brunus and sold in European business sectors for around 50 years (Maurer, 2004). During the 1960s, Novo Industry A/S introduced alkaline protease in the market produced by *Bacillus licheniformis* under the brand Alkalase (Gupta et al., 2002a). Furthermore, in 1985, Novo produced another multienzyme preparation containing cellulase, soluble protease, and different chemicals, which caused transformative changes in the market. In general, an ideal enzyme for detergent cleaning ought to be successful at low levels (0.4%–0.8%) in the cleanser solutions. Today, proteases are the leaders of the modern enzyme market worldwide (Maurer, 2004; Gupta et al., 2002a).

Enzymes discover broad applications in textile processing, basically for desizing and biopolishing of different types of fabrics including cotton-based ones. They have acquired considerable attention because of their nonharmful and eco-friendly nature. These enzymes offer high proficiency and specific activity, improving them better substitutes than most strong acids, bases, and oxidizing agents utilized for textile processing. Changing trends fads and rising expendable income of grown-ups across the world are trends driving the extension of the textile enzyme market. The increasing focus toward textile processing in textile mills and laundries is powering the steps of the market. Advances made in jeans desizing and synthetic fabric finishing will shape the development direction over the next years. Industry players are centered around reducing the general cost of production of textile enzymes to support their take-up. Besides, developing impulse by governments toward low-pollution textile processing in emerging economies will open new roads in the worldwide textile enzymes market. The global textile enzymes market stood at US$201.5 million in 2017 and is projected to clock a CAGR of 4.0% during 2018–2026 (https://www.transparencymarketresearch.com).

Presently enzymes have accomplished the status of a household commodity. Recent improvements in the field of enzymology, for example, genetically engineering development for recombinant compounds, look for biodegradable transporters for immobilization, extremozymes and cross-linked enzymes produced stable enzymes, are more reasonable for commercial natural synthesis and their resolution (Zlinski and Aldmann, 1997). Current enzyme technology, related to multidisciplinary scientific information and process innovation, is important for the advancement of new, clean, and cost-effective making ideas for food specialties (e.g., bread, cheddar, cocktails, vinegar, organic product juice, and so on), fine-chemical (e.g., amino acids, nutrients), bulk substance commodities (bioethanol, biodiesel, xylitol, 2, 3-propanediol, and so on), and drug and nutraceutical products (Bruggink et al., 1998; Herrera, 2004). Enzymes also have an application for a wide range of scientific purposes particularly in food diagnostics and as sensors in electrochemical responses. Fig. 6.1 presents the major industrial enzymes and their wide range of applications.

India imports around 70% of the total enzymes consumed in the country, most of which goes to the textile, detergent, starch, and pharmaceutical industries. Based on available information, the estimated consumption of industrial enzymes in India in various sectors has been summarized in Fig. 6.2.

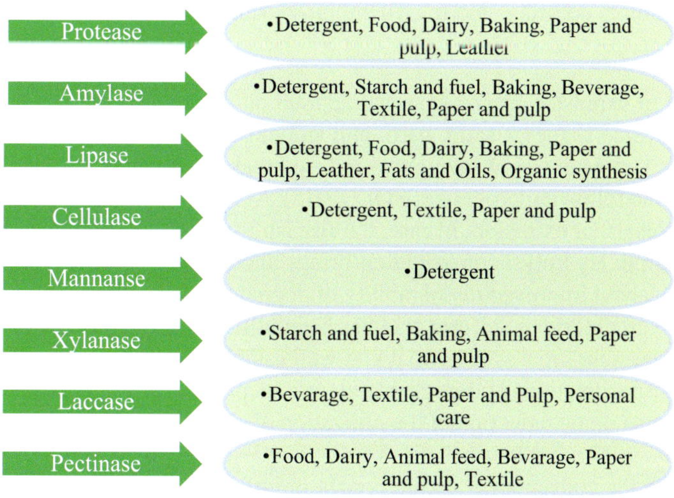

FIG. 6.1

Major industrial enzymes and their applications.

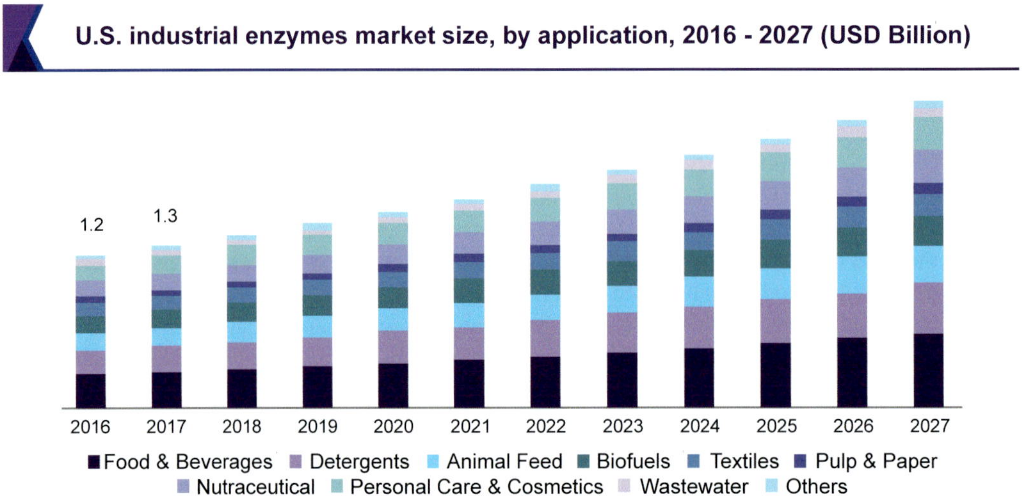

FIG. 6.2

World sales of industrial enzymes.

Source: www.grandviewresearch.com.

6.3 Applications of enzymes in textile industry

6.3.1 Applications in textile processing

The textile industry uses enzymes at various phases of textile processing like desizing, scouring, bleaching, stone washing, coloring, and in effluent bioremediation.

6.3.1.1 Desizing

It is the interaction of expulsion of size material applied on wrap strings of a fabric to work with the process of weaving. Size shapes a solid, hard, and smooth covering on warp yarns to make them to withstand the cyclic strains during weaving and reduce breakage. In general, the expulsion of impurities can be done by the following processes: SOLUBILIZATION, EMULSIFICATION, COMPOUND breakdown by hydrolysis, or oxidation. In case, the size isn't solvent, it is eliminated by one or the other hydrolysis or oxidation. Hydrolysis or oxidation brings about lowering down of molecular weight of starch, which brings about improved dissolvability of degraded starch. Conventionally, acid hydrolysis of starch utilizing hydrochloric or sulfuric acid has been utilized for the process of desizing. However, the amylase-based desizing process is normal and well established in the various industries. Amylase enzymes of various specifications and workable at various pH ranges are available. The utilization of amylase enzymes helps in the reduction of the utilization of water, chemicals, and energy during desizing (Saravanan et al., 2012). Another benefit of using amylase enzyme is that it doesn't harm the fiber when compared with the conventional textile processing of using acids. Regardless of the enzyme or acids, the starch is changed into sugar and its derivatives during desizing. In spite of the fact that sugar is nontoxic, it is nonrecyclable and the effluents cause numerous ecological issues. Henceforth, the enzyme-based desizing process should be adjusted to resolve the above issue. Toward this direction, attempts have been made to desize using amyl glucosidase enzyme rather than a conventional amylase to hydrolyze starch into single glucose units. After that, the glucose oxidase enzyme is utilized to create hydrogen peroxide from the glucose produced during desizing (Eren et al., 2009).

6.3.1.2 Scouring

In this process, impurities on the outside of the fabrics like wax, gelatins, fats, and different impurities are eliminated. This gives the fabric better wetting capacity, which consequently assists the fabric to get bleached and colored appropriately. Chemicals like caustic soda are utilized for this process. Consequently, it eliminates impurities as well as reduces the strength and weight of the fabric. Also, it discharges harsh chemicals in the wastewater. By using enzymes in the new scouring process known as bioscouring, all impurities are taken out and it keeps the cellulose structure of the fabric safe. Therefore, the fabric has higher wetting and infiltration capacities. Conventionally, cotton textile is treated with sodium hydroxide for scouring purposes at high temperature and pressure. An alkaline pectinase enzyme-based process was created as an option to alkaline scouring which should likewise be possible at moderate temperature. The enzyme delicately acts on cotton fabric and causes less harm, less pilling, and uniform color take-up contrasted with conventional alkaline scouring. Apart from pectinase enzymes, different enzymes, for example, cellulase, protease, cutinase, and others have been proposed by the scientists for the scouring of cotton. The enzyme is substrate (pectin) specific in nature and is not able to eliminate noncellulosic impurities totally because of poor access to the substrates at the point when utilized on modern scale. To make the enzyme-based scouring process industrially adoptable, it is proposed to utilize a mix of enzymes, for example, pectinase+cellulase, pectinase+protease, pectinase+cutinase (Agrawal et al., 2008), pectinase+lipase (Kalantzi et al., 2010), or pectinase+xylanase (Battan et al., 2012). Attempts have additionally been made to combine enzyme-based scouring and activator assisted in the single bath to eliminate the noncellulosic impurities from the cotton textiles (Hebeish et al., 2009).

6.3.1.3 Bleaching

This process requires a lot of water, energy, and chemicals. Bleaching is the process of expulsion of regular coloring matter (Yellowish dim shading) present in the material by utilizing, H_2O_2, NaCl, CaCl, SO_2, etc. Bleaching is utilized and followed by washing with fresh water to eliminate the traces of chlorine. Actually, the radical reactions that work with the action of bleaching agents with the fibers, cause decrease in degree of polymerization and thus severe damage to fibers (Basto et al., 2007). Furthermore, large amounts of water are required to remove hydrogen peroxide from fabrics which would cause problems in dyeing. At present, hydrogen peroxide is the favored bleaching agent for the fading of fabrics. At the point when hydrogen peroxide bleaching of cotton is done preceding dyeing of cotton with reactive dyes, the presence of spent peroxide will make uneven dyeing and promote hydrolysis of the reactive dyes. In order to counter the above process, ordinarily fabrics are treated with chemicals to totally clean up the peroxide prior to coloring with reactive dyes. The hydrogen peroxide bleached fabrics need to go through a few washes with water and treatment with sodium thiosulphate for eliminating the peroxide. This step requires a higher amount of water, energy, and time. Biotechnology has come into play to lessen the washes with the introduction of catalase enzymes for eliminating the spent hydrogen peroxide. Catalase enzyme can viably eliminate the leftover hydrogen peroxide instead of chemicals. The catalase-based process requires less water as far as less steps and it is additionally eco-accommodating. The catalase-treated fabrics show uniform color take-up and great shading values (Amorim et al., 2002).

6.3.1.4 Finishing

Finishing completes with the various treatments, mechanical and chemical, performed on fibers, yarns, or fabrics, to improve look and surface (Bajaj, 2002). Potential pollutants released due to finishing processes are formaldehyde, resins, antistatics, softenings, crosslinking agents, and biocides (Le Marechal et al., 2012). To counter this problem new method comes into play which is biopolishing. It is a finishing process that improves fabric quality by principally bringing down the fluffiness and pilling property of cellulosic fiber. The target of the process is the removal of microfibrils of cotton through the activity of cellulase enzyme. Biopolishing treatment brings the fabric a cleaner surface, a cooler vibe, brilliance, and softer feel (Mojsov, 2011). Denim is heavy-grade cotton. In this, the color is basically adsorbed on the outside of the fiber. That is why fading can be accomplished without considering loss of strength. In traditional processes, sodium hypochlorite or potassium permanganate was utilized called as pumice stones (Pederson and Schneider, 1998). However, these techniques have certain inherent drawbacks like pumice stones cause black-staining, wear, and tear of machines and the stones are required in extremely enormous amount. These detriments lead to give rise in the process of utilization of enzymes. Cellulase enzyme is utilized in denim washing. Cellulase works by releasing the indigo color on the denim in the process known as "Bio-Stonewashing." The utilization of less pumice stones brings about less harm to piece of clothing, machine, and less pumice dust (Campos et al., 2001).

6.3.2 Applications in bioremediation of effluents from textile industry

The textile effluents are generally high in color, introducing a conspicuous issue when released into open waters. The recalcitrant nature of present reactive dyes has prompted the obligation to meet severe environmental legislations and guidelines. The requirement for a practical strategy to eliminate the colorants from wastewater discharged by the textile industry has been recognized, and a few procedures had been researched (Pearce et al., 2006). The removal of dyed wastewaters can be achieved through physical, chemical, and biological advancements and the mix of these. The significant strategies for the treatment of textile effluents include membrane filtration, coagulation/flocculation, precipitation, buoyancy, adsorption and ultrasonic mineralization, and chemical processes like electrolysis, chemical reduction, and progressed chemical oxidation (Gogate and Pandit, 2004; Tan et al., 2010). A wide range of chemical, physical, and natural methods studied on the removal of dyes from industrial effluents have their own benefits, negative impacts, and limitations. Some are very costly, and others have some operational issues, for example, high sludge development particularly with chemical methods (Çınar et al., 2008). Besides, conventional wastewater treatment systems that depend on adsorption process, aerobic biodegradation presents challenges with respect to expulsion of synthetic dyes utilized in textile industries (Lourenco et al., 2001; Çınar et al., 2008). Some water treatment methods like coagulation and flocculation with lime, alum, and Fe(II) salts produce enormous amounts of sludge that impose removal issues (Daneshvar et al., 2003; Selcuk, 2005). The expulsion of dyes by adsorption, utilizing activated carbon and some different absorbents, and the photochemical oxidation of dyes utilizing UV radiations and using sunlight with oxidation agents are also not feasible (Shen et al., 2001; Muruganandham and Swaminathan, 2004). Thus, such methodologies can be relied upon to raise some secondary pollution issues, because of unnecessary chemicals used. Many researchers have centered their research goal toward the decoloration of textile effluents that contain dyestuffs and recalcitrant materials utilizing some biological procedures,

Table 6.1 Major enzymes used in the textile industries.

Processes	Enzymes
Uses in textile processing steps	
Desizing	Amylase, Pullulanase
Bioscouring	Pectinases, Cutinases, Proteases, Xylanases, Lipases,
Biobleaching	Lipase, Protease, Cellulase, Pectinase, Glucose oxidase
Stonewashing	Cellulase, Laccase
Uses in bioremediation of effluents from textile industries	
Effluent treatment	Laccase

for example, anaerobic, oxygen consuming, and combined anaerobic/aerobic cycles (Georgiou et al., 2003; Pearce et al., 2003; Alam et al., 2009). Biological treatment systems offer better alternatives that can effectively eliminate dyes from huge volumes of modern effluents at extremely minimal rate (Robinson et al., 2001). Many factors like nature of dyestuffs, the composition of wastewater, operational and chemical expenses, ecological fate, and handling expenses of generated waste byproducts determine the specialized and economic feasibility of procedures. Some biotechnological techniques had been prescribed as suitable ways to deal with the pollution issue in an eco-proficient way (Pearce et al., 2003; Asgher et al., 2008). The most well-known route is the utilization of bacteria or fungi, frequently in blend with physicochemical processes (Robinson et al., 2001). White rot fungi (WRF) are among the important organisms, which have been explored widely for the textile dyed wastewater degradation and decolorization purposes. These organisms have an extraordinary potential to change an expansive scope of recalcitrant organic compounds. Until now, several WRF strains have been used to viably degrade different synthetic dyes and dye-based effluents (Bilal et al., 2017). The ligninolytic enzymes of WRF seem to be promising enzymes involved in dye metabolism (Iqbal and Asgher, 2013), and dyes having structural similarity with lignin are more susceptible to degradation by these chemicals. The potential biotechnological and ecological importance has prompted an extensive expansion in the interest of ligninolytic enzymes. (Asgher et al., 2008). Thus, different enzymes are utilized at various stages to make the processes more effective and nature-friendly (Table 6.1). The textile industry addresses the second most polluting industry making drastic impacts on the environment. The reasoning for the use of enzymes in fabric processing is to discover organic solutions for harsh treatments, which could not create harmful waste materials in the environment (Araujo et al., 2008).

6.4 Fungal enzymes in textile industries

Enzymes are biocatalysts, which can accelerate the chemical processes. Enzymes initiate like other inorganic catalysts like acids, bases, metals, and metal oxides. The molecule that an enzyme follows upon is known as its substrate, which is converted into a product. Enzymes were originally attempted to classify according to their function. The International Commission on Enzymes (EC) was set up in 1956 by the International Union of Biochemistry (IUB) with the International Union of Pure and

Applied Chemistry (IUPAC) suggested many enzymes that had been found. The EC characterization framework is classified into six categories:

EC1 Oxidoreductases: catalyze oxidation/reduction reactions.
EC 2 Transferases: move a functional group.
EC 3 Hydrolases: catalyze the hydrolysis of different bonds.
EC 4 Lyases: separate different bonds by means other than hydrolysis and oxidation.
EC 5 Isomerases: catalyze isomerization changes inside a single molecule.
EC 6 Ligases: Join two molecules with covalent bonds.

In textile industry, chiefly hydrolases and oxidoreductases are used for different enzymatic applications. The group of hydrolases enzyme includes amylases, cellulases, pectinases, proteases, and lipases. In addition, hydrolase enzymes mainly act as hydrolysis.

6.4.1 Amylases (EC 3.2.1.1)

Being the main enzyme found in 1833, amylase has been all around examined and utilized in various areas. Amylases catalyze the breakdown of starch to glucose units and discover applications in almost all ventures (de Souza and de Oliveira Magalhães, 2010). Amylases are of four types, viz., α amylase, β amylase, gluco-amylase, and pullulanase based on the method of action on starch and their finished products (Table 6.2).

Alpha-amylases are endoenzymes with a molecular weight between 10 and 210 kDa, breaking interior α-1,4-glycosidic bonds of linear starch molecules to yield glucose, maltose, straight oligosaccharides α -limit dextrin (El-Enshasy et al., 2013). β-Amylases are exoenzymes that divide α-1, 4-glycosidic linkages of starch yielding maltose and β limit dextrin. Glucoamylases are exoenzymes breaking the α-1,4-glycosidic linkages of amylose and amylopectin to glucose subunits, enabling complete starch hydrolysis (Struyf et al., 2017). Pullulanase proteins can be produced both as *exo-* and *endo*-enzyme equipped for dividing both the α-1,4-glycosidic linkages and β-1,6-glycosidic linkages of partially processed amylopectin and are regularly applied in combination with glucoamylases in industries (Hii et al., 2012).

Amylases are produced from a variety of plant and animal sources. However, bacterial and fungal amylases have gained much access in the industrial area (Rana et al., 2013). The ease of extracting of enzymes from organisms, their simple accessibility, and plasticity are a few elements adding to its preference. The amylases in textile industries are accessible in various forms to be utilized in gentle, high, and psychrophilic temperatures and hence microorganisms from various specialties are investigated to get such enzymes. Amylases are generally produced by *Bacillus* sp., *Clostridium thermos sulfurogenes*, and *Geobacillus,* giving variations of alkali stable and heat tolerant enzymes (Takasaki, 1976; Sundarram and Murthy, 2014; Swamy and Seenayya, 1996; Emanuilova and Kambourova, 1992). The fungal sources comprise species of *Aspergillus*, *Penicillium, Aureobasidium pullulans* (Seo et al., 2004), etc.

6.4.2 Cellulases(EC 3.2.1.4)

Cellulases catalyze the hydrolytic cleavage of cellulose into small sugar segments like glucose units by cutting the β-1, 4-glucosidic bonds of the parent particle. The debasement of cellulose is normally done by the activity of three kinds of cellulases viz. endoglucanase (E.C. 3.2.1.4), exoglucanase

Table 6.2 Different types of amylases, their structure, and mode of action.

Type of amylase	Structure	Method of action
α amylase	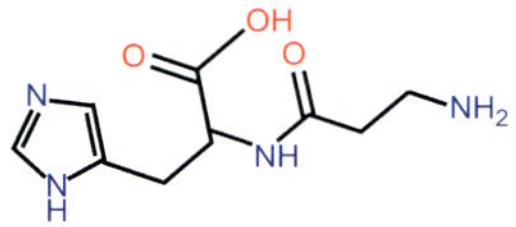	It can break down long-chain carbohydrates such as starch, amylose Into maltotriose and maltose or amylopectin into maltose, glucose, and limit dextrin.
β amylase	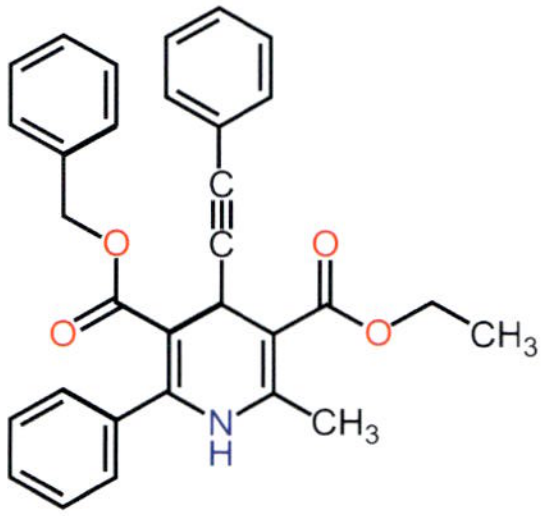	It catalyzes the enzymatic breakdown of the second α-1,4 glycosidic bond of nonreducing sugars, thereby cleaving off maltose at a time
Glucoamylase	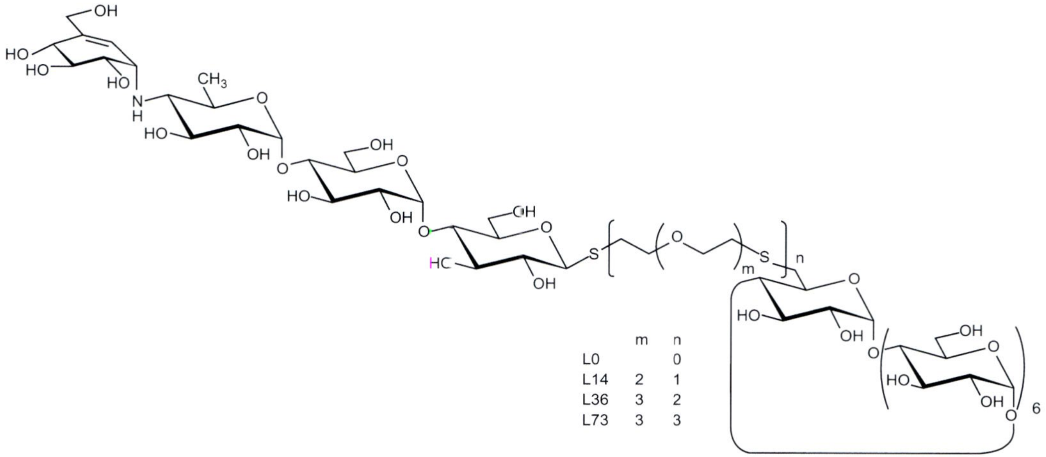	Acts preferably on α-1,4-than β-1,6 glycosidic linkages present in starch and other related polysaccharides

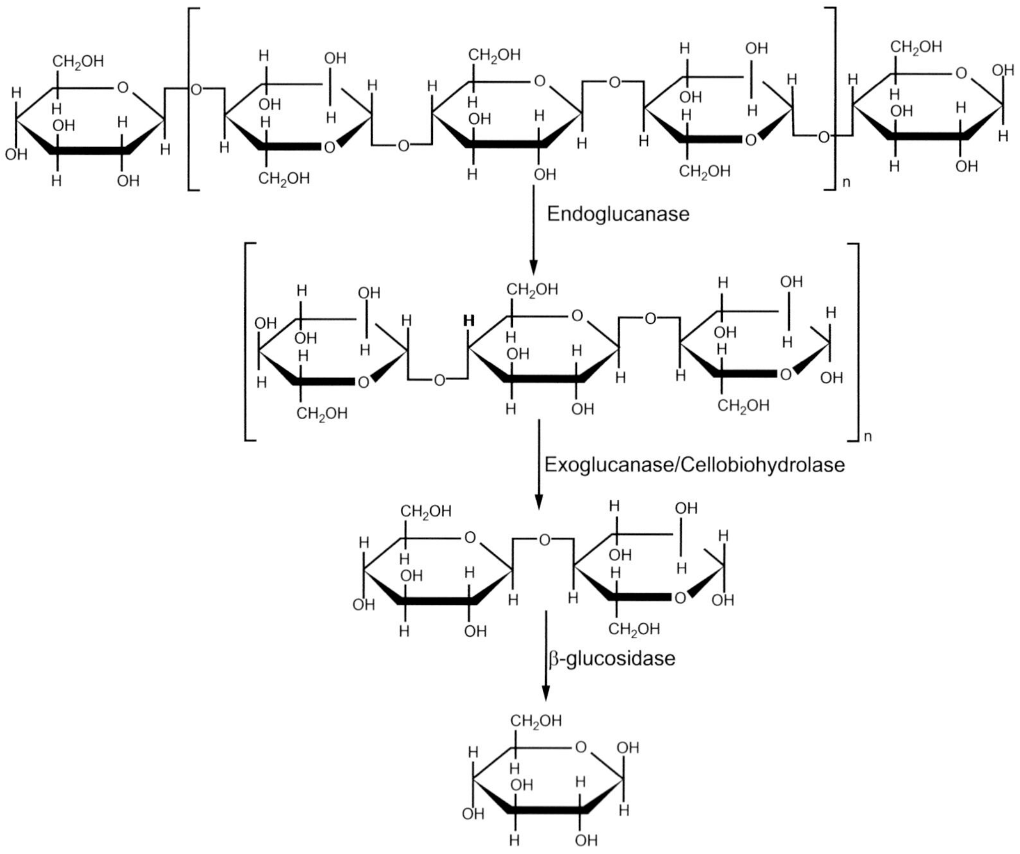

FIG. 6.3

Structure of different types of cellulases and their mode of action.

(E.C. 3.2.1.176) and (E.C. 3.2.1.91) and β-glucosidase (E.C. 3.2.1.21) (Juturu and Wu, 2014). Endoglucanases generally give irregular cuts on the internal bonds of glycan chain to create more limited cello-oligosaccharides which can be additionally cleaved by exoglucanase to yield cellobiose. The β-glucosidases hence follow up on the cellobiose to deliver glucose to be used by organisms (Jørgensen et al., 2007) (Fig. 6.3).

Cellulose degradation is achieved predominantly by aerobic bacteria, fungi, and yeast (Payne et al., 2015). Nonetheless, anaerobic bacteria are not that viable in cellulase enzyme production (de Lourdes Moreno et al., 2013). Bacterial cellulases are discovered to be attached to their cell wall while fungal cellulases are discharged outside, in this way expanding the ease of extraction (Juturu and Wu, 2014). Production of cellulase is recorded also in plants and in various invertebrate taxa that incorporate bugs, shellfish, annelids, mollusks, mussels, and nematodes. Fungal cellulases are discovered to be better work in harsh conditions and are industrially produced by mostly *Trichoderma* species compared with *Aspergillus* and *Humicola* species (Muhammad et al., 2016).

6.4.3 Proteases (EC 3.4.2.1)

According to worldwide market analyses, proteases addressed the largest share of industrial enzymes and would surpass US$3 billion by 2024 (https://www.gminsights.com/industry-analysis/proteases-market). Proteases are proteolytic compounds produced by number of microscopic bacteria and fungi and they play a basic part in numerous industries. Bacterial proteases are more vital for their thermo-resistance than the fungal partners. They are divided into *exo*-and *endo*-peptidases dependent on their activity at or away from the ends, respectively. Based on the functional group in their dynamic site, they are classified as serine proteases, aspartic proteases, cysteine proteases, and metalloproteases (Rao et al., 1998). Generally, serine proteases are alkaline and neutral proteases of low molecular weight (18.5–35 kDa) synthesized by organisms and animals and are generally utilized as detergent additives (Ellaiah et al., 2002). A detail on the mechanism of its action and reactant job is reviewed (Hedstrom, 2002). Bacteria, for example, *Bacillus* sp., *Clostridium,* and *Pseudomonas* add to the significant degree of production of this enzyme (Gupta et al., 2002b). Cysteine proteases are members of the papain family and are described by Cys-His-Asn set of three at the dynamic site. The cysteine residue attacks the carbon of the reactive peptide bond and fills in as a proton contributor, which brings about the cleavage of the thioester bond to create carboxylic moiety from the parent protein (Coulombe et al., 1996). Aspartic proteases address the acidic proteases and are delivered by viruses, bacteria, fungi, plants, and animals (Hasiao et al., 2014). The high stability and action of aspartic proteases represent their utilization in the food processing industry, for example, protein haze of wine, production of cheese, flavor upgrade of food, etc. (Theron and Divol, 2014). The metallo-proteases require metallic particles for their activity and hence are classified so. Aside from their modern importance, these serine proteases play an important role in physiological and patho-physiological cycles as signaling molecules in humans and are imminent medication targets (Turk, 2006; Patel, 2017). A few instances of proteases physiological functions are in blood coagulation, fibrinolysis, supplement initiation, chemical production, transformation, and proper digestion (Neurath and Walsh, 1976). Cysteine proteases additionally play a significant part in immunomodulation and antigen show, hemoglobin hydrolysis, digestion, parasitic intrusion, and processing surface proteins (Verma et al., 2016). Proteases are generally produced by microorganisms like *Bacillus, Aspergillus, Penicillium, Pseudomonas, Streptomyces,* etc.

6.4.4 Laccases (EC 1.10.3.2)

Laccases are multi-copper enzymes interesting in their capacity to oxidize pesticides, steroids, phenolic as well as nonphenolics toxins, etc. (Gianfreda et al., 1999). They are essentially oxidoreductases and have a wide scope of substrates including polyphenols, methoxy-substituted monophenols, aliphatic and sweet-smelling amines (Karaki et al., 2016; de Freitas et al., 2017). They are fit for ring cleavage, breakage of polymers mediated by the oxidation of substrate electrons alongside a generation of water. Laccase mediated remediation of industrial effluents from textiles, dairy, and pharmaceutical industries is generally practiced (Becker et al., 2017; Asif et al., 2018). Laccases are synthesized by plants, insects, and bacteria, however, fungi, for example, *Deuteromycetes, Ascomycetes*, and *Basidiomycetes* are noticeable laccase producers (Abd El Monssef et al., 2016). Studies also demonstrate that bacterial sources of laccases are discovered to be promising in dye degradation (Chauhan et al., 2017). Regularly laccases are produced in numerous isoforms than single isoforms and give 95% degradation of colors, as verified in various reports (Luo et al., 2018) (Fig. 6.4).

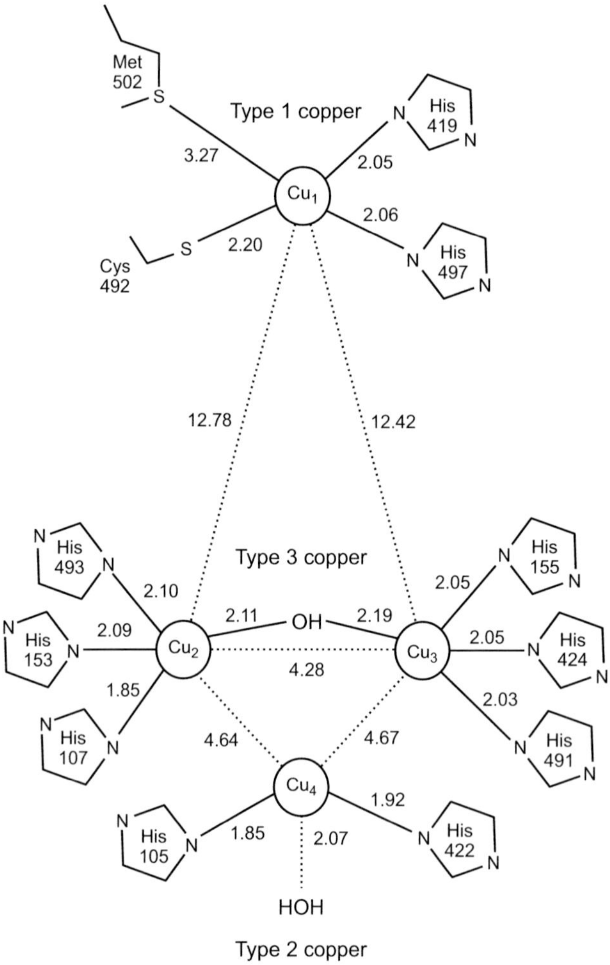

FIG. 6.4

Structure of laccase.

6.4.5 Catalases (EC 1.11.1.21)

Catalases, which are more suitably known as hydroperoxides, play an important part in catalyzing the degradation of H_2O_2 to H_2O and O_2. They are produced by various microorganisms, including bacteria and fungi (Zámocký et al., 2012). Their ideal temperature is from 20°C to 50°C and works best and effectively at neutral pH. Catalases produced by animals are basically less expensive; therefore, the production of catalase from microorganisms would possibly be economically feasible and worthwhile when cheap technology and recombinant strains are utilized (Yumoto et al., 2000) (Fig. 6.5).

After desizing and scouring, bleaching of H_2O_2 can occur, however, it takes place before dying. Catalases are also used to decompose an excess of H_2O_2 and obviously, this hinders the requirement for a

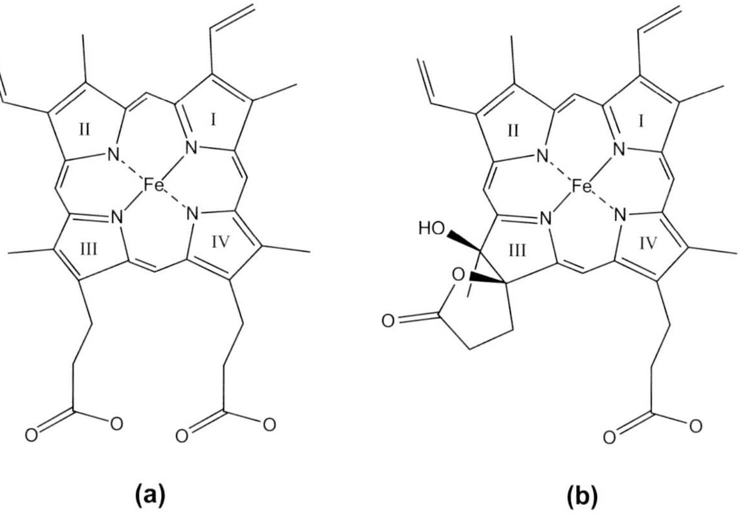

FIG. 6.5

Structure of catalase.

reducing agent and reduces the requirement for rinsing water, bringing about lower polluted wastewater and lower utilization of water (UlAleem, 2013). By using immobilized catalysts, the cost of enzymes for degrading hydrogen peroxide in bleaching effluents could easily be reduced. This also permits the recovery of enzymes and the reuse of treated bleaching effluents for dyeing (Araujo et al., 2008).

6.4.6 Pectinases (EC 3.2.1.15)

Gelatin and pectic substances are the polysaccharides that are found in the center lamella and cell wall of the plants. Pectinases are enzymes that degrade these pectic substances. This enzyme is predominantly produced by plants and microbes (bacteria and fungi) including saprophytes to degrade the cell wall of plants (Pedrolli et al., 2009; Yadav et al., 2018). The pectinase is additionally divided into three fundamental groups, that is, pectinesterases (PEs), polygalacturonases (PGs), and polygalacturonatelyases (PGLs). Bacteria and fungi produce the pectinesterases that catalyze the hydrolysis of pectin methyl esters to shape pectic acid on plants like banana, lemon, orange, and tomato. It typically acts up on the methyl ester site of galacturonate present next to nonesterified galacturonate (Muthu, 2014). The molecular weight of pectinesterase obtained from microorganisms and plants falls in the range of 30–50kDa. Generally, the ideal temperature is discovered to be 40–60°C and pH is found to be in the range of 4–7 for PEs except for *Erwinia*, which shows the ideal activity in the alkaline range (Kohli et al., 2015). PG is the enzyme that hydrolyzes the α-1,4glycosidic linkages through *exo*-and *endo*-dividing systems in pectin (Palanivelu, 2006). Endo PGs are most normally produced by bacteria, fungi, and yeast, and have a molecular weight of 3080kDa. This enzyme is found to be active in the acidic range, i.e., 2.5–6.0, and in the temperature range of 30–50°C, in spite of the fact that exo PGs have been reported to be synthesized by plants such as apple, carrot, and peach, and *Aspergillus niger* and *Erwinia* sp. This enzyme is found to have a molecular weight in the range of 30–50kDa (Araujo et al., 2008). PGLs are enzymes that cut the pectin chain through the β-end, which leads to the development of a double bond between C4 and C5 present at the reducing end and frees CO_2. The endo PGL cleaves arbitrarily, though exo PGL cuts toward the end of polygalacturonate to yield unsaturated galacturonic acid (Hoondal et al., 2002). The molecular weight of the PGLs falls in the range of 30–50 kDa, with optimal activity in the range of pH 8–10. However, PGLs got from *Erwinia* and *Bacillus licheniformis* are discovered to be active at pH 6 and 10, respectively. The enzyme PGL is regularly discovered to be active inside a range of 30–40°C, though PGL obtained from thermophiles is accounted to have an ideal temperature between 50°C and 75°C. The pectatelyases have been broadly investigated for bioscouring (Tierny et al., 1994).

The crude cotton contains different noncellulosic impurities, for example, waxes, hemicelluloses, and mineral salts, that accumulate in the cell wall and fingernail skin of the filaments. These impurities add to the hydrophobic nature of raw cotton, which interferes with chemical processing, for example, the coloring and finishing of cotton (Sawada et al., 1998). In this way, before the dyeing of the cotton yarn, pre-treatment is needed to eliminate the materials that hinder the binding of the color. This interaction is known as scouring, which upgrades the wettability of the fabric, and for which sodium hydroxide is utilized (Araujo et al., 2008). Although these synthetic compounds have been known to lessen the strength of cellulose, the weight of the fabric is also decreased. Moreover, this treatment increases COD, BOD, and the salt content of wastewater generated after the treatment. Bioscouring overcomes the issues related with chemicals. During processing by the enzyme, the cellulose stays intact, which prevents the loss of strength and weight of the material (Duran and Duran, 2000). Bioscouring has different advantages over conventional scouring. This process is generally performed at neutral pH; thus, the total water consumption is decreased generally to a large extent as compared with the conventional strategy, and the delicacy of the cotton fiber is also maintained (Muthu, 2014). Various enzymes, for example, cellulases, cutinases, lipases, and pectinases have been studied exclusively and in combination for the bioscouring of cotton, but pectinases are reported to be exceptionally effective (Karapinar and Sariisik, 2004). Table 6.3 summarizes the various enzymes of fungal origin used in textile industries of varied processes.

Table 6.3 Use of major fungal enzymes in textile industries.

Enzymes	Fungal species	Applications	References
Amylases	*Fusarium oxysporum*, *Pyrococcus furiosus*	Sizing agents for denim	Colomera and Kuilderd (2015), Laderman et al. (1993)
Proteases	*Aspergillus candidus*, *Fusarium eumartii*	Diffusion of the dye into the fibers	Periolatto et al. (2011); Srilakshmi et al. (2015)
Lipases	*Thermobifida fusca*, *Fusarium solani*	To enhance its dyeability with a basic dye	El-Shemy et al. (2016); Zumstein et al. (2017)
Laccases	*Trametes hirsuta*, *Pleurotus* sp.	Used as an enzymatic rinse process after reactive dyeing, the oxidative splitting of hydrolyzed reactive dyes on the fiber, and biobleaching of important industrial dyes	Osuji et al. (2014); Pereira et al. (2005)
Cellulases	*Trichoderma reesei*	Biofinishing, biopolishing	Ali et al. (2012); McMullan et al. (2001)
Pectinases	*Aspergillus niger*, *Penicillium notatum*	In textile processing and bioscouring of cotton fibers, desizing	Mojsov (2012)

6.5 Manufacturers of textile enzymes & entrepreneurship potentials

The global textile enzymes market size is projected to reach USD 182.7 million by 2026, from USD 178.3 million in 2020, at a CAGR of 2.3% during 2021–2026. In India, the major industrial enzyme producers are Novozymes India, Biocon India Ltd., Advanced Enzyme Technologies Ltd., Anil Bioplus Ltd., Concord biotech Ltd., and Precise Laboratories. However, Biocon is leading the Indian enzymes sector with Rs 85 crore (2005–06) followed by Novozymes. The other leading companies in this sector include Advanced Enzymes Technologies, Rossari Biotech, Maps India, Zytex India, Lumis Biotech, and Textan Chemicals. The total revenue generated by the main five Indian enzyme producers is given in Table 6.4.

The Biocon India Group is the largest biotechnology conglomerate in India today with highly automated and modern integrated manufacturing and research facilities set on a 30-acre site on the outskirts of the high tech city of Bangalore in South India. It has acquired numerous honors worldwide for its solid R&D-based products toward product specificity with high yield and better reproducibility. The organization has an enhanced product range from microbial-derived enzymes to generic mass drug(s)/medicine(s). The organization is, nonetheless, presently focusing on biopharmaceuticals and is quick to develop some oral medication plans for insulin and rheumatoid joint pain. According to a very recent report, the organization is divesting its industrial enzyme business for Rs 4670 m to Novozymes (The Novozymes Report, 2019).

Advanced Biochemicals Ltd., Pune, has also played a critical part in the improvement of the enzyme business in India. The organization is committed to the production of various microbial enzymes utilized in drugs, nutraceuticals, textiles, and refining. The solid R&D arrangement of the organization won to launch new novel enzymes with more prominent proficiency, less specificity, and quick transformation rates with high reproducibility. The organization is engaged in the production of alpha-acetolactate decarboxylase, alpha amylase, cellulase, and alkaline protease. The organization is setting up an assembling office at Indore with speculation of Rs 150 million. The organization has saved another Rs 350 million for its extension.

Mumbai-based Rossari Biotech Ltd. is offering the enzyme solutions for textile auxiliaries. The organization has also recently employed more up-to-date areas like food, feed, and paper starch-based enzymes.

Table 6.4 Top five Indian enzymes manufacturers.

Company	Revenue (millions)
Novozymes	1000
Biocon	950
Advanced Enzyme Technologies	693
Rossari Biotech	660
Zytex	150
Others	497
Total Bio-Industries	3950

Source: Biospectrum, Industrial Enzymes, 2005.

Table 6.5 Potential enzyme distributors/suppliers in India.

Distributors/suppliers	Business type	Products offered
Texnzymes India, Mumbai, Maharashtra	Manufacturers	Textile auxiliaries and enzymes
Americos Industries Inc., Ahmedabad, Gujarat	Manufacturers And exporters	Textile auxiliaries and enzymes
United Biochemicals Private Limited, Mumbai, Maharashtra	Manufacturers And exporters	Enzymes and biochemicals for textiles
Zytex India Private Limited Mumbai, Maharashtra	Manufacturers And exporters	Industrial enzymes
Protos Trading Private Limited, New Delhi	Importers/Buyers	Textile enzymes

One of the pioneers of industrial biotechnology in India is Zytex, Gujrat. With over five decades of business experience in enzyme and biotech products, Zytex set up its first manufacturing facility in 1996 at Silvassa, India. In 2004, a new export-oriented plant was set up in Surat and in 2009 a new state-of-the-art fermentation facility in Baroda. Zytex has a world-class R&D center & application support lab approved by Department of Science and Industrial Research (DSIR), Govt. of India.

Today, the Indian biotech sector has achieved a critical amount in manufacturing and research services. Indian biotechnology is presently ready to use its logical abilities and specialized experience to have a worldwide effect on a solid advancement-led platform. The mechanical capacity of a firm to produce a few items is just a part of the requirements for the commercial accomplishment of the business. Apart from innovative capacities, the firm should have the option to situate its products in the market. The market, hence, serves as the connection between buyers' necessities and the pattern of industrial response (Negi and Tewari, 1999). Table 6.5 summarizes the potential enzyme wholesalers/providers in India.

6.6 Conclusion

There has been a developing interest in demand of enzymes with novel properties. Enzymes from marine microorganisms have exceptional properties and have proven their modern industrial applications (Chandrasekaran, 1997). Other uncommon enzymes are thermo-tolerant enzymes, which have shown enhanced stability considerably under highly tough conditions. These enzymes give exceptional benefits like more stability during the substrate conversion, which brings about more transformation cycles, accordingly saving energy and time (Haki and Rakshit, 2003). Genetic engineering is of central significance for the advancement of novel genetically altered or "designer" enzymes. India is yet to satisfy its enzyme demand by bringing in to the tune of 70% from US, Canada, and China. However, if great attention is given on research and entrepreneurship around in-house production of enzymes, it will lead to more employment generation and income for the country, which can facilitate the better approaches for saving the foreign trade reserves. Investigating the increasing demand for enzymes in the coming way, a country wide production development program should be initiated through governmental interventions and private sector funds. This will lead to a noteworthy enthusiasm for structure-function relationships in novel enzyme exploration, screening of high titer-yielding strains, improvement in production yields embracing statistical methodologies, and improvement in downstream recovery of enzymes. The Government of India through the Department of Biotechnology (DBT), the Department of Science and Technology (DST), and the Council of Scientific and Industrial Research (CSIR) have been concentrating on these areas and funding projects related to industrial enzymes. This will assist the enzyme industry growth and compete with worldwide players.

References

Abd El Monssef, R.A., Hassan, E.A., Ramadan, E.M., 2016. Production of laccase enzyme for their potential application to decolorize fungal pigments on aging paper and parchment. Ann. Agric. Sci. 61 (1), 145–154.

Agrawal, P.B., Nierstrasz, V.A., Warmoeskerken, M.M.C.G., 2008. Role of mechanical action in low temperature cotton scouring with F. solani pisi cutinase and pectate lyase. Enzym. Microb. Technol. 42 (6), 473–482.

Alam, M.Z., Mansor, M.F., Jalal, K.C.A., 2009. Optimization of decolorization of methylene blue by lignin peroxidase enzyme produced from sewage sludge with Phanerocheate chrysosporium. J. Hazard. Mater. 162 (2), 708–715.

Ali, H., Hashem, M., Shaker, N., Ramadan, M., El-Sadek, B., Hady, M.A., 2012. Cellulase enzyme in bio-finishing of cotton-based fabrics: effects of process parameters. Res. J. Text. Appar. 16, 57–65.

Amorim, A.M., Gasques, M.D., Andreaus, J., Scharf, M., 2002. The application of catalase for the elimination of hydrogen peroxide residues after bleaching of cotton fabrics. An. Acad. Bras. Cienc. 74 (3), 433–436.

Araujo, R., Casal, M., Cavaco-Paulo, A., 2008. Application of enzymes for textile fibres processing. Biocatal. Biotransform. 26, 332–349.

Asgher, M., Bhatti, H.N., Ashraf, M., Legge, R.L., 2008. Recent developments in biodegradation of industrial pollutants by white rot fungi and their enzyme system. Biodegradation 19 (6), 771–783.

Asif, M.B., Hai, F.I., Kang, J., van de Merwe, J.P., Leusch, F.D.L., Price, W.E., et al., 2018. Biocatalytic degradation of pharmaceuticals, personal care products, industrial chemicals, steroid hormones and pesticides in a membrane distillation-enzymatic bioreactor. Bioresour. Technol. 247, 528–536.

Bailey, J.E., Ollis, D.F., 1997. Biochemical Engineering Fundamentals. TP 248. 3. B34.

Bajaj, P., 2002. Finishing of textile materials. J. Appl. Polym. Sci. 83 (3), 631–659.

Basto, C., Tzanov, T., Cavaco-Paulo, A., 2007. Combined ultrasound-laccase assisted bleaching of cotton. Ultrason. Sonochem. 14 (3), 350–354.

Battan, B., Dhiman, S.S., Ahlawat, S., Mahajan, R., Sharma, J., 2012. Application of thermostable xylanase of *Bacillus pumilus* in textile processing. Indian J. Microbiol. 52 (2), 222–229.

Becker, D., Rodriguez-Mozaz, S., Insa, S., Schoevaart, R., Barceló, D., de Cazes, M., et al., 2017. Removal of endocrine disrupting chemicals in wastewater by enzymatic treatment with fungal laccases. Org. Process. Res. Dev. 21 (4), 480–491.

Bilal, M., Asgher, M., Parra-Saldivar, R., Hu, H., Wang, W., Zhang, X., Iqbal, H.M.N., 2017. Immobilized ligninolytic enzymes: an innovative and environmental responsive technology to tackle dye-based industrial pollutants—a review. Sci. Total Environ. 576, 646–659.

Bruggink, A., Roos, E.C., De Vroom, E., 1998. Penicillin acylase in the industrial production of beta-lactam antibiotics. Org. Process. Res. Dev. 2, 128–133.

Buchholz, K., Poulson, P.B., 2000. Overview of history of applied biocatalysis. In: Straathof, A.J.J., Adlercreutz, P. (Eds.), Applied Biocatalysis. Harwood Academic Publishers, Amsterdam, pp. 1–15.

Campos, R., Kandelbauer, A., Robra, K.H., Cavaco-Paulo, A., Gubitz, G.M., 2001. Indigo degradation with purified laccases from Trametes hirsuta and Sclerotium rolfsii. J. Biotechnol. 89, 131–139.

Chand, S., Mishra, P., 2003. Research and application of microbial enzymes—India's contribution. Adv. Biochem. Eng. Biotechnol. 85, 95–124. Springer Berlin/Heidelberg. Biotechnology in India.
Chandrasekaran, M., 1997. Industrial enzymes from marine microorganisms: the Indian scenario. J. Mar. Biotechnol. 5, 86–89.
Chauhan, P.S., Goradia, B., Saxena, A., 2017. Bacterial laccase: recent update on production, properties and industrial applications. 3Biotech 7 (5), 323.
Çınar, Ö., Yaşar, S., Kertmen, M., Demiröz, K., Yigit, N.Ö., Kitis, M., 2008. Effect of cycle time on biodegradation of azo dye in sequencing batch reactor. Process. Saf. Environ. Prot. 86 (6), 455–460.
Colomera, A., Kuilderd, H., 2015. Biotechnological washing of denim jeans. Denim J., 357–403.
Coulombe, R., Grochulski, P., Sivaraman, J., Ménard, R., Mort, J.S., Cygler, M., 1996. Structure of human procathepsin L reveals the molecular basis of inhibition by the prosegment. EMBO J. 15 (20), 5492–5503.
Daneshvar, N., Ashassi-Sorkhabi, H., Tizpar, A., 2003. Decolorization of orange II by electrocoagulation method. Sep. Purif. Technol. 31 (2), 153–162.
Duran, N., Duran, M., 2000. Enzyme applications in the textile industry. Rev. Prog. Color. Relat. Top. 30, 41–44.
El-Enshasy, H.A., Abdel Fattah, Y.R., Othman, N.Z., 2013. Bioprocessing Technologies in Biorefinery for Sustainable Production of Fuels, Chemicals, and Polymers. vol. 111 John Wiley & Sons, Inc.
Ellaiah, P., Srinivasulu, B., Adinarayana, K., 2002. A review on microbial alkaline proteases. J. Sci. Ind. Res. 61, 690.
El-Shemy, N.S., El-Hawary, N.S., El-Sayed, H., 2016. Basic and reactive-dyeable polyester fabrics using lipase enzymes. J. Chem. Eng. Process Technol. 7, 271.
Emanuilova, E.I., Kambourova, M.S., 1992. Effect of carbon source and dissolved oxygen level on cell growth and pullulanase production by Bacillus stearothermophilus G-82. World J. Microbiol. Biotechnol. 8 (1), 21–23.
Eren, H.A., Anis, P., Davulcu, A., 2009. Enzymatic one-bath desizing—bleaching—dyeing process for cotton fabrics. Text. Res. J. 79 (12), 1091–1098.
de Freitas, E.N., Bubna, G.A., Brugnari, T., Kato, C.G., Nolli, M., Rauen, T.G., et al., 2017. Removal of bisphenol A by laccases from Pleurotus ostreatus and Pleurotus pulmonarius and evaluation of ecotoxicity of degradation products. Chem. Eng. J. 330, 1361–1369.
Georgiou, D., Aivazidis, A., Hatiras, J., Gimouhopoulos, K., 2003. Treatment of cotton textile wastewater using lime and ferrous sulfate. Water Res. 37 (9), 2248–2250.
Gianfreda, L., Xu, F., Bollag, J.-M., 1999. Laccases: a useful group of oxidoreductive enzymes. Bioremediat. J. 3 (1), 1–26.
Global Enzymes Market, 2023. www.deloitte.com. www.ficci.com/.
Gogate, P.R., Pandit, A.B., 2004. A review of imperative technologies for wastewater treatment I: oxidation technologies at ambient conditions. Adv. Environ. Res. 8 (3), 501–551.
Gupta, R., Beg, Q.K., Lorentz, P., 2002a. Bacterial alkaline proteases: molecular approaches and industrial application. Appl. Microbiol. Biotechnol. 59, 15–32.
Gupta, R., Beg, Q.K., Lorenz, P., 2002b. Bacterial alkaline proteases: molecular approaches and industrial applications. Appl. Microbiol. Biotechnol. 59 (1), 15.
Haki, G.D., Rakshit, S.K., 2003. Developments in industrially important thermostable enzymes: a review. Bioresour. Technol. 89, 17–34.
Hasiao, N. W, Chen, Y., Kuan, Y.C., Lee, Y.C., Lee, S.K., Chan, H.H., Kao, C.H., 2014. Purification and characterization of an aspartic protease from the Rhizopus oryzae protease extract, Peptidase R. Electron. J. Biotechnol. 17 (2), 89–94.
Hebeish, A., Ramadan, M., Hashem, M., Shaker, N., Abdel-Hady, M., 2009. New development for combined bioscouring and bleaching of cotton-based fabrics. Res. J. Text. Appar. 17 (1), 94–103.
Hedstrom, L., 2002. Serine protease mechanism and specificity. Chem. Rev. 102 (12), 4501–4524.
Herrera, S., 2004. Industrial biotechnology- a chance at redemption. Nat. Biotechnol. 22, 671–675.
Hii, S.L., Tan, J.S., Ling, T.C., Ariff, A.B., 2012. Pullulanase, role in starch hydrolysis and potential industrial applications. Enzyme Res. 2012, 1–14.

Hoondal, G., Tiwari, R., Tewari, R., Dahiya, N.B.Q.K., Beg, Q., 2002. Microbial alkaline pectinases and their industrial applications, a review. Appl. Microbiol. Biotechnol. 59, 409–418.

Iqbal, H.M.N., Asgher, M., 2013. Characterization and decolorization applicability of xerogel matrix immobilized manganese peroxidase produced from Trametes versicolor IBL-04. Protein Pept. Lett. 20 (5), 591–600.

Jørgensen, H., Kristensen, J.B., Felby, C., 2007. Enzymatic conversion of lignocellulose into fermentable sugars: challenges and opportunities. Biofuels Bioprod. Biorefin. 1 (2), 119–134.

Juturu, V., Wu, J.C., 2014. Microbial cellulases: engineering, production and applications. Renew. Sust. Energ. Rev. 33, 188–203.

Kalantzi, S., Mamma, D., Kalogeris, E., Kekos, D., 2010. Improved properties of cotton fabrics treated with lipase and its combination with pectinase. Fibres Text. East. Eur. 18 (5), 82.

Karaki, N., Aljawish, A., Humeau, C., Muniglia, L., Jasniewski, J., 2016. Enzymatic modification of polysaccharides: mechanisms, properties, and potential applications: a review. Enzym. Microb. Technol. 90, 1–18.

Karapinar, E., Sariisik, M.O., 2004. Scouring of cotton with cellulases pectinases and proteases. Fibres Text. East. Eur. 12, 79–82.

Kirk, O., Borchert, T.V., Fuglsang, C.C., 2002. Industrial enzyme applications. Curr. Opin. *Biotechnol.* 13, 345–351.

Kohli, P., Kalia, M., Gupta, R., 2015. Pectin methylesterases: a review. J. Bioprocess. Biotech. 5, 1.

Laderman, K.A., Asada, K., Uemori, T., Mukai, H., Taguchi, Y., Kato, I., Anfinsen, C.B., 1993. Alphaamylase from the hyperthermophilic archaebacterium Pyrococcus furiosus. Cloning and sequencing of the gene and expression in Escherichia coli. J. Biol. Chem. 268, 24402–24407.

Le Marechal, A.M., Križanec, B., Vajnhandl, S., Valh, J.V., 2012. Textile finishing industry as an important source of organic pollutants. In: Organic Pollutants Ten Years after the Stockholm Convention Environmental and Analytical Update.

de Lourdes Moreno, M., Pérez, D., García, M.T., Mellado, E., 2013. Halophilic bacteria as a source of novel hydrolytic enzymes. Life 3 (1), 38–51.

Lourenco, N.D., Novais, J.M., Pinheiro, H.M., 2001. Effect of some operational parameters on textile dye biodegradation in a sequential batch reactor. J. Biotechnol. 89 (2), 163–174.

Luo, Q., Chen, Y., Xia, J., Wang, K.Q., Cai, Y.J., Liao, X.R., et al., 2018. Functional expression enhancement of Bacillus pumilus CotA-laccase mutant WLF through site-directed mutagenesis. Enzym. Microb. Technol. 109, 11–19.

Mani, S., 2006. Growth of new technology-based industries in India, the contrasting experiences of biotechnology and information technology industries. Int. J. Technol. Glob. 2, 200–216.

Maurer, K.H., 2004. Detergent proteases. Curr. Opin. Biotechnol. 15, 330–334.

McMullan, G., Meehan, C., Connely, M., 2001. Microbial decolourisation and degradation of textile dyes. Appl. Microbiol. Biotechnol. 56, 81–87.

Mojsov, K., 2011. Application of Enzymes in the Textile Industry: A Review.

Mojsov, K., 2012. Biotechnological applications of pectinases in textile processing and bioscouring of cotton fibers. In: II International Conference Industrial Engineering and Environmental Protection Conference Proceedings, Zrenjanin, Serbia, pp. 314–322.

Muhammad, I., Zahid, A., Muhammad, I., Muhammad, J.A., Hassan, A., 2016. Cellulase production from species of fungi and bacteria from agricultural wastes and its utilization in industry: a review. Adv. Enzyme Res. 4, 44–55.

Muruganandham, M., Swaminathan, M., 2004. Photochemical oxidation of reactive azo dye with UV-H_2O_2 process. Dyes Pigments 62 (3), 269–275.

Muthu, S.S. (Ed.), 2014. Roadmap to Sustainable Textiles and Clothing: Eco-Friendly Raw Materials Technologies and Processing Methods. Springer, Hong Kong.

Negi, Y.S., Tewari, S.C., 1999. Economics of fermented products. In: Joshi, V.K., Pandey, A. (Eds.), Biotechnology: Food Fermentation. vol. 1. Educational Publishers and Distributors, Kerala, India.

Neurath, H., Walsh, K.A., 1976. Role of proteolytic enzymes in biological regulation (a review). Proc. Natl. Acad. Sci. U. S. A. 73 (11), 3825–3832.

Osuji, A.C., Eze, S.O.O., Osayi, E.E., Chilaka, F.C., 2014. Biobleaching of industrial important dyes with peroxidase partially purified from garlic. Sci. World J. 2014, 183163.

Palanivelu, P., 2006. Polygalacturonases: active site analyses and mechanism of action. Indian J. Biotechnol. 5, 148–162.

Patel, S., 2017. A critical review on serine protease: key immune manipulator and pathology mediator. Allergol Immunopathol (Madr) 45 (6), 579–591.

Payen, A., Persoz, J.F., 1833. Mémoiresur la diastase, les principauxproduits de sesréactions, et leurs Applications aux arts industriels. Ann. Chim. Phys., 73–92. 2me Série 53.

Payne, C.M., Knott, B.C., Mayes, H.B., Hansson, H., Himmel, M.E., Sandgren, M., et al., 2015. Fungal cellulases. Chem. Rev. 115 (3), 1308–1448.

Pearce, C.I., Christie, R., Boothman, C., von Canstein, H., Guthrie, J.T., Lloyd, J.R., 2006. Reactive azo dye reduction by Shewanella strain J18 143. Biotechnol. Bioeng. 95 (4), 692–703.

Pearce, C.I., Lloyd, J.R., Guthrie, J.T., 2003. The removal of colour from textile wastewater using whole bacterial cells: a review. Dyes Pigments 58 (3), 179–196.

Pederson, A.H., Schneider, P.N.N., 1998. US Pat. 5795855 A. US Patent.

Pedrolli, D.B., Monteiro, A.C., Gomes, E., Carmona, E.C., 2009. Pectin and pectinases: production characterization and industrial application of microbial pectinolytic enzymes. Open Biotechnol. J. 3, 9–18.

Pereira, L., Bastos, C., Tzanov, T., Cavaco-Paulo, A., Gübitz, G.M., 2005. Environmentally friendly bleaching of cotton using laccases. Environ. Chem. Lett. 3, 66–69.

Periolatto, M., Ferrero, F., Giansetti, M., Mossotti, R., Innocenti, R., 2011. Influence of protease on dyeing of wool with acid dyes. Open Chem. 9, 157–164.

Rana, N., Walia, A., Gaur, A., 2013. α-Amylases from microbial sources and its potential applications in various industries. Natl. Acad. Sci. Lett. 36 (1), 9–17.

Rao, M.B., Tanksale, A.M., Ghatge, M.S., Deshpande, V.V., 1998. Molecular and biotechnological aspects of microbial proteases. Microbiol. Mol. Biol. Rev. 62 (3), 597.

Robinson, T., McMullan, G., Marchant, R., Nigam, P., 2001. Remediation of dyes in textile effluent: a critical review on current treatment technologies with a proposed alternative. Bioresour. Technol. 77 (3), 247–255.

Saravanan, D., Sivasaravanan, S., Sudharshan Prabhu, M., Vasanthi, N.S., Senthil Raja, K., Das, A., Ramachandran, T., 2012. One-step process for desizing and bleaching of cotton fabrics using the combination of amylase and glucose oxidase enzymes. J. Appl. Polym. Sci. 123, 2445–2450.

Sawada, K., Tokino, S., Ueda, M., Wang, X.Y., 1998. Bioscouring of cotton with pectinase enzyme. J. Soc. Dye. Colour. 114, 333–336.

Selcuk, H., 2005. Decolorization and detoxification of textile wastewater by ozonation and coagulation processes. Dyes Pigments 64 (3), 217–222.

Seo, H.P., Son, C.W., Chung, C.H., Jung, D.I., Kim, S.K., Gross, R.A., et al., 2004. Production of high molecular weight pullulan by Aureobasidium pullulans HP-2001 with soybean pomace as a nitrogen source. Bioresour. Technol. 95 (3), 293–299.

Shen, Z., Wang, W., Jia, J., Ye, J., Feng, X., Peng, A., 2001. Degradation of dye solution by an activated carbon fiber electrode electrolysis system. J. Hazard. Mater. 84 (1), 107–116.

de Souza, P.M., de Oliveira Magalhães, P., 2010. Application of microbial α-amylase in industry—a review. Braz. J. Microbiol. 41 (4), 850–861.

Srilakshmi, J., Madhavi, J., Lavanya, S., Ammani, K., 2015. Commercial potential of fungal protease: past, present and future prospects. J. Pharm. Chem. Biol. Sci. 2 (4), 218–234.

Struyf, N., Verspreet, J., Verstrepen, K.J., Courtin, C.M., 2017. Investigating the impact of α-amylase,α-glucosidase and glucoamylase action on yeast-mediated bread dough fermentation and bread sugar levels. J. Cereal Sci. 75, 35–44.

Sundarram, A., Murthy, T.P.K., 2014. α-Amylase production and applications: a review. J. Appl. Environ. Microbiol. 2, 166.

Swamy, M.V., Seenayya, G., 1996. Thermostable pullulanase andα-amylaseactivity from Clostridium thermos sulfurogenes SV9—optmization of culture conditions for enzyme production. Process Biochem. 31 (2), 157–162.

Takasaki, Y., 1976. Productions and utilizations of β-amylase and pullulanase from Bacillus cereus var. mycoides. Agric. Biol. Chem. 40, 1515.

Tan, L.S., Jain, K., Rozaini, C.A., 2010. Adsorption of textile dye from aqueous solution on pretreated mangrove bark, an agricultural waste: equilibrium and kinetic studies. J. Appl. Sci. Environ. Sanit. 5 (3), 283–294.

The Novozymes Report, 2019. report2020.novozymes.com.

Theron, L.W., Divol, B., 2014. Microbial aspartic proteases: current and potential applications in industry. Appl. Microbiol. Biotechnol. 98 (21), 8853–8868.

Tierny, Y., Bechet, M., Joncquiert, J.C., Dubourguier, H.C., Guillaume, J.B., 1994. Molecular cloning and expression in Escherichia coli of genes encoding pectate lyase and pectin methylesterase activities from Bacteroides thetaiotaomicron. J. Appl. Bacteriol. 76, 592–602.

Turk, B., 2006. Targeting proteases: successes, failures and future prospects. Nat. Rev. Drug Discov. 5 (9), 785–799.

UlAleem, A., 2013. An Investigation of Alternatives to Reductive Clearing in the Dyeing of Polyester. (Doctoral Dissertation Heriot Watt University).

Verma, S., Dixit, R., Pandey, K.C., 2016. Cysteine proteases: modes of activation and future prospects as pharmacological targets. Front. Pharmacol. 7 (125), 107.

Yadav, A.N., Verma, P., Kumar, V., Sangwan, P., Mishra, S., Panjiar, N., Gupta, V.K., Saxena, A.K., 2018. Biodiversity of the genus Penicillium in different habitats. In: Gupta, V.K., Rodriguez-Couto, S. (Eds.), New and Future Developments in Microbial Biotechnology and Bioengineering, Penicillium System Properties and Applications. Elsevier, Amsterdam, pp. 3–18, https://doi.org/10.1016/B978-0-444-63501-3.00001-6.

Yumoto, I., Ichihashi, D., Iwata, H., Istokovics, A., Ichise, N., Matsuyama, H., Okuyama, H., Kawasaki, K., 2000. Purification and characterization of a catalase from the facultatively psychrophilic bacterium Vibrio rumoiensis S-1T exhibiting high catalase activity. J. Bacteriol. 182, 1903–1909.

Zámocký, M., Gasselhuber, B., Furtmüller, P.G., Obinger, C., 2012. Molecular evolution of hydrogen peroxide degrading enzymes. Arch. Biochem. Biophys. 525, 131–144.

Zlinski, T., Aldmann, H., 1997. Cross linked enzyme crystals (clecs): efficient and stable biocatalysts for preparative organic chemistry. Angew. Chem. Int. Ed. Eng. 36, 22–724.

Zumstein, M.T., Kohler, H.P.E., McNeill, K., Sander, M., 2017. High-throughput analysis of enzymatic hydrolysis of biodegradable polyesters by monitoring cohydrolysis of a polyester-embedded fuorogenic probe. Environ. Sci. Technol. 51 (8), 4358–4367.

CHAPTER

Fungi in nutraceutical and baking purposes

7

Sabyasachi Banerjee[a], Subhasis Banerjee[a], Santanu Banerjee[b], Avik Das[b], and Sankhadip Bose[c]

[a]*Department of Pharmaceutical Chemistry, Gupta College of Technological Sciences, Asansol, West Bengal, India*
[b]*Department of Pharmacology, Gupta College of Technological Sciences, Asansol, West Bengal, India,* [c]*School of Pharmacy, The Neotia University, Sarisa, West Bengal, India*

7.1 Introduction

The fungi as food and feed are exceptionally nutritive since they comprise nonessential and essential amino acids. The utilization of fungi in dietary sources as well as in fermented beverages is popular since ancient times. Archeological evidence traces the relation of eatable wild mushroom with inhabitants of Chile, right around 13,000 years afore time (Rojas and Mansur, 1995); howsoever, it was in China where the utilization of wild fungi was first dependably noted, a few hundred years before the birth of the Christ (Aaronson, 2000). The first records of the fermentation process and the utilization of fermented food products were found in Babylon and Sumeria (Elander and Lowe, 1994). In spite of this deep-rooted practice of the utilization of fungi and its utilization in preparation of other food products, their entire potency wasn't explored until the latter half of twentieth century, when it was supported by the appearance of the brilliant period of industrial microbiology. Since then, this various community addressed by yeasts, filamentous fungi, and mushrooms, has been exploited in a bunch of food items for both human and livestock consumptions.

The fungal realm has certain common benefits as far as their dietary supremacy over the remainder of the vegetarian platter. Those are (i) decent protein content (20%–30% of the dry matter) having all the vital amino acids (yeasts are particularly enriched in lysine), consequently able to substituting meat; (ii) chitinous wall to act as a wellspring of dietary fibers; (iii) higher vitamin B contents; (iv) lower fat content; and (v) basically liberated of cholesterol. The benefit of mushroom propagation is additionally increased by their lower cost of production, since the majority of them could be cultivated on agro wastes or other industrial waste products. All horticultural production pro creates huge waste in light of the fact that such a tiny portion of each harvest is really utilized (7% in sugarcane, 5% in coconut and palm, 2% in sisal plantation, etc.). These could easily be dealt with by cultivation of mushrooms. For instance, oyster mushroom species (*Pleurotus cystidiosus*, *P. ostreatus*, *P. sajor-caju*) develop promptly on cotton wastes. Also, albeit the *Volvariella volvacea* (straw mushroom) is generally grown in Southeastern Asia on rice straws, this too could be grown on cotton waste. The capability of certain Pleurotus sp. to develop on numerous lignocelluloses agricultural wastes has been utilized in both the bioremediation and the production of a consequent cash crop as mushroom.

Entrepreneurship with Microorganisms. https://doi.org/10.1016/B978-0-443-19049-0.00016-5

Also, the harvested mushrooms (spent compost) can be utilized as a valuable animal feed because of their higher protein contents and soil conditioner with its higher nutrient and polymeric components which upgrade soil structure and are surprisingly utilized to digest contaminations (like polychlorinated phenols) on landfill waste sites by virtue of its populace of microorganisms capable of digesting the natural phenolic contents of lignin (Chiu et al., 1998). Additionally, the multispecies amenability and subsequent ease of upgradation to larger-scale cultures have substantially aided the commercial utilization of fungi. These factors have contributed in bringing together various research groups and organizations who have taken an interest in sequential "Eurofung" projects subsidized by the European Union to accelerate the developments of the industry. As an outcome, the yearly average yield of mushrooms has ascended to a stunning 6161 thousand tons, and the trade has thrived as an intercontinental one (Carlile et al., 2001). Howsoever, not the entirety of the wild assortments is propagated on a business scale. The decision of strain is made based on the production yield and regulatory issues, particularly for fungi utilized in the food industries. Host strains are normally chosen from among those, which have accomplished the alleged generally recognized as safe (GRAS) status, by the U.S. Food and Drug Administration (FDA).

In this chapter, the significant uses of the fungal realm (in particular mushroom/macro-fungi, filamentous fungi, and yeast) in foods and animal feed have been investigated predominantly under two broad categories: (i) direct utilization of either freshly fruiting body or processed mycelia and (ii) a more indirect fermentation-based methodology where the fungal enzymes and the secondary metabolites have been utilized in the processing of a wider range of foods and health products. Still other significant aspects of the study might relate to the expanded utilization of these organisms in nutraceutical and pharmaceutical industries.

7.2 Utilization of Fungi as nutraceutical

Over ongoing years, the utilization of fungal food had expanded on a worldwide premise with an elevation in public concern about dietary and health-related issues. Particularly, those belonging to the vegan community have turned to eating either newly cooked mushrooms or prepared foods, beverages, and dietary supplements of fungal origin.

7.2.1 Utilization of fruiting body

The fruiting body of the mushrooms has been taken freshly or cooked and utilized as a delicacy. Fungi could be cultivated technically through the fermentative method and through phases of media preparation, inoculation, and incubation. The media might be in the form of substrates accessible from cheap-valued sources, for example, agro-biomass and industrial wastes, however, transformed into higher value-added food and pharmaceutical items. In this manner, utilization of fungi is significant from economical as well as environmental viewpoints.

Albeit a large number of species of edible mushrooms exist in the wild, <20 species are utilized broadly as food, and just 8 to 10 species are consistently produced to any huge extent. The most usually consumed species is *Agaricus bisporus*, sold as portobello mushroom when larger or button mushroom when smaller, and utilized in soups, salads, and numerous different dishes. Numerous Asian fungi are presently economically cultivated and have acquired immense notoriety in the West. They could be benefited freshly from supermarkets and markets, inclusive of oyster mushrooms (*P. ostreatus*), straw mushrooms (*V. volvacea*), enokitakes (*Flammulina* sp.), and shiitakes (*Lentinula edodes*). They are frequently utilized for the preparation of diverse types of dishes. There are numerous different fungi, for example, milk mushrooms, black trumpets, chanterelles, morels, truffles, and porcini mushrooms (also termed as "*Boletus edulis*" or "king boletes"); all of them are having higher market price. The most regularly utilized propagated table macro-fungi are enrolled in Table 7.1, alongside their nutritional and known therapeutic properties. They could be utilized as a vegetarian substitute for protein source with nonvegetarian sources. Fruiting body of the fungi could be propagated through horticulture.

Table 7.1 Few medicinal and nutritional benefits of commonly used consumable macro-fungi.

Fungi	Medicinal values	Nutritional values	References
Lentinula edodes	Tyrosinase present in L. *edodes* in general diminishes blood pressure. An active polysaccharide lentinam and (1–3) β-D-glucan, decreases malignant growth as well as cholesterol and upregulates TH1 response	High in protein including all necessary amino acids; notable natural resource of vitamin D; presence of choline and adenine helps in forestalling the events of liver and vascular sclerosis effectively	Murata et al., 2002; Rossi et al., 1993
Flammulina velutipes	Induce antibody productions via modulating TH-cell differentiations and functions	Aheterogalactan, namely, Mannofucogalactan procured from *F. velutipes* is exhibited to possess nutritional benefits	Carbonero et al., 2008; Ko et al., 1995
Volvariella volvacea	A fungal immunomodulatory enzyme FIP-Vvo induces TH1-specific cytokines (IL-2, IFN-γ, LT), TH2-specific cytokine (IL-4)	Rich source of antioxidants because of higher β-carotene content	Cheung et al., 2003; Hsu et al., 1997; She et al., 1998
Tuber melanosporum	Its having antiviral, anticarcinogenic, and anticholesterolaemic effects; it also possesses prophylactic action with respect to hypertension and coronary heart disease	Its the most sought-after delicacy with incredible monetary worth	Breene, 1990
Pleurotus ostreatus	Exhibits antiviral, hematological, antitumor, antibacterial, antibiotic, immunomodulation, and hypocholesterolemic activities	Rich in carbohydrates, protein, fiber, vitamins, and minerals. Among the volatile constituents, 1-octen-3-ol is accepted to be the principal component which used as a flavoring agent	Cohen et al., 2002
Auricularia polytricha	Fruiting body makes a new immunomodulatory protein (APP) that increases the preparation of both tumor necrosis factor-α (TNF-α), and nitric oxide (NO) which suggests that APP stimulates immune response in the host. APP also activates murine splenocytes, notably increases their proliferation and γ-interferon (IFN-γ) secretions	Rich in Mg, P, Se, and K; higher dietary fiber content >50% of net weight. Helps to relieve constipation	Kim et al., 2004; Sheu et al., 2004

(Continued)

Table 7.1 Few medicinal and nutritional benefits of commonly used consumable macro-fungi—cont'd

Fungi	Medicinal values	Nutritional values	References
Ganoderma lucidum	A proteoglycan, namely, GLIS, procured from the fruiting body, is a B-cells stimulator. This constituent stimulates the activation of B lymphocyte, differentiation, proliferation, and immunoglobulin production	Utilized in dietary preparations and to make support. The protein content is 7.3% of the dry weight. Glucose and metal (like Mg, K, Ca, and Ge) contents are 11% and 10.2% of dry mass, respectively	Bao et al., 2002; Zhang et al., 2002
Agaricus bisporus	Helps in diminishing blood serum cholesterol	Source of vitamin B and minerals like potassium, sodium, selenium, and phosphorus	Beelman et al., 2003
Morchella esculenta	Methanolic extract exhibits antioxidant action, scavenging impacts on radicals, and chelating effects with Fe^{2+} ions; comprises galactomannan which induces macrophage activity	Used in many Provencal cuisines	Duncan et al., 2002
Tremella fuciformis	Potent hypocholesterolemic activity	Due to higher fibers it is exoteric among cholesterol-affected people	Cheung, 1996
Morchella elata	According to the Chinese medicinal system, *M. elata is* used to treat high blood pressure, tuberculosis, and the common cold	Source of Vitamin D2	Mattila et al., 2000
Morchella semilibera	Ethanolic extract shows 85% of antioxidant action	Used in the preparation of palatable dishes	Carbonero et al., 2008

Lentinus edodes, also known as Shiitake mushroom, is the second most propagated mushroom around the globe, solely after Agaricus, the "Paris mushroom." Other than Japan and China, Shiitake is likewise widely propagated in Thailand, Taiwan, Korea, and Singapore, along with the United States, Holland, and Canada. Proteins present in Shiitake mushroom contain a full supplement of essential amino acids and tend to be utilized widely in the vegetarian diet. Its active metabolite, Lentinan, a polysaccharide, has been appeared to lessen cancer and cholesterol. The Shiitake mushroom is as regular in Asian nations as *A. bisporus* is in the West. Its propagation strategy is like that of *P. ostreatus*. *Volvariella volvacea* is generally termed as "paddy mushroom" and is utilized in numerous Chinese recipes. It is industrially propagated on a blend of raw cotton waste and rice bran and harvested in the button or egg stage before the pileus arises. The fungus will in general develop on decaying vegetation and wood in the wild. *Flammulina velutipes*, also called winter mushroom or Enoki, is minuscule and delicate mushroom. It's developed on sawdust medium in an enormous container. It would appear to be an improbable contender for cultivation due to its smaller size, however, is generally sold in supermarkets. Japan is to be accepted as the origin of the cultivation of this species. *P. ostreatus* or oyster mushroom is a saprobic fungus that could generally be found, grown on deceased trees, in nature. *Tuber melanosporum* or truffle, a fungus belongs to the order Tuberales, is an underground European fungus and has been gathered, since at least 3600 years. The flesh of all truffles is almost white when youthful; as the truffle develops, the flesh gets more obscure with a marbling of lighter tissue. The aroma and taste of commercially procured truffles are so intense that they are utilized as flavoring rather than a separate dish. Another mushroom *Ganoderma lucidum* also called Reishior Lingzhi isn't exactly a tasteful mushroom, anyway utilized as a consumable in dietary formulations, yet is included as one of the most respected ingredients in conventional oriental medicine. It is usually propagated for its medicinal as well as tonic values. The earlier records, the species *Auricularia polytricha* also called "wood ear" dated back to around 200–300 BCE. It's now propagated all through the Asia and South Pacific. The cultivation method of these species is similar as of Shiitake mushroom. It is developed over logs and furthermore on a blend of sawdust and cotton wastes. Possessing gelatinous jelly-like behavior of the basidiocarps, the Jelly Fungi, which comprise leaflike folds, *Tremella fuciformis* is also called silver ear or snow ear fungus, is generally eaten in the eastern zones, and is regarded as a Chinese delicacy. It has been utilized as a herb to cure many diseases from the long time. Chinese people believe that it can cure tuberculosis, hypertension, and the common cold. The strategy of cultivating this species is similar to that of Auricularia and Shiitake since it's a wood-possessing species. True morels are a genus of consumable mushrooms closely associated with anatomically simpler cup fungi. Though morels are commonly sold canned or dried, they could be bought fresh. During the preparation of fresh morels to eat, soaking might ruin their delicate flavor. Because of their natural porousness, morels might comprise trace amounts of soil that can't be cleaned out. Truly outstanding and least difficult approaches to enjoy morels is by tenderly sauteing them in butter, cracking pepper on top, and sprinkling with salt. Morels are mainstream in various assortments: *Morchella esculenta* is better acquainted as "The common morel." At the point when youthful, this species has white ridges and dark brownish pits and is called white morel. At the time it become mature, both the ridges as well as pits turned yellowish-brown, and it turns to a "yellow morel."

7.3 Fungi in baking industries

Fungi constitute a decent amount of foods and food additives exist in the markets as human food or animal feed.

7.3.1 Single cell protein

Single cell protein (SCP) is the name aggregately given to an assortment of microbial items manufactured by the fermentation process. They could be utilized to ferment a portion of the tremendous amounts of waste materials, like straws; woods and wood processing wastes; foods, cannery, and food processing wastes; and surplus from alcohol productions or from animal and human excreta. These factors combinedly might see SCP arise as the potential protein source for domestic livestock. Generally, they are obtained in a diluted form comprising <5% solids, that are further concentrated by strategies like filtration, precipitation, coagulation, and centrifugation. Expulsion of water is important to stabilize the materials for storage in many cases; SCP should be dried to about 10% moisture, or consolidated and acidified in order to restrain spoilage from occurring, or fed promptly after being produced; nonetheless, these dehydration strategies aren't at present economical. Another significant part of exploitation is the productions of protein from hydrocarbon wastes of the petroleum industries by utilizing filamentous fungi. These organisms have the simplicity of harvesting yet bother from slower growth rates. In comparison, yeasts have better production and utility values and as such have established their dominance over other members of the community. The significant uses of yeasts have been featured as follows.

7.3.2 Baker's yeast

The preparation of baker's yeast is the biggest domestic utilization of a microorganism for food purposes. It is a strain of *Saccharomyces cerevisiae*. The strain of the yeast is deliberately chosen for its capacity to generate plentiful gas rapidly, its feasibility during ordinary storage, and its capability to deliver desirable flavor. The organisms are blended in with bread doughs to achieve vigorous sugar fermentation. The carbon dioxide (CO_2) produced during the fermentation is liable for the rising or leavening of the dough.

7.3.3 Utilization of yeast cells in foods and fodders

The terms "food yeast" and "fodder yeast" are limited to a specific kind of single cell biomass. Dr. Biloraudand Professor Jacquot first proposed the definition of food yeast in 1957, which was hence embraced by the IUPAC and afterward by the European Economic Community in 1975. The statement goes as "Food yeast is a kind of yeast which has been stained and dried; it ought to have no diastase activity, and hasn't been submitted to extraction process nor received any additive". Yeasts function as food additives for domestic livestock and could be securely named as fodder yeasts. Basically, these organisms desired for human and animal utilization should conform to certain dietary properties of protein content, vitamin content, amino acid composition, great digestibility, and inexistence of toxic substances.

Notwithstanding these essential standards, the strains selected should easily be amenable to larger-scale production and ought to be cost-effective to contend at standard with other more traditional items in the market. The following aspects should thusly be clearly characterized for a given strain used to produce food and fodder yeasts, the development conditions, the decision of potent substrates, and the related downstream treatments.

7.4 Processed fungal foods as an alternative to SCPs

Other than these, few well-known processed food items are available in the market, the most striking of which is the myco-protein "Quorn." Myco-protein is the terminology authored by the UK Foods Standards Committee to serve as the generic name for a food item resulting from the continuous fermentation of a selected strain *Fusarium venenatum* (initially called *F. graminearum*). Initially, the item was dried and powdered, available to be purchased as higher protein SCP flour, yet for its organoleptic characteristics of the hyphal mass it has been produced as a meat substitute under the brand name "Quorn" as a high-technology item. The mycelium is grown in an enormous large air-lift fermenter in a persistent-culture mode. Its filamentous structure empowers it to reproduce the fibrous nature of meat; combined with the inherent nutritive value of fungal biomass, the item is a low-calorie, low-fat, and cholesterol-free health food.

7.4.1 Use in fermentation-based food industries

Albeit the development of macro-fungi had thrived in recent years, utilization of fresh mycelia as food hasn't actually achieved the impetus to be popularized on a global scale. Maybe, more imperative is the role of these organisms, particularly yeast in industrial food preparation and processing. Fungal cell factories are generally utilized in brewing, winemaking, and bread-preparing industries because of their intrinsic ability to secrete a wider titter of enzymes into the growth medium. The accompanying sections will introduce and expound on these indirect yet essential aspects of fungal utilization.

7.4.2 Production of alcoholic beverages

The yeast *S. cerevisiae* is generally utilized for the manufacturing of numerous alcoholic beverages. They have been classified into 3classifications; those produced utilizing fruit juices, those produced using starchy materials, and those produced utilizing other plant materials (Carlile et al., 2001). The restrictions imposed by alcohol concentration on the fermentation by yeast are in the ranges of 10%–12% or 15%–16% as on account of "sake" production, and the alcoholic concentration is increased by distillation. Alcoholic beverages are produced utilizing fruit juices inclusive of wine, perry, and cider. Wines are manufactured by fermenting white and red grapes, where the yeast changes over the fruit sugars into liquor. *Botrytis cinerea*, a grapevine pathogen, is utilized to manufacture a specific wine "Sauternes" or "Edel faule" in Germany. The pathogen assaults the growing grapes and concentrates the sugars inside and eliminates any residual acidity; the grapes are then reaped and fermented utilizing the conventional processes. The resulting wine is very sweet and of high quality. Cider is prepared to utilize apples. Ciders could be classified into sweet (lower in tannin and acid), bittersweet (high in tannin, lower in acid), sharp (high in acid, low in tannin), and bitter sharp (higher in both acid and tannin). There is lesser sugar in apples than in grapes, diminishing the liquor concentration of cider. Perry is produced utilizing Perry pears or dessert pears. Again, the similar fermentation method utilized for wine and cider is used to prepare Perry.

With respect to alcoholic beverages prepared from starchy compounds, the most widely utilized item is a cereal grain (barley, rye, wheat) Nonetheless, other starchy compounds like roots, seeds, and tubers are likewise utilized. Before fermentation continues, the starchy compounds must be debased into less difficult sugars. In the beer industries, malting is utilized to accomplish that objective. Malting happens under aerobic condition in which typically barley grains are steeped in the water for as long as a day and afterward permitted to germinate under moist aerobic conditions. The germination cycle generates enzymes that degenerate the polysaccharides as well as proteins into simpler sugars and amino acids. The compound is fermented into beer. The production of "sake" utilizes rice as its starting material in Japan. Rice is rich in starch and accordingly utilizable by the yeast. Therefore, a fungus, namely, *Aspergillus oryzae*, is inoculated to be grown on the surface of the rice. The fungus changes over the starch into simpler sugars, which could then be able to utilized by the yeast in order to produce sake. "Sake" is inoculated with *A. oryzae*to produce "koji." Further rice mash is lactic acid fermented utilizing bacteria and yeasts. The mash and "koji" are blended and fermented for approximately 20 days, till the alcohol concentration has reached approximately 18%. The product is filtered, pasteurized, and stored before utilization.

Preparation of nonalcoholic beverages, for example, "coco" (*Candida* sp., *Leuconostoc mesenteroides*), coffee (*Saccharomyces marscianus*, *Leuconostoc mesenteroides*, *Fusarium* sp., *Flavobacterium* sp.) (Soccol et al., 2008), and black tea (Pasha and Reddy, 2005), is likewise reported.

7.4.3 Preparation of bakery and cheese products

Bakery products comprise a blend of flour (generally from cereals particularly wheat), along with water, sugar, and salt fermented by yeast. Flour is blended in with the residual stuffs and incubated at almost 25°C. The yeast leavens the sugar to form alcohol and CO_2. The acquitted gas causes bubbles by elastic augmentation of gluten protein present in the flour. Over baking, the alcohol gets evaporated. The duration of fermenting, the extent of gluten in flour, the components of grain, and the temperature decide the flavor and texture of the bread.

In the preparation of cheese, the availability of apparent fungal mycelium is a section of the moldy cheeses favorites among connoisseurs. Camembert and Roquefort or blue cheese are the two most familiar examples. These cheeses are prepared from two species of Penicillium, *P. roqueforti* is used to prepare Roquefort cheese and *P. camemberti* for Camembert cheese. Cheese making is considered one of the earliest forms of food production, dating back approximately 4000 years. The process involves the addition of molds and various microorganisms to enhance the flavor of the cheese. The processes are generally alluded to as mold-ripened.

7.5 Production of other food products/condiments/additives

A lot of researches have been performed on some fermented food products, so that the activity of the fungus engaged with the process could be established. Some of the more common ones comprise tofu, shoyu, miso, and tempeh. Howsoever, the microorganisms (inclusive of bacteria) engaged with most of the fermented foods (~500 in numbers) are unknown. In contrast to western societies, in which fermented foods are generally prepared by yeasts, eastern societies have used various distinctive mycelial fungi. Soya sauce (soy or shoyu sauce) is perhaps the most recognizable Asian food products prepared by cooked soybeans blended with wheat flour, squeezed into cakes, and inoculated with *A. oryzae*. The molded cake koji, blended with water and salt, is alluded to as moromi. The moromi is inoculated with bacterium *Peiococcussoyae*, and yeasts with bacteria *Torulopsis* sp., and *Saccharomyces rouxii*, in order to ferment the blend for around 6 months to transform into the soy sauce. Shoyu, found in China >2500 years prior, is referred to as a flavoring and flavor-boosting element as a meatless seasoning in the West. Tempeh produced from fermented items of legume seeds with the fungus *Rhizopus oligosporus* is found to be originated in Indonesia. The fungus inoculating inside the bubbled beans digest complex carbohydrate and various organic components that might produce gas. Miso, a Japanese term for fermented soybean paste, is not generally devoured without itself, yet as a base for soup or utilized as a flavoring agent. Miso fermentation comprises washed or polished rice, which is steamed and inoculated with organism *A. oryzae*, which results in rice koji. The proteins and carbohydrates of the already inoculated rice are digested by the fungus and changed over to amino acids and sugars. The rice koji is thereafter inoculated with yeast and bacteria for fermentation.

The entirety of the above processes is really completed by a large group of fungal enzymes, which are being delivered into the particular substrates by the concerned fungi. These enzymes along with their brief area of applications in the food processing industries are listed in Tables 7.2 and 7.3.

Table 7.2 Enzymes associated with their area of applications.

Area of applications	Enzymes	References
Butter and its oils	Lipase, catalase, glucose oxidase	Gupta et al., 2003
Animal feed	Glucoamylases, amylase, glucanase, pentosanases, cellulases, xylanases, phytases, proteinases	Wang et al., 2006
Cheese	Lipase, rennet, proteinases	Freitas and Malcata, 2000
Bread	Amyloglucosidases, amylases, cellulases, glucose oxidase, glucanases,hemicellulases, pentosanases, lipases,proteinases	Taniwaki et al., 2001
Biscuits	Cellulases, amylases, hemicellulases, pentosanases, proteinases	
Alcohol	Amyloglucosidase, amylase, β-glucanases, cellobiase, cellulases, proteinases, pectinase	Sharma et al., 2002
Brewing	Decarboxylase, acetolactase, amylases, cellulase, amyloglucosidase, glucanase, pentosanase, lipase, proteinase, xylanase	Stroh, 1998
Coffee	Hemicellulases, cellulase, galactomannanase, pectinase	Soccol et al., 2008
Egg processing	Lipase, proteinase, phospholipase, glucose oxidase, catalase	Singh et al., 2007
Fish	Proteinase	Prasad, 2001
Confectionery	Invertase, amylase, pectinase, proteinase	Stroh, 1998
Fats	Glucose oxidase, esterase, lipases	Bobek et al., 1994
Dairy products	Proteinase, lactase, sulfhydryl oxidase, lysozyme, lactoperoxidase, peroxidase, catalase	Archer, 2000; Beauchemin et al., 1999; Rode et al., 1999
Flavors	Peptidase, glucanase, proteinase, lipases, esterase, amylase	Shahani et al., 1976
Fruit, cloudy juices	Pectinases, amylases, proteinases, cellulases	Brandelli et al., 2005
Dibittering	Naringinase, peptidase	Knoss et al., 1998
Fruit extracts	Anthocyanase	Albershein, 1966
Vegetable and fruit processing	macerating enzymes, cellulases, pectinases	Brandelli et al., 2005
Fructose	Inulinase, glucose isomerase, amylase, cellulase, amyloglucosidase, glucanases, isomerase, hemicellulases, lipase, phospholipase, proteases, pectinases	Sørensen et al., 2004
Wine	Amyloglucosidase, amylase, cellulase, hemicellulase, glucanase, pectinases, glucose oxidase, proteases, catalase, anthocyanase, pentosanase	Okamura et al., 2001
Tea	Glucanase, cellulase, pectinase, tannase	Pasha and Reddy, 2005
Malt extract	Amyloglucosidase, amylase, cellulase, proteinase, glucanase, xylanase	Feng et al., 2007
Animal fats/oil	Lipases, esterases, proteinase	Beldman et al., 1984
Botanical extraction	Amyloglucosidase, amylase, cellulase, hemicellulase, glucanase, pectinases, proteases	Knoss et al., 1998
Protein	Cellulase, amylase, glucanase, pectinase, hemicellulase, protease	Semenova et al., 2006
Starch	Amyloglucosidase, amylase, cellulase, hemicellulase, glucanase, isomerase, phospholipase, lipase, pectinases, proteases	Albershein, 1966
Fruit pulps	Amylase, pectinase, amyloglucosidase, glucanase, cellulase, hemicellulase, protease, pectinase	Brandelli et al., 2005
Fruit juice	Amyloglucosidase, amylase, cellulase	Semenova et al., 2006
Fruit extraction	Amyloglucosidase, amylase, cellulase, pentosanase, pectinase, limonoate, naringinase, dehydrogenase	Albershein, 1966

Table 7.3 Enzymes as additives utilized in food and feed bioprocessing (Ghorai et al., 2009).

Enzyme	Applications
α-Amylase	Fermentation, starch syrups, ethanol, animal feed
β-Amylase	Maltose syrup, brewing
β-Glucanase	Brewing industry
Cellulase	Animal feed
β-Glucosidase	Transform isoflavone phytoestrogen in the soymilk
Glucoamylase	Manufacturing dextrose as well as high-fructose syrup
α-Galactosidase	Increases sucrose yield; also used in beet sugar industries
Invertase	Preparation of invert syrup from beet sugar
Hemicellulase/xylanase/pentosanase	Baking and/or fruit juice manufacturing
Pectinase	Fruit processing
Naringinase	Debitter citrus peel
Lactase	Eliminating lactose from dairy products
Proteases	Baked foods, distilled spirits, protein processing
Pullulanase	Antistaling agent in baked foods

7.6 Use of enzymes in food and feed bioprocessing

Enzymes have been utilized in food producing since the starting of humanity in manufacturing cheese and indirectly by yeasts (Schmid et al., 2001). In food industry, notwithstanding cheese preparation, enzymes were already utilized in 1930s in fruit juice preparation. These enzymes generally clarify the juice. They are termed as pectinases, which comprise various distinctive enzyme activities. The significant use of microbial enzymes among food industries started in 1960 in starch industries. The conventional acid hydrolysis of starch was totally replaced by glucoamylases and α-amylases that could transform starch to glucose with >95% of yield. Starch industry turned into the second biggest utilizer of enzymes just after the detergent industries. Intensive experiments to utilize enzymes in animal feeding began in mid-1980s. The first commercial achievement was the addition of β-glucanase in barley-based feeding diets.

Enzymes were additionally examined later in wheat-based diets. Xylanase enzymes were discovered to be the most feasible ones for this situation. Xylanases are these days regularly utilized in feed formulations. Typically, preparation of a feed enzyme is a multi-enzyme cocktail comprising xylanases, glucanases, proteinases, and amylases. The addition of enzymes decreases viscosity, which enhances nutrients absorption, releases nutrients either by hydrolysis of nondegradable fibers or by releasing the nutrients obstructed by these fibers, and lessens the measure of feces. Moreover, in poultries, enzymes are utilized in turkey and pig feeds. Enzymes additionally have diverse applications in beverage industries. Chymosin is generally utilized in cheese preparation for coagulating milk protein. Another enzyme utilized in milk industries is lactase or β-galactosidase, which divides milk-sugar lactose into galactose and glucose. This method is utilized for milk products that are devoured by lactose-intolerant masses.

Enzymes are additionally utilized in fruit juice preparation. Amalgamation of xylanase, pectinase, and cellulase enhances the release of juice from the pulp. Usage of amylases and pectinases results in juice clarification. Brewing is an enzyme-based method, whereas the malting is a method that raises enzyme levels in the grains. In case of mashing method, the enzymes are released which hydrolyzes starch to form fermentable sugars such as maltose, a disaccharide glucose. Additional enzymes could be employed to assist starch hydrolysis (commonly α-amylases), solve filtration issues caused by β-glucans exist in malt (β-glucanases), hydrolyze proteins (like neutral proteinase), and control haze during maturating, filtrating, and storing (α-amylase, β-glucanase, and papain). Similarly, enzymes are widely utilized in wine preparation to procure a better color and flavor.

7.7 Fungal enzymes used in feed

Enzymes have been utilized for quite a long time to improve the usage of poultry and swine diets. A few recent investigations have inspected the utilization of exogenous enzyme items in high-forage diets fed to the growing cattle (Beauchemin et al., 1997; Beauchemin et al., 1995; McAllister et al., 1999; Michal et al., 1996; Pritchard et al., 1996; Wang et al., 1999; ZoBell et al., 2000) (Table 7.2). For example, amylase, phytase, xylanase, and β-glucanase are included in the cereal-based diet plans such as monogastrics to expand the usage of starch, dietary phosphorus, arabinoxylans, and β-glucans, respectively. Researches have shown that supplementing dairy cows and feedlot cattles diet with fiber-degraded enzymes can possibly improve feed usage and animal performances (Beauchemin and Rode, 1996) (Table 7.3). Ruminant-feeding enzyme additives, fundamentally cellulases and xylanases, are concentrated extractives resulted from fungal or bacterial fermentations that have explicit enzymatic actions. Progresses in animal performances because of the utilization of enzyme additives could be attributed basically to improving ruminal fiber digestion which results in expanded digestible energy intake(Arambel et al., 1987). This methodology offers opportunities for utilizing enzymes to promote nutrient digestions, utilizations, as well as animal productivities and simultaneously diminish animal fecal materials and pollutions. Enzymes sprayed over feeds immediately before feeding provides expanded administration adaptability and bypasses any negative interaction that ensiling process might have on silage enzyme execution. Feeds treated with enzymes as such might improve digesting ability through various mechanisms inclusive of direct hydrolysis, enhancements in palatability, changes in the gut viscosity, integral activities with ruminant enzymes, and change in digestion site.

Protease enzymes might improve the digestibility of cereal grains, since the digestion of starch is partly a component of protein-starch matrix inside the seed. Steam-flaked sorghum treated with an enzyme blend improves weight gaining and feed effectiveness in steers by around 10% (Boyles et al., 1992). Fiber-degrading enzymes might likewise assist with improving the digestibility of cereal grains along with fibrous seed coat. Cellulase or xylanase enzymes sprayed over a barley silage and barley diet also improve weight gaining and feed efficacy in steers (Beauchemin and Rode, 1996).

Fungal direct-fed microbials have already been well known in addition to ruminant diet for a long time. Generally, three kinds of additives are available. First, a few items comprise and ensure "live" yeast. The greater part of these items comprises different strains of *S. cerevisiae* (Martin and Nisbet, 1992; Savage, 1987). Second, other additives comprise *S. cerevisiae* as well as culture extracts, however, make no assurance for live organisms. Third, there're fungal additive dependant on *A. oryzae* fermentation final products that likewise make no claims for providing live microbes.

Direct-fed microbial items are available in a diversity of forms inclusive of pastes, powders, boluses, and capsules. In certain applications, direct-fed microbial items might be blended with feeds or administered with drinking water.

7.8 Commercial utilization of recombinant fungi enzymes

Recombinant fungi are one of the principal resources of enzymes for commercial applications. The industrial enzyme market came to 1.6 billion dollars in 1998 for the accompanying application sectors: 45% in food, 34% in detergents, 11% in textiles, 3% in leather, and 1.2% in paper and pulp (Stroh, 1998). This doesn't include therapeutic and diagnostic enzymes. The market for those nonpharmaceutical enzymes came to 2 billion dollars in 2000. More than 60% of those enzymes utilized in food, detergent, and starch processing industries are recombinant items (Cowan, 1996), albeit the quantity of heterologous fungal enzymes endorsed for food applications isn't huge (Table 7.4). Because of the lower yields procured with nonfungal enzymes, numerous recombinant food-grade enzymes are of fungal origin (Archer, 2000). There is one special case in which the donor strain isn't another fungus, for example, calf rennin (chymosin), which is utilized for cheese preparation. Preparation of the bovine enzyme in recombinant *Aspergillus niger* var. awamori added up to around 1 g per liter after the nitrosoguanidine mutagenesis and determination for 2-deoxyglucose resistances (Dunn-Coleman et al., 1991). Furthermore, improvements were performed by parasexual recombination, which results in a strain produced 1.5 g per liter from parents produced 1.2 g per liter (Bodie et al., 1994). A recombinant strain of *A. oryzae* produces an aspartic proteinase from the *Rhizomucor miehei*, which has already been permitted by US-FDA for cheese preparation (Pariza and Johnson, 2001). Microbial lipases have a gigantic potency in sectors like food technologies, biomedical sciences, and chemical and detergent industries. In food industries, lipases are generally utilized for the production of baking foods, fruit juices, alluring flavors in cheese, and inter-esterification of oils and fats to prepare modified acylglycerols. There're three fungal recombinant lipases presently utilized in the food industries, those from *Thermomyces lanuginosus*, *R. miehei*, and *Fusarium oxysporum*, which are all produced in *A. oryzae* (Pariza and Johnson, 2001). The application of hydrolytic lipases in the detergent manufacturing

Table 7.4 Recombinant enzymes and source fungi.

Enzyme	Fungi
Catalase	*Aspergillus niger*
Cellulase	*A. oryzae*, *Trichoderma reesei*
β-Galactosidase	*A. oryzae*
Glucose oxidase	*A. niger*
β-Glucanase	*T. reesei*
Lipase	*A. oryzae*
Xylanase	*A. oryzae*, *A. niger*, *T. reesei*
Phytase	*A. oryzae*, *A. niger*
Protease	*A. oryzae*
Chymosin	*A. niger* var.

industry is another significant area of commercial utilization. Enzymes used in detergent usually make up almost 32% of the absolute lipase sales. Lipase to be used in detergents should be thermostable and stays dynamic in an alkaline environment of a common machine wash. In 1995, two bacterial lipases Luma fast and Lipomax were introduced from *Pseudomonas mendocina* and *P. alcaligenes*, respectively, by Genencor International which were discovered to be appropriate for the purposes (Jaeger and Reetz, 1998). Biotechnology has great possibilities to expand quality and supply feedstocks for paper and pulp, to diminish productivity costs, and to prepare novel higher value products. Novel enzyme technologies could diminish environmental hazards and change fiber properties. Wild as well as recombinant hydrolases and oxidoreductases have higher potency for eco-friendly paper pulp bleaching (from pulp). The ultimate lignin content of the flax pulp is much more diminished if a peroxide stage is incorporated. This characteristic has been utilized by cloning these genes (feruloyl esterase from *A. niger*, Mn^{2+}-oxidizing peroxidases from *Phanerochaete chrysosporium* and *Pleurotus eryngii*) into a totally chlorine free (TCF) sequence that also included a peroxide stage (Sigoillot et al., 2005).

7.9 Secondary metabolites used in food and feed from fungi

Secondary metabolites are components prepared by an organism that aren't needed for primary metabolic processes. Fungi generate an abundant array of secondary metabolites, some of which are significant in the industry. They are generally utilized to promote the coloration of food products. Fungi generate a series of components which alter the color of food. For example, *Monascus purpureus* has traditionally been utilized for red wine production since longer time (Went, 1895). The polyketides pigments are insoluble in acidic conditions. *β*-Carotene is procured by a series of Mucorales(Ende and Stegwee, 1971). This could be added to a diversity of foods. Concerning with the possibly poisonous or allergic attributes of some artificial colors has prompted to a closer study of colors from natural sources. Fermentation of rice by *M. purpureus* to produce ang-kak (red rice) or koji has been utilized as a conventional Chinese food and medicine since 800 CE (Li et al., 1998). The aqua-soluble red pigments, viz., rubropunctamine and monascorubramine are generated by reacting orange pigments such as rubropunctatin and monascorubrin, respectively, with the amino acids that exist in the fermentation media. The fungus is utilized for producing red rice, soybean cheese, wine, fish, and meat, and is approved for use in foods in China as well as Japan. *Phaffia rhodozyma*, a yeast has turned into the most significant microbial resource for the production of carotenoid astaxanthin. This pigment is accountable for the orange to the pinkish color of salmonid flesh and the reddish color of boiled crustacean shells. Feeding pen-reared salmonids with this yeast in a diet plan cites pigmentation of white muscles. *Blakeslea trispora* has been utilized for the preparation of *β*-carotene industrially in Russia for quite a long time (Nelis and De Leenheer, 1991). Fermentation, a fungal-mated culture, is utilized with a preferred proportion of negative and positive mating strains. The *β*-carotene accumulation is emphatically associated with sexual interactions between two mating types (i) during mating a hormone-alike substance principally trisporic acid is prepared, which stimulates pigment productivity, and (ii) a group of harmful secondary metabolites like mycotoxins frequently contaminate animal feed (Bennett and Klich, 2003). These components are possessed with harmful properties toward humans as well as other animals, causing a wider scope of acute and chronic impacts altogether known as mycotoxicoses. Mixing of contaminated feed with the uncontaminated feed is a typical practice to decrease mycotoxin contaminations. Howsoever, this is under tough guideline by the federal organizations. The US-FDA commonly establishes impediments for concentrations of mycotoxins in human as well as animal foods. Limitations are labeled utilizing various terms, inclusive of "guidance levels" for fumonisins, "action levels" for aflatoxins, and "advisory levels" for vomitoxin.

7.10 Pharmaceutical and nutraceutical by-products from Fungi

Mushrooms with medicinal effects are utilized for nutraceutical as well as pharmaceutical products (Kidd, 2000; Wasser and Weis, 1999). Because of their higher compatibility and tolerance with the radiotherapy and chemotherapy, the products procured from mushrooms are utilized for cancer therapies. The fruit bodies of mushroom and their extracts are successfully utilized and are likewise economically feasible alternative because of quicker growth of the fruiting body or the mycelial stage. Few prominent mushrooms with pharmacological activities are *Coriolus versicolor*, *Agaricus brasiliensis*, *L. edodes*, *G. lucidum*, *Grifola frondosa*, *P. ostreatus*, etc. (Chu et al., 2002; Rossi et al., 1993). The significant bioactive components found in mushroom include lentinan, schizophyllan, and PSK, utilized in cancer chemotherapy with noteworthy outcomes. Other biologically active components found in the mushrooms include nematicide, immunosuppressive, antiviral, antimicrobial, and hypocholesterolemic agents. Black tea fermented with *Dabaryomyces hansenii* results in accumulating leading vitamins, like A, B1, B2, B12, and C adequate amounts to satisfy recommended dietary allowance. It additionally brings about a decrease of tannins and caffeine in significant amounts. Moreover, the theophylline accrued because of fermentation confers a broncho dilatory impact to the tea (Pasha and Reddy, 2005).

7.11 Symbiotic fungus termitomyces

Termitomyces, agarics paleotropical genus is captivating to both entomologists and mycologists, which grows only in alignment with termites along with their nests and is reliant upon the organic matters brought by the insects from their feeding on trees. In spite of the fact that *Termitomyces* are saprobic, they are harmonious with termites and comprise significant wild edible species. The fungus assists the termites with debasing the plant-derived materials (e.g., dry grass, leaf litter, and wood) on which they live. *Termitomyces* are having 20 edible species, recorded from Asia and Africa. These fungi are routinely gathered and furthermore sold. *Termitomyces titanicus* is the biggest edible fungus around the globe as indicated by Guinness Book of Records, as having a cap diameter of about 1 m, albeit different species *Termitomyces microcarpus* (seldomly over steps 2 cm) are a lot more modest (Bignell, 2000; Natarajan, 1977). There are 23consumable species of *Termitomyces* are accounted for from 35 nations. The genus is profoundly regarded and numerous species are broadly consumed with higher nutritional benefits. The mushrooms are procured all through the Africa and are utilized generally in Asia, yet aren't well documented. Notable species incorporate *Termitomyces clypeatus*, *T. striatus*, and *T. microporus*. The species have medicinal properties and are regarded as good for brain and memory (Wei and Yao, 2003).

Another significant consumable fungus and enzymes producer *T. clypeatus* comprises 32% carbohydrate, 31% protein, and 10%–14% ascorbic acid (Ogundana and Fagade, 1982). Several enzymes of higher therapeutic values have been reported from *T. clypeatus* (Rouland-Lefèvre, 2000). *T. clypeatus* is known to be a potent manufacturer of various enzymes in culture media (Khowala et al., 1992; Khowala and Sengupta, 1992). It also has been discovered to be a potent manufacturer of a broader spectrum of extracellular glucosidases (sucrase, cellulase, cellobiose, etc.), able for hydrolyzing polysaccharides, for instance, cellulose, hemicellulose, and starch. Various enzymes, for example, 1,4-β-D-xylosidase, endo 1,4-β-D-xylanase, α-L-arabinofuranosidase, α-amylase, acetyl esterase, and amyloglucosidase were likewise purified from the fungus. All the enzymes are having different industrial applications. Several other enzymes procured from this fungus are enrolled alongside their utilization (Table 7.3).

7.12 Bioprocessing of food by *T. clypeatus*

7.12.1 Softening and leavening of bread

A method for preparing a novel enzymatic production effective for enhanced leavening of the bakery items was developed from *T. clypeatus* (Sengupta et al., 1999a). The enzymes amylases, hemicellulases, and cellulases procured from culture media were included in flour notwithstanding other routine ingredients like sugar, salt, yeast, and additives, and the bread produced was much soften and large in volume and size because of better leavening. The method expanded the falling number (softening index) of the bread by greater than 3 to 4 times. Different enzymes are utilized in baked items and utilized for the specific purpose of softening and leavening to increase the palatability of the products and furthermore for inhibiting staling, which is of impressive significance for expanding the shelf life of bakery items.

7.12.2 Clarification of noncitrus fruit juice

An enzyme prepared from *T. clypeatus* comprising a blend of cellulase, pectinase, hemicellulase, xylanase, and arabinose was effective for clarifying noncitrus fruit juice. The enzyme blend at 40°C was mixed with the apple juice and incubated for 2–4 h. The subsequent suspension was filtered to procure a clear juice. A method for preparation of an enzyme composition comprising a blend of xylanase and pectinase effective for clarification of noncitrus fruit juice was developed and patented (Sengupta et al., 1999b). Pectinase is utilized in extraction, filtration, clarification, and depectinization of fruit juices as well as wines by collapsing the cell wall enzymatically, and for maceration of vegetables and fruits, removal of the inner wall of garlic, lotus seed, peanut, and almond (Kashyap et al., 2000).

7.13 Conclusion and future prospects

In the course of recent many years, there has been a solid upsurge of the fungal community mostly in the circles of feed, food, and therapeutics. Since the fab majority of these mycorrhizal and saprotrophic organisms could genetically be modified easily, research activities like EUROFUNG in progress have taken up the strain improvement programs with state-of-the-art technologies to achieve a unified goal of developing "secretion giants" out of those modest higher potential organisms. In such manner, the enhancement of production method deserves equal significance. Albeit most current techniques depend upon submerged fermentation (SmF), the utilization of the conventional solid-state fermentation (SSF) method sought to be investigated more thoroughly. A few outcomes have exhibited that the SSF method results in improved degrees of different secreted fungal hydrolases (Pandey et al., 2000). Up to this point, the greater part of the exploration in the field of SSF has been centered around measure and fermenter design (Weber et al., 1999), treating the organism required as a black box. Howsoever, with the appearance of the "omics" age and ultramodernization of farming strategies and detection instruments, it will really be feasible to look at the reactions that appeared by these organisms toward evolving habitat, growth parameters, and nutritional status. Future utilization won't be limited exclusively to food production yet will include the employment of fungal community to cater for an ever-increasing number of differentiated human needs. It has effectively seen successful endeavors of this industry in food additives, nutraceuticals, and condiments. This will expand even more with expanded necessities of treating human diseases, bioremediation, and biofuel production. At present, the mushroom industry has been organized in a perspective on providing impressive returns to the farmers. Accordingly, both developed and developing nations have turned to mushroom cultivations as one of the additional promising choices for expanding both rural income as well as foreign currency earning. Thus, the industry is under a cycle of decentralization to an accomodable degree. Howsoever, to augment benefits for the respective nations, the farmers and industry proprietors should be more philanthropic. Cumulatively, this industry bears the extraordinary potential to thrive as quite possibly the most expansive trade ever.

References

Aaronson, S., 2000. Fungi. In: Kiple, K.F., Ornelas, K.C. (Eds.), The Cambridge World History of Food, pp. 313–336.

Albershein, P., 1966. Pectinlyase from fungi. Methods Enzymol. 8, 628–631.

Arambel, M.J., Weidmeier, R.D., Walters, J.L., 1987. Influence of donor animal adaptation to added yeast culture and/or *Aspergillusoryzae* fermentation extract on in vitro rumen fermentation. Nutr. Rep. Int. 35, 433–437.

Archer, D., 2000. Filamentous fungi as microbial cell factories for food use. Curr. Opin. Biotechnol. 11, 478–483.

Bao, X.F., Wang, X.S., Dong, Q., et al., 2002. Structural features of immunologically active polysaccharides from *Ganoderma lucidum*. Phytochemistry 59, 175–181.

Beauchemin, K.A., Jones, S.D.M., Rode, L.M., Sewalt, V.J.H., 1997. Effects of fibrolytic enzymes in corn or barley diets on performance and carcass characteristics of feedlot cattle. Can. J. Anim. Sci. 77 (4), 645–653.

Beauchemin, K.A., Rode, L.M., 1996. Use of feed enzymes in ruminant nutrition. In: Rode, L.M. (Ed.), Animal Science Research and Development—Meeting Future Challenges. Minister of Supply and Services, Ottawa, ON, pp. 103–130.

Beauchemin, K.A., Rode, L.M., Sewalt, V.J.H., 1995. Fibrolytic enzymes increase fiber digestibility and growth rate of steers fed dry forages. Can. J. Anim. Sci. 75 (4), 641–644.

Beauchemin, K.A., Yang, W.Z., Rode, L.M., 1999. Effects of grain source and enzyme additive on site and extent of nutrient digestion in dairy cows. J. Dairy Sci. 82, 378–390.

Beelman, R.B., Royse, D., Chikthimmah, N., 2003. Bioactive components in button mushroom *Agaricusbisporus* (j.lge) imbach (Agaricomycetideae) of nutritional, medicinal, or biological importance (review). Int. J. Med. Mushrooms 5, 321–337.

Beldman, G., Pilnik, W., Rombouts, F.M., Voragen, A.G.J., 1984. Application of cellulase and pectinase from fungal origin for the liquifaction and saccharification of biomass. Enzym. Microb. Technol. 6, 503–507.

Bennett, J.W., Klich, M., 2003. Mycotoxins. Clin. Microbiol. Rev. 3, 497–516.

Bignell, D.E., 2000. In: Abe, T., Bignell, D.E., Higashi, M. (Eds.), Termites: Evolution Sociality, Symbioses, Ecology. Kluwer Academic Publishers, Dordrecht, The Netherlands, pp. 189–208.

Bobek, P., Ozdin, L., Kuniak, L., 1994. Mechanism of hypocholesterolemic effect of oyster mushroom (*Pleurotus ostreatus*) in rats: reduction of cholesterol absorption and increase of plasma cholesterol removal. Z. Ernahrungswiss. 33, 44–50.

Bodie, E.A., Armstrong, G.L., Dunn-Coleman, N.S., 1994. Strain improvement of chymosin-producing strains of *Aspergillusniger* var. awamori using parasexual recombination. Enzym. Microb. Technol. 16 (5), 376–382.

Boyles, D.W., Richardson, C.R., Robinson, K.D., Cobb, C.W., 1992. Feedlot performance of 25 steers fed steam-flaked grain sorghum with added enzymes. In: Proceedings Western Section American 26 Society of Animal Science. vol. 43, pp. 502–505.

Brandelli, A., Geimba, M.P., Mantovani, C.F., 2005. Enzymatic clarification of fruit juices by fungal pectin lyase. Food Biotechnol. 19, 173–181.

Breene, W.M., 1990. Nutritional and medicinal value of specialty mushrooms. J. Food Prot. 53, 883–894.

Carbonero, E.R., Gorin, P.A.J., Iacomini, M., et al., 2008. Characterization of a heterogalactan: some nutritional values of the edible mushroom *Flammulinavelutipes*. Food Chem. 108, 329–333.

Carlile, M.J., Watkinson, S.C., Gooday, G.W., 2001. The Fungi. Academic Press, London; San Diego, CA.

Cheung, C.K.P., 1996. The hypocholesterolemic effect of two edible mushrooms: *Aurzculariaauricula* (tree-ear) and *Tremellafuciformis* (white jelly-leaf) in hypercholesterolemic rat. Nutr. Res. 16, 1721–1725.

Cheung, L.M., Cheung, P.C., Ooi, V.E., 2003. Antioxidant activity and total phenolics of edible mushroom extracts. Food Chem. 81 (2), 249–255.

Chiu, S.W., Ching, M.L., Fong, K.L., Moore, D., 1998. Spent oyster mushroom substrate performs better than many mushroom mycelia in removing the biocide pentachlorophenol. Mycol. Res. 102, 1553–1562.

Chu, K.K., Ho, S.S., Chow, A.H., 2002. Coriolusversicolor: a medicinal mushroom with promising immunotherapeutic values. J. Clin. Pharmacol. 42 (9), 976–984.

Cohen, R., Persky, L., Hadar, Y.P.L., 2002. Biotechnological applications and potential of wood-degrading mushrooms of the genus Pleurotus. Appl. Microbiol. Biotechnol. 58, 582–594.

Cowan, D., 1996. Industrial enzyme technology. Trends Biotechnol. 14, 177–178.

Duncan, C.J., Pasco, D.S., Pugh, N., Ross, S.A., 2002. Isolation of a galactomannan that enhances macrophage activation from the edible fungus *Morchellaesculenta*. J. Agric. Food Chem. 50, 5683–5685.

Dunn-Coleman, N.S., Bloebaum, P., Berka, R.M., Bodie, E., Robinson, N., Armstrong, G., Ward, M., Przetak, M., Carter, G.L., LaCost, R., Wilson, L.J., 1991. Commercial levels of chymosin production by aspergillus. Bio/Technology 9 (10), 976–981.

Elander, R.P., Lowe, D.A., 1994. Fungal biotechnology: an overview. In: Handbook of Applied Mycology. Marcel Dekker, New York, NY, pp. 1–34.

Ende, H.V.D., Stegwee, D., 1971. Physiology of sex in Mucorales. Bot. Rev. 37, 22–36.

Feng, X.M., Larsen, T.O., Schnürer, J., 2007. Production of volatile compounds by *Rhizopusoligosporus* during soybean and barley tempeh fermentation. Food Biotechnol. 19, 173–181.

Freitas, C., Malcata, F.X., 2000. Microbiology and biochemistry of cheeses with appélationd' Origineprotegée and manufactured in the Iberian Peninsula from ovine and caprine milks. J. Dairy Sci. 83, 584–602.

Ghorai, S., Banik, S.P., Verma, D., Chowdhury, S., Mukherjee, S., Khowala, S., 2009. Fungal biotechnology in food and feed processing. Food Res. Int. 42 (5–6), 577–587.

Gupta, R., Rathi, P., Bradoo, S., 2003. Lipase mediated upgradation of dietary fats and oils. Crit. Rev. Food Sci. Nutr. 43, 635–644.

Hsu, C.I., Hsu, H.C., Kao, C.L., et al., 1997. Fip-vvo, a new fungal immunomodulatory protein isolated from *Volvariellavolvacea*. Biochem. J. 323, 557–565.

Jaeger, K.E., Reetz, M.T., 1998. Microbial lipases form versatile tools for biotechnology. Trends Biotechnol. 16 (9), 396–403.

Kashyap, D.R., Chandra, S., Kaul, A., Tewari, R., 2000. Production, purification and characterization of pectinase from a Bacillus sp. DT7. World J. Microbiol. Biotechnol. 16, 277–282.

Khowala, S., Ghosh, A.K., Sengupta, S., 1992. Saccharification of xylan by an amyloglucosidase of *Termitomycesclypeatus* and synergism in the presence of xylanase. Appl. Microbiol. Biotechnol. 37, 287–292.

Khowala, S., Sengupta, S., 1992. Secretion of β-glucosidase by *Termitomycesclypeatus*. Enzym. Microb. Technol. 14, 144–149.

Kidd, P.M., 2000. The use of mushroom glucans and proteoglycans in cancer treatment. Altern. Med. Rev. 5 (1), 4–27.

Kim, T.I., Park, S.J., Choi, C.H., et al., 2004. Effect of ear mushroom (Auricularia) on functional constipation. Korean J. Gastroenterol. 44, 34–41.

Knoss, W., Reuter, B., Alkorta, I., et al., 1998. Industrial applications of pectic enzymes: a review. Process Biochem. 33, 21–28.

Ko, J.L., Hsu, C.I., Lin, R.H., Kao, C.L., Lin, J.Y., 1995. A new fungal immunomodulatory protein, FIP-fve isolated from the edible mushroom, *Flammulinavelutipes* and its complete amino acid sequence. Eur. J. Biochem. 228 (2), 244–249.

Li, C., Zhu, Y., Wang, Y., Zhu, J.S., Chang, J., Kritchevsky, D., 1998. *Monascuspurpureus*-fermented rice (red yeast rice): a natural food product that lowers blood cholesterol in animal models of hypercholesterolemia. Nutr. Res. 18 (1), 71–81.

Martin, S.A., Nisbet, D.J., 1992. Effect of direct-fed microbials on rumen microbial fermentation. J. Dairy Sci. 75 (6), 1736–1744.

Mattila, P., Piironen, V., Suonpaa, K., 2000. Functional properties of edible mushrooms. Nutrition 16, 694–696.

McAllister, T.A., Oosting, S.J., Popp, J.D., Mir, Z., Yanke, L.J., Hristov, A.N., Treacher, R.J., Cheng, K.J., 1999. Effect of exogenous enzymes on digestibility of barley silage and growth performance of feedlot cattle. Can. J. Anim. Sci. 79 (3), 353–360.

Michal, J.J., Johnson, K.A., Treacher, R.J., 1996. The impact of direct fed fibrolytic enzymes on the growth rate and feed efficiency of growing beef steers and heifers. J. Anim. Sci. 74 (Suppl. 1), 296–299.

Murata, Y., Shimamura, T., Tagami, T., Takatsuki, F., Hamuro, J., 2002. The skewing to Th1 induced by lentinan is directed through the distinctive cytokine production by macrophages with elevated intracellular glutathione content. Int. Immunopharmacol. 2 (5), 673–689.

Natarajan, K., 1977. A new species of Termitomyces from India. Curr. Sci. 46, 679–680.

Nelis, H., De Leenheer, A.P., 1991. Microbial sources of carotenoid pigments used in foods and feeds. J. Appl. Bacteriol. 70 (3), 181–191.

Ogundana, S.K., Fagade, O.E., 1982. Nutritive value of some Nigerian edible mushrooms. Food Chem. 8, 263–268.

Okamura, T., Ogata, T., Minamimoto, N., et al., 2001. Characteristics of wine produced by mushroom fermentation. Biosci. Biotechnol. Biochem. 65, 1596–1600.

Pandey, A., Soccol, C.R., Mitchell, D., 2000. New developments in solid state fermentation: I-bioprocesses and products. Process Biochem. 35 (10), 1153–1169.

Pariza, M.W., Johnson, E.A., 2001. Evaluating the safety of microbial enzyme preparations used in food processing: update for a new century. Regul. Toxicol. Pharmacol. 33 (2), 173–186.

Pasha, C., Reddy, G., 2005. Nutritional and medicinal improvement of black tea by yeast fermentation. Food Chem. 89, 449–453.

Prasad, C., 2001. Improving mental health through nutrition: the future. Nutr. Neurosci. 4, 251–272.

Pritchard, G., Hunt, C., Allen, A., Treacher, R., 1996. Effect of direct-fed fibrolytic enzymes on digestion and growth performance in beef cattle. J. Anim. Sci. 74 (Suppl. 1), 296–299.

Rode, L.M., Yang, W.Z., Beauchemin, K.A., 1999. Fibrolytic enzyme supplements for dairy cows in early lactation. J. Dairy Sci. 82, 2121–2126.

Rojas, C., Mansur, E., 1995. Ecuador: Informaciones Generales Sobre Productos Non Madererosen Ecuador. In: Memoria, Consulta de Expertos SobreProductos For estales no Madereros Para America Latinayel Caribe, SerieForestal. FAO Regional Office for Latin America and the Caribbean, Santiago, Chile, pp. 208–223.

Rossi, V., Jovicevic, L., Nistico, V., et al., 1993. *In vitro* antitumor activity of Lentinus edodes. Pharmacol. Res. 27, 109–110.

Rouland-Lefèvre, C., 2000. In: Abe, T., Bignell, D.E., Higashi, M. (Eds.), Termites: Evolution, Sociality, Symbioses, Ecology. Kluwer Academic Publishers, Dordrecht, The Netherlands, pp. 289–306.

Savage, D.C., 1987. Microorganisms associated with epithelial surfaces and stability of the indigenous gastrointestinal microflora. Food/Nahrung 31 (5–6), 383–395.

Schmid, A., Dordick, J.S., Hauer, B., Kiener, A., Wubbolts, M., Witholt, B., 2001. Industrial biocatalysis today and tomorrow. Nature 409 (6817), 258–268.

Semenova, M.V., Sinitsyna, O.A., Morozova, V.V., et al., 2006. Use of a preparation from fungal pectin lyase in the food industry. Appl. Biochem. Microbiol. 42, 598–602.

Sengupta, S., Ghosh, A.K., Naskar, A.K., et al., 1999b. A Process for the Preparation of an Enzyme Composition Containing a Mixture of Pectinase and Xylanase Useful for Clarification of Non-Citrus Fruit Juice. Indian Patent NF/686/99.

Sengupta, S., Sengupta, D., Naskar, A.K., Jana, M.L., 1999a. A Process for the Preparation of a Novel Enzymatic Formulation Useful for Improved Leavening of Bakery Products. Indian Patent NF/403/99.

Shahani, K.M., Arnold, R.G., Kilara, A., Dwivedi, B.K., 1976. Role of microbial enzymes in flavor development in foods. Biotechnol. Bioeng. 18, 891–907.

Sharma, S., Pandey, M., Saharan, B., 2002. Fermentation of starch to ethanol by an amylolytic yeast *Saccharomyces diastaticus* SM-10. Indian J. Exp. Biol. 40, 325–328.

She, Q.B., Ng, T.B., Liu, W.K., 1998. A novel lectin with potent immunomodulatory activity isolated from both fruiting bodies and cultured mycelia of the edible mushroom *Volvariellavolvacea*. Biochem. Biophys. Res. Commun. 247, 106–111.

Sheu, F., Chien, P.J., Chien, A.L., et al., 2004. Isolation and characterization of an immunomodulatory protein (APP) from the Jew's ear mushroom *Auriculariapolytricha*. Food Chem. 87, 593–600.

Sigoillot, C., Camarero, S., Vidal, T., Record, E., Asther, M., Pérez-Boada, M., Martínez, M.J., Sigoillot, J.C., Asther, M., Colom, J.F., Martínez, Á.T., 2005. Comparison of different fungal enzymes for bleaching high-quality paper pulps. J. Biotechnol. 115 (4), 333–343.

Singh, S., Wakeling, L., Gamlath, S., 2007. Retention of essential amino acids during extrusion of protein and reducing sugars. J. Agric. Food Chem. 55, 8779–8786.

Soccol, C.R., Dalla Santa, H.S., Rubel, R., et al., 2008. Mushrooms—a promising source to produce nutraceuticals and pharmaceutical by products. In: Koutinas, A.A., Pandey, A., Larroche, C., Larroche, A. (Eds.), Current Topics on Bioprocesses in Food Industry 1. Asiatech Publishers, Inc, New Delhi, pp. 439–448.

Sørensen, J.F., Kragh, K.M., Sibbesen, O., et al., 2004. Potential role of glycosidase inhibitors in industrial biotechnological applications. Biochim. Biophys. Acta 1696, 275–287.

Stroh, W.H., 1998. Industrial enzymes market. Gen. Enzyme News 18, 11–38.

Taniwaki, M.H., Silva, N., Banhe, A.A., Iamanaka, B.T., 2001. Comparison of culture media, simplate, and petrifilm for enumeration of yeasts and molds in food. J. Food Prot. 64, 1592–1596.

Wang, X.J., Bai, J.G., Liang, Y.X., 2006. Optimization of multienzyme production by two mixed strains in solid-state fermentation. Appl. Microbiol. Biotechnol. 73, 533–540.

Wang, Y., McAllister, T.A., Wilde, R.E., et al., 1999. Effects of monensin, exogenous fibrolytic enzymes and tween 80 on performance of feedlot cattle. Can. J. Anim. Sci. 79, 587.

Wasser, S.P., Weis, A.L., 1999. Medicinal properties of substances occurring in higher basidiomycetes mushrooms: current perspectives. Int. J. Med. Mushrooms 1 (1).

Weber, F.J., Tramper, J., Rinzema, A., 1999. A simplified material and energy balance approach for process development and scale-up of *Coniothyriumminitans* conidia production by solid-state cultivation in a packed-bed reactor. Biotechnol. Bioeng. 65 (4), 447–458.

Wei, T.Z., Yao, Y.J., 2003. Literature review of Termitomyces species in China. Fungal Sci. 22, 39–54.

Went, F.A.F.C., 1895. *Monascuspurpureus* le champignon de l' ang-quacune nouvelle thele bole. Ann. Sci. Nat. 8, 1–17.

Zhang, J., Tang, Q., Zimmerman-Kordmann, M., et al., 2002. Activation of B lymphocytes by GLIS, a bioactive proteoglycan from *Ganoderma lucidum*. Life Sci. 71, 623–638.

ZoBell, D.R., Wiedmeier, R.D., Olson, K.C., Treacher, R., 2000. The effect of an exogenous enzyme treatment on production and carcass characteristics of growing and finishing steers. Anim. Feed Sci. Technol. 87 (3–4), 279–285.

CHAPTER

Precision fermentation of sustainable products in the food industry

8

C.S. Siva Prasath[a], C. Aswini Sivadas[b], C. Honey Chandran[a], and T.V. Suchithra[b]
[a]Synthite Industries Pvt. Ltd., Kadayiruppu, Kerala, India, [b]National Institute of Technology, Calicut, Kerala, India

8.1 Introduction

Fermentation is an ancient biotechnological process used to produce products with desirable properties and extend the shelf life of foods. Yeast, bacteria, and molds are the common microbes in food processing. Bacteria are vital agents in food fermentations. Among them, lactic acid bacteria like *Lactobacillus*, *Pediococcus*, *Streptococcus*, *Oenococcus*, etc., are mainly employed in food fermentations (Ray and Joshi, 2014). *Lactobacillus* is used as a preservative agent in meat either naturally or added. *Lactobacillus* reduces the meat's pH and nitrate upon fermentation, enabling preservation potency (Velioglu and Murat, 2009). *Lactobacillus*, *Pediococci,* and *Leuconostocs* are widely used to ferment vegetables, for example, sauerkraut (fermented cabbages). *Acetobacter* species widely produce fruit vinegar (cider vinegar). *Bacillus* species, another set of bacteria, contributes alkaline fermentation, usually utilized in protein-rich foods like soybeans and legumes. Most industrially important yeast strains are under the *Saccharomyces* family, including *S. cerevisiae*, *S. bulderi*, *Saccharomyces cerevisiae* var. *ellipsoideus*, and *Schizosaccharomyces pombe*. Commercially important molds include *Aspergillus, Penicillium, Ceratocystis,* etc., and possess a significant role in producing enzymes, citric acid, and flavor (Ray and Joshi, 2014). During the metabolic process, fermentation favors the oxidation of carbohydrates into organic acids and alcohols. It will act as a preservative. The developmental trail of fermentation is shown in Fig. 8.1 (Ray and Joshi, 2014). Storage stability is one of the main attributes contributed by fermentation. The fermented product can maintain stability for up to 1 year under proper storage conditions. Fermentation is considered the cheapest and most natural preservation method (Das et al., 2016).

8.2 Precision fermentation

Fermentation is now widely used to produce alcoholic beverages, bread and pastry, dairy products, pickled vegetables, soy sauce, and so on. Modern biotechnology retains great promise for intensifying the scope of fermentation to invent novel foods and thus expand the sustainability of food production. Innovative approaches in genomics and synthetic biology comprise precision and biomass fermentation to synthesize specific compounds for the food, nutraceutical, and pharmaceutical industries. Besides the particular products produced by fermentation, it also generates many by-products, which lowers

Entrepreneurship with Microorganisms. https://doi.org/10.1016/B978-0-443-19049-0.00020-7

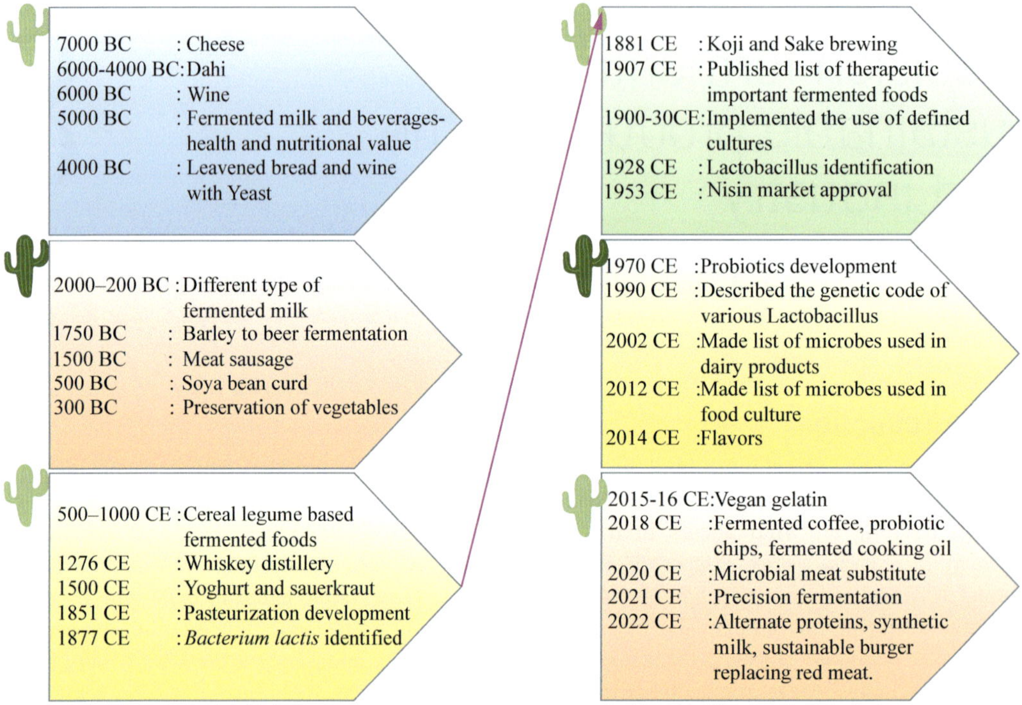

FIG. 8.1

The developmental trail of fermentation.

the production efficiency and also creates difficulty in downstream processing. In precision fermentation, the creation of microbial cell factories (MCFs) reduces the generation of by-products. Precision fermentation or synthetic biology is a potential substitute for traditional fermentation. The combined effect of genomics and fermentation possesses significant potential for generating novel food/related products (Teng et al., 2021).

Precision fermentation can improve traditional food fermentations by employing MCFs, which significantly generate functional food components with higher purity and yield. It also enables the production of high-value functional food ingredients, including enzymes, lipids, carbohydrates, vitamins, flavors, colorants, antioxidants, and preservatives (Chai et al., 2022).

8.3 Microbial cell factories

Microbial cell factories are genetically engineered microbes via strain engineering, employed to enhance productivity or metabolic yields. MCFs used to generate industrially valuable compounds comprising fine chemicals, biofuels, food additives, natural products, and pharmaceuticals by utilizing low-cost and renewable substrates in an economical and eco-friendly approach (Davy et al., 2017). *Saccharomyces cerevisiae* is considered a key model for eukaryotic systems and serves as a frame

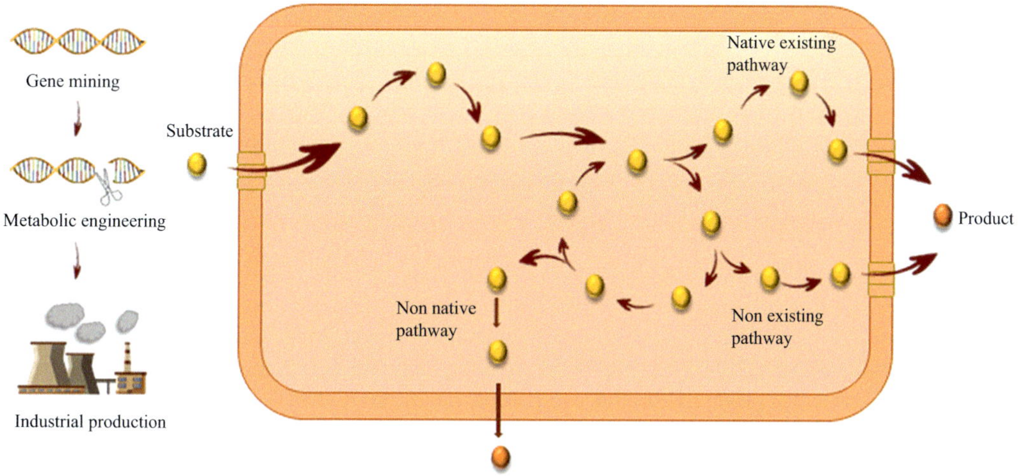

FIG. 8.2

Designing of MCFs.

for the diverse production of natural compounds (Otto et al., 2022). Metabolic engineering regulates the gene to produce chemicals for pharmaceuticals and other valuable materials efficiently. Propanol, butanol, and isobutanol are among the products produced by MCFs, as well as dicarboxylic acids like succinic, fumaric, malic, and adipic acids, diols like 1, 2-propanediol,1, 3-propanediol, 1,4, and 2, 3-butanediol, diamines like putrescine and cadaverine, and polymers like polyhydroxyalkanoates, polylactic acid, spider silk protein, and poly-gamma-glutamic acid. MCFs are the alternative source for petrochemical-derived products (Corchero et al., 2013). There are reports regarding the industrial production of three natural products (artemisinin, resveratrol, and carotenoids) through MCFs (Liu et al., 2017). The diagrammatic representation of the design of MCFs, optimization, and industrial production can be seen in Fig. 8.2. The process includes exploring biosynthetic pathways of natural plant products and artificial pathway recombination or generating MCFs.

8.4 Flavors in the food industry

Flavors and fragrances are important for food and beverages. Presently, most of these compounds are produced by chemical processes and extraction. The formation of an undesirable racemic mixture is a drawback of this process. The substances are thought to be poisonous or unhealthy by consumers. The main source of natural flavors is plants, especially their oils. Dependence is the primary drawback of plant sources. It depends on a variety of environmental conditions that are difficult to manage (Janssens et al., 1992). The use of biotechnology enables us to overcome all the hurdles in the chemical process in the flavor production. For instance, the microorganisms can produce various flavors through fermentation at a large scale. The heat-induced food processing with a combination of fermentation and enzymatic treatment is a sophisticated approach in food processing to generate flavors under mild conditions, which requires optimized conditions of flavors and enzymatic reactions (Dias et al., 2017).

An array of potentially valuable flavors and fragrances can be synthesized by utilizing the relevant de novo pathway in the MCFs (Bachmann et al., 2015). The agricultural wastes such as lignocellulosic waste molecules which are rich in cellulose, lignin, and hemicellulose are one of the potentials, cheaper feedstocks for producing the flavor compounds.

Fungus *Ceratocystis fimbriata* can produce a fruity aroma when it grows on sugarcane bagasse with synthetic media. When the amino acid-like leucine or valine is added to this media it will give a banana flavor. The production of different flavor compounds by fortifying with special nutrients in the synthetic media and the alteration of culture conditions have been proven already. Nowadays chemical industry has been using biotechnology practices like enzymatic or microbial processes to manufacture more than 100 flavor molecules. Few of the microbial flavors are present only in low concentrations in the fermentation broth, which in turn causes high purification cost. Hence, their market price is more costly than synthetic ones. However, due to the growing demand for natural aroma makes the scope of the biotechnologically produced flavors are getting much more attraction (Bel-Rhlid et al., 2018). Being an emerging field of molecular biology, the next-generation sequencing is also evolving as a promising method for enhancing the yield of industrially important flavors.

Evolutionary engineering is to identify the genes responsible for flavor production and construct them into more suitable phenotypes, which will allow further improvement of the fermentation process via alterations in the growth conditions and product improvements (Braga et al., 2018). Precision fermentation uses specially designed cell factories to produce specific food ingredients. These developed ingredients have specific characteristics and powerful attributes of plant-based products and cultivated meat. Recently alternative proteins are produced through precision fermentation. The production of flavors through the biotechnological route has several advantages over traditional methodologies. Due to the flavor's natural designation and increased market accessibility, more people are drawn to it (Hadj Saadoun et al., 2021).

8.5 Industrial process overview

8.5.1 Bioconversion

The de novo synthesis or biotransformation is the main biotechnological means for the flavor production of the de novo pathway, the synthesis of flavors is a complex process and yields less; the de novo synthesis involves complex processes and less yield whereas biotransformation, a single reaction that catalyzes enzymatically to produce products; hence, biotransformation is the potentially viable method for industrial production of flavors (Braga et al., 2018). The selection of microbe/s of interest is very crucial in the upstream processing. The microbes should utilize the substrate and undergo different reactions such as oxidation, reduction, hydrolytic, and dehydration reactions to form new flavors (Scharpf et al., 1986) which in turn results in specific products from specific substrates. Various microbes use ferulic acid from agricultural waste as the substrate for vanillin production. A few products manufactured by bioconversion are shown in Fig. 8.3.

The first commercially available fermentation-derived vanillin is Rhovanil, a product of a Belgian chemical company named Solvay. It was obtained by bioconversion of ferulic acid, an identified substrate for vanillin production. Ferulic acid is abundantly available in nature and shares a structural similarity to vanillin. Vanillin production occurs during the degradation pathway of ferulic acid (Fig. 8.4) (Gallage and Møller, 2015).

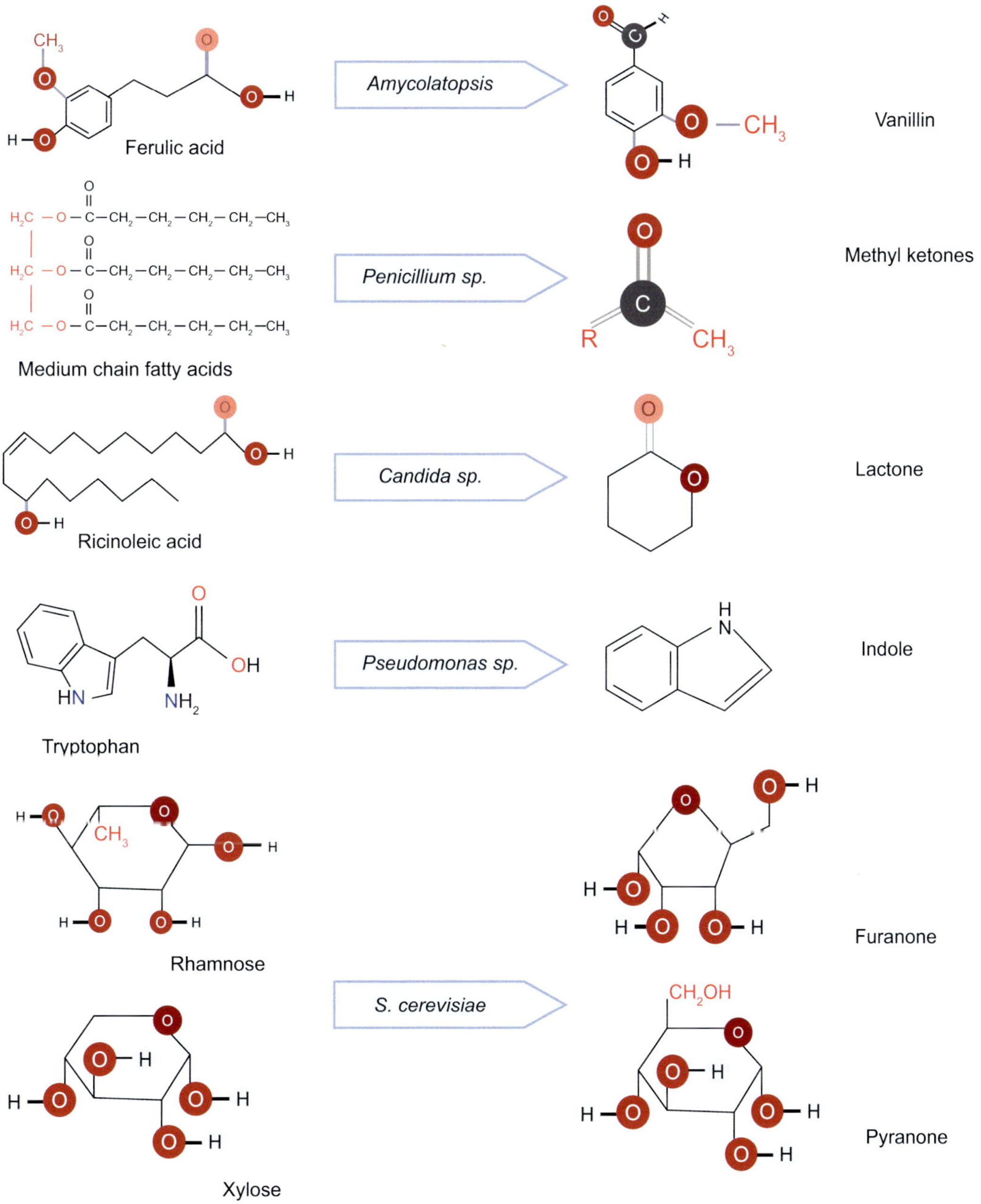

FIG. 8.3

Bioconversion process.

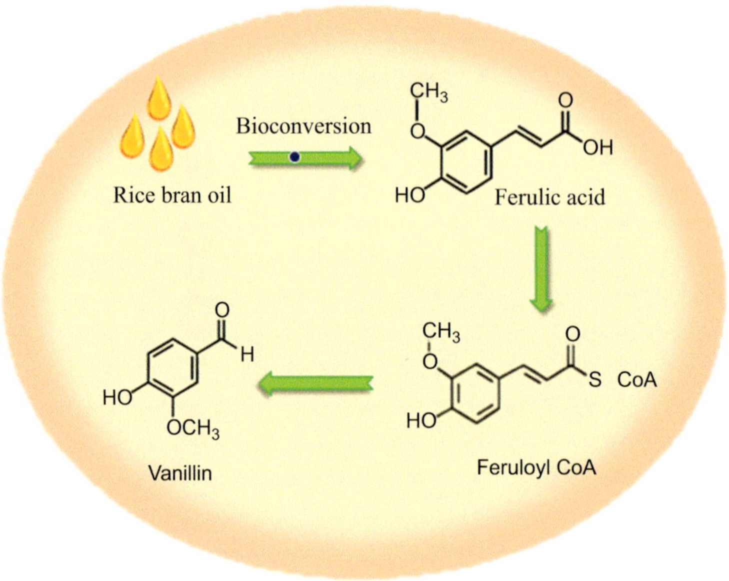

FIG. 8.4

Vanillin production through bioconversion.

Vanillin is purified by extraction, vacuum distillation, ultrafiltration, and multistage re-crystallization from aqueous solutions. Pervaporation is used to recover volatile solutes on hydrophobic membranes and desorb them into the vapor phase (Converti et al., 2010). Omega ingredients are commercially manufactured in different lactone flavors from coconut oil and castor oil using yeast through bioconversion. Terpenes are produced naturally in plant cells in low quantities. Isbiobionic is a Dutch-based company that produces terpene in *Rhodobacter sphaeroides* by altering the metabolic pathway (US 2020/0010822 A1, 2020). Isobionic manufactures are Natural Valencene (orange flavors with sweet and woody notes), Natural β-Bisabolene (like lemon, bergamot, and oregano), Natural Nootkatone (grapefruit taste and woody notes), and Santalol (alternative to sandalwood oil) through classical mutation and strain engineering technology.

8.6 Sweeteners through fermentation

Sugar is an inevitable part of our diet. Global demand for zero-calorie natural-sourced sweeteners has increased significantly over the last decades. Increased interest in plant-derived, natural, low-calorie, or zero-calorie sweeteners indicates that the public has become more aware of the adverse health impacts of excessive sugar consumption (More et al., 2021). Alternative sweeteners are prominent in the food and pharmaceutical sectors because of their potential applicability. Besides industrial production, most fruits and vegetables possess natural sweeteners such as maltitol, lactitol, sorbitol, xylitol, erythritol, and isomaltose (Philippe et al., 2014), but the extraction from such sources is still expensive. Microorganism-based alternative sweetener production is regarded as a sustainable method for many industrial applications since the process may also operate in a controlled condition. There are several benefits to the biotechnological production of natural sweeteners. Ideally, cell culture can produce the compound of interest from a critical plant.

Moreover, plant cell culture is expensive, time-consuming, and low yielding for the production of specialized natural products; thus, plant tissue culture may not be a viable method for the large-scale production of natural product sweeteners. Implementing metabolic engineering of plant or microbial cells is for producing the sweetener even without a productive cell culture. By developing safe, natural, zero-calorie sweeteners with enjoyable taste profiles, we can help mitigate the effects of excessive sugar consumption on a generation of human beings, both in terms of increased quality of life and decreased cost of healthcare (Philippe et al., 2014). We can generate value from sugar by multiplying its sweetening power, all while providing stable and sustainable farming opportunities and supply markets, and finally the improvised methodologies that enable the biotechnological production of natural sweeteners will have significant scientific and industrial implications for producing other small molecules as well.

Erythritol is four carbon sugar alcohol (or polyol) produced by fermenting glucose and sucrose by *Trichosporonoides megachiliensis (Trichosporonceae)*. It has a sweetness of approximately 60%–80% of sucrose. It is also manufactured from wheat or corn starch by enzymatic hydrolysis by yeast fermentation (More et al., 2021). Xylitol is another naturally occurring five-carbon sugar equivalent to sucrose in sweetness. The production of xylitol involves the extraction of xylan from plant cell walls or algae by acid hydrolysis, which further undergoes hydrogenation to yield the desired product. Xylitol is a good sweetener for diabetics and other purposes and possesses a low glycemic index. Generally, it is synthesized via a chemical process, and fermentation is considered an alternate method for the chemical process. Xylitol is an intermediate in xylose metabolism (Rafiqul and Sakinah, 2013). Yeast is the best xylitol producer among the other microorganisms. *Yarrowia lipolytica* is involved in the biotransformation of xylose into xylitol with more than 9% yield. It is the safest organism because of its GRAS (Generally Recognized as Safe; the United States Federal food, drug, and cosmetic acts) status. The initial purification stages are centrifugation, charcoal treatment, and precipitation. The final form of xylitol is made through crystallization (Prabhu et al., 2020).

Stevia (*Stevia rebaudiana*) is a herb that is used extensively in various areas of the world (without documentation of long-term use and effects) as a noncaloric sugar substitute (Džoljić, 2020). Nowadays, scientists focus on alternative sweeteners and their application in food technology. Since stevia glycosides (SvGls)—the natural plant sweeteners are about 400 times sweeter than sucrose, the stevia plant is one of the most significant sources of intense natural sweeteners. Stevia extracts are a potent tool for lowering the amount of sugar in food and beverages, and they can be acceptable alternatives to the full-sugar version of these products. The sweetness of stevia comes from its leaves, which produce steviol glycosides. Developing effective methods of SvGls production is of high research interest because of their benefits to human health (e.g., antidiabetic properties, lowering blood pressure, and good radical scavenging activity) and high taste quality. As a result, traditional and biotechnological methods have been developed to improve the yield of SvGls biosynthesis (Libik-konieczny, 2021), (Spakman, 2015). Biotech companies such as Evolva explored recombinant SvGls production in *Saccharomyces cerevisiae*, showing that eukaryotic cells are preferable as host cells for manufacturing rare steviol glycosides such as Reb M or Reb D (Libik-konieczny, 2021). The Rebaudioside M possesses an off-white color with a sweet taste. DSM Food and Pure Circle took the patent for manufacturing stevia Reb M through precision fermentation (Fig. 8.5). This Reb M is identical to molecules, which are present in the stevia plant. The microbes are specially programmed for this production of Reb M. The steviol glycoside

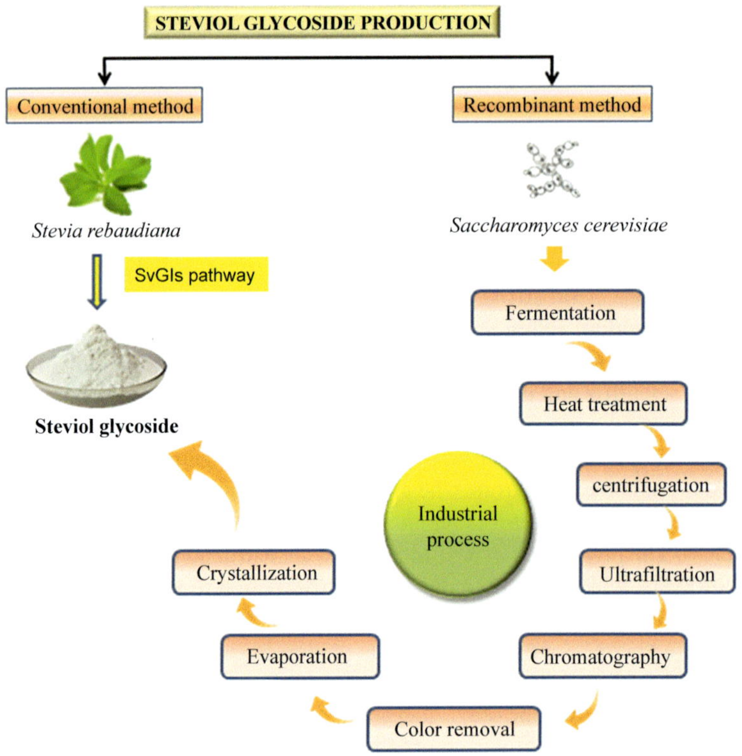

FIG. 8.5

Overview of the production of Reb M through precision fermentation.

production pathway was introduced into the microbes through metabolic pathway engineering (Xiao and Wu, 2015). Brazzein has an intensity in sweetener and is manufactured by Sweegan, United States of America. It is 500–2000 times sweetener than regular sugar with zero calories.

8.7 Antioxidants of fermented origin

The increasing awareness of healthy food habits and lifestyles has driven the scientific community to pay significant attention to free radicals and antioxidants. Free radicals are unstable, highly reactive molecules harming nucleic acids, lipids, proteins, and other biologically relevant molecules. Antioxidant defense systems must balance their daily production as they are continuously delivered. An unbalanced antioxidant system will lead to medical conditions like cardiovascular and neurodegenerative diseases, atherosclerosis, cancer, inflammatory state, and many others. Dietary intake of antioxidants is a must need for protecting our body from free radicals (Verni et al., 2019). Antioxidants are a valuable group of food preservatives, and they are in either natural or synthetic forms. Natural antioxidants have received significant attention, including flavonoids, phenolic acids, carotenoids, tocopherols, and some protein-derived compounds, such as amino acids and bioactive peptides (Verni et al., 2019). Antioxidant dietary supplements have a significant role in protecting human health. Natural sources of antioxidants, including fermented foods, have received immense attention because of the enhanced concern about artificial antioxidant consumption (Melini et al., 2019). There are several reports regarding the in vitro antioxidant capacity of dairy products such as yogurt and fermented milk (antioxidant activity is higher than normal milk), contributed by the discharge of specific bioactive peptides or the proteolysis of milk proteins (α-lactalbumin, β-lactoglobulin, and α-casein). Milk, cereals, fruits, vegetables, meat, fish, and dairy products have been reported to have higher antioxidant activity after fermentation. Fermented milk and cereals also contain antihypertensive peptides (Bachmann et al., 2015). Compounds such as carotenoids or vitamins are also important antioxidants. Fermentation processes can increase antioxidant molecules, modify the profile and types of bioactive antioxidant compounds, or even produce new compounds with biological activities of interest (Martí-Quijal et al., 2021).

Dihydroquercetin and flavonoids are found in plants. Dihydroquercetin is also known as taxifolin. Taxifolinacts as an antioxidant, anticancer, antimicrobial, and antiAlzheimer compound (Sunil and Xu, 2019). Commercial production of an active form of dihydroquercetin is a complicated process. The manufacturing of dihydroquercetin includes the processing of wood, extraction, preparation, and production of nanosuspensions and nanoemulsions, crystallization, and lyophilization (Orlova et al., 2022). Blue California is a leading company manufacturing dihydroquercetin through biotechnological routes by the combination of fermentation and enzymatic processes. The Flavanone 3-Hydroxylase (F3H) gene was cloned in *E. coli* for the sustainable production of particular enzymes. The carbon source act as the substrate for the production of enzymes (Fig. 8.6) (Park et al., 2020). Eriodictyol is a

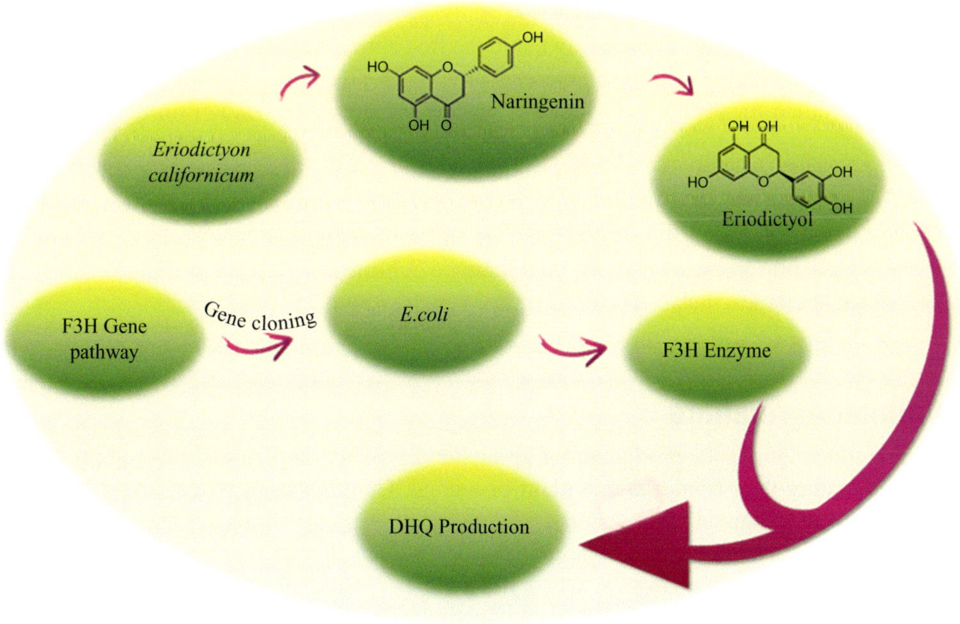

FIG. 8.6

Production flow of Dihydroquercetin.

flavonoid derived from yerba santa used as the key substrate for this production. Eriodictyol reacts with the derived enzyme from *E. coli* to produce dihydroquercetin.

Menaquinone 7 is named as Vitamin K2. Vitamin K2 is responsible for blood coagulation, possesses antiinflammatory, anticancer, and cardiovascular health, and has a protective role in brain damage. Vitamin K is present in meat and dairy products, fermented cheese, yogurt, and the traditional Japanese food natto (Ravishankar et al., 2015). There are reports regarding the production of Menaquinone 7 by *Bacillus subtilis* with soy as substrate. Optimization of fermentation parameters could increase the yield of Menaquinone 7 (Sato et al., 2001). In India, Syngergia Life Scineces Pvt. Ltd., commercially manufacturing Menaquinone 7 by using *Bacillus licheniformis*. Conagen, United States of America commercialized Salidroside (with 99% purity) by precision fermentation technology. Salidroside possesses strong antioxidant properties and thus has a promising role in reducing inflammation.

8.8 Alternative protein via fermentation

Several companies marketed plant-based meat and its derivatives. Nowadays, consumers are looking for an alternative protein to compromise the nutrients of meat-derived protein and products. Livestock farming requires a large area of land, water, and energy sources. Precision fermentation is the alternative for these addressed problems. The microorganism is programmed in a specific way to express a particular protein. These fermented products have been recently approved by customers globally (Thomas and Bryant, 2021). PerfectDay made animal-free milk protein through precision fermentation. β-lactoglobulin is an ideal alternative for dairy products. This β–lactoglobulin is used in ice cream and other dairy products. Perfect Day made the first animal-free whey protein. Clara Foods Co, United States of America, produced soluble egg protein by precision fermentation. *Pichia pastoris* is genetically modified for nonanimal soluble egg white protein glycosylation formation. The *Pichia pastoris* is able to synthesize ovomucoid, which is the major egg protein, because of its modification. Soy leg hemoglobin is used as an ingredient in plant-based meat. Impossible foods, United States of America made soy leghemoglobin by precision fermentation. The soy leghemoglobin protein was incorporated into *Pichia pastoris* for production. Soy leghemoglobin contains the iron molecule heme, which releases a volatile aroma on cooking, generating a rich-umami meat flavor (Fig. 8.7). In addition to the flavor characteristics, it also offers nutritional benefits (Food Frontier, 2022).

8.9 Cellular agriculture

Cellular agriculture defines the production of agricultural products and animal-derived products from particular cells rather than from a whole plant or animal. Cultured meat is real animal meat manufactured by cultivating animal cells directly. This production process removes the demand for farm animals for food. Cultivated meat is obtained from the same cell types and is arranged in a similar way as animal tissues, thus copying the sensory as well as nutritional profiles of conventional meat. Decades of awareness in cell culture, tissue engineering, fermentation, and chemical and bioprocess engineering preceded the field of cultivated meat. Several companies and academic laboratories are engaged in intense research to establish a new system for manufacturing meat products at industrial scales. Cell-based seafood production is the alternative to solve the industrial aquaculture challenge

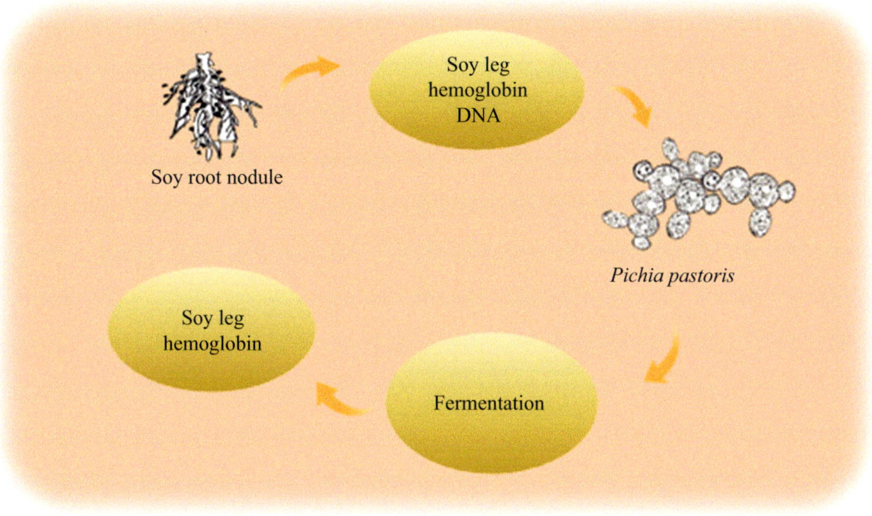

FIG. 8.7

Process of creating soy leghemoglobin via precision fermentation.

(Rubio et al., 2019). GOOD Meat United States of America made real meat by cultivating cells. The Cellular Agriculture Society, Miami, is committed to researching the production of plant and animal-derived products without plants and animals. ORF Genetics from Ice land, Laurus Bio Pvt. Ltd. from India, and Peprotech from Israel initiated research on cultivated cell line technology for plant and animal-derived products. The Perfect Day, the United States of America joined hands with the Cellular Agricultural Society for alternative protein production. Blue Nalu, United States of America, working on cellular aquaculture for trusted seafood products. Fig. 8.8 depicts various technological aspects of in vitro meat production (Table 8.1).

8.10 National and international food regulation

The main objective of this agency is to protect people from consuming unhealthy foods and create awareness about what you have. In 2006, the Food Safety and Standers Authority of India (FSSAI) was established under food safety and standards. FSSAI regulates the manufacturing, storage, and distribution of foods and food products in India. The FSSAI standards regulations 2011 (Ministry of Health and Family Welfare (FSSAI) notification, 2011) explains flavoring foods and enhancers. It describes (i) Natural flavors and natural flavoring means which one is obtained by a physical process from vegetables, (ii) Natural-identical flavoring substance means which are obtained by chemical process from aromatic raw materials, and (iii) Artificial flavoring means those substances which have not been identified in natural products. The labeling procedures for the fermentation of milk products, such as fermented/cultured/sour cream, are mentioned in this act. In this article, in 2011, fermented meat product means meat that is preserved by approved microbes. FSSAI proposes labeling for genetically modified organism products should mention as "Contain GMO/Ingredients derived from GMO." The

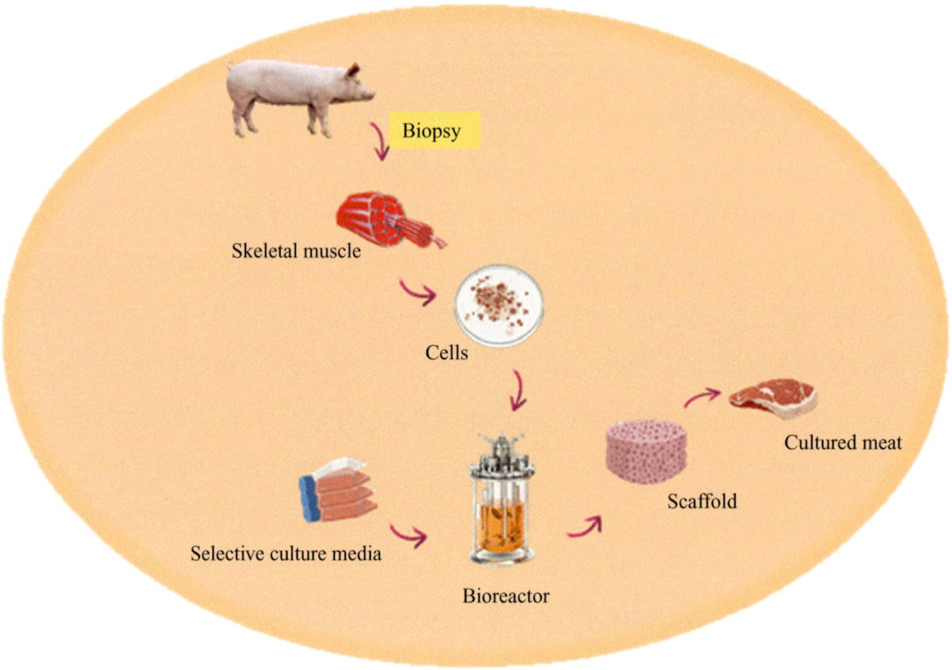

FIG. 8.8

Overview of in vitro meat production.

United States Food and Drug Administration (FDA or USFDA) is responsible for the United States of America which ensures food safety. The food additives and ingredients used in food should get premarket approval from the GRAS committee. Article 21CFR101.22 (CFR, 2012), Title 21, volume 2, says that fermentation flavors can label as Natural flavor or Natural flavoring. The list of microbial-derived additives and ingredients used in food is mentioned in Article 21CFR172 (Food Preservatives, 2023) and 173 (Govinfo, 2023). European Food Safety Authority (EFSA) evaluates the food safety additive and ingredients used in food. The list of fermentation products approved in the Europe region is mentioned in the (EU) No 231/2012. In (EC) No 1334/2008 (EC, 2008) explains that microbiologically derived flavors can be labeled as natural flavoring substances.

Table 8.1 List of some industrially important compounds obtained through precision fermentation.

Categories	Compounds	Main microorganism	Company
Aromatic compounds	Vanillin	*S. cerevisiae*	Evolva
		S. cerevisiae	IFF
		Aspergillus niger	Kraft General Foods
		Streptomyces setonii	Givaudan
		Amycolatopsis	Solvay

Table 8.1 List of some industrially important compounds obtained through precision fermentation—cont'd

Categories	Compounds	Main microorganism	Company
	Resveratrol	*S. cervisiae*	Evolva
		Yarrowia lipolytica	Du pont
	Aromatic amino acids	*E. coli*	Ajinomoto
		E. coli	Du Pont
	Para-Hydroxybenoic acid	*E. coli*	Generic electrics
	Terephtalic acid	*E. coli*	Amyris
		E. coli	Genomatica
	p-Coumaric acid	*E. coli*	Dupont
	2-Phenyllactic acid	*Pseudomonas gladioli*	IFF
	3-Phenyllactic acid	*Brevibacterium lactofermentum*	Ajinomoto
	Phenylacetaldehyde	*Lactococcus lactis*	Nestec
	Trans-cinnamic acid	*E. coli*	Dupont
	3,4-Dimethoxycinnamic acid	*E. coli*	Symrise
	Gallic acid	*Aspergillus niger*	ZunyiBeiyuan Chemicals
	Protocatechuic acid	*A. lactofermentum*	Ajinomoto
	Nootkatone	*Rhodobacter sphaeroides*	Iso Bionics
	Sandalwood oil	*S. cervisiae*	Iso Bionics
	Gamma Decalactone	*Yarrowia lipolytica*	Blue California Flavors and Fragrances
Sweeteners	Steviol Reb M	*Yarrowia lipolytica*	DSM Food Specialties
	Brazzein	*Pichia pastoris*	Tate & Lyle Sugars
Alternative protein	Mycoprotein	*Fusarium*	3F Bio
	Whole cut meats	*Fusarium*	Atlastfood co
	Egg proteins	*Pichia pastoris*	Clara foods
	Edible protein	*Fusariumn ovum*	Nature's fynd
	Milk protein	*Trichoderma reesei*	Perfect day
	Beta lactoglobulin	*E. coli*	Glycom
	Soy leghemoglobin	*Pichia pastoris*	Impossible foods

8.11 Conclusion

The precision fermentation era in the fermentation industry can switch traditional food fermentations by facilitating the consistent production of valuable foodstuffs. Synthetic biology and precision fermentation will reduce the dependence on conventional agriculture, animal husbandry, and demand for input, power, land, and water; it helps to provide sustainable food and thus validate food security. Because of enhanced global health awareness, consumers are conscious about their diet and health and are interested in natural, clean-label, and sustainably and ethically sourced food products, which are considered the dynamic force for the fermentation industry. The recent trend in precision fermentation will create a remarkable bloom in the food, nutraceutical, and pharmaceutical industries.

References

Bachmann, H., Pronk, J.T., Kleerebezem, M., Teusink, B., 2015. Evolutionary engineering to enhance starter culture performance in food fermentations. Curr. Opin. Biotechnol. 32, 1–7.

Bel-Rhlid, R., Berger, R.G., Blank, I., 2018. Bio-mediated generation of food flavors—towards sustainable flavor production inspired by nature. Trends Food Sci. Technol. 78, 134–143.

Braga, A., Guerreiro, C., Belo, I., 2018. Generation of flavors and fragrances through biotransformation and De novo synthesis. Food Bioprocess Technol. 11 (12), 2217–2228. https://doi.org/10.1007/s11947-018-2180-8.

CFR, 101.22, 2012. Foods: Labeling of Spices, Flavorings, Colorings. and Chemical Preservatives 2, 72–77. https://www.gpo.gov/fdsys/granule/CFR-2012-title21-vol2/CFR-2012-title21-vol2-sec101-22.

Chai, K.F., Ng, K.R., Samarasiri, M., Chen, W.N., 2022. Precision fermentation to advance fungal food fermentations. Curr. Opin. Food Sci. 47, 100881.

Converti, A., Aliakbarian, B., Domínguez, J.M., Vázquez, G.B., Perego, P., 2010. Microbial production of biovanillin. Braz. J. Microbiol. 41 (3), 519–530.

Corchero, J.L., Gasser, B., Resina, D., Smith, W., Parrilli, E., Vázquez, F., Abasolo, I., Giuliani, M., Jäntti, J., Ferrer, P., Saloheimo, M., Mattanovich, D., Schwartz, S., Tutino, M.L., Villaverde, A., 2013. Unconventional microbial systems for the cost-efficient production of high-quality protein therapeutics. Biotechnol. Adv. 31 (2), 140–153.

Das, R., Pandey, H., Das, B., Sarkar, S., 2016. Fermentation and its application in vegetable preservation: a review. Int. J. Food Ferment. Technol. 6 (2), 207–217.

Davy, A.M., Kildegaard, H.F., Andersen, M.R., 2017. Cell factory engineering. Cell Syst. 4 (3), 262–275.

Dias, F.M.S., Gomez, J.G.C., Silva, L.F., 2017. Exploring the microbial production of aromatic fine chemicals to overcome the barriers of traditional methods. Adv. Appl. Sci. Res. 8 (1), 94–109.

Džoljić, J., 2020. Stevia Plant—Alternative Sweetner in Food Technology. October.

EC, 2008. B Regulation (EC) No 1334/2008 of the European Parliament And Of The Council, MachineryDirective: 1.

Food Frontier, 2022. What is Soy Leghemoglobin? Issue www.foodfrontier.org, pp. 1–4.

Food Preservatives, 2023. Part 172—FOOD ADDITIVES PER- MITTED FOR DIRECT ADDITION TO FOOD FOR HUMAN CONSUMP- Subpart A—General Provisions Subpart B—Food Preservatives.

Gallage, N.J., Møller, B.L., 2015. Vanillin-bioconversion and bioengineering of the most popular plant flavor and its de novo biosynthesis in the vanilla orchid. Mol. Plant 8 (1), 40–57.

Govinfo, 2023. Part 173, Secondary Direct Food Additives Permitted In Food For Human Consump- Subpart A—Polymer Substances and Polymer Adjuvants for. https://www.govinfo.gov/content/pkg/CFR-2014-title21-vol3/pdf/CFR-2014-title21-vol3-part173.pdf.

Hadj Saadoun, J., Bertani, G., Levante, A., Vezzosi, F., Ricci, A., Bernini, V., Lazzi, C., 2021. Fermentation of agri-food waste: a promising route for the production of aroma compounds. Foods 10 (4).

Janssens, L., De Pooter, H.L., Schamp, N.M., Vandamme, E.J., 1992. Production of flavours by microorganisms. Process Biochem. 27 (4), 195–215.

Libik-konieczny, M., 2021. Synthesis and production of steviol glycosides: recent research trends and perspectives. Appl. Microbiol. Biotechnol., 3883–3900.

Liu, X., Ding, W., Jiang, H., 2017. Engineering microbial cell factories for the production of plant natural products: from design principles to industrial-scale production. Microb. Cell Factories 16 (1), 1–9.

Martí-Quijal, F.J., Khubber, S., Remize, F., Tomasevic, I., Roselló-Soto, E., Barba, F.J., 2021. Obtaining antioxidants and natural preservatives from food by-products through fermentation: a review. Fermentation 7 (3), 1–11.

Melini, F., Melini, V., Luziatelli, F., Ficca, A.G., Ruzzi, M., 2019. Health-promoting components in fermented foods: an up-to-date systematic review. Nutrients 11 (5), 1–24.

Ministry of Health and Family Welfare (Food Safety and Standards Authority of India) Notification, 2011. The Gazette of India: Extraordinary. pp. 65–114.

More, T.A., Shaikh, Z., Ali, A., 2021. Artificial sweeteners and their health implications: a review. Biosci. Biotechnol. Res. Asia 18 (June), 227–237.

Orlova, S.V., Tatarinov, V.V., Nikitina, E.A., Sheremeta, A.V., Ivlev, V.A., Vasil'ev, V.G., Paliy, K.V., Goryainov, S.V., 2022. Bioavailability and safety of dihydroquercetin (review). Pharm. Chem. J. 55 (11), 1133–1137.

Otto, M., Liu, D., Siewers, V., 2022. Saccharomyces cerevisiae as a heterologous host for natural products. In: Clifton, N.J. (Ed.), Methods in Molecular Biology. Vol. 2489. Springer, pp. 333–367. https://doi.org/10.1007/978-1-0716-2273-5_18.

Park, S.Y., Yang, D., Ha, S.H., Lee, S.Y., 2020. Biosynthesis of dihydroquercetin in Escherichia coli from glycerol. BioRxiv, 1–22. 2020.11.27.401000.

Philippe, R.N., De Mey, M., Anderson, J., Ajikumar, P.K., 2014. Biotechnological production of natural zero-calorie sweeteners. Curr. Opin. Biotechnol. 26, 155–161.

Prabhu, A.A., Thomas, D.J., Ledesma-Amaro, R., Leeke, G.A., Medina, A., Verheecke-Vaessen, C., Coulon, F., Agrawal, D., Kumar, V., 2020. Biovalorisation of crude glycerol and xylose into xylitol by oleaginous yeast Yarrowia lipolytica. Microb. Cell Factories 19 (1), 1–18.

Rafiqul, I.S.M., Sakinah, A.M.M., 2013. Processes for the production of xylitol-a review. Food Rev. Int. 29 (2), 127–156.

Ravishankar, B., Dound, Y.A., Mehta, D.S., Ashok, B.K., De Souza, A., Pan, M.H., Ho, C.T., Badmaev, V., Vaidya, A.D.B., 2015. Safety assessment of menaquinone-7 for use in human nutrition. J. Food Drug Anal. 23 (1), 99–108.

Ray, R.C., Joshi, V., 2014. Fermented foods: past, present and future. In: Ray, R.C., Montet, D. (Eds.), Microorganisms and Fermentation of Traditional Foods. CRC Press, Taylor & Francis Group, pp. 1–36. Issue August 21.

Rubio, N., Datar, I., Stachura, D., Kaplan, D., Krueger, K., Ahmed, S., 2019. Cell-based fish: a novel approach to seafood production and an opportunity for cellular agriculture. Front. Sustain. Food Syst. 3 (June), 1–13.

Sato, T., Yamada, Y., Ohtani, Y., Mitsui, N., Murasawa, H., Araki, S., 2001. Production of menaquinone (vitamin K2)-7 by Bacillus subtilis. J. Biosci. Bioeng. 91 (1), 16–20.

Scharpf, L.G., Seitz, E.W., Morris, J.A., Farbood, M.I., 1986. Generation of Flavor and Odor Compounds through Fermentation Processes. ACS Publications, pp. 323–346.

Spakman, D., 2015. How to broaden the applicability of the high potency sweeteners steviol glycosides: from enzymatic glycosylation to recombinant production. Doctoral dissertation, Faculty of Science and Engineering.

Sunil, C., Xu, B., 2019. An insight into the health-promoting effects of taxifolin (dihydroquercetin). Phytochemistry 166, 112066.

Teng, T.S., Chin, Y.L., Chai, K.F., Chen, W.N., 2021. Fermentation for future food systems. EMBO Rep. 22 (5), 1–6.

Thomas, O.Z., Bryant, C., 2021. Don't Have a Cow, man: consumer acceptance of animal-free dairy products in five countries. Front. Sustain. Food Syst. 5 (June), 1–14.

Velioglu, I.Y., Murat, H., 2009. Fermented meat products. In: Yılmaz, I. (Ed.), Quality of Meat and Meat Products. Transworld Research Network, Indie, pp. 1–16.

Verni, M., Verardo, V., Rizzello, C.G. (2019). How fermentation affects the antioxidant properties of cereals and legumes. Foods (vol. 8, Issue 9). MDPI Multidisciplinary Digital Publishing Institute.

Xiao, J., Wu, J., 2015. (12) Patent Application Publication (10) Pub. No.: US 2017/0215756A1. Wo2015145143, Us20170107208a1, pp. 1–25.

CHAPTER

9

Exploitation of mycometabolites in weed management: Global scenario and future application

Ajay Kumar Singh and Akhilesh Kumar Pandey

Mycological Research Laboratory, Department of Biological Science, Rani Durgawati University, Jabalpur, Madhya Pradesh, India

9.1 Introduction

The management of weed in current agriculture practices has been depended on chemical herbicides, but almost no new herbicides mode of actions has been explored. Another drawback of using chemical herbicide is developing resistance varieties of weeds. Resistant weeds are developed through the intensive uses of the same synthetic herbicides for long times. Hence, these synthetic herbicides need to be replaced with environmentally friendly biological herbicides (bioherbicides). Various fungi with herbicidal potential have been discovered, but very few have become commercial realities or viable alternatives due to biological, technological, and commercial constraints. Herbicidal metabolites produced by fungi play an important role in host-pathogen interactions and better alternate for mycoherbicide constraint.

Mycoherbicides developed from fungal phytotoxins have been used for controlling weeds. It has been considered that mycoherbicides can be better supplementary tools in weed control. In pursuit of a better environmental sustainability, industries are now moving away from the conventional and chemically manufactured compounds and are focusing on the development of biobased naturally harvestable compounds to reduced carbon footprints, ecological, and environmental impacts, as well as to reduce manufacturing costs. It can be achieved by harnessing the versatile potential of fungi. Fungi are known as important microorganisms for industrialization. Many fungal secondary metabolites that include pigments, antibiotics, vitamins, amino acids and organic acids have been discovered, leading to a productive growth of bio-commercialization. These metabolites consist of a wide array of chemical structures. They can be important factors of pathogenicity or virulence, can have different behaviors with respect to the host varying from strictly host-specific to completely nonspecific compounds, and can act with different mechanisms affecting several sites in the host.

The exploitation of mycometabolites in agricultural weed management is safer to the user and the environment. They have formulated and applied in the same manner as chemical herbicides. Plant pathogenic fungi that are basically classified as necrotrophs, hemibiotrophs, and biotrophs constitute

Entrepreneurship with Microorganisms. https://doi.org/10.1016/B978-0-443-19049-0.00008-6

one of the main infectious agents in weeds, causing alterations during developmental and later stages, taking nutrients from the plants they invade and therefore resulting in huge damages to plants. Fungi infect weeds using different strategies: biotrophs exploit plants' resources, while keeping the host alive; necrotrophs kill the host in order to thrive off dead or dying tissue; and hemibiotrophs have an asymptomatic phase followed by a necrotrophic stage (Horbach et al., 2011).

This paper reviews on the application of biochemical (natural or biorational) herbicides based on specific fungal weed pathogens and it believes to assist the decreasing harmful impact of the chemicals.

9.2 Mycometabolites-entrepreneurs approach

Nonetheless, only after the turn of the 19th century (1900s) had the scientific community begun to realize the potential of fungi to produce beneficial primary and secondary metabolites and compounds, ranging from alcohol, organic acids, antibiotics, vitamins, pigments, immunosuppressant and immunomodulatory agents, and economically important proteins and enzymes (Royse, 2003), thus having the industrial importance that could revolutionize biotechnology. The initiative to harvest biological compounds made by fungi has taken hold in the minds of both biotechnologists and entrepreneurs for decades now, since Alexander Fleming's seminal discovery of penicillin (Fleming, 1929), an antimicrobial compound produced as a secondary metabolite (extrolite) by the fungus *Penicillium chrysogenum* (previously *Penicillium notatum*). *Tolipocladium inflatum* producing cyclosporine used as immunosuppressive agents in modern medicine (Stucker and Ackermann, 2011). In effect, mycometabolites has established its very own niche in the ever-growing field of industrial biotechnology.

9.3 Mycometabolites as natural herbicides

The rationale behind this approach is the use of toxic secondary metabolites of microbes or phytotoxins, which are the safest pesticides for the environment and people. Phytotoxins are usually isolated from in vitro cultures of the pathogen grown on either solid or liquid media (Strange, 2007). Phytotoxins are largely represented by low molecular weight secondary metabolites capable of deranging the vital activity of plant cells or causing their death at low concentration. The effect of phytotoxins on plants is characterized by the appearance of specific symptoms, wilting, and general growth suppression, and chlorosis, necrosis, and spotting of aerial portions are the most common. The weed pathogenic fungi grown in suitable media for a particular incubation time to extract herbicidal compound (Banowetz et al., 2008). It has become increasingly evident that phytotoxins are important disease determinants. Various approaches have been developed to improve the amount and quality of phytotoxins, which is synthesized by fungal bio-control agents (Dayan et al., 2000). Most of the information available on the phytotoxicity of fungal products is not useful in evaluating their potential as herbicides. In the few cases in which a molecular target site has been established, it has generally been one that has not yet been exploited by the herbicide industry (Duke et al., 1996). Some fungal phytotoxins also vary in host specificity, ranging from host specificity to having no specificity whatever (Strobel et al., 1992). The phytotoxin tentoxin (a cyclic tetrapeptide) that is produced by several *Alternaria* species and causes severe chlorosis in many of the problem species associated with soybeans and maize without affecting either crop (Duke and Lydon, 1987). Non-host-specific toxins are of considerably more interest because they often have the potential for killing a range of weeds without phytotoxicity to crops (Duke et al., 1991). It is certain that biodegradable, microbially derived herbicides will be on the market within the next decade (Duke et al., 2000).

Both pathogenic and nonpathogenic fungi are known to synthesize array of phytotoxic metabolites. Some of them have been evaluated and patented, and few of them have been commercialized as herbicides. We have significantly tested herbicidal activity of partially purified filtrate (CFCF) of *C. gloeosporioides* f. sp. *parthenii* FGCC #18, *C. dematium* FGCC#20, *F. oxysporum* FGCC#39, *F. solani* FGCC#86, *S. rolfsii* FGCC#19, *Aspergillus flavus* FGCC#14, and *Curvularia lunata* FGCC#41 have been observed against *Parthenium hysterophorus* (Thapar et al., 2002; Pandey et al., 2003, 2004) and *Lantana camara* (Saxena and Pandey, 2000; Pandey et al., 2005; Singh, 2007; Singh and Pandey, 2019). Saxena and Pandey (2001) reported extremely high biological activity in CFCF of *Alternaria alternata* FGCC#25 against *L. camara*. Pandey et al. (2007) reported extremely high biological activity in CFCF of *Helminthosporium* sp. FGCC#74 against *Hyptis suaveolens*. Singh (2007) reported herbicidal compounds from some selected fungi against different noxious weeds of Madhya Pradesh. *Alternaria eichhorniae* metabolites Perlenginones and Alteichin have shown good potential agent Water hyacinth weed.

Fungal secondary metabolites divided into four main chemical classes, viz. polyketides, terpenoids, shikimic acid-derived compounds and nonribosomal peptides. Whole-genome analysis of fungi revealed that ascomycetes have more genes of secondary metabolism than archeo ascomycetes, basidiomycetes, chytridiomycetes, and hemiascomycetes and zygomycetes have nothing (Collemare et al., 2008). Ascomycete genomes code for on average 16 polyketide synthases (PKS), 10 nonribosomal protein synthases (NRPS), two tryptophan synthetases (TS), and two dimethylallyl tryptophan synthetases (DMATS) with crucial importance in SM synthesis. These types of SM genes encode signature enzymes that can be enriched in secondary metabolism gene clusters and responsible for main synthesis steps of metabolites. PKS-NRPSs have been identified only in ascomycetes, with an average of three genes per species. Whole-genomic analysis has identified 12–15 PKS genes in *F. graminearum* (Sieber et al., 2014), where six have been linked to metabolites.

9.4 Culturing conditions for production of mycometabolites

It is possible to isolate phytotoxins from infected plant tissues and germinating conidia of fungi, but this approach in not productive because of the low content of the target compounds. Therefore, in order to isolate phytotoxin in amounts sufficient for studies of chemical and biological properties, the fungi are cultured in liquid nutrient media (the average yield ranges from 1 to 50 mg per 1 L culture liquid). In several cases, it is possible to isolate phytotoxin in settings that involve solid-phase fermentation on natural substrates. Phytotoxin formation is sensitive to a few diverse factors (e.g., the composition of the medium, its acidity, and the duration and conditions of culturing), most of which are not identified in advance as being able to affect the process. Distinct strains of the same species may vary considerably in their capacity for phytotoxin production, including in the responses to one and the same factor (out of those just mentioned above). Microorganism strains are genetically unstable (this is most obvious in the case of imperfect fungi), and their storage or reinoculation may adversely affect the ability to produce toxins (Kale and Bennett, 1992).

In the course of fungi culturing for production of herbicidal compound, it is affected by temperature, pH, aeration, the degree of illumination, media, and other parameters. The formation of certain herbicide (such as cercosporin) requires light (Daub and Ehrenshaft, 2000). Toxin formation by fungi may be stimulated under the conditions of cold stress. Thus, a strain of Fusarium sporotrichiella actively forms T-2 toxin, if the fungus is cultured on millet for 1 week, first at 18°C and then at 5°C (Greenhalgh et al., 1988). According to Shchekochikhina (Shchekochikhina, 1975), culturing of *Bipolaris sorokiniana* in liquid Czapek medium demonstrates that phytotoxins accumulation in the culture liquid occurs more rapidly at moderate (7°C–17°C) than normal (20°C–27°C) temperature.

9.5 Composition of nutrient media

Because phytotoxin is not used in industry, it is problematic to locate reports describing optimization of media for the purpose of increasing the toxin yield. Usually, insignificant amounts (of less than 100 mg) of phytotoxin are required for preliminary assessment of their biological activity. If the production level is insufficient, the yield of phytotoxin may be increased by selecting appropriate carbon and nitrogen sources and finding the right combination (in terms of the concentrations and the molar ratio). Certain trace elements may affect considerably toxin production by fungi. For this reason, it is recommended to use artificial media (rich in vitamins and trace elements), during early stages of studies, to ensure phytotoxin isolation (Stierle et al., 1992). It was demonstrated that optimization of the composition of the Czapek medium (by changing the contents of glucose, sodium nitrate, and potassium phosphate) used for culturing patulin producers made it possible to increase the yield of the toxin two- to threefold (Berestetskiy et al., 1976). The contents of glucose and peptone, optimum for ascochytin biosynthesis by *Ascochyta fabae* (grown in liquid nutrient medium), amounted, respectively, to 2%–5% and 0.25% (Oku and Nakanishi, 1963). Strains of *Fusarium oxysporum* form fusaric acid only in the presence of Zn ions (0.08–0.40 mg/L) (Gäumann, 1957).

9.6 Medium acidity

During culturing, the concentration of hydrogen ions in the medium may affect the activity of fungal enzymes and the stability of secondary metabolites. Although the formation of aflatoxin by the fungus *Aspergillus flavus* does not show a clear-cut dependence on the initial pH of the medium, maximum accumulation of the toxin was observed at pH 4.0 or lower (Calvo et al., 2002).

The formation of aspergillomarasmine A by *Pyrenophora teres* grown in modified de Vries medium was augmented considerably upon its neutralization; pH was increased by (1) addition of phosphate buffer to the liquid medium or (2) sterile titration of the medium by alkali in the course of culturing (Weierganag et al., 2002).

9.7 Incubation time for phytotoxin production

Some of the fungal phytotoxins are detected in the medium early on, in a matter of days after the germination of the conidia (Kohmoto, 1992). Maximum accumulation of the phytotoxin of *Bipolaris euphorbiae* in the culture liquid was observed after 6 or 7 days, coinciding with the peak of biomass gain, minimum pH (4.2), and near-complete consumption of the carbon source (Barbosa et al., 2002). In static cultures of *Fusarium oxysporum* f. sp. *lycopersici*, lycomarasmin is detected on day 7 (10 mg/L), and its concentration increases to 200–300 mg/L by day 44 (Gäumann, 1957). Some reports note that maximum phytotoxin accumulation in cultures is more frequently observed prior to the onset of spore formation (Berestetskii and Parkhomenko, 2004).

9.8 Bioassay of phytotoxins

Bioassays are used for identifying phytotoxin in culture liquid and assessing phytotoxic activity (of extracts or pure substances). In selecting bioassays, the biology and ecology of the fungus are considered for further selection. If a phytotoxin originates in soil fungi or causative agents of root rot diseases, the bioassay involves plant seedlings and the extent of growth suppression of roots treated with serial dilutions of the culture filtrate or the pure toxin is calculated using untreated roots as controls. If the symptoms of the disease caused by a phytotoxin in leaves, a solution of the substance tested or the culture filtrate approx. 5–20 μL is applied onto the leaf punctured by a needle (as a rule, symptoms of toxicosis are not as pronounced in intact leaves); the results (chloroses, necrotic spots) are read after 24–72 h. In order to reduce the requisite time for obtaining the results, isolated leaves (or parts thereof) may be placed into a moist chamber or onto the surface of water agar (Stierle et al., 1992).

9.9 Economics for development herbicide

Fungal metabolites-based herbicides have been regarded to be cost-effective. Zorner et al. (1993) have estimated that the development cost regarding mycoherbicide was about 1.5–2 million, which is much lower than that required for development of effective chemical herbicide formulation (about US $ 10–20 million). Bowers (1982) estimated that the cost of application of Collego was about $ 29/ha.

9.10 Limitations in commercializing mycometabolites

Fungal metabolites, both primary and secondary, are a subset of a larger repertoire of natural compounds that act as intermediary in the interplay between fungi and the ecology, hence opening up a potential industrial avenue to explore. Indeed, mycological biotechnology, as it stands now, is an important and ever-growing economic driver as industries begin to realize the potential and the cost-effectiveness of bio-production, with around 40% of the total drugs approved for commercial market being of natural products or biologically modified natural metabolites. Nevertheless, the industry is not without its technical limitations. Several significant technical hindrances to successful cultivation and large-scale bio-production of useful microbial (including fungal) natural products include relatively high capital and subsequent cultivation cost, stringent need for sterile conditions, and expensive media components and carbon source. Barig and colleagues (Barig et al., 2011) reported in 2011 that the main issue riddling most microbial fermentation plants is the stringent need for sterile conditions, as up-scaled, industrial fermenters are typically exposed to a working environment that could be extremely difficult to keep sterile, unlike the more manageable sterile conditions in the laboratory. Because of the nature of microbial cultures, especially when nonfastidious growth media are involved (for instance, glucose used as carbon source), undesirable growth of contaminating colonies from other microbial organism within the fermentation tanks may jeopardize the output performance of the cultivated industrial strain of interest, and may reduce the quality of the metabolites that would be harvested and purified. Furthermore, high cultivation cost is directly proportional to the complexity of the media components needed for the cultivation process and is inversely proportional to the size of the market for the metabolic product being harvested—higher cost effectiveness could be associated with small, niche market for metabolic products of less economic demands, and vice versa. In addition to this, when a medium requires a large-scale preparation, certain key ingredients such as glucose (carbon source) would become costly, leading to a reduced industrial cost-effectiveness (Barig et al., 2011; Jamal et al., 2008). It is reported that growth media components carry a strong impact on global industrial bioprocesses and can account for 30% of the total production cost (Mattanovich et al., 2014).

9.11 Potential improvements

There are several improvements that researchers around the globe are currently attempting to overcome the current industrial limitations. Undergoing studies are being done especially based on the idea of utilizing cheaper raw biomaterials as growth media or as an alternative carbon source, such as using sugar cane bagasse and molasses as substrates in solid-state fermenters (Veana et al., 2014) or using plant oil to replace glucose as carbon source (Darvishi et al., 2009). Barig and colleagues are also extensively experimenting with various physicochemical parameters, such as culture viability under low pH, and minimized up-scaled sterile conditions (Barig et al., 2011), to explore unique combinations of growth conditions that would optimize fungal growth and at the same time, limit or even hinder the unwelcome growth of contaminant organisms.

The application of fungal phytotoxic metabolite could be a replacement for synthetic herbicide, which is more economical in controlling weeds than the synthetic herbicides. We are confident that the future of fungal metabolites-based herbicide will accelerate agriculture production and serve as an alternative to the chemical herbicides because of its safety in the environment.

Thus, the study of natural substances with herbicidal activity has great prospects. The known spectrum of the fungi species producing biologically active substances (BAS) of this type is yet very narrow. Therefore, the search for new BAS producers and the study of their properties are very topical. In order to attain the goal, it is necessary to solve the following tasks:

- Search for phytopathogenic fungi of weeds from different ecological zones.
- Determination of the toxicity of the active strains of weed pathogenic fungi under laboratory conditions and the isolation of toxins responsible for biological activity.
- Physio-chemical characteristics of new fungal phytotoxic metabolites, the structure and mechanism of the action of purified substances.
- Scaling-up production of fungal isolates with sufficient herbicidal activity and extraction of secondary metabolites for further testing.
- Testing of identified fungal secondary metabolites with herbicidal and insecticidal activity on agriculturally important crops.

9.12 Future prospect

Fungal natural compounds present a great potential as natural herbicides. Today, future research is driving modern agriculture towards crop production systems that are healthier, safer and friendlier to the environment as consumers demand pesticide-free products and environmentally safe cultural practices. Considering these perspectives, the richest sources of natural compounds present in nature and their ecology, assume higher and higher importance and can increase the possibility of finding natural herbicides with new scaffolds and modes of action, fundamental factors overcoming resistance in weeds to conventional, synthetic herbicides. Being the result of co-evolution of the producing organism and its biotic environment, these compounds can have high target selectivity, with potentially reduced risks for humans and nontarget organisms. Furthermore, they can have a shorter environmental half-life than synthetic compounds, thus reducing potential environmental impact. We could see how important it is to research deeper into this metabolic treasure trove, to improve our industrial yield and bring benefits to the society as a whole, as well as reducing human's carbon footprints.

9.13 Conclusion

Mycoherbicide based on metabolites for weed control is nowadays gaining momentum. New mycoherbicides will find a place in irrigated tropical, subtropical agro-ecosystems, forestry, waste lands, as well as in managing parasite weeds or resistant weed control. Research on synergy test of pathogens and pesticides for inclusion in IPM, developmental technology, fungal toxins and application of biotechnology, especially genetic engineering, is required. In the search for alternative solutions to weed control, the interest in application of bioactive secondary metabolites has increased. The success of natural products or natural product-derived products in weed management is weak compared to that for insecticide and fungicide. The biochemical or biorational herbicide available for organic farmers and those who wish to reduce synthetic herbicide use are ineffective and costly to use. In this review, we found that original culture filtrates were applied in the foliar spray bioassays showed good results. It is likely that if these culture filtrates are used in a concentrated form, these will be more toxic to weed species.

The great structural diversity of fungal phytotoxins with high potency and unique mechanisms of action (compared to synthetic herbicides) make fungal toxins highly attractive for discovering herbicidal activity. Even if natural phytotoxins are not necessarily suitable for direct use as a commercial herbicide, the identification of mechanisms is very important for new herbicide developments. Newly developed herbicides with environmentally friendly component could be used more safely in integrated pest management systems. Further studies are required to isolate the active herbicidal constituents from these fungal culture filtrates. The problem of the development of ecologically safe pesticides with new mechanisms of action is very urgent. The main reasons for that are as follows: first, the appearance of weed forms resistant to permitted pesticides, second, the strict requirements to the weedicides applied in terms of their safety for people and the environment. That is why the attention of researchers has been increasingly attracted by a compromise option: the isolation and characterization of pathogenicity factors of biocontrol agents, mainly toxins, in order to create new pesticides on their chemical basis. For a successful research and development process leading to a commercial product, a wide range of criteria (biological, environmental, toxicological, regulatory, and commercial) must be satisfied from the beginning. Among the major challenges to be faced by the candidate products to reach the market are the sustainable use of raw materials, the standardization of chemically complex extracts, and the regulatory requirements and approval. The unique set of secondary metabolites produced by fungi may play an important role in weed management as new products directly, as novel chemical frameworks for synthesis and/or for identifying original modes of action. The tremendous promising herbicidal potential of many of these natural products reported for noxious weeds in this current review will prompt a continued interest in developing fungal metabolites as natural safe herbicides.

Acknowledgments

We are thankful to the Head, Dept of Biological Sciences. Financial assistance received from Council of Scientific and Industrial Research (CSIR) is also thankfully acknowledged.

References

Banowetz, G.M., Azevedo, M.D., Armstrong, D.J., Halgren, A.B., Mills, D.I., 2008. Germination-arrest factor (GAF): biological properties of a novel, naturally occurring herbicide produced by selected isolates of rhizosphere bacteria. Biol. Control 46, 380–390.

Barbosa, A.M., Souza, C.G.M., Dekker, R.F.H., Fonseca, R.C., Ferreira, D.T., 2002. Phytotoxin produced by *Bipolaris euphorbiae in vitro* is effective against the weed *Euphorbia heterophylla*. Braz. Arch. Biol. Technol. 45 (2), 233–240.

Barig, S., Alisch, R., Nieland, S., Wuttke, A., Gräser, Y., Huddar, M., Stahmann, K.P., 2011. Monoseptic growth of fungal lipase producers under minimized sterile conditions: cultivation of *Phialemonium curvatum* in 350 L scale. Eng. Life Sci. 11 (4), 387–394.

Berestetskii, A.O., Parkhomenko, N.V., 2004. A review of fungal phytotoxins: from basic studies to practical use. Mikol. Fitopatol. 38 (2), 78–88.

Berestetskiy, O.A., Patyka, V.F., Nadkernichnyi, S.P., 1976. Voprosy ekologii i fiziologii mikroorganizmov, ispol'zuemykh v sel'skom khozyaistve (Problems of Ecology and Physiology of Microorganisms Used in Agriculture). VNII Sel'skokhoz. Mikrobiol, Leningrad, pp. 56–60.

Bowers, R.C., 1982. Commercialization of microbial biological control agents. In: Charudattan, R., Walker, H.L. (Eds.), Biological Control of Weeds with Plant Pathogens. Wiley, New York, pp. 157–173.

Calvo, A.M., Wilson, R.A., Bok, J.W., Keller, N.P., 2002. Relationship between secondary metabolism and fungal development. Microbiol. Mol. Biol. Rev. 66 (3), 447–459.

Collemare, J., Billard, A., Böhnert, H.U., Lebrun, M.-H., 2008. Biosynthesis of secondary metabolites in the rice blast fungus *Magnaporthe grisea*: the role of hybrid PKS-NRPS in pathogenicity. Mycol. Res. 112, 207–215.

Darvishi, F., Nahvi, I., Zarkesh-Esfahani, H., Momenbeik, F., 2009. Effect of plant oils upon lipase and citric acid production in Yarrowia lipolytica yeast. Biomed. Res. Int. 2009, 562943. https://doi.org/10.1155/2009/562943.

Daub, M.E., Ehrenshaft, M., 2000. The photoactivated Cercospora toxin cercosporin: contributions to plant disease and fundamental biology. Annu. Rev. Phytopathol. 38, 461–490.

Dayan, F.E., Rimando, A.M., Duke, S.O., 2000. Investigating the mode of action of natural phytotoxins. J. Chem. Ecol. 26 (9), 2079–2094.

Duke, S.O., Lydon, J., 1987. Herbicides from natural compounds. Weed Technol. 1, 122–128.

Duke, S.O., Abbas, H.K., Boyette, C.D., 1991. Microbial compounds with the potential for herbicidal use. In: 20. Brighton Crop Protection Conference, Weeds, pp. 155–164.

Duke, S.O., Abbas, H.K., Amagasa, T., Tanaka, T., 1996. Phytotoxins of microbial origin with potential for use as herbicide. In: Copping, L.G. (Ed.), Crop Protection Agents from Nature: Natural Products and Analogues. The Royal Society of Chemistry, Cambridge, UK, pp. 82–113.

Duke, S.O., Dayan, F.E., Romagni, J.G., Rimando, A.M., 2000. Natural products as sources of herbicides: current status and future trends. Weed Res. 40, 99–111.

Fleming, A., 1929. On the antibacterial action of cultures of a Penicillium, with special reference to their use in the isolation of B. influenzae. Br. J. Exp. Pathol. 10 (3), 226.

Gäumann, E., 1957. Fusaric acid as a wilt toxin. Phytopathology 47 (3), 342–357.

Greenhalgh, R., Blackwell, B.A., Savard, M., Miller, J.D., Taylor, A., 1988. Secondary metabolites produced by *Fusarium sporotrichioides* DAOM 165006 in liquid culture. J. Agric. Food Chem. 36, 216–219.

Horbach, R., Navarro-Quesada, A.R., Knogge, W., Deising, H.B., 2011. When and how to kill a plant cell: infection strategies of plant pathogenic fungi. J. Plant Physiol. 168, 51–62.

Jamal, P., Alam, M.Z., Salleh, N.U., 2008. Media optimization for bioproteins production from cheaper carbon source. J. Eng. Sci. Technol. 3 (2), 124–130.

Kale, S., Bennett, J.W., 1992. Mycotoxins in ecological systems. In: Bhatnagar, D., Lillehoj, E.B., Arora, D.K. (Eds.), Handbook in Applied Mycology. vol. 5. Marcel Dekker, New York, pp. 311–331.

Kohmoto, K., 1992. Determination of host-selective toxins. In: Linkens, H.F., Jackson, J.F. (Eds.), Modern Methods of Plant Analysis. Plant Toxin Analysis. vol. 13. Springer, Berlin, pp. 51–73.

Mattanovich, D., Sauer, M., Gasser, B., 2014. Yeast biotechnology: teaching the old dog new tricks. Microb. Cell Factories 13 (1), 1.

Oku, H., Nakanishi, T., 1963. A toxic metabolite from *Ascochyta fabae* having antibiotic activity. Phytopathology 53 (1963), 1321–1325.

Pandey, A.K., Shrivastava, G.M., Singh, A.K., Verma, S.K., 2003. Herbicidal potential of secondary metabolites of *Sclerotium rolfsii* against *Parthenium*: a preliminary observation. J. Basic Appl. Mycol. 2 (1), 27–30.

Pandey, A.K., Singh, A.K., Quereshi, S., Agrawal, D., 2004. Herbicidal activities of secondary metabolites of *Strepomyces* sp. against *Hyptis suaveolens*. J. Basic Appl. Mycol. 3, 95–97.

Pandey, A.K., Singh, A.K., Quereshi, S., Pandey, C., 2005. Herbicidal activities of secondary metabolites of *Aspergillus* sp. against *Lantana camara*. J. Basic Appl. Mycol. 4 (I and II), 65 67.

Pandey, A.K., Sharma, R., Quereshi, S., Singh, A.K., 2007. Herbicidal potential of cell free culture filtrate of *Helminthosporium* sp FGCC#74 against *Hyptis suaveolens*. J. Basic Appl. Mycol. 6 (I and II), 89–93.

Royse, D.J., 2003. Fungi. In: Encyclopedia of Food and Culture. Retrieved April 19, 2016 from Encyclopedia.com. http://www.encyclopedia.com/doc/1G2-3403400278.htm.

Saxena, S., Pandey, A.K., 2000. Preliminary evaluation of fungal metabolites as natural herbicides for the management of *Lantana camara*. Indian Phytopathol. 53 (4), 490–491.

Saxena, S., Pandey, A.K., 2001. Microbial metabolites as eco-friendly agrochemicals for the next millennium. Appl. Microbiol. Biotechnol. 55 (4), 395–403.

Shchekochikhina, R.I., 1975. Study of phytotoxic properties of some pathogenic fungi. Mikol. Fitopatol. 9 (6), 518–523.

Sieber, C.M.K., Lee, W., Wong, P., Münsterkötter, M., Mewes, H.W., Schmeitzl, C., et al., 2014. The *Fusarium graminearum* genome reveals more secondary metabolite gene clusters and hints of horizontal gene transfer. PLoS ONE 9 (10), e110311.

Singh, A.K., 2007. Isolation and Characterization of Herbicidal Compounds from some Selected fungi. Ph.D Thesis. Bioscience, R.D. University, Jabalpur.

Singh, A.K., Pandey, A.K., 2019. Fungal metabolites as a natural source of herbicide: a novel approach of weed management. J. Appl. Nat. Sci. 11, 158–163.

Stierle, A., Strobel, G., Stierle, D., Sugawara, F., 1992. Analytical methods for phytotoxins. In: Linskens, H.F., Jackson, J.F. (Eds.), Modern Methods of Plant Analysis New Series, Vol. 13, Plant Toxin Analysis. Springer-Verlag Berlin Heidelberg, pp. 1–32.

Strange, R.N., 2007. Phytotoxins produced by microbial plant pathogens. Nat. Prod. Rep. 24, 127–144.
Strobel, G.A., Sugawara, F., Hershenhorn, J., 1992. Pathogens and their products affecting weedy plants. Phytoparasitica 20, 307–323.
Stucker, F., Ackermann, D., 2011. Immunosuppressive drugs-how they work, their side effects and interactions. Ther. Umsch., Rev. Ther. 68 (12), 679–686.
Thapar, R., Singh, A.K., Pandey, A., Pandey, A.K., 2002. Bioactivity of CFCF of *Curvularia lunata* in *Parthenium hysterophorus* L. J. Basic Appl. Mycol. 1, 126–129.
Veana, F., Martínez-Hernández, J.L., Aguilar, C.N., Rodríguez-Herrera, R., Michelena, G., 2014. Utilization of molasses and sugar cane bagasse for production of fungal invertase in solid state fermentation using *Aspergillus niger* GH1. Braz. J. Microbiol. 45 (2), 373–377.
Weiergang, I., Jorgensen, H.J.L., Møller, I.M., Friis, P., Smedegaard-Petersen, V., 2002. Optimization of *in vitro* growth of *Pyrenophora teres* for production of the phytotoxin aspergillomarasmine A. Physiol. Mol. Plant Pathol. 60, 131–140.
Zorner, P.S., Evans, S.L., Savage, S.D., 1993. Perspectives on providing a realistic technical foundation for the commercialization of bioherbicides. In: Duke, S.O., Menn, J.J., Plimmer, J.R. (Eds.), Pest Control with Enhanced Environmental Safety. ACS Symp. Ser 524. American Chemical Society, Washington, DC, pp. 79–86.

CHAPTER

10 Fungi as a tool for decontaminating the range of soil contaminants

Akshita Maheshwari, Sonal Srivastava, and Suchi Srivastava
Microbial Technology Division, CSIR-National Botanical Research Institute, Lucknow, India

10.1 Introduction

The industrial revolution was the response to the explosive rise of population. Natural resource exploitation for meeting the energy demand of the population resulted in prolonged habitat contamination with pollutants. These pollutants are unwanted substances present in the natural environment causing undesired or deleterious effects on the abiotic and biotic system in a visible manner. Urbanization, industrialization, and land use activities result in the accumulation of toxic environmental pollutants like polycyclic aromatic hydrocarbons, polychlorinated dibenzo furans/dioxins, heavy metals, pesticides, pHthalates, etc. These contaminants enter the food web and eventually get biomagnified (Panigrahi et al., 2019). According to the World Health Organization report on drinking water, safe water is not available to 2.2 billion people and contaminated water is actively used by 144 million people globally (WHO, 2021). Pollution is accountable to 16% of global death and loss of 5% gross domestic product (Reddy and Behera, 2006). Poor management practices have left a legacy of resource contamination threatening ecology, economy, food and water safety, health, and biodiversity.

Bioremediation is a sustainable strategy for the removal and degradation of pollutants from soil and water. Robust growth, enzymatic activity, and versatility of fungi make it a supreme candidate for the biological remediation of toxins and pollutants from the environment referred to as mycoremediation (Reddy and Behera, 2006; Harms et al., 2011). Unlike physical and chemical methods of treatment, it is inexpensive and easy with no toxic by-products. Thus mycoremediation is an economic feasible and eco-friendly approach to diminish contamination and toxicity of pollutants. Contaminants like heavy metals, dyes, polycyclic hydrocarbons, petroleum, industrial and agricultural waste, etc., are decontaminated from the environment by using fungal metabolic processes. These fungal agents use these toxicants as energy source and rendering them harmless (Gupta et al., 2017). In this chapter, we have discussed various pollutants existing in the environment and the strategies for the exploitation of fungi for sustainable remediation and detoxification of those pollutants.

10.2 Bioremediation

Biological processes used for decontamination of the environment, preventing pollution, and restoring the natural ecosystem by exploiting plant (phytoremediation) and microbial (microremediation) metabolic abilities are called bioremediation. It involves mineralization, conversion to biomass,

Entrepreneurship with Microorganisms. https://doi.org/10.1016/B978-0-443-19049-0.00004-9

biodegradation, bioaccumulation, etc., to remediate the contaminants. Bioremediation is an eco-friendly, economic, and rapid process as compared to conventional physical and chemical methods of remediation; however, integrated efforts are more efficient. Bioremediation can permanently eliminate or biotransform detrimental pollutants to residues like water, cell biomass, carbon dioxide, etc., eliminating the liability of disposal and treatment. It prevents entry and biomagnifications of contaminants in the ecosystems. Bioremediation of hazardous waste can be done in situ without disrupting the natural ecosystem along with prevention of transport. It can also be carried out ex situ by excavating and treating the contaminants in bioreactors, however, it will be more expensive and less advantageous (Sardrood et al., 2013).

Natural and anthropogenic contaminants like heavy metals, pesticides, oil spills, industrial waste, etc. can be removed, transformed, immobilized, or detoxified using various mechanisms like methylation, oxidation-reduction, complexation, and chelation using siderophores and biosurfactants. These processes possess low-cost technology with high degradation potential of contaminants; being resistant to extreme environments they are found to be environmentally compatible to convert external polymers to simpler molecules. Thus, bioremediation is useful to metabolize and remove impurities of the recalcitrant compounds in an economic and eco-friendly manner (Sardrood et al., 2013; Kanaujiya et al., 2019).

Microorganisms are ubiquitous showing diverse properties and have the ability to perform catalytic mechanisms even in extreme conditions. Although they are not omnivorous for the degradation of all compounds and highly specific, and sometimes slow, they can be modified genetically and biochemically using biotechnological and molecular tools to enhance their mechanisms, activity, and occurrence, making it economically and functionally viable. Cooperation and coordination of metabolic potential, interaction ability, and availability of microbes is very important as microbes can work synergistically or antagonistically with the other population requiring specific suitable conditions and might disrupt the ecosystem (Gupta et al., 2017). Thus, exploration of microbial populations, their capabilities, and interaction with contaminants and environment developing new technology as the most attractive feature of bioremediation is the study of choice in current perspectives (Fig. 10.1).

10.3 Mycoremediation

Mycoremediation is the use of fungal technology for removing the contaminants active in the environment. Fungi being vigorous omnipresent organisms possess versatile features to metabolize toxic pollutants in order to degrade them in nontoxic forms. They are an essential ecological component of the food web for decomposition and degradation because of distinct metabolic and morphological capabilities. They are the only organism on the planet that can degrade lignin in wood as they possess a robust hyphal network exuding strong enzymes, bioactive compounds, and acids. They can degrade starch, cellulose, pectins, hemicelluloses, other polymers, and substrates like chitin, oils, fats, and keratins (Panigrahi et al., 2019; Sardrood et al., 2013). Fungi have the ability to convert, degrade and mineralize toxic compounds, reactive and radioactive substances, ions, organic and inorganic contaminants and other complex hydrocarbons, xenobiotics, etc., hazardous to the environment.

Fungi can penetrate deep into soil, colonize, tolerate high toxin concentration, translocate water, minerals, and nutrients and thus can survive and grow well in stressed conditions like low pH and poor nutrient status. The hyphal growth is an evolutionary notable feature that enables mycoremediation by penetrating inner layers of substrates and searching for new noncolonized sources for energy (Akhtar

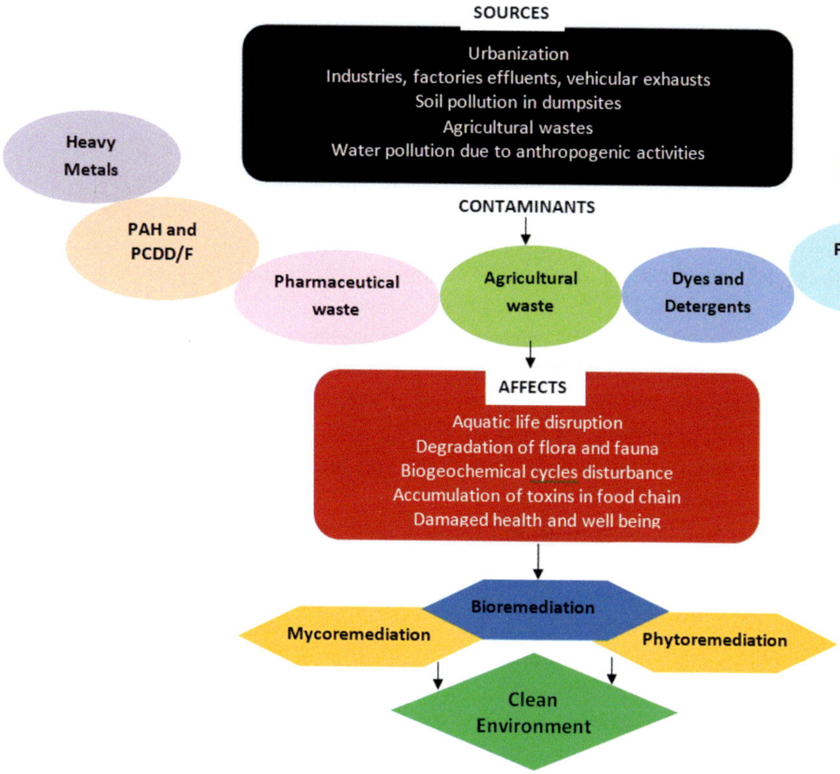

FIG. 10.1

Sources and effects of contamination and their remediation.

and Mannan, 2020). Ample nutrients and essential elements enhance the rate of degradation and bioactivity of fungi for remediation but fungi can perform well in deficient conditions as well. Short life span of fungi and high sporulation rate make them multiply more and faster for application in remediation and reclamation processes, e.g. utilization of yeast to remediate oil spills as it multiplies several multitudes (Kumar and Kaur, 2018; Sardrood et al., 2013).

Successful mycoremediation requires essential screening procedures to select appropriate species for a targeted pollutant depending on pH, nutrient and water availability, oxygen levels, competitive organisms, moisture, temperature, soil conditions, availability of organic matter, etc., for adequate fungal growth and enzyme production. Other factors include contaminant distribution, toxicity, bond of contaminants with soil and water, chemical nature, concentration, abundance, etc. Diversity of contaminants requires specific fungi depending on the metabolic activity for their bioremediation (Khatoon et al., 2021). Genetic engineering and the upscale of fungi make it a powerful remediation weapon in such conditions.

10.3.1 Heavy metal

Heavy metals are dense metals with high atomic numbers and specific density greater than 5 g/cm^3. Major heavy metals are arsenic (As), lead (Pb), copper (Cu), chromium (Cr), cadmium (Cd), nickel

(Ni), cobalt (Co), silver (Ag), iron (Fe) and zinc (Zn) (Jaishankar et al., 2014). In the environment, they are involved in various physiological and biochemical metabolic functions but beyond a threshold limits they possess adverse effects on environment (Järup, 2003). They threaten biodiversity and health by entering water and soil through natural processes like soil erosion, weathering, volcanic eruptions, etc., and anthropogenic effluents from agriculture and industries, mining, electronic waste, etc. Toxic concentrations of these elements in the soil eventually accumulate in plants and thus enter the food cycle, this subsequent transfer leads to biomagnification. Heavy metals enter in human body through consumption of contaminated food and water, causing cancer, mutation, and deleterious effects on metabolic systems of body like liver kidney failure, malfunctioning in bone marrow, and central nervous system (Morais et al., 2012; Khan et al., 2019; Masindi and Muedi, 2018). Conventional methods like ion exchange, redox reaction, electrochemical processes, osmosis, precipitation, membrane filtration, coagulation-flocculation, and other physical and chemical processes consume a huge energy and are inefficient and expensive. They alter the properties of soil and water, produce hazardous by-products, and do not remove the metals completely (Majeed, 2021). Thus bioremediation by intracellular accumulation, bioleaching, biomineralization, redox reactions, and biosorption might be the answer to overcome these limitations (Lloyd et al., 2014).

Heavy metals cause severe toxic effects in the terrestrial, aquatic ecosystem like animals, birds, and plants when present in excess by hindering various physiological activities (Poli et al., 2009). They oxidize different biomolecules like protein and lipids producing reactive oxygen species causing oxidative stress (Gupta et al., 2012). In plants, metals like mercury, arsenic, and lead hinders the growth and developmental processes causing stunting, deformities of leaves, fruits, and eventually yields (Chibuike and Obiora, 2014). Heavy metals like Hg, As, and Cr forms harmful thiol and methyl derivatives and Pb, Fe, and Cu cause free radical imbalance leading to oxidative stress in plants. Iron leads to corrosion, lipid peroxidation, organ penetration, and saturation, while Cr and Zn also imbalances ion channels and hinders cell membrane permeability, disintegrates organelles along with DNA, and damages protein. Pb and Cd hinder supply of magnesium and thus impair chlorophyll synthesis, and they also inhibit enzymes like protochlorophyllide reductase, delta-aminolevulinic dehydrase, etc. (Chibuike and Obiora, 2014; Gupta and Diwan, 2017; John et al., 2008). In soil, these metals hinders metabolic processes like respiration, mineralization, nitrification, etc. (Singh et al., 2018).

Fungi have the ability to mycoremediate metal ions by binding and making a metal ion complex. They have relatively higher tolerance for heavy metal due to robust and large surface area as compared to plants and other microorganisms. Several fungi evolve metal-resistant genes to adapt and bioremediate heavy metal stress (Das and Dash, 2017). Determination of specific fungal species to target pollutants like lead, arsenic, etc., is important. Some fungi participating in heavy metal bioremediation are *Rhizopus* sp., *Mucor* sp., *Penicillium* sp., *Phanerochaete* sp., *Paxillus* sp., *Absidia* sp., etc. (Singh et al., 2018). Mycoremediation of heavy metals is highly dependent on environmental conditions. *Rhodotorula mucilaginosa*, a pigmented yeast, can bioaccumulate silver ions and adsorb lead present in contaminated soil (Gomes et al., 2002). *Aspergillus* species was found to bioremediate toxic metals like arsenic, lead, cadmium chromium, and nickel from the contaminated sites (Mishra and Malik, 2014).

10.3.2 Polyaromatic hydrocarbons

The polycyclic aromatic hydrocarbons (PAH) are hazardous organic compounds constituting fused benzene rings in multiple arrangements. They consist of simple and complex arrangements of carbon and hydrogen thus called polyaromatic hydrocarbons (Singh and Haritash, 2019). Some major examples are chrysene, pyrene, 2-methyl naphthalene, anthracene, nanthrene, and acenaphthene. Based on aromatic ring number, there can be low- or high-molecular-weight PAH (Abdel-Hamid et al., 2013). Pyrene, chrysene, and benzo(*a*)pyrene are high-molecular-weight PAH consisting of more than three benzene rings while anthracene, naphthalene, and phenanthrene are low-molecular-weight PAHs (Kumar et al., 2021; García-Alonso et al., 2008). These contaminants alter the porosity, particle size, and water holding capacity of the soil. These compounds cause toxicity and also affect microbial populations in soil. They arise in the environment due to incomplete combustion of petroleum, coal, wood, and other organic materials and present in soil and water as undegraded impurities (Morelli et al., 2013; Anasonye et al., 2014). They are ubiquitous in soil, air, and water due to forest fires, incineration, vehicles, heating systems, etc., and causes prolonged damage to the ecosystem (Abdel-Shafy and Mansour, 2016). These contaminants are heat and corrosion resistant with high melting point and high boiling point but low aqueous solubility and vapor pressure. Solubility of PAH is high in organic solvents due to their lipophilic nature (Gitipour et al., 2018). They are stable xenobiotics in the environment thus difficult to degrade in comparison to other contaminants. Use of microbes has been observed in degradation of these toxins successfully.

PAH generally formed due to incomplete combustion at high temperature and low oxygen. There are three types of PAH present in the environment. These sources are (a) pyrogenic: These PAH arise by anthropogenic activities like combustion of wood, fossil fuel, tires, agriculture waste, thermal cracking of petroleum, and destructive distillation of coal in industries and natural processes like volcanic eruptions forest fires, etc., at high temperatures about 350°C to more than 1200°C and generally found in urban regions. (b) Petrogenic sources are mainly petroleum sources due to oil spills, oil seeps and gas leaks. They are formed at lower temperatures. These arise due to maturation of oil and coals from oil spills, storage tank leaks, release of motor oil, and gasoline during transportation. (c) Biogenic sources are naturally occurring surface waxes, woods, etc., synthesized by plants and microbes and left undegraded in vegetative matter (Decesaro et al., 2017; Smith et al., 2006; Cameselle and Gouveia, 2019). PAH are used in various commercial forms in agriculture, pharmaceuticals, plastics, lubricants, and various other chemicals. Acenaphthene, fluoranthene, and fluorene are used to manufacture dyes, pigments plastics, pharmaceuticals, pesticides, and other agrochemicals. Anthracene acts as a diluent in preservation of woods, while phenanthrene is used in manufacturing processes of resins.

PAH are omnipresent and arise from small and large anthropogenic and natural processes. Thus, their association with soil is dependent on various factors like size of particle, carbon, porosity, and sometimes pH of the soil (Mishra and Singh, 2014). Acidic pH of the soil indicated higher humic and fumic acid content, which can bind to the PAH as compared to basic soils (Saba et al., 2011). It was found that PAH binds easily to clay and silt, which are finer in size, as they provide higher surface area thus more active binding sites for adsorption (Smith et al., 2006; Magi et al., 2002). Stable and persistent toxicity arises due to less movement of adsorbed compounds due to their hydrophobic nature in less porous fine soils, which leads to their long-term persistence and difficult to remediate effects, as microbial growth of some species requires aeration (Mao and Guan, 2016). Low-molecular-weight PAH is loosely bound to the soil as compared to high-molecular-weight PAHs (García-Alonso et al., 2008).

Carcinogenic nature of PAH is recognized by the International Agency on Research of Cancer (Trakoli, 2012). They are mutagens causing lung cancer by inhalation of smoke from various sources along with cyclic presence of PAH in the food (Kiamarsi et al., 2019). PAH in soil majorly arise from crops and airborne fallouts and get accumulated in adipose tissues of humans due to their lipophilic nature (Gitipour et al., 2018). PAH can be detoxified and mineralized by various fungi in the presence of optimal environmental conditions. Various other fungi like *Aspergillus, Phanerochaete chrysosporium, B. adusta, Pleurotus* sp., and *Ipex lacteus* were major fungus documented to biodegrade PAH in contaminated soils (Haritash and Kaushik, 2016). Fungi have proven to be better in degrading PAH than bacteria, viz., C12 and C18 n-alkanes were degraded 80%–85% by *Penicillium frequentans* and *Candida parapsilosis,* while bacteria can degrade 15% (Omar and Rehm, 1988). *Ipex lacteus*-treated soil contaminated with fluorine and fluoranthene has been shown to have reduced PAH through mycoremediation (Bhatt et al., 2002).

10.3.3 Mycoremediation of agricultural wastes

Intensification of agriculture for high yield goals has led to rigorous use of pesticides like insecticides, herbicides, nematicides, and various fertilizers. These compounds cause severe damage to the ecosystem as their residues leach into the ground water and persist for a long time due to less degradable complex structures (Gupta et al., 2017). They also get biomagnified into food webs via. Consumption of the residual chemicals in food and water affecting animal and human health. With the changing perspective of people and shifting toward organic agriculture, the demand of these products is still unfazed in the global market and is projected to reach 39.15 billion dollars of herbicide in 2022 (Akhtar and Mannan, 2020).

Agriculture intensification has led to misuse of pesticides, herbicides, fertilizers, and various chemical-based products to satisfy the greed of high yield, which has led to toxicity of the residuals of these chemicals in soil and groundwater. Endosulfan a highly carcinogenic pesticide can be degraded by *Aspergillus tamari, A. niger, A. terreus, Botryosphaeria laricina, Cladosporium* sp., *Fusarium* sp., *Trichoderma* sp., *Phanerochaete* sp., and *Mucor* sp. by utilizing it as a carbon source for food and energy (Bhalerao and Puranik, 2007; Silambarasan and Abraham, 2013).

10.3.4 Mycoremediation of dyes

Dyes are substances that chemically bond with the substrate to impart color. They are released into the environment in thousands of liters from textile and printing industries and become recalcitrant due to chemical stability, fade resistance. Although dyes do not possess extreme threat to the environment some exceptions are always there. Azo, cationic, and metalized anthraquinone are some of the classes of dyes known to be toxic; however, some dyes like congo red, aromatic amine benzidine have been proven to be carcinogenic (Akhtar and Mannan, 2020; Li et al., 2010). Highly toxic congo red produced from the paper industry was found to be degraded by *Aspergillus flavus* in the nearby soil (Bhattacharya and Das, 2011). Nonspecific fungal enzymes like laccase and peroxidase were easily able to degrade the dyes from the environment (Novotný et al., 2004). *Phanerochaete chrysosporium* is a multifunctioning fungi which can degrade dyes like azo dye, congo red, orange II, azure B, etc., along with various PAHs PCDD/Fs and phthalates (Sardrood et al., 2013; Angel, 2002). *Trametes versicolor* was found to have high activity of laccase and manganese peroxidase and was able to degrade azo dye and anthraquinones (Yang et al., 2017). *Bjerkandera adusta* can decolorize highly reactive dyes and secretes enzymes like peroxidase for degrading various dyes by destructing chromophoric assemblies in these dyes (Sodaneath et al., 2017). In various studies, *Trametes trogii, Myrothecium roridum*

I'M 6482, *Pycnoporus sanguineus, Phanerochaete chrysosporium, Penicillium ochrochloron,* and *P. chrysogenum* were found to degrade various industrial and azo dyes (Durruty et al., 2015; Jasińska et al., 2015; Shedbalkar and Jadhav, 2011; Yan et al., 2014). These fungi can also degrade heavy metals and other industrial effluents and thus can be used in large-scale bioreactors for treatment of these products before release into the environment.

10.3.5 Mycoremediation of microplastics and phthalates

Microplastics are plastic particles used in daily products or industrial product waste released into the environment due to improper management and degradation. Their size is 0.1–5 mm. These are found everywhere terrestrial, landfills, and finally washed off to aquatic ecosystems produced by natural processes like mechanical abrasion or UV radiation or by intentional processes like plastic pellets in industries, microbeads in pharmaceuticals or consumer care products (Alimi et al., 2018). This plastic debris can cause damage to animals, alter biodiversity, disrupt habitat, reduce recreational aesthetics, transmit or propagate disease causing organisms, might cause climate change, disrupt food chain, and alter the biophysical and geochemical properties of soil (Laskar and Kumar, 2019). Disposal of plastics typically requires more virgin plastic than expected due to reduced quality and only less than 1/5th of all the plastic waste produced is recycled, while other options, such as incineration produces dioxins like toxic products after combustion (Ajibade et al., 2021). Microplastics produced from the plastic waste are found to be everywhere and have been proven to be remediated by fungi by the process of bio-fragmentation and mineralization into simpler compounds. High-density polyethylene treated with *Aspergillus terreus* was found to produce straight and branched carboxylic acids, alkanes as well as phthalate esters and their derivatives, indicating fragmentation of the plastic compounds (Balasubramanian et al., 2014; Sangale et al., 2019). Another phenomenon of decrease in crystallization was observed in certain plastic compounds when kept in incubation with *Curvularia lunata*, indicating degradation due to increase in oxidation as oxidation/degradation occurs in amorphous regions it helps fungal enzymes to access the crystalline regions of the compound (Raut et al., 2015). In another study, polyurethane was found to be degraded by saprophytic fungi, *Marasmius oreades*, and *Agaricus bisporus* (Brunner et al., 2018).

Di (2-ethyl hexyl) phthalate (DEHP) is one of the most abundant contaminants in soil and water, mostly present in plastics. Phthalates are plasticizers used to increase the durability and flexibility of vinyl, toys, plastic devices, capacitors, etc., and they are the long lasting pollutants existing in the environment. They are the reason behind widespread plastic pollution and ultimately enter the food chain harming terrestrial as well aquatic health. In humans, they intercept the respiratory and hormonal functioning. Conventionally radiation, hydrolysis, photocatalytic degradation can be used but mycoremediation might be more hassle free in the degradation of phthalates (Benjamin et al., 2015). Fungi consortium of *Aspergillus parasiticus*, *Penicillium funiculosum,* and *Fusarium subglutinans* degraded upto 70% of DEHP of plastic blood bags in 2 weeks. In further studies, it was found that *A. parasiticus* and *F. subglutinans* could also degrade poly vinyl chloride pipes, and therefore, it can be commercialized to degrade plastic in large-scale bioreactors (Benjamin et al., 2015; Pradeep and Benjamin, 2012). Fungal enzyme cutinase from *F. oxysporum* could rapidly and efficiently degrade various DEHP and convert them into nontoxic compounds while esterase another fungal enzyme produced toxic byproducts during conversion of DEHP (Pradeep et al., 2013).

10.3.6 Mycoremediation of petroleum and oil spills

Petroleum and oil spills are proven hazardous pollutants unintentionally or intentionally released into the environment harming soil and oceanic life. They make the resources unusable and carcinogenic and disrupt the food chain. Fungi can be a rapid, economic, and more effective means of bioremediation of the spillage (Kumar and Kaur, 2018). Enzymes secreted by fungi like oxidoreductases, laccases, and other extremozymes have proven to detoxify and bioremediate hydrocarbons caused due to oil spills. Marine ecosystem is most prone to these oil spills and natural oil seepage causing hydrocarbon contamination. Extremozymes like alkane hydroxylases and cytochrome 450 proteins like nonheme di-iron (Alk A, B) and Cyt P153 were found to bioremediate petroleum and oil spill contaminants in marine ecosystem (Rothschild and Mancinelli, 2001; Bolhuis and Cretoiu, 2016; Donato et al., 2018).

Various fungal genera viz. *Aspergillus, Fusarium, Candida, Geotrichum, Mucor, Penicillium, Cephalosporium, Cladosporium, Saccharomyces, Trichoderma, Rhodosporidium, Hansenula, Trichosporon, Sporobolomyces, Cunninghamella, Aureobasidium, Rhodotorula,* etc., have been studied and identified to bioremediate the petroleum and oil spills in the environment (Bartha and Atlas, 1977; Atlas et al., 1992; Obire and Nwaubeta, 2010). In the aquatic system, several yeast have been identified belonging to genera *Saccharomyces, Candida, Rhodotorula,* and *Sporobolomyces,* which successfully degraded oil. Along with that some filamentous fungi like *A. niger, A. terreus, Fusarium* sp., *Penicillium chrysogenum, P. glabrum, Trichoderma harzianum, Nigrospora* sp., *Botryodiplodia theobromae,* etc., were also able to degrade oils by penetrating oil and increasing the surface area for biodegradation in synergy with bacteria from the environment (Obire et al., 2008). *Penicillium* sp. 06 strain was effectively used to degrade PAH in petroleum-contaminated soil by oxidation of phenanthrene by 89%, and acenaphthene, fluoranthene, and fuorene by 75%. *Punctularia strigosozona*ta overexpressed hydrophobins to degrade hydrocarbons in fuel oil. Fungi *Daedaleopsis* sp. was observed to bioremediate petroleum products. *Aspergillus niger*, *Candida krusei*, *C. glabrata*, and *Saccharomyces cerevisiae* were also found to degrade crude oil (Zheng and Obbard, 2003; Young et al., 2015; Pourfakhraei et al., 2018; Al-Tamimi et al., 2021; El-Borai et al., 2016).

10.3.7 Mycoremediation of pharmaceutical wastes

Pharmaceuticals are biologically active compounds made with the intention to affect the function of the body system. These are easily absorbable hydrophilic compounds excreted into the environment as a mixture of metabolites, which might become toxic to the ecosystem and can be biomagnified in the food chain (Akerman-Sanchez and Rojas-Jimenez, 2021). There are about 3000 variants of chemicals required in the pharmaceutical industry to produce various drugs. Antibiotics, antiinflammatory, analgesics, antidiabetics, contraceptives, antihistamines, synthetic hormones, antidepressants, antihypertensive, diuretics, and beta blockers like ibuprofen, naproxen, paracetamol, diclofenac, chloramphenicol, tetracyclines, atenolol, ranitidine, benzodiazepines, etc., are found to be present in high concentration in the environment due to their low degradation rate. These pharmaceuticals have been considered as one of the main chemical contaminants due to their exponential increase in consumption. They can also hinder carbon cycling, nitrogen cycle, and soil respiration by killing the ecologically important microorganisms (Akerman-Sanchez and Rojas-Jimenez, 2021; Verlicchi et al., 2012). Drug-resistant microbes can be used in degradation of these compounds.

A vast range of fungal genera is observed to remediate pharmaceuticals from the environment. Strains of fungi *Aspergillus niger, A. niveus, A. fumigates* were found to remove pharmaceuticals from diverse concentrations of industrial wastes. Ascomycetes group of fungi-like strains of *Penicillium decumbens* and *P. lignorum* were used to treat pharmaceutical contaminated industrial sewage and substantial results were found in decrement of phenol, color, and chemical oxygen demand (Mohammad et al., 2006; Angayarkanni et al., 2003). Aquatic fungus, *Mucor hiemalis* was found to remove acetaminophen and diclofenac from water bodies contaminated by pharmaceutical effluents (Esterhuizen-Londt et al., 2016, 2017). *Pleurotus ostreatus* eliminated the antibiotic oxytetracycline by absorbing it in a liquid culture (Migliore et al., 2012). Ligninolytic and nonligninolytic enzymes secreting fungi help in degradation of pharmaceuticals. *Leptosphaerulina* sp. produce versatile enzymes like laccase and peroxidase to degrade isoxazolyl-penicillins like oxacillin (Migliore et al., 2012; Copete-Pertuz et al., 2018).

Trametes versicolor was found to be highly efficient white rot fungus to remediate pharmaceuticals in various studies. It absorbed and removed naproxen in only 6 h, while in another study, degrading an analgesic, ketoprofen using ligninolytic enzymes and cytochrome 450 (Marco-Urrea et al., 2010a,b). *T versicolor* was also found to eliminate pharmaceuticals like diazepam, codeine, metoprolol, carbamazepine, and sulfamethazine (Asif et al., 2017; García-Galán et al., 2011). Role of ligninolytic enzyme and cytochrome 450 was also observed in degradation of fluoroquinolones antibiotics, like ofloxacin and ciprofloxacin (Prieto et al., 2011). Similar role was seen in *Irpex lacteus* to remove ofloxacin and ciprofloxacin along with flumequine antibiotics and it produce enzyme manganese peroxidase to degrade the antibacterial activity of ofloxacin and norfloxacin (Čvančarová et al., 2013, 2014). *Bjerkandera adusta* MUT 2295 strain was also used to treat industrial wastewater sludge contaminated with pharmaceuticals (Anastasi et al., 2010). Oxidative degradation of various drugs was done by *P. chrysosporium* in a bioreactor (Rodarte-Morales et al., 2012). Mycelia of *Lentinula edodes* secretes various oxidizing enzymes for oxidative degradation of piroxicam (Muszyńska et al., 2019). It was also found to degrade endocrine disruptors like 17 alpha ethinyl estradiol and synthetic testosterone and absorbed antifungal drugs like clotrimazole and bifonazole (Muszyńska et al., 2018; Kryczyk-Poprawa et al., 2019; Singh et al., 2021).

10.4 Mechanism and processes of mycoremediation

Fungi are ubiquitous, broadly distributed organisms, which have the ability to successfully adapt to extreme, unfavorable, and stressed conditions as well. Mycoremediation involves varied mechanisms and processes like breakdown by fungal enzymes, mobilization, immobilization, biosorption, bioaccumulation, and biotransformation (Deshmukh et al., 2016). They convert the toxic heavy metals to nontoxic forms by the process of biotransformation and volatilization. Combination of fungi and other microbes also helps in mineralization processes.

10.4.1 Fungal enzymes

Fungi secrete extracellular enzymes for breaking larger complex compounds like cellulose, hemicelluloses, lignin, pectin, and starch. Different enzymes known to be produced by fungi are cellulase, amylase, lipase, protease, laccase, catalase, peroxidase, etc., which have unique potential for dissociating/hydrolyzing substrate polymers like cellulose, starch, lipids, proteins, etc., for managing waste. Mycoremediation of contaminants releases various secondary metabolites, exoenzymes, into the substrate leading to degradation and production of some breakdown products, which can be used as substrates by internal enzymes. Environmental conditions and species of fungi diversify the enzymatic reactions, which can occur even in low concentration of substrate contaminants. Fungi produce different types of ligninolytic enzymes, which degrade organic pollutants and other xenobiotic compounds like dyes, pesticides, etc., along with lignin to remediate them from contaminated sites. These enzymes produce water soluble polar products which can be metabolized further by the fungus itself or other microflora (Natalia, 2017). Among all the fungal groups, various extracellular ligninolytic enzymes secreted by white rot fungi completely degrade the pollutants by nonspecific oxidation mechanism (Dashtban et al., 2009). These enzymes are categorized as laccases, manganese peroxidases, lignin peroxidase, oxidase, dehydrogenase, and other versatile peroxidases. White rot fungi reported for mycoremediation are *Agaricus bisporus, Pleurotus* sp., *Trametes versicolor, Bjerkandera adusta, Lentinula edodes,* etc., (Khatoon et al., 2021). Laccase are eco-friendly enzymes, which have the ability to degrade phenols, PAH, etc., by reduction of oxygen and water production (Li et al., 2010; Couto and Toca-Herrera, 2007; Riva, 2006; Dias et al., 2007; Wu et al., 2008). Certain lignin and manganese peroxidases are heme proteins that have protoporphyrin groups. Manganese peroxidase has specific manganese binding sites. These glycoproteins are also having heme ionic groups (Dias et al., 2007; Piontek et al., 1993; Sundaramoorthy et al., 1994; Plácido and Capareda, 2015).

Fungi also produce nonligninolytic and other hydrolytic enzymes like peroxidases and proteases (Alves et al., 2002). The cytochrome *p*450 monooxygenase configures epoxide hydrolase by incorporating single oxygen molecule compounds and then further transforming it to trans-dihydrodiols (Jerina, 1983). Hydrolases disrupt bonds of hydrocarbons, carbamates, organophosphates and other chemicals. Certain hydrolases catalyzed DDT and heptachlor degradation via alcoholysis and condensation (Williams et al., 2013; Vasileva-Tonkova and Galabova, 2003; Williams, 1977; Lal and Saxena, 1982). Genera of *Aspergillus, Penicillium, Mucor, Trichoderma, Chrysosporium,* and *Rhizopus* are reported to produce proteases. These proteases participate in breakdown of peptide bonds to remediate pharmaceuticals, food, and leather contaminants (Bensmail et al., 2015; Germano et al., 2003; Andrade et al., 2002; Dienes et al., 2007; Souza et al., 2015; Singh, 2003). Lipases catalyzes the reactions like esterification, hydrolysis, aminolysis, and alcoholysis and helps to bioremediate chemicals, detergents, PAH, cosmetics, etc., *Candida rugosa, C. Antarctica, A. niger,* and *Penicillium* are the fungi involved in lipase degradation (Prasad and Manjunath, 2011; Sharma et al., 2010; Liu et al., 2015; Sharma et al., 2016; Amoah et al., 2016; Wolski et al., 2008). Cellulases are the extracellular enzymes hydrolyzing various compounds. *Aspergillus, Penicillium, Fusarium, Cladosporium, Chaetomium, Trichoderma,* and *Stachybotrys* are the genera that produce these enzymes (Wood, 1985; Sajith et al., 2016; Aneja and Mehrotra, 1980). Fungal enzymes like catalase, laccase, and peroxidase were found to bioremediate heavy metals. *Aspergillus, Penicillium, Rhizopus,* and *Acremonium* are some of the genus that use catalase for bioremediation of heavy metals like cadmium, nickel, and copper (Agarwala et al., 1961; Kaplan, 1965; Mohammadian et al., 2017; Ling, 1997) (Fig. 10.2).

Recent studies are performed for development of specific enzymes through recombinant gene technology and protein engineering techniques from various fungal species for eco-friendly bioremediation. The mechanism for remediation of contaminants via fungi directly or indirectly involves oxidation of organic compounds. Role of fungal enzymes in mycoremediation of various pollutants is shown in Table 10.1. Enzymatic degradation by fungi using various ligninolytic and nonligninolytic enzymes is the most economical and eco-friendly way to remove PAH from the sites of contamination. Enzymes like laccase and peroxidase can act on the compounds, structurally similar to lignin, and partially degrade them to quinones, phthalic acid, etc. (Pozdnyakova, 2012; Brodkorb and Legge, 1992).

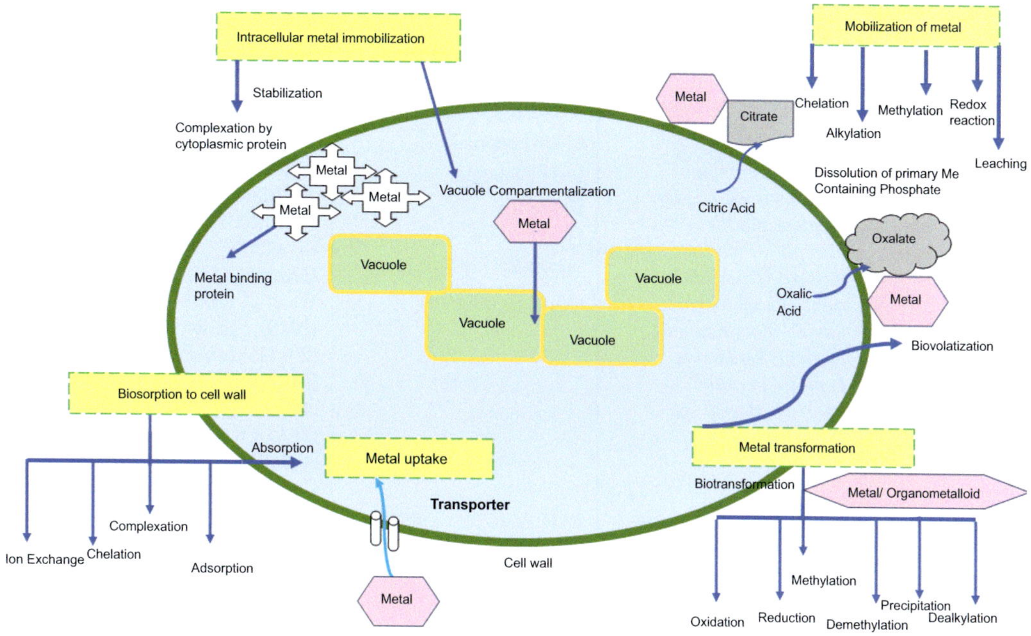

FIG. 10.2

Mechanism of mycoremediation of heavy metals.

Nonligninolytic enzymes like dehydrogenase, hydrolases along with cytochrome P450 monooxygenase were utilized by *Dentipellis* sp. to degrade PAH (Bhattacharya et al., 2013; Park et al., 2019).

Ligninolytic fungi belonging to basidiomycetes and ascomycetes have the ability to degrade various compound like PAH, wood preservatives, dyes, PCCD/Fs, etc., due to enzyme nonspecificity, generation of free hydroxyl radicals and electron donation to form quinones and acids (Akhtar and Mannan, 2020; Natalia, 2017; Burlacu et al., 2018; Tuomela and Hatakka, 2011). *Trametes* sp. were found to degrade different pesticides like atrazine and paraquat using ligninolytic enzymes (Chan-Cupul et al., 2016; Camacho-Morales et al., 2017). Eutrophication leading to algal bloom can also be remediated by various algicidal fungi like *Trametes versicolor, Mucor hiemalis,* and *Trichoderma* sp. producing algicidal enzymes like laccase, cellulase, peroxidase, etc. (Mohamed et al., 2014). *Coprinus* sp. was producing peroxidase for phenol treatment (Ikehata et al., 2005).

Ligninolytic enzymes like laccase, peroxidase are extracellular in nature. These enzymes break down PAH complexes in contaminated soils to simpler soluble and bioavailable forms of byproducts. These byproducts are less toxic in nature and can be used as substrate by other organisms or can be further degraded by nonligninolytic cytochrome P450 monooxygenases (Harms et al., 2011; Pozdnyakova, 2012; Sack et al., 2006). Ligninolytic fungi completely mineralize and degrade PAH by epoxide hydrolases, ligninolytic enzymes and cytochrome P450 monooxygenases. *Pleurotus ostreatus* and *Phanerochaete chrysosporium* degrades benzo(*a*)pyrene via both ligninolytic and nonligninolytic enzymatic reactions for complete mineralization. Similarly, *Polyporus* sp. and *Irpex lacteus* degrades various PAH in the same manner (Bezalel et al., 1997; Cajthaml et al., 2002, 2006; Hadibarata et al., 2012). In *Pleurotus* sp.

Table 10.1 Ligninolytic fungal enzymes for mycoremediation of contaminants.

Fungal enzyme	Fungi species	Contaminant remediation	References
Laccase	*Yarrowia lipolytica, Pleurotus eryngii, P. pastoris, Gandoderma lucidum* and *Pleurotus ostreatus, Pichia pastoris, Trametes trogii, Phanerochaete flavido-alba, A.niger Coriolus versicolor, Leptosphaerulina* sp. *Trichoderma* sp., *Pleurotus* sp.	Pharmaceuticals PAH Pesticide Herbicide Dyes Phthalate Organic pollutants	Brodkorb and Legge (1992); Piontek et al. (1993); Rothschild and Mancinelli (2001); Novotný et al. (2004); Colao et al. (2006); Riva (2006); Couto and Toca-Herrera (2007); Dias et al. (2007); Wu et al. (2008); Nasrin et al. (2010); Shedbalkar and Jadhav (2011); Mohamed et al. (2014); Bleve et al. (2014); Benghazi et al. (2014); Rivera-Hoyos et al. (2015); Kalyani et al. (2015); Yang et al. (2017); Amjad et al. (2017); Copete-Pertuz et al. (2018)
Lignin Peroxidase	*Panustigrinus, Pleurotus pulmonarius, Irpexlacteus, Trametes versicolor, Phanerochaete chrysosporium, Coriolus versicolor, Gloeophyllum striatum Phanerochaete chrysosporium*	PAH Pesticide, herbicide, dyes, organic pollutants	Khindaria et al. (1996); Lau et al. (2003); Novotný et al. (2004); Leontievsky et al. (2004); Yu et al. (2005); Asgher et al. (2006); Ford et al. (2007); Yadav et al. (2009)
Manganese peroxidase	*P. tigrinus, P. pulmonarius, Ceriporiopsis subvermispora, Pulmonarius lacteus, Bjerkandera adusta, Phanerochaete chrysosporium, Nematoloma frowardii, Trametes versicolor, Irpex lacteus*	Heavy metals PAH and pesticide Herbicide Dyes Pharmaceutical Phthalate and microplastics Organic pollutants	Piontek et al. (1993); Sundaramoorthy et al. (1994); Khindaria et al. (1996); Angel (2002); Lau et al. (2003); Novotný et al. (2004); Leontievsky et al. (2004); Yu et al. (2005); Sack et al. (2006); Asgher et al. (2006); Dias et al. (2007); Ford et al. (2007); Valentín et al. (2007); Chiu et al. (2009); Yadav et al. (2009); Wang et al. (2009); Čvančarová et al. (2013, 2014); Plácido and Capareda (2015); Yang et al. (2017)
Catalase	*Curvularia* sp., *Pythyme* sp., *A. flavus, Rhizopus* sp., *Aspergillus* sp., *Acremonium* sp., *A. foetidus, Penicillium* sp., *F radiculosa*	Heavy metals	Agarwala et al. (1961); Kaplan (1965); Ling (1997); Akhtar et al. (2013); Chakraborty et al. (2013); Kurniati et al. (2014); Deshmukh et al. (2016); Mohammadian et al. (2017); Akgul et al. (2018)

Table 10.1 Ligninolytic fungal enzymes for mycoremediation of contaminants—cont'd

Fungal enzyme	Fungi species	Contaminant remediation	References
Lipase	*Curvularia* sp., *Mucor* sp., *Trichoderma* sp., *Fusarium* sp., *Lasiodiplodia* sp., *Rhizopus* sp., *Drechslera* sp., *Candida rugosa, C. antarctica, A niger* and *Penicillium* sp.	PAHs	Margesin et al. (1999); Alves et al. (2002); Wolski et al. (2008); Sharma et al. (2010); Prasad and Manjunath (2011); Lladó et al. (2013); Balaji et al. (2014); Liu et al. (2015); Sharma et al. (2016); Amoah et al. (2016)
Peroxidase	*A. foetidus, Geotrichum candidum, Pycnoporus cinnabarinus, Ganoderma* sp., *T. versicolor, B. adjusta, A. niger, T. viride, Penicillium* sp., *Ceriporiametamorphosa, Chrysosporium, Pleurotus* sp.,	Pesticide Organic pollutants Dye decolorization	Agarwala et al. (1961); Brodkorb and Legge (1992); Novotný et al. (2004); Ikehata et al. (2005); Sack et al. (2006); Pozdnyakova (2012); dos Santos et al. (2013); Mohamed et al. (2014); Ma et al. (2014); Radhika et al. (2014); Baratto et al. (2015); Copete-Pertuz et al. (2018)

enhanced activity of laccase by the presence of copper has been reported for lignin degradation (Amjad et al., 2017; Bhattacharya et al., 2014). Lignin peroxidase activation occurs by providing electron substitutes for the oxidation of aromatic rings (Khindaria et al., 1996).

Nonligninolytic fungi mostly oxidize PAH using cytochrome P450 monooxygenases present intracellularly in fungi. *Aspergillus niger, Cunninghamella elegans,* and *Chrusporium pannorun* degrades phenanthrene by oxidation using cytochrome P450 monooxygenases (Al-Hawash et al., 2018). Lipase, catalase, and dehydrogenase are used as an indicator of PAH degradation in soil (Ogbolosingha et al., 2015). *Cunninghamella elegans* was studied to transform fluorine, fluoranthene, and anthracene to various simpler by-products like 9-fluorenol, 9-fluorenone, 3 fluoranthene-b glucopyranoside, b-glucopyranoside, fluoranthene trans-2,3-dihydrodiol, 3–8- hydroxy fluoranthene, 1 anthryl-sulfate, and anthracene trans-1-2-dihydrodiol metabolites (Pothuluri et al., 1993, 1996; Cerniglia and Sutherland, 2001).

10.4.2 Mobilization

Inactive biological sources of toxic substrate are activated by various active metabolic mechanisms to move or accumulate the contaminants within the metabolic processes for their breakdown (Khatoon et al., 2021; Mohammadian et al., 2017). Mobilization of metals is influenced by its concentration, pH, and ability to form complexes. These processes involve.

(a) Redox transformations: Activating microbes by oxidation and reduction or both processes of mobilization of organometallic compounds, metals, metalloids, etc.
(b) Chelation by siderophores: Low molecular weight iron chelating secondary metabolites facilitates uptake of iron and associated metals like magnesium, manganese, chromium, and

gallium by ligand exchange mechanism. Iron accumulating siderophores can trace and bind various other heavy metals as well. Siderophores were found to solubilize heavy metals like Pb, Zn, and Cd and precipitate them in the fungal biomass.

(c) Methylation: Different metalloids are formed by assimilation of enzymatically transferred methyl groups to a metal. These metalloids show varied volatility and solubility and eventually reduce the toxicity. Major methyl group acceptors in secondary metabolic processes are carbon, nitrogen, sulfur, and oxygen (Gadd and Sayer, 2000).

(d) Alkylation: Transformation of alkyl group among different molecules making it active and available for mobilization.

(e) Leaching: Formation of low molecular weight organic acids after breakdown donates protons and anions to form complexes (Gadd, 2004). They also help to combat metal stress by reducing degradation of IAA induced by metals (Diels et al., 1999; Neubauer et al., 2000). *Phanerochaete chrysosporium* detoxifies heavy metals from the soil by forming oxalic acid mediated cadmium metal complexes thus enhancing its bioavailability (Xu et al., 2015).

10.4.3 Immobilization

Restriction of mobility and physical interaction of toxic contaminants by modification of physical and chemical characteristics by microbes is called immobilization. It can occur through precipitation and/or sequestration in an active or passive process (Gadd and Sayer, 2000). Accumulation can be intracellular or extracellular and can occur on the cell surface. Mycoremediation through immobilization involves two mechanisms, i.e., stabilization and solidification. It is a temperature and moisture dependent process. In this method the toxic contaminated substrate is mixed with a stabilizer in presence of water, forming a matrix of solidified mass which is then further processed with chemicals to precipitate the contaminants like metals as hydroxides (Gadd, 2004). Immobilization of metals is facilitated by the presence of ligands on the cell surface like hydroxyl, carboxyl, amine, phosphate, etc., giving the fungi enhanced metal binding capacity by forming complexes. Polysaccharides and glycoproteins in the cell wall also act as active metal chelating sites (Ghosh et al., 2021; Ahluwalia and Goyal, 2007). Metal speciation, ionic radius, accessibility, electro-negativity, and electrostatic interaction influence the binding ability (Rahman and Sathasivam, 2015). Other extracellular polymeric substances (EPS) having negatively charged ionizable functional groups released by microbes also have positively charged metal binding potential (Gupta and Diwan, 2017). Fungi release oxalic acids and form immobile metal oxalate complexes like calcium, cobalt, cadmium and zinc oxalate. Fungal spp. such as *Bjerkandera fumosa, Trametes versicolor*, and *Phlebia radiata* formed oxalate crystals (Gadd, 2004; Franceschi and Nakata, 2005; Jarosz-Wilkolazka and Gadd, 2003).

10.4.4 Biosorption

Biosorption is a passive, reversible, physiochemical process which involves mechanisms like absorption, adsorption, precipitation, reduction, ion exchange, chelation, and surface bonds with functional groups to uptake the pollutants from toxic sources. It is a metabolically independent quick process involving a solid phase (biosorbent) and liquid phase (solvent). Fungal biomass acts as biosorbent constituting cell wall materials forming electrostatic interactions between ions, metals like copper, cadmium, cobalt, and cell wall of fungi. This ion exchange mechanism supports uptake of toxic materials for successful mycoremediation (Gadd and Sayer, 2000; Veglio' and Beolchini, 1997; Dhankhar and

Hooda, 2011). Biosorption of metal by yeast mold and mushrooms is very common. The process of sorption is affected by pH, carbon, aeration, temperature, etc. Biosorption of heavy metals viz. Pb and Cu were increased at high pH using *P. purpurogenum* and *P. canescens* (Say et al., 2003; Ianis et al., 2006). Similarly, increased aeration and level of carbon sources like glucose mediated the enhanced biosorption of Cd and Pb by *Saccharomyces cerevisiae* (Ying et al., 2009; Damodaran et al., 2015). *Trichoderma* sp. was found to bioremediate most of the heavy metals like Cd, Cu, Zn, Pb, Cr, and Ni by producing large biomass for biosorption of these metals (Siddiquee et al., 2015).

10.4.5 Biotransformation

Biotransformation is the active process of modification of the contaminants into simpler nontoxic components. It involves the mechanism of bioprecipitation, oxidation, reduction, methylation, and demethylation to catalyze the surrounding environment and to interact with the substrate. Mineralization of organic compounds to carbon dioxide leads to production of excess bicarbonates due to alkaline pH in the cell causing precipitation of the compounds in nontoxic forms. Reduction of metals by catalysis of enzymes is used in producing electron acceptors from heavy metals. Transformation of organic and inorganic compound to volatile forms using biological processes is called biovolatilization (Singh et al., 2019; Singh, 2006). Biotransformation of metals, organometals, and metalloids improves their mobility and reduces their toxicity. Methylation of metals occurs by 3 different pathways like methylcobalamine, S-adenosylmethionine (SAM), and *N*-methyltetrahydrofolate pathway (Mason, 2012; Mukhopadhyay et al., 2002). Fungi provide a large surface to volume ratio thus is useful in bioremediation by biosorption, methylation, accumulation, volatilization, and precipitation of arsenic. Fungi methylate As to their organic forms like mono-, di-, and tri-methyl arsenate (MMA, DMA, TMA) and resulted in volatilization. Fungal strains *Trichoderma* sp., *Rhizopus, Fusarium, Penicillium, Candida* sp. *and Scopularis,* and *Neospora* sp. oxidized and methylated the arsenic into its volatile organic forms (Cullen and Reimer, 1989; Srivastava et al., 2011; Singh et al., 2015; Zeng et al., 2015). *Fusarium* converts arsenate to volatile TMA by a long process, which involves enzyme mediated arsenate reduction, methylation, reoxidation, and further reduction with SAM. Selenium gets methylated by SAM pathway while mercury and tin use cobalamine pathway (Mason, 2012; Mukhopadhyay et al., 2002. Biotransformation of substrate by the process of esterification, hydroxylation, deoxygenation, dehydrogenation, etc., by fungi can transform toxic pesticides to simpler forms like *Aspergillus niger* could hydroxylate and oxygenate phenoxybenzoic acid to less toxic compounds such as gallic acid and phenols (Deng et al., 2015). *Pleurotus ostreatus*, white rot basidiomycete fungi, could biotransform aldrin via epoxidation and hydroxylation processes (Purnomo et al., 2017).

10.4.6 Bioaccumulation

Bioaccumulation in contradiction to biosorption, is an active metal influx mechanism, but is slower and requires favorable conditions. *Trichoderma* sp., *Penicillium janthinellum, Fusarium oxysporum* strains were reported to accumulate Cu, Zn, As, and Cd under in vitro conditions (Zeng et al., 2010; Su et al., 2010; Tripathi et al., 2017). Mycorrhizal fungi, *Funneliformis geosporum* not only improve soil health by bioaccumulating metal and minerals from the soil but also fight with heavy metal toxicity (Abu-Elsaoud et al., 2017). Mushrooms act as effective heavy metal accumulators for As, Cu, Pb, Hg, Cd, and other radioactive elements like selenium and cesium. They absorb the metals through their mycelia and then channel them to their fruiting bodies (Zhu et al., 2011).

10.4.7 Bioaugmentation

Bioaugmentation is dependent on availability of contaminant, microbes, and enzymatic activity to catabolize (Kiamarsi et al., 2019). *Irpex lacteus* and *Pleurotus ostreatus* were found to remove PAH by 27%–36% and 55%–67% in the soil (Byss et al., 2008). *Absidia cylindrospora* degraded ~90% of fluorine in a soil slurry in 288 h (Garon et al., 2004). Similarly, lignin peroxidase producing *Gloeophyllum striatum* MTCC-1117 degraded coal (Yadav et al., 2009) and 77% PAH is reported to be degraded by *Scopulariopsis brecicaulis* (Mao and Guan, 2016).

10.4.8 Other known mechanisms

Biosurfactants, an ecologically safe, hydrophobic or hydrophilic surface active compounds produced by fungi lowers the surface tension between the molecules and are used in bioremediation (Lang, 2002; Satpute et al., 2010). Addition of nutrients to biostimulate the contaminant degradation by microbes is another process. Carbon, nitrogen phosphates, and other organic biostimulants can be added to provide optimal; conditions to synthesize degradative enzymes and fungal growth. Addition of nitrogen and phosphorus as NH_4Cl and NaH_2PO_4 compounds biostimulated removal of PAH by 20% more efficiency in contaminated desert soils (Abed et al., 2015). Nutrient addition mediated enhanced PAH biodegradation in the top soil and 62% increase in biodegradation efficiency of PAH and 32% of total hydrocarbons reduced with surfactant addition was reported in crude oil contaminated soil by addition of nutrients (Breedveld and Sparrevik, 2000; Zucchi et al., 2003; Chaîneau et al., 2005).

Degradation of other major hydrocarbon complexes like, polychlorinated dibenzo dioxins/furans (PCDD/Fs) by litter decomposing fungi, *Stropharia rugosoannulata,* has been recorded as first instance of PCDD/F degradation in solid state cultivation (Valentín et al., 2013). In a study *Pleurotus pulmonarius* could degrade PCDD/Fs coupled with solid state fermentation technique in contaminated soil slurry and reduced to negligible and hexa-, hepta-, and octa-CDD/F by more than 80%, 97%, and 90%, respectively. Major stages of degradation were mycelium colonization and fruiting body stage with 71% and 26% removal efficiency (Kaewlaoyoong et al., 2020) (Table 10.2).

10.5 Role of environmental factors on mycoremediation

Fungal growth and its ability to biodegrade are dependent on various abiotic factors as fungi perpetuate under specific environments. Temperature, pH, humidity, aeration, light, and the presence of other compounds are major conditions, which are needed to be optimized for bioremediation. When microbial growth and activity is facilitated by optimal environment conditions its application for degradation enhances. *Fusarium* sp. and *Trichosporon* spp. could be grown in a wide range of pH 4 to 9, while *Sepedonium* grows in pH range of 4 to 10. Though *Geotrichum* sp. can tolerate a wide pH range, it grows best at pH 3, while *Rhizopus* sp. and *Aspergillus* at pH 5.5, *Alternaria* sp., and *Phytophthora* at pH 6.5 and *Fusarium* sp. showed maximum growth at pH 7 under dark conditions (Hassan et al., 2017). Production of biomass is necessary for better remediation. High aeration helps in enhancing the biomass and exopolysaccharide in *Ganoderma lucidum* (Agudelo-Escobar et al., 2017). Nutrient exhaustion can sometimes reduce the ability of fungus to bioremediate, which can be overcome by biphasic treatment. In nutrient sufficient condition white rot fungus, *Phanerochaete chrysosporium* degrade PAH by cytochrome P450 monooxygenase and under nutrient deficient conditions action of

Table 10.2 Fungal mechanisms to mycoremediate heavy metals.

Mechanism	Fungi	Metal	References
Biosorption	*Circinella* sp., *Cunninghamella* sp., *Rhizophus* sp., *Penicillium ap., Hymenoscyphus ericae, Trichoderma* sp., *P. purpurogenum, P. canescens, Saccharomyces cerevisiae*	Pb, Ni, Zn, Cd, Cr, Cu, Fe, As	Fourest and Roux (1992); Say et al. (2003); El-Morsy (2004); Ianis et al. (2006); Ying et al. (2009); Bhattacharyya et al. (2009); Luo et al. (2010); Alpat et al. (2010); Mohsenzadeh and Shahrokhi (2014); Damodaran et al. (2015); Binsadiq (2015); Siddiquee et al. (2015)
Bioadsorption	*Apergiilus letulus, Pisolithus* sp.	Pb, Cu,Cr, Ni, Zn	Tam (1995); Mishra and Malik (2014)
Bioaccumulation	*Mucor circinelloides, Trichoderma* sp., *Penicillium* sp., *Paxillus involutus, Phanerochaete chrysosporium, Pleurotus ostreatus, Fusarium oxysporum, Mushroom, Funneliformis geosporum*	Pb, Zn. Cu, Ni, Cd As, Hg	Blaudez et al. (2000); Zeng et al. (2010); Su et al. (2010); Zhu et al. (2011); Mohsenzadeh and Shahrokhi (2014); Xu et al., 2014; Fazli et al. (2015); Tripathi et al., 2017; Abu-Elsaoud et al. (2017); Zhang et al. (2017); Kapahi and Sachdeva (2017); Hassan et al. (2017)
Mobilization	*Phanerochaete chrysosporium*	Cd	Xu et al. (2015)
Immobilization	*Bjerkandera fumosa Trametes versicolor, Phlebia radiata*	Oxalates of Ca, Cd, Zn	Jarosz-Wilkolazka and Gadd (2003); Gadd (2004); Franceschi and Nakata (2005)
Biotransformation	*Trichoderma* sp., *Rhizopus* sp., *Fusarium* sp., *Penicillium* sp., *Candida* sp., *Scopularis* sp., *Neospora* sp.	As, Se, Sn, Hg	Cullen and Reimer (1989); Mukhopadhyay et al. (2002); Srivastava et al. (2011); Mason (2012); Singh et al. (2015); Zeng et al. (2015)

lignolytic enzymes making it a more efficient system (Bhattacharya et al., 2013). Some robust fungi like *Pleurotus ostreatus* and *P. eryngii* can degrade PAH along with heavy metals by activating antioxidant enzymes (Vaseem et al., 2017). Biosorption ability of *Circinella* sp. and *Cunninghamella echinulata* to remediate nickel was found to be influenced by pH temperature, biomass and time (Alpat et al., 2008; El-Morsy, 2004). *Penicillium* sp., *Rhizopus* sp., *Candida* sp., and *Mucor hiemalis* was studied to be influenced by same parameters for bioaccumulation of heavy metals (Ahmad, 2005; Dugal and Gangawane, 2012; Zafar et al., 2007; Rodríguez et al., 2018; Akhtar et al., 2008; Godlewska-Żyłkiewicz et al., 2019).

Similarly, remediation of Pb, Zn, Cr, Ni, Cd, and Cu by *Aspergillus* sp. is also dependent on pH, temperature, time, and metal concentration. They are also reported to bioremediate heavy metals from saline soils. In another study, *Aspergillus* sp. was also able to remove Cr and nickel by 92% and 90%, respectively, at 7.0 pH (Thippeswamy et al., 2012; Congeevaram et al., 2007; Price, 2000; Joshi et al., 2011; Siokwu and Anyanwn, 2012). The mushroom species *Pleurotus ostreatus* was found to remove manganese from surfactant containing contaminated water and metals like Co, Pb, Zn, Ni, Cr, and Cu from coal mining areas. They were also found to increase antioxidant enzymes to tolerate these toxic metals (Vaseem et al., 2017; Wu et al., 2016). Thus we can conclude that mycoremediation is a green, economic way to fight deleterious heavy metals with indigenous fungi and improve the soil and water quality. Optimum temperature range for fungi belonging to genus *Beauveria, Paecilomyces,* and *Metarhizium* is needed for proper mycelia growth (Hallsworth and Magan, 1999; Piątkowski and Krzyżewska, 2013). Marine fungus *Cochliobolus lunatus* was able to grow in low nutrient high salt concentration and also degrade PAH chrysene (Bhatt et al., 2014).

10.6 Omics in mycoremediation

Metabolic characters responsible for bioremediation are governed by various specific genes in the microorganisms. These specific genes can be engineered for manipulation of the metabolic processes in targeted organisms for enhancing decontamination mechanism. Various tools can be used for developing genetically engineered fungi making them more efficient due to presence of specific characters, enzyme, genes to degrade its targeted contaminant like heavy metal PAH, pesticide, etc. (Marihal and Jagadeesh, 2013).

Modulation of gene expression is a recent strategy for detoxification of heavy metals. *Odiodendron maius* has proven as a model organism to demonstrate heavy metal tolerance by omics and heterologous expressions studies (Chiapello et al., 2015). In *Gigaspora margarita,* metallothionein gene was upregulated when exposed to copper, while in *G. intraradices*, gint ABC1, an ABC transporter gene upregulated on cadmium and copper exposure (Lanfranco et al., 2002; González-Guerrero et al., 2010). Transgenic yeast expressing low affinity cation transporter encoding gene *LCT1* (TC 9.A.20.1.1) from wheat has been shown to hyperaccumulate Cd (Clemens et al., 1998). Yeast expressing AtNramp cDNA, metal transporters from *Arabidopsis thaliana* showed accumulation of cadmium and iron (Thomine et al., 2000). Overexpression of gene *acrA*, (a plasma membrane arsenic efflux pump), along with its fusion with *egfp* in *Aspergillus niger* act as a biosensor for monitoring bioremediation (Choe et al., 2012). In *Fibroporia radiculosa,* specific genes encoding regulatory mechanisms like Cu transporting ATPase pump and Cu homeostasis gene (*CutC*) were upregulated during early Cu degradation (Tang et al., 2013). In another study, it was found that the ATPase pump in *F. radiculosa* expels copper from the cell to adapt to copper in the early decay stage in ammoniacal copper citrate (1.2%) treated wood, where ATPase pump overexpressed (Jenkins, 2012). Copper resistance P-type ATPase pump, catalase, aryl alcohol oxidase, and oxalate decarboxylase upregulated in *F. radiculosa* decay the alkaline copper treated wood (Akgul et al., 2018).

Symbiotic association of Mycorrhiza with plants influences expression of plant genes at heavy metal stress. In *Saccharomyces cerevisiae,* various functional complementation studies of ericoid mycorrhiza genes encoding transporters, pumps, enzymes, and other proteins has been done (Daghino et al., 2016). Under copper stress, detoxification by metallothionein through yeast complementation assay were observed while *Rhizophagus irregularis-Salix purpurea* symbiotic association responded to regulation of aquaporin genes *PIP1*, 2 and *TIP* 1,2 of *Salix purpurea* (Kohler et al., 2004; Almeida-Rodríguez et al., 2016). *Hymenoscyphus ericae,* ericoid mycorrhizal fungus, has shown to influence uptake of heavy metals like As by the plant *Calluna vulgaris* in arsenic contaminated soils (Sharples et al., 2000). Various genes responsible for polyamines were also found to be upregulated in zinc- and copper-contaminated conditions (Prabhavathi and Rajam, 2007). In white poplar-AMF symbiosis various metallothionein and polyamine genes (*PaMT* 1, 2,3, *PaADC*, *PaSPDS* 1 and 2) were induced in heavy metal contaminated soil (Cicatelli et al., 2010). In *Glomus intraradices* Sy167, various genes viz. metallothionein, Zn transporter, glutathione S-transferase, and heat shock proteins were expressed during heavy metal stress for tolerance (Hildebrandt et al., 2007). These genes upregulated in *Medicago truncatula* (MtPT4) roots in zinc-contaminated soil (Nguyen et al., 2019). Various phosphate transporters based on mycorrhizal association showed enhance expression in AMF, symbiosis with *Lycopersicum esculentum* (*LePT3* and *LePT4*) and *Solanum tuberosum* (*StPT3*) (Nagy et al., 2005).

Arbuscular mycorrhizal fungi (AMF) colonization with *Rhizophagus irregularis* and *Funneliformis mosseae* was used to phyto remediate ammonia, methyl tetra butyl ether, and benzene contaminated ground water (Fester, 2013). A binding affinity to cellulose has been engineered by expression of *Trichoderma reesei* genes that encodes cellulose binding domain from cellobiohydrolase I and II on cell surface of *Saccharomyces cerevisiae* (Nam et al., 2002). Phenotype Microarray technique was used to study oxidative stress pathways in *Penicillium griseofulvum* in presence of beta-hexachloro cycloalkane and toluene (Ceci et al., 2015). PAH degrading microbial consortium using two strains of genetically engineered *Aspergillus niger* expressed lignin and manganese peroxidase (LiP+5 strain and MnP +7 strain) genes were derived from *Phanerochaete chrysosporium* for prolonged survival, tolerance, and efficient degradation of low- and high-molecular-weight PAH (Zafra et al., 2017). A complete transcriptomic and genomic analysis of PAH degradation by *Dentipellis* sp. KUC8613 indicated the use of nonligninolytic enzymes, upregulation of PAH specific and general P450s genes. 3D model based on homology along with ligand docking stimulation to compare versatility of fungal P450 enzyme system (CYP63A2) showed a large sized active site in the fungal enzyme (Syed et al., 2013). This concept was used in another fungal system (CYP1A1- P450), which enlarges site of substrate binding and enhance the 2,3,7,8 tetrachloro-dibenzo-p-dioxin metabolizing ability (Sakaki et al., 2013).

A promoter (prom1) from *Cochliobolus heterostrophus* and terminator (*trp* C) from *Aspergillus nidulans* were used to transform and express *opd* gene in *Gliocladium virens* for degradation of organophosphates derived contaminants, paraoxon, and di-isopropyl-fluorophosphate (Dave et al., 1994). A complete metabolic genome analysis of *A. niger* showed it had 1100 unique enzyme encoding genes which were helpful in its survival in diverse contaminated conditions (Sun et al., 2007). In *P. pastoris, Ganoderma lucidum*, and *Pleurotus ostreatus,* laccase-synthesizing genes GILCC 1 and POXA1B were expressed (Rivera-Hoyos et al., 2015). Recombinant laccase constructs from *Pleurotus eryngii* (ERY 4) laccase genes to obtain chimerical enzymes, 4NC3 in *S. cerevisiae* showed multiple substrate affinity with temperature and pH stability (Bleve et al., 2014). A similar study using modified thermal asymmetric interlaced PCR was done to obtain highly efficient recombinant laccase, yeast laccase gene 1557 bp (YILac) for hydrolyzing wood by yeast *Yarrowia lipolytica*. Its expression in *Pichia pastoris* showed degradation of phenolic compounds in biomass of acid treated wood (Kalyani et al., 2015).

Various dyes like azo dyes and anthraquinone dyes were bioremediated by transgenic *Pichia pastoris* fungus transferred with cDNA lcc1, a fungal laccase, isolated from *Trametes trogii* (Colao et al., 2006). *Penicillium oxalicum* SAR-3 strain having a broad spectrum azo dye catabolism ability possess various genes coding for ABC transporters and peroxidases for the degradation of azo dyes along with stress response genes. Forward suppression subtractive hybridization (SSH) of cDNA library showed about 183 unique expressed sequence tags *in P. oxalicum* SAR-3 in response to Acid RED azo dye (Saroj et al., 2014). Expression of laccase gene of *Phanerochaete flavido-alba* in *Aspergillus niger* along with recombinant enzyme Lac-LPFS (active) showed biotransformation and decoloration of synthetic dyes used in textiles (Benghazi et al., 2014).

Genetically engineered *Fusarium solani* strains with improved dehalogenase activity due to parasexual cycle lead to degradation capability of DDT, dichloro diphenyl trichloroethane, in the recombinant fungi (Mitra et al., 2001). Degradation and tolerance of the pesticide dichlorvos, 2,2-dichlorovinyl dimethyl phosphate by *Trichderma atroviride* was found to be associated with alteration and functioning of Hex1-related transcriptome and other ABC transporters (Zhang et al., 2015).

10.7 Fungal interactions for enhanced mycoremediation

Plant microbe interaction enhances metabolic capacities for remediation by improving metal availability and mobility to the plants by releasing chelating agents and acidifying the redox reactions by soil microbes (Hong-Bo et al., 2010). *Trichoderma* sp. and arbuscular mycorrhizal fungi (AMF) facilitates uptake by plant and thus helps in phytoremediation processes. Siderophore production by microbes helps sequestration of iron and nickel enhancing plant growth and removing toxicity in soil (Dimkpa et al., 2008). AMF symbiosis enables metallophytes like plants belonging to *Brassicaceae* and *Caryophyllaceae* to grow under heavy metal stress at high nutrient demand conditions (Demars and Boerner, 1996). *Paxillus pinus* symbiosis with AMF under cadmium toxicity presents a phenolic defense system as well as GSH protection pathway by mycorrhiza (Schützendübel and Polle, 2002). *Trichoderma atroviride* association with *Brassica juncea* enhanced the translocation and uptake of Ni, Cd, and Zn. *T. harzianum* promoted *Salix fragilis* growth in heavy metal contaminated soil, while detoxified potassium cyanide and hyperaccumulation of arsenic enhanced the growth of *Pteris vittata* (Lynch and Moffat, 2005; Adams et al., 2007).

Degradation of total petroleum hydrocarbons was more than twice for *Pharbitis nil* along with its microbial community (27.63%–64.72%) as compared to control (10.2%–35.6%) (Zhang et al., 2010). Endophytic fungi infected grass species *Festuca arundinacea* and *F. pratensis* degraded total petroleum hydrocarbons in soil contaminated with aged petroleum. These grasses had more root and shoot biomass as compared to infected ones (Soleimani et al., 2010). Mutual symbiosis between host plants like maize and dark septate endophytic colonies of *E. piscipula* act as an efficient strategy to alleviate stress due to heavy metal. It also improves root and shoot growth (Li et al., 2011). Culture of watermelon seeds in endophytic fungus *Ceratosasidium stevensii* treated soil showed degradation of phenolics along with increase in stem and leaf length (Yi et al., 2014). *Penicillium janthinellum* LK5, a gibberellin-producing endophytic fungal strain, reduces electrolyte and lipid peroxidation in infected *Solanum lycopersicum* plants along with increasing reduced glutathione and catalase to sustain cadmium induced oxidative stress and membrane injury (Khan et al., 2014). AMF and plant symbiosis improves plant growth due to improved phosphorus and water uptake and reduced enzymes for oxidative stress (peroxidase and superoxide dismutase) activities (Neagoe et al., 2013). Thus contaminant stress effects on plant growth can be counteracted by such symbiosis.

Trichoderma H8 and *Aspergillus* G16 strains associated with *Acacia auriculiformis* enhanced growth of mustard (*Brassica juncea* cross var. *foliosa Bailey*) soils contaminated with cadmium and nickel. Along with that 167% more fresh weight was observed in *Trichoderma*, 44% more yield in *Aspergillus* and 178% higher yield in both *Trichoderma* and *Aspergillus* inoculated plants (Jiang et al., 2008). Yeast, from the roots of canola plants showed host resistance to various metals. Endophytic

Cryptococcus sp. CBSB78 inoculation in *B. alboglabra* enhanced Cd, Pb, and Zn extraction in multi metal contaminated soils (Deng et al., 2012). *Enterobacter* sp. when compared with yeast, *Rhodotorula* sp. inoculated *Brassica* seedlings showed better Zn, Cd, Pb, and Cu extraction efficacy in field conditions (Wang et al., 2013). Phytoremediation ability of *Brassica juncea* was enhanced when the associated fungi *Enterobacter* sp. was transformed with bifunctional glutathione synthase (*gcsgs*) gene. Seedlings inoculated with CBSB1-GCSGS showed increased dry weight, fresh weight, and shoot length by 160%, 123%, and 67%, respectively (Qiu et al., 2014). *Lasiodiplodia* sp., an endophytic fungus resistant to cadmium, zinc, and lead obtained from *Portulaca oleracea* extracted Cd in canola (*Brassica napus*) from the Cd- and Pb-contaminated soils with increment in biomass (Deng et al., 2014b). Self-fusion protoplast engineered fungi showed enhanced cadmium resistance. The fusant inoculated in canola showed 62% more dry weight than wild *Mucor* sp. in Cd- and Pb-contaminated soils (Deng et al., 2013). Interspecific fusion of inactivated protoplasts of three stable fusants F1, F2, and F3 were constructed from *Mucor* sp., resistant to cadmium, *Fusarium* sp. resistant to zinc. These bioremediated Zn, Cd, and Pb by enhancing metal translocation and extraction efficiency in canola (Deng et al., 2014a).These mechanisms are dependent on plant and fungal species, availability of metal, environment, and soil conditions.

Lichens have the ability to accumulate and retain heavy metals by sequestering them as oxalate crystals or making complexes with lichen acids. Cell wall of lichen restrains various heavy metals as well as they get transported by chelation and sequestration mechanisms across the cell membrane to maintain homeostasis of metal. Algae and fungi cell walls are efficient to segregate metal for detoxification, while the cell does other mechanisms of producing organic acids and secondary metabolites. *Acarospora rugulosa* was able to accumulate copper by 16% by forming copper oxalate in copper tolerant lichen (Chisholm et al., 1987). Zinc oxalate production and lead immobilization were observed in *Diploschistes muscorum* (Sarret et al., 1998). Various organic acids like oxalic, citric, and malic acid performed bioremediation in *Lecanora polytropa* (Pawlik-Skowrońska et al., 2006). Lichen has shown a capability to remove colors of textile synthetic dyes. *Permelia perlata*, a lichen, was used to remove color, detoxify, and biodegrade the dye solvent Red 24 (SR24) and also has the ability to remove the color of other structurally different dyes (Saratale et al., 2009; Kulkarni et al., 2014).

Fungi and bacterial co-cultures are explored for bioremediation of contaminants like PAHs, petroleum hydrocarbons, dyes, etc. Soil inoculated with bacterial fungal co-culture showed increased PAH degradation as well as reduced mutagenicity of organic soils extracts as compared to axenic culture inoculated soils (Boonchan et al., 2000). Co-culture of *Aspergillus terreus*, *Rhodococcus* spp., and *Penicillium* sp. showed total biodegradation of anthracene pyrene and phenanthrene PAHs along with reduced ecotoxicity of soil (Kim and Lee, 2007). A microbe consortium having *Trametes gibbosa, Pseudomonas, Gordonia, Acinetobacter, Chryseobacterium, stenotrophomonas,* and *Alcaligenes* was able to eradicate 90% of the diesel fuel in a culture within 10 days (Zanaroli et al., 2010). Co inoculation of *Cunninghamella echinulata,* and *Vibrio rumoiensis* into mangrove sediments contaminated with hydrocarbon bioremediated total petroleum hydrocarbons (TPH) in a more efficient manner than single microorganism (Li and Li, 2011). Yeast- bacterium consortium was able to bioaugment TPH and PAH thus detoxifying highly contaminated oilfield soils (Qiao et al., 2014). Another yeast bacteria coculture was able to clean up polychlorinated biphenyl by enhancing its degradation by 69.9%, while bacteria and fungi alone could degrade it by 57% and 27.8%, respectively (Chen et al., 2015). Bacterial-fungal consortium was able to decolorize and degrade textile dyes in wastewater treatments and prevention of bacterial contamination during the process of bioremediation of textile effluents by fungi, as they enhance the process of decolorization of red dye (Qu et al., 2010; Zhou et al., 2014). Chlorpyrifos degrading *Serratia* sp. and fungal strain, *Thichosporon* degraded chlorpyrifos and intermediate 3,5,6-trichloro-2 pyridinol and results in complete mineralization of chlorpyrifos, when inoculated together (Xu et al., 2007). *Bacillus licheniformis,* when degraded β-cypermethrin, produced a toxic intermediate, 3-phenoxybenzoic acid which was further degraded by *Aspergillus oryzae.* Thus their co-inoculation gave more improved results for degradation of β-cypermethrin (Zhao et al., 2016).

10.8 Transgenic plants using fungal genes for different contaminants

Fungi mediated bioremediation through modulation of certain genes offer utilization of these novel genes for transgenic development both for phyto and bioremediation purposes. Mechanisms involving decontamination of pollutants present in the fungi can be transferred via. Various tools to create super plants with improved abilities to remove recalcitrant compounds. Number of researchers has engineered certain specific genes sourced from various fungi into plants to produce transgenic plants for phytoremediation.

Bioremediation ability to uptake, remove, and translocate metals involves various genes which can be transferred to plants. Yeast metallothioneins (MT) gene, responsible for metal binding proteins, was expressed in tobacco plants which improved its tolerance to Cd (Misra and Gedamu, 1989). Later, *CUP1* gene was transferred from yeast to cauliflower improving Cd tolerance by 16 times, and in tobacco, it enhanced phytoextraction of copper along with cadmium (Thomas et al., 2003; Hasegawa et al., 1997). An ABC transporter *YCF 1* from yeast transferred to *Arabidopsis thaliana* showed improved lead and cadmium tolerance in the transgenic plants as the gene is responsible for vacuolar immobilization of metals by conjugation with glutathione (Song et al., 2003). Ferric reductase encoding genes *FRE1* and *FRE2* transferred from yeast to tobacco showed 1.5 fold improved iron in the transgenic plant (Samuelsen et al., 1998).

Due to plants inability to volatilize or methylated arsenic, several genetic engineering approaches to develop transgenic *Oryza sativa* and *Arabidopsis thaliana* using arsenic methyltransferase (*arsM*) from microbial sources was found to induce methylation. Expression of *asrM* gene from *R. palustris* in rice, methylate As and made it tenfold volatile resulting in reduced toxicity. In rice low As was found in transgenic line with *asrM*, as nontoxic forms like monomethylarsonic acid and dimethyl arsonic acid present in grain, root, and shoot. Soil fungus, *Westerdykella aurantiaca,* has also been used as novel *arsM* (*Waars*M) gene with S-adenosylmethionine binding motifs in rice (Meng et al., 2011; Verma et al., 2016).

Manganese peroxidase gene from fungi *Coriolus versicolor* was expressed in tobacco plants. These transgenic tobacco plants were successful in removing pentachlorophenol. In later studies hybrid *Populus* spp. was introduced with 1098 bp full cDNA encoding manganese peroxidase from *Coriolus versicolor* with a double 35S promoter control from cauliflower mosaic virus was found to remove bisphenol from the medium (Iimura et al., 2002, 2007). Similarly, *Coriolos versicolor* gene responsible for laccase enzyme was transformed to tobacco plants (Nasrin et al., 2010). Although various biotechnological advancements might be useful in developing promising, efficient, and economic methods of bioremediation, the ethical debate to use the genetically modified fungi and transgenic plants is yet to be resolved for effective remediation of contaminants in the environment.

10.9 Conclusion and future prospects

Metallic, organic, and inorganic toxic environmental contaminants are rising with the increase in anthropogenic and industrial activities into the natural resources. Mycoremediation has proven as an effective, eco-friendly, economic, and sustainable approach for achieving a clean green environment. Diverse fungal species have been successfully used to remediate pollutants like heavy metals, PAH, pharmaceutical, agricultural, and other industrial waste into nontoxic products successfully due to their robust mycelia, versatile metabolism, and powerful extracellular enzymes.

There is a need to develop better technology to enhance and economize the practice using various engineering and genetic advancements. The potential of fungi should be exploited and interacted with different branches of science to produce more efficient, substrate-specific, and faster remediation technology to relieve the aquatic and terrestrial resources from the environmental stress and develop a safe habitat.

References

Abdel-Hamid, A.M., Solbiati, J.O., Cann, I.K., 2013. Insights into lignin degradation and its potential industrial applications. Adv. Appl. Microbiol. 82, 1–28.

Abdel-Shafy, H.I., Mansour, M.S.M., 2016. A review on polycyclic aromatic hydrocarbons: source, environmental impact, effect on human health and remediation. Egypt. J. Pet. 25, 107–123.

Abed, R.M.M., Al-Kharusi, S., Al-Hinai, M., 2015. Effect of biostimulation, temperature and salinity on respiration activities and bacterial community composition in an oil polluted desert soil. Int. Biodeterior. Biodegradation 98, 43–52.

Abu-Elsaoud, A.M., Nafady, N.A., Abdel-Azeem, A.M., 2017. Arbuscular mycorrhizal strategy for zinc mycoremediation and diminished translocation to shoots and grains in wheat. PLoS One 12 (11), e0188220.

Adams, P., De-Leij, F.A.A., Lynch, J.M., 2007. *Trichoderma harzianum* Rifai 1295-22 mediates growth promotion of crack willow (*Salix fragilis*) saplings in both clean and metal-contaminated soil. Microb. Ecol. 54, 306–313.

Agarwala, S.C., Kumar, A., Sharma, C.P., 1961. Effect of excess supply of heavy metals on barley during germination, with special reference to catalase and peroxidase. Nature 191, 726–727.

Agudelo-Escobar, L.M., Gutiérrez-López, Y., Urrego-Restrepo, S., 2017. Effects of aeration, agitation and pH on the production of mycelial biomass and exopolysaccharide from the filamentous fungus *Ganoderma lucidum*. DYNA 84, 72–79.

Ahluwalia, S.S., Goyal, D., 2007. Microbial and plant derived biomass for removal of heavy metals from wastewater. Bioresour. Technol. 98, 2243–2257.

Ahmad, F., 2005. Heavy metal biosorption potential of *Aspergillus* and *Rhizopus* sp. isolated from wastewater treated soil. J. Appl. Sci. Environ. Mgt., 9.

Ajibade, F.O., Adelodun, B., Lasisi, K.H., Fadare, O.O., Ajibade, T.F., Nwogwu, N.A., Sulaymon, I.D., Ugya, A.Y., Wang, H.C., Wang, A., 2021. Environmental pollution and their socioeconomic impacts. In: Microbe Mediated Remediation of Environmental Contaminants. Elsevier, pp. 321–354.

Akerman-Sanchez, G., Rojas-Jimenez, K., 2021. Fungi for the bioremediation of pharmaceutical-derived pollutants: a bioengineering approach to water treatment. Environ. Adv. 4, 100071.

Akgul, A., Tang, J.D., Diehl, S.V., 2018. Gene expression analysis of wood decay fungus *Fibroporia radiculosa* grown in acq-treated wood. Wood Fiber Sci. 50, 12.

Akhtar, S., Mahmood-Ul-Hassan, M., Ahmad, R., Suthor, V., Yasin, M., 2013. Metal tolerance potential of filamentous fungi isolated from soils irrigated with untreated municipal effluent. Soil Environ. 32, 55–62.

Akhtar, N., Mannan, M.A., 2020. Mycoremediation: expunging environmental pollutants. Biotechnol. Rep. 26, e00452.

Akhtar, K., Waheed Akhtar, M., Khalid, A.M., 2008. Removal and recovery of zirconium from its aqueous solution by *Candida tropicalis*. J. Hazard. Mater. 156, 108–117.

Al-Hawash, A.B., Alkooranee, J.T., Zhang, X., Ma, F., 2018. Fungal degradation of polycyclic aromatic hydrocarbons. Int. J. Pure Appl. Biosci. 6, 8–24.

Alimi, O.S., Farner Budarz, J., Hernandez, L.M., Tufenkji, N., 2018. Microplastics and nanoplastics in aquatic environments: aggregation, deposition, and enhanced contaminant transport. Environ. Sci. Technol. 52, 1704–1724.

Almeida-Rodríguez, A.M., Gómes, M.P., Loubert-Hudon, A., Joly, S., Labrecque, M., 2016. Symbiotic association between *Salix purpurea* L. and *Rhizophagus irregularis*: modulation of plant responses under copper stress. Tree Physiol. 36, 407–420.

Alpat, S., Alpat, S.K., Çadirci, B.H., Özbayrak, Ö., Yasa, I., 2010. Effects of biosorption parameter: kinetics, isotherm and thermodynamics for Ni(II) biosorption from aqueous solution by *Circinella sp*. Electron. J. Biotechnol. 13, 4–5.

Alpat, S., Alpat, S., Cadirci, B., Yasa, I., Telefoncu, A., 2008. A novel microbial biosensor based on *Circinella sp*. modified carbon paste electrode and its voltammetric application. Sensors Actuators B Chem. 134, 175–181.

Al-Tamimi, W.H., Burghal Burghal, A.A., Abu-Mejdad Abu-Mejdad, N.M.J., 2021. Production of biosurfactant from *Candida cruzi* isolated from produce water oil fields in Basrah for microbial enhance oil recovery. J. Pet. Res. Stud. 8, 1–14.

Alves, M.H., Campos-Takaki, G.M., Porto, A.L.F., Milanez, A.I., 2002. Screening *of Mucor* spp. for the production of amylase, lipase, polygalacturonase and protease. Braz. J. Microbiol. 33, 325–330.

Amjad, A.L., Di, G.U., Mahar, A., Ping, W., Feng, S.H., Ronghua, L.I., Zhang, Z., 2017. Mycoremediation of potentially toxic trace elements—a biological tool for soil cleanup: a review. Pedosphere 27 (2), 205–222.

Amoah, J., Ho, S.-H., Hama, S., Yoshida, A., Nakanishi, A., Hasunuma, T., Ogino, C., Kondo, A., 2016. Lipase cocktail for efficient conversion of oils containing phospholipids to biodiesel. Bioresour. Technol. 211, 224–230.

Anasonye, F., Winquist, E., Kluczek-Turpeinen, B., Räsänen, M., Salonen, K., Steffen, K.T., Tuomela, M., 2014. Fungal enzyme production and biodegradation of polychlorinated dibenzo-p-dioxins and dibenzofurans in contaminated sawmill soil. Chemosphere 110, 85–90.

Anastasi, A., Spina, F., Prigione, V., Tigini, V., Varese, G.C., 2010. Textile wastewater treatment: scale-up of a bioprocess using the fungus *Bjerkandera adusta*. J. Biotechnol. 150, 52.

Andrade, V.S., Sarubbo, L.A., Fukushima, K., Miyaji, M., Nishimura, K., de Campos-Takaki, G.M., 2002. Production of extracellular proteases by *Mucor circinelloides* using D-glucose as carbon source/substrate. Braz. J. Microbiol. 33.

Aneja, K.R., Mehrotra, R.S., 1980. Comparative cellulolytic ability of microfungi inhabiting various types of litter. Proc. Indian Natl. Sci. Acad. B. Biol. Sci. 46, 566–571.

Angayarkanni, J., Palaniswamy, M., Swaminathan, K., 2003. Biotreatment of distillery effluent using *Aspergillus niveus*. Bull. Environ. Contam. Toxicol. 70, 268–277.

Angel, T.M., 2002. Molecular biology and structure-function of lignin degrading heme peroxidases. Enzym. Microb. Technol. 30, 425–444.

Asgher, M., Asad, M.J., Legge, R.L., 2006. Enhanced lignin peroxidase synthesis by *Phanerochaete chrysosporium* in solid state bioprocessing of a lignocellulosic substrate. World J. Microbiol. Biotechnol. 22, 449–453.

Asif, M.B., Hai, F.I., Singh, L., Price, W.E., Nghiem, L.D., 2017. Degradation of pharmaceuticals and personal care products by white-rot fungi—a critical review. Curr. Pollut. Rep. 3 (2), 88–103.

Atlas, R.M., Atlas, R.M., Bartha, R., 1992. Hydrocarbon biodegradation and oil spill bioremediation. Adv. Microb. Ecol., 287–338.

Balaji, V., Arulazhagan, P., Ebenezer, P., 2014. Enzymatic bioremediation of polyaromatic hydrocarbons by fungal consortia enriched from petroleum contaminated soil and oil seeds. J. Environ. Biol. 35, 521–529.

Balasubramanian, V., Natarajan, K., Rajeshkannan, V., Perumal, P., 2014. Enhancement of in vitro high-density polyethylene (HDPE) degradation by physical, chemical, and biological treatments. Environ. Sci. Pollut. Res. 21, 12549–12562.

Baratto, M.C., Juarez-Moreno, K., Pogni, R., Basosi, R., Vazquez-Duhalt, R., 2015. EPR and LC-MS studies on the mechanism of industrial dye decolorization by versatile peroxidase from *Bjerkandera adusta*. Environ. Sci. Pollut. Res. Int. 22, 8683–8692.

Bartha, R., Atlas, R.M., 1977. The microbiology of aquatic oil spills. Adv. Appl. Microbiol., 225–266.

Benghazi, L., Record, E., Suárez, A., Gomez-Vidal, J.A., Martínez, J., de la Rubia, T., 2014. Production of the *Phanerochaete flavido-alba* laccase in *Aspergillus niger* for synthetic dyes decolorization and biotransformation. World J. Microbiol. Biotechnol. 30, 201–211.

Benjamin, S., Pradeep, S., Josh, M.S., Kumar, S., Masai, E., 2015. A monograph on the remediation of hazardous phthalates. J. Hazard. Mater. 298, 58–72.

Bensmail, S., Mechakra, A., Fazouane-Naimi, F., 2015. Optimization of milk-clotting protease production by a local isolate of *Aspergillus niger* ffb1 in solid-state fermentation. J. Microbiol. Biotechnol. Food Sci. 04, 467–472.

Bezalel, L., Hadar, Y., Cerniglia, C.E., 1997. Enzymatic mechanisms involved in phenanthrene degradation by the white rot fungus *Pleurotus ostreatus*. Appl. Environ. Microbiol. 63, 2495–2501.

Bhalerao, T.S., Puranik, P.R., 2007. Biodegradation of organochlorine pesticide, endosulfan, by a fungal soil isolate, *Aspergillus niger*. Int. Biodeterior. Biodegradation 59, 315–321.

Bhatt, M., Cajthaml, T., Šašek, V., 2002. Mycoremediation of PAH-contaminated soil. Folia Microbiol. 47, 255–258.

Bhatt, J.K., Ghevariya, C.M., Dudhagara, D.R., Rajpara, R.K., Dave, B.P., 2014. Application of response surface methodology for rapid chrysene biodegradation by newly isolated marine-derived fungus *Cochliobolus lunatus* strain CHR4D. J. Microbiol. 52, 908–917.

Bhattacharya, S., Das, A., 2011. Mycoremediation of Congo red dye by filamentous fungi. Braz. J. Microbiol. 42 (4), 1526–1536.

Bhattacharya, S., Das, A., Prashanthi, K., Palaniswamy, M., Angayarkanni, J., 2014. Mycoremediation of benzo[a] pyrene by *Pleurotus ostreatus* in the presence of heavy metals and mediators. Biotechnology 4, 205–211.

Bhattacharya, S.S., Syed, K., Shann, J., Yadav, J.S., 2013. A novel P450-initiated biphasic process for sustainable biodegradation of benzo [a] pyrene in soil under nutrient-sufficient conditions by the white rot fungus *Phanerochaete chrysosporium*. J. Hazard. Mater. 261, 675–683.

Bhattacharyya, S., Pal, T.K., Basumajumdar, A., 2009. Modulation of enzyme activities of a lead-adapted strain of *Rhizopus arrhizus* during bioaccumulation of lead. Folia Microbiol. 54, 505–508.

Binsadiq, A.R.H., 2015. Fungal absorption and tolerance of heavy metals. J. Agric. Sci. Technol. B, 77–80.

Blaudez, D., Jacob, C., Turnau, K., Colpaert, J.V., Ahonen-Jonnarth, U., Finlay, R., Botton, B., Chalot, M., 2000. Differential responses of ectomycorrhizal fungi to heavy metals *in vitro*. Mycol. Res. 104, 1366–1371.

Bleve, G., Lezzi, C., Spagnolo, S., Rampino, P., Perrotta, C., Mita, G., Grieco, F., 2014. Construction of a laccase chimerical gene: recombinant protein characterization and gene expression via yeast surface display. Appl. Biochem. Biotechnol. 172, 2916–2931.

Bolhuis, H., Cretoiu, M.S., 2016. What is so special about marine microorganisms? Introduction to the marine microbiome-from diversity to biotechnological potential. In: The Marine Microbiome, pp. 3–20.

Boonchan, S., Britz, M.L., Stanley, G.A., 2000. Degradation and mineralization of high-molecular-weight polycyclic aromatic hydrocarbons by defined fungal-bacterial cocultures. Appl. Environ. Microbiol. 66, 1007–1019.

Breedveld, G.D., Sparrevik, M., 2000. Nutrient-limited biodegradation of PAH in various soil strata at a creosote contaminated site. Biodegradation 11, 391–399.

Brodkorb, T.S., Legge, R.L., 1992. Enhanced biodegradation of phenanthrene in oil tar-contaminated soils supplemented with *Phanerochaete chrysosporium*. Appl. Environ. Microbiol. 58 (9), 3117–3121.

Brunner, I., Fischer, M., Rüthi, J., Stierli, B., Frey, B., 2018. Ability of fungi isolated from plastic debris floating in the shoreline of a lake to degrade plastics. PLoS One 13, 2020–2047.

Burlacu, A., Israel-Roming, F., Cornea, C.P., 2018. Depolymerization of Kraft lignin with laccase and peroxidase: a review. Sci. Bull. Ser. F Biotechnol. 22, 172–179.

Byss, M., Elhottová, D., Tříska, J., Baldrian, P., 2008. Fungal bioremediation of the creosote-contaminated soil: influence of *Pleurotus ostreatus* and *Irpex lacteus* on polycyclic aromatic hydrocarbons removal and soil microbial community composition in the laboratory-scale study. Chemosphere 73, 1518–1523.

Cajthaml, T., Erbanová, P., Šašek, V., Moeder, M., 2006. Breakdown products on metabolic pathway of degradation of benz[a]anthracene by a ligninolytic fungus. Chemosphere 64, 560–564.

Cajthaml, T., Möder, M., Kačer, P., Šašek, V., Popp, P., 2002. Study of fungal degradation products of polycyclic aromatic hydrocarbons using gas chromatography with ion trap mass spectrometry detection. J. Chromatogr. A 974, 213–222.

Camacho-Morales, R.L., Guillén-Navarro, K., Sánchez, J.E., 2017. Degradation of the herbicide paraquat by macromycetes isolated from southeastern Mexico. 3Biotech 7 (5), 1.

Cameselle, C., Gouveia, S., 2019. Phytoremediation of mixed contaminated soil enhanced with electric current. J. Hazard. Mater. 361, 95–102.

Ceci, A., Pierro, L., Riccardi, C., Pinzari, F., Maggi, O., Persiani, A.M., Gadd, G.M., Papini, M.P., 2015. Biotransformation of β-hexachlorocyclohexane by the saprotrophic soil fungus *Penicillium griseofulvum*. Chemosphere 137, 101–107.

Cerniglia, C.E., Sutherland, J.B., 2001. Bioremediation of polycyclic aromatic hydrocarbons by ligninolytic and non-ligninolytic fungi. In: Fungi in Bioremediation. British Mycological society symposium series 2001, 23, pp. 136–187.

Chaîneau, C.H., Rougeux, G., Yéprémian, C., Oudot, J., 2005. Effects of nutrient concentration on the biodegradation of crude oil and associated microbial populations in the soil. Soil Biol. Biochem. 37, 1490–1497.

Chakraborty, S., Mukherjee, A., Das, T.K., 2013. Biochemical characterization of a lead-tolerant strain of *Aspergillus foetidus*: an implication of bioremediation of lead from liquid media. Int. Biodeterior. Biodegradation 84, 134–142.

Chan-Cupul, W., Heredia-Abarca, G., Rodríguez-Vázquez, R., 2016. Atrazine degradation by fungal co-culture enzyme extracts under different soil conditions. J. Environ. Sci. Health B 51 (5), 298–308.

Chen, F., Hao, S., Qu, J., Ma, J., Zhang, S., 2015. Enhanced biodegradation of polychlorinated biphenyls by defined bacteria-yeast consortium. Ann. Microbiol. 65, 1847–1854.

Chiapello, M., Martino, E., Perotto, S., 2015. Common and metal-specific proteomic responses to cadmium and zinc in the metal tolerant ericoid mycorrhizal fungus *Oidiodendron maius* Zn. Metallomics 7, 805–815.

Chibuike, G.U., Obiora, S.C., 2014. Heavy metal polluted soils: effect on plants and bioremediation methods. Appl. Environ. Soil Sci., 1–12.

Chisholm, J.E., Jones, G.C., Purvis, O.W., 1987. Hydrated copper oxalate, moolooite, in lichens. Mineral. Mag. 51, 715–718.

Chiu, S.-W., Gao, T., Chan, C.S.-S., Ho, C.K.-M., 2009. Removal of spilled petroleum in industrial soils by spent compost of mushroom *Pleurotus pulmonarius*. Chemosphere 75, 837–842.

Choe, S.-I., Gravelat, F.N., Al Abdallah, Q., Lee, M.J., Gibbs, B.F., Sheppard, D.C., 2012. Role of *Aspergillus niger* acrA in arsenic resistance and its use as the basis for an arsenic biosensor. Appl. Environ. Microbiol. 78, 3855–3863.

Cicatelli, A., Lingua, G., Todeschini, V., Biondi, S., Torrigiani, P., Castiglione, S., 2010. Arbuscular mycorrhizal fungi restore normal growth in a white poplar clone grown on heavy metal-contaminated soil, and this is associated with upregulation of foliar metallothionein and polyamine biosynthetic gene expression. Ann. Bot. 106, 791–802.

Clemens, S., Antosiewicz, D.M., Ward, J.M., Schachtman, D.P., Schroeder, J.I., 1998. The plant cDNA LCT1 mediates the uptake of calcium and cadmium in yeast. PNAS 95, 12043–12048.

Colao, M.C., Lupino, S., Garzillo, A.M., Buonocore, V., Ruzzi, M., 2006. Heterologous expression of lcc1 gene from *Trametes trogii* in *Pichia pastoris* and characterization of the recombinant enzyme. Microb. Cell Factories 5, 1–11.

Congeevaram, S., Dhanarani, S., Park, J., Dexilin, M., Thamaraiselvi, K., 2007. Biosorption of chromium and nickel by heavy metal resistant fungal and bacterial isolates. J. Hazard. Mater. 146, 270–277.

Copete-Pertuz, L.S., Plácido, J., Serna-Galvis, E.A., Torres-Palma, R.A., Mora, A., 2018. Elimination of Isoxazolyl-Penicillins antibiotics in waters by the ligninolytic native Colombian strain *Leptosphaerulina* sp. considerations on biodegradation process and antimicrobial activity removal. Sci. Total Environ. 630, 1195–1204.

Couto, S.R., Toca-Herrera, J.L., 2007. Laccase production at reactor scale by filamentous fungi. Biotechnol. Adv. 25, 558–569.

Cullen, W.R., Reimer, K.J., 1989. Arsenic speciation in the environment. Chem. Rev. 9, 713–764.

Čvančarová, M., Moeder, M., Filipová, A., Cajthaml, T., 2014. Biotransformation of fluoroquinolone antibiotics by ligninolytic fungi—metabolites, enzymes and residual antibacterial activity. Chemosphere 136, 311–320.

Čvančarová, M., Moeder, M., Filipová, A., Reemtsma, T., Cajthaml, T., 2013. Biotransformation of the antibiotic agent flumequine by ligninolytic fungi and residual antibacterial activity of the transformation mixtures. Environ. Sci. Technol. 47 (24), 14128–14136.

Daghino, S., Martino, E., Perotto, S., 2016. Model systems to unravel the molecular mechanisms of heavy metal tolerance in the ericoid mycorrhizal symbiosis. Mycorrhiza 26, 263–274.

Damodaran, D., Shetty, V.K., Balakrishnan, R.M., 2015. Interaction of heavy metals in multimetal biosorption by *Galerina vittiformis* from soil. Biorem. J. 19, 56–68.

Das, S., Dash, H.R., 2017. Handbook of Metal-Microbe Interactions and Bioremediation. CRC Press.

Dashtban, M., Schraft, H., Qin, W., 2009. Fungal bioconversion of lignocellulosic residues; opportunities & perspectives. Int. J. Biol. Sci., 578–595.

Dave, K.I., Lauriano, C., Xu, B., Wild, J.R., Kenerley, C.M., 1994. Expression of organophosphate hydrolase in the filamentous fungus Gliocladium virens. Appl. Microbiol. Biotechnol. 41, 352–358.

Decesaro, A., Rampel, A., Machado, T.S., Thomé, A., Reddy, K., Margarites, A.C., Colla, L.M., 2017. Bioremediation of soil contaminated with diesel and biodiesel fuel using biostimulation with microalgae biomass. J. Environ. Eng. 143, 04016091.

Demars, B.G., Boerner, R.E.J., 1996. Vesicular arbuscular mycorrhizal development in the Brassicaceae in relation to plant life span. Flora 191, 179–189.

Deng, Z., Cao, L., Zhang, R., Wang, W., Shi, Y., Tan, H., Wang, Z., Cao, L., 2014b. Enhanced phytoremediation of multi-metal contaminated soils by interspecific fusion between the protoplasts of endophytic *Mucor* sp. CBRF59 and *Fusarium* sp. CBRF14. Soil Biol. Biochem. 77, 31–40.

Deng, W., Lin, D., Yao, K., Yuan, H., Wang, Z., Li, J., Zou, L., Han, X., Zhou, K., He, L., Hu, X., 2015. Characterization of a novel β-cypermethrin-degrading *Aspergillus niger* YAT strain and the biochemical degradation pathway of β-cypermethrin. Appl. Microbiol. Biotechnol. 99, 8187–8198.

Deng, Z., Wang, W., Tan, H., Cao, L., 2012. Characterization of heavy metal-resistant endophytic yeast *Cryptococcus sp.* CBSB78 from rapes (*Brassica chinensis*) and its potential in promoting the growth of Brassica spp. in metal-contaminated soils. Water Air Soil Pollut. 223, 5321–5329.

Deng, Z., Zhang, R., Shi, Y., Hu, L., Tan, H., Cao, L., 2013. Enhancement of phytoremediation of Cd- and Pb-contaminated soils by self-fusion of protoplasts from endophytic fungus *Mucor sp.* CBRF59. Chemosphere 91, 41–47.

Deng, Z., Zhang, R., Shi, Y., Hu, L., Tan, H., Cao, L., 2014a. Characterization of Cd-, Pb-, Zn-resistant endophytic *Lasiodiplodia sp.* MXSF31 from metal accumulating *Portulaca oleracea* and its potential in promoting the growth of rape in metal-contaminated soils. Environ. Sci. Pollut. Res. 21, 2346–2357.

Deshmukh, R., Khardenavis, A.A., Purohit, H.J., 2016. Diverse metabolic capacities of fungi for bioremediation. Indian J. Microbiol. 56, 247–264.

Dhankhar, R., Hooda, A., 2011. Fungal biosorption—an alternative to meet the challenges of heavy metal pollution in aqueous solutions. Environ. Technol. 32, 467–491.

Dias, A.A., Sampaio, A., Bezerra, R.M., 2007. Environmental applications of fungal and plant systems: decolourisation of textile wastewater and related dyestuffs. Environ. Bioremediat. Technol., 445–463.

Diels, L., De Smet, M., Hooyberghs, L., Corbisier, P., 1999. Heavy metals bioremediation of soil. Mol. Biotechnol. 12, 149–158.

Dienes, D., Börjesson, J., Hägglund, P., Tjerneld, F., Lidén, G., Réczey, K., Stålbrand, H., 2007. Identification of a trypsin-like serine protease from *Trichoderma reesei* QM9414. Enzym. Microb. Technol. 40, 1087–1094.

Dimkpa, C.O., Svatoš, A., Dabrowska, P., Schmidt, A., Boland, W., Kothe, E., 2008. Involvement of siderophores in the reduction of metal-induced inhibition of auxin synthesis in *Streptomyces* spp. Chemosphere 74, 19–25.

Donato, P.D., Di Donato, P., Buono, A., Poli, A., Finore, I., Abbamondi, G., Nicolaus, B., Lama, L., 2018. Exploring marine environments for the identification of extremophiles and their enzymes for sustainable and green bioprocesses. Sustainability 11, 149.

Dugal, S., Gangawane, M., 2012. Metal tolerance and potential of *Penicillium* sp for use in mycoremediation. J. Chem. Pharm. Res. 4, 2362–2366.

Durruty, I., Fasce, D., González, J.F., Wolski, E.A., 2015. A kinetic study of textile dyeing wastewater degradation by *Penicillium chrysogenum*. Bioprocess Biosyst. Eng. 38, 1019–1031.

El-Borai, A., Eltayeb, K., Mostafa, A., El-Assar, S., 2016. Biodegradation of industrial oil-polluted wastewater in Egypt by bacterial consortium immobilized in different types of carriers. Pol. J. Environ. Stud. 25, 1901–1909.

El-Morsy, E.-S.M., 2004. *Cunninghamella echinulata* a new biosorbent of metal ions from polluted water in Egypt. Mycologia 96, 1183.

Esterhuizen-Londt, M., Hendel, A.L., Pflugmacher, S., 2017. Mycoremediation of diclofenac using *Mucor hiemalis*. Toxicol. Environ. Chem. 99, 795–808.

Esterhuizen-Londt, M., Schwartz, K., Pflugmacher, S., 2016. Using aquatic fungi for pharmaceutical bioremediation: uptake of acetaminophen by *Mucor hiemalis* does not result in an enzymatic oxidative stress response. Fungal Biol. 120, 1249–1257.

Fazli, M.M., Soleimani, N., Mehrasbi, M., Darabian, S., Mohammadi, J., Ramazani, A., 2015. Highly cadmium tolerant fungi: their tolerance and removal potential. J. Environ. Health Sci. Eng. 13, 19.

Fester, T., 2013. Arbuscular mycorrhizal fungi in a wetland constructed for benzene-, methyltert-butyl ether- and ammonia-contaminated groundwater bioremediation. Microb. Biotechnol. 6, 80–84.

Ford, C.I., Walter, M., Northcott, G.L., Di, H.J., Cameron, K.C., Trower, T., 2007. Fungal inoculum properties: extracellular enzyme expression and pentachlorophenol removal in highly contaminated field soils. J. Environ. Qual. 36, 1599–1608.

Fourest, E., Roux, J.-C., 1992. Heavy metal biosorption by fungal mycelial by-products: mechanisms and influence of pH. Appl. Microbiol. Biotechnol. 37, 399–403.

Franceschi, V.R., Nakata, P.A., 2005. Calcium oxalate in plants: formation and function. Annu. Rev. Plant Biol. 56, 41–71.

Gadd, G.M., 2004. Microbial influence on metal mobility and application for bioremediation. Geoderma 122, 109–119.

Gadd, G.M., Sayer, J.A., 2000. Influence of fungi on the environmental mobility of metals and metalloids. In: Environmental Microbe-Metal Interactions, pp. 237–256.

García-Alonso, S., Pérez-Pastor, R.M., Sevillano-Castaño, M.L., Escolano, O., García-Frutos, F.J., 2008. Influence of particle size on the quality of pah concentration measurements in a contaminated soil. Polycycl. Aromat. Compd. 28, 67–83.

García-Galán, M.J., Rodríguez-Rodríguez, C.E., Vicent, T., Caminal, G., Díaz-Cruz, M.S., Barceló, D., 2011. Biodegradation of sulfamethazine by *Trametes versicolor*: removal from sewage sludge and identification of intermediate products by UPLC-QqTOF-MS. Sci. Total Environ. 409, 5505–5512.

Garon, D., Sage, L., Wouessidjewe, D., Seigle-Murandi, F., 2004. Enhanced degradation of fluorene in soil slurry by *Absidia cylindrospora* and maltosyl-cyclodextrin. Chemosphere 56, 159–166.

Germano, S., Pandey, A., Osaku, C.A., Rocha, S.N., Soccol, C.R., 2003. Characterization and stability of proteases from *Penicillium* sp. produced by solid-state fermentation. Enzym. Microb. Technol. 32, 246–251.

Ghosh, S., Selvakumar, G., Ajilda, A.A.K., Webster, T.J., 2021. Microbial biosorbents for heavy metal removal. New trends in removal of heavy metals from industrial wastewater. Appl. Microbiol. Biotechnol., 213–262.

Gitipour, S., Sorial, G.A., Ghasemi, S., Bazyari, M., 2018. Treatment technologies for PAH-contaminated sites: a critical review. Environ. Monit. Assess. 190, 1–17.

Godlewska-Żyłkiewicz, B., Sawicka, S., Karpińska, J., 2019. Removal of platinum and palladium from wastewater by means of biosorption on fungi *Aspergillus sp.* and yeast *Saccharomyces* sp. Water 11, 1522.

Gomes, N.C.M., Rosa, C.A., Pimentel, P.F., Mendonça-Hagler, L.C.S., 2002. Uptake of free and complexed silver ions by different strains of *Rhodotorula mucilaginosa*. Braz. J. Microbiol. 33, 62–66.

González-Guerrero, M., Benabdellah, K., Valderas, A., Azcón-Aguilar, C., Ferrol, N., 2010. GintABC1 encodes a putative ABC transporter of the MRP subfamily induced by Cu, Cd, and oxidative stress in *Glomus intraradices*. Mycorrhiza 20, 137–146.

Gupta, P., Diwan, B., 2017. Bacterial exopolysaccharide mediated heavy metal removal: a review on biosynthesis, mechanism and remediation strategies. Biotechnol. Rep. 13, 58–71.

Gupta, S.C., Hevia, D., Patchva, S., Park, B., Koh, W., Aggarwal, B.B., 2012. Upsides and downsides of reactive oxygen species for cancer: the roles of reactive oxygen species in tumorigenesis, prevention, and therapy. Antioxid. Redox Signal. 16 (11), 1295–1322.

Gupta, S., Wali, A., Gupta, M., Annepu, S.K., 2017. Fungi: An Effective Tool for Bioremediation. Plant-Microbe Interactions in Agro-Ecological Perspectives. Springer, pp. 593–606.

Hadibarata, T., Khudhair, A.B., Salim, M.R., 2012. Breakdown products in the metabolic pathway of anthracene degradation by a ligninolytic fungus *Polyporus sp.* S133. Water Air Soil Pollut. 223, 2201–2208.

Hallsworth, J.E., Magan, N., 1999. Water and temperature relations of growth of the entomogenous fungi *Beauveria bassiana, Metarhizium anisopliae* and *Paecilomyces farinosus*. J. Invertebr. Pathol. 74, 261–266.

Haritash, A.K., Kaushik, C.P., 2016. Degradation of low molecular weight polycyclic aromatic hydrocarbons by microorganisms isolated from contaminated soil. J. Hazard. Mater. 6, 472–782,

Harms, H., Schlosser, D., Wick, L.Y., 2011. Untapped potential. exploiting fungi in bioremediation of hazardous chemicals. Nat. Rev Microbiol. 9, 177–192.

Hasegawa, I., Terada, E., Sunairi, M., Wakita, H., Shinmachi, F., Noguchi, A., Nakajima, M., Yazaki, J., 1997. Genetic improvement of heavy metal tolerance in plants by transfer of the yeast metallothionein gene (CUP1). In: Plant Nutrition Sustainable Food Production and Environment. Springer (Dordrecht 391–395).

Hassan, N., Rehman, S.-U., Bano, H., 2017. Environmental factors affecting growth of pathogenic fungi causing fruit rot in tomato (*Lycopersicon Esculentum*). Int. J. Eng. Res. Technol. 2, 578–584.

Hildebrandt, U., Regvar, M., Bothe, H., 2007. Arbuscular mycorrhiza and heavy metal tolerance. Phytochemistry 68, 139–146.

Hong-Bo, S., Li-Ye, C., Cheng-Jiang, R., Hua, L., Dong-Gang, G., Wei-Xiang, L., 2010. Understanding molecular mechanisms for improving phytoremediation of heavy metal-contaminated soils. Crit. Rev. Biotechnol. 30, 23–30.

Ianis, M., Tsekova, K., Vasileva, S., 2006. Copper biosorption by *Penicillium cyclopium*: equilibrium and modelling study. Biotechnol. Biotechnol. Equip. 20, 195–201.

Iimura, Y., Ikeda, S., Sonoki, T., Hayakawa, T., Kajita, S., Kimbara, K., Tatsumi, K., Katayama, Y., 2002. Expression of a gene for Mn-peroxidase from *Coriolus versicolor* in transgenic tobacco generates potential tools for phytoremediation. Appl. Microbiol. Biotechnol. 59 (2), 246–251.

Iimura, Y., Yoshizumi, M., Sonoki, T., Uesugi, M., Tatsumi, K., Horiuchi, K.-I., Kajita, S., Katayama, Y., 2007. Hybrid aspen with a transgene for fungal manganese peroxidase is a potential contributor to phytoremediation of the environment contaminated with bisphenol A. J. Wood Sci. 53, 541–544.

Ikehata, K., Buchanan, I., Pickard, M., Smith, D., 2005. Purification, characterization and evaluation of extracellular peroxidase from two species for aqueous phenol treatment. Bioresour. Technol. 96, 1758–1770.

Jaishankar, M., Tseten, T., Anbalagan, N., Mathew, B.B., Beeregowda, K.N., 2014. Toxicity, mechanism and health effects of some heavy metals. Interdiscip. Toxicol. 7, 60.

Jarosz-Wilkolazka, A., Gadd, G.M., 2003. Oxalate production by wood-rotting fungi growing in toxic metal-amended medium. Chemosphere 52, 541–547.

Järup, L., 2003. Hazards of heavy metal contamination. Br. Med. Bull. 68, 167–182.

Jasińska, A., Paraszkiewicz, K., Sip, A., Długoński, J., 2015. Malachite green decolorization by the filamentous fungus *Myrothecium roridum*- mechanistic study and process optimization. Bioresour. Technol. 194, 43–48.

Jenkins, K.M., 2012. Evaluating the Mechanism of Oxalate Synthesis of *Fibroporia Radiculosa* Isolates Adapting to Copper-Tolerance. A thesis submitted to the faculty of Mississippi State University in the department of Forest Products, 146 p.

Jerina, D.M., 1983. The 1982 Bernard B. brodie award lecture. Metabolism of aromatic hydrocarbons by the cytochrome P-450 system and epoxide hydrolase. Drug Metab. Dispos. 11, 1–4.

Jiang, M., Cao, L., Zhang, R., 2008. Effects of acacia (*Acacia auriculaeformis* A. Cunn)-associated fungi on mustard (*Brassica juncea* (L.) Coss. var. foliosa bailey) growth in Cd- and Ni-contaminated soils. Lett. Appl. Microbiol. 47, 561–565.

John, R., Ahmad, P., Gadgil, K., Sharma, S., 2008. Effect of cadmium and lead on growth, biochemical parameters and uptake in *Lemna polyrrhiza* L. Plant Soil Environ. 54, 262–270.

Joshi, C., Mathur, P., Khare, S.K., 2011. Degradation of phorbol esters by *Pseudomonas aeruginosa* PseA during solid-state fermentation of deoiled *Jatropha curcas* seed cake. Bioresour. Technol. 102, 4815–4819.

Kaewlaoyoong, A., Cheng, C.Y., Lin, C., Chen, J.R., Huang, W.Y., Sriprom, P., 2020. White rot fungus *Pleurotus pulmonarius* enhanced bioremediation of highly PCDD/F-contaminated field soil via solid state fermentation. Sci. Total Environ. 738, 139670.

Kalyani, D., Tiwari, M.K., Li, J., Kim, S.C., Kalia, V.C., Kang, Y.C., Lee, J.-K., 2015. A highly efficient recombinant laccase from the yeast *Yarrowia lipolytica* and its application in the hydrolysis of biomass. PLoS One 10, e0120156.

Kanaujiya, D.K., Paul, T., Sinharoy, A., Pakshirajan, K., 2019. Biological treatment processes for the removal of organic micropollutants from wastewater: a review. Curr. Pollut. Rep. 5, 112–128.

Kapahi, M., Sachdeva, S., 2017. Mycoremediation potential of *Pleurotus* species for heavy metals: a review. Bioresour. Bioprocess. 4, 32.

Kaplan, J.G., 1965. Action of non-penetrating heavy metals on the catalase activity of yeast cells. Nature 205, 76–77.

Khan, I., Ali, M., Aftab, M., Shakir, S., Qayyum, S., Haleem, K.S., Tauseef, I., 2019. Mycoremediation: a treatment for heavy metal-polluted soil using indigenous metallotolerant fungi. Environ. Monit. Assess. 191 (10), 1–5.

Khan, A.L., Waqas, M., Hussain, J., Al-Harrasi, A., Lee, I.-J., 2014. Fungal endophyte *Penicillium janthinellum* LK5 can reduce cadmium toxicity in *Solanum lycopersicum* (Sitiens and Rhe). Biol. Fertil. Soils 50, 75–85.

Khatoon, H., Rai, J.P.N., Jillani, A., 2021. Role of fungi in bioremediation of contaminated soil. In: Fungi Bio-Prospects in Sustainable Agriculture, Environment and Nano-Technology. Elsevier, pp. 121–156.

Khindaria, A., Yamazaki, I., Aust, S.D., 1996. Stabilization of the veratryl alcohol cation radical by lignin peroxidase. Biochemistry 35, 6418–6424.

Kiamarsi, Z., Soleimani, M., Nezami, A., Kafi, M., 2019. Biodegradation of n-alkanes and polycyclic aromatic hydrocarbons using novel indigenous bacteria isolated from contaminated soils. Int. J. Environ. Sci. Technol. 16, 6805–6816.

Kim, J.-D., Lee, C.-G., 2007. Microbial degradation of polycyclic aromatic hydrocarbons in soil by bacterium-fungus co-cultures. Biotechnol. Bioprocess Eng. 12, 410–416.

Kohler, A., Blaudez, D., Chalot, M., Martin, F., 2004. Cloning and expression of multiple metallothioneins from hybrid poplar. New Phytol. 164, 83–93.

Kryczyk-Poprawa, A., Żmudzki, P., Maślanka, A., Piotrowska, J., Opoka, W., Muszyńska, B., 2019. Mycoremediation of azole antifungal agents using in vitro cultures of *Lentinula edodes*. 3Biotech 9 (6), 207.

Kulkarni, A.N., Kadam, A.A., Kachole, M.S., Govindwar, S.P., 2014. Lichen *Permelia perlata*: a novel system for biodegradation and detoxification of disperse dye Solvent Red 24. J. Hazard. Mater. 276, 461–468.

Kumar, M., Bolan, N.S., Hoang, S.A., et al., 2021. Remediation of soils and sediments polluted with polycyclic aromatic hydrocarbons: to immobilize, mobilize, or degrade? J. Hazard. Mater. 420, 126534.

Kumar, R., Kaur, A., 2018. Oil spill removal by mycoremediation. In: Microbial Action on Hydrocarbons. Springer, pp. 505–526.

Kurniati, E., Arfarita, N., Imai, T., Higuchi, T., Kanno, A., Yamamoto, K., Sekine, M., 2014. Potential bioremediation of mercury-contaminated substrate using filamentous fungi isolated from forest soil. J. Environ. Sci. 26, 1223–1231.

Lal, R., Saxena, D.M., 1982. Accumulation, metabolism, and effects of organochlorine insecticides on microorganisms. Microbiol. Rev. 46, 95–127.

Lanfranco, L., Bolchi, A., Ros, E.C., Ottonello, S., Bonfante, P., 2002. Differential expression of a metallothionein gene during the presymbiotic versus the symbiotic phase of an arbuscular mycorrhizal fungus. Plant Physiol. 130, 58–67.

Lang, S., 2002. Biological amphiphiles (microbial biosurfactants). Curr. Opin. Colloid Interface Sci. 7, 12–20.

Laskar, N., Kumar, U., 2019. Plastics and microplastics: a threat to environment. Environ. Technol. Innov. 14, 100352.

Lau, K.L., Tsang, Y.Y., Chiu, S.W., 2003. Use of spent mushroom compost to bioremediate PAH-contaminated samples. Chemosphere 52, 1539–1546.

Leontievsky, A.A., Myasoedova, N.M., Baskunov, B.P., Evans, C.S., Golovleva, L.A., 2004. Transformation of 2,4,6-trichlorophenol by the white rot fungi *Panus tigrinus* and *Coriolus versicolor*. Biodegradation 11, 331–340.

Li, Y.Y., Li, B., 2011. Study on fungi-bacteria consortium bioremediation of petroleum contaminated mangrove sediments amended with mixed biosurfactants. Adv. Mater. Res. 183-185, 1163–1167.

Li, X., Lin, X., Zhang, J., Wu, Y., Yin, R., Feng, Y., Wang, Y., 2010. Degradation of polycyclic aromatic hydrocarbons by crude extracts from spent mushroom substrate and its possible mechanisms. Curr. Microbiol. 60, 336–342.

Li, T., Liu, M.J., Zhang, X.T., Zhang, H.B., Sha, T., Zhao, Z.W., 2011. Improved tolerance of maize (*Zea mays* L.) to heavy metals by colonization of a dark septate endophyte (DSE) *Exophiala pisciphila*. Sci. Total Environ. 409, 1069–1074.

Ling, T.C., 1997. Heavy metals biosorption by powdered *Rhizopus oligosporus* biomass.

Liu, G., Hu, S., Li, L., Hou, Y., 2015. Purification and characterization of a lipase with high thermostability and polar organic solvent-tolerance from *Aspergillus niger* AN0512. Lipids 50, 1155–1163.

Lladó, S., Covino, S., Solanas, A.M., Viñas, M., Petruccioli, M., D'annibale, A., 2013. Comparative assessment of bioremediation approaches to highly recalcitrant PAH degradation in a real industrial polluted soil. J. Hazard. Mater. 248-249, 407–414.

Lloyd, J.R., Anderson, R.T., Macaskie, L.E., 2014. Bioremediation of metals and radionuclides. Bioremediation, 293–317.

Luo, J.-M., Jin-ming, L.U.O., Xiao, X., LUO, s.-l., 2010. Biosorption of cadmium(II) from aqueous solutions by industrial fungus *Rhizopus cohnii*. Trans. Nonferrous Metals Soc. China 20, 1104–1111.

Lynch, J.M., Moffat, A.J., 2005. Bioremediation—prospects for the future application of innovative applied biological research. Ann. Appl. Biol. 146, 217–221.

Ma, L., Zhuo, R., Liu, H., Yu, D., Jiang, M., Zhang, X., Yang, Y., 2014. Efficient decolorization and detoxification of the sulfonated azo dye reactive Orange 16 and simulated textile wastewater containing reactive Orange 16 by the white-rot fungus *Ganoderma sp.* En3 isolated from the forest of Tzu-chin mountain in China. Biochem. Eng. J. 82, 1–9.

Magi, E., Bianco, R., Ianni, C., Di Carro, M., 2002. Distribution of polycyclic aromatic hydrocarbons in the sediments of the Adriatic Sea. Environ. Pollut. 119, 91–98.

Majeed, A., 2021. Heavy metal pollution in agricultural soils consequences and bioremediation approaches. In: Bioremediation Science from Theory to Practice. CRC Press, pp. 227–239.

Mao, J., Guan, W., 2016. Fungal degradation of polycyclic aromatic hydrocarbons (PAHs) by *Scopulariopsis brevicaulis* and its application in bioremediation of PAH-contaminated soil. Acta Agric. Scand. Sect. B Soil Plant Sci. 66, 399–405.

Marco-Urrea, E., Pérez-Trujillo, M., Blánquez, P., Vicent, T., Caminal, G., 2010b. Biodegradation of the analgesic naproxen by *Trametes versicolor* and identification of intermediates using HPLC-DAD-MS and NMR. Bioresour. Technol. 101 (7), 2159–2166.

Marco-Urrea, E., Pérez-Trujillo, M., Cruz-Morató, C., Caminal, G., Vicent, T., 2010a. White-rot fungus-mediated degradation of the analgesic ketoprofen and identification of intermediates by HPLC–DAD–MS and NMR. Chemosphere 78 (4), 474–481.

Margesin, R., Zimmerbauer, A., Schinner, F., 1999. Soil lipase activity-a useful indicator of oil biodegradation. Biotechnol. Tech. 13, 859–863.

Marihal, A.K., Jagadeesh, K.S., 2013. Plant–microbe interaction: a potential tool for enhanced bioremediation. In: Plant Microbe Symbiosis: Fundamentals and Advances. Springer, pp. 395–410.

Masindi, V., Muedi, K.L., 2018. Environmental contamination by heavy metals. In: Heavy Metals. InTech.

Mason, R.P., 2012. The methylation of metals and metalloids in aquatic systems. In: Methylation from DNA, RNA and Histones to Diseases and Treatment. InTech.

Meng, X., Qin, J., Wang, L., Duan, G., Sun, G., Wu, H., Chu, C., Ling, H., Rosen, B.P., Zhu, Y., 2011. Arsenic biotransformation and volatilization in transgenic rice. New Phytol. 191, 49–56.

Migliore, L., Fiori, M., Spadoni, A., Galli, E., 2012. Biodegradation of oxytetracycline by *Pleurotus ostreatus* mycelium: a mycoremediation technique. J. Hazard. Mater. 215–216, 227–232.

Mishra, A., Malik, A., 2014. Novel fungal consortium for bioremediation of metals and dyes from mixed waste stream. Bioresour. Technol. 171, 217–226.

Mishra, S., Singh, S.N., 2014. Biodegradation of benzo(a)pyrene mediated by catabolic enzymes of bacteria. Int. J. Environ. Sci. Technol. 11, 1571–1580.

Misra, S., Gedamu, L., 1989. Heavy metal tolerant transgenic *Brassica napus* L. and *Nicotiana tabacum* L. plants. Theor. Appl. Genet. 78, 161–168.

Mitra, J., Mukherjee, P.K., Kale, S.P., Murthy, N.B., 2001. Bioremediation of DDT in soil by genetically improved strains of soil fungus *Fusarium solani*. Biodegradation 12, 235–245.

Mohamed, Z.A., Hashem, M., Alamri, S.A., 2014. Growth inhibition of the cyanobacterium *Microcystis aeruginosa* and degradation of its microcystin toxins by the fungus *Trichoderma citrinoviride*. Toxicon 86, 51–58.

Mohammad, P., Azarmidokht, H., Fatollah, M., Mahboubeh, B., 2006. Application of response surface methodology for optimization of important parameters in decolorizing treated distillery wastewater using *Aspergillus fumigatus* UB2 60. Int. Biodeterior. Biodegradation 57, 195–199.

Mohammadian, E., Babai Ahari, A., Arzanlou, M., Oustan, S., Khazaei, S.H., 2017. Tolerance to heavy metals in filamentous fungi isolated from contaminated mining soils in the Zanjan Province, Iran. Chemosphere 185, 290–296.

Mohsenzadeh, F., Shahrokhi, F., 2014. Biological removing of cadmium from contaminated media by fungal biomass of *Trichoderma* species. J. Environ. Health Sci. Eng. 12, 102.

Morais, e Costa, F.G., de Lourdes Pereir, M., 2012. Heavy metals and human health. In: Environmental Health. Emerging Issues and Practice. InTech.

Morelli, I.S., Saparrat, M.C.N., Del Panno, M.T., Coppotelli, B.M., Arrambari, A., 2013. Bioremediation of PAH-contaminated soil by fungi. Soil Biol., 159–179.

Mukhopadhyay, R., Rosen, B.P., Phung, L.T., Silver, S., 2002. Microbial arsenic: from geocycles to genes and enzymes. FEMS Microbiol. Rev. 26, 311–325.

Muszyńska, B., Dąbrowska, M., Starek, M., Żmudzki, P., Lazur, J., Pytko-Polończyk, J., Opoka, W., 2019. *Lentinula edodes* mycelium as effective agent for piroxicam mycoremediation. Front. Microbiol. 10, 313.

Muszyńska, B., Żmudzki, P., Lazur, J., Kała, K., Sułkowska-Ziaja, K., Opoka, W., 2018. Analysis of the biodegradation of synthetic testosterone and 17α-ethynylestradiol using the edible mushroom *Lentinula edodes*. 3Biotech 8 (10), 1–5.

Nagy, R., Karandashov, V., Chague, V., Kalinkevich, K., M'barek, T., Xu, G., Jakobsen, I., Levy, A.A., Amrhein, N., Bucher, M., 2005. The characterization of novel mycorrhiza-specific phosphate transporters from *Lycopersicon esculentum* and *Solanum tuberosum* uncovers functional redundancy in symbiotic phosphate transport in solanaceous species. Plant J. 42, 236–250.

Nam, J.-M., Fujita, Y., Arai, T., Kondo, A., Morikawa, Y., Okada, H., Ueda, M., Tanaka, A., 2002. Construction of engineered yeast with the ability of binding to cellulose. J. Mol. Catal. B Enzym. 17, 197–202.

Nasrin, Z., Yoshikawa, M., Nakamura, Y., et al., 2010. Overexpression of a fungal laccase gene induces nondehiscent anthers and morphological changes in flowers of transgenic tobacco. J. Wood Sci. 56, 460–469.

Natalia, P., 2017. Ligninolytic fungi: their degradative potential and the prospect for the development of environmentally significant biotechnologies. In: 2nd International Conference on Environmental Health & Global Climate Change. Occup Med Health Aff 5:2 (Suppl).

Neagoe, A., Iordache, V., Bergman, H., Kothe, E., 2013. Patterns of effects of arbuscular mycorrhizal fungi on plants grown in contaminated soil. J. Plant Nutr. Soil Sci. 176, 273–286.

Neubauer, U., Furrer, G., Kayser, A., Schulin, R., 2000. Siderophores, NTA, and citrate: potential soil amendments to enhance heavy metal mobility in phytoremediation. Int. J. Phytoremed. 2, 353–368.

Nguyen, T.D., Cavagnaro, T.R., Watts-Williams, S.J., 2019. The effects of soil phosphorus and zinc availability on plant responses to mycorrhizal fungi: a physiological and molecular assessment. Sci. Rep. 9, 1–13.

Novotný, Č., Svobodová, K., Erbanová, P., Cajthaml, T., Kasinath, A., Lang, E., Šašek, V., 2004. Ligninolytic fungi in bioremediation: extracellular enzyme production and degradation rate. Soil Biol. Biochem. 36, 1545–1551.

Obire, O., Anyanwu, Okigbo, R.N., 2008. Saprophytic and crude oil-degrading fungi from cow dung and poultry droppings as bioremediating agents. Int. J. Agric. Technol. 4 (2), 81–89.

Obire, O., Nwaubeta, O., 2010. Biodegradation of refined petroleum hydrocarbons in soil. J. Appl. Sci. Environ. Manag. 5, 43–46.

Ogbolosingha, A.J., Essien, E.B., Ohiri, R.C., 2015. Variation of lipase, catalase and dehydrogenase activities during bioremediation of crude oil polluted soil. J. Environ. Earth Sci. 5, 128–141.

Omar, S.H., Rehm, H.-J., 1988. Degradation of n-alkanes by *Candida parapsilosis* and *Penicillium frequentans* immobilized on granular clay and aquifer sand. Appl. Microbiol. Biotechnol. 28, 103–108.

Panigrahi, S., Velraj, P., Rao, T.S., 2019. Functional microbial diversity in contaminated environment and application in bioremediation. In: Microbial Diversity in the Genomic Era. Elsevier, pp. 359–385.

Park, H., Min, B., Jang, Y., et al., 2019. Comprehensive genomic and transcriptomic analysis of polycyclic aromatic hydrocarbon degradation by a mycoremediation fungus, *Dentipellis* sp. KUC8613. Appl. Microbiol. Biotechnol. 103, 8145–8155.

Pawlik-Skowrońska, B., Purvis, O.W., Pirszel, J., Skowroński, T., 2006. Cellular mechanisms of cu-tolerance in the epilithic lichen *Lecanora polytropa* growing at a copper mine. Lichenologist 38, 267–275.

Piątkowski, J., Krzyżewska, A., 2013. Influence of some physical factors on the growth and sporulation of entomopathogenic fungi. Acta Mycol 42, 255–265.

Piontek, K., Glumoff, T., Winterhalter, K., 1993. Low pH crystal structure of glycosylated lignin peroxidase from *Phanerochaete chrysosporium* at 2.5 Å resolution. FEBS Lett. 315, 119–124.

Plácido, J., Capareda, S., 2015. Ligninolytic enzymes: a biotechnological alternative for bioethanol production. Bioresour. Bioprocess. 2, 23.

Poli, A., Salerno, A., Laezza, G., di Donato, P., Dumontet, S., Nicolaus, B., 2009. Heavy metal resistance of some thermophiles: potential use of alpha-amylase from *Anoxybacillus amylolyticus* as a microbial enzymatic bioassay. Res. Microbiol. 160, 99–106.

Pothuluri, J.V., Evans, F.E., Heinze, T.M., Cerniglia, C.E., 1996. Formation of sulfate and glucoside conjugates of benzo[e]pyrene by *Cunninghamella elegans*. Appl. Microbiol. Biotechnol. 45, 677–683.

Pothuluri, J.V., Freeman, J.P., Evans, F.E., Cerniglia, C.E., 1993. Biotransformation of fluorene by the fungus *Cunninghamella elegans*. Appl. Environ. Microbiol. 59, 1977–1980.

Pourfakhraei, E., Badraghi, J., Mamashli, F., Nazari, M., Saboury, A.A., 2018. Biodegradation of asphaltene and petroleum compounds by a highly potent *Daedaleopsis* sp. J. Basic Microbiol. 58, 609–622.

Pozdnyakova, N.N., 2012. Involvement of the ligninolytic system of white-rot and litter-decomposing fungi in the degradation of polycyclic aromatic hydrocarbons. Biotechnol. Res. Int., 1–20.

Prabhavathi, V.R., Rajam, M.V., 2007. Polyamine accumulation in transgenic eggplant enhances tolerance to multiple abiotic stresses and fungal resistance. Plant Biotech. 24, 273–282.

Pradeep, S., Benjamin, S., 2012. Mycelial fungi completely remediate di(2-ethylhexyl)phthalate, the hazardous plasticizer in PVC blood storage bag. J. Hazard. Mater. 235-236, 69–77.

Pradeep, S., Faseela, P., Josh, M.K.S., Balachandran, S., Devi, R.S., Benjamin, S., 2013. Fungal biodegradation of phthalate plasticizer in situ. Biodegradation 24, 257–267.

Prasad, M.P., Manjunath, K., 2011. Comparative Study on Biodegradation of Lipid-Rich Wastewater Using Lipase Producing Bacterial Species. vol. 10 NISCAIR-CSIR, India, pp. 121–124.

Price, M.S., 2000. Characterization of *Aspergillus niger* for Removal of Copper and Zinc From Swine Watstewater.

Prieto, A., Möder, M., Rodil, R., Adrian, L., Marco-Urrea, E., 2011. Degradation of the antibiotics norfloxacin and ciprofloxacin by a white-rot fungus and identification of degradation products. Bioresour. Technol. 102 (23), 10987–10995.

Purnomo, A.S., Nawfa, R., Martak, F., Shimizu, K., Kamei, I., 2017. Erratum to: biodegradation of aldrin and dieldrin by the white-rot fungus *Pleurotus ostreatus*. Curr. Microbiol. 74, 889.

Qiao, J., Zhang, C., Luo, S., Chen, W., 2014. Bioremediation of highly contaminated oilfield soil: bioaugmentation for enhancing aromatic compounds removal. Front. Environ. Sci. Eng. 8, 293–304.

Qiu, Z., Tan, H., Zhou, S., Cao, L., 2014. Enhanced phytoremediation of toxic metals by inoculating endophytic *Enterobacter* sp. CBSB1 expressing bifunctional glutathione synthase. J. Hazard. Mater. 267, 17–20.

Qu, Y., Shi, S., Ma, F., Yan, B., 2010. Decolorization of reactive dark blue K-R by the synergism of fungus and bacterium using response surface methodology. Bioresour. Technol. 101, 8016–8023.

Radhika, R., Jebapriya, G.R., Gnanadoss, J.J., 2014. Decolourization of synthetic textile dyes using the edible mushroom fungi *Pleurotus*. Pak. J. Biol. Sci. 17, 248–253.

Rahman, M.S., Sathasivam, K.V., 2015. Heavy metal adsorption onto *Kappaphycus* sp. from aqueous solutions: the use of error functions for validation of isotherm and kinetics models. Biomed. Res. Int.

Raut, S., Raut, S., Sharma, M., Srivastav, C., Adhikari, B., Sen, S.K., 2015. Enhancing degradation of low density polyethylene films by *Curvularia lunata* SG1 using particle swarm optimization strategy. Indian J. Microbiol. 55, 258–268.

Reddy, V.R., Behera, B., 2006. Impact of water pollution on rural communities: an economic analysis. Ecol. Econ. 58, 520–537.

Riva, S., 2006. Laccases: blue enzymes for green chemistry. Trends Biotechnol. 24, 219–226.

Rivera-Hoyos, C.M., Morales-Álvarez, E.D., Poveda-Cuevas, S.A., Reyes-Guzmán, E.A., Poutou-Piñales, R.A., Reyes-Montaño, E.A., Pedroza-Rodríguez, A.M., Rodríguez-Vázquez, R., Cardozo-Bernal, Á.M., 2015. Computational analysis and low-scale constitutive expression of laccases synthetic genes GlLCC1 from *Ganoderma lucidum* and POXA 1B from Pleurotus ostreatus in Pichia pastoris. PLoS One 10, e0116524.

Rodarte-Morales, A.I., Feijoo, G., Moreira, M.T., Lema, J.M., 2012. Biotransformation of three pharmaceutical active compounds by the fungus *Phanerochaete chrysosporium* in a fed batch stirred reactor under air and oxygen supply. Biodegradation 23 (1), 145–156.

Rodríguez, I.A., Cárdenas-González, J.F., Juárez, V.M.M., Pérez, A.R., de Guadalupe Moctezuma Zarate, M., NCP, C., 2018. Biosorption of heavy metals by *Candida albicans*. In: Advances in Bioremediation and Phytoremediation. IntechOpen.

Rothschild, L.J., Mancinelli, R.L., 2001. Life in extreme environments. Nature 409, 1092–1101.

Saba, B., Rafique, U., Hashmi, I., 2011. Adsorption kinetics of anthracene and phenanthrene in different soils of Attock refinery limited (ARL) Rawalpindi, Pakistan. Desalin. Water Treat. 30, 333–338.

Sack, U., Hofrichter, M., Fritsche, W., 2006. Degradation of polycyclic aromatic hydrocarbons by manganese peroxidase of *Nematoloma frowardii*. FEMS Microbiol. Lett. 152, 227–234.

Sajith, S., Priji, P., Sreedevi, S., Benjamin, S., 2016. An overview on fungal cellulases with an industrial perspective. J. Nutr. Food Sci. 6, 461.

Sakaki, T., Yamamoto, K., Ikushiro, S., 2013. Possibility of application of cytochrome P450 to bioremediation of dioxins. Biotechnol. Appl. Biochem. 60, 65–70.

Samuelsen, A.I., Martin, R.C., Mok, D.W.S., Mok, M.C., 1998. Expression of the yeast FRE genes in transgenic tobacco. Plant Physiol. 118, 51–58.

Sangale, M.K., Shahnawaz, M., Ade, A.B., 2019. Gas chromatography-mass spectra analysis and deleterious potential of fungal based polythene-degradation products. Sci. Rep. 9, 1–6.

dos Santos, B.G., Araujo, C.V., Castoldi, R., Maciel, G., Inacio, F., de Souza, C.M., Bracht, A., Peralta, R., 2013. Ligninolytic enzymes from white-rot fungi andapplication in the removal of synthetic dyes. In: Fungal Enzymes. CRC Press, Boca Raton, pp. 258–279.

Saratale, R.G., Saratale, G.D., Chang, J.S., Govindwar, S.P., 2009. Decolorization and biodegradation of textile dye navy blue HER by *Trichosporon beigelii* NCIM-3326. J. Hazard. Mater. 166, 1421–1428.

Sardrood, B.P., Goltapeh, E.M., Varma, A., 2013. An introduction to bioremediation. Soil Biol., 3–27.

Saroj, S., Kumar, K., Prasad, M., Singh, R.P., 2014. Differential expression of peroxidase and ABC transporter as the key regulatory components for degradation of azo dyes by *Penicillium oxalicum* SAR-3. Funct. Integr. Genomics 14, 631–642.

Sarret, G., Manceau, A., Cuny, D., Van Haluwyn, C., Déruelle, S., Hazemann, J.-L., Soldo, Y., Eybert-Bérard, L., Menthonnex, J.-J., 1998. Mechanisms of lichen resistance to metallic pollution. Environ. Sci. Technol. 32, 3325–3330.

Satpute, S.K., Banat, I.M., Dhakephalkar, P.K., Banpurkar, A.G., Chopade, B.A., 2010. Biosurfactants, bioemulsifiers and exopolysaccharides from marine microorganisms. Biotechnol. Adv. 28, 436–450.

Say, R., Yilmaz, N., Denizli, A., 2003. Removal of heavy metal ions using the fungus *Penicillium canescens*. Adsorpt. Sci. Technol. 21, 643–650.

Schützendübel, A., Polle, A., 2002. Plant responses to abiotic stresses: heavy metal-induced oxidative stress and protection by mycorrhization. J. Exp. Bot. 53, 1351–1365.

Sharma, A.K., Sharma, V., Saxena, J., 2016. A review on properties of fungal lipases. Int. J. Curr. Microbiol. App. Sci. 5, 123–130.

Sharma, D., Sharma, B., Shukla, A.K., 2010. Biotechnological approach of microbial lipase: a review. Biotechnology 10, 23–40.

Sharples, J.M., Meharg, A.A., Chambers, S.M., Cairney, J.W.G., 2000. Mechanism of arsenate resistance in the ericoid mycorrhizal fungus *Hymenoscyphus ericae*. Plant Physiol. 124, 1327–1334.

Shedbalkar, U., Jadhav, J.P., 2011. Detoxification of malachite green and textile industrial effluent by *Penicillium ochrochloron*. Biotechnol. Bioprocess Eng. 16, 196–204.

Siddiquee, S., Rovina, K., Azad, S.A., 2015. Heavy metal contaminants removal from wastewater using the potential filamentous fungi biomass: a review. J. Microb. Biochem. Technol. 7, 384–393.

Silambarasan, S., Abraham, J., 2013. Mycoremediation of endosulfan and its metabolites in aqueous medium and soil by *Botryosphaeria laricina* JAS6 and *Aspergillus tamarii* JAS9. PLoS One 8.

Singh, C.J., 2003. Optimization of an extracellular protease of *Chrysosporium keratinophilum* and its potential in bioremediation of keratinic wastes. Mycopathologia 156, 151–156.

Singh, H., 2006. Mycoremediation: Fungal Bioremediation. John Wiley & Sons.

Singh, S.K., Haritash, A.K., 2019. Polycyclic aromatic hydrocarbons: soil pollution and remediation. Int. J. Environ. Sci. Technol. 16, 6489–6512.

Singh, A., Mukhopadhyay, K., Sachan, S.G., 2019. Biotransformation of eugenol to vanillin by a novel strain *bacillus safensis* SMS1003. Biocatal. Biotransform. 37, 291–303.

Singh, G., Singh, A., Singh, P., Gupta, A., Shukla, R., Mishra, V.K., 2021. Sources, fate, and impact of pharmaceutical and personal care products in the environment and their different treatment technologies. In: Microbe Mediated Remediation of Environmental Contaminants. Woodhead Publishing, pp. 391–407.

Singh, P.C., Srivastava, S., Shukla, D., Bist, V., Tripathi, P., Anand, V., Arkvanshi, S.K., Kaur, J., Srivastava, S., 2018. Mycoremediation mechanisms for heavy metal resistance/tolerance in plants. Fungal Biol., 351–381.

Singh, M., Srivastava, P.K., Verma, P.C., Kharwar, R.N., Singh, N., Tripathi, R.D., 2015. Soil fungi for mycoremediation of arsenic pollution in agriculture soils. J. Appl. Microbiol. 119, 1278–1290.

Siokwu, S., Anyanwn, C.U., 2012. Tolerance for heavy metals by filamentous fungi isolated from a sewage oxidation pond. Afr. J. Microbiol. Res. 6, 2038–2043.

Smith, M.J., Flowers, T.H., Duncan, H.J., Alder, J., 2006. Effects of polycyclic aromatic hydrocarbons on germination and subsequent growth of grasses and legumes in freshly contaminated soil and soil with aged PAHs residues. Environ. Pollut. 141, 519–525.

Sodaneath, H., Lee, J.I., SO, Y., Jung, H., Choi, J.H., Ryu, H.W., Cho, K.S., 2017. Decolorization of textile dyes in an air-lift bioreactor inoculated with *Bjerkandera adusta* OBR105. J. Environ. Sci. Health A 52, 1099–1111.

Soleimani, M., Afyuni, M., Hajabbasi, M.A., Nourbakhsh, F., Sabzalian, M.R., Christensen, J.H., 2010. Phytoremediation of an aged petroleum contaminated soil using endophyte infected and non-infected grasses. Chemosphere 81, 1084–1090.

Song, W.-Y., Sohn, E.J., Martinoia, E., Lee, Y.J., Yang, Y.-Y., Jasinski, M., Forestier, C., Hwang, I., Lee, Y., 2003. Engineering tolerance and accumulation of lead and cadmium in transgenic plants. Nat. Biotechnol. 21, 914–919.

Souza, P.M., Bittencourt, M.L., Caprara, C.C., Freitas, M.D., Almeida, R.P., Silveira, D., Fonseca, Y.M., Ferreira, E.X., Pessoa, A., Magalhães, P.O., 2015. A biotechnology perspective of fungal proteases. Braz. J. Microbiol. 46 (2), 337e346.

Srivastava, P.K., Vaish, A., Dwivedi, S., Chakrabarty, D., Singh, N., Tripathi, R.D., 2011. Biological removal of arsenic pollution by soil fungi. Sci. Total Environ. 409, 2430–2442.

Su, S., Zeng, X., Bai, L., Jiang, X., Li, L., 2010. Bioaccumulation and biovolatilisation of pentavalent arsenic by *Penicillin janthinellum, Fusarium oxysporum and Trichoderma asperellum* under laboratory conditions. Curr. Microbiol. 61, 261–266.

Sun, J., Lu, X., Rinas, U., Zeng, A., 2007. Metabolic peculiarities of *Aspergillus niger* disclosed by comparative metabolic genomics. Genome Biol. 8, R182.

Sundaramoorthy, M., Kishi, K., Gold, M.H., Poulos, T.L., 1994. The crystal structure of manganese peroxidase from *Phanerochaete chrysosporium* at 2.06-a resolution. J. Biol. Chem. 269, 32759–32767.

Syed, K., Porollo, A., Lam, Y.W., Grimmett, P.E., Yadav, J.S., 2013. CYP63A2, a catalytically versatile fungal p450 monooxygenase capable of oxidizing higher-molecular-weight polycyclic aromatic hydrocarbons, alkylphenols, and alkanes. Appl. Environ. Microbiol. 79, 2692–2702.

Tam, P.C.F., 1995. Heavy metal tolerance by ectomycorrhizal fungi and metal amelioration by *Pisolithus tinctorius*. Mycorrhiza 5, 181–187.

Tang, J.D., Parker, L.A., Perkins, A.D., Sonstegard, T.S., Schroeder, S.G., Nicholas, D.D., Diehl, S.V., 2013. Gene expression analysis of copper tolerance and wood decay in the brown rot fungus *Fibroporia radiculosa*. Appl. Environ. Microbiol. 79, 1523–1533.

Thippeswamy, B., Shivakumar, C.K., Krishnappa, M., 2012. Bioaccumulation potential of *Aspergillus niger* and *Aspergillus flavus* for removal of heavy metals from paper mill effluent. J. Environ. Biol. 33, 1063–1068.

Thomas, J.C., Davies, E.C., Malick, F.K., Endreszl, C., Williams, C.R., Abbas, M., Petrella, S., Swisher, K., Perron, M., Edwards, R., Ostenkowski, P., 2003. Yeast metallothionein in transgenic tobacco promotes copper uptake from contaminated soils. Biotechnol. Prog. 19, 273–280.

Thomine, S., Wang, R., Ward, J.M., Crawford, N.M., Schroeder, J.I., 2000. Cadmium and iron transport by members of a plant metal transporter family in *Arabidopsis* with homology to Nramp genes. PNAS 97, 4991–4996.

Trakoli, A., 2012. IARC monographs on the evaluation of carcinogenic risks to humans. Volume 99: some aromatic amines, organic dyes, and related exposures. International agency for research on cancer. Occup. Med. 62, 232.

Tripathi, P., Singh, P.C., Mishra, A., Srivastava, S., Chauhan, R., Awasthi, S., Mishra, S., Dwivedi, S., Tripathi, P., Kalra, A., Tripathi, R.D., 2017. Arsenic tolerant *Trichoderma* sp. reduces arsenic induced stress in chickpea (*Cicer arietinum*). Environ. Pollut. 223, 137–145.

Tuomela, M., Hatakka, A., 2011. Oxidative fungal enzymes for bioremediation. In: Comprehensive Biotechnology. Elsevier, pp. 183–196.

Valentín, L., Lu-Chau, T.A., López, C., Feijoo, G., Moreira, M.T., Lema, J.M., 2007. Biodegradation of dibenzothiophene, fluoranthene, pyrene and chrysene in a soil slurry reactor by the white-rot fungus *Bjerkandera sp.* BOS55. Process Biochem. 42, 641–648.

Valentín, L., Oesch-Kuisma, H., Steffen, K.T., Kähkönen, M.A., Hatakka, A., Tuomela, M., 2013. Mycoremediation of wood and soil from an old sawmill area contaminated for decades. J. Hazard. Mater. 260, 668–675.

Vaseem, H., Singh, V.K., Singh, M.P., 2017. Heavy metal pollution due to coal washery effluent and its decontamination using a macrofungus, *Pleurotus ostreatus*. Ecotoxicol. Environ. Saf. 145, 42–49.

Vasileva-Tonkova, E., Galabova, D., 2003. Hydrolytic enzymes and surfactants of bacterial isolates from lubricant-contaminated wastewater. Z. Naturforsch. C 58, 87–92.

Veglio', F., Beolchini, F., 1997. Removal of metals by biosorption: a review. Hydrometallurgy 44, 301–316.

Verlicchi, P., Al Aukidy, M., Zambello, E., 2012. Occurrence of pharmaceutical compounds in urban wastewater: removal, mass load and environmental risk after a secondary treatment-a review. Sci. Total Environ. 429, 123–155.

Verma, S., Verma, P.K., Meher, A.K., Dwivedi, S., Bansiwal, A.K., Pande, V., Srivastava, P.K., Verma, P.C., Tripathi, R.D., Chakrabarty, D., 2016. A novel arsenic methyltransferase gene of *Westerdykella aurantiaca* isolated from arsenic contaminated soil: phylogenetic, physiological, and biochemical studies and its role in arsenic bioremediation. Metallomics 8, 344–353.

Wang, W., Deng, Z., Tan, H., Cao, L., 2013. Effects of Cd, Pb, Zn, Cu-resistant endophytic *Enterobacter sp* CBSB1 and *Rhodotorula* sp. CBSB79 on the growth and phytoextraction of Brassica plants in multimetal contaminated soils. Int. J. Phytoremed. 15, 488–497.

Wang, C., Sun, H., Li, J., Li, Y., Zhang, Q., 2009. Enzyme activities during degradation of polycyclic aromatic hydrocarbons by white rot fungus Phanerochaete *chrysosporium* in soils. Chemosphere 77, 733–738.

WHO, 2021. Drinking-Water. https://www.who.int/news-room/fact-sheets/detail/drinking-water. (Accessed 25 July 2021).

Williams, P.P., 1977. Metabolism of synthetic organic pesticides by anaerobic microorganisms. Residue Rev. 66, 63–135.

Williams, E.S., Spencer Williams, E., Mahler, B.J., Van Metre, P.C., 2013. Cancer risk from incidental ingestion exposures to pahs associated with coal-tar-sealed pavement. Environ. Sci. Technol. 47, 1101–1109.

Wolski, E., Menusi, E., Mazutti, M., et al., 2008. Response surface methodology for optimization of lipase production by an immobilized newly isolated Penicillium sp. Ind. Eng. Chem. Res. 47, 9651–9657.

Wood, T.M., 1985. Properties of cellulolytic enzyme systems. Biochem. Soc. Trans. 13, 407–410.

Wu, Y., Teng, Y., Li, Z., Liao, X., Luo, Y., 2008. Potential role of polycyclic aromatic hydrocarbons (PAHs) oxidation by fungal laccase in the remediation of an aged contaminated soil. Soil Biol. Biochem. 40, 789–796.

Wu, M., Xu, Y., Ding, W., Li, Y., Xu, H., 2016. Mycoremediation of manganese and phenanthrene by *Pleurotus eryngii* mycelium enhanced by tween 80 and saponin. Appl. Microbiol. Biotechnol. 100 (16), 7249–7261.

Xu, P., Leng, Y., Zeng, G., et al., 2015. Cadmium induced oxalic acid secretion and its role in metal uptake and detoxification mechanisms in *Phanerochaete chrysosporium*. Appl. Microbiol. Biotechnol. 99, 435–443.

Xu, G., Li, Y., Zheng, W., Peng, X., Li, W., Yan, Y., 2007. Mineralization of chlorpyrifos by co-culture of *Serratia* and *Trichosporon* spp. Biotechnol. Lett. 29, 1469–1473.

Xu, X., Yao, W., Xiao, D., Heinz, T.F., 2014. Spin and pseudospins in layered transition metal dichalcogenides. Nat. Phys. 10, 343–350.

Yadav, M., Yadav, P., Yadav, K.D.S., 2009. Purification, characterization, and coal depolymerizing activity of lignin peroxidase from *Gloeophyllum sepiarium* MTCC-1170. Biochemistry (Mosc) 74, 1125–1131.

Yan, J., Niu, J., Chen, D., Chen, Y., Irbis, C., 2014. Screening of *Trametes* strains for efficient decolorization of malachite green at high temperatures and ionic concentrations. Int. Biodeterior. Biodegradation 87, 109–115.

Yang, S.O., Sodaneath, H., Lee, J.I., Jung, H., Choi, J.H., Ryu, H.W., Cho, K.S., 2017. Decolorization of acid, disperse and reactive dyes by *Trametes versicolor* CBR43. J. Environ. Sci. Health A 52, 862–872.

Yi, X., Chuan-Chao, D., Xing-Xiang, W., Fu-Yan, L., Hong-Wei, W., Xiao-Gang, L., 2014. Effect of the endophyte *Ceratobasidium stevensii* on 4-HBA degradation and watermelon seed germination. Afr. J. Microbiol. Res. 8, 1535–1543.

Ying, Z.G., Lin, L., He, L., Yuanhao, H., 2009. Microorganism-plant combined bioremediation on heavy metal contaminated soil in the industrial district. In: 2009 International conference on environmental science and information application technology.

Young, D., Rice, J., Martin, R., Lindquist, E., Lipzen, A., Grigoriev, I., Hibbett, D., 2015. Degradation of bunker C fuel oil by white-rot fungi in sawdust cultures suggests potential applications in bioremediation. PLoS One 10, 0130381.

Yu, G., Wen, X., Qian, Y., 2005. Production of the ligninolytic enzymes by immobilized *Phanerochaete chrysosporium* in an air atmosphere. World J. Microbiol. Biotechnol. 21, 323–327.

Zafar, S., Aqil, F., Ahmad, I., 2007. Metal tolerance and biosorption potential of filamentous fungi isolated from metal contaminated agricultural soil. Bioresour. Technol. 98, 2557–2561.

Zafra, G., Absalón, Á.E., Anducho-Reyes, M.Á., Fernandez, F.J., Cortés-Espinosa, D.V., 2017. Construction of PAH-degrading mixed microbial consortia by induced selection in soil. Chemosphere 172, 120–126.

Zanaroli, G., Di Toro, S., Todaro, D., Varese, G.C., Bertolotto, A., Fava, F., 2010. Characterization of two diesel fuel degrading microbial consortia enriched from a non-acclimated, complex source of microorganisms. Microb. Cell Factories 9, 10.

Zeng, X., Su, S., Feng, Q., et al., 2015. Arsenic speciation transformation and arsenite influx and efflux across the cell membrane of fungi investigated using HPLC–HG–AFS and in-situ XANES. Chemosphere 119, 1163–1168.

Zeng, X., Su, S., Jiang, X., Li, L., Bai, L., Zhang, Y., 2010. Capability of pentavalent arsenic bioaccumulation and biovolatilization of three fungal strains under laboratory conditions. Clean: Soil, Air, Water 38, 238–241.

Zhang, T., Tang, J., Sun, J., Yu, C., Liu, Z., Chen, J., 2015. Hex1-related transcriptome of *Trichoderma atroviride* reveals expression patterns of ABC transporters associated with tolerance to dichlorvos. Biotechnol. Lett. 37, 1421–1429.

Zhang, X., Yang, H., Cui, Z., 2017. *Mucor circinelloides*: efficiency of bioremediation response to heavy metal pollution. Toxicol. Res. 6, 442–447.

Zhang, Z., Zhou, Q., Peng, S., Cai, Z., 2010. Remediation of petroleum contaminated soils by joint action of *Pharbitis nil* L. and its microbial community. Sci. Total Environ. 408, 5600–5605.

Zhao, J., Chi, Y., Xu, Y., Jia, D., Yao, K., 2016. Co-metabolic degradation of β-cypermethrin and 3-phenoxybenzoic acid by co-culture of *Bacillus licheniformis* B-1 and *Aspergillus oryzae* M-4. PLoS One 11, e0166796.

Zheng, Z., Obbard, J.P., 2003. Oxidation of polycyclic aromatic hydrocarbons by fungal isolates from an oil contaminated refinery soil. Environ. Sci. Pollut. Res. 10, 173–176.

Zhou, D., Zhang, X., Du, Y., Dong, S., Xu, Z., Yan, L., 2014. Insights into the synergistic effect of fungi and bacteria for reactive red decolorization. J. Spectrosc. 2014, 1–4.

Zhu, F., Qu, L., Fan, W., Qiao, M., Hao, H., Wang, X., 2011. Assessment of heavy metals in some wild edible mushrooms collected from Yunnan Province, China. Environ. Monit. Assess. 179, 191–199.

Zucchi, M., Angiolini, L., Borin, S., Brusetti, L., Dietrich, N., Gigliotti, C., Barbieri, P., Sorlini, C., Daffonchio, D., 2003. Response of bacterial community during bioremediation of an oil-polluted soil. J. Appl. Microbiol. 94, 248–257.

CHAPTER

11

Exploitation of microbial consortia for formulating biofungicides, biopesticides, and biofertilizers for plant growth promotion

J. Verma[a], C. Kumar[b], M. Sharma[a], Amritesh C. Shukla[c], and S. Saxena[a]

[a]Department of Biotechnology, Babasaheb Bhimrao Ambedkar University, Lucknow, Uttar Pradesh, India, [b]Amity Institute of Organic Agriculture, Amity University Uttar Pradesh, Noida, Uttar Pradesh, India, [c]Department of Botany, University of Lucknow, Lucknow, Uttar Pradesh, India

11.1 Introduction

Climate change is a foremost challenge that affects the ecosystem worldwide due to emissions of carbon dioxide and other greenhouse gases from human activities, which needs to be focused for a better and sustainable development in the current scenario [Intergovernmental Panel on Climate Change (IPCC), 2001]. Currently, many strategies are at play to avoid stress, such as water conservation, better breeding, and the development of drought-tolerant engineered crops; however, these methods are costly and challenging to execute (Luo et al., 2019). In addition to these strategies, organisms found in ecological systems and other beneficial microbes capture nitrogen and incorporate it into plants via nitrogen fixation and other plant growth-promoting activities (Hayat et al., 2010). Resistance to various biotic and abiotic stresses is provided by several beneficial microorganisms (Enebe and Babalola, 2018). Bacterial genera that have been well-documented and successful in encouraging plant growth include *Azospirillum*, *Rhizobium*, *Bacillus*, *Pseudomonas*, *Serratia*, *Stenotrophomonas*, and *Streptomyces*. Similarly, microorganisms like *Ampelomyces*, *Coniothyrium*, and *Trichoderma* are utilized to promote plant growth all over the world (Glick et al., 2007; Patel et al., 2017; Niu et al., 2018). *Trichoderma* is a model form of bacterium that has a greater impact on plant growth and yield. Farmers are now advised that microbial inoculants are available on the market in the form of bioagents, biofertilizers, phytostimulants, and biopesticides. Besides this, some reports of exopolysaccharide-producing microorganisms like *Bacillus*, and *Azospirillum*, strains are being used in developed countries for drought and salt stress management (Khan and Bano, 2019).

This chapter highlights the beneficial role of biofungicides, biopesticides, and biofertilizers in sustainable agriculture as a viable and environmentally acceptable method for plant growth promotion.

Entrepreneurship with Microorganisms. https://doi.org/10.1016/B978-0-443-19049-0.00019-0

11.2 Problems in agriculture and need for formulations

Drought and salinity stress are two of the most critical limiting abiotic factors impacting agricultural yield, and they have become a global problem. It is important to remember that the salt stress problem in crops is continuously increasing due to poor management of irrigation water. Adding salt to water reduces its osmotic capacity, providing less water for the roots and thus exposing the plant to secondary osmotic stress. This means that all physiological responses associated with drought stress can be abolished by mitigating salt stress. In some cases, plants show increased susceptibility to bacteria under stress such as salinity (Ullah et al., 2021). Both host and pathogen produce reactive oxygen species during infection. This condition can limit bacterial colonization if the plant produces reactive oxygen species, and may cause disease if bacteria produce reactive oxygen species (Rouhier and Jacquot, 2008). Drought resistance in plants is frequently associated with high osmotic potential between root hairs and rhizospheric soils (Bray, 1997; Wang et al., 2003), although drought tolerance in microorganisms is not correlate with an increase in osmotic potential (Rodriguez et al., 2009). Despite increasing their osmotic capacity in response to drought, nonsymbiotic plants wither earlier (6–10 days) than symbiotic plants. As a result, managing abiotic pressures for example drought and salinization is a significant concern in agriculture, and Fig. 11.1 depicts the contribution of biotic and abiotic factors to plant growth. Precipitation, temperature, and salinization are affecting people and plants to a higher extent. Many abiotic stresses such as drought, salinity, alkalinity, and temperature affect plant ecosystems negatively (Enebe and Babalola, 2018).

Beneficial bacteria have been utilized for many years by scientists to combat weed infestation (Babalola, 2010), nutrient deficiency and heavy metal contamination (Sheng, 2005), drought stress (Zahir et al., 2008), and salt stress (Egamberdieva, 2008). The potential of PGPR to induce tolerance to salinity and precipitation is known as "inducing systemic tolerance" (Yang et al., 2009). However, beneficial bacteria play a vital part in preventing soil diseases, stimulating plant growth, and transforming plants (Doran et al., 1996). These organisms help the crop avoid heat stress in different ways, including exopolysaccharides (EPS), plant hormones (auxin, gibberellic acid, and cytokinin), 1-aminocyclopropane-1-carboxylic acid (ACC) deaminase, cycling of ammonia, phosphorus, potassium, and zinc, nitrogen fixation, improved nutrition, osmolyte accumulation, synthesis of antioxidants, modulating activities of stress response genes (Yadav and Yadav, 2018) (Fig. 11.2). Recently, *Pseudomonas fluorescens*, *Enterobacter hormaechei*, and *Pseudomonas migulae* were investigated as drought tolerance isolates that can stimulate seed germination, seedling along with the high production of ACC deaminase, and exopolysaccharide and serve as effective inoculants in sustainable agriculture (Niu et al., 2018). In seeking environmental remediation, this chapter defines the fundamental role of fungicides, biopesticides, and biofertilizers in sustainable agriculture development.

11.3 Fungicides

11.3.1 What are fungicides?

Fungicides are pesticides that are nonorganic agrochemicals, organic mineral salt solutions, and biological organisms. Fungicides contain fungicidal compounds that are fungicidal, producing fungicidal compounds to kill fungi and their spores or inhibit their growth. Fungicides are used to control plant diseases such as rust, mold, and blight. They can also be used to control mold and fungus in other bios settings. Fungicides work in many ways, but most interfere with fungal diseases or interfere with the energy produced in fungal diseases (Palmieri et al., 2022).

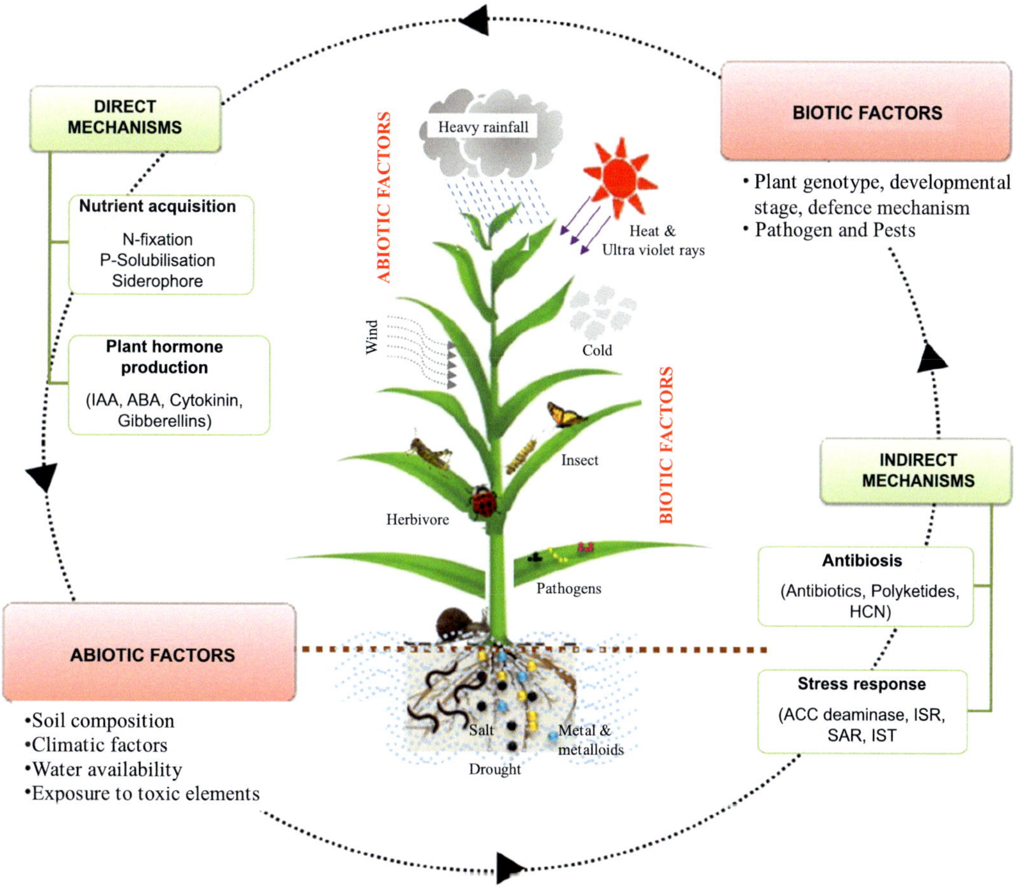

FIG. 11.1

The effect of biotic and abiotic factors on plant growth promotion, as well as the cyclic (direct and indirect) mechanism of growth promotion factors in plant systems.

11.3.2 What are biofungicides and why do we need bio-fungicides?

Biofungicides are preparations made from living organisms used to kill phytopathogenic fungi. Using a biofungicide is based on natural processes where helpful bacteria, frequently obtained from the soil, inhibit the spread of plant diseases. These microbes produce many antibiotics, cause disease, and compete with other microbes to cause local disease or disease in plants. Disease management is crucial for the majority of crops because diseased plants have a considerable negative economic impact on production and quality. In general, there are three main reasons for the use of fungicides:

(a) Improve agricultural sustainability (Sharma et al., 2020).
(b) Disease control during planting and development.
(c) Increase product yield and reduce defects. Crop diseases can reduce crop yields as leaves affected by the disease are less suitable for photosynthesis (Burgess et al., 2023).

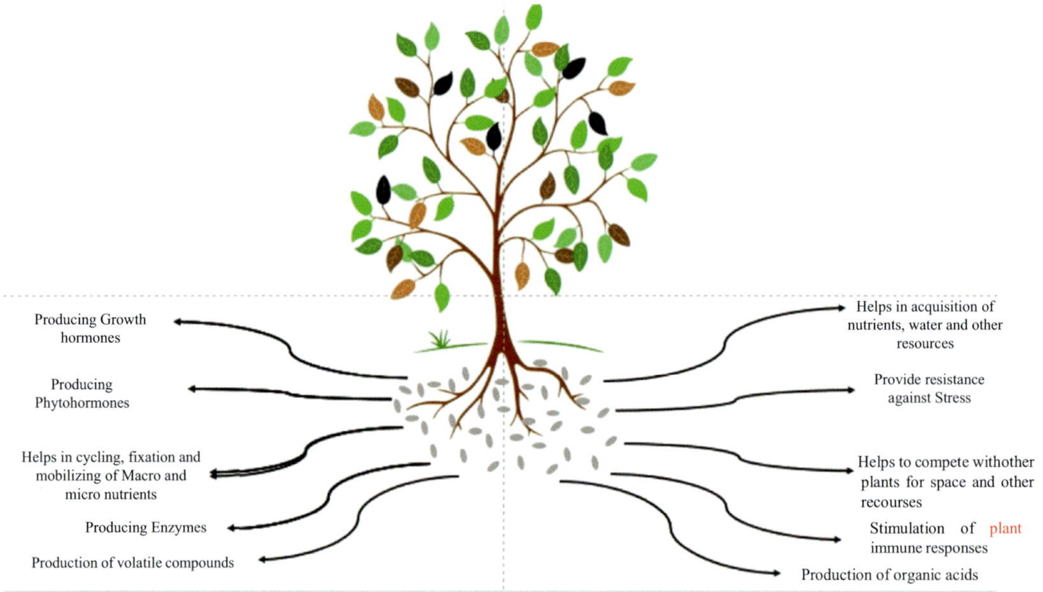

FIG. 11.2

PGPR mechanisms as biofungicides.

(d) Improving the shelf life and quality of crops and products. Some of the worst damage occurs in crops after harvest (Abbey et al., 2018).
(e) To decrease the usage of hazardous chemicals to produce good quality produce.

11.3.3 Current status of fungicides

In the current scenario, pesticide-resistant plant pathogenic fungi have a great demand due to enormous crop destruction and the development of novel fungicides for plant growth promotion. Various antifungal strategies are used worldwide for biofungicide development. Among these, peptides and antifungal proteins (APS) are crucial for sustainable agriculture. Recently, Toth et al. (2020) reported *Neosartorya fischeri* as a crop preservative by inhibiting the growth of ascomycetes in vitro analysis. APs display extraordinary stability and resistance to the degradation of protease enzymes to their beta-protein and tertiary structure of protein and significantly impede the development of several plant pathogens like *Penicillium* sp., *Aspergillus* sp., etc. (Galgoczy et al., 2010).

11.3.4 Bioformulation and development of biofungicides

Biofungicides are microorganisms like bacteria, fungi, mycorrhiza, and viruses, however, most efficient, and commercial biofungicides are made from pure cultures or consortia of bacteria. These biofungicides are a solution to maintaining agricultural yield and environmental quality. The design and application of biofungicides against fungi in cultivated plants must be carefully considered for them to have a positive environmental impact. Fungicides have many of the advantages of fungicides, such as being biodegradable, inexpensive, and leaving no visible residue. Agricultural fungicides are secondary metabolites produced during microbial fermentation that can suppress or eradicate plant pathogens at low concentrations and control crop yield and growth. For example, kasugamycin, jinggangmycin, polyoximycin, and streptomycin. Currently, agricultural fungicides are the powerhouse in biocides. Table 11.1 shows examples of famous biopesticides all over the world. Biofungicides when applied through seed/seedling treatment, foliar spry, and fertigation, have various benefits in various modes of action (Fenibo et al., 2021) like Rhizosphere competence, Parasitism, Antibiosis, and providing metabolic changes (Arthurs and Dara, 2019).

11.4 Plant growth promoting rhizobacteria (PGPR)

11.4.1 Plant growth promotion management

These bacteria also function as rhizobacteria, which aid in the growth of plants, in addition to having biocidal or biocontrol activities. Microbes that colonize the roots of plants help them grow and increase production. Some biofungicides, when used for planting or seed treatment, can promote root growth, and increase drought resistance in some plants. Growth in plants is enhanced by directly impacting the plant and the soil microbial community. Biocides can improve nutrient supply and quality (copper, phosphorus, iron, and manganese) in plants (Hashem et al., 2019). Microbes that are used as biofungicides can also produce extracellular enzymes, hormones, and organic acids which benefit plants in their day-to-day activities which is represented in Fig. 11.2.

Table 11.1 Microbes used as biofungicides in all over the world.

S. no.	Microbes used as biofungicides	References
1.	*Streptomyces NEAU-S7GS2*	Liu et al. (2019)
2.	*Ulocladium oudemansii U3*	Thomidis et al. (2015)
3.	*Bacillus subtilis QST 713*	EFSA (2021)
4.	*Bacillus subtilis GB03*	Hashem et al. (2019)
5.	*Coniothyrium minitans*	Whipps et al. (2008)
6.	*Bacillus velezensis P1*	Nievierowski et al. (2023)
7.	*Burkholderia sp. F25*	Yu et al. (2022)
8.	*Pseudomonas sp.*	Win et al. (2022)
9.	*Trichoderma harzianum*	Sheir et al. (2015)
10.	*Gliocladium catenulatum JII446*	Khakimov et al. (2020)
11.	*Bacillus subtilis*	Twizeyimana et al. (2023)
12.	*Pichia guilliermondii*	Thambugala et al. (2020)
13.	*Ampelomyces quisqualis*	Kiss et al. (2004)
14.	*Trichoderma virens*	Srivastava et al. (2016)
15.	*Pseudomonas chlororaphis*	Twizeyimana et al. (2023)
16.	*Pseudomonas fluorescens*	Castaldi et al. (2021)
17.	*Trichoderma viride*	Waghunde et al. (2016)
18.	*T. harzianum (T39)*	Abbey et al. (2018)

The rhizosphere is a zone of life where plants grow their roots. Among all layers of soil, rhizosphere, and rhizoplane have the maximum microbial count/microbiotas population, with a diversity greater than 30,000 species identified and a lot more to be discovered. The diversity of the rhizosphere determines the soil quality/soil productivity. Various forms of sustainable agriculture like organic, biodynamic farming, etc. deploy the soil microbiota like PGPR, VAM, AMF, etc. for fulfilling plant requirements for nutrients, enzymes, hormones, biocontrol, etc., which makes the situation of soil microbiotas diversity directly proportional to crop productivity in sustainable agriculture. As a result of the green revolution, a wide variety of agrochemicals are being utilized to increase productivity as a result soil microbiota's diversity and population have been declining from crop to crop. Inoculation of specific consortia of microbes like N_2-fixing cyanobacteria, PGPR, mycorrhizae, stress-tolerant endophytes, and other beneficial microbes, have the potential to maintain soil health and even have the potential to repair the damage caused by conventional farming (Fig. 11.3).

Endophytic microorganisms are good friends of plants. They are present in tissues through phosphate dissolution, siderophores, the formation of indole-3-acetic acid (IAA), ACC deaminase (which reduces the level of ethylene and detoxifies pollutants in plants to stimulate plant growth), biological nitrogen fixation, etc. Bhattacharyya and Jha (2012) presented almost similar findings on plant growth-stimulating

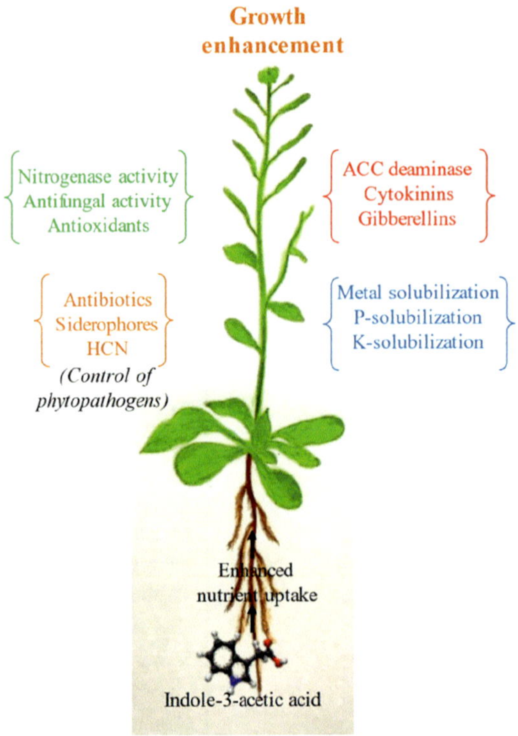

FIG. 11.3

Enzymes and growth hormones accountable for plant growth promotion in different parts of a plant.

rhizobacteria. The inhibition of ethylene stress by ACC deaminase, which is only effective in phytosanitary applications, can be reduced by endophytic bacteria, as recently demonstrated (Rashid et al., 2012; Glick, 2014). The deaminase enzyme is involved in the deamination of ACC to ammonia and α-ketobutyrate, thereby supporting the plant's tolerance to stress (El-Tarabily et al., 2008; Glick, 2014). The effectiveness of eight endophytes isolated from tomato seeds and chili plants on biocontrol activity and qualities that promote plant growth was described in a recent work (Amaresan et al., 2012). Similarly, in a tomato seed inoculation experiment, Karthik et al. (2017) discovered that endophytic bacteria had a favorable impact on seed germination, growth of seedlings, and biomass production. Following immunization with the ACC deaminase virus, Maxton et al. (2017) evaluated the impact of salinity and drought stress on the distinct morphological and physiological development of peppers. Forni et al. (2017) provided a thorough analysis of the strategies employed by plant growth-promoting organisms to mitigate the negative effects of salinity, drought, and other abiotic stresses on growth and development and boost agricultural yields. In the aforesaid context, we will construct GFP-containing plant growth-promoting endophytes and inoculate them in surface sterilized tomato and chili seeds which will provide inherent resistance to tomato and chili plants against pathogens.

11.5 Biopesticides

11.5.1 Role of biopesticides in plant growth promotion

Biopesticides are naturally derived chemical compounds obtained from living organisms to protect and manage plants from severe pests and pathogens (EPA, 2023). Living organisms include microbes (bacteria, fungi, and viruses), plants, and animals. Biopesticides mode of action is nontoxic; it is eco-friendly and hence used all over the world (Essiedu et al., 2020). Its products are successfully utilized for integrating pest management (IPM) (Chang et al., 2003; Essiedu et al., 2020).

It is also beneficial for the growth of plants. Although it has a direct effect by reducing the number of bacteria, many studies have shown that it has a direct effect by colonizing legumes, increasing strength, and therefore improving plant growth (Qi et al., 2016). Numerous bacteria including *Azotobacter*, *Caulobacter*, *Klebsiella*, *Pseudomonas*, and *Rhizobium*, support its cultivation (Kumar et al., 2016). *Bacillus* species (*Bt* and *B. sphaericus*) were shown to solubilize phosphates, allowing the plant to flourish. Because the mineral is abundant in soil, this is advantageous. However, due to its undissolved state, its uptake is constrained (Seshadri et al., 2007). Similarly, iron dissolution is complicated, and the plant utilizes the siderophore-iron complex for iron absorption. Microorganisms create siderophores, which bind to iron and help in its absorption.

According to studies, catechol-type siderophore synthesized by *Bt* strain ATCC33679 has a good affinity for iron and makes it usable in plants (Wilson et al., 2006). Plant growth was improved when *Bacillus thuringiensis*-KR1 and *Bradyrhizobium japonicum* increased bean nodule number, volume, weight, and shoot weight (Mishra et al., 2009). Another study found that combining *Bt* with another pathogen, *Rhizobium leguminosarum* promoted root nodule and pea dry weight formation. *Bt* colonization has been observed in lentils (Mishra et al., 2009; Pindi et al., 2014). Similar to this, drought resistance and oxidative metabolism significantly improved in lavender plants co-inoculated with mycorrhiza (Armada et al., 2016). In a recent study, *Bt* A5-BRSC was utilized for 2 years as a coal-based biofertilizer on *Abelmoschus esculentus*. Evaluates its ability to support plant growth, including nutritional and morphological factors. These conclusive studies demonstrate that utilizing *Bt* is helpful as an insecticide and as a biofertilizer that aids plant development (Bandopadhyay, 2020).

11.5.2 Global status of biopesticides

In many areas of the world, emphasis has turned to higher plant products as chemotherapeutics. Botanical pesticides are now widely utilized across the world. As botanical pesticides, pyrethroids and neem products are produced, and various essential oils from higher plants are utilized as antimicrobials (Dubey et al., 2008). Biopesticides have a US$ 3 billion global market value and accounting 5% of the worldwide pesticides market according to Marrone (2014), the value represents 5% of the worldwide pesticide market. The market share of biopesticides should be comparable to synthetic pesticides from 2040 to 2050, with a CAGR of over 15% (Olson, 2015; Damalas and Koutroubas, 2018). Business Market, Inc. estimates that the market for synthetic insecticides and biopesticides worldwide was worth 61.2 billion USD in 2017 and will be worth $79.3 billion by 2022 (Lehr, 2014; Chen, 2018). China, the United States, and Brazil are the three nations that use the most pesticides, according to data from the United Nations Food and Agriculture Organization (FAO) for the years 2017–18. FAO also reported that Asia accounted for 52.2%, United States 32.4%, Europe 11.8%, Africa 2%, and Oceania 1.6% of global pesticide use from 2015 to 2018 (FAOSTAT, 2021). The use of bacteria to make antiinflammatory drugs in China began in the 1960s. Over 32 viruses have been found beneficial for pest control in agriculture, farms, forests, and buildings (Sun, 2015). Matsukemin, a cypovirus product comprising both live and inactive *Bt*, was the first microbial control product approved in Japan in 1974. There are now 25 microbial insecticides on the market, with less than 2% of them being used in Japan. The Ministry of Agriculture of China authorized 57 products derived from 11 viruses in 2014, amounting to 1600 tons of viral insecticidal formulations generated every year. This amounts to around 0.2% of China's overall pesticide production. This microbial biopesticide is sold in limited quantities in specific markets to treat arthropods (Lacey et al., 2015).

When compared to synthetic chemical pesticides, the application of biopesticides is currently limited due to several drawbacks, including high manufacturing costs, poor storage stability, vulnerability to environmental conditions, effectiveness difficulties, quality control, short shelf lives, and knowledge gaps (Arthurs and Dara, 2019). Some of these issues can be addressed by formulation changes, which have been beneficial in enhancing and maintaining biopesticide activity (Gašić and Tanović, 2013). The isolated toxin from the *Bt* strain is used as an efficient microbial pesticide in most biochemical biopesticides used across the world. Farmers can purchase commercially available biopesticides. According to current data, around 175 registered biopesticides worldwide, with 700 active ingredient products available for usage. Only 12 biopesticides have been registered in India, comprising bacterial, viral, fungal, and plant items.

11.5.3 Scope and importance of biopesticides

If their potential is wholly realized, biopesticides might replace synthetic pesticides soon without impacting production or yield considerably. The use of biopesticides in agriculture, like other programs such as biostimulants and biofertilizers, balances the social, economic, and environmental quality control that can enable sustainability in agriculture.

11.5.4 Classification/types of biopesticides

Due to their great effectiveness, species specificity, and ecological sustainability, pest management methods frequently use pesticides derived from microorganisms and their compounds. These biochemicals are produced by microbes like bacteria, actinomycetes, viruses, and fungi.

These microorganisms contain molecules/compounds active against herbivorous insects or disease species. Biopesticides target crop-damaging insects, while bioherbicides use organisms such as fungi to control weeds. Biopesticides can act as growth regulators, metabolic toxicants, gastrointestinal disruptors, neuromuscular toxicants, and general multisite inhibitors (Sparks and Nauen, 2015; Dar et al., 2021). Given their importance, their economic growth potential and economic value are high. They are divided into several groups according to their origin and role.

11.5.4.1 Bacterial biopesticides and mechanisms of action

Genera *Bacillus* and *Pseudomonas* with various species and subspecies have been identified as biopesticides. They are primarily used for the treatment of insect infestations and plant diseases. The most well-known of these insecticides are made from different *B. thuringiensis* Berliner strains. Among them, *B. thuringiensis* subspecies, kurtaki and aizawa are effective against Lepidopteran larvae; *B. thuringiensis israelensis*, which is effective against mosquitoes, houseflies (simuliidae) and fungal sandflies; *B. thuringiensis tenebrionis* and Co., especially against the adult Colorado potato beetle (*Leptinotarsa decemlineata*); and *B. thuringiensis japonensis* species Buibui versus soil beetle (Carlton, 1993; Copping and Menn, 2000). Four different categories of bacteria are utilized as biopesticides: crystalline spore-forming (*B. thuringiensis*), obligate pathogens (*Bacillus popilliae*), prospective pathogens (*Serratia marcescens*), and facultative pathogens (*Pseudomonas aeruginosa*). For business, bacteria cause most infectious diseases. According to Roh et al. (2007), *B. thuringiensis* and *B. sphaericus* are the most widely utilized bacteria since they are distinct strains, safe, and efficient. The successful application of *Bacillus thuringiensis* (*Bt*) and other microbial strains has resulted in the discovery of numerous new strains and strains that seem to be advantageous to toxins and pathogens that may be advantageous to the biopesticide industry; some of these have also been converted into the products. The most prevalent entomopathogens are *Pseudomonas*, *Yersinia*, *Chromobacterium*, and others are used as biocontrol agent.

The *B. thuringiensis* Cry family of crystallization proteins are generated in parasporal crystals by the Cry gene. *Bt* produces crystal proteins that can kill certain types of pests such as Lepidoptera. Identification of insect targets by binding *Bt* crystal proteins to insect receptors (Kumar, 2012). The toxicity of the insecticidal crystal protein (delta-endotoxin) of *Bt* and its subspecies was evaluated by Thurley et al. (1985) and Aronson and Shai (2001). To be lethal, endotoxins must be absorbed by the larvae. When the larvae eat this poison, they destroy the intestinal tract and cause intestinal paralysis. The larvae then stop feeding and die due to a combination of starvation and damage to the midgut epithelium. This is because neurotoxins can cause brain damage (Betz et al., 2000; Darboux et al., 2001; Zhu et al., 2000).

11.5.4.2 Fungal biopesticides and their mechanism of action

Metarhizium anisopliae is an entomopathogenic fungus commonly used in pest control and found worldwide (Tulloch, 1976). Many entomopathogenic fungal species and their derivatives are used as microbial antibiotics. These are *Beauveria*, *Metarhizium*, *Verticillium*, *Lecanicillium*, *Hirsutella*, *Paecilomyces*, and other fungi (Roberts and St Leger, 2004). Fegan et al. (1993) found dung beetles in the soil in some places. Environmental factors including the right temperature and moisture content promote filamentous growth and the development of conidia, infectious spores that, upon contact, infect soil organisms. Future research and development efforts should focus on *M. anisopliae* since it has the potential to be exploited as a biocontrol agent, particularly against species of malaria vectors

(Mnyone et al., 2010). These entomopathogenic organisms are considered safe and effective for the use of insecticides. Brazil processes 100,000 hectares of sugarcane each year, and *M. anisopliae* is commonly employed there (Faria and Magalhães, 2001).

The yield and productivity of crops are significantly influenced by arbuscular mycorrhizal fungi (AMFs) (Mishra et al., 2018; Ellouze et al., 2018). By strengthening their defense systems, they make plants resistant to diseases. The composition of AMF varies with soil type, crop, fertilizer, and tillage application, and its prevalence decreases (Vannette and Hunter, 2009). Plants have developed several ways to fight biotic stress. They can produce or release phytochemicals like gossypol, nicotine, pyrethrins, and many other compounds that can protect plants from herbivores. AMF colonization of agricultural plants aids in determining the optimal defense for the host by altering gene expression patterns and, possibly indirectly, affecting crop nutrition (Nuruzzaman et al., 2016; Kremer, 2019).

Fungicides and Fungicidal Mechanisms: These biopesticides often block or prevent translation, which includes binding to the prokaryotic 50S ribosome to prevent peptide transfer and chain elongation (e.g., blasticidin) (Parker et al., 2019; Svidritskiy et al., 2013). Kasugamycin similarly prevented aminoacyl tRNAs from attaching to the 30S and 70S ribosomal subunit complexes, hence inhibiting translation (Schuwirth et al., 2006). Streptomycin and mildiomycin bind to the 30S ribosomal subunit causing abnormal protein synthesis (nonfunctional) and block peptidyl transferase activity (Arena et al., 1995; Feduchi et al., 1985). According to Gwinn (2018), these substances can cause cell death (natamycin), modify the permeability of the plasma membrane, enhance toxicity (amino acids and electrolytes), block chitin synthase activity (polyoxin), and hinder the synthesis of glucose by Trehalase (validamycin). When pesticides enter the bloodstream, they produce γ-aminobutyric acid (GABA), which opens GABA-gated chloride channels, hyperpolarizes the brain, and inhibits electrical neurotransmission (abamectin and emamectin) (Arena et al., 1995; Feduchi et al., 1985). Polynactins cause potassium ion leakage from mitochondria. Antibiotics prevent plant phosphorylation by increasing ammonia by inhibiting glutamine synthase (Feduchi et al., 1985).

11.5.4.3 Viral biopesticides and mechanisms of action

Baculoviruses are another important class of microbial antibiotics. The *Bt* toxin and the baculovirus inclusion body combine to generate the baculovirus, also known as Color Btrus. Baculoviruses have double-stranded DNA and are found in class insects of the phylum Arthropoda. Baculoviruses are generally less virulent and have been used successfully as antibiotics against many major pathogens. Baculovirus causes death only in the larval stage of Lepidoptera, the main group from which it is isolated. Larvae must ingest baculovirus to become infected. The drug enters the cells in the body through the environment and spreads throughout. However, in some lines, the infection may be limited to the midline or the fatty body (Williams and Faulkner, 1997). Granular virus (GV) and nuclear polyhedrosis virus (NPV) are the two categories of baculoviruses. In NPV, the body typically contains a large number of organisms, but in GV, the body typically only contains one. Baculoviruses are closed, which means they exist within a protein matrix. In baculovirus biology, the existence of the inclusion body is crucial for the virus' survival (Cory et al., 2000). Azadirachtin extracted from neem plant and used for insect pest management in agriculture (Adhikari et al., 2020).

11.5.4.4 Nematode biopesticides and mechanisms of action

Entomopathogenic nematodes also defend plants against major crop pests and diseases. Several attempts have been made to biocontrol field pest populations in agriculture (McSorley, 1999). Covering plants, planting crops, and mixing organic matter into the soil can help control nematodes.

Entomopathogenic nematodes used as biocontrol agents are mainly from *Heterorhabditis* and *Steinernema* (Nematoda: Rhabditida) and are related to the symbiotic symbiotic organisms Photorhabdus and Xenorhabdus (Thanwisai et al., 2022). They do not harm animals, the environment, or nontarget creatures (Shannag and Capinera, 2000; Duncan and McCoy, 1996; Shapiro and McCoy, 2000). The ease with which they are synthesized in large quantities has contributed to their commercial development as biocontrol agents by in vivo or in vitro methods and their tolerance (Peters, 1996). Insect-parasitic nematodes enter the soil and destroy harmful bacteria, killing them within 48 hours. After the death of the host, the disease stage of the nematode matures into an adult, and a new generation of larvae is formed. Infected larvae in the third stage begin the cycle of infection in nematodes. These unfed larvae infect susceptible insects by entering through body openings such as the anus, mouth, and stomata (Grewal et al., 1997). The nematodes infect the host's hemocoel after entering, and they build up their symbiotic bacteria in the stomach. After that, the bacterium causes sepsis, which kills the host in 24–48 hours. The bacteria quickly suppress the larvae's infection while destroying the host tissues. Nematodes can complete one to two generations inside the host (cadaver) (Bird and Akhurst, 1983).

11.5.4.5 Protozoan biopesticides and their mechanisms of action

In nature, insect-protozoan infections are prevalent and significantly affect insects (Fuxa and Tanada, 1987). *Nosema* spp. like Microsporidia, it is usually specific and transient and causes a long-term infection. Most of the pathogenic protozoa have biological activity. They can only grow in one host, and some species need intermediate hosts, such as Microsporidia.

Their similarity is that they are persistent and circulating in the participants and cause serious harm to the growth and health of insects. Some strains have achieved significant results under the use of heavy microbial antibiotics (Senthil-Nathan, 2015).

11.5.5 Other biopesticides

11.5.5.1 Biochemical pesticides

Biochemical pesticides are conventional chemicals used to control pests by nontoxic methods of plant origin. Biochemical pesticides are also classified according to whether they control pests using pheromones (semiochemicals), plant extracts/oils, or insect breeding methods. Plants have developed and produced many chemicals that help kill microbes during infection and invasion. Steroids, alkaloids, phenylpropanoids, phenols, terpenes, and nitrogenous compounds are examples of secondary metabolites (Weber et al., 2019; Du et al., 2016). Cigarettes contain nicotine, which is toxic to many herbivorous insects. Pesticides produced from it are called "green pesticides" and have high efficacy and nontoxicity. Smoking contains secondary metabolites such as nicotine and solanesol, which are potent inhibitors of *Bacillus subtilis*, *Staphylococcus aureus*, and *Micrococcus lysozyme* (Du et al., 2016).

Mechanism of action: Microbicides affect plasma membrane integrity and meristems by interfering with metabolic processes such as sporulation, while pesticides like nicotine and azadirachtin act by obstructing insect growth regulators, binding to Na^+ channels, or interfering with respiratory enzymes (Gomiero, 2018; Elgar et al., 2018).

11.5.5.2 Feather bug pheromones

Pheromones are chemicals secreted by insects to elicit certain behavioral responses in other insects. They play a variety of roles and are given names based on how they react, including alarm, aggregation, and sexual pheromones. Some pheromones serve as sexual attractants, helping individuals locate and mate, while others encourage the hunting, mating, and gathering of other congeners Witzgall et al. (2004) provided a list of more than 1600 pheromones and sexual attractants because pheromones have grown in importance as a tool for monitoring and managing agricultural pests. In IPM these are compounds created by insects used for pest control. These substances work well to interfere with the mating process, preventing successful mating and lowering the number of insects. Pheromone sinks become disorganized when they disperse into the environment thanks to the role of insects used in the process as pheromone distributors. Since they do not really kill insects but instead alter their behavior by influencing their factory system, insect pheromones are not true insecticides.

Mechanism of action: Sensation The insect's antenna absorbs pheromones, which are then diffused through the pores in the epidermis into the sensilla. Once inside, pheromone-binding proteins transport them to the chemo-sensing membrane of the hydrophilic sensilla. Pheromones or pheromone-PBP complexes bind to specific receptor proteins that convert chemical signals into electrical signals, sending second messages connected to neural circuits.

11.5.6 Formulation and development of biopesticides

11.5.6.1 Dry formulation

Dustable Powders: The dust formulations have an active ingredient concentration generally of ten percent. They are formulated on the delicate ground by the sorption of a solid powder of mineral solids (talc, clay, etc.), with a 50–100 mm particle size. UV protection products, and adhesive components (i.e., sticks) for enhancing adsorption and anticoagulation are inert constituents in the dust compositions (Slavica and Brankica, 2013).

Granules: Between 2% and 20% of a granule's weight is made up of active compounds, which are either covered on the outside or absorbed by the granule. Granules are primarily used in soils, weeds, and nematodes in order of root absorption to control insects. The coarse-size granules comprise 100–600 μm of kaoline, silica, starch, polymers, residues from groundnut plants, dry fertilizer, etc. After contact with soil moisture, specific granules release their active components. According to Tadros (2006) and Slavica and Brankica (2013), granules can be coated with polymers or resins to change the active components' effectiveness.

Seed Dressing: A kind of formulation of biopesticide produced by dissolving in powder form and accompanying inert active ingredient transport to enable the adhesion of the end product to seed coatings. Seed dressing is a mixing of powder formulation with seeds. The product is designed to attach to it and contain coloring chemicals that notify the manufacturer that the red pigment is safe for seeds that have been treated (Woods, 2003).

Wettable powder: Active chemicals are mixed with solvents and disinfectants, coordinating agents, heat-resistant materials, and nontoxic materials to make a dust-resistant powder. These are also carefully applied dry soil formulations following water suspension. Due to their dustiness, strict safety measures are generally performed that might create significant health concerns for the producers and their usage. Furthermore, wettable powders have long storage stability and excellent water miscibility and may be used in ordinary sprinklers (Brar et al., 2006; Knowles, 2008).

Water-dispersible granules: They are made to remain suspended in liquid, get around damp dust issues, retain the lack of dust, and store readily (Knowles, 2008).

11.5.6.2 Liquid formulations

Emulsions: Emulsion formulations are made to bind water as inert or conventional emulsions (water in oil or oil in water). However, in the case of water-in-oil emulsions, losses from evaporation and spray entrainment are reduced since the oil is in the outer phase of the formulation. Most importantly, the right choice of emulsifier for stabilization should prevent damage due to instability (Brar et al., 2006).

Suspension Concentrate: The finely ground active ingredient that is dissolved in the liquid phase (often water) to create a formulation. Because the product does not dissolve in the liquid phase, it is frequently necessary to agitate the product before use to equally distribute the particles. The range of particle sizes is between 1 and 10 m, and the smaller size of the particles improves the material's ability to penetrate plant tissue and stimulate biological activity. It is a form of production because of its safety for workers and the environment (Knowles, 2008).

Suspo-Emulsion: It is a complex emulsion and suspension formulation as it requires particulate removal of the ingredients to form a homogeneous emulsion composition so that the final product remains in a stable state of freezing. It is prepared by measuring the storage stability of the product (Knowles, 2008).

Oil Dispersion: This formula is made in the same way as a concentrated suspension. Careful parameter selection will avoid instability problems (Vernner and Bauer, 2007).

Capsule Suspension: Gelatin, starch, cellulose, and other polymers are used to create capsules that enclose and shield biological agents from harsh environments. The active ingredients are packaged in stable microencapsulated solutions that must be diluted with water before use. The encapsulation process is the concept of interfacial polymerization leading to small and potent formulations, particularly fungal biopesticides (Winder et al., 2003).

Ultra Low Volume Liquid: Active component formulations shouldn't be diluted with water before usage. It is portable and can be used as a biocontrol agent's active component (Woods, 2003). Because they are very effective, species-specific, and eco-friendly, pest management systems all over the world use bacteria and their products to generate pesticides. Bacteria such as bacteria, viruses, and fungi produce these chemicals. These bacteria contain molecules/compounds active against herbivorous insects or disease species. Biopesticides target crop-damaging insects, while bioherbicides use organisms such as fungi to control weeds. The modes of action of biopesticides include growth regulators, gastrointestinal disrupters, metabolic toxicants, neuromuscular toxins, and nonspecific multisite inhibitors. Given their importance, their economic growth potential and economic value are high. They are divided into several groups according to their origin and role.

11.6 Biofertilizers

11.6.1 What are biofertilizers and their discriminative aspects?

In this industrialization era, the population of the world currently reached 7.75 billion and will increase rapidly day-by-day, which creates a huge problem for food, feed, and other agricultural problems. Therefore, development of sustainable agriculture we need to increase the nutrient requirement as well as decrement of chemical fertilizers (Selim, 2020). Although Chemical fertilizers increase crop productivity they also create a hazardous effect on soil health. In this context, we used naturally available fertilizers (i.e., microbial-based fertilizers) which enhanced the crop productivity in sustainable agriculture and also the fertility of soil called biofertilizer (Raiz et al., 2020). Due to increasing health awareness, biofertilizers are being widely used to avoid toxicity and to mitigate ecological imbalance. The production of dangerous chemicals by chemical fertilizer companies endanger human health and contributes to ecological imbalance (Fasusi et al., 2021).

Further, biofertilizers are cost-effective also as high cost is involved in the production of chemical fertilizers. Not only that biofertilizers are capable of increasing crop productivity and soil fertility and therefore have the capability of meeting the demand for food production as per the increasing population of the world. Further application of biofertilizers promotes green technology which is essential for sustainable development (Bhardwaj et al., 2014).

Living cells found in microbial biofertilizers interact with the rhizosphere or endosphere, increasing soil fertility and promoting nutrient yield. To fulfill the expanding demand for food supply, biofertilizers can be employed to boost crop yield. Preparing biofertilizers using mycorrhizal fungi and PGPR for rhizosphere control is becoming popular in developing countries (Raklami et al., 2019). New technologies have been developed to modify plant growth-promoting bacteria with nanoparticles made from organic and inorganic materials (Shang et al., 2019; Mittal et al., 2020). Therefore, the application of biofertilizers is a viable microbial technology for sustainable agriculture and rhizosphere management. To boost crop yield and enhance soil quality, biofertilizers made from many kinds of organisms are employed. As depicted in Fig. 11.4, biofertilizers have a significant impact on growth through nitrogen fixation, phytohormone production, siderophore generation, phosphate solubilization, and crop disease eradication.

11.6.2 Importance of biofertilizers in plant growth promotion

In the management of soil nutrients, such as nitrogen fixation, phosphate dissolution, herbicides, and plant growth enhancement, biofertilizers are essential (Nosheen et al., 2021). They are also utilized to improve soil fertility and serve as biocontrol agents in a variety of crops, along with crop rotation and cropping patterns. *Azospirillum*, *Azotobacter*, *Cyanobacteria*, *Rhizobium*, P and K-solubilizing microbes, and mycorrhiza are beneficial PGPRs that have been shown to increase crop production even under no-tillage or minimum tillage (Bhardwaj et al., 2014). Recently a biological process was used to convert complex substrate into biogas and digest it under different microbial activities such as hydrolysis, methanogenesis, acetogenesis, and acidogenesis. This technology is also known as Anaerobic digestion which turned into biofertilizers and has high potential for energy recovery and quality product development (Wainaina et al., 2019). The classification of biofertilizers is shown in Fig. 11.5.

11.6.3 Universal status of biofertilizers

Farmers are adopting more efficient ways of managing soil nutrients as they become more concerned with sustainable agriculture, which is driving the growth of the global biofertilizer market. Biofertilizers are favored over chemical fertilizers because they are less expensive, more environmentally friendly, and provide long-term agricultural productivity. Rising demand for organic foods and environmentally friendly farming practices is gaining attraction for biofertilizers in developed and developing nations.

In 2016, the market for biofertilizers was valued at US$ 668.47 million, and by 2021, it was anticipated to increase at a CAGR of 13.3%. By 2020, the market for biofertilizers was anticipated to be worth US$2.3 billion. With countries like India, China, Argentina, Canada, Europe, and the United States joining in, the worldwide biofertilizer market is growing (Malusà et al., 2016; Markets and Markets, 2019).

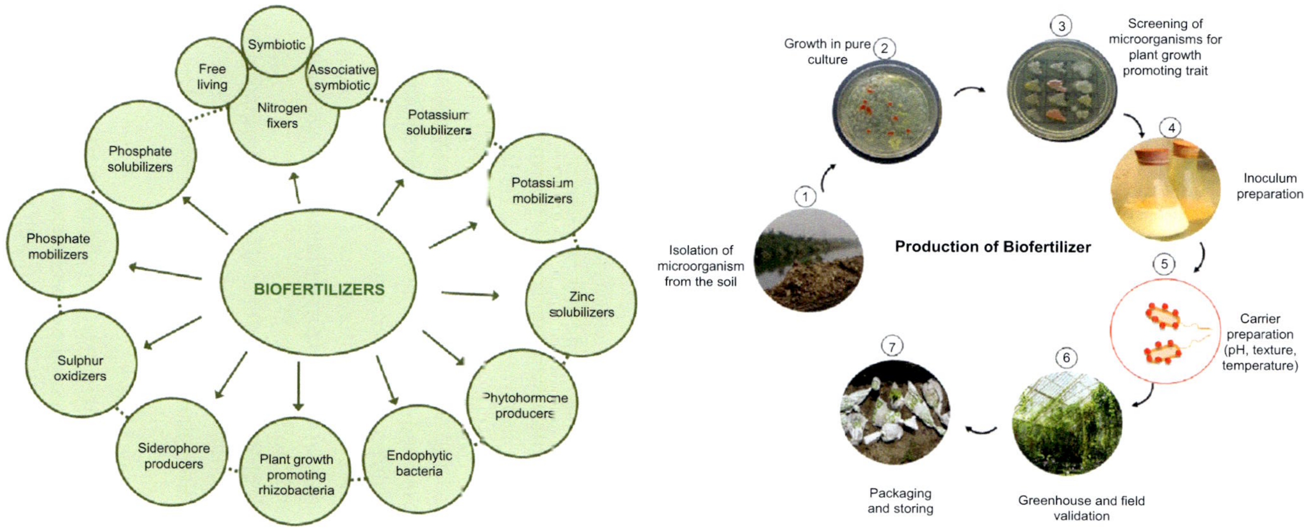

FIG. 11.4

Application and production of biofertilizers in different fields as growth promoting agent.

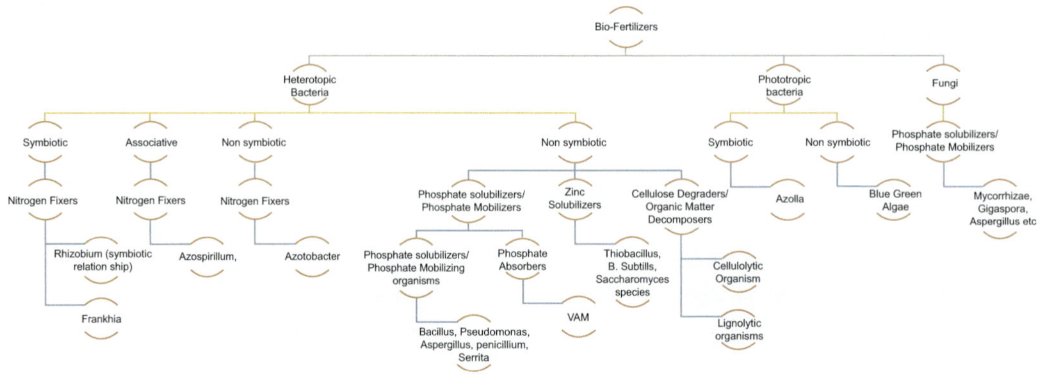

FIG. 11.5

Classification of biofertilizers.

These nations have acknowledged the advantages of biofertilizers and are working to promote their use, as seen by the robust biofertilizer markets in these nations (Masso et al., 2015). In addition, these countries have programs to assist in the production of biofertilizers. These strategies include privatization, commercialization, tax breaks, consultancy, economic development, and financial aid (Mukhongo et al., 2016; Chianu et al., 2011).

11.6.4 Types of biofertilizers

To restore nutrients, biofertilizers use organic methods such as atmospheric nitrogen fixation, phosphorus dissolution, and the production of chemicals that promote plant development. Based on microorganisms, nature, and purpose, biofertilizers are categorized into many categories (Itelima et al., 2018). Some important microbes used as biofertilizers are given in Table 11.2.

a. ***N_2 biofertilizer*:** This group fixes N_2 in a symbiotic manner. N_2 biofertilizer helps to improve N_2 levels in the soil. Soils undergo several N-transformation cycles such as nitrification, denitrification, ammonification, and volatilization under different environmental conditions in soils. The choice of N_2 biofertilizer to be used relies on the type of crop planted because different biofertilizers have distinct advantages in different soils. Crops are planted with *Rhizobium*, nonlegume crops with *Azotobacterium* or *Azospirillum*, sugarcane with *Acetobacter*, and lowland rice with blue-green algae and azolla (Table 11.2).

b. ***P-biofertilizer*:** Due to a lack of P, plant growth is affected. Phosphorus biofertilizer helps soil improve phosphorus levels and improve soil conditions. P-biofertilizer is not dependent on soil-grown products. All crops utilize phosphate including *Rhizobium*, *Azotobacter*, *Azospirillum*, or *Acetobacter* (Table 11.2).

c. ***Nitrogen fixers and nutrient solubilizers***

Nitrogen is an essential nutrient for crop production and is necessary for plant growth and development. Bacteria that fix nitrogen can be classified as symbiotic or nonsymbiotic. Typically, symbionts develop a symbiotic (mutual) relationship with the host legumes and are bacteria from the Rhizobiaceae family. When free-living and endophytic microorganisms perform N_2-fixation, they are known as nonsymbiotic. Examples are *Azotobacter*, *Azospirillum*, and *cyanobacteria*. Biofertilizers based on nutrient-solubilizing microbial strains are a potential technique for increasing minerals'

Table 11.2 Microbes used as biofertilizers and their mode of action.

S. no.	Element	Function	Nature	Examples
01.	Nitrogen	Fixing	Symbiotic	*Bradyrhizobium*, *Rhizobium*, *Azorhizobium*, *Sinorhizobium Mesorhizobium*, *Allorhizobium*, etc.
			Asymbiotic, Free-living, nonphotosynthetic	*Azospirillum*, *Azotobacter*, *Cyanobacteria*, etc.
			Asymbiotic, Free-living, photosynthetic	Heterocyst: *Anabaena*, *Nostoc*, *Aulosira*, *Calothrix*, *Tolypothrix Cylindrospermum*, and *Stigonema*; Nonheterocystous: Unicellular (*Aphanothece*, *Chroococcidiopsis*, *Dermocapsa*) and filamentous (*Oscillatoria*, *Schizothrix*, *Trichodesmium*)
			Associative	*Azospirillum* sp.
02.	Phosphorus	Solubilizing	Bacteria	*Aspergillus niger*, *Bacillus flexus*, *Sinorhizobium meliloti*, *Penicillium variable*, etc.
			Fungi	*Aspergillus*
				Awamori and *Penicillium* sp.
		Mobilizing	Bacteria	*Glomus deserticola*, *G. mosseae*, *G. intraradices*, etc.
			Arbuscular mycorrhiza	*Acaulospora* sp., *Gigaspora* sp., *Glomus* sp., *Sclerocystis* sp., *Scutellospora* sp., etc.
			Ectomyccorhiza	*Boletus* sp., *Laccaria* sp., *Pisolithus* sp., etc.
			Ericoid mycorrhiza	*Pezizella ericae*
			Orchid mycorrhiza	*Rhizoctonia solani*
03.	Zinc	Solubilizing		*Bacillus aryabhattai*, *Thiobacillus thiooxidans*, etc.
04.	Potassium	Solubilizing		*Acidithiobacillus*, *Burkholderia*, *Pseudomonas*, etc.
05.	Sulfur	Oxidizing		*Alcaligenes*, *Bacillus*, *Xanthobacter*, etc.
06.	Silicate	Solubilizers		*Bacillus* sp.
07.		Biocontrol		*Agrobacterium radiobacter*, *Bacillus subtilis*, *Pseudomonas aureofaciens*, *Pseudomonas fluorescens*, etc.
08.		Plant-growth promoting rhizobacteria	*Pseudomonas*	*Pseudomonas fluorescens*
				Achromobacter, *Actinoplanes*, *Azospirillum*, *Agrobacterium*, *Alcaligenes*, *Arthrobacter*, *Azotobacter*, *Bacillus*, *Cellulomonas*, *Pseudomonas* sp., *Rhizobium*, *Bradyrhizobium*, *Enterobacter*, *Erwinia*, *Flavobacterium*, *Xanthomonas*, etc.

bioavailability in agricultural soils while also having an ecologically favorable impact on the usage of mineral-rich fertilizers. The nutritional role of biofertilizers is summarized in Fig. 11.6.

d. *Zinc, Phosphorus, and Potassium solubilizers*

Zn solubilizers: It has been demonstrated that the PGPRs *Azospirillum* sp., *Pseudomonas*, *Bacillus aryabhattai*, *Rhizobium* sp., *Trichoderma* sp., and *Thiobacillus thiooxidans* boost the growth and zinc content of infected plants. Numerous research (Kamran et al., 2017; Vyas and Meena, 2018) have demonstrated that the bacteria *Bacillus megaterium*, *Burkholderia cenocepacia*, *Gluconacetobacter diazotrophicus*, *Serratia liquefaciens*, and *S. marcescens* may solubilize Zn depending on soil response. The pH of the soil affects the mechanism for dissolving zinc. The first of the two processes take place in acidic soils and follows cation exchange. The second is the formation of $ZnCaCO_3$ due to the chemical adsorption of Zn on $CaCO_3$. Zinc solubilization is interconnected with the synthesis of organic acids and siderophores (gluconate or 2-ketogluconate) (Saravanan et al., 2011).

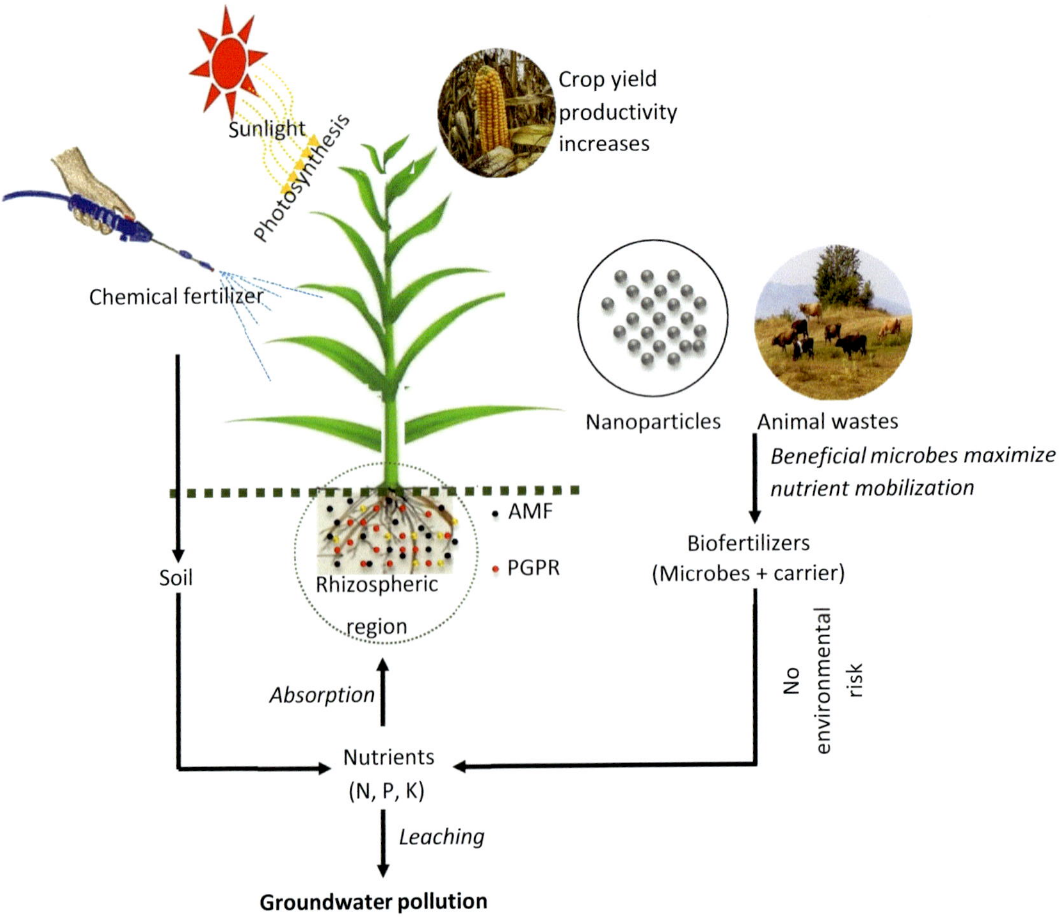

FIG. 11.6

Nutritional role of biofertilizers in crop yield productivity

P solubilizers: Bacteria and fungi, both may produce phosphate compounds that are both water-soluble and insoluble. The effectiveness of organophosphate mineralization is influenced by the physicochemical and biological characteristics of molecules including nucleic acids, phospholipids, and sugar phosphates. Nucleic acids, phospholipids, and sugar phosphate molecules mineralize faster than sulfuric acid, polyphosphate, and phosphonate molecules. According to Shrivastava et al. (2018), some organophosphorus compounds in soil are highly mineralized and accessible to microorganisms. They are frequently seen in association with clay particles.

Some soil organisms of the genera *Aspergillus*, *Bacillus*, *Mucor*, *Rhizopus*, *Penicillium*, and *Pseudomonas* can generate phosphatase, an enzyme that breaks down phosphate or phosphoric anhydride bonds and dephosphorylates organophosphorus compounds (Mengel et al., 2001). It has been demonstrated that microbial phosphatase works better in mineralizing organophosphorus chemicals than phosphatase produced by plants (Tarafdar et al., 2001). In addition, phytase, an enzyme that removes phosphorus from phytate, has been produced by *Aspergillus niger* (Neira-Vielma et al., 2018). Plants can therefore take up phosphorus from organophosphorus chemicals directly. Microbial enzymes called phosphate and C-P lyases break the C-P bond in organic phosphonates. These enzymes are isolated from *Bacillus licheniformis*, *Bacillus stearothermophilus*, *Thermus caldophilus*, and *Thermus thermophilus* (Nalini et al., 2015; Rodríguez et al., 2006). Experiment results show that specific *Enterobacter* sp., *Arthrobacter* sp., and *Azotobacter* sp. strains that produce substantial amounts of exopolysaccharides may solubilize tricalcium phosphate.

Rock phosphates, tricalcium phosphate $Ca_3(PO_4)_2$, dicalcium phosphate $CaHPO_4$, and hydroxyapatite $Ca_5(PO_4)_3$ are insoluble phosphorus compounds. $Ca_3(PO_4)_2$ and $Ca_5(PO_4)_3$ are degraded by *Bacillus flexus*, *Sinorhizobium meliloti*, and *B. megaterium* (Ibarra-Galeana et al., 2017). Other microbial species can also dissolve insoluble P compounds that include *A. niger*, *Paecilomyces mazei* AA1, *Penicillium variegata*, and *Yarrowia lipolytica* (Shrivastava et al., 2018). During the P-solubilizing mechanism, low molecular weight organic acids like malate, succinate, fumarate, citric acid, glutamate, and 2-ketoglutarate are released, as are mineral solvent compounds, extracellular enzymes, and phosphate. Substrate degradation (biological phosphate mineralization) is an example of phosphorus dissolution mechanisms (Choudhary et al., 2018). Mineral phosphorus, organic acids cause reduction in the pH of the rhizosphere, chelates cations involved in phosphorus precipitation, metal ions and (Ca-P, Al-P, Fe-P) forms soluble complexes with insoluble phosphorus compounds and competes with P for the soil adsorption site (Shrivastava et al., 2018). Organic acids produced by microorganisms during metabolism dissolve P-minerals directly by substituting acid anions with phosphate or phosphorus with chelated cations such as Fe, Al, and Ca. Gluconic acid, oxalic acid, citric acid, lactic acid, tartaric acid, and aspartic acid are examples of organic acids.

K solubilizers: By producing different organic acids (such as malic, citric, coumaric, formic, oxalic, succinic, and tartaric acids), polysaccharides, acidolysis, complexolysis, chelation coupling, and exchange reactions, bacteria, fungi, and actinomycetes are active K solubilizers (Etesami et al., 2017; Kumar et al., 2016). Bacteria including, *Acidithiobacillus ferrooxidans*, *Aminobacter* sp., *Arthrobacter* sp., *Bacillus circulans*, *Bacillus edaphicus*, *Burkholderia* sp., *Cladosporium* sp., *E. hormaechei*, *Paenibacillus* sp., *Paenibacillus glucanolyticus*, *Paenibacillus mucilaginosus*, *Sphingomonas* sp., etc. were found to exhibit K-lytic activity (Meena et al., 2016). The process of potassium dissolving involves obtaining potassium in the soil's usable form by building metal-organic complexes with Si^{4+} ions. pH drops cause feldspar dissolution, H^+ decreases and releases cationic acids from mineral structures, releasing carboxylic acids and other functional groups. Potassium is also released from feldspar breakdown and capsular polysaccharide synthesis. Biofilms have also been shown to dissolve potassium in biotite and anorthite (Das and Pradhan, 2016). Some approaches for developing types and mechanisms of biofertilizers are given in Fig. 11.7.

11.6.5 Formulation and commercial development of biofertilizers

The process of integrating a specified microbial strain with a carrier is defined as formulation (Bargaz et al., 2018). The active component is formulated in an appropriate carrier, usually with additives responsible for microbial stability and protection during storage and transit (Namasivayam et al., 2014). Proper formulations allow for innovative strategies to introduce the bacteria of interest for enhanced activity to achieve the best results in the inoculated host plants. The formulation must be stable during production and delivery. In addition, according to Nehra and Choudhary (2015), it should be easy to manage and use by the farmer, provide the best quality and appearance, and keep plant health under control by protecting it from external conditions. Biofertilizers are classified into several varieties

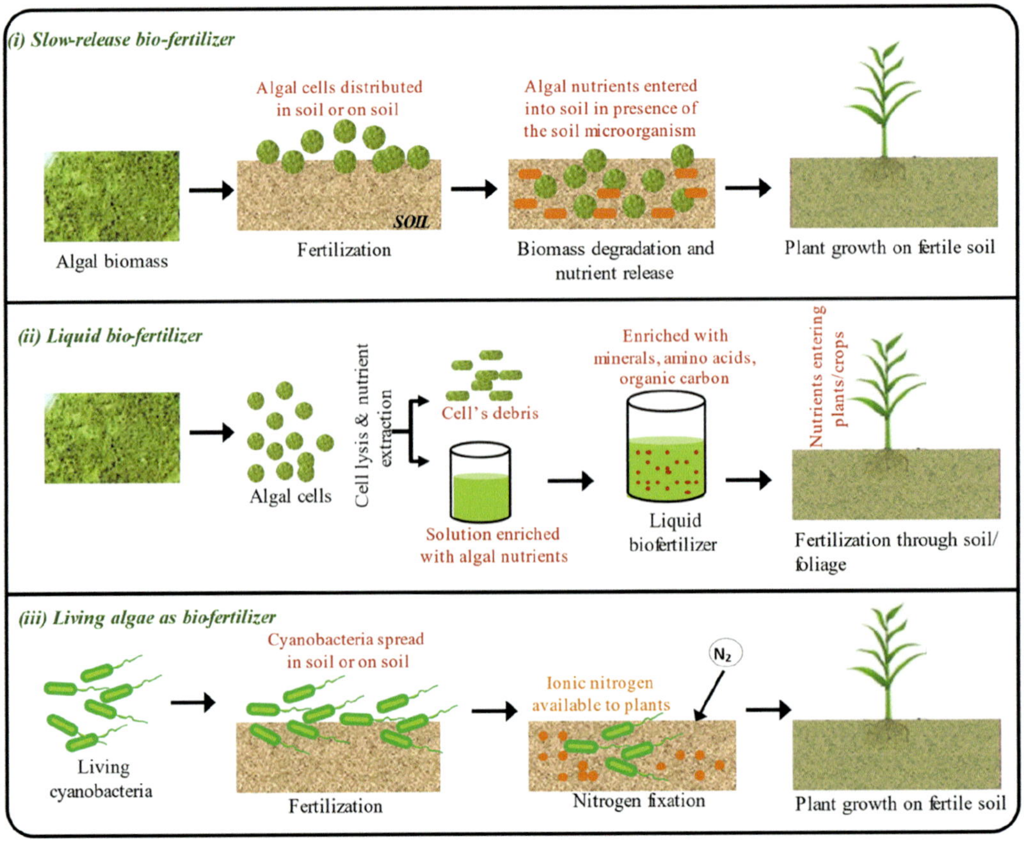

FIG. 11.7

Biofertilizer types and biofertilizer development in the field.

based on their physical features and the materials used. Formulations based on solid carriers, liquid formulations, polymer-entrapped formulations, and fluid-bed dry formulations (Fig. 11.8).

11.6.5.1 Solid-carrier bioformulation

A carrier is a substance that can effectively transport microorganisms within and regulate their physical activity under specific circumstances. It goes beyond the drawbacks of carrier-based formulations, namely their short shelf life, sensitivity to temperature changes, contamination, and viable cell count. Microbial organisms are required to deliver the vehicle in solid carrier-based biofertilizers and are capable of exchanging material for the production of hydrolyzing enzymes, volatile organic compounds, phytohormones, etc., and applied to formulation and better bio-inoculants development (Gupta et al., 2022).

11.6.5.2 Liquid bioformulation

These biofertilizers are further classified as aqueous or flowable suspensions based on culture in broth, mineral/organic oil, or oil-in-water suspension (Bharti et al., 2017). According to Yadav and Chandra (2014), liquid formulations typically contain 35%–65% liquid (water or oil), 10%–40% microorganisms, 1%–33% suspension components, 1% dispersants, and 3%–8% surfactants.

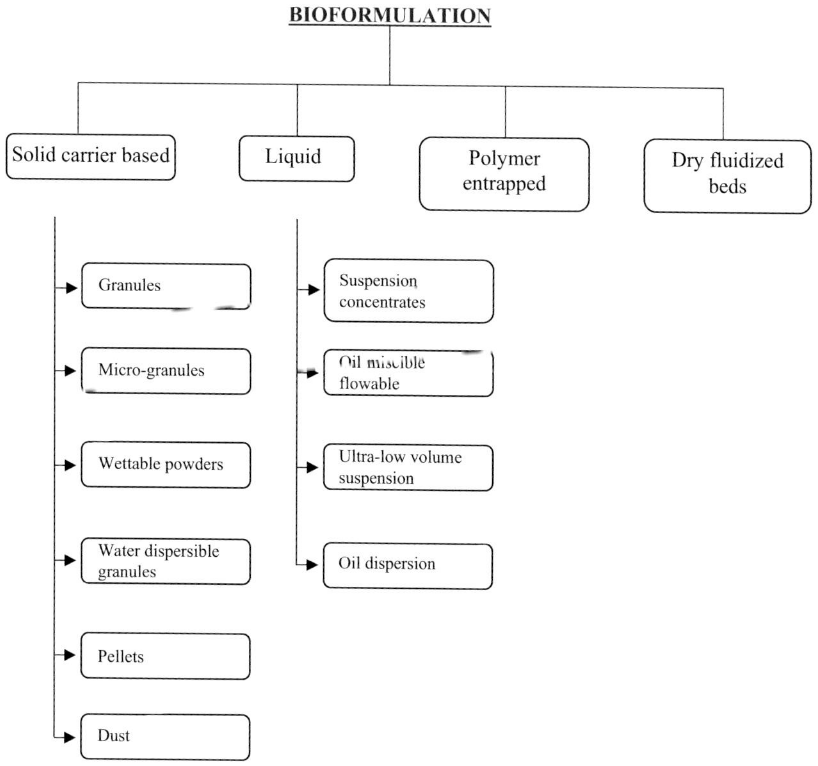

FIG. 11.8

A flow chart of bioformulation based on their physical features and material used.

11.7 Future prospectus

In the upcoming years, organic farming will be executed with new techniques, new formulations in sustainable agriculture development, and biofungicides, biopesticides, and biofertilizers emphasized as key elements in modern crop production. Rhizosphere microorganisms are recognized as a promoting agent in organic farming but very few microorganisms are used to formulate biofertilizer as well as biopesticides development. To sustain their plantations, farmers require safer pesticides to make biopesticides rather than chemical agents a great alternative. However, there are numerous difficulties in the adoption, production, and research of biopesticides and biofertilizer. Further research on manufacturing, delivery, and formulation must be carried out to enable the marketing of bio-control microbes. The combination of public-private sectors can improve the production, development, and marketing in developing nations of environmentally benign alternatives to chemical pesticides.

Moreover, help from programs financed by governmental authorities, business investors, and pesticide industries also has to be provided. Strict regulatory procedures for maintaining the availability of biopesticides in developing nations at affordable costs are a crucial concern. In the future, *Bt*-based insects and their insecticidal crystalline proteins will continue to enhance environmental management techniques, especially as transgenic crops become more extensively used. Discoveries of toxins and new recombinant DNA and proteomic approaches to modify how the toxin is given to the target insects are being developed. Numerous natural species have been used as insecticides in many agricultural systems across the world. Crop fields of over two million acres, which have shown to be viable and reasonably inexpensive, may be protected against bio-control microbes There are, therefore, likely to be approached to attempts to enhance the use and production of biopesticides. Numerous hurdles to several bio-control microbes are therefore still present. A lot of work has been carried out till now however, it is not enough, to cross these hurdles massive research grants with cutting-edge technology shall be launched with aspiring insinuations which may change the situation in the coming future and can bring some commercial formulations which are easily acceptable, less toxic, highly efficient and nonpolluting.

11.8 Conclusion

Keeping pests away in whatever way possible has been necessary to avoid famine and increase food availability but, pest control blew up in the global market and gave birth to the pesticide era. The chemicals were synthesized from organic compounds and used widely for the first time. This was a huge lead forward, as the crop yield shot up and the developing world got its so-called green revolution. Agrochemicals flooded the market and high-yielding seeds are being used. These agrochemicals have saved millions of lives in various nations, which are on the verge of food crisis. In a short period, countries prospered as more people were fed. It seems that they have found a magical solution but, that joy didn't last for a long period, lately however this method of farming gained a lot of criticism throughout the world. Most farmers are affected by pesticide problems and their side effects. Therefore, a public-private sector approach is needed to develop, produce, and market environmentally friendly alternatives to pesticides, particularly for developing countries.

As a result, bio-control agents like biofungicides, biopesticides, and biofertilizers are gaining more attention due to the reason the global population is gaining knowledge on the advantages and toxic effects of harmful and toxic agrochemicals. This growing demand has driven the scientific community to discover more and more organisms, that produce chemicals/compounds/metabolites that are biocidal eco-friendly and can give constant sustainable production without damaging the biodiversity, which leads footsteps towards sustainable agriculture like organic, biodynamic, etc. have gained popularity; however, they have their disadvantages like low yield, pest issues, etc. There is a growing knowledge in the world population and their demand to use eco-friendly products, whose carbon footprint is very less.

As a result, the production, creation, and presentation of scientific research to assist the work of biocontrol agents (e.g., biofungicides, biopesticides, etc.) have increasingly begun. More research is required to incorporate these biochemicals into production systems and enhance the production and use of biopesticides in developing countries, even though numerous methods and various strains of bacteria and other microbes have been identified that gave the best results. While developing a stringent control process to regulate the quality and accessibility of biopesticides/biopesticides, public finances, commercial enterprises, and pesticide corporations should be encouraged to invest in the biopesticide/biofungicide industry. Biofungicides are inexpensive and accessible to all farmers. Therefore, all farmers can use this product with high efficiency, usability, and biocompatibility. At the same time, alternative, effective, and sustainable uses of biofungicides and biopesticides for the benefit of humanity must be constantly reviewed.

References

Abbey, J.A., Percival, D., Abbey, L., Asiedu, S.K., Prithiviraj, Schilder, A., 2018. Biofungicides as alternative to synthetic fungicide control of grey mould (Botrytiscinerea) prospects and challenges. Biocontrol. Sci. Technol. https://doi.org/10.1080/09583157.2018.1548574.

Adhikari, K., Niraula, D., Shrestha, J., 2020. Use of neem (*Azadirachta indica* A. Juss) as a biopesticide in agriculture: a review. J. Agric. Appl. Biol. 1 (2), 100–117.

Amaresan, N., Jayakumar, V., Kumar, K., Thajuddin, N., 2012. Isolation and characterization of plant growth promoting endophytic bacteria and their effect on tomato (Lycopersicon esculentum) and chilli (Capsicum annuum) seedling growth. Ann Microbiol. 62, 805–810.

Arena, J.P., Liu, K.K., Paress, P.S., Frazier, E.G., Cully, D.F., Mrozik, H., Schaeffer, J.M., 1995. The mechanism of action of avermectins in *Caenorhabditis elegans*: correlation between activation of glutamate-sensitive chloride current, membrane binding, and biological activity. J. Parasitol., 286–294.

Armada, E., Probanza, A., Roldán, A., Azcón, R., 2016. Native plant growth promoting bacteria *Bacillus thuringiensis* and mixed or individual mycorrhizal species improved drought tolerance and oxidative metabolism in *Lavandula dentata* plants. J. Plant Physiol. 192, 1–12.

Aronson, A.I., Shai, Y., 2001. Why *Bacillus thuringiensis* insecticidal toxins are so effective: unique features of their mode of action. FEMS Microbiol. Lett. 195 (1), 1–8.

Arthurs, S., Dara, S.K., 2019. Microbial biopesticides for invertebrate pests and their markets in the United States. J. Invertebr. Pathol. 165, 13–21.

Babalola, O.O., 2010. Beneficial bacteria of agricultural importance. Biotechnol. Lett. 32, 1559–1570.

Bandopadhyay, S., 2020. Application of plant growth promoting *Bacillus thuringiensis* as biofertilizer on *Abelmoschus esculentus* plants under field condition. J. Pure Appl. Microbiol. 14 (2), 1287–1294.

Bargaz, A., Lyamlouli, K., Chtouki, M., Zeroual, Y., Dhiba, D., 2018. Soil microbial resources for improving fertilizers efficiency in an integrated plant nutrient management system. Front. Microbiol. 9, 1606.

Betz, F.S., Hammond, B.G., Fuchs, R.L., 2000. Safety and advantages of *Bacillus thuringiensis*-protected plants to control insect pests. Regul. Toxicol. Pharmacol. 32 (2), 156–173.

Bhardwaj, D., Ansar, M.W., Sahoo, R.K., Tuteja, N., 2014. Biofertilizers function as key player in sustainable agriculture by improving soil fertility, plant tolerance and crop productivity. Microb. Cell Factories 13, 66.

Bharti, N., Sharma, S.K., Saini, S., Verma, A., Nimonkar, Y., Prakash, O., 2017. Microbial plant probiotics: problems in application and formulation. In: Probiotics and Plant Health. Springer, Singapore, pp. 317–335.

Bhattacharyya, P.N., Jha, D.K., 2012. Plant growth-promoting rhizobacteria (PGPR): emergence in agriculture. World J. Microbiol. Biotechnol. 28, 1327–1350.

Bird, A.F., Akhurst, R.J., 1983. The nature of the intestinal vesicle in nematodes of the family Steinernematidae. Int. J. Parasitol. 13 (6), 599–606.

Brar, S.K., Verma, M., Tyagi, R.D., Valéro, J.R., 2006. Recent advances in downstream processing and formulations of *Bacillus thuringiensis* based biopesticides. Process Biochem. 41 (2), 323–342.

Bray, 1997. Plant responses to water deficit. Trends Plant Sci. 2 (2), 48–54. https://doi.org/10.1016/S1360-1385(97)82562-9.

Burgess, A.J., et al., 2023. Improving crop yield potential: underlying biological processes and future prospects. Food Energy Secur. 12, e435. https://doi.org/10.1002/fes3.435.

Carlton, B.C., 1993. Genetics of Bt insecticidal crystal proteins and strategies for the construction of improved strains. In: Duke, S.O., Menn, J.J., Plimmer, J.R. (Eds), Pest Control With Enhanced Environmental Safety. ACS Symposium Series, vol. 524. American Chemical Society, Washington, DC, pp. 326–337.

Castaldi, S., Masi, M., Sautua, F., Cimmino, A., et al., 2021. Pseudomonas fluorescens showing antifungal activity against Macrophomina phaseolina, a severe pathogenic fungus of soybean, produces phenazine as the main active metabolite. Biomolecules 11 (1728). https://doi.org/10.3390/biom11111728.

Chang, J.H., Choi, J.Y., Jin, B.R., Roh, J.Y., Olszewski, J.A., Seo, S.J., O'Reilly, D.R., Je, Y.H., 2003. An improved baculovirus insecticide producing occlusion bodies that contain *Bacillus thuringiensis* insect toxin. J. Invertebr. Pathol. 84, 30–37.

Chen, J., 2018. Biopesticides: Global Markets to 2022. BCC Research, Wellesley, MA (Report Code CHM029G).

Chianu, J.N., Nkonya, E.M., Mairura, F.S., Chianu, J.N., Akinnifesi, F.K., 2011. Biological nitrogen fixation and socioeconomic factors for legume production in Sub-Saharan Africa. In: Sustainable Agriculture. vol. 2. Springer, Dordrecht, pp. 309–329.

Choudhary, M., Ghasal, P.C., Yadav, R.P., Meena, V.S., Mondal, T., Bisht, J.K., 2018. Towards plant-beneficiary rhizobacteria and agricultural sustainability. In: Role of Rhizospheric Microbes in Soil. Springer, Singapore, pp. 1–46.

Copping, L.G., Menn, J.J., 2000. Biopesticides: a review of their action, applications and efficacy. Pest Manag. Sci. 56 (8), 651–676.

Cory, J.S., Hirst, M.L., Sterling, P.H., Speight, M.R., 2000. Narrow host range nucleopolyhedrovirus for control of the browntail moth (Lepidoptera: Lymantriidae). Environ. Entomol. 29 (3), 661–667.

Damalas, C.A., Koutroubas, S.D., 2018. Current status and recent developments in biopesticide use. Agriculture 8 (1), 13. https://doi.org/10.3390/agriculture8010013.

Dar, S.A., Wani, S.H., Mir, M., Showkat, A., Dolkar, T., Dawa, T., 2021. Biopesticides: mode of action, efficacy and scope in pest management. J. Adv. Res. Biochem. Pharmacol. 4, 1–8.

Darboux, I., Nielsen-LeRoux, C., Charles, J.F., Pauron, D., 2001. The receptor of *Bacillus sphaericus* binary toxin in *Culex pipiens* (Diptera: Culicidae) midgut: molecular cloning and expression. Insect Biochem. Mol. Biol. 31 (10), 981–990.

Das, I., Pradhan, M., 2016. Potassium-solubilizing microorganisnms and their role in enhancing soil fertility and health. In: Meena, V.S. (Ed.), Potassium Solubilizing Microorganism for Sustainable Agriculture. Springer India, pp. 281–291, https://doi.org/10.1007/978-81-322-2776-2.

Doran, J.W., Sarrantonio, M., Liebig, M.A., 1996. Relative effectiveness of organic and inorganic nutrient sources in improving yield, seed quality and nutrient uptake of canola. Adv. Agron. 56, 2–54.

Du, Y., Hou, X., Yan, N., Dong, W., Mao, X., Zhang, Z., 2016. Chemical basis of the fungicidal activity of tobacco extracts against *Valsa mali*. Molecules 21 (12), 1743.

Dubey, K., Srivastava, B., Kumar, A., 2008. Current status of plant products as botanical pesticides in storage pest management. Current status of botanicals in storage pest management. J. Biopestici. 1 (2), 182–186.

Duncan, L.W., McCoy, C.W., 1996. Vertical distribution in soil, persistence, and efficacy against citrus root weevil (Coleoptera: Curculionidae) of two species of entomogenous nematodes (Rhabditida: Steinernematidae; Heterorhabditidae). Environ. Entomol. 25 (1), 174–178.

EFSA, 2021. Conclusion on the peer review of the pesticide risk assessment of the active substance *Bacillus amyloliquefaciens* strain QST713(formerly *Bacillus subtilis* strain QST 713). EFSA J. 19, 6381.

Egamberdieva, D., 2008. Plant growth promoting properties of rhizobacteria isolated from wheat and pea grown in loamy sand soil. Turk. J. Biol. 32 (1), 9–15.

Elgar, M.A., Zhang, D., Wang, Q., Wittwer, B., Pham, H.T., Johnson, T.L., Freelance, C.B., Coquilleau, M., 2018. Focus: ecology and evolution: insect antennal morphology: the evolution of diverse solutions to odorant perception. Yale J. Biol. Med. 91 (4), 457.

Ellouze, W., Hamel, C., Singh, A.K., Mishra, V., DePauw, R.M., Knox, R.E., 2018. Abundance of the arbuscular mycorrhizal fungal taxa associated with the roots and rhizosphere soil of different durum wheat cultivars in the Canadian prairies. Can. J. Microbiol. 64 (8), 527–536.

El-Tarabily, K.A., Nassar, A.H., Sivasithamparam, K., 2008. Promotion of growth of bean (Phaseolus vulgaris L.) in a calcareous soil by a phosphate-solubilizing, rhizosphere-competent isolate of Micromonospora endolithica. Appl. Soil Ecol. 39, 161–171.

Enebe, M.C., Babalola, O.O., 2018. The influence of plant growth-promoting rhizobacteria in plant tolerance to abiotic stress: a survival strategy. Appl. Microbiol. Biotechnol. 102, 7821–7835.

EPA, 2023. Ingredients Used in Pesticide Products: Pesticides. What Are Biopesticides? Available online: https://www.epa.gov/ingredients-used-pesticide-products/what-are-biopesticides.

Essiedu, J.A., Adepoju, F.O., Ivantsova, M.N., 2020. Benefits and limitations in using biopesticides: a review. In: Proceedings of the VII International Young Researchers' Conference—Physics, Technology, Innovations (PTI-2020), Ekaterinburg, Russia, 18–22 May. vol. 2313, p. 080002.

Etesami, H., Emami, S., Alikhani, H.A., 2017. Potassium solubilizing bacteria (KSB): mechanisms, promotion of plant growth, and future prospects—a review. J. Soil Sci. Plant Nutr. 17 (4), 897–911.

FAOSTAT, 2021. Pesticides. Food and Agriculture Organization, Rome. Available online: http://www.fao.org/faostat/en/#data/RP. (Accessed 10 May 2021).

Faria, M.D., Magalhães, B.P., 2001. O uso de fungos entomopatogênicos no Brasil. Biotecnol. Ciênc. Desenvolv. 22 (1), 18–21.

Fasusi, O.A., Cruz, C., Babalola, O.O., 2021. Agricultural sustainability: microbial biofertilizers in rhizosphere management. Agriculture 11, 163.

Feduchi, E., Cosin, M., Carrasco, L., 1985. Mildiomycin: a nucleoside antibiotic that inhibits protein synthesis. J. Antibiot. 38 (3), 415–419.

Fegan, M., Manners, J.M., Maclean, D.J., Irwin, J.A.G., Samuels, K.D.Z., Holdom, D.G., Li, D.P., 1993. Random amplified polymorphic DNA markers reveal a high degree of genetic diversity in the entomopathogenic fungus *Metarhizium anisopliae* var. *anisopliae*. Microbiology 139 (9), 2075–2081.

Fenibo, E.O., Ijoma, G.N., Matambo, T., 2021. Biopesticides in sustainable agriculture: a critical sustainable development driver governed by green chemistry principles. Front. Sustain. Food Syst. 5, 619058. https://doi.org/10.3389/fsufs.2021.619058.

Forni, C., Duca, D., Glick, B., 2017. Mechanisms of plant response to salt and drought stress and their alteration by bacteria. Plant Soil 410, 335–356.

Fuxa, J.R., Tanada, Y. (Eds.), 1987. Epidemiological Concepts Applied to Insect Epizootiology Epizootiology of Insect Diseases. John Wiley & Sons.

Galgoczy, L., Kovacs, L., Vagvolgyi, C.S., 2010. Defensin-like antifungal proteins secreted by filamentous fungi. In: Méndez-Vilas, A. (Ed.), Current Research, Technology and Education Topics in Applied Microbiology and Microbial Biotechnology, Microbiology Book (Bajadoz: Formatex), vol. 1, Issue 2. pp. 550–559.

Gašić, S., Tanović, B., 2013. Biopesticide formulations, possibility of application and future trends. Pestic. fitomed. 28 (2), 97–102.
Glick, B.R., 2014. Bacteria with ACC deaminase can promote plant growth and help to feed the world. Microbiol. Res. 169, 30–39.
Glick, B.R., et al., 2007. Promotion of plant growth by ACC deaminase-producing soil bacteria. Eur. J. Plant Pathol. 119 (3), 329–339.
Gomiero, T., 2018. Food quality assessment in organic vs. conventional agricultural produce: findings and issues. Appl. Soil Ecol. 123, 714–728.
Grewal, P.S., Lewis, E.E., Gaugler, R., 1997. Response of infective stage parasites (Nematoda: Steinernematidae) to volatile cues from infected hosts. J. Chem. Ecol. 23 (2), 503–515.
Gupta, A., Bano, A., Rai, S., Sharma, S., Pathak, N., 2022. Selection of carrier materials to formulate bioinoculant package for promoting seed germination. Lett. Appl. NanoBioScience 12 (3), 1–13.
Gwinn, K.D., 2018. Bioactive natural products in plant disease control. In: Studies in Natural Products Chemistry. vol. 56. Elsevier, pp. 229–246.
Hashem, A., Tabassum, B., Allah, E.F.A., 2019. Bacillus subtilis: a plant-growth promoting rhizobacterium that also impacts biotic stress. Saudi J. Biol. Sci. 26 (6), 1291–1297. https://doi.org/10.1016/j.sjbs.2019.05.004.
Hayat, R., et al., 2010. Soil beneficial bacteria and their role in plant growth promotion: a review. Ann. Microbiol. 60 (4), 579–598.
Ibarra-Galeana, J.A., Castro-Martínez, C., Fierro-Coronado, R.A., Armenta-Bojórquez, A.D., Maldonado-Mendoza, I.E., 2017. Characterization of phosphate-solubilizing bacteria exhibiting the potential for growth promotion and phosphorus nutrition improvement in maize (*Zea mays* L.) in calcareous soils of Sinaloa, Mexico. Ann. Microbiol. 67 (12), 801–811.
Intergovernmental Panel on Climate Change (IPCC), 2001. Intergovernmental Panel on Climate Change. Climate Change 2001: Third Assessment Report. vol. I Cambridge University Press, Cambridge, UK, p. 2001.
Itelima, J.U., Bang, W.J., Onyimba, I.A., Sila, M.D., Egbere, O.J., 2018. A review: biofertilizers—a key player in enhancing soil fertility and crop productivity. J. Microbiol. Biotechnol. Rep. 2, 22–28.
Kamran, S., Shahid, I., Baig, D.N., Rizwan, M., Malik, K.A., Mehnaz, S., 2017. Contribution of zinc solubilizing bacteria in growth promotion and zinc content of wheat. Front. Microbiol. 8, 2593.
Karthik, M., Pushpakanth, P., Krishnamoorthy, R., Senthilkumar, M., 2017. Endophytic bacteria associated with banana cultivars and their inoculation effect on plant growth. J. Hortic. Sci. Biotechnol. 1–9.
Khakimov, A.A., Omonlikov, A.U., Utaganov, S.B.U., 2020. Current status and prospects of the use of biofungicides against plant diseases. GSC Biol. Pharm. Sci. 13 (03), 119–126.
Khan, N., Bano, A., 2019. Exopolysaccharide producing rhizobacteria and their impact on growth and drought tolerance of wheat grown under rainfed conditions. PLoS One 14 (9), e0222302.
Kiss, L., Russell, J.C., Szentiványi, O., Xu, X., Jeffries, P., 2004. Biology and biocontrol potential of Ampelomyces mycoparasites, natural antagonists of powdery mildew fungi. Biocontrol Sci. Technol. 14, 635–651. https://doi.org/10.1080/09583150410001683600l.
Knowles, A., 2008. Recent developments of safer formulations of agrochemicals. Environmentalist 28 (1), 35–44.
Kremer, R.J., 2019. Bioherbicides and nanotechnology: current status and future trends. In: NanoBiopesticides Today and Future Perspectives. Academic Press, pp. 353–366.
Kumar, S., 2012. Biopesticides: a need for food and environmental safety. J. Biofertil. Biopestic. 3 (4), 1–3.
Kumar, A., Singh, M., Singh, P.P., Singh, S.K., Singh, P.K., Pandey, K.D., 2016. Isolation of plant growth promoting rhizobacteria and their impact on growth and curcumin content in *Curcuma longa* L. Biocatal. Agric. Biotechnol. 8, 1–7.
Lacey, L.A., Grzywacz, D., Shapiro-Ilan, D.I., Frutos, R., Brownbridge, M., Goettel, M.S., 2015. Insect pathogens as biological control agents: back to the future. J. Invertebr. Pathol. 132, 1–41.
Lehr, P., 2014. Global Markets for Biopesticides. Report Code CHM029E BCC Research.

Liu, D., Yan, R., Fu, Y., Wang, X., Zhang, J., Xiang, W., 2019. Antifungal, plant growth-promoting, and genomic properties of an endophytic actinobacterium Streptomyces sp. NEAU-S7GS2. Front. Microbiol. 10 (2077). https://doi.org/10.3389/fmicb.2019.02077.

Luo, L., Xia, H., Lu, B.-R., 2019. Editorial: crop breeding for drought resistance. Front. Plant Sci. 10, 314.

Malusà, E., Pinzari, F., Canfora, L., 2016. Efficacy of biofertilizers: challenges to improve crop production. In: Microbial Inoculants in Sustainable Agricultural Productivity. Springer, New Delhi, pp. 17–40.

Markets and Markets, 2019. Biofertilizer Market by Form (Liquid, Carrier-Based), Mode of Application (Soil Treatment, Seed Treatment), Crop Type, Type (Nitrogen-Fixing, Phosphates Solubilizing and Mobilizing, Potash Solubilizing and Mobilizing), Region-Global Forecast to 2025. https://www.marketsandmarkets.com/Market-Reports/compound-biofertilizers-customized-fertilizers-market-856.html. (Accessed 1 September 2020).

Marrone, P.G., 2014. The market and potential for biopesticides. In: Gross, A.D., Coats, J.R., Duke, S.O., Seiber, J.N. (Eds.), Biopesticides: State of the Art and Future Opportunities. American Chemical Society, Washington DC, pp. 245–258. https://doi.org/10.1021/bk-2014-1172.ch016.

Masso, C., Ochieng, J.R.A., Vanlauwe, B., 2015. Worldwide contrast in application of bio-fertilizers for sustainable agriculture: lessons for sub-Saharan Africa. J. Biol. Agric. Healthc. 5 (12), 34–50.

Maxton, A., Singh, P., Masih, S.A., 2017. ACC deaminase producing bacteria mediated drought and salt tolerance in Capsicum annum. J. Plant Nutr., 1–26.

McSorley, R., 1999. Nonchemical management of plant-parasitic nematodes. In: The IPM Practitioner: The Newsletter of Integrated Pest Management (USA).

Meena, V.S., Kumar, A., Meena, R.K., 2016. Potassium-Solubulizing Microorganism in Evergreen Agriculture: An Overview Agroforestry and Fodder Production Management View Project. Springer Link, pp. 1–20, https://doi.org/10.1007/978-81-322-2776-2_1.

Mengel, K., Kirkby, E.A., Kosegarten, H., Appel, T., 2001. Principles of Plant Nutrition. Kluwer Academic Publishers, Dordrecht. https://doi.org/10.1007/978-94-010-1009-2.

Mishra, P.K., Mishra, S., Selvakumar, G., Bisht, J.K., Kundu, S., Gupta, H.S., 2009. Coinoculation of *Bacillus thuringeinsis*-KR1 with *Rhizobium leguminosarum* enhances plant growth and nodulation of pea (*Pisum sativum* L.) and lentil (*Lens culinaris* L.). World J. Microbiol. Biotechnol. 25 (5), 753–761.

Mishra, V., Ellouze, W., Howard, R.J., 2018. Utility of arbuscular mycorrhizal fungi for improved production and disease mitigation in organic and hydroponic greenhouse crops. J. Hortic. 5 (237). https://doi.org/10.4172/2376-0354.1000237.

Mittal, D., Kaur, G., Singh, P., Yadav, K., Ali, S.A., 2020. Nanoparticle-based sustainable agriculture and food science: recent advances and future outlook. Front. Nanotechnol. 2, 579954.

Mnyone, L.L., Koenraadt, C.J., Lyimo, I.N., Mpingwa, M.W., Takken, W., Russell, T.L., 2010. Anopheline and culicine mosquitoes are not repelled by surfaces treated with the entomopathogenic fungi *Metarhizium anisopliae* and Beauveria bassiana. Parasit. Vectors 3 (1), 1–6.

Mukhongo, R.W., Tumuhairwe, J.B., Ebanyat, P., AbdelGadir, A.H., Thuita, M., Masso, C., 2016. Production and Use of Arbuscular Mycorrhizal Fungi Inoculum in Sub-Saharan Africa: Challenges and Ways of Improving.

Nalini, P., Ellaiah, P., Prabhakar, T., Girijasankar, G., 2015. Microbial alkaline phosphatases in bioprocessing. Int. J. Curr. Microbiol. App. Sci. 4, 384–396.

Namasivayam, S.K.R., Saikia, S.L., Bharani, A.R.S., 2014. Evaluation of persistence and plant growth promoting effect of bioencapsulated formulation of suitable bacterial bio-fertilizers. Biosci. Biotechnol. Res. Asia 11 (2), 407–415.

Nehra, V., Choudhary, M., 2015. A review on plant growth promoting rhizobacteria acting as bioinoculants and their biological approach towards the production of sustainable agriculture. J. Appl. Nat. Sci. 7 (1), 540–556.

Neira-Vielma, A.A., Aguilar, C.N., Ilyina, A., Contreras-Esquivel, J.C., das Graça Carneiro-da-Cunha, M., Michelena-Álvarez, G., Martínez-Hernández, J.L., 2018. Purification and biochemical characterization of an

Aspergillus niger phytase produced by solid-state fermentation using triticale residues as substrate. Biotechnol. Rep. 17, 49–54.

Nievierowski, T.H., et al., 2023. A Bacillus-based biofungicide agent prevents ochratoxins occurrence in grapes and impacts the volatile profile throughout the Chardonnay winemaking stages. Int. J. Food Microbiol. 389. https://doi.org/10.1016/j.ijfoodmicro.2023.110107.

Niu, X., Song, L., Xiao, Y., Ge, W., 2018. Drought-tolerant plant growth-promoting rhizobacteria associated with foxtail millet in a semi-arid agroecosystem and their potential in alleviating drought stress. Front. Microbiol. 8, 2580.

Nosheen, S., Ajmal, I., Song, Y., 2021. Microbes as biofertilizers, a potential approach for sustainable crop production. Sustainability 13, 1868.

Nuruzzaman, M.D., Rahman, M.M., Liu, Y., Naidu, R., 2016. Nanoencapsulation, nano-guard for pesticides: a new window for safe application. J. Agric. Food Chem. 64 (7), 1447–1483.

Olson, S., 2015. An analysis of the biopesticide market now and where it is going. Outlooks Pest Manag. 26 (5), 203–206.

Palmieri, D., Ianiri, G., Del Grosso, C., Barone, G., De Curtis, F., Castoria, R., Lima, G., 2022. Advances and perspectives in the use of biocontrol agents against fungal plant diseases. Horticulturae 8, 577. https://doi.org/10.3390/horticulturae8070577.

Parker, K.M., Barragán Borrero, V., van Leeuwen, D.M., Lever, M.A., Mateescu, B., Sander, M., 2019. Environmental fate of RNA interference pesticides: adsorption and degradation of double-stranded RNA molecules in agricultural soils. Environ. Sci. Technol. 53 (6), 3027–3036.

Patel, S., Jinal, H.N., Amaresan, N., 2017. Isolation and characterization of drought resistance bacteria for plant growth promoting properties and their effect on chilli (*Capsicum annum*) seedling under salt stress. Biocatal. Agric. Biotechnol. 12, 85–89.

Peters, A., 1996. The natural host range of Steinernema and Heterorhabditis spp. and their impact on insect populations. Biocontrol Sci. Technol. 6 (3), 389–402.

Pindi, P.K., Sultana, T., Vootla, P.K., 2014. Plant growth regulation of Bt-cotton through *Bacillus* species. 3 Biotech 4 (3), 305–315.

Qi, J., Aiuchi, D., Tani, M., Asano, S.I., Koike, M., 2016. Potential of entomopathogenic *Bacillus thuringiensis* as plant growth promoting rhizobacteria and biological control agents for tomato *Fusarium* wilt. Int. J. Environ. Agric. Res. 2 (6), 55.

Raiz, U., et al., 2020. Bio-fertilizers: eco-friendly approach for plant and soil environment. Bioremediat. Biotechnol. 189–213.

Raklami, A., Bechtaoui, N., Tahiri, A., Anli, M., Meddich, A., Oufdou, K., 2019. Use of rhizobacteria and mycorrhizae consortium in the open field as a strategy for improving crop nutrition, productivity and soil fertility. Front. Microbiol. 10, 1106.

Rashid, S., Charles, T.C., Glick, B.R., 2012. Isolation and characterization of new plant growth-promoting bacterial endophytes. Appl. Soil Ecol. 61, 217–224.

Roberts, D.W., St Leger, R.J., 2004. *Metarhizium* spp., cosmopolitan insect-pathogenic fungi: mycological aspects. Adv. Appl. Microbiol. 54 (1), 1–70.

Rodríguez, H., Fraga, R., Gonzalez, T., Bashan, Y., 2006. Genetics of phosphate solubilization and its potential applications for improving plant growth-promoting bacteria. Plant Soil 287 (1), 15–21.

Rodriguez, S.J., Suarez, R., Caballero, M.J., Itturiaga, G., 2009. Trehalose accumulation in *Azospirillum brasilense* improves drought tolerance and biomass in maize plant. FEMS Microbiol. Lett. 296, 52–59.

Roh, J.Y., Choi, J.Y., Li, M.S., Jin, B.R., Je, Y.H., 2007. *Bacillus thuringiensis* as a specific, safe, and effective tool for insect pest control. J. Microbiol. Biotechnol. 17 (4), 547–559.

Rouhier, N., Jacquot, J.-P., 2008. Silicon alleviates salt stress by modulating antioxidant enzyme activities in *Dianthus caryophyllus* 'Tula'. New Phytol. 180, 738–741.

Saravanan, V.S., Kumar, M.R., Sa, T.M., 2011. Microbial zinc solubilization and their role on plants. In: Bacteria in Agrobiology: Plant Nutrient Management. Springer, Berlin, Heidelberg, pp. 47–63.

Schuwirth, B.S., Day, J.M., Hau, C.W., Janssen, G.R., Dahlberg, A.E., Cate, J.H.D., Vila- Sanjurjo, A., 2006. Structural analysis of kasugamycin inhibition of translation. Nat. Struct. Mol. Biol. 13 (10), 879–886.

Selim, M.M., 2020. Introduction to the integrated nutrient management strategies and their contribution to yield and soil properties. Hindawi, Int. J. Agron. 2020, 1–14.

Senthil-Nathan, S., 2015. A review of biopesticides and their mode of action against insect pests. In: Environmental Sustainability, pp. 49–63.

Seshadri, S., Ignacimuthu, S., Vadivelu, M., Lakshminarasimhan, C., 2007. Inorganic phosphate solubilization by two insect pathogenic *Bacillus* sp. In: First International Meeting on Microbial Phosphate Solubilization. Springer, Dordrecht, pp. 351–355.

Shang, Y., Hasan, M.K., Ahammed, G.J., Li, M., Yin, H., Zhou, J., 2019. Applications of nanotechnology in plant growth and crop protection: a review. Molecules 24, 2558.

Shannag, H.K., Capinera, J.L., 2000. Interference of Steinernema carpocapsae (Nematoda: Steinernematidae) with *Cardiochiles diaphaniae* (Hymenoptera: Braconidae), a parasitoid of melonworm and pickleworm (Lepidoptera: Pyralidae). Environ. Entomol. 29 (3), 612–617.

Shapiro, D.I., McCoy, C.W., 2000. Virulence of entomopathogenic nematodes to *Diaprepes abbreviatus* (Coleoptera: Curculionidae) in the laboratory. J. Econ. Entomol. 93 (4), 1090–1095.

Sharma, A., Jain, A., Gupta, P., Chowdary, V., 2020. Machine Learning Applications for Precision Agriculture: A Comprehensive Review. Digital Object Identifier. https://doi.org/10.1109/ACCESS.2020.3048415.

Sheir, S.K., et al., 2015. Biohazards of the biofungicide, Trichoderma harzianum on the crayfish, Procambarus clarkii: histological and biochemical implications. Egyptian. J. Basic Appl. Sci. 2 (2), 87–97. https://doi.org/10.1016/j.ejbas.2015.02.001.

Sheng, X.F., 2005. Growth promotion and increased potassium uptake of cotton and rape by a potassium releasing strain of Bacillus edaphicus. Soil Biol. Biochem. 37, 1918–1922.

Shrivastava, M., Srivastava, P.C., D'souza, S.F., 2018. Phosphate-solubilizing microbes: diversity and phosphates solubilization mechanism. In: Role of Rhizospheric Microbes in soil. Springer, Singapore, pp. 137–165.

Slavica, G., Brankica, G., 2013. Biopesticide formulations, possibility of application and future trends. Pestic. I Fitomedicina 28 (2), 97–102. https://doi.org/10.2298/PIF1302097G.

Sparks, T.C., Nauen, R., 2015. IRAC: Mode of action classification and insecticide resistance management. Pest. Biochem. Physiol. 121, 122–128. https://doi.org/10.1016/j.pestbp.2014.11.014

Srivastava, M., et al., 2016. Trichoderma- a potential and effective bio fungicide and alternative source against notable phytopathogens. a review. Afr. J. Agric. Res. 11 (5), 310–316. https://doi.org/10.5897/AJAR2015.9568.

Sun, X., 2015. History and current status of development and use of viral insecticides in China. Viruses 7 (1), 306–319.

Svidritskiy, E., Ling, C., Ermolenko, D.N., Korostelev, A.A., 2013. Blasticidin S inhibits translation by trapping deformed tRNA on the ribosome. Proc. Natl. Acad. Sci. 110 (30), 12283–12288.

Tadros, T.F., 2006. Applied Surfactants: Principles and Applications. John Wiley & Sons, ISBN: 978-3-527-30629-9.

Tarafdar, J.C., Yadav, R.S., Meena, S.C., 2001. Comparative efficiency of acid phosphatase originated from plant and fungal sources. J. Plant Nutr. Soil Sci. 164, 279–282.

Thambugala, K.M., Daranagama, D.A., Phillips, A.J.L., Kannangara, S.D., Promputtha, I., 2020. Fungi vs. fungi in biocontrol: an overview of fungal antagonists applied against fungal plant pathogens. Front. Cell. Infect. Microbiol. 10 (604923). https://doi.org/10.3389/fcimb.2020.604923.

Thanwisai, A., Muangpat, P., Meesil, W., Janthu, P., Dumidae, A., Subkrasae, C., Ardpairin, J., Tandhavanant, S., Yoshino, T.P., Vitta, A., 2022. Entomopathogenic nematodes and their symbiotic bacteria from the national parks of Thailand and larvicidal property of symbiotic bacteria against Aedes aegypti and Culex quinquefasciatus. Biology 11, 1658. https://doi.org/10.3390/biology11111165.

Thomidis, T., Pantazis, S., Navrozidisaand, E., Karagiannidis, N., 2015. Biological control of fruit rots on strawberry and grape by BOTRY-ZenT New Zealand. J. Crop Hortic. Sci. 43 (1), 68–72.

Thurley, P., Chilcott, C.N., Kalmakoff, J., Pillai, J.S., 1985. Characterization of proteolytic activity associated with *Bacillus thuringiensis* var. *darmstadiensis* crystals. FEMS Microbiol. Lett. 27 (2), 221–225.

Toth, L., Váradi, G., Boros, É., Borics, A., Ficze, H., Nagy, I., Tóth, G.K., Rákhely, G., Marx, F., Galgóczy, L., 2020. Biofungicidal potential of *Neosartorya* (*Aspergillus*) *fischeri* antifungal protein NFAP and novel synthetic-core peptides. Front. Microbiol. 11, 820.

Tulloch, M., 1976. *Metarhizium* spp. isolates from Madagascar: morphology and effect of high temperature on growth and infectivity to the migratory locust, *Locusta migratoria*, The genus *Metarhizium*. Trans. Br. Mycol. Soc. 66, 407–411.

Twizeyimana, M., Hammer, P.E., Gachango, E., Craig, K., Espejo, B., Biggs, M.B., Kremer, J., Ingham, D.J., 2023. Diverse environmental bacteria displaying activity against *Phakopsora pachyrhizi*, the cause of soyabean rust. Front. Plant Sci. 14, 1080116. https://doi.org/10.3389/fpls.2023.1080116.

Ullah, A., Bano, A., Khan, N., 2021. Climate change and salinity effects on crops and chemical communication between plants and plant growth-promoting microorganisms under Stress. Front. Sustain. Food Syst. 5, 618092.

Vannette, R.L., Hunter, M.D., 2009. Mycorrhizal fungi as mediators of defence against insect pests in agricultural systems. Agric. For. Entomol. 11, 351–358.

Vernner, R., Bauer, P., 2007. Q-TEO, a formulation concept that overcomes the incompability between water and oil. Pfalzenschutz-Nachrichten Bayer 60 (1), 7–26.

Vyas, D., Meena, R.H., 2018. Role of biofertilizers in integrated plant nutrient system (IPNS). Int. Res. J. Nat. Appl. Sci. 5, 77–94.

Waghunde, R.R., Shelake, R.M., Sabalpara, A.N., 2016. Trichoderma: a significant fungus for agriculture and environment. Afr. J. Agric. Res. 11, 1952–1965.

Wainaina, S., Lukitawesa, Awasthi, M.K., Taherzadeh, M.J., 2019. Bioengineering of anaerobic digestion for volatile fatty acids, hydrogen or methane production: a critical review. Bioengineered 10 (1), 437–458.

Wang, W., Vinocur, B., Altman, A., 2003. Plant responses to drought, salinity and extreme temperatures: towards genetic engineering for stress tolerance. Planta 218, 1–14.

Weber, S.S., Kaminski, K.P., Perret, J.L., Leroy, P., Mazurov, A., Peitsch, M.C., Ivanov, N.V., Hoeng, J., 2019. Antiparasitic properties of leaf extracts derived from selected *Nicotiana* species and *Nicotiana tabacum* varieties. Food Chem. Toxicol. 132, 110660.

Whipps, J.M., Sreenivasaprasad, S., Muthumeenakshi, S., Rogers, C.W., Challen, M.P., 2008. Use of Coniothyrium minitans as a biocontrol agent and some molecular aspects of sclerotial mycoparasitism. Eur. J. Plant Pathol. 121, 323–330. https://doi.org/10.1007/s10658-007-9238-1.

Williams, G.V., Faulkner, P., 1997. Cytological changes and viral morphogenesis during baculovirus infection. In: Miller, L.K. (Ed.), The Baculoviruses. Plenum, New York.

Wilson, M.K., Abergel, R.J., Raymond, K.N., Arceneaux, J.E., Byers, B.R., 2006. Siderophores of *Bacillus anthracis*, *Bacillus cereus*, and *Bacillus thuringiensis*. Biochem. Biophys. Res. Commun. 348 (1), 320–325.

Win, K.T., Kobayashi, M., Tanaka, F., Takeuchi, K., Oo, A.Z., Jiang, C.J., 2022. Identification of *Pseudomonas* strains for the biological control of soyabean red crown root rot. Sci. Rep. 12, 1–10.

Winder, R.S., Wheeler, J.J., Conder, N., Otvos, I.S., Nevill, R., Duan, L., 2003. Microencapsulation: a strategy for formulation of inoculum. Biocontrol Sci. Technol. 13 (2), 155–169.

Witzgall, P., Lindblom, T., Bengtsson, M., Tóth, M., 2004. The Pherolist (online) http://www.pherolist.slu.se. (Accessed 30 October 2006).

Woods, T.S., 2003. Biopesticide formulations, possibility of application and future trends. In: Encyclopedia of Agrochemicals, pp. 30–56.

Yadav, A.K., Chandra, K., 2014. Mass production and quality control of microbial inoculants. Proc. Indian Natl. Sci. Acad. 80, 483–489.

Yadav, A.N., Yadav, N., 2018. Stress-adaptive microbes for plant growth promotion and alleviation of drought stress in plants. Acta Sci. Agric. 2 (6), 85–88.

Yang, J., Kloepper, J.W., Ryu, C.-M., 2009. Rhizosphere bacteria help plants tolerate abiotic stress. Trends Plant Sci. 14 (1), 1–4.

Yu, H., Chen, W., Bhatt, K., Zhou, Z., Zhu, X., Liu, S., et al., 2022. A novel bacterial strain *Burkholderia* sp. F25 capable of degrading diffusible signal factor signal shows strong biocontrol potential. Front. Plant Sci. 13, 1071693.

Zahir, Z.A., Munir, A., Asghar, H.N., Shaharoona, B., Arshad, M., 2008. Effectiveness of rhizobacteria containing ACC deaminase for growth promotion of peas (*Pisum sativum*) under drought conditions. J. Microbiol. Biotechnol. 18, 958–963.

Zhu, Y.C., Kramer, K.J., Oppert, B., Dowdy, A.K., 2000. cDNAs of aminopeptidase-like protein genes from *Plodia interpunctella* strains with different susceptibilities to *Bacillus thuringiensis* toxins. Insect Biochem. Mol. Biol. 30 (3), 215–224.

CHAPTER 12

Biofungicides and plant growth promoters: Advantages and opportunities in entrepreneurship

A.K. Rana, K. Kaur, and P. Vyas
Department of Microbiology, Punjab Agricultural University, Ludhiana, Punjab, India

12.1 Introduction

Agriculture is the backbone of Indian economy, which is facing major problems pertaining to yield, quality, and profit due to the damage caused by fungal diseases. The use of agrochemicals, in particular, controlling plant diseases, has increased crop productivity, but their excessive use has caused a disastrous effect on environment and human health. It has been found only 0.1% of used agrochemical reaches the target and the remaining 99.5% enters the environment causing harm to other nontarget organisms, as well as humans. This has resulted in the search for alternatives to agrochemicals. One of the safe approaches is the use of naturally occurring beneficial microorganisms in agriculture. Microorganisms are known to show different types of associations ranging from antibiosis, commensalism, parasitism, and symbiosis in nature. The use of these associations for biological control of plant pathogen is being studied. Some of the microorganisms have been found to show biocontrol potential. For example, *Bacillus* and *Pseudomonas* spp. have been found to act as biocontrol agents because of their ability to produce antimicrobial metabolites (Choudhary et al., 2014). Fungicides of biological origin comprising of bacteria, fungi, and actinomycetes are referred to as biofungicides. They can also be the secondary metabolites produced in animal or plant tissues.

Some of the other reasons for increased phytopathogen infestation are the high diversity and greater host range of pathogens, easy wind dispersal, water, farm machinery, climate changes, and large-scale genetically uniform intensive monoculture crops (Pan et al., 2010). Biocontrol agents have been found to show more transient and localized action to the rhizosphere. Their use guarantees higher efficiency, a healthier environment, user and consumer safety, and cost-effectiveness. It decreases the risk of development of pest resistance and overall lead to the improvement of crop health and yield (Urja and Saraf, 2014). In the last decade, biofungicides have successfully gained a lot of interest particularly due to the rising demand of organic foods (Kumar and Singh, 2014). Plant extracts, including neem (*Azadirachta indica*), wild tobacco (*Calotropis procera*), wood ash, and dried chilies, are being used by several farmers across Asia and Africa for controlling and repelling some insect pests (Anukwuorji et al., 2013).

Entrepreneurship with Microorganisms. https://doi.org/10.1016/B978-0-443-19049-0.00007-4

The substitution of chemical fertilizers with plant growth-promoting microbes (PGPM) has resulted in increased plant growth, yield, and control of diseases in some cases (Joko et al., 2012). The direct effects shown on plants by PGPM are usually because of the production and release of certain compounds that are required by plants during their growth phases, including phytohormones. Absorption of a number of essential nutrients/elements from the environment is also promoted by this group of organisms (Ahemad and Kibret, 2014). PGPM have also been found to show some indirect effects on the plants that includes their defensive actions against organisms of phytopathogenic nature (Beneduzi et al., 2012). Through various mechanisms, these organisms can act as a biocontrol agent, nevertheless, of their action on plant growth promotion, through auxin production (Miransari and Smith, 2014), ethylene levels decrement (Liu and Zhang, 2015), or by nitrogen fixation in association with roots of the plants (Reed et al., 2011). Bacteria from a wide range of genera have been reported to show various plant growth promotion activities of which *Bacillus* sp. and *Pseudomonas* spp. are of principal use (Raaijmakers et al., 2010).

12.2 Classification of biofungicides

According to the United States Environmental Protection Agency (EPA), biofungicides are classified into three groups (Fig. 12.1).

12.2.1 Microbial fungicides

Microorganisms such as bacteria, fungi, viruses, or protozoans consist of microbial fungicides (Kumar and Singh, 2014). It has been reported that fungicides of microbial origin can control fungi of different kinds having separate active ingredient that is relatively specific to its target pests. It has been noted that microbial fungicides act by competing with fungal organisms. Continuous monitoring of fungicides of microbial origin is need as to ensure they do not develop the capacity to cause any harm to non-target organisms, including humans (Gupta and Dikshit, 2010).

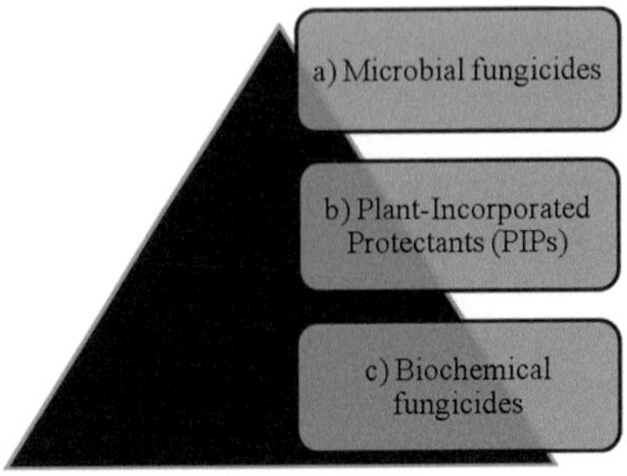

FIG. 12.1

Classification of biofungicides.

12.2.2 Plant-incorporated pesticides (PIPs)

Fungicidal substances produced by plants are known as PIPs. It was found that genetic modification in a crop plant is the reason for the production of plant-incorporated pesticides, for example, Bt cotton. These have shown to be more effective and economically good approach in producing more food, feed, and forages in an eco-friendly and safer manner in the developing countries (Gupta and Dikshit, 2010).

12.2.3 Biochemical fungicides

These occur naturally in plants and animals having an ability to control pests by nontoxic mechanisms. These include substances that interrupt the growth on reproduction, for example, plant growth regulators or substances that repel or attract pests, including insect sex pheromones. Plant extract and botanical oils are other biochemical fungicides (Kilic and Akay, 2008).

12.2.4 Bioactive component in biofungicides

A large number of secondary metabolites, including alkaloids, amino acids, flavonoids, polyacetylenes, saponins, sugars, terpenoids, tannis, etc., are produced by plants. These are biologically active substances and are also referred to as phytochemicals. Several examples of these phytochemical biopesticides and their use in the integrated pest management (IPM) have been described in detail (Singh et al., 2012). It has been observed that the plants during their long evolution have synthesized a varied number of chemicals to protect themselves from colonization by insects and other herbivores. These chemicals protect plants from approaching insects, disrupt feeding, and oviposition on the plants with even having the ability to have toxic and harmful effect on the developmental stages of many insects (Koul and Dhaliwal, 2000).

12.2.5 Phytoelaxins as plant protectants

The concept of phytoelaxins was enlightened by Muller and Borger almost 70 years back, while studying potato tuber infected by different strains of casual organism of "late blight of potato" "*Phytophthora infestans*" (Müller and Börger, 1940). It was reported that the infection caused by pathogenic fungus initiated some hypersensitive reactions resulting in the production of some "plant secondary metabolite." This disrupted the further infection by the fungus of another strain of the same genus *Phytophthora* on the same plant. This "principle" was named as "phytoelaxins" by Muller and his coworkers as it has protected the plants from a secondary infection (Deverall, 1982). Therefore, phytoelaxins may be defined as products naturally secreted and accumulated in response to pathogenic organism or abiotic stress agents like heavy metal toxicity, UV radiations, and wounds on tissue (Naoumkina et al., 2007). Phytoelaxins target different pathogenic organisms by various mechanisms, including inhibition of glucose uptake, disruption of plasma membrane, or mitochondria (Table 12.1). The inducers of phytoelaxins can be divided into two types as elicitor and elicitin. The oligosaccharide from fungal cell origin (like hepatosaccharide from soja cell wall) consists of the elicitors. The another type, elicitin, is the general glycolipids that are secreted by the fungal cells (Cordelier et al., 2003).

Table 12.1 Mechanisms used by phytoalexins for the inhibition of fungal pathogens.

Phytoelaxin	Host plant	Mechanisms	Target organism	Reference
Kievitone or Phaseolin	*Phaseolus mungo*	Inhibits glucose uptake by fungal cell	*Rhizoctonia solani*	Van Etten and Bateman, 1971
Phytuberin, rishitin, Anhydro-β-rotunol, solavetivone	*Solanum tuberosum*	Disrupts motility and swelling of the zoospores, cell membrane rupture	*Phytophthora* sp.	Harris and Dennis, 1977
Stilbene, resveratrol, pterostilbene	*Vitis vinifera*	Mitochondrial, Plasma membrane disruption, Uncoupled ETS and photosynthesis prohibition	*Botrytis cinerea*	Adrian and Jeandet, 2012
Camalexin	*Phaseolus vulgaris*	By apoptotic mechanisms it induces fungal programmed cell death (PCD)	*Botrytis cinerea*	Shlezinger et al., 2011

12.3 An emerging source of biofungicides: Endophytes

The useful microorganisms living inside the host plant without causing any visible symptom of disease or syndrome, as well as helping in promoting the plant growth during harsh environmental conditions, are referred to as endophytes. Endophytes have proved to be a stable and reliable source for exploring the bioactive compounds produced with the plants. An extensive research is required to give a huge opportunity to the endophyte biologists to explore and identify endophytic fungal and actinobacterial flora with antifungal properties. The safer the biofungicides, better accepted in the society, scientific community and particularly in agriculture sector (Table 12.2).

12.4 Plant growth promoters

The soil surrounding the plant roots known as the rhizosphere is the store house of various microorganisms. A number of chemicals are synthesized and secreted by plant roots that attract a varied number of microbial communities or populations able to promote plant growth. Ever since the concept of plant growth promotion has come into light, various scientists have been trying to solve this mystery of soil harboring million of populations of microorganisms. These microbes present in soil directly affect soil quality and plant growth. This has attracted attention of agriculturist working on developing an environmentally friendly and sustainable method for plant growth promotion in agriculture (Debasis et al., 2019). These plant growth-promoting microorganisms (PGPM) carry out various mechanisms to affect plant growth positively (Fig. 12.2).

Table 12.2 Endophytic microbes showing antifungal activity.

Endophyte	Plant source	Antifungal properties against	Reference
Alternaria sp. and *Ovulariopsis* sp.	*Datura stramonium*	*Aspergillus niger*, *Fusarium* sp., *Colletotrichum gloeosporioides*, *Phytophthora nicotianae*, *Scopulariopsis* sp., *Trichoderma viride*, *Verticillium* sp.	Li et al., 2005
Acremonium strictum, *Alternaria alternata*, *Beauveria bassiana*, *Phoma* sp., *Trichoderma koningii*	*Zea mays* roots	*Fusarium oxysporum*, *F. pallidoroseum*, *F. verticillioides*, *Cladosporium herbarum*	Orole and Adejumo, 2009
Alternaria sp., *Dothideomycetes* sp., *Chaetomium* sp., *Thielavia subthermophila*	*Tylophora indica*	*Sclerotinia sclerotiorum*, *Fusarium oxysporum*	Kumar et al., 2011
Chaetomium globosum	*Withania somnifera*	*Sclerotinia sclerotiorum*	Kumar et al., 2013
Fusarium proliferatum, *Nigrospora oryzae*, *Guignardia cammillae*, *Alternaria destruens*	*Jatropha curcas*	*Sclerotinia sclerotiorum*	Kumar and Kaushik, 2013
Phytophthora infestans, *Fusarium oxysporum*	*Triticum durum*	*Aspergillus* sp., *Penicillium* sp., *Alternaria* sp., *Cladosporium* sp., *Chaetomium* sp., *Phoma* sp.	Sadrati et al., 2013
Penicillium sp., *Curvularia* sp., *Cladosporium* sp.	Moso bamboo (*Phyllostachys edulis*) (seeds)	*Arthrinium sacchari*, *Arthrinium phaeospermum*, *Curvularia eragrostidis*, *Pleosporaherbarum*	Shen et al., 2014
Trichothecium sp.	*Phyllanthus amarus*	*Penicillium expansum* (blue mold of apples)	Taware et al., 2014
Alternaria sp., *Biscogniauxia mediterranea*, *Cladosporium funiculosum*, *Paraconiothyrium* sp.	*Opuntia humifusa*	*Colletotrichum fragariae*, *C. gloeosporioides*, *C. acutatum*	Silva-Hughesa et al., 2015
Rhexocercosporidium sp., *F. solani*	*Sophora tonkinensis* Gapnep	*Alternaria panax*, *F. solani*, *C. gloeosporioides*	Yao et al., 2017
Glomerellacingulate, *Colletotrichum gloeosporioides*, *C. truncatum*, *Dothideomycetes* sp., *Lasiodiplodia pseudotheobromae*	*Houttuynia cordata*	*Fusarium oxysporum*, *Rhizoctonia* sp., *Sclerotium rolfsii*, *Trichoderma harzianum*, *Alternaria brassicicola*, *Phytopthorapalmivora*	Aramsirirujiwet et al., 2016
Colletotrichum boninense, *Fusarium chlamydosporum*, *Myiarchus yucatanensis*, *Cladosporium* sp.	*Monarda citriodora* (leaf, roots, and flowers)	*Fusarium solani*, *Sclerotinia* sp., *Colletotrichum capsici*, *Aspergillus flavus*, *A. fumigatus*	Katoch and Pull, 2017
Trichoderma longibrachiatum strain BHU-BOT-RYRL17, *Syncephalastrumracemosum* strain AQGSS 12, *Trichoderma longibrachiatum* voucher 50	*Markhamia tomentosa*	*Fusarium oxysporum*, *Sclerotinia sclerotiorum*, *Rhizoctonia solani*, *Botrytis cinerea*	Ibrahima et al., 2017
Penicillium simplicissimum, *Leptosphaeria* sp., *Talaromyces flavus*, *Acremonium* sp.	Cotton roots (*Gossypium hirsutum*)	*V. dahlia* (*Verticillium* wilt disease)	Yuan et al., 2017
Penicillium sp. (ARDS-2.3), *Aspergillus oryzae* (ARHS-1.1)	*Asparagus racemosus* Wild	*Botrytis cinerea* (gray mold), *Sclerotinia sclerotiorum* (stem rot), *Rhizoctonia solani* (root rot), *Fusarium oxysporum* (wilt)	Chowdhary and Kaushik, 2018
Aspergillus sp., *Curvularia* sp., *Fusarium oxysporum*	*Dendrobium lindleyi*	*Fusarium* sp., *Sclerotium* sp., *Colletotrichum* sp., *Curvularia* sp., *Phytophthora* sp.	Bungtongdee et al., 2019

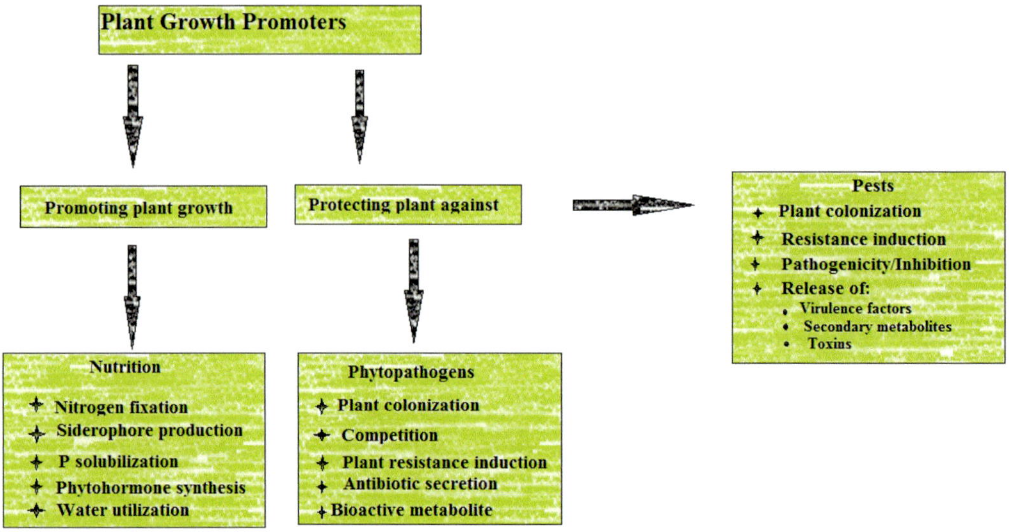

FIG. 12.2

Mechanisms of plant growth-promoting microorganisms for plant growth and protection.

12.4.1 Plant growth-promoting microorganisms as biopesticides

The pesticides that incorporate microorganisms for biocontrol activities are termed as biopesticides and consist of bacteria, fungi, actinomycetes, viruses, nematode, protozoa, and yeast. Among the biopesticides utilized currently, microbial biopesticides form the largest group of broad-spectrum biopesticides, showing pest specificity; i.e., they do not target non-pest species and are environment friendly. It has been estimated that there are 1500 naturally occurring insect-specific microorganisms, out of which 100 are insecticidal, 200 are microbial biopesticides, which have availability in 30 countries affiliated by the Organization for Economic Co-operation and Development (OECD). A total of 53 microbial biopesticides have been registered in the USA, 22 in Canada, and 21 in European union (EU) though the products registered in Asia are varying. Technological expansion has spiked the interests of consumers and producers around the globe toward developing microbial products. Bacterial biopesticides can be divided into four types based on utilization of bacteria strain: crystalliferous spore formers (*Bacillus thuringiensis*); obligate pathogens (*Bacilus popilliae*); potential pathogens (*Serratia marcesens*) along with facultative pathogens (*Pseudomonas aeruginosa*). Spore-forming category has been extensively used among the four as they are safer and effective. *Bacillus thuringiensis* and *B. sphaericus*are the most widely used bacteria for this purpose. Most of the mechanism of this process is still not understood as their mechanism has been found to be complex, which can be indicated by the presence of novel receptors and signal transduction pathways induced in the host after intoxication. The production of these bacteria is carried out in larger portions either by solid or liquid fermentation. These bacteria during sporulation show the production of an intracellular protein toxic along with a parasporal crystalline toxin at the time of sporulation. Mosquito control is done by the means of *Bacillus sphaericus*-based products. There are some other strains including *Agrobacterium* *radiobacter, B. popilliae, B. subtilis, P. cepacia, P. chlororaphis, P. flourescens, P. solanacearum* and *P. syringae* available which have a comparatively less effect on pest management. Mosquito control can be done by using *Bacillus sphaericus*-based products. Some species of bacteria have minor effect on pest management though there are certain commercial products based on *Agrobacterium radiobacter, B. popilliae, B. subtilis, P. cepacia, P. chlororaphis, P. flourescens, P. solanacearum,* and *P. syringae* available (Dulmage, 1981).

12.4.2 Plant growth-promoting microorganisms as bioinsecticides

Microorganisms or their products having the capability to attack and kill pest or insects are termed as microbial insecticides. Among the bacteria, *Bacillus thuringiensis* (Bt) has shown notable results from 30 distinct isolates exhibiting insecticidal activity (Li et al., 2003; Charudattan, 2001). After the phase of robust vegetative growth, sporulation of the cells takes place giving rise to nongrowing, heat-resistant spores. Concurrently, a proteinaceous delta-endotoxin crystal is formed, while sporulating, most of the isolates belonging to this group of bacteria are involved in production of toxic protein crystals. The most extensive toxin is the delta-endotoxin (protoxin) that gets activated inside the midgut of the insect with the help of proteolytic enzymes. The toxins released disrupt the epithelial membrane of the midgut, which causes imbalance in the osmotic balance of the cells and consequently paralyzing and causing death of insect. Some of the commercial formulations of Bt include Thuricide, Dipel, Agritol, Biotrol, and Larvatrol, along with many others having viable spore and delta-endotoxins (the variety Bt kurstaki does not produce beta-exotoxin). They are effective against a number of important pests of agriculture and forestry. In Israel, Bt formulations of variety kurstaki are utilized to keep avocado pests in check. Also, biocontrol of *Spodoptera littoralis* in alfalfa and *Heliothisarmigera* in tomatoes is effectual in bioorganic agriculture. There are some Bt varieties reported to be more powerful against many lepidopterous insects. Three entomopathogenic fungi for insect and nematode control like *Beauveria bassiana, Lecanicillium muscarium, and Isaria (Paecilomyces) fumosorosea* have been reported to control whitefly, thrips along with some other soft-bodied insects in greenhouse crops. *Metarhizium anisopliae* has been approved for control of black vine weevil in soft fruit and tree nurseries. *Paecilomyces lilacinus* is reported to control root knot nematodes.

12.4.3 Plant growth-promoting microorganisms as biofertilizers

Various formulations of agriculturally significant microbes (bacteria or fungi) with favorable physiological and behavioral characteristics utilized for crop nutrient management are termed as biofertilizers. The major characteristics for quality of different biofertilizers are well entrenched and constant efforts are being made for regularizing the development and commercialization of the biofertilizer products. For success in commercializing of the biofertilizers, it is important to assure the quality of the product being introduced in the market. While formulating the quality standards of biofertilizers, viable cell count is considered a vital parameter and the capability of microbes to fix nitrogen, solubilizing phosphorous, potassium, and zinc is crucial efficiency characters that are to be kept in mind. The microbes constituting the biofertilizer enhance the nutrients of the plant by replacing the nutrition required or by making nutrients easily available for the plants or by increasing access to nutrients (Table 12.3). These microbes play a crucial role in integrated nutrient management (INM) system. They provide the plants with not only the nutrients but also helping them in maintenance of

Table 12.3 Some examples of plant growth-promoting bacteria with their plant growth-promoting traits.

Strain	Traits	Biocontrol action	Environmental stress/others	References
Bacillus cereus	HCN production, siderophore production, in vitro growth experiments	Against *Rhizoctonia solani, Fomes lamenensis, Corticiuminvisum*	–	Vandana et al., 2018
Pseudomonas rhizophila S211	ACCD, putative dioxygenases, auxin, pyroverdin, exopolysaccharide levan and rhamnolipid BS	–	Potential in pesticide bioremediation	Hassen et al., 2018
Pseudomonas spp	PGPR, ACCD, fatty acid methyl ester analysis and antioxidant activity	–	drought tolerant: ~1.0 MPa (30% polyethylene glycol 8000)	Chandra et al., 2018
Bacillus cereus	PGPR, phosphate solubilizing bacteria, IAA production and root colonization in *Zea mays*	–	salt tolerant	Mohan et al., 2017
Bacillus, Pseudomonas and *Arthrobacter*	SPME & GC-MS analysis and biocontrol	Against *Phytophthora cinnamomi*	–	Méndez-Bravo et al., 2018
Rhizobium, Pseudomonas, KRB, ZSB	Potassium, zinc solubilization ACCD and exopolysaccharide (EPS) production	–	Stress tolerant	Kumari et al., 2018
Pseudomonas fluorescence K-34, *Pseudomonas fluorescence* 1773/K and *Pseudomonas trivalis* BIHB 745	Siderophore production, phosphate solubilization, cell wall degrading ability, IAA and ecofriendly multifunctional biofertilizer in agriculture	–	–	Li et al., 2003

soil health. They are economical, safe for environment, natural formulations that provide an alternate source of nutrition to the plants, which consequently helps to increase the income of a farmer as it provides increased yield along with reduced cost for input (Banayo et al., 2012). Biofertilizers are being accepted and appreciated all around the globe but especially among the mid- and small-scale farmers from Asia and South America (Njira, 2013). Asia being the biggest producer and consumer of biofertilizers still has a few undone sectors, which might strike a major biosafety concern for the environmentalists. Foremost hurdle is the distinction in the guidelines for large-scale production and marketing in developed, developing, and underdeveloped countries. These differences in the regulations make it difficult for intercontinental trade to be conducted. An ever-increasing demand for healthier organic food creates a very exceptional opportunity for development of agro-based economies especially those in Asia. The rise in demand makes the regulation bodies of government, industrial, and academic sectors realize the need to have strict regulations for production of higher quality products providing the favorable results as promised. The regulations and registration body work are observed to be easier in countries like China, India, Korea, Japan, and Taiwan, which has enormously contributed toward the acceptance of biofertilizers. Nevertheless, adoptions of biofertilizers turn out to be successful only when the formulations chosen are of a better quality and appropriate awareness and knowledge is provided to the consumer (Pandey et al., 2016). In Taiwan, the central and the state governmental authorities are supporting to popularize biofertilizers such as P-solubilizing bacteria and mycorrhizal inoculants for horticultural crops. During previous studies, the number of inoculants isolated was sufficient to inoculate agricultural land estimated to be 65,091 Ha in size. A notable increase in the annual income of a farmer was observed after usage of biofertilizers (Tabassum et al., 2017; Marrone, 2019).

12.5 Advantages of biofungicides and plant growth promoters

Biofungicides and plant growth promoter offer a number of advantages over the chemical fungicides and fertilizers. The final product of biofungicides is residues free; therefore, they prove to be very effective in reducing the number of chemical treatments, particularly, during the postharvesting time where they have a limited or no requirement. This helps particularly in the case of crops having prolonged harvesting where plants are needed to be treated close to harvest. They exhibit beneficial impacts on preharvest treatment targeting postharvest pathogens (e.g. *Aureobasidium* spp., *Bacillus* spp.). None of the existing microbial biofungicides have shown any phytotoxicity on crops; therefore, they can serve as suitable tools at plant stages where the tissues are soft and tender or to treat flowers. The genera registered as biofungicides have not shown any negative impacts on *Saccharomyces cerevisiae* also as they prevent infection by gray mold or sour rot without influencing the fermentation during the vinification process; thus, they are being recommended to be used on grapes. Microbial biofungicides are found to be nontoxic for pollinators and bees have been used to deliver them to flowers in some of the cases (Charudattan, 2001).

Biofungicides exhibit a complex mechanism of action such as production of metabolites and enzymes, induction of resistance, competition for space and nutrients, and mycoparasitism; therefore, they have lesser chances of developing resistance to pathogenic species. For practicing organic production, they have proved to be a useful tool because of their natural origin. Investigating from an eco-friendly view, biofungicides are fully biodegradable, as well as renewable resources, and biomagnification of their metabolites has never been reported (Pertot et al., 2016). There are few, ineffective, or nonexistent chemical solutions for some fastidious plant diseases. Thus, biological control is considered as an alternative or supplemental way in agriculture for increasing crop productivity (Gerhardson, 2002).

Plant growth-promoting microorganisms exhibit different biocontrol activities through synthesis of bacterial allo chemicals, including iron-chelating siderophores, antibiotics, biocidal volatiles, lytic enzymes, and detoxication enzymes (Glick, 1995). The application of nonpathogenic strains of *Streptomyces* for controlling scab disease in potato (*Solanum tuberosum* L.) caused by *S. scabies* is one of the applications of plant growth promoters. Among all biopesticides, 30% of total sales are represented by microorganism-based products and new related products are timely brought into the market (Thakore, 2006).

12.6 Opportunities in entrepreneurship

The global demand for food and fiber has been escalating. This is where green revolution played an important role in intensifying agriculture to meet these demands. Nonetheless, this has not only led to ground water pollution but has also degraded the food and environment. Significant amount of money is lost in farm produce all around the world, and one of the crucial factors contributing to it are plant diseases. The ever-increasing world population demands a stable availability of marketable produce which makes it vital to control these plant diseases. In attempt to manage the plant diseases, an overwhelming amount of chemicals have been utilized, which led to a derogatory effect on quality of environment and as a result gave rise to a trend for the utilization of living forms resistant to chemicals. Witnessing the constant change in agricultural sector, the only techniques permissible for managing the diseases are by incorporating biocontrol agents in agricultural practice in this way any kind of harmful imbalance in the environment and ecosystem can be resisted. A jump-start for commercializing this technology can be provided by increasing the shelf-life of the formulations, as well as broadening the spectrum of action providing consistency in performance in field conditions (Galindo et al., 2013).

A worldwide rise in trends for the identification of safe and environment friendly alternatives for utilization in agricultural practices can be observed. Despite the fact, that many researchers are aware of the trends still various nonchemical agriculture products occupy a very minimal number of shares in the market. Nowadays, the production and consumption of biological pesticides adds up to a minimal amount of percentage in the business but an exceptional rate of growth has been observed in the biopesticide market, suggesting the masses concerned are accepting this class of innovative solution gradually. The global market for bio fungicides is approximated to be 1.6 billion USD in 2020 and is speculated to grow at a CAGR 16.1% from 2020, amounting to be 3.4 USD by 2025. This growing market of bio fungicides is driven by the factors like increasing demand of organic food products, government rigorously regulating against the utilization of chemical-based plant protection products along with their increasing health concern. The biofungicide market has distinctions based on various factors leading to different formulations and different bioproducts. It is speculated that with the rise of biopesticide market a slight downfall will be observed for synthetic pesticides as the former has a lot of factors favoring it. Some of them are withdrawal of synthetic pesticides from market due to their harmful nature for humans, as well as environment, awareness among farm workers of their work place safety and consumers demanding safer and healthier products. Additionally, the acquisition of integrated pest management magnifies the need for safe control measures like biofungicides with the growing biopesticide market, many opportunities are on the horizon for entrepreneurs for commercializing biofungicides/biopesticides. The ever-increasing demands of entrepreneurs can be fulfilled by helping university students pursuing the interests with a concern toward plant protection. They can be taught required skills and knowledge like introduction to biofungicides, their importance, development, intellectual rights (patent), quality control, and commercialization for developing them and opening an opportunity for even the teachers and professors. The main intention of technopreneur class is to illustrate the process of biofungicide development. Various steps involved start from isolating potential local biocontrol agents, inclusive of the endophytes, entomopathogen fungi, parasitoid, and predator and are performed at both laboratory and field till the commercialization of the product. Teaching endeavors have contributed in finding various isolates of endophytic fungi from cocoa plants

(Daha et al., 2012). Certain isolates have been utilized in research on biopesticide development against under MP3EI research scheme funded by the Directorate General of Higher Education, Ministry of Education and Culture, Republic of Indonesia (Retchelderfer, 1984). Another most important factor is formulation technology as it could potentially affect the economic practicality of a biofungicide/biopesticide product. In order to stabilize, the product and increase the viability making it a consistent product on the field requires favorable formulations as inconsistency in performance of the product is one of the major hindrances faced in product development. It has been observed that finding a favorable formulation is a time-consuming process (Greaves, 1993; Urquhart and Punja, 1997). With several formulations available in the market granules can be preferred for the protection provided to prevent desiccation and providing basic nutrition, while powder is popular due to the ease of application involved suitable for seed treatment (Amin et al., 2011; Amin, 2013). Intellectual property rights (patent) are another important objective to teach the students importance of intellectual right in the research, development, and innovation in university and to teach them how to apply for intellectual rights.

Commercialization is the most essential factor of developing and launching a bioproduct, so it is crucial for the students to know how to make a business plan and execute it successfully. The success is dependent on the comprehensive and thorough understanding of the respective field. So, the subject of technopreneur ship is another opportunity arises out of the biofungicide market and should be offered in different universities with special facilities to be used for the development of biopesticides and biofertilizers (Prasad, 2014). So, the main entrepreneurial opportunities emerging from plant growth-promoting microorganisms in the form of biopesticides and biofertilizers.

North America has been reported to be the largest shareholder in the market as major companies are based in this region, and is the largest producer of fruits and vegetables. The acquisition of integrated pest management and organic farming practice in these regions along with rising concerns regarding food safety worldwide have also escalated the growth of the market. Some leading companies involved in biofungicides are BASF, SE (Germany), Bayer AG (Germany), Syngenta AG (Switzerland), FMC Corporation (US), Nufarm (Australia), Novozymes (Denmark), Marrone Bio Innovations (US), Koppert Biological Systems (Netherlands), Isagro S.P.A (Italy), T. Stanes & Company Limited (India), BioWorks (US), The Stockton Group (Israel), Valent Biosciences (US), Agri Life (India), Certis U.S.A (US), Andermatt Biocontrol AG (Switzerland), Lesattre (France), Rizobacter (Argentina), Vegalab S.A (US), Biobest Group NV (Belgium), and Biolchim (Italy). Biosave (*Pseudomanos syringae)* registered in the USA for postharvest application to citrus fruits extending to cherries, potatoes, and sweet potatoes as well. Shemar (*Metschikowia fructicola*) registered in Israel for the pre- and postharvest application on a variety of fruits and vegetable like grape, peaches, apricots, citrus, pepper, and strawberry (Marrone, 2008).

12.7 Nanobiofungicides: An emerging opportunity

Synthesis of nanometals by biomediated ways using organisms like bacteria, fungi, algae, yeast, actinomycetes, and virus has been prevalent lately (Prasad et al., 2014; Aziz et al., 2015; Lee et al., 2013). It started by using copper nanoparticles, by dissolving them in water and using them as fungicides to control diseases in grape and other fruit trees (Bhattacharyya et al., 2015; Rai and Ingle, 2012). It has been known since long those fungi are associated with plant diseases and consequently play a role in loss of economy. Some of the most effective pathogens include *Fusarium* sp., *Phoma* sp., *Aspergillus* sp., *Phytopthora* sp., and *Phyllosticta* sp. (Parveen and Rao, 2015). Germination potential of seeds was observed to increase when silver nanoparticles were coted over a seed (Jo et al., 2009). Biosynthesized silver nanoparticles have shown antifungal properties when applied against plant pathogenic fungi (Gopinath and Velusamy, 2013; Mishra et al., 2014). The trend of utilization of nanopesticides appears to be escalating. Polymer-based nanoformulations, inorganic nanoparticles like silica and titanium dioxide, and nanoemulsions are various formulations that are overriding currently (Nair et al., 2010). These formulations when applied tend to discharge the active substance in a slower manner in comparison with existing formulations which in turn protects the seeds from degradation. Plant pathogen control by the means of nanoencapsulated fungicides requires smaller quantity and works effectively without deteriorating environmental conditions around. Various nanopesticides that are biosafe are under development and have molecular communication with the plants, soil, and environment depending on the dose dispersed in the soil (Aziz et al., 2015; Abd-Elsalam, 2013; Ingle et al., 2014). It has been speculated that a considerable number of nanopesticide formulations are said to be released into agrochemical market in coming days to maintain sustainability in agricultural practices (Adil et al., 2015). Nanomaterials or nanobiofungicides have notably shown to manage most of the pathogens (Yadav et al., 2015; Singh et al., 2015; Lamsal et al., 2011). It might have been advised to formulate nanoparticles with pesticide colloidal suspension or in powdered form with nano- or microscale which have proven to improve stability of active organic composite (UV, thermal, hydrolysis, etc.), foliar settling, reduction in foliar leaching systemic action, synergism, specificity, and variety of forms. The mechanisms mentioned collectively result in minimal application of pesticide. Silver nanoparticles can be synthesized efficiently by environment friendly ways notably by using extracellular processes (Kah et al., 2013).

12.8 Impediments to commercialization

Accessibility of microorganisms either as a product or a formulation plays a vital role to successfully suppress pests and diseases, in turn making it easy for technology to be transferred from laboratory to land. Some important commercialized products based on biofungicides and plant growth promoters are listed in Table 12.4 and 12.5.

However, the impediments faced while developing or utilizing the biofungicides and biofertilizers are reflections of the factors that hinder the development all around the globe. These include:

- Shortfall of appropriate screening protocol vital for selecting favorable strains of microbes.
- Absence of adequate knowledge on the microbial ecology of the bacterial strains, as well as the plant pathogens.
- Augmentation of technology for large-scale production.
- Not having a consistent performance along with a below par shelf-life.
- Lacking patent protection.
- Exorbitant prices for registration.
- Lacking to spread awareness, training, and education.
- Absence of interdisciplinary approach.
- Impediments due to technology.

Table 12.4 Some examples of commercialized products based on plant growth promoters and biofungicides (Marrone, 2019).

Product name	Composition	Benefit/target pathogen/ disease	Appropriate for
Actinovate G	*Streptomyces lydicus* strain WYEC 108	Soil-borne diseases like powdery mildew, botrytis, etc.	Fruits and vegetables
Bio-phospho	*Bacillus subtilis*	Phosphate solubilization	Wheat, maize, and rice
Biopromotor Rhizobium	*Rhizobium* sp.	Nitrogen Fixation	Legumes
Cyanonuticon	*Cyanobacterial* strains	Carbon status and soil health	Rice and wheat
NPK liquid	*Azotobacter chroocomccum, Pseudomonas striata, Bacillus* sp.	nitrogen fixing, P-solubilizing, k-solubilizing	All crops
Potash solubilizing liquid	*Bacillus* sp.	K-solubilizing	Cereals, millets, fiber, vegetables
Kodiakr	*Bacillus subtilis*	*Fusarium* and *Rhizoctonia* soil pathogens	Peanuts, soyabean, small grain, corn and vegetable crops
AzoBac	*Azospirillum* CM1404	Bioremediation, nitrogen fixing, root stimulator	Maize, sugarcane, wheat and grass pastures
Biozote-MAX	Plant growth-promoting rhizobacteria	Phytohormones IAA and GA production	Major crops
Rice-Biofert	*Azolla*	Microbial metabolite	Rice
Blight Ban A506	*Pseudomonas fluorescens* A 506	*Erwinia amylovora* and russet—inducing bacteria	Almond, Apple, Apricot, Blue berry, Cherry, Peach, Pear, Potato, strawberry, Tomato
Conquer	*Pseudomonas fluorescens* 276	*P. tolassii*	Mushrooms
Cedomon	*Pseudomonas chloroaphis*	Leaf stripe, net blotch, Fusarium sp., spot blotch, leaf spot, and others	Cereals
Rhizogold plus	a mix-culture of Rhizobium and ACC deaminase containing PGPR	Reduce salinity stress	Cereals
Subtilex	*Bacillus subtilis* MB1600	*Fusarium* spp., *Rhizoctonia* spp., and *Pythium* spp.	Ornamental and vegetable crops
Pix plus plant regulator	*Bacillus cereus* BPO1 technical, *Bacillus cereus* strain UW85	Growth regulator	Cotton
Bio-save 10LP, 110	*Pseudomonas syringae*	*Botrytis cinerea, Penicillium spp., Geotrichumcandidum*	Pome fruit, Citrus, Cherries, and Potato
MeloCon	*Purpureocillium lilacinus*	Bionematicide	Citrus fruits
KlamiC and others	*Pochonia chlamydosporia*	Bionematicide	Horticultural crops
DiTera	*Myrothecium verrucaria*	Bionematicide	Fruits and vegetables

Table 12.5 Examples of some commercialized biofungicide products.

Product name	Composition	Pathogen attacked	Action
Messenger	*Erwinia amylovora* (HrpN harpin protein)	Multi-spectrum	Insecticide, fungicide and nematicide
Rotstop	*Phelbia gigantea*	Effective against pine and spruce rust	Biofungicide
Galltrol-A, Dygall Norbac 84-C*, Nogall	*A. radiobacter*	Crown galls	Antagonist
Serenade, Epic, Kodiak, MBI 600, Cillus	*B. subtilis*	Effective against root rot caused by *Rhizoctonia, Fusarium, Alternaria, Aspergillus,* and *Pythium.* Also effective against some foliar diseases	Fungicide and antagonist
Ballad, Sonata AS, Yield shield	*B. pumilus*	Effective against rust, downy and powdery mildews	Fungicide
Actinovate SP	*Streptomyces lydicus*	Effective against soil borne diseases of turf, nursery crops	Fungicide
RootShield, Bio Trek 22 g, Supresivit	*Trichoderma harzianum*	Effective against variety of soil pathogens	Mycoparasitic, Antagonistic
Protus WG	*Talaromyces flavus* V117b	Effective against fungal pathogens of tomato, cucumber, strawberry and rape oilseeds	Antagonist
RootShield WP, PlantShield HC	*Trichoderma harzianum* T-22	Acts as a biocontrol agent for beans, corns ornamental flowers	Biofungicide
BIO-TAM 2.0	*Trichoderma asperellum* and *Trichoderma gamsii*	Acts as Biocontrol in crops like potato, wheat, rice and maize	Biofungicide
SoilGard	*Gliocladium virens*	Effective BCA for Lentils against fungal pathogens	Biofungicide
Contans WG	*Coniothyrium minitans*	Targets *Sclerotinia*	Biofungicide
Howler	*Pseudomonas chlororaphis* strain AFS009	Targets fungal pathogens causing plant diseases	Biofungicide
Actinovate, ActinoGrow	*Streptomyces lydicus*	Biocontrol of soil pathogens such as damping off fungus, root rot, *Rhizoctonia* and turf brown patch	Biofungicide
Serifel	*Bacillus amyloliquefaciens* MBI600	Protects tomato against tomato spotted wilt virus and potato virus	Biofungicide
Taegro 2 WP	*Bacillus subtilis* var. *amyloliquefaciens* FZB24	Biocontrol for seedling diseases caused by *Fusarium, Phytopthora* and *Rhizoctonia*	Biofungicide
LifeGard WG	*Bacillus mycoides*	Biocontrol against plant pathogens	Biofungicide
ENV503	*Bacillus amyloliquefaciens* ENV503 (genetically identical to *B. subtilis* GB03)	Protection against pathogenic microbes	Biofungicide
Serenade, Cease	*Bacillus subtilis* (renamed *amyloliquefaciens*) 713	Biocontrol against pests, pathogens, nematodes	Biofungicide
Prevont	*Bacillus subtilis* IAB/BS03	Biocontrol against plant diseases like powdery mildew, late blight, gray mold	Biofungicide
DoubleNickel 55	*Bacillus amyloliquefaciens* D747 (similar in lipopeptides to Serenade)	Protects fruits against pathogens	Biofungicide

12.9 Policies for promoting commercialization

The present scenario of crop production, biocontrol, and plant growth promotion are two vital factors to keep in check. The research in this field is restricted to the laboratories making it difficult to be able to fully exploit it. Moreover, negligible attention has been paid toward the production of commercial products of bioagents, as well as plant growth promoters. The absence of knowledge makes it worse as the products that have been commercialized already are not being utilized by the unaware farmers. So for commercializing biofungicides and plant growth promoters following points could be kept in mind:

(i) Increasing the popularity of biocontrol agents, as well as the plant growth promoters.
- Attracting attention of media for promotion.
- Demonstrating on the fields.
- Farmer's day.
- Adopting a biovillage.
- Conduct of periodic training sessions for commercial farmers and producers to improvise their yield.

(ii) Making links with industries
- Make sure the availability of technical support to the entrepreneurs on quality control and registration.
- Consistent surveillance is vital for the maintenance of quality.
- Availability of consistent support for research should be kept in check to help in standardizing dosage, storage and delivery systems.
- Positive policy support by government should be extended to help in encouraging usage of plant growth promoters and biocontrol agents for plant protection.

12.10 Conclusions

A number of biofungicides along with natural plant growth promoters have provided an alternative and good approach for increasing crop yield and controlling diseases rather than using synthetic chemicals. These have been reported to exert major impacts on the livelihoods of farmers by boosting agricultural yields and food security. Several studies have revealed that these can curb the damage caused by arthropod and other agricultural pests, this, opens several opportunities for developing a business in this field. The production of biofungicides and plant growth promoters at industrial scale can immensely contribute to the economics of developing countries through improved export earning, creating employment and, thus, helping in alleviating the poor. This can also prove that the fact that how biofungicides market is anticipated to show a vibrant growth on account of environmental awareness and education spreading among the farmers of India.

References

Abd-Elsalam, K.A., 2013. Nano platforms for plant pathogenic fungi management. Appl. Phys. 100, 829–834.

Adil, S.F., Assal, M.E., Khan, M., 2015. Biogenic synthesis of metallic nanoparticles and prospects toward green chemistry. Dalton Trans. 44, 9709–9717.

Adrian, M., Jeandet, P., 2012. Effects of resveratrol on the ultrastructure of *Botrytis cinerea* conidia and biological significance in plant/pathogen interactions. Fitoterapia 83, 1345–1350.

Ahemad, M., Kibret, M., 2014. Mechanisms and applications of plant growth promoting rhizobacteria: current perspective. J. King Saudi Univ. Sci. 26 (1), 1–20.

Amin, N., 2013. Teaching of biopesticide development as a techno prenuership opportunity in plant protection. J. Biol. Agric. Health 3, 2224–3208.

Amin, N., Daha, L., Nasruddin, A., 2011. Fungal Endophyte as Biopesticide and Biofertilizer in Form of Tablet. Patent No. 00201100098. (In Indonesian Language).

Anukwuorji, C.A., Anuagasi, C.L., Okigbo, R.N., 2013. Occurrence and control of fungal pathogens of potato (*Ipomoea batatas* L. Lam) with plant extracts. Int. J. PharmTech Res. 2, 278–289.

Aramsirirujiwet, Y., Gumlangmak, C., Kitpreechavanich, V., 2016. Studies on antagonistic effect against plant pathogenic fungi from endophytic fungi isolated from *Houttuynia cordata*Thunband screening for siderophore and indole-3-acetic acid production. KKU Res. J. 21 (1), 55–66.

Aziz, N., Faraz, M., Pandey, R., Sakir, M., Fatma, T., Varma, A., Barman, I., Prasad, R., 2015. Facile algaederived route to biogenic silver nanoparticles: synthesis, antibacterial and photocatalytic properties. Langmuir 31, 11605–11612.

Banayo, N.P., Cruz, P.C., Aguilar, E.A., Badayos, R.B., Haefele, S.M., 2012. Evaluation of biofertilizers in irrigated rice: effects on grain yield at different fertilizer rates. Agriculture 2 (1), 73–86.

Beneduzi, A., Ambrosini, A., Passaglia, L.M.P., 2012. Plant growth-promoting rhizobacteria (PGPR): their potential as antagonists and biocontrol agents. Genet. Mol. Biol. 35 (4), 1044–1051.

Bhattacharyya, A., Antoney, P.U., Raja Naika, H., Reddy, S.J., Adeyemi, M.M., Omkar, 2015. Nanotechnology and butterflies: a mini review. J. Appl. Biosci. 41, 27–32.

Bungtongdee, N., Sopalun, K., Laosripaiboon, W., Iamtham, S., 2019. The chemical composition, antifungal, antioxidant and antimutagenicity properties of bioactive compounds from fungal endophytes associated with Thai orchids. J. Phytopathol. 167, 56–64.

Chandra, D., Srivastava, R., Glick, B.R., Sharma, A.K., 2018. Drought-tolerant *Pseudomonas* spp. improve the growth performance of finger millet (*Eleusine coracana* (L.) Gaertn.) under non-stressed and drought-stressed conditions. Pedosphere 28 (2), 227–240.

Charudattan, R., 2001. Biological control of weeds by means of plant pathogens: significance for integrated weed management in modern agro-ecology. BioControl 46 (2), 229–260.

Choudhary, D.K., Verma, S.K., Patel, A.K., Dayaram, 2014. Formation and development of biofungicide. Int. Res. J. Nat. Sci. 2, 14–22.

Chowdhary, K., Kaushik, N., 2018. Diversity and antifungal activity of fungal endophytes of *Asparagus racemosus* Willd. Agric. Res. 8, 27–35.

Cordelier, S., Ruffray, P., Fritg, B., Kauffmann, S., 2003. Biological and molecular comparison between localized and systemic acquired resistance induced in tobacco by *Phytophthora megasperma* glycoprotein elicitin. Plant Mol. Biol. 51, 109–118.

Daha, L., Amin, N., Nurariaty, A., 2012. Endophytic fungi in form of capsule to control of Cocoa Pod Borer Conophomorpa cramerella on cocoa plant. In: Final Report of National Research Priority in Masterplan of acseleration and Extention of Indonesian Economic Development, pp. 2011–2025.

Debasis, M., Snežana, A., Panneerselvam, P., Manisha, S.A., Tanja, V., Ganeshamurthy, A.N., Devvret, V., Poonam, R.T., Divya, J., 2019. Plant growth promoting microorganisms (PGPMs) helping in sustainable agriculture: current perspective. IJAVMS 7 (2), 50–74.

Deverall, B.J., 1982. Introduction. In: Bailey, J.A., Mansfield, J.W. (Eds.), Phytoalexins. Blackie, Glasgow/London, pp. 19–20.

Dulmage, H.D., 1981. Insecticidal activity of isolates of Bacillus thuringiensis and their potential for pest control. In: Burges, H.D. (Ed.), Microbial Control of Pests and Diseases. Academic Press, London, pp. 193–222.

Galindo, E., Serrano-Carreón, L., Gutiérrez, C.R., Allende, R., Balderas, K., Patiño, M., Trejo, M., Wong, M.A., Rayo, E., Isauro, D., Jurado, C., 2013. The challenges of introducing a new biofungicide to the market: a case study. Electron. J. Biotechnol. 16 (3), 5–7.

Gerhardson, B., 2002. Biological substitutes for pesticides. Trends Biotechnol. 20, 338–343.

Glick, B.R., 1995. The enhancement of plant growth by free-living bacteria. Can. J. Microbiol. 41 (2), 109–117.

Gopinath, V., Velusamy, P., 2013. Extracellular biosynthesis of silver nanoparticles using *Bacillus* sp. GP-23 and evaluation of their antifungal activity towards Fusarium oxysporum. Spectrochim. Acta A Mol. Biomol. Spectrosc. 106, 170–174.

Greaves, M.P., 1993. Formulation of microbial herbicides to improve performance in the field. In: Proceedings of 8th EWRS Symposium Quantitative Approaches in Weed and Herbicide Research and their Practical Application. Braunschweig, Germany, pp. 219–225.

Gupta, S., Dikshit, A.K., 2010. Biopesticides: an ecofriendly approach for pest control. J. Biopestic. 3, 186–188.

Harris, J.E., Dennis, C., 1977. The effect of post-infectional potato tuber metabolites and surfactants on zoospores of oomycetes. Physiol. Plant Pathol. 11, 163–169.

Hassen, W., Neifar, M., Cherif, H., Najjari, A., Chouchane, H., Driouich, R.C., Salah, A., Naili, F., Mosbah, A., Souissi, Y., Raddadi, N., 2018. *Pseudomonas rhizophila* S211, a new plant growthpromoting rhizobacterium with potential in pesticide bioremediation. Front. Microbiol. 9 (34), 1–17.

Ibrahima, M., Kaushik, N., Sowemimoa, A., Chhipab, H., Koekemoerc, T., Venterc, M., Odukoyaa, O.A., 2017. Antifungal and antiproliferative activities of endophytic fungi isolated from the leaves of *Markhamia tomentosa*. Pharm. Biol. 55 (1), 590–595.

Ingle, A., Duran, N., Rai, M., 2014. Bioactivity, mechanism of action and cytotoxicity of copper-based nanoparticles: a review. Appl. Microbiol. Biotechnol. 98, 1001–1009.

Jo, Y.K., Kim, B.H., Jung, G.H., 2009. Antifungal activity of silver ions and nanoparticles on phytopathogenic fungi. Plant Dis. 93, 1037–1043.

Joko, T., Koentjoro, M.P., Somowiyarjo, S., Rohman, M.S., Liana, A., Ogawa, N., 2012. Response of rhizobacterial communities in watermelon to infection with cucumber green mottle mosaic virus as revealed by cultivation-dependent RISA. Arch. Phytopathol. Plant Protect. 45 (15), 1810–1818.

Kah, M., Beulke, S., Tiede, K., 2013. Nano-pesticides: state of knowledge, environmental fate and exposure modelling. Crit. Rev. Environ. Sci. Technol. 43 (16), 1823–1867.

Katoch, M., Pull, S., 2017. Endophytic fungi associated with *Monarda citriodora*, an aromatic and medicinal plant and their biocontrol potential. Pharm. Biol. 55 (1), 1528–1535.

Kilic, A., Akay, M.T., 2008. A three generation study with genetically modified Bt corn in rats: biochemical and histopathological investigation. Food Chem. Toxicol. 46, 1164–1170.

Koul, O., Dhaliwal, G.S., 2000. Phytochemical Biopesticides (Advances in Biopesticides Research). CRC Press, Boca Raton, Florida.

Kumar, S., Kaushik, N., 2013. Endophytic fungi isolated from oil-seed crop *Jatropha curcas* produces oil and exhibit antifungal activity. PLoS One 8 (2), e56202.

Kumar, S., Kaushik, N., Edrada-Ebel, R., Ebel, R., Proksch, P., 2011. Isolation, characterization, and bioactivity of endophytic fungi of *Tylophora indica*. World J. Microbiol. Biotechnol. 27, 571–577.

Kumar, S., Kaushik, N., Proksch, P., 2013. Identification of antifungal principle in the solvent extract of an endophytic fungus *Chaetomium globosum* from *Withaniasomnifera*. Springerplus 2 (1), 37.

Kumar, S., Singh, A., 2014. Biopesticides for integrated crop management: environmental and regulatory aspects. J. Biofertil. Biopestic. 5, 121–129.

Kumari, M.E.R., Gopal, A.V., Lakshmipathy, R., 2018. Effect of stress tolerant plant growth promoting Rhizobacteria on growth of blackgram under stress condition. Int. J. Curr. Microbiol. App. Sci. 7 (1), 1479–1487.

Lamsal, K., Kim, S.W., Jung, J.H., Kim, Y.S., Kim, K.S., Lee, Y.S., 2011. Application of silver nanoparticles for the control of *Colletotrichum* species in vitro and pepper anthracnose disease in field. Mycobiology 39, 194–199.

Lee, K.J., Park, S.H., Govarthanan, M., 2013. Synthesis of silver nanoparticles using cow milk and their antifungal activity against phytopathogens. Mater. Lett. 105, 128–131.

Li, Y., Sun, Z., Zhuang, X., Xu, L., Chen, S., Li, M., 2003. Research progress on microbial herbicides. Crop Prot. 22 (2), 247–252.

Li, S., Zhang, Z., Cain, A., Wang, B., Long, M., Taylor, J., 2005. Antifungal activity of camptothecin, trifolin, and hyperoside isolated from *Camptotheca acuminata*. J. Agric. Food Chem. 53, 32–37.

Liu, X.M., Zhang, H., 2015. The effects of bacterial volatile emissions on plant abiotic stress tolerance. Front. Plant Sci. 6, 774–777.

Marrone, P.G., 2008. Integrated Pest Management (Concept, Tactics, Strategies and Case Studies). Cambridge University Press.

Marrone, P.G., 2019. Pesticidal natural products—status and future potential. Pest Manag. Sci. 75 (9), 2325–2340.

Méndez-Bravo, A., Cortazar-Murillo, E.M., Guevara Avendaño, E., Ceballos-Luna, O., Rodríguez-Haas, B., Kiel Martínez, A.L., Hernández-Cristóbal, O., Guerrero-Analco, J.A., Reverchon, F., 2018. Plant growth-promoting rhizobacteria associated with avocado display antagonistic activity against Phytophthora cinnamomi through volatile emissions. PLoS One 13 (3), 1–18.

Miransari, M., Smith, D.L., 2014. Plant hormones and seed germination. Environ. Exp. Bot. 99, 110–121.

Mishra, S., Singh, B.R., Singh, A., 2014. Biofabricated silver nanoparticles act as a strong fungicide against *Bipolaris sorokiniana* causing spot blotch disease in wheat. PLoS One 9 (5), e97881.

Mohan, V., Devi, K.S., Anushya, A., Revathy, G., Kuzhalvaimozhi, G.V., Vijayalakshmi, K.S., 2017. Screening of salt tolerant and growth promotion efficacy of phosphate solubilizing bacteria. JAIR 5 (12), 168–172.

Müller, K.O., Börger, H., 1940. Experimentelle Untersuchungenüber die Phytophthora Resistenz der Kartoffel. Arb. Biol. Reichsanst. Land Forstwirtsch. 23, 189–231.

Nair, R., Varghese, S.H., Nair, B.G., Maekawa, T., Yoshida, Y., Kumar, D.S., 2010. Nanoparticulate material delivery to plants. Plant Sci. 179, 154–163.

Naoumkina, M., Farag, M.A., Sumner, L.W., Tang, Y.H., Liu, C.J., Dixon, R.A., 2007. Different mechanisms for phytoalexin induction by pathogen and wound signals in *Medicago truncatula*. Proc. Natl. Acad. Sci. 104, 17909–17915.

Njira, K.O., 2013. Microbial contributions in alleviating decline in soil fertility. Microbiol. Res. J. Int. 3, 724–742.

Orole, O.O., Adejumo, T.O., 2009. Activity of fungal endophytes against four maize wilt pathogens. Afr. J. Microbiol. Res. 3 (12), 969–973.

Pan, Z., Li, X., Yang, B.X., Andrade, D., Xue, J., Mckinney, N., 2010. Prediction of plant diseases through modelling and monitoring airborne pathogen dispersal. In: Hemming, D. (Ed.), CAB Reviews: Perspecttive in Agriculture, Veterinary Science. Nutrition and Natural Resources CABI, UK, pp. 191–202.

Pandey, S., Verma, A., Chakraborty, D., 2016. Potential use of rhizobacteria as biofertilizer and its role in increasing tolerance to drought stress. In: Recent Trends in Biofertilizers, pp. 115–140.

Parveen, A., Rao, S., 2015. Effect of nanosilver on seed germination and seedling growth in *Pennisetum glaucum*. J. Clust. Sci. 26, 693–701.

Pertot, I., Prodorutti, D., Colombini, A., Pasini, L., 2016. *Trichoderma atroviride* SC1 prevents *Phaeomoniella chlamydospora* and *Phaeoacremonium aleophilum* infection of grapevine plants during the grafting process in nurseries. BioControl 61 (3), 257–267.

Prasad, R., 2014. Synthesis of silver nanoparticles in photosynthetic plants. J. Nanopart. https://doi.org/10.1155/2014/963961.

Prasad, R., Kumar, V., Prasad, K.S., 2014. Nanotechnology in sustainable agriculture: present concerns and future aspects. Afr. J. Biotechnol. 13 (6), 705–713.

Raaijmakers, J.M., de Bruijn, I., Nybroe, O., Ongena, M., 2010. Natural functions of lipopeptides from *Bacillus* and *Pseudomonas*: more than surfactants and antibiotics. FEMS Microbiol. Rev. 34 (6), 1037–1062.

Rai, M., Ingle, A., 2012. Role of nanotechnology in agriculture with special reference to management of insect pests. Appl. Microbiol. Biotechnol. 94, 287–293.

Reed, S.C., Cleveland, C.C., Townsend, A.R., 2011. Functional ecology of free living nitrogen fixation: a contemporary perspective. Annu. Rev. Ecol. Evol. Syst. 42, 489–512.

Retchelderfer, K., 1984. Factors affecting the economic feasibility of the biological control of weeds. In: Delfosse, E.S. (Ed.), Proceedings of VI International Symposium on Biological Control of Weeds. Vancouver, Canada, pp. 135–144.

Sadrati, N., Daoud, H., Zerroug, A., Dahamna, S., Bouharati, S., 2013. Screening of antimicrobial and antioxidant secondary metabolites from endophytic fungi isolated from wheat (*Triticum durum*). J. Plant Protect. Res. 53 (2), 129–136.

Shen, X.-Y., Cheng, Y.-L., Cai, C.-J., Fan, L., Gao, J., et al., 2014. Diversity and antimicrobial activity of culturable endophytic fungi isolated from Moso bamboo seeds. PLoS One 9 (4), e95838.

Shlezinger, N., Minz, A., Gur, Y., Hatam, I., Dagdas, Y.F., Talbot, N.J., Sharon, A., 2011. Anti-apoptotic machinery protects the necrotrophic fungus *Botrytis cinerea* from host-induced apoptotic-like cell death during plant infection. PLoS Pathog. 7, 1002185.

Silva-Hughesa, A.F., Wedgeb, D.E., Cantrellb, C.L., Carvalhoa, C.R., Panb, Z., Moraesc, R.M., Madoxxe, V.L., Rosaa, L.H., 2015. Diversity and antifungal activity of the endophytic fungi associated with the native medicinal cactus *Opuntia humifusa* (Cactaceae) from the United States. Microbiol. Res. 175, 67–77.

Singh, A., Khare, A., Singh, A.P., 2012. Use of vegetable oils as biopesticides in grain protection- a review. J. Biofertil. Biopestic. 3, 114–117.

Singh, A., Singh, N.B., Hussain, I., Singh, H., Singh, S.C., 2015. Plant-nanoparticle interaction: an approach to improve agricultural practices and plant productivity. Int. J. Pharm. Sci. Invent. 4, 25–40.

Tabassum, B., Khan, A., Tariq, M., Ramzan, M., Khan, M.S., Shahid, N., Aaliya, K., 2017. Bottlenecks in commercialisation and future prospects of PGPR. Appl. Soil Ecol. 121, 102–117.

Taware, R., Abnave, P., Patil, D., Rajamohananan, P.R., Raja, R., Soundararajan, G., Kundu, G.C., Ahmad, A., 2014. Isolation, purification and characterization of Trichothecinol-a produced by endophytic fungus *Trichothecium* sp. and its antifungal, anticancer and antimetastatic activities. Sustainable Chem. Processes 2, 8.

Thakore, Y., 2006. The biopesticide market for global agricultural use. Ind. Biotechnol. 2 (3), 194–208.

Urja, P., Saraf, M., 2014. In vitro evaluation of PGPR strains for their iocontrol potential against fungal pathogens. In: Kharwar, R.N., Upadhay, R.S., Dubey, N.K., Roicha, R. (Eds.), Microbial Diversity and Biotechnology in Food Security. SperingerVerley, Berlin, pp. 293–305.

Urquhart, E.J., Punja, Z.K., 1997. Epiphytic growth and survival of *Tilletiopsis pallescens*, a potential biological control agent of *Sphaerotheca fuliginea*, on cucumber leaves. Can. J. Bot. 75 (6), 892–901.

Van Etten, H.D., Bateman, D.F., 1971. Studies on the mode of action of the phytoalexin phaseolin. Phytopathology 61, 1363–1372.

Vandana, K.U., Chopra, A., Choudhury, A., Adapa, D., Mazumder, P.B., 2018. Genetic diversity and antagonistic activity of plant growth promoting bacteria, isolated from tea-rhizosphere: a culture dependent study. Biomed. Res. 29 (4), 1–12.

Yadav, A., Kon, K., Kratosova, G., Duran, N., Ingle, A.P., Rai, M., 2015. Fungi as an efficient mycosystem for the synthesis of metal nanoparticles: progress and key aspects of research. Biotechnol. Lett. 37, 2099–2120.

Yao, Y.Q., Lan, F., Qiao, Y.M., Wei, J.G., Huang, R.S., Li, L.B., 2017. Endophytic fungi harbored in the root of *Sophora tonkinensis* Gapnep: diversity and biocontrol potential against phytopathogens. Microbiol. Open 6, e437.

Yuan, Y., Feng, H., Wang, L., Li, Z., Shi, Y., Zhao, L., et al., 2017. Potential of endophytic fungi isolated from cotton roots for biological control against *Verticillium* wilt disease. PLoS One 12 (1), e0170557.

CHAPTER

Potential use of fungi as biofertilizer in sustainable agriculture

13

Kena P. Anshuman

Department of Microbiology, Sir P.P. Institute of Science, MK Bhavnagar University, Bhavnagar, Gujarat, India

13.1 Introduction

Fungi are eukaryotic, spore-bearing organisms. It lacks chlorophyll. Generally, fungi are found in air, water, soil, marine environment, dead matter as well as in extreme environments such as deserts and high salt concentrations. Fungi have velvety, furry, and powdery growth with different colors like gray, red, pink, orange, yellow, green, black, brown, etc. This coloration is due to various pigments such as quinones, anthraquinones, rubropuntamine, rubropunctatin, ankaflavin, monascin, and β-carotene produced by various fungi (Mukherjee and Mishra, 2017). Thus, fungi have varieties of morphological appearances depending on the species. Thus, fungi attain great practical and scientific interest in microbiology, biotechnology, agriculture, industrial level, etc.

13.2 Structure

Fungi comprise mold, yeast, and mushroom. Mold is filamentous and multicellular while yeast is unicellular. The yeast and mold filament is surrounded by a true cell wall. Some fungi are "dimorphic," having two forms, for instance, switch between a yeast phase and a hyphal phase (Dube, 2013). For example, some pathogenic fungi of humans and other animals have unicellular yeast-like form in their host, but when growing in soil or on laboratory medium they have filamentous mold form.

Fungal cell wall is made up of chitin, polymer of *N*-acetylglucosamine unit, which contains 80%–90% of carbohydrate and remaining protein and lipids. The cell membrane has a unique sterol and ergosterol. Most fungal cell walls lack cellulose (except Oomycetes). The cell wall compositions, cell size, and shapes are variable by age and environmental factors (Dube, 2013).

Yeast is larger than bacteria, ranging from 1 to 5 μm in width and 5–30 μm or more in length. They are commonly oval and egg shaped. Some are spherical too. Mold is a filamentous structure. Thallus of mold is made up of two parts (1) the Mycelium and (2) Spores. The mycelium is a complex of several filaments called "hyphae." New hyphae generally arise from spore which on germination form the germ tube and these germ tubes elongate and branch to form hyphae.

Hyphae are septate, nonseptate, or septate with multinucleated cells. Each has more than one nucleus in each compartment. Vegetative mycelia penetrate into the medium and derive nutrition. Whereas reproductive mycelium produces spores and usually extends from the medium into the air.

Entrepreneurship with Microorganisms. https://doi.org/10.1016/B978-0-443-19049-0.00005-0

Fungi reproduce asexually or sexually. Asexual reproduction by fragmentation, budding, fission, or by asexual spores, for instance, sporangiospores and conidiospores and sexual reproduction by three principal events, for instance, plasmogamy, karyogamy, and meiosis (Dube, 2013).

13.3 Physiology

Fungi is heterotrophic and derives nutrition from external sources. They live as saprophytes, parasites, or as symbionts. Symbiotic relation of fungi found with lichen and mycorrhizae. In lichen, they are a partner to algae and in mycorrhizae association with the root system. Some fungi can use organic nitrogen. Fungi could not use inorganic carbon compounds; they must be in an organic form such as glucose. Some are facultative and grow in both aerobic and anaerobic conditions. Fungi grow best at acidic pH, optimum range is 3.8–5.6. The temperature range for saprophytic species is from 22°C to 30°C and for pathogenic species range is 30°C–37°C. Some fungi can grow at or near 2°C and cause spoilage of vegetables and meat in cold storage. Thus, fungi can use wide varieties of nutrition for growth.

Fungi are found in every environment on Earth. It plays a great role in the ecosystem and imparts in the biogeochemical cycle. Fungi easily grow on food materials such as jam, jelly, bread, or cold storage materials where sugar is available.

Fungi are pathogenic and nonpathogenic. Pathogenic fungi cause disease in humans, animals, and plants, while nonpathogenic fungi have vast applications in microbiology, biotechnology, marine, and in agriculture as it produces antibiotics, siderophores, various enzymes, etc.

13.4 Classification of fungi

Many mycologists used various fungal characteristics such as thallus characters, reproductive structure, ultrastructure, and more recently molecular evidence to classify fungi. The fungi were divided by Saccardo, 1886 into four classes, for instance, Phycomycetes, Ascomycetes, basidiomycetes, and Deuteromycetes (Saccardo, 1886). Later, Sparrow (1958) recognized for classes within the aquatic phycomycetes, these are Chytridiomycetes, Hyphochytridiomycetes, Oomycetes, and Plasmodiophoromycetes (Sparrow, 1958). Further two other subclasses were zygomycetes and trichomycetes (Sumbali and Mehrotra, 2009). International Code of Botanical Nomenclature Based on the Evolutionary System in 1996 include Chytridiomycota, Zygomycota, Ascomycota, and Basidiomycota. Recently, Schuessler et al. (2001) include the fifth phylum in the kingdom of fungi, for instance, Mycota or Eumycota (Schuessler et al., 2001). Thus, five phyla are Chytridiomycota, Glomeromycota, Zygomycota, Ascomycota, and Basidiomycota (Sumbali and Mehrotra, 2009).

13.4.1 Chytridiomycota

The chytridiomycota is the simplest aquatic fungi, termed as "Chytrids." Generally saprophytic lives on plant and animal matter in freshwater, mud, or soil. Parasitic lives on aquatic plants, animals, and also in insects. Chytridiomycota reproduces asexually and sexually. Asexual reproduction by means of zoospores with single, posterior, and whiplash flagellum. Sexual reproduction is either by planogametic copulation or by gametangial copulation or by somatogamy, for example, *Allomyces macrogynus* and *Rhizophydium* (Sumbali and Mehrotra, 2009).

13.4.2 Glomeromycota

Glomeromycota is economically important mycorrhizal fungi, forming important associations with the roots of herbaceous plants and tropical trees. Earlier, Arbuscular Mycorrhizal (AM) fungi were considered a member of Zygomycota. But ribosomal RNA analysis shows that Arbuscular Mycorrhizal fungi are not related to Zygomycota. So, they have been assigned to a new group, for instance, Glomeromycota (Schuessler et al., 2001). An important feature of this Arbuscular Mycorrhizal Fungi (AMF) is the production of large, swollen "Vesicles" as food storage reserves and Arbuscular Mycorrhizal fungal hyphae penetrate in host root cell wall to form dichotomous tree-like branching structure called "Arbuscules," for example, *Acaulospora, Entrophospora, Archaeospora, Glomus, Paraglomus, Gigaspora, Scutellospora,* etc.

13.4.3 Zygomycota

These are terrestrial fungi. Generally found in soil, dung, plant parts, vertebrates, and invertebrates. Saprophytic species are found on decaying plants and animal matter in the soil. Few are parasitic to plants, insects, and animals (including humans). Sexual reproduction forms thick-walled Zygote called "Zygospore." Two gametangia direct contact with one another and fuse to form large Zygospores, an important feature of these fungi, for example, *Rhizopus, mucor, phycomycetes, thermo mucor, pilobolus,* etc. (Sumbali and Mehrotra, 2009).

13.4.4 Ascomycota

This is the largest phylum of fungi, commonly known as "Sac fungi." Generally, they are found in freshwater, marine, and terrestrial habitat. It degrades many organic compounds including lignin, cellulose, and collagen. It includes yeast, powdery mildew, molds, cup fungi, the morels, and truffles.

The pair of nuclei from the female gametangium migrate into specialized hyphae called "Ascogenous hyphae." After the maturation of ascogenous hyphae, nuclear fusion occurs at the tip of hyphae in the ascus mother cell. The diploid zygote nucleus undergoes meiosis and forms four haploid nuclei, dividing mitotically again to produce ascospores. Typically, eight ascospores are formed within the ascus, so it is called "Sac fungi." Most of the members of ascomycetes produce fruiting bodies called "ascocarp," enclosing the asci, when ascospores mature it releases from asci with great force. Under favorable conditions, ascospores germinate, produce germ tubes, and form new mycelium. Asexual reproduction is by budding, fission, fragmentation, chlamydospores, and conidia. Ascomycota may be unicellular, bicellular, or multicellular. Multicellular may be divided by septa in one to three planes, for example, *Aspergillus, Penicillium, Alternaria, Fusarium,* and *Trichothecium* (Sumbali and Mehrotra, 2009).

13.4.5 Basidiomycota

Basidiomycota commonly known as basidiomycetes or "Club fungi." It contains about 30,000 species, which form 37% of described species of true fungi (Kirk et al., 2001). These include diverse forms such as mushrooms, puffball, boletes, fairy club, birds nest fungi, jelly fungi, bracket fungi, rust, and smut fungi. They are saprophytic. Some are parasitic and destroy wood and herbaceous plants. Many members of basidiomycetes are edible, especially, button mushrooms, oyster mushrooms, milky mushrooms, etc. They form the characteristic structure "Basidium" at the tip of hyphae, involved in sexual reproduction. It is of club shape so-called Club Fungi. Two or more basidiospores are produced in basidium. Basidia may be held within fruiting bodies called "basidiocarps." Basidiocarps can reach 60 cm in diameter and may have spores, each lined with basidia that produce basidiospores. Thus, a single fruiting body can produce millions of spores, for example, *Cryptococcus, Ustilaginomycetes, Urediniomycetes,* and *Polyporus*.

Thus, fungi are used as a "Model organisms" for researchers due to their important role in various fields. There are 97,330 species of fungi (Hawksworth, 2004; Kirk et al., 2008). As of 2020, around 1, 48, 000 species of fungi have been described (Cheek et al., 2020). Research suggested that fungal species richness was much higher than plants and demonstrated that tree species diversity was a good predictor of macrofungal diversity (Schmit et al., 2005). Fungi have various applications in many fields and one of the potential applications is "Biofertilizer" to sustain agriculture. Today chemical fertilizers cause hazardous effects on all living beings. So, the recent trend to develop biofertilizer is in vogue to overcome this problem.

13.5 What is biofertilizer?

Biofertilizers (BF) are biologically origin made up of living cells of different types of microorganisms which have the ability to convert nonusable forms to usable forms through biological processes. Microorganisms transform soil nutrients such as phosphorus, zinc, copper, iron, and sulfur from nonusable to usable form through biological processes. Such biofertilizers are low cost, nonbulky and play an important role in plant nutrients, and most importantly they are nonhazardous to plants and animals including human beings. Biofertilizer is also referred to as "Microbial inoculant" or "Bioinoculant."

Generally, live or latent cells of the efficient strain of various microorganisms were used as biofertilizer. It includes bacteria, fungi, actinomycetes, algae, etc. And algae are found in specific situations. Biomass of these organisms in soil is found under specific situations only. Soil harbors numerous microorganisms, and the presence of these microorganisms make the soil a living and active system. And these microorganisms play an important role in the life cycle of plants and animals through decomposition, nutrient (N_2) fixation, solubilization (P), and secretion of important compounds and serve as plant growth promoter in stress adaptation (Paul and Nair, 2008; Subba Rao, 1986).

In recent years, the use of biofertilizers has become a hope for many countries, from an economical and environmental perspective. Therefore, developing countries like India can solve the problem of the high cost of fertilizers and help in saving the economy of countries. Thus, using biofertilizers has attained special significance in modern agriculture.

The organisms are used as a broad spectrum for making biofertilizers in concentrated from 10^7 to 10^9 per gram (Vyas et al., 1998). Biofertilizers have many advantages to plants and soil, which are usually poor in organic matters. It increases the yield and growth of plants, decomposes organic matter, solubilizes phosphate, produces hormones, antimetabolites, and most importantly it is eco-friendly too (Ateia et al., 2009; Bhattacharjee and Dey, 2014).

Biofertilizers are available in many forms, such as liquid biofertilizers, carrier-based biofertilizers, granular biofertilizers, and encapsulated biofertilizers. It can be used according to requirements and availability. Carriers such as peat, lignite, wood charcoal, and rice husk are used. And binders like gum aerobic, vegetable oil, carboxymethylcellulose (CMC), etc. are used (Bhattacharjee and Dey, 2014).

1. Liquid base biofertilizers are generally used in seed treatment, root dipping, and soil applications.
2. Carrier-based biofertilizers are used with carriers such as peat, coal, lignite for seed/soil inoculants.
3. Granular biofertilizers, here granules are made up of peat prill/small marble, calcite, or silica grains. Granules are coated or impregnated with target organisms. It is used in soil application, more effective in stressful environments.
4. Encapsulated biofertilizers, here sticker solution of gum aerabic/Carboxymethylcellulose (CMC) are used for seed dressing.

13.6 Application methods of biofertilizer

Biofertilizer is used in various ways in liquid or solid form to seeds, soil, roots, and foliage. Different crops require different treatments. So, different techniques are used, and it also varies accordingly. It can be used as seed inoculation/dressing up, root treatment, or as a foliar spray.

13.6.1 Seed inoculation

When the soil has adverse conditions such as dryness, acidity, alkalinity, or excess of fertilizers, to protect a seed, adopt special methods of inoculation, for instance, seed inoculation. Here seeds are coated with microorganisms and then sawn in the soil. Otherwise, seed damage or could not survive in unfavorable conditions. Seed inoculated with *Rhizobium* sp. enhanced the population of other organisms such as *Azotobacter* and *Azospirillum* in the root region of wheat cultivars grown in unsterile soil (Kavimandan, 1993). Carrier base culture of microorganisms suspends in 10% sugar gur (Jaggery) solution in water. Also, a high amount of gum aerobic or carboxymethylcellulose (CMC) is added to the inoculum slurry before mixing with seeds. Finally, the pelleting agent is mixed when inoculated seeds are moist. It helps to coat seeds evenly. The commonly used pelleting agents are calcium carbonate, rock phosphate, charcoal powder, gypsum, bentonite, etc.

Under normal conditions, cell count should be 10^5–10^6 per seed used. And for 10 kg of seeds 200 g of microorganisms for small scale or for large scale 400–600 g of microorganisms can be used (Bhattacharjee and Dey, 2014). The gur solution is sprinkled on seeds as the binder. Seeds are mixed for uniform coating then seeds are spread on the polyethylene sheet. Sometimes seed coat toxicity affects the survival of *Rhizobia*, so to avoid it, soak seeds in water for a few hours, remove the water-soluble toxic substances, and later coat seeds with carrier base culture (Sumbali and Mehrotra, 2009).

Thus, biofertilizer as seed inoculants during sowing is a regular practice now in vogue and increases crop yield. In addition, farmers can also restore seeds too.

13.6.2 Granular inoculation

Granular biofertilizer is prepared by compost, *Rhizobia, Azotobacter,* etc. Generally, 25 kg well decomposed sieved compost is filled in a 50 kg polyethylene bag. For one hectare drilling, 5 kg compost is required. It is moistened by the addition of 4–5 L of water in the evening and the mouth of the bag is tied, then the bag is kept in tin or types of roofs in the sunshine for 25–30 days, generally this process is done in the month of April–May. The moistened compost is ready to use (Subba Rao, 1986).

Artificially prepared culture of *Rhizobia* to leguminous seed used before sowing. This is referred to as "Legume inoculation." Peat as carrier use with 10^8 cells per gram for legume inoculation. Generally, 1 kg carrier and 300 mL of culture broth are used for preparation [8]. In India peat-like materials are available in Nilgiri Valley. In Australia, peat is harvested, dried, and ground to pass a 200-mesh sieve. Peat is generally acidic in nature so it is neutralized by adding sufficient $CaCO_3$. *Rhizobial* culture is mixed with unsterilized peat, a mechanical mix. Then peat is sieved through a coarse sieve to remove lumbs and then allow to mature for 4 days at 26 °C in a tray covered with polyethylene and packed.

If microbial culture is added to sterilized peat then a required quantity of broth is added through a syringe to provide moisture (60%), then the puncture is sealed by adhesive tape. The content was mixed by rolling in hand, incubated at 26 °C for 2 weeks and then stored at 4 °C. For sterilization of peat, generally gamma rays are used at a dose of 5.0×10^6 rads. Standard *Rhizobial* culture in peat is 10^8–10^9 viable cells/grown in Yeast Extract Mannitol (YEM) broth is advisable (Subba Rao, 1986).

13.6.3 Liquid inoculation

The liquid inoculation method at sowing has become widely used in Australia [10]. Dissolving peat-based culture in water and applying this suspension to the soil surface. Gradually the culture reaches below the root zone (2–3 cm). This is applied to a 21 days crop of soybean and chickpea at the rate of 2.5 g/L (20 L/ha) and it increases nodulation (Vyas et al., 1998). This liquid-based culture can also apply to root treatment and as a foliar spray.

1. ***Root treatment:*** The roots of transplantable crops such as vegetables are dipped into aqueous suspension of carrier cultures (20–30 min) and then sawn in the field.
2. ***Foliar spray*:** This method is used for seed or soil inoculation but sometimes adversely affects. Seed and soil contain several chemicals which are toxic to biofertilizers. When seeds germinate, they excrete several phenols, which produce a toxic effect on spore. So, keeping in view biofertilizer on foliage and standing crops are more advisable (Vyas et al., 1998).

Liquid biofertilizer is more important than carrier-based biofertilizer because of their longer shelf life, no contamination, no loss of properties, easy to use, high export potential, cheap and easy to produce. Therefore, nowadays ready to use liquid biofertilizer of effective microorganisms is more popular in the market. And liquid biofertilizers replace chemical fertilizers (Maheswari and Kalaiyarasi, 2015).

13.7 Fungi as biofertilizer

There are different types of biofertilizers available by use of bacteria, fungi, algae, mycorrhizae, etc. These are used for composting, nitrogen fixation, phosphate solubilization, siderophore production, as antagonist, etc. However nowadays one of the potential uses of fungi as biofertilizer is in vogue.

13.7.1 Composting

Enriching the soil by the addition of organic materials of various origins is the most common practice since long. It helps to rehabilitate soil and improves the quality of soil as soil physical properties are affected by organic fertilizers (Celik et al., 2004). Compost can be used as fertilizer. The formation of compost varies with raw materials and processes. Generally, compost contains farm manure, sewage biosolids, yard waste, dead leaves, etc. It is prepared by microbial action on special organic wastes such as leaves, roots, crop residues, straw, weeds, kitchen waste, sugar factory waste, and marine waste. Aerobic or anaerobic soil microorganisms decompose this organic matter (OM) into the final product known as "Compost." So, organic matters are the key constituent to form compost and cultivate soil. It also contains several chemical substances and acts as precursors for humus formation in soil.

The organic matters decompose under medium high temperature in heaps or pits with adequate moisture. Within 3–6 months amorphous, brawn to black humified compost was obtained. The C:N ratio is one of the important parameters in composting for optimum production. Other factors such as particle size, blending or proportion of raw materials, moisture, aeration, and temperature play important roles in microbial activities (Vyas et al., 1998).

Many fungi decompose these organic matters for energy and growth. Various enzymes released from fungi help in the decomposition of compost. Many cellulolytic and lignolytic fungi are isolated from compost. Mesophilic cellulolytic fungi such as *Aspergillus niger, Aspergillus* sp. (R), *Trichoderma viride*, and *Penicillium* sp. are more rapido Jowar stalk, wheat straw, and Jamun leaves. It is reduced by 1 month and also improves the quality of compost (Gaur et al., 1982). As fungi easily grow in sugar and moisture-rich environments. And compost rich in carbohydrates, protein, organic acids, alkaloids, and other miscellaneous substances such as antibiotics, auxins, vitamins, enzymes, and pigments. It keeps recycling nature. Fungi such as *Trichinella spiralis, Paecilomyces fusisporus, Trichoderma viride,* and *Aspergillus awamori* are used in composting processes. It added 300/tons of material. The initial stage of composting moisture should be maintained at 100%. Rock phosphate was also added (1%) to narrow down the crude protein ratio of the substrate. It makes the soil rich in nutrients as compared to traditional one. As fungi increase composting rates. Thus, fungi play a key role in composting (Vyas et al., 1998).

There are many types of compost such as vermicompost, phosphor-compost, oil cakes, and poultry waste compost. This compost is prepared by three different methods, that is, Indore method, Bangalore method, and Nadep method (Vyas et al., 1998).

Saprophytic fungi from class Basidiomycota are the primary wood decomposer fungi (Hoppe et al., 2016). Also, white rot fungi can play a vital role (pivotal role) in dead wood decomposition due to having lignocellulolytic enzymes that can easily degrade lignin (Van der Wal et al., 2007). Brown rot basidiomycetes also degrade lignin by nonenzymatic mechanism (Martinez et al., 2005), and Ascomycete's fungi also degrade cellulose and hemicelluloses and provide spongy texture to wood (Bani et al., 2018).

Generally, fungal spore suspension is used for composting (Vyas et al., 1998). According to Sikora and Enkiri (2000) suggested that compost combined with sufficient fertilizer is more beneficial. It helps in nitrogen uptake and reduces the accumulation of heavy metal, salt, phosphate, etc., in soil.

(Sikora and Enkiri, 2000). Research suggested that when additives like fly ash, phosphogypsum, jaggery, lime, and polyethylene glycol are used on green waste compost it influences microbial growth, enzyme activities, and degradation of organic matters. Compost containing jaggery and polyethylene glycol was more superior than other additives and it improves the quality of finished compost (Gabhane et al., 2012).

13.7.2 Nitrogen fixation

Nitrogen is the most essential plant nutrient. It is required in huge amounts by the plants. The large percentage of nitrogen is in the form of gas or molecular N_2, which is unavailable to crop plants. So, it must be converted into an available form or in combined nitrogen (NH_4^+). Only a few bacteria can use molecular N_2 directly, whereas plants, animals, and other organisms depend on fixed forms of N_2 (Sumbali and Mehrotra, 2009). This conversion is known as "Nitrogen fixation." Many microorganisms responsible to fix nitrogen are known as "Nitrogen fixer" or "Diazotroph" (Vyas et al., 1998). Nitrogen moves from the atmosphere through many organisms and then back into the atmosphere called "Nitrogen Cycle." This is run by decomposers and various nitrifying organisms, whose nitrogen fixation is one of the important reactions to make it available to plants.

Many bacteria, fungi, cyanobacteria, lichens, etc., can be widely used as nitrogen fixer. They fix nitrogen by symbiotic, nonsymbiotic, semisymbiotic, associative, or independently. It may be aerobic, anaerobic, or microaerophilic.

Almost half of the world population uses rice in their diet. Rice grown in flooded areas where moist conditions are present. Algae require the same condition for growth, the algae fix nitrogen and increase rice yield. Algae such as Azolla, Anabaena, Nostoc, and Calothrix are found in wetland rice. Azolla is a green manure used in Vietnam, China, Philippine, and India. Azolla can accumulate 40–120 kg/ha nitrogen within 30 days. And symbiosis of Azolla and Anabaena also increases rice yield (Tuan and Thuyet, 1979; Liu, 1979). Research also suggested that nitrogen fixing fungi in association with bacteria increase N_2 fixation activity rather than grow alone (Kononkov et al., 1979).

Fungal and bacterial mutual interaction plays a significant role in wood decomposition, especially nitrogen fixing bacteria (NFB) and wood inhabiting fungi (WIF) (Hoppe et al., 2014). As per report basidiomycetes fungi with more biomass decay wood faster in the presence of yeast and N_2-fixing bacteria than in their absence (Blanchette and Shaw, 1978). Generally, fungal communities are influencing bacterial communities. Some pairs of organisms occur frequently while some pairs less frequently occur. Thus, such type of biotic interaction plays a key role in the ecosystem (Odriozola et al., 2021).

Some mycorrhizae are also associated with associative mechanisms. As endosymbiont bacteria exist in AMF (Bertaux et al., 2003). AMF and NFB association in the pieces, *Piptadenia gonoacantha (mart.) macbr.* (Belong to the piptadenia group) showed that AMF and NFB combination increase plant growth and nodulation. As phosphorus is important for nodule formation and AMF affect nodulation. Nitrogen fixation requires energy which are provided by plants. However, the great phosphorus deficiency in tropical soil limits the maximum development of the symbiosis. Thus, increased phosphorus absorption by AMF increases fixation (de Oliveira Junior et al., 2017; Moreira and Siqueira, 2006).

13.7.3 Phosphate solubilization

Next to nitrogen, phosphorus is another key compound in the soil. Phosphorus is a macronutrient required by plants for growth and development. It helps in plant metabolism and in soil microbiological processes including photosynthesis (Sharma et al., 2013). It consists of nucleic acid, phytin, phospholipids, coenzymes, and many inorganic molecules including ATP. Phosphorus supply in the early stage of plant growth is important for early maturity of crops, particularly, for cereals. And it is also essential for seed formation and germination. Phosphorus is found in soil, manure, plant, animal debris, etc., in organic and inorganic forms. Plant absorbs phosphate only in the soluble form such as orthophosphate ions. About 98% of Indian soils are poor in phosphate availability. The phosphorus cycle is the microbial process of mineralization, solubilization, and immobilization in soil. Many microorganisms including fungi play a significant role in this process and make phosphate available.

Fungi convert insoluble phosphate compounds such as rock phosphate, bone meal, basic slag, etc. into soluble form. Fungi secrete many organic acids such as lactic, citric, gluconic, succinic, malic, fumaric acids, etc. (Magnuson and Lasure, 2004; Gaur, 1990). These organic acids can act as chelating agents as well as having acidifying effects. These form complexes with calcium, copper, manganese, iron, aluminum, and nickel salt of phosphate and make it soluble. In addition, fungi mineralize organic phosphorus by enzymatic activities and by production of growth promoting substances that induce plant growth. Fungi such as *A. niger, Aspergillus awamori, Penicillium digitatum, Mucor* spp., and *Schwanniomyces occidentalis* are widely found with phosphate solubilization process. It reduces pH up to 3.1–3.9. *A. niger* can solubilize ferric phosphate by producing H_2S gas. And these H_2S gases reduce ferric phosphate to ferrous sulfide and make phosphate soluble. Some rhizosphere fungi can also solubilize phosphate by releasing organic acids (Whitelaw, 1999; Rashid et al., 2014; Kanse et al., 2015). These fungi are involved in phosphorus cycles and chelate or solubilize phosphate.

Some fungi produce enzymes such as phytase and phosphatase also help in phosphate solubilization. Various fungi like *Aspergillus candidus, A. fumigatus, A. niger, A. parasiticus, A. rugulosus, A. terreus, Penicillium rubrum, P. simplicissimum, Pseudeurotium zonatum, Trichoderm harzianum,* and *T. viride* produces enzymes such as phytase and phosphatase, which helps in phosphate solubilization (Aseri et al., 2009; Narsian and Patel, 2000; Tarafdar et al., 2003). Phospho—compost is another use as a biofertilizer. Microorganisms mixed with farm waste, cattle dung, soil, compost, chopped grasses, crop residue, etc. with rock phosphate. All these are made into a slurry form to provide moisture. This slurry is allowed to decompose for about 60–90 days. Such compost is better than other sources of phosphate for soil reclamation (Vyas et al., 1998).

Weathering of stones and rocks is a continuous process in the ecosystem through physical, chemical, and biological processes (Goudie and Parker, 1999), where microorganisms play an important role in weathering of such stones and rocks. The byproduct of their metabolism is responsible for weathering. Fungi such as lichen produce organic acid, that is, phenolic acid, which is an important solubilizing nutrient from the inorganic substrate (Neaman et al., 2005).

In addition, biofilm is observed in many environments and such biofilm gives promising results in phosphate solubilization of rock phosphate (Seneviratne and Jayasinghearachchi, 2005; Jayasinghearachchi and Seneviratne, 2006). According to Seneviratne and Indrasena (2006) research on eppawala rock phosphate, Sri Lanka with lichen fungi such as *X. mexicana* fused with *B. elkanii* SEMA 5090; the findings showed that biofilm releases the highest amount of phosphate than fungi alone.

Phosphate solubilization does not affect stress environments such as high salinity (1%–3% w/v NaCl) with *T. funiculosum* SLS8, some strains of fungi *Penicillium, Aspergillus, Eupenicillium* spp. can also solubilize phosphate at various salinity levels. Carbon and nitrogen sources and environmental factors also affect phosphate solubilization (Narsian and Patel, 2000; Rinu and Pandey, 2010; Srinivasan et al., 2012). Thus, fungi solubilize insoluble phosphate in various ways. So, it is advisable to use fungi as biofertilizer to overcome macronutrients such as phosphate from soil to sustain agriculture as well as the ecosystem.

13.7.4 Siderophore

The plant and microbes in soil require micronutrients—metals such as nickel, copper, zinc, and iron. These metal elements are often inadequate in the environment. Many microorganisms produce low molecular weight iron-chelating compounds to combat low iron-stress environment known as "Siderophore." It solubilizes and transports iron into cells (Neilands and Leong, 1986; Lankford, 1973).

Lesser iron available, more siderophore production. No system analogs to siderophore have been known for any other metal ion. Thus, making iron unique in requiring such specific ligands said to be "Virtually specific for Fe (III)." Microorganisms growing in aerobic conditions need iron for many functions such as reduction of oxygen for ATP synthesis, reduction of ribotide precursors of DNA, formation of heme, and for optimum growth (at least 1 μM iron require) (Neilands, 1995). Thus, since 1942, when Waring and Werkman documented the microbial requirement of iron, much emphasis has been laid on the role that iron plays in microbial life ranging from respiration to nucleic acid synthesis (Waring and Werkman, 1942). Free iron available in the environment is 10^{-17} M. This is below what required for the optimum growth of microbes (10^{-8} to 10^{-6} M). Microorganisms have developed three sophisticated strategies to make iron available (i) chelation, (ii) reduction of Fe (III) to soluble Fe (II) (de Silva et al., 1996), and (iii) utilization of iron chelated by siderophores from other organisms (Matzanke et al., 1997). Based on their chemical nature, siderophores are classified as Hydroxamates, Catecholates, Carboxylates, Mixed type, and Marinobactins.

Soil is a rich source of microorganisms including fungi. Most fungi produce siderophores under aerobic and iron-stress conditions, except, *Saccharomyces* sp. (Lesuisse and Labbe, 1994; Neilands, 1995). Common soil fungi such as *Aspergillus* and *Penicillium* predominantly produce ferrichrome. Most of the fungi secrete hydroxamate-type siderophores including fusarinines, coprogens, and ferrichromes. Many strains simultaneously produce more than one type of hydroxamate siderophore (Renshaw et al., 2002). Carboxylates are novel siderophores produced exclusively by fungi of class Zygomycota and few bacteria. These coordinate iron with carboxyl and hydroxyl groups (Drechsel et al., 1991). Rhizoferrin, a siderophore produced by zygomycetes, contains two citric acid residues linked to putrescine through two amide bonds. Both constituents—citric acid and putrescine—are common products of primary metabolism and their combination seems to be an easy way of forming a siderophore (Thieken and Winkelmann, 1992).

The putative siderophore producing strains such as *Cunninghamella elegans*, *Penicillium* sp., *P. chrysogenum*, and *P. funiculosum* isolated from the water column; *Monilia* sp. and *Paecilomyces variotii* isolated from mangrove plants surface; and *P. citrinum, Rhizopus* sp., *Syncephalastrum* sp., and *S. racemosum* isolated from sediment (Vala et al., 2000). As might be possible to infer from studies with terrestrial fungi, the four zygomycetes, that is, *C. elegans, Rhizopus* sp., *Syncephalastrum* sp., and *S. racemosum*, were synthesized carboxylate type siderophores (Vala et al., 2000). And, *Paecilomyces variotii* produced both hydroxamate and carboxylate type siderophores. So, it is suggested that carboxylate siderophores may also be produced by fungi outside the Zygomycota (Vala et al., 2000; Holinsworth and Martin, 2009). According to the comparative study of Baakza et al. (2004), fungi belonging to Zygomycota, for example, *C. elegans, Rhizopus* sp., and *Syncephalastrum racemosum* and seven varieties of Ascomycota, for example, *Aspergillus flavus, A. niger, A. ochraceous, A. versicolor, P. chrysogenum, P. citrinum*, and *P. funiculosum* were studied for siderophore production. And for each fungus, siderophore production by a marine isolate was compared to siderophore production by a terrestrial isolate. With the exception of one marine strain of *Aspergillus flavus*, all studied fungi were shown to secrete siderophores of carboxylate type. Further, marine strains produce more siderophores compared to terrestrial ones (Baakza et al., 2004).

Another study of *Aspergillus* from different sources, for instance, marine sources, such as *Aspergillus* sp., *A. nidulans, A. niger, A. ochraceus*, and *A. versicolor*, and five from terrestrial sources, for instance, *A. duricaulis, A. fumigatus, A. niger, A. ochraceus,* and *A. versicolor* was found to secrete hydroxamate-type siderophores (Vala et al., 2006).

Some Ectomycorrhizal fungi found in agricultural soil also produce hydroxamate-type siderophores (Haselwandter, 1995). Endophytic fungi isolated from orchid plants also produce hydroxamate-type siderophores, for example, *Cymbidium aloifolium.*

Thus, siderophore producing fungi could be exploited as biofertilizers too. This is also another way to fertile soil for sustaining agriculture.

13.7.5 Antagonistic activity

Many fungi used to control other pathogenic fungi are known as "Antagonistic fungi." Fungal biocontrol activity was observed in 1943 by Waksman and Horning and they found many fungi distributed in nature and constantly active to kill other soil pathogens (Waksman and Horning, 1943).

These antagonistic fungi are used for seed, soil, foliar, and fruit treatment. Fungi attack harmful insect, weed, or plant disease by secretion of enzymes, toxins, antibiotics, etc. Some common fungi such as *Aspergillus, Penicillium, Fusarium, Trichoderma,* and yeast show antagonistic activity. *Trichoderma* is one of the important fungi that shows antagonistic activity against disease-causing fungal pathogens (Adnan et al., 2019). The Phylloplane fungi of guava such *as A. niger, F. oxysporum*, and *P. citrinum* were shown to have a strong antagonistic effect on *Colletotrichum gloeosporioides* and *Pestalotia psidii* (Pandy et al., 1993). Yeast like *Candida oleophila* and *Debaryomyces hansenii* are effective on citrus fruit (Agrios, 2005). Antagonist effect was also observed with the endophyte group of organisms. These organisms are present inside the plant tissue and protected from environmental stress. For example, *Fusarium* sp. is found in plants to protect against wilt disease (Postma and Luttikholt, 1996).

Antagonists are more effective when applied at the flowering stage than at the fruit maturity stage. Time of application, moisture, and soil condition also affect on effectiveness (Agrios, 2005). Thus, fungi can be exploited to develop antagonism against other phytopathogens in crop plants.

Further, crop rotation, tillage, recycling of crop residue, and soil treatment can be useful to improve soil conditions. Thus, fungi with their vast applications are used as biofertilizers to sustain the environment and agriculture.

References

Adnan, M., Islam, W., Shabbir, A., Khan, K.A., Ghramh, H.A., Haung, Z., Chen, H.Y.H., Lu, G.D., 2019. Plant defence against fungal pathogens by antagonistic fungi with *Trichoderma* in focus. Microb. Pathog. 129, 7–18.

Agrios, G.N., 2005. Control of plant disease. In: Agrios, G.N. (Ed.), Plant Pathology, fifth ed. Elsevier Academic Press, pp. 293–353, https://doi.org/10.1016/C2009-0-02037-6. ISBN: 978-0-12-044565-3.

Aseri, G.K., Jain, N., Tarafdar, J.C., 2009. Hydrolysis of organic phosphate forms by phosphatases and phytase producing fungi of arid and semi-arid soils of India. Am. Eurasian J. Agric. Environ. Sci. 5, 564–570.

Ateia, E.M., Osman, Y.A.H., Meawad, A.E.A., 2009. Effect of organic fertilization on yield and active constituents of *Thymus vulgaris L.* Under North Sinai conditions. Res. J. Agric. Biol. Sci. 5 (4), 555–565.
Baakza, A., Vala, A.K., Dave, B.P., Dube, H.C., 2004. A comparative study of siderophore production by fungi from marine and terrestrial habitats. J. Exp. Mar. Biol. Ecol. 311, 1–9.
Bani, A., Piolia, S., Ventura, M., Panzacchi, P., Borruso, L., Tognetti, R., Tonon, G., Brusetti, L., 2018. The role of microbial community in the decomposition of leaf litter and deadwood. Appl. Soil Ecol. 126, 75–84.
Bertaux, J., Schmid, M., Prevost-Boure, N.C., Churin, J.L., Hartmann, A., Garbaye, J., Frey-Klett, P., 2003. In situ identification of intracellular bacteria related to *Paenibacillus* spp. in the mycelium of the ectomycorrhizal fungus *Laccaria bicolor* S238N. Appl. Environ. Microbiol. 69, 4243–4248.
Bhattacharjee, R., Dey, U., 2014. Biofertilizer, a way towards organic agriculture: a review. Afr. J. Microbiol. Res. 8 (24), 2332–2342.
Blanchette, R.A., Shaw, C.G., 1978. Associations among bacteria, yeast and basidiomycetes during wool decay. Phytopa 68 (4), 631–637.
Celik, I., Ortas, I., Kilic, S., 2004. Effect of compost, mycorrhiza, manure and fertilizer on some physical properties of a chromoxerert soil. Soil Tillage Res. 78 (1), 59–67.
Cheek, M., Lughadha, E.N., Kirk, P., Lindon, H., Carretero, J., Looney, B., Douglas, B., Haelewaters, D., Gaja, E., Llewellyn, T., Ainsworth, A.M., Gafforov, Y., Hyde, K., Crous, P., Hughes, M., Walker, B.E., Forzza, R.C., Wong, K.M., Niskanen, T., 2020. New scientific discoveries, plants and fungi. Plants, People, Planet 2 (5), 371–388.
de Oliveira Junior, J.Q., da Conceicao, J.E., de Faria, S.M., 2017. Nitrogen-fixing bacteria and arbuscular mycorrhizal fungi in *Piptadeniagonoacantha (Mart.) Macbr*. Braz. J. Microbiol. 48 (1), 95–100.
de Silva, D.M., Askwith, C.C., Kaplan, J., 1996. Molecular mechanisms of iron uptake in eukaryotes. Physiol. Rev. 76, 31–47.
Drechsel, H.J., Metzger, S., Freund, S., Jung, G., Boelaert, J.R., Winkelmann, G., 1991. Rhizoferrin—a novel siderophore from the fungus *Rhizopusmicrosporus* var. *rhizopodiformis*. Biometals 4, 238–243.
Dube, H.C., 2013. An Introduction to Fungi, fourth ed. Scientific publishers, India.
Gabhane, J., Prince William, S.P.M., Bidyadhar, R., Bhilawe, P., Anand, D., Vaidya, A.N., Wate, S.R., 2012. Additives aided composting of green waste: effect on organic matter degradation, compost maturity and quality of finished compost. Bioresour. Technol. 114, 382–388.
Gaur, A.C., 1990. Phosphate Solubilizing Microorganisms as Biofertilizer. Omega Scientific publications, New Delhi, p. 176.
Gaur, A.C., Sadasivam, K.V., Mathur, R.S., Magu, S.P., 1982. Role of mesophilic fungi in composting. Agric. Waste 4 (6), 453–460.
Goudie, A.S., Parker, A.G., 1999. Experimental simulation of rapid rock block disintegration by sodium chloride in a foggy coastal desert. J. Arid Environ. 40, 347–355.
Hawksworth, D.L., 2004. Fungal diversity and its implications for genetic resources collections. Stud. Mycol. 50 (1), 9–18.
Holinsworth, B., Martin, J.D., 2009. Siderophore production by marine-derived fungi. Biometals 22 (4), 625–632.
Hoppe, B., Kahl, T., Karasch, P., Wubet, T., Bauhus, J., Buscot, F., Kruger, D., 2014. Network analysis reveals ecological links between N-fixing bacteria and wood-decaying fungi. PloS One 9 (3), e91389.
Hoppe, B., Purahong, W., Wubet, T., Kahl, T., Bauhus, J., Arnstadt, T., Hofrichter, M., Buscot, F., Kruger, D., 2016. Linking molecular deadwood-inhabiting fungal diversity and community dynamics to ecosystem functions and processes in Central European forests. Fungal Divers. 77, 367–379.
Jayasinghearachchi, H.S., Seneviratne, G., 2006. Fungal solubilization of rock phosphate is enhanced by forming fungal- rhizobial biofilms. Soil Biol. Biochem. 38, 405–408.
Kanse, O.S., Whitelaw-Weckert, M., Kadam, T.A., Bhosale, H.J., 2015. Phosphate solubilization by stress-tolerant soil fungus *Talaromycesfuniculosus* SLS8 isolated from the Neemrhizosphere. Ann. Microbiol. 65, 85–93.
Kavimandan, S.K., 1993. Use of biofertilizer in wheat. In: Wheat Workshop. vol. 5.

Kirk, P.M., Cannon, P.F., David, J.C., Stalpers, J., 2001. Ainsworth and Bisby's Dictionary of the Fungi, ninth ed. CAB International, Wallingford, UK.

Kirk, P.M., et al., 2008. Ainsworth and Bisoy's Dictionary of the Fungi, 10th ed. C A B International, Oxon, UK.

Kononkov, E.P., Umarov, M.M., Mirchink, T.G., 1979. Nitrogen-fixing fungal associations with bacteria. Mikrobiologiia 48, 734–737.

Lankford, E., 1973. Bacterial assimilation of iron. Crit. Rev. Microbiol. 2, 273–331.

Lesuisse, E., Labbe, P., 1994. Reductive iron assimilation in *Saccharomyces cerevisiae*. In: Winkelmann, G., Winge, D.R. (Eds.), Metal Ions in Fungi. Marcel Dekker, New York, pp. 149–178.

Liu, C., 1979. Use of Azolla in rice production in China. In: International Rice Research Institute (Ed.), Nitrogen & Rice. International Rice Research Institute, Los Banson, Philippines, pp. 375–394.

Magnuson, J.K., Lasure, L.L., 2004. Organic acid production by filamentous fungi. In: Tkacz, J.S., Lange, L. (Eds.), Adv Fungal Biotechnolindagric Med. Springer US, Boston, MA, pp. 307–34010.

Maheswari, U.N., Kalaiyarasi, M., 2015. Comparative study of liquid bio-fertilizer and carrier-based bio-fertilizer on green leafy vegetables. Int. J. Pharm. Sci. Rev. Res. 33 (1), 229–232. Arti no. 42.

Martinez, A.T., Speranza, M., Ruiz-Duenas, F.J., Ferreira, P., Camarero, S., Guillen, F., Martinez, M.J., Gutierrez, A., del Rio, J.C., 2005. Biodegradation of lignocellulosics: microbial, chemical, and enzymatic aspects of the fungal attack of lignin. Int. Microbiol. 8, 195–204.

Matzanke, B.F., Bohnke, R., Mollmann, U., Reissbrodt, R., Schunemann, V., Trautwein, A.X., 1997. Iron uptake and intracellular metal transfer in mycobacteria mediated by xenosiderophores. Biometals 10, 193–203.

Moreira, F.M.S., Siqueira, J.O., 2006. Microbiologia e Bioquimica do Solo, second ed. Editora UFLA, Lavras.

Mukherjee, G., Mishra, T., 2017. Fungal pigments: an overview. In: Developments in Fungal Biology & Applied Mycology, pp. 525–541, https://doi.org/10.1007/978-981-10-4768-8_26.

Narsian, V., Patel, H., 2000. *Aspergillus aculeatus* as a rock phosphate solubilizer. Soil Biol. Biochem. 32, 559–565.

Neaman, A., Chorover, J., Brantley, S.L., 2005. Implication of the evolution of organic acid moieties for basalt weathering over ecological time. Am. J. Sci. 305, 147–185.

Neilands, J.B., 1995. Siderophores: structure and function of microbial iron transport compounds. J. Biol. Chem. 270 (45), 26723–26726.

Neilands, J.B., Leong, S.A., 1986. Siderophores in relation to plant growth and disease. Annu. Rev. Plant Physiol. 37, 187–208.

Odriozola, I., Abrego, N., Tlaskal, V., Zrustova, P., Morais, D., Vetrovsky, T., Ovaskainen, O., Baldrian, P., 2021. Fungal communities are important determinants of bacterial community composition in deadwood. Am. Soc. Microbiol. 6 (1), e01017-20. https://doi.org/10.1128/mSystems.01017-20.

Pandy, P.R., Arora, D.K., Dube, R.C., 1993. Antagonistic interaction between fungal pathogens and phylloplane fungi of guava. Mycopathologia 124, 31–39.

Paul, D., Nair, S., 2008. Stress adaptation in a plant growth promoting Rhizobacterium (PGPR) with increasing salinity in the coastal agricultural soils. J. Basic Microbiol. 48, 1–7.

Postma, J., Luttikholt, A.J.G., 1996. Colonization of carnation stems by a nonpathogenic isolate of *Fusarium oxysporum* and its effect on wilt caused by *Fusarium oxysporumf. sp. dianthi*. Can. J. Bot. 74, 1841–1851.

Rashid, K.I., Ahmmed, S.J., Mahmood-Mukhtar, Z.F., 2014. Study the antibacterial activity of Bauhinia variegata Linn. plant leaf extracts against some species of pathogenic bacteria. J. Al-Nahrain Univ. 17 (1), 55–59.

Renshaw, J.C., Robson, G.D., Trinci, A.P.J., Wiebe, M.G., Livens, F.R., Collinson, D., Taylor, R.J., 2002. Fungal siderophores: structures, functions and applications. Mycol. Res. 106 (10), 1123–1142. https://doi.org/10.1017/S0953756202006548.

Rinu, K., Pandey, A., 2010. Temperature-dependent phosphate solubilization by cold and pH-tolerant species of *Aspergillus* isolated from Himalayan soil. Mycoscience 51 (4), 263–271.

Saccardo, P.A., 1886. In: Fungi Italici Autographic Delineation: Discomycetes. Sumpt. Auctoris, p. 14.

Schmit, J.P., Mueller, G.M., Huang, Y., Leacock, P.R., Mata, J.L., Wu, Q., Haung, Y., 2005. Assessment of tree species richness as a surrogate for macrofungal species richness. Biol. Conserv. 121 (1), 99–110.

Schuessler, A., Gehrig, H., Schwarzott, D., Walker, C., 2001. Analysis of partial Glomeles SSU rRNA gene sequences: implications for primer design and phylogeny. Mycol. Res. 105, 5–15.

Seneviratne, G., Indrasena, I.K., 2006. Nitrogen fixation in Lichens important for improved rock weathering. J. Biosci. 31 (5), 639–643.

Seneviratne, G., Jayasinghearachchi, H.S., 2005. A rhizobial biofilm with nitrogenase activity alters nutrient availability in a soil. Soil Biol. Biochem. 37, 1975–1978.

Sharma, S.B., Sayyed, R.Z., Trivedi, M.H., Gobi, T.A., 2013. Phosphate solubilizing microbes: sustainable approach for managing phosphorus deficiency in agricultural soils. Springerplus 2, 587.

Sikora, L.J., Enkiri, N.K., 2000. Efficiency of compost—fertilizer blends compared with fertilizer alone. Soil Sci. 165 (5), 444–451.

Sparrow, F.K., 1958. Interrelationships and phylogeny of the aquatic phycomycetes. Mycologia 50 (6), 797–813.

Srinivasan, R., Yandigeri, M.S., Kashyap, S., Alagawadi, A.R., 2012. Effect of salt on survival and P-solubilization potential of phosphate solubilizing microorganisms from salt affected soils. Saudi J. Biol. Sci. 19, 427–434.

Subba Rao, N.S., 1986. Soil Microorganisms and Plant Growth, second ed. Oxford and IBH publishing Co, New Delhi.

Sumbali, G., Mehrotra, R.S., 2009. Principles of Microbiology. Tata McGraw-Hill education Pvt Ltd, New Delhi.

Tarafdar, J.C., Bareja, M., Panwar, J., 2003. Efficiency of some phosphatase producing soil-fungi. Indian J. Microbiol. 43, 27–32.

Thieken, A., Winkelmann, G., 1992. Rhizoferrin: a complexone type siderophore of the Mucorales and entomophthorales (Zygomycetes). FEMS Microbiol. Lett. 73, 37–41.

Tuan, D.T., Thuyet, T.R., 1979. The use of Azolla in rice production in Vietnam. In: International Rice Research Institute (Ed.), Nitrogen & Rice. International Rice Research Institute, Los Banson, Philippines, pp. 395–504.

Vala, A.K., Vaidya, S.Y., Dube, H.C., 2000. Siderophore production by facultative marine fungi. Ind. J. Mar. Sci. 29, 339–340.

Vala, A.K., Dave, B.P., Dube, H.C., 2006. Chemical characterization and quantification of siderophores produced by marine and terrestrial *Aspergilli*. Can. J. Microbiol. 52, 603–607.

Van der Wal, A., De Boer, W., Smant, W., Van Veen, J.A., 2007. Initial decay of woody fragments in soil is influenced by size, vertical position, nitrogen availability and soil origin. Plant Soil 301, 189–201.

Vyas, S.C., Vyas, S., Vyas, S., Modi, H.A., 1998. Biofertilizers and Organic Farming. Aktaprakashan, Nadiad, Gujarat, India.

Waksman, S.A., Horning, E.S., 1943. Distribution of antagonistic fungi in nature & their antibiotic action. Mycologia 3 (1), 47.

Waring, W.S., Werkman, C.N., 1942. Growth of bacteria in an iron-free medium. Arch. Biochem. 1, 303–310.

Whitelaw, M.A., 1999. Growth promotion of plants inoculated with phosphate-solubilizing fungi. Adv. Agron. 69, 99–151.

CHAPTER

14

Nutraceutical metabolites, value addition and industrial products for developing entrepreneurship through edible fleshy fungi

Rakesh Pandey[a], Vaibhav Sharan Pandey[a,b], and Vashist Narayan Pandey[a]

[a]*Experimental Botany and Nutraceutical Lab, Department of Botany, DDU Gorakhpur University, Gorakhpur, Uttar Pradesh, India,* [b]*Plant Resource Centre, Department of Botany, SVM Mahila Mahavidyalaya, Arya Nagar, Gorakhpur, Uttar Pradesh India*

Abbreviations

ABTS	2,2′-azino-bis(3-ethylbenzothiazoline-6-sulfonic acid)
ACE	angiotensin converting enzyme
AkP	alkaline serine protease
Bcl-2	B-cell lymphoma 2 protein
Bcl-XL	B-cell lymphoma—extra-large protein
CA46	Burkitt's lymphoma cell line
CDH	cellobiose dehydrogenase
DPPH	2,2-diphenylpicrylhydrazyl
GAE	gallic acid equivalents
HepG2	cancer cells
HL-60	promyelocytic leukemia cell line
IC_{50}	half maximal inhibitory concentration
IFN-γ	interferon-γ
iNOS	inducible nitric oxide synthase
JAK-STAT1	Janus kinase/signal transducer and activator of transcription-1
LPS	lipopolysaccharide
MAPK	mitogen activated protein kinases
MDA	malondialdehyde
NF-κB	nuclear factor kappa B
NO	nitric oxide
NPEP	*Pleurotus eryngii* polysaccharide-2
OCTN1	organic cation transporter novel-type 1
RAW264.7	macrophage-like cell line
SCFA	short chain fatty acids
TLR4	toll-like receptor 4

Entrepreneurship with Microorganisms. https://doi.org/10.1016/B978-0-443-19049-0.00010-4

U937 cell lines	pro-monocytic human myeloid leukemia cell line
VEGFR-1	vascular endothelial growth factor receptor-1
VEGFR-2	vascular endothelial growth factor receptor-2
WPEP	*Pleurotus eryngii* polysaccharide-1

14.1 Introduction

Edible fleshy fungi (EFF) are alternative underutilized source of food and medicine next to plants. EFF can be defined as group of edible fungi high in proteins, carbohydrates, and fibers, rich in unsaturated fatty acids, vitamins and minerals, low in calories, fat, and sodium and free from cholesterol. The proteins of EFF contain all nine essential amino acids essential for human beings. Among these essential amino acids, lysine is present abundantly, while methionine and tryptophan are present in low quantity (Rahi and Malik, 2016). Some prominently edible forms are species of *Agaricus*, *Auricularia*, *Boletus*, *Calocybe*, *Cantharellus*, *Craterellus*, *Flammulina*, *Ganoderma*, *Grifola*, *Hericium*, *Laccaria*, *Lentinula*, *Lycoperdon*, *Morchella*, *Pleurotus*, *Russula*, *Volvariella*, *Termitomyces,* etc. (Pandey et al., 1993; Chaurasia, 2004; Srivastava et al., 2011; Li et al., 2021). A total of 2006 species are considered as safely consumed EFF species. The edibility of EFF was probably decided on a trial-and-error basis by early civilizations. Fungal species that was palatable without causing trouble was considered as edible and those which caused some adverse reactions or taste were considered as nonedible. Processing of edible fungi may also change their aroma, texture, and palatability; however, safety and suitability must be the main criteria for edibility of any fleshy fungi (Li et al., 2021). Edibility of some EFF with their parts used is shown in Table 14.1. Asia has largest number of EFF species followed by Europe and North America. Out of 2006 EFF, only a few are cultivated and consumed on a wide scale throughout the world. EFF are cultivated mainly for food, medicine, and other industrial purposes. On a global scale, *Lentinula edodes*, *Auricularia* spp., *Pleurotus ostreatus*, *Agaricus bisporus*, *Flammulina velutipes*, *Pleurotus eryngii*, and *Volvariella volvacea* cover the major share in cultivation and production (Singh et al., 2020; Li et al., 2021). EFF are well known for their primary and secondary metabolites. These metabolites are responsible for their nutritional, medicinal, and flavor qualities. The metabolites of EFF are prized for their immunomodulatory, anticancer, antioxidant, neuroprotective, hepatoprotective role, etc., and in this respect, they can be considered as "nutraceuticals," "myco-nutraceuticals," or "mycoceuticals." EFF are generally grown on dead organic matters and are full of diverse range of ligninolytic and cellulolytic enzymes. These enzymes are valuable tools in food processing and functional food development. Also, global food security issues, search for new alternative sources for food, necessity for new effective therapeutic agents, and income generation for local people provide a large-scale opportunity for industrial entrepreneurship. The aim of present work is to describe nutraceutical metabolites of EFF, value addition, and industrial products for developing entrepreneurship.

14.2 Nutraceuticals

The term "Nutraceutical" was introduced in 1989 by Stephen De Felice, Founder and Chairman of the Foundation for Innovation in Medicine at Cranford, New Jersey, USA. It is a combination of two words "nutrition" and "pharmaceuticals" demonstrating health benefits of foods or isolated substances from the foods (Gupta et al., 2012; Rahi and Malik, 2016). The concept of nutraceutical and new substances

Table 14.1 Edibility of fleshy fungi collected.

S. no.	Fungi	Growth stage	Parts used	Edibility	Reference
01.	*Agaricus arvensis* Schaeff.	Elongation stage	Stipe + Pileus	Excellent	Pandey and Srivastava (2004)
02.	*A. bisporus* (Lange). Imbach.	Elongation stage	Stipe + Pileus	Very good	Chaurasia (2004)
03.	*A. bitorquis* (Quél.) Sacc.	Elongation stage	Stipe + Pileus	Excellent	Pandey et al. (1993)
04.	*A. brunnescens* Peck	Elongation stage	Stipe + Pileus	Excellent	Pandey et al. (1993)
05.	*A. campestris* L.	Elongation stage	Stipe + Pileus	Excellent	Ram et al. (2010)
06.	*Auricularia polytricha* (Mont.) Sacc.	Mature stage	Full fruit body	Good	Pandey and Srivastava (2004)
07.	*Calocybe indica* Purkay. & A. Chandra	Mature stage	Stipe + Pileus	Good	Ram et al. (2010)
08.	*Cantharellus cibarius* Fr.	Mature stage	Stipe + Pileus	Good	Pandey and Srivastava (2004)
09.	*Craterellus tubaeformis* (Fr.) Quél./*Cantharellus infundibuliformis* (Scop.) Fr./Cantharellus tubaeformis Fr.	Mature stage	Stipe + Pileus	Very good	Pandey and Srivastava (2004)
10.	*Flammulina velutipes* (Curtis) Singer	Mature stage	Stipe + Pileus	Good	Pandey et al. (1993)
11.	*Lentinula edodes* (Berk.) Pegler/*Lentinus edodes* (Berk.) Singer	Elongation stage	Stipe + Pileus	Very good	Pandey et al. (1993)
12.	*Lentinus sajor-caju* (Fr.) Fr./*Pleurotus sajor-caju* (Fr.) Singer	Mature stage	Stipe + Pileus	Very good	Pandey et al. (1993)
13.	*Lycoperdon giganteum* Batsch ex. Pers./Clavatia gigantea (Batsch) Llyod	Young stage	Button	Edible while young	Pandey et al. (1993)
14.	*L. perlatum* Pers.	Young stage	Button	Edible while young	Pandey et al. (1993)
15.	*L. pyriforme* (Schaeff.) Pers./Apioperdon pyriforme (Schaeff.) Vizzini	Young stage	Button	Excellent when flesh is white and hard	Pandey et al. (1993)
16.	*Morchella deliciosa* Fr.	Mature stage	Stipe + Pileus	Excellent	Pandey et al. (1993)
17.	*M. esculenta* (L.) Pers./Morchella conica Pers.	Mature stage	Stipe + Pileus	Very good	Chaurasia (2004), Pandey and Srivastava (2004)
18.	*Pleurotus eryngii* (Dc ex Fr.) Quel	Mature stage	Pileus	Very good	Pandey et al. (1993)
19.	*P. flabellatus* (Berk.) Br. Sacc.	Mature stage	Stipe + Pileus	Good	Pandey et al. (1993)
20.	*P. ostreatus* (Jacq. Fr. Kummer)	Mature stage	Stipe + Pileus	Excellent	Pandey et al. (1993)

Continued

Table 14.1 Edibility of fleshy fungi collected—cont'd

S. no.	Fungi	Growth stage	Parts used	Edibility	Reference
21.	*P. florida* Eger/*Pleurotus pulmonarius* (Fr.) Quel	Mature stage	Stipe + Pileus	Very good	Pandey et al. (1993)
22.	*Termitomyces clypeatus* R. Heim	Elongation stage	Stipe + Pileus	Excellent	Pandey et al. (1993), Chaurasia (2004), Srivastava et al. (2011), Srivastava et al. (2012)
23.	*T. heimii* Natarajan	Elongation stage	Stipe + Pileus	Excellent	Pandey et al. (1993), Chaurasia (2004), Srivastava et al. (2011), Srivastava et al. (2012)
24.	*T. mammiformis* R. Heim	Elongation stage	Stipe + Pileus	Very good	Pandey et al. (1993), Chaurasia (2004), Srivastava et al. (2011), Srivastava et al. (2012)
25.	*T. robustus* (Beeli) R. Heim	Elongation stage	Stipe + Pileus	Very good	Pandey et al. (1993), Srivastava et al. (2011), Srivastava et al. (2012)
26.	*Volvariella diplasia* (Berk& Broome) Singer	Button stage	Stipe + Pileus	Good	Pandey et al. (1993)
27.	*V. esculenta* (Massee) Singer	Button stage	Stipe + Pileus	Very good	Pandey et al. (1993)
28.	*V. volvacea* (Bull: Fr.) Singer	Button stage	Stipe + Pileus	Excellent	Pandey et al. (1993)

in food has given a new dimension that can improve biofunctions, health, and health benefits in total. Nowadays, the present world has focused toward plants and fungal food as source of bioenhancers. These products have been devised variously as functional food, dietary supplements, vitamins, minerals, nutraceuticals, and so on (Rahi and Malik, 2016).

Fleshy fungi, wild or cultivated, have been used for the production of food ingredients, feed, enzymes, and other metabolites of nutritional and medicinal importance for generations. Their use has been mentioned for the preparation of various folk medicine, e.g., Somras, an ancient traditional medicine consumed for joy and happiness by the angels mentioned in the Vedas. Wide range of primary and secondary metabolites in EFF has armored them with powerful nutraceutical potential to combat with many modern day's diseases. EFF have polysaccharides, proteins, peptides, terpenes, phenolic compounds, and polyunsaturated fatty acids that have potential for disease treatment and prevention (Ma et al., 2018). The metabolites from EFF and their nutraceutical activities are shown in Table 14.2.

14.2.1 Primary nutraceuticals

Edible fleshy fungi are known for their unique range of phytomolecules in biological world. The spectrum of metabolites ranges from primary metabolites including proteins, carbohydrates, lipids, organic acids, nucleic acids, vitamins, and minerals to secondary metabolites derived from the primary ones.

14.2.1.1 Carbohydrates

EFF are a reservoir of different types of polysaccharides with known therapeutic effects. The most important EFF polysaccharides that are commercialized belong to β-glucan family. In addition to β-glucans EFF also have α-glucans and α,β-glucans. β-glucans are polymer of D-glucose units joined through β-glycosidic bonds. These polysaccharides may have linear or branched structure. Some of them are named after the source of their identification and isolation like lentinan (*L. edodes*; branched polymer), pleuran (*P. ostreatus*; branched polymer), schizophyllan (*Schizophyllum commune*; branched polymer), grifolan (*Grifola frondosa*; branched polymer), and Krestin (*Trametes versicolor/Coriolus versicolor*; branched polymer) (Maity et al., 2021; Novak and Vetvicka, 2008; Tsukagoshi et al., 1984). These and other polysaccharides fractions have shown immunomodulatory, antitumor, and antioxidant activity. The polysaccharide fraction from *P. eryngii* residues was shown to have antiaging and antioxidant potential. The monosaccharide composition of polysacharide revealed five different monosaccharides with high proportion of xylose (34.91%) and arabinose (23.31%) (Zhang et al., 2021). Two polysaccharide fractions WPEP and NPEP from *P. eryngii*, chiefly composed of glucose subunits, have shown anti-inflammatory potential against Lipopolysaccharide challenged RAW264.7 cells. The antiinflammatory effect was due to suppression of MAPK and NF-κB signaling via inhibition of p38 phosphorylation (Ma et al., 2020). The carbohydrate-rich aqueous extract from *Termitomyces clypeatus* demonstrated antioxidant and anticancer activity against all the tested cancer cell lines especially against U937 cell lines (Mondal et al., 2016).

Lentinan, a β-1,3 glucan having β-1,6 glucose branches from *L. edodes*, has been reported to have chemotherapeutic properties against several types of cancers, viz. liver, gastric, lung, breast, etc. (Trivedi et al., 2022). It has shown immunomodulatory and antioxidant activity in lipopolysaccharide primed bovine mammary epithelial cells (Meng et al., 2022). A polysaccharide from *L. edodes* mycelium demonstrated protective effects on high glucose challenged INS-1 cell models. The polysaccharide reduced oxidative stress, MDA level, cytotoxicity and apoptosis. Cell signaling studies

Table 14.2 Metabolites of edible fleshy fungi and their actions.

S no.	Name of fungal species	Metabolites	Mechanism of action/activities	References
01.	*Agaricus bisporus* (Lange). Imbach.	Mannitol, oxalic acid, malic acid, gallic acid, α-tocopherol, γ-tocopherol, δ-tocopherol, ergosterol	Antioxidant, antimicrobial	Stojković et al. (2014)
02.	*A. blazei* Murrill	Ergosterol, blazein, agaritine, blazeispirols, agarol	Anticancer	Misgiati et al. (2021), Itoh et al. (2008), Akiyama et al. (2011), Hirotani et al. (2002), Shimizu et al. (2016)
03.	*A. brasiliensis* Wasser, M. Didukh, Amazonas & Stamets	Lovastatin, GABA, cinnamic acid, oxalic acid, fumaric acid, p-coumaric acid, mannitol, trehalose, α-tocopherol, γ-tocopherol, ergosterol	Antioxidant, antibacterial	Lo et al. (2012), Stojković et al. (2014)
04.	*Auricularia auricula-judae* (Bull.) Quel.	β-glucans, melanin, catechin, chlorogenic acid, epicatechin, rutin and quercetin	Antioxidant, immunomodulatory, hypoglycemic, antitumor	Liu et al. (2021)
05.	*Auricularia nigricans* (Sw.) Birkebak, Looney & sánchez-García/*Auricularia polytricha* (Mont.) Sacc.	GABA, ceramide, cerevisterol, 9-hydroxycerevisterol, cerebroside	Antinociceptive	Lo et al. (2012), Koyama et al. (2002)
06.	*Boletus aereus* Bull.	Mannitol, trehalose, tocopherol, ascorbic acid, p-hydroxybenzoic acid, p-coumaric acid, cinnamic acid	Antioxidant	Heleno et al. (2011)
07.	*Boletus edulis* Bull.	Lovastatin, GABA, ergothioneine, boledulins A, B, C, protocatechuic acid, homogentisic acid, pyrogallol, gallic acid, p-catechin, dihydroxybenzoic acid	Cytotoxic, anticancer, antimicrobial, antibiofilm activity	Lo et al. (2012), Feng et al. (2011), Garcia et al. (2022)
08.	*Calocybe indica* Purkay. & A. Chandra	3-Octanol, 1-octen-3-ol, n-octanol, 3-octanone, 1-octen-3-one, t-linalool oxide, polysaccharide fraction	Flavor compounds, antioxidant, immunostimulatory	Subbiah and Balan (2015), Ghosh et al. (2021)
09.	*Cantharellus cibarius* Fr.	Cibaric acid, ergosterol, glycerol tridehydrocrepenynate, Glycerol 1,3-dilinoleate	Antimicrobial, insecticidal	Daniewski et al. (2012)

10.	*Craterellus tubaeformis* (Fr.) Quél./*Cantharellus infundibuliformis* (Scop.) Fr.	Butanal, pentanal, 1-octene, octane, 3-cyclohepten-1-one, α-pinene, 1-octen-3-one	Aroma volatile compounds	Aisala et al. (2019)
11.	*Cyclocybe cylindracea* (DC.) Vizzini & Angelini/*Agrocybe cylindracea* (DC.) Maire	Lovastatin, GABA, ergothioneine, polysaccharides fraction, fructose, mannitol, trehalose, glutamic acid	Antiobesity, antiinflammatory, taste	Lo et al. (2012), Zhu et al. (2022), Mau and Tseng (1998)
12.	*Flammulina velutipes* (Curtis) Singer	Flamvelutpenoids, flammufuranones, flammuspirones, sterpurols, sterpuric acid, flammulinolides, enokipodins	Antibacterial, HMG-CoA reductase inhibition, cytotoxic, antioxidant	Fukushima-Sakuno (2020), Wang et al. (2012), Tao et al. (2016)
13.	*Ganoderma lucidum* (Curtis) P. Karst.	Lovastatin, GABA, ganoderal A, lucialdehyde A, ganoderic aldehyde TR, ganoderol A, B, ganoderic acids, lucideric acids, lucidenic lactone, lucidones, lingzhiol, ganofuran B	Anti-tumor, immunomodulation, antioxidant, anti-neurodegerative	Lo et al. (2012), Sharma et al. (2019)
14.	*Grifola frondosa* (Dicks.) Gray	Grifolaone A, riboflavin, folate, pantothenic acid, niacin, coumarin	Antimicrobial, nutritional, intestinal flora modulatory activity	He et al. (2016), Sato et al. (2017), Wang et al. (2021)
15.	*Hericium erinaceus* (Bull.) Pers.	Erinacines, hericerins, hericenones, resorcinols, erinaceolactones, ergosterol	Anticarcinogenic, antibiotic, neuroprotective	Friedman (2015)
16.	*Inonotus obliquus* (Fr.) Pilát	Lanosterol, inotodiol, trametenolic acid, inonotusoxide A, B, inonotusoriol A, D, E, inonotusic acid, inotolactone A, B, chagabusone-A, inonobilin, hispidin, interfunginn, phelligridin	Anti-tumor, antioxidant	Zhao and Zheng (2021), Zheng et al. (2010)
17.	*Lentinula edodes* (Berk.) Pegler/*Lentinus edodes* (Berk.) Singer	Lovastatin, GABA, eritadenine, LEPS1 proteoglycan, lentinan, lentinamycin	Anti-inflammatory, antioxidant, antiapoptotic, chemotherapeutic, antimicrobial	Lo et al. (2012), Kaya and Cam (2022), Zhang et al. (2023), Meng et al. (2022), Trivedi et al. (2022), Fukushima-Sakuno (2020)
18.	*Lentinus sajor-caju* (Fr.) Fr./*P. sajor-caju* (Fr.) Singer/	β-1,3-glucanoligosaccharide, linoleic acid, chlorogenic acid, vanillic acid, myricetin, 4-hydroxymethylbenzoic acid	Osteoblastogenesis, antioxidant, antimicrobial	Yodthong et al. (2020), Krümmel et al. (2022)

Continued

Table 14.2 Metabolites of edible fleshy fungi and their actions—cont'd

S no.	Name of fungal species	Metabolites	Mechanism of action/activities	References
19.	*Morchella esculenta* (L.) Pers./*Morchella conica* Pers.	Lovastatin, GABA, linoleic acid, ethyl linoleate, β-sitosterol; ergosta-5,7,22-triene-3β-ol; 6,22-diene-3-hydroxy-5,8-epidioxy ergosta; 2-(2-hydroxypropanamido) benzamide; (2,4-dichlorophenyl)-2,4-dichlorobenzoate; 1-O-octadecanoyl-sn-glycerol; (3β,5α,22E)-Ergosta-7,22,24(28)-trien-3-ol; galactomannan; 5-dihydroergosterol; ergosterol peroxide; ergosterol; cerevisterol; trilinolein; Methyl myristate; 1-linoleoylglycerol, ceramide	Lung cancer, immunostimulatory, antioxidant, suppression of NF-κB activation	Lo et al. (2012), Yang et al. (2019), Lee et al. (2018), Sunil and Xu (2022), Kim et al. (2011a)
20.	*Pleurotus citrinopileatus* Singer	Lovastatin, GABA	HMG-CoA reductase inhibitor, Neurological disorders	Lo et al. (2012), Möhler (2012)
21.	*P. eryngii* (Dc ex Fr.) Quel	Strophasterol E and F, steroids pleurocins A and B, eringiacetal B, ergostane type steroids, bisabolane type sesquiterpenes, catechin, epicatechin, chlorogenic acid, vanillic acid, ferulic acid, sinapic acid, β-carotene, α-tocopherol and δ-tocopherol	Inhibition of NO production (macrophage inhibition); aromatase inhibitory activity, antioxidant, antiinflammatory	Kikuchi et al. (2019), Kikuchi et al. (2017a,b), Kikuchi et al. (2016), Kikuchi et al. (2018), Lin et al. (2014)
22.	*P. ostreatus* (Jacq. Fr. Kummer)	Lovastatin, GABA, ergothioneine, hydrophobin Vmh3-1, polysaccharide fractions	HMG-CoA reductase inhibitor, neurological disorders, antitumor activity	Lo et al. (2012), Kulkarni et al. (2022), Corrêa et al. (2016)
23.	*P. pulmonarius* (Fr.) Quel /*P. florida* Eger	Ergothioneine, lovastatin, β-glucan	Antiinflammatory	Chilanti et al. (2022), Corrêa et al. (2016)
24.	*Suillus bellini* (Inzenga) Kuntze	Suillin, ergosterol, D-mannitol, D-sorbitol, fumaric acid, sucrose, glutamic acid, isoleucine, leucine, threonine, tyrosine, valine, alanine	Essential dietary supplements and flavor enhancing agents, anticancer	Venditti et al. (2017), Liu et al. (2009)
25.	*Termitomyces clypeatus* R. Heim	Cellulase, amylase, amyloglucosidase, aqueous fraction rich in xylitol, raffinose, sorbitol, arabitol, inositol, ribitol, ascorbic acid, thermostable β-glucosidase, serine protease (AkP), cellobiose dehydrogenase, phosphoketolase, laccase	Cellulolytic, amylolytic, xylanolytic, antioxidant, antitumor, transglycosylation activity by β-glucosidase, apoptosis, pentose sugar utilization, ligninolytic	Jonathan and Adeoyo (2011), Ghosh et al. (1997), Mondal et al. (2016), Pal et al. (2010), Majumder et al. (2016), Saha et al. (2008), Sarkar and Roy (2014), Bose et al. (2007)

26.	*T. heimii* Natarajan	β-glucan with (1 → 6)-linked β-D-glucopyronosyl backbone, heteroglycan, tannic acid, gallic acid, protocatechuic acid, gentisic acid, coumaric acid, ergosterol, ergosta-5,8-dien-3-ol	Antioxidant, anti-hepatocarcinoma activity	Manna et al. (2015), Maity et al. (2020), Puttaraju et al. (2006), Ray et al. (2022)
27.	*T. mammiformis* R. Heim	Tannic acid, gallic acid, protocatechuic acid, gentisic acid, high mineral content	Antioxidant	Puttaraju et al. (2006), Kansci et al. (2003)
28.	*T. robustus* (Beeli) R. Heim	Water solube and insoluble β-glucans, water soluble fucoglucan	Immunostimulatory properties	Bhanja et al. (2012), Mondal et al. (2008)
29.	*Tricholoma matsutake* (S. Ito & S. Imai) Singer	Matsutakone, matsutoic acid, 3-octanol, 1-octen-3-ol, €-2-octen-1-ol, 3-octanone, octanoic acid, junipene	Acetylcholinesterase inhibitory activity, nutritional	Zhao et al. (2017), Cho et al. (2006)
30.	*Tremella fuciformis* Berk.	Polysaccharide fraction, 4-hydroxybenzoic acid, gentisic acid, 4-coumaric acid	Anti-colitis, antioxidant, antiinflammatory	Xu et al. (2021), Li et al. (2014)
31.	*Tuber aestivum* (Wulfen) Spreng	Trehalose, myo-inositol, mannitol, choline, malic acid, citric acid, fumaric, succinic, acetic acid, ergosterol, brassicasterol, palmitic acid, stearic acid, linoleic acid, propanedioic acid, bis (trimethylsilyl) ester, 2,6-dimethyl-4-nitrosophenol, salutaridinol, naringenin-7-O-neohesperidoside	Nutritional, antioxidant, anti-angiogenic, antiinflammatory	Mannina et al. (2004), Marathe et al. (2020)
32.	*Volvariella diplasia* (Berk& Broome) Singer	Hot aqueous extract polysaccharide fraction-II (glucose only); water soluble polysaccharide fraction I (D-glucose, D-mannose, D-galactose)	Nutritional	Ghosh et al. (2008)
33.	*V. volvacea* (Bull: Fr.) Singer	Lovastatin, GABA, ergothioneine, limonene, octa-1,5-dien-3-ol, 3-octanol, 1-octen-3-ol, 1-octanol, 2-octen-1-ol, mannitol, trehalose	Nutritional and flavoring agents	Lo et al. (2012), Mau et al. (1997)

demonstrated downregulation of p38 MAPK and NF-κB pathways while activation of Nrf2 signaling cascade (Cao et al., 2019). Similarly, polysaccharides present in aqueous extract of *L. edodes* demonstrated HMG-CoA reductase inhibition and modulated expression of genes participating in cholesterol metabolism. The active principle was found to be α- and β-glucans and fucomannogalactans (Gil-Ramírez et al., 2016). Glucanoligosaccharide from *Pleurotus sajor-caju* has demonstrated osteogenic activity on MC3T3-E1 model cell lines. Higher bone resorption to formation ratio causes osteoporosis, which leads to bone fragility and fracture. Enzymatic treatment of β-glucan from *P. sajor-caju* yields glucanoligosaccharide that enhances the expression of bone morphogenetic protein-2 (BMP-2), runt-related transcription factor-2 (Runx2), osteocalcin (OCN), alkaline phosphatase (ALP), and collagen type 1 (COL1) genes and promote osteogenic activity (Yodthong et al., 2020). *L. edodes* spent substrate yielded a proteoglycan (LEPS1) with antiinflammatory properties. The protein was connected to polysaccharide fraction with O-glycosidic bond. LEPS1 suppressed the inflammatory markers NO, TNF-α, IL-1β, and IL-6 in LPS primed RAW264.7 macrophages through downregulating JAK-STAT1 and p38 MAPK pathways (Zhang et al., 2023). *Termitomyces* are harbors of novel linear and branched polysaccharides and a water soluble fucoglucan was characterized in *T. robustus* having L-fucose and D-glucose in 1:4 ratio (Mondal et al., 2008).

A selenium polysaccharide (Se-POP-3) was isolated and characterized from selenium enriched *P. ostreatus* and its anticancer efficacy was studied against gastrointestinal cancer. Se-POP-3 was found effective against gastric and colon cancer and induced apoptosis in them. Interestingly, it was non-toxic against normal cell (NCM460) at the effective concentration. It can be further explored in developing anticancer functional food formulations for gastrointestinal cancers (Zhang et al., 2022). Similarly, sulfation of polysaccharides extracted by subcritical water at 180°C improves its anticoagulant properties with less cytotoxic effect and can be an alternate option in developing anticoagulant formulations. Subcritical water extraction is an efficient technique for extraction of metabolites, and in this study at 180°C, the polysaccharide yield was 20.35% (Rizkyana et al., 2022).

14.2.1.2 Proteins

EFF are good source of vegetarian proteins comparable to legumes and milk. The protein content in EFF ranges from 19% to 35% on a dry weight basis. The commonly consumed *A. bisporus* (14.1%–27.1%), *P. ostreatus* (21.1%–34.5%), and *L. edodes* (17.1%–24.6%) can be a good choice for fungal protein supplements in addition to their dietary fiber content. Amino acids profiles of edible fungal proteins indicates presence of all essential amino acids in EFF proteins making them foods of nutraceutical importance (Wani et al., 2010; Corrêa et al., 2016; Wang and Zhao, 2023). Also, these proteins and peptides exert therapeutic physiological effects like antihypertensive, antioxidant, hepatoprotective, antimicrobial, and immunomodulatory effects, (Zhou et al., 2020). Antihypertensive activity was demonstrated by peptide LSMGSASLSP present in aqueous extract of *Hypsizygus marmoreus* fruiting bodies. The peptide sequence with a molecular weight of 567.3 Da was a potent inhibitor of angiotensin converting enzyme (ACE) with an IC_{50} of 0.19 mg/mL (Kang et al., 2013). Similarly, antihypertensive peptide WALKGYK was isolated from aqueous extract of *Tricholoma matsutake* fruiting bodies. It was found to be a noncompetitive inhibitor (IC_{50} 0.40 μM) of ACE (Geng et al., 2016). *Termitomyces heimii* from Nagaland, India, was found to be a good source of fungal proteins as it had 60.53 ± 0.01 g protein/100 g dry weights, and it can be a source of novel bioactive peptides of therapeutic potential. Sufficient quantity of dietary fibers and sugars were also present in *T. heimii* as demonstrated by the biochemical analysis, and it can be processed as fungal biofunctional food (Ao and Deb, 2019). A novel bioactive protein present in aqueous extract of *P. eryngii* was demonstrated to have anti-inflammatory potential in LPS primed RAW264.7 cells. The 40 kDa protein decreased NO, IL-1β, IL-6, and iNOS expression by downregulating NF-κB and MAPK signaling pathways (Yuan et al., 2017). The serine protease (AkP) from *T. clypeatus* has demonstrated antiproliferative efficacy against HepG2 cancer cells by promoting cleavage of cell surface proteoglycans. Caspase-3 activation and p53 upregulation were responsible for apoptotic activity (Majumder et al., 2016). *G. frondosa* fruiting bodies possess a novel 83 kDa heterodimer protein that activates murine splenocytes and natural killer cells (NK cells) and enhance the secretion of IFN-γ. It promotes TLR4-dependent activation of bone marrow-derived dendritic cells (BMDCs) and demonstrates tumoricidal activity in mice models (Tsao et al., 2013).

Hydrophobin proteins are a class of biosurfactant globular proteins with 100–150 amino acids in their primary amino acid sequence. These proteins are divided into Class I and Class II category on the basis of solubility pattern, hydropathy values, and formation of different structure during self-assembly at the interface (Berger and Sallada, 2019; Dokouhaki et al., 2021) These proteins were first characterized from *Schizophyllum commune* and, since then, different types of hydrophobins were isolated from edible fungi such as *F. velutipes* (Kim et al., 2016a), *P. ostreatus* (Kulkarni et al., 2020, 2022), *G. frondosa* (Wang et al., 2010), and *Pleurotus floridanus* (Rafeeq et al., 2021). High surface activity of these proteins can be exploited in drug delivery systems, protein engineering, and stabilization of food dispersions (Berger and Sallada, 2019; Dokouhaki et al., 2021; Green et al., 2013). The hydrophobin protein Vmh3-1 extracted from *P. ostreatus* is an amphipathic protein with emulsifying properties. The Vmh3-1 has a higher proportion of β-sheets (51%) with a very low α-helix (2%) configuration. The protein resisted self-aggregation after vigorous shaking and this property could have immense industrial applications in drug delivery, in making biomaterials for implants and in food technology (Kulkarni et al., 2022).

14.2.1.3 Lipids, fats, and fatty acids

EFF are low fat (0.1%–16.3%) containing cholesterol less food material. In their fatty acid composition, they contain more unsaturated fatty acids than saturated fatty acids. The unsaturated fatty acids especially polyunsaturated fatty acids (PUFA) have many demonstrated health benefits. EFF are rich vegetarian source of linoleic acid (ω-6), linolenic acid (ω-3), and oleic acid (ω 9). Linoleic and linolenic are PUFA while oleic acid is monounsaturated fatty acid (MUFA). Besides, palmitic acid and stearic acid were also reported in EFF (Sande et al., 2019). Fatty acid analysis in wild EFF reported high amount of *cis*-linoleic acid in total fatty acid percentage in the fruiting bodies of *P. ostreatus* (65.29), *Lactarius salmonicolor* (59.44), *F. velutipes* (40.87), *Polyporus squamosus* (38.91), *Boletus reticulatus* (36.60), *and Russula anthracina* (22.39). In another study, palmitic acid was the dominant fatty acid in *Hericium erinaceus* (Ergönül et al., 2013; Saini et al., 2021).

14.2.1.4 Vitamins

EFF are prominent source of ergocalciferol, i.e., vitamin D_2, the predominant form of vitamin D present in fungi. Ergosterol after UV exposure is converted to pre-vitamin D_2 which is converted to vitamin D_2 in a heat dependent process. Thus, sun light exposure to mushroom has potential to increase its vitamin D_2 content (Cardwell et al., 2018). *G. frondosa*, *Cantharellus* spp., and *Morchella* spp. procured from retail suppliers in different cities of the USA have high content of vitamin D_2 28.1, 5.30, and 5.15 μg/100 g fresh weight, respectively (Phillips et al., 2011). Ultrasound Assisted Extraction (UAE) of Vitamin D_2 and its precursor ergosterol was performed in edible mushroom *A. bisporus* and *P. ostreatus*.

UAE enhanced extraction of vitamin D_2 and ergosterol. Application of 40kHz frequency/240W/40°C for 30min in ethanol was found to be optimum condition for fast and effective recovery of these metabolites for food industry (Nzekoue et al., 2022). Apart from vitamin D, EFF also have vitamin B and vitamin C. Wild edible mushrooms *Macrocybe gigantea* J124 (33 ± 0.400 mg/100 g dry wt.), *Lactifluus leptomerus* J201 (34.00 ± 0.570 mg/100 g dry wt.), and *Ramaria thindii* J 470 (32.00 ± 0.288 mg/100 g dry wt.) from North-Eastern India are found to be good in vitamin C content (Khumlianlal et al., 2022). *Pleurotus* species are rich in folic acid content like *P. flabellatus* (1.22 mg/100 g dry wt.), *P. florida* (1.41 mg/100 g dry wt.), *P. sajor-caju* (1.23 mg/100 g dry wt.), and *P. eous* (1.35 mg/100 g dry wt.). Thiamin (B_1), Riboflavin (B_2), Niacin (B_3), and ascorbic acid are also present abundantly in *Pleurotus* species (Raman et al., 2021).

14.2.1.5 Minerals

Edible fungi are good natural resources of minerals and an analysis of mineral content of six *Termitomyces* species from Cameroon revealed high ash content (14.39 ± 0.50 g/100 g dry wt.) in *T. mammiformis,* which was higher than other commonly consumed fungal species *P. ostreatus* and *P. eous* (Kansci et al., 2003). Mineral composition of nine edible fungi revealed abundant quantity of macro (phosphorus, potassium) and micronutrients (iron, copper, and zinc) in *Agaricus brasiliensis, L. edodes, P. eryngii, P. ostreatus, and P. djamor*. Intake of 100 g (dry wt) *A. brasiliensis* would alone sufficient to contribute daily recommended requirements of minerals phosphorus, iron, copper, and zinc (Bach et al., 2017). Selenium is essential for the functioning of selenocysteine enzymes and species of *Boletus* are particularly rich in selenium. *Boletus edulis* possess approximately 20 μg selenium/ g dry weight, maximum amount of which reaches up to 70 μg/g dry wt. *Lycoperdon* spp. and *Macrolepiota* spp. are also prominent source of fungal selenium (Falandysz, 2008).

14.2.2 Secondary nutraceuticals

EFF have different types of secondary metabolites, viz., ergosterol, GABA, lovastatin, ergothioneine, eritadenine, tocopherols, blazeispirols, melanin, ganoderols, lingzhiol, erinacines, lentinamycin, chlorogenic acid, vanillic acid, myricetin, 4-hydroxymethylbenzoic acid, etc., present in different proportions in mycelium and fruiting bodies. These secondary metabolites may be specific to a particular genus or occur in a more general way. Many of these compounds provide aroma and flavor to EFF and enhance their acceptability and palatability among consumers. Extraction of these metabolites depends on the processing, stages of collection, and extraction techniques implied in the process.

14.2.2.1 Alkaloids

Alkaloids are nitrogen containing compounds generally derived from amino acids. EFF are also source of diverse range of alkaloidal compounds. *Ganoderma lucidum* fruiting bodies possess four polycyclic alkaloids, namely, lucidimine A, B, C, and D, of which lucidimine B was found to be antiproliferative against MCF-7 cancer cell line. It promoted DNA fragmentation and blocked cell cycle in S phase (Zhao et al., 2015; Chen and Lan, 2018). Eight pyrrole alkaloids including pyrrolefronine were isolated from hydroethanolic (95%) extract of *G. frondosa* fruiting bodies. These pyrrole alkaloids demonstrated strong inhibitory action against α-glucosidase indicating importance of pyrrole ring structure in α-glucosidase inhibition (Chen et al., 2018).

Thermal stability of nutraceutical compounds is necessary for making functional foods which require thermal treatment for their formation, processing or packaging. Eritadenine, (2*R*,3*R*)-4-(6-aminopurin-9-yl)-2,3-dihydroxybutanoic acid, is another molecule of nutraceutical importance present in edible fungus L. *edodes*. It competitively inhibited angiotensin converting enzyme (ACE) with IC_{50} 0.091 μM, which was comparable to the standard drug captopril (Afrin et al., 2016). Eritadenine also demonstrated hypocholesterolimic activity in rat models (Shimada et al., 2003a,b; Takashima et al., 1973). Extraction of eritadenine by pressurized liquid extraction method at 142°C for 5 min in aqueous solvent for single extraction cycle yielded higher amount (659 ± 40 mg/100 g dry wt.) than other conventional method of extraction. The thermal stability of eritadenine makes it a suitable nutraceutical agent for thermally treated functional foods (Kaya and Cam, 2022). Also, the extraction of eritadenine in aqueous medium can be useful in making nutraceutical soups and beverages (Kaya and Cam, 2022; Sánchez-Minutti et al., 2019). Application of different processing methods can change the eritadenine concentration in *L. edodes*. Drying of mushroom at 4°C and 60°C enhanced eritadenine concentration by 52% and 90.7%, respectively (Sánchez-Minutti et al., 2019). The aqueous extract of freeze-dried *L. edodes* powder also has antibacterial and antifungal properties. Phytochemical studies revealed the presence of 34 compounds including alkaloids muscarine, choline, and eritadenine (Rao et al., 2009).

14.2.2.2 Polyphenols

Different types of polyphenolic compounds are present in EFF like gallic acid, protocatechuic acid, catechol, gentisic acid, *p*-hydroxybenzoic acid, *p*-coumaric acid, ferulic acid, rutin, sinapic acid, catechin, vanillic acid, syringic acid, rosmarinic acid, quercetin, apigenin, and kaempferol. Polyphenolic compounds act as powerful antioxidant molecules (Bach et al., 2019; Kaewnarin et al., 2016; Nowacka et al., 2014). *A. brasiliensis* is rich in gallic acid (491.89 ± 1.59 μg/g dry wt.), catechol (148.83 ± 0.81 μg/g dry wt.), p-hydroxybenzoic acid (332.76 ± 2.98 ± μg/g dry wt.), and ferulic acid (752.54 ± 1.90 μg/g dry wt.). High quantity of phenolic compounds in *A. brasiliensis* imparts it with high antioxidant potential (DPPH 50.64 ± 0.37 μmol TE/g; ABTS 128.60 ± 2.02 μmol TE/g). Catechol is present in *F. velutipes* and *A. bisporus*, while gallic acid is reported from *A. bisporus* and *L. edodes* (Bach et al., 2019). Phenolic profile of wild edibles from Thailand reported high gallic acid, protocatechuic acid, catechin, vanillic acid, syringic acid, and apigenin content in water and methanolic extract of *Rugiboletus extremiorientalis*. Rosmarinic acid was found in methanolic extract only. *Russula emetica* was rich in protocatechuic acid and catechin. Quercetin was reported in high quantity (1762.0 μg/100 g dry wt.) in methanolic extract of *R. emetic,* while kaempferol was detected in *Phlebopus portentosus*. The methanolic extract of *R. extremiorientalis* was found to be a potent inhibitor of enzyme tyrosinase, a key copper-containing enzyme in melanin synthesis. Overproduction of melanin can be responsible for several skin-related disorders (Kaewnarin et al., 2016). Analysis of wild edible fungi from Poland revealed the presence of phenolic compounds in *Armillaria mellea* (protocatechuic acid), *Calvatia excipuliformis* (salicylic acid), *Craterellus cornucopioides* (protocatechuic acid, 4-OH-benzoic acid, vanillic acid), *Laccaria amnethystea* (4-OH-benzoic acid), *Pholiota mutabilis* (protocatechuic acid, 4-OH-benzoic acid, *p*-coumaric acid), *Coprinus micaceus* (4-OH-benzoic acid, *p*-coumaric acid), *Lycoperdon perlatum* (4-OH-benzoic acid, *p*-coumaric acid), and *Xerocomus badius* (protocatechuic acid) by Nowacka et al. (2014). Similarly, water extract of *T. heimii* and *T. mammiformis* fruiting bodies collected from Himachal Pradesh, India, also showed very high antioxidant index with high phenolic content, i.e., 37 mg/g and 19.2 mg/g of sample, respectively

(Puttaraju et al., 2006). The methanolic extract from *T. heimii* also showed prominent antioxidant activity with IC_{50} value 107.2 ± 0.05 μg/mL and total phenolic content 17.4 ± 0.05 g GAE/100 g dry wt. (Ao and Deb, 2019).

14.2.2.3 Terpenes and terpenoids

EFF are potent source of terpenes and terpenoids. Sesquiterpenes flammufuranone A and B, flammuspirone A-J, and flamvelutpenoids E and F were isolated from *F. velutipes* fruiting bodies. Compounds flammuspirone A and flammuspirone C inhibited HMG-CoA reductase with IC_{50} 114.7 and 77.6 μM, respectively (Tao et al., 2016). Lanostane triterpenoid Ganosidone A with its eight derivatives methyl ganoderate A, methyl ganoderate H, lucidumol A, lucidumol C, ganoderic acid A, ganodermanon triol, ganolucidic acid A, and ganolucidic acid E were isolated from the methanolic extract of *G. lucidum* fruiting bodies and evaluated for their anti-inflammatory activity. Lucidumol A and ganodermanon triol inhibited nitric oxide (NO) production in lipopolysaccharide challenged RAW 264.7 murine macrophage cells without showing cytotoxicity at a concentration of 25 μM (Koo et al., 2021). Eight new lanostane triterpenoids including ganoluciduone A, ganoluciduone B, ganolucidoid A, and ganolucidoid B with seven previously defined lanostane triterpenoids from the ethyl acetate fraction of methanolic extract of *G. lucidum* fruiting bodies were tested for antiinflammatory efficacy and ganoluciduone B showed moderate activity with 45.5% inhibition of NO production (Su et al., 2020). Similarly, ethyl acetate fraction from *G. lucidum* fruiting bodies also yielded triterpenoids ethyl lucidenates A, ganodermanondiol, lucidumol B, and methyl lucidenates A. Ethyl lucidenates A demonstrated cytotoxic activity against HL-60 (IC_{50} 25.98 μg/mL) and CA46 (IC_{50} 20.42 μg/mL) cancer cell lines (Li et al., 2013).

14.2.2.4 Sterols

EFF are notably rich in sterols and ergosterol is the main sterol present in them. These sterols can be in free form or esterified with fatty acids. Free ergosterol estimation in commonly consumed EFF reported high ergosterol in *A. bisporus* (485 ± 51 mg/100 g dry wt.), *L. edodes* (351 ± 46 mg/100 g dry wt.), *P. eryngii* (265 ± 34 mg/100 g dry wt.), and *P. ostreatus* (254 ± 38 mg/100 g dry wt.) by Hammann et al. (2016). A study was conducted to assess sterol composition of edible mushrooms available in the retail markets of different cities in the United States. It revealed the presence of ergosterol and its derivatives ergosta-5,7-dienol (22,23-dihydroergosterol), ergosta-7,22-dienol, and ergosta-7-enol in *A. bisporus*, *F. velutipes*, *L. edodes*, *G. frondosa*, *P. ostreatus*, and *Cantharellus*. Highest amount of ergosterol was reported for *L. edodes* and *G. frondosa*. Brassicasterol and campesterol were found in the samples of *Morchella* spp. (Phillips et al., 2011).

Cancer is uncontrolled division of cells under the proliferative signals developing resistance to cell death. Proliferative signaling, evasion of growth suppressors, metastasis, replicative immortality, angiogenesis, and cell death resistance are hallmarks of cancer (Hanahan and Weinberg, 2011). Hyperestrogenic signaling is responsible for the development of breast cancer in most of the cases. Frequent application of estrogen receptor inhibitors develops serious side effects. Aromatase is the enzyme responsible for the conversion of androgens into estrogens. Therefore, aromatase inhibitors provide alternative route to treat breast cancer (Johnston and Dowsett, 2003). Ergostane-type sterols from the edible fungus *P. erygii* have been evaluated for their aromatase inhibitory activity and demonstrated comparable activity to standard aminoglutethimide. These sterols have the potential to be used in breast cancer medication (Kikuchi et al., 2017b). The ethyl acetate extract from

T. heimii demonstrated antiproliferative potential against hepatocarcinoma with IC_{50} 263.53 (8.09) μg/mL. The molecular docking studies supported the anticancer efficacy of *T. heimii* extract as compounds ergosterol and ergosta-5,8-dien-3-ol showed higher interaction for Bcl-XL, while lanosterol and eburicol had stronger affinity for Bcl-2. The in silico docking also demonstrated strong affinity of ergosterol and ergosta-5,8-dien-3-ol for VEGFR-1 and VEGFR-2, the angiogenic receptors, and acted antagonistic to angiogenic proliferation (Ray et al., 2022). Eight compounds including six new compounds were characterized in the methanolic extract of *Morchella esculenta* fruiting body. These compounds belong to fatty acids and sterols category. Compound 1 (1-O-octadecanoyl-sn-glycerol), 3 ((3β,5α,8α,22E,24S)-5,8-epidioxyergosta-6,9(11),22-trien-3-ol), and 5 ((3β,5α,22E)-Ergosta-7,22,24(28)-trien-3-ol) were found effective against lung cancer cell lines by inducing apoptosis in them (Lee et al., 2018).

14.2.2.5 Fatty acid derivatives

Low-cost natural substitute of synthetic drugs in food sources are in demand owing to their less side effects and health consciousness of public. Lovastatin, a fatty acid ester, is a prodrug and is used in cardiovascular diseases as an inhibitor of HMG-CoA reductase, the rate limiting enzyme in cholesterol biosynthesis (Sirtori, 1990; Frishman and Rapier, 1989). The substrate for HMG-CoA reductase is 3-hydroxy-3-methylglutaryl coenzyme A (HMG-CoA) to produce mevalonate and lovastatin competitively inhibits this enzyme (Frishman and Rapier, 1989). Lovastatin is first described from fungus *Aspergillus terreus* as a potent natural inhibitor of HMG-CoA (Frishman and Rapier, 1989; Tobert, 2003). Since then, many fungal sources were demonstrated to have lovastatin either in fruiting body or in mycelium. The presence of lovastatin was demonstrated in the fruiting body of *Agrocybe cylindracea* (583.0 ± 21.3 mg/kg dry wt.), *L. edodes* (316.7 ± 3.6 mg/kg dry wt.), and *Hypsizygus marmoreus* (628.1 ± 8.5 mg/kg dry wt.) in good amount; however, the mycelium of *M. esculenta* (1438.4 ± 56.3 mg/kg dry wt.), *P. citrinopileatus* (930.9 ± 28.4 mg/kg dry wt.), and *G. lucidum* (908.1 ± 49.5 mg/kg dry wt.) had more lovastatin than fruiting body (Lo et al., 2012).

14.2.2.6 Protein and amino acid derivatives

The physiologically active Gamma-aminobutyric acid (GABA) is a potent nonprotein amino acid that is formed by the decarboxylation of L glutamate in a reaction catalyzed by the enzyme glutamate decarboxylase (GAD) (Diana et al., 2014). GABA works as inhibitory neurotransmitter in the central nervous system and can be used in many physiological disorders like neurological (Möhler, 2012), renal failure (Kim et al., 2004), and hypertension (Shimada et al., 2009). Efforts have been made for the formulation of GABA enriched functional food, e.g., GABA enriched fermented sheep's milk (Ramos and Poveda, 2022), tomato juice (Nakatni et al., 2022), probiotic yogurts (Garavand et al., 2022; El-Fattah et al., 2018), etc. Edible fungi are also a prominent source of GABA. Quantification of GABA in fruiting body and mycelium of some edibles reported high GABA content in *A. brasiliensis*, *A. cylindracea*, *L. edodes*, *Tremella fuciformis*, *V. volvacea*, *B. edulis*, *M. esculenta*, *P. citrinopileatus*, *P. ostreatus,* and *Termitomyces albuminosus* (Lo et al., 2012).

Ergothioneine (tri-methylbetaine of 2-thiol-L-histidine) is a water-soluble derivative of amino acid histidine and transported into tissues through organic cation transporter novel-type 1 (OCTN1) transporter (Cheah and Halliwell, 2012, 2021). Ergothioeine is a powerful antioxidant biomolecule and has its role in various pathological conditions like neurodegeration and dementia (Wu et al., 2021), sleep difficulties (Katsube et al., 2022), diabetic nephropathy (Dare et al., 2022), and age-related diseases (Apparoo et al., 2022). Ergothioneine estimation in some edible mushroom resulted in high quantity in *V. volvacea* fruiting body (537.3 ± 1.7 mg/kg dry wt.) and moderate quantity in *B. edulis* (258.0 ± 5.6 mg/kg dry wt.) mycelium (Lo et al., 2012).

14.2.3 Enzymes of industrial applications

EFF are source of not only therapeutic metabolites and nutrients but also serve as a source of diverse range of enzymes with industrial applications. β-glucosidase (EC 3.2.1.21) is an enzyme of cellulolytic enzyme system with the ability to hydrolyze glycosidic bonds in different substrates. β-glucosidase has its application in beverage and other food processing industries (Singh et al., 2016). A thermostable β-glucosidase having a tendency of enzyme aggregate formation was isolated and characterized in *T. clypeatus*. It has low molecular weight (6.688 kDa) and can form co-aggregates with sucrose in the fungus. The enzyme has transglycosylation activity and can be used in the formation of complex bioactive compounds of industrial uses without chemical hazards (Pal et al., 2010). β-glucosidase isolated from *Lycoperdon pyriforme* fruiting bodies demonstrated resistance to metal ions and pH stability over a wide range of pH from 3.0–9.0. Its maximum activity was found at pH 4.0 and 50°C temperature making it suitable for industrial purposes (Akatin, 2013).

T. clypeatus is a good source of amylase and cellulase enzymes and use of carboxymethyl-cellulose in culture medium as a carbon source would enhance the production of these enzymes. These enzymes are used in cellulolytic and amylolytic biodegradable processes (Jonathan and Adeoyo, 2011). Similarly, this fungus is also a source of important enzyme amyloglucosidase with xylanolytic activity. The xylanolytic activity may be attributed to high amount of serine and threonine present in the amino acid composition of this enzyme. The enzyme has the ability to hydrolyze glucose from both starch and xylan (Ghosh et al., 1997).

The extracellular cellobiose dehydrogenase (EC 1.1.99.18) enzyme (CDH) is a flavocytochrome protein secreted by many fungal species for biomass degradation. This enzyme has separate catalytic and electron transferring domains that makes it suitable to be used in making biosensors, biofuel cells, bioanodes, biosupercapacitors, biocatalyst, and bioremediation (Scheiblbrandner and Ludwig, 2020). *T. clypeatus* produces cellobiose dehydrogenase in significant amount in culture medium enriched with cellulose at pH 6.5 and is a good source of this enzyme for industrial applications (Saha et al., 2008). Further, the use of tamarind kernel powder in culture media augmented production of CDH (Saha et al., 2014). The computational study of CDH from *T. clypeatus* revealed the presence of a third carbohydrate binding domain that enhances enzyme activity and makes it more useful than from any other sources (Banerjee et al., 2019; Lakhundi et al., 2015). *G. frondosa*, *T. versicolor*, and *V. volvacea* are some other sources of cellobiose dehydrogenase (Chen et al., 2017; Yoshida et al., 2002).

T. clypeatus is also an important source of enzyme phosphoketolase, which is deployed in Pentose Phosphoketolase pathway (PPK). This enzyme is central to the utilization of hemicellulosic pentose sugar arabinose and xylose abundantly present in biomasses and can be utilized in the production of low-cost bioethanol (Sarkar and Roy, 2014).

P. eryngii is armed with a number of enzymes used by the ligninolytic system. These are extracellular laccases, aryl-alcohol oxidases (AAO), Mn-dependent peroxidase, and versatile peroxidase (VP). These enzymes can be applied in food, beverages, paper and pulp production, bioremediation, production of aromatic compounds, utilization of agricultural wastes, and byproducts and in textile industry (Stajić et al., 2009). Versatile peroxidase (EC 1.11.1.16) can be applied as a tool to produce homo and heteropolymer of low- and high-molecular-weight biomolecules. The environment-friendly biotransformation of existing biomolecules into novel compounds provides an opportunity to synthesize more specific molecules of industrial application with improved or new properties. VP from *P. eryngii* exhibited formation of lignan homopolymer of 8 units with secoisolariciresinol and hydroxymatairesinol. Similarly, peptide sequences YIGSR and VYV also formed homopolymer of 11 units in the presence of VP. Interestingly, VP catalyzed formation of heteropolymer of peptide sequence YIGSR with both lignans. VP can also be employed in the formation of feruloylated arabinoxylan gels (Salvachúa et al., 2013). *P. eryngii* was used to produce ligninolytic enzymes on apricot and pomegranate agroindustrial wastes and apricot wastes were found to be a good substratum than pomegranate owing to the high lignin content. Both these wastes can be used as a replacement of conventional substrates in solid state fermentation process. In another study, grape waste from juice industry was used as a substrate for solid state fermentation to produce ligninolytic enzymes by *P. eryngii*. The study demonstrated the efficacy of grape wastes for the production of low-cost extracellular enzymes (Akpinar and Urek, 2012, 2014). The solid-state culture is an economic and acceptable process for large-scale production of ligninolytic enzymes (Iwashita, 2002).

Laccases are part of ligninolytic enzyme system and participate in oxidation reaction of a wide range of phenolic substrates like syringic acid, vanillic acid, phenol, naphthol, ferulic acid and hydroquinone to form free radicals. These radicals are central to the formation of homodimers, heterodimers, oligomers, and polymers (Mikolasch and Schauer, 2009). The coupling of a laccase substrate to a non-laccase substrate could result in C-O, C-S, or C-N coupled heterodimers and trimers depending upon the groups involved in the reaction (Mikolasch and Schauer, 2009). Laccases are potent fungal biotechnological tool due to their wide range of substrates and biological actions. Therefore, new isozymes of this enzyme were searched and characterized from different fungal sources. The heterologously expressed laccase isozymes LACC6, LACC9, and LACC10 from *P. ostreatus* HAUCC 162 were isolated and evaluated for their dye removal capacity. The isozyme LACC6 significantly removed Malachite Green (91.5%), Ramazol Brilliant Blue R (84.9%), Bromophenol Blue (79.1%), and Methyl Orange (73.1%) within 24 h. It can withstand high metal and organic solvent concentration and can be utilized in decolorization of dyes in bioremediation (Zhuo et al., 2019). Earlier studies reported synergistic induction of laccase production in the presence of metal ions and aromatic compounds in the same strain. There was 20.1-fold increase in lacc 6 transcription after treatment with 2 mM Cu^{2+} + 0.5 mM ferulic acid (Zhuo et al., 2017). Laccases can also be used for the removal of hazardous environmental pollutants like Chlorophenols, nitrophenols, and sulfonamide antibiotics. Toxicity, prevalence, and resistance from degradation make these compounds hazardous to the human health. LACC6, LACC9, and LACC10 laccase isozymes from *P. ostreatus* efficiently degraded these compounds in a laccase-syringic acid system. Syringic acid was used as a laccase mediator (Zhuo et al., 2018). Laccases can also be employed in food processing industries, synthesis of antioxidant and antimicrobial compounds, transforming food polymers and developing functional foods (Backes et al., 2021).

14.2.4 Secondary metabolites extraction

The concentration of fungal metabolites depends upon the species strains, process of extraction, extraction solvents, growth medium, supplements, and different processing methods of edibles. The ergothioneine and lovastatin content of different *Pleurotus* species strains were evaluated by growing fungus in culture medium supplemented with different agro-industrial residues (Chilanti et al., 2022). For ergothioneine production, it was found that basidiocarps grown in coffee grounds had higher concentration in *Pleurotus* cf. *pulmonarius* 41D strain as compared to other strains. However, in organic grape waste medium it was *P. pulmonarius* 26C strain that had higher ergothioneine content. Among strains grown in *Pinus* sp. sawdust highest ergothioneine content was recorded in *P. pulmonarius* PS-2001. In case of mycelium, highest ergothioneine was occurred in *P. pulmonarius* 122H.5 and *Pinus* sp. sawdust medium was more effective than other two medium used in the study confirming the role of growth medium, strains, and duration of culture in ergothioneine production (Chilanti et al., 2022). The effect of extraction solvent on lovastatin content in the fruiting body of different *Pleurotus* spp. strains was also evaluated and it was found that acetonitrile was more effective than ethyl acetate and methanol in extracting lovastatin. The highest lovastatin content (10.2 ± 0.1 mg/100 g) was found in basidiocarp of *P.* cf. *pulmonarius* 41D grown in organic grape waste (Chilanti et al., 2022).

14.3 Prebiotics

Prebiotics are substances that are resistant to intestinal digestion, absorption and to promote the growth and activity of beneficial microorganisms present in the colon. A substance must fulfill three criteria to be recognized as prebiotics; 1- resistant to gastric pH, enzymatic hydrolysis and absorption, 2-fermentable by intestinal microorganisms and 3- encourage growth and activity of intestinal health-promoting microflora, selectively (Gibson et al., 2004). Inulin and β-glucans are two well-known groups of prebiotics as they fulfilled the above-mentioned criteria (Gibson et al., 2004; Ruthes et al., 2021; Moumita and Das, 2022). In a recent study, prebiotics potential of common Indian edible mushrooms *A. bisporus*, *P. ostreatus*, *P. florida*, *P. sajor-caju,* and L. *edodes* has been evaluated for making synbiotic formulations. The study reveals highest β-glucans, inulin, and eritadenine content in the extracts from *P. florida* indicating its highest prebiotic potential. Also, it has been found that *P. florida* extract has maximum compatibility with probiotic bacterium *Enterococcus faecium* and can be explored for the formation of synbiotic microcapsules and its incorporation into nutraceutical foods (Moumita and Das, 2022). Alteration of gut microbiome has a possible link with the development of pathological conditions as suggested by several recently published scientific literatures (HIV-associated metabolic disorders—Chachage, 2021; COVID-19—Zuo et al., 2021; SARS-CoV-2-linked colorectal cancer—Mozaffari et al., 2022; obesity—Benahmed et al., 2021; eye diseases—Zysset-Burri et al., 2023; diabetes—Lau et al., 2021; atherosclerosis—Anto and Blesso, 2022). Gut microbiome remodeling (GMR) has been associated with the function of prebiotics in alleviating certain diseased conditions, e.g., colitis (Chen et al., 2022), tumor (Li et al., 2020), atopic dermatitis (Lee et al., 2021), and stress (Cruz-Pereira et al., 2022). Changing the composition of gut microbiome can modulate the metabolites dynamics and may help to overcome pathophysiological conditions. The microbes residing in gut produce short chain fatty acids having 2–6 carbon in their aliphatic skeleton like acetic (C2), propionic (C3), butyric (C4), valeric (C5), and caproic (C6) acids which exert different therapeutic effects. However, acetic, propionic and butyric acids are the key therapeutic modulators produced by these microbes (Ohira et al., 2017; Yang et al., 2022). Polysaccharides from *G. lucidum* fruiting body increased the production of key short-chain fatty acids and enhanced the Bacteroidetes to Firmicutes ratio after 48 h of fermentation in in vitro simulation model. It also reduced branched short-chain fatty acids like isobutyric and isovaleric acids. *G. lucidum* polysaccharides beneficially modulated microbial diversity with a remarkable increase in *Bacteroides ovatus* and *Bacteroides uniformis* while decreasing *Clostridium symbiosum* and *Peptostreptococcus anaerobius* (Yang et al., 2022). *G. lucidum* polysaccharides from fruiting body also showed its positive effect on gut microbiota in C57BL/6J mice in vivo models when administered 750mg/ kg dose for 15days. The polysaccharide promoted *Lactobacillus johnsonii*, *Bifidobacterium* sp., *Lactococcus lactis*, *Roseburia intestinalis*, *Akkermansia muciniphila*, *Paraprevotella clara*, *Bacteroides xylanolyticus* and downregulated *Dehalobacterium* spp., *Pseudobutyrivibrio* spp., and *Robinsoniella peoriensis*. These beneficial bacteria help in polysaccharide degradation and produce SCFA with several known health effects (Khan et al., 2018). Therefore, it is clear that the polysaccharides from edible fleshy fungi not only support the growth of beneficial microbiota but also change the composition of gut microbiome in favor of human health. The hydroalcoholic extract from *L. edodes* induced growth of probiotic strain *L. acidophilus,* while *P. pulmonarius* hydroalcoholic extract promoted *L. plantarum*. These extracts rich in polysaccharides may act as better prebiotics than conventional prebiotics inulin and fructooligosacharides (Sawangwan et al., 2018). Similarly, glucan-rich water soluble (branched 1,3-1,6-β-D-glucan), alkali soluble (linear 1,3-α-D-glucan), and insoluble (branched 1,3-1,6-β-D-glucan) fraction were characterized from *P. eryngii* and *P. ostreatus* strains for their prebiotic potential for *Bifidobacterium*, *Enterococcus*, and *Lactobacillus* probiotic strains. All these fractions differentially stimulated growth of probiotic strains and may be used in making synbiotic combinations (Synytsya et al., 2009). Diet supplemented with 1% (1g/100g) *A. bisporus* powder influenced gut microbial community in C57BL/6 mice and especially increased the population of *Prevotella* responsible for propionic and succinic acid production (Tian et al., 2018). The propionate and succinate have a role in intestinal gluconeogenesis. Reduced hepatic glucose and glycogen levels were found after WB (white button) feeding. Application of WB feeding increased members from the order Lactobacillales, Bacteroidales, and Campylobacterales. Higher expression of *G6pase*, *Glut2*, *Pepck*, *Sglt3a, Sglt3b,* and SCFA receptor *Ffar3* mRNA after white button feeding indicated microbial-inducted glucose homeostasis through intestinal gluconeogenesis (Tian et al., 2018). In another in vivo study, the 5% supplementation of *P. sajor-caju* was also found effective in modulation of colonic microbial community in Zucker rat obese models. The diet supplementation provoked the growth of SCFAs-producing bacteria from *Bifidobacterium*, *Roseburia*, *Faecalibaculum*, etc., and downregulated pathogenic strains. It also decreased plasma bile acids viz. cholic acid, β muricholic acid, glycocholic acid, taurocholic acid, and tauromuricholic acid and increased chenodeoxycholic acid, α-muricholic acid, deoxycholic acid, hyodeoxycholic acid, ursodeoxycholic acid, and taurodeoxycholic acid. The positive effect of *P. sajor-caju* on SCFA-producing microbes could be exploited in inflammatory hepatic conditions and other ailments (Maheshwari et al., 2021).

14.4 Value addition and industrial products

EFF can be used in nutritional enrichment of available foods and in the development of industrial products.

14.4.1 Nutritional enrichment

Grain flours are consumed for their nutritional values in the form of bread, cakes, pastry, and other purposes. The commonly consumed cereal grain flours are wheat, rice, and corn. Fortification of edible grain flours with edible fungi can enhance the nutritional composition and therapeutic potential of flours to form functional foods. Colonization of brown rice, corn, and wheat by the mycelia of macrofungi *Agaricus blazei*, *Auricularia fuscosuccinea,* and *Pleurotus albidus* improves nutritional, functional and physico-chemical characteristics of grain flours. Wheat grain colonization with *P. albidus* through solid state culture produced high protein (18.34 g/100 g) flour with high ergosterol content (0.60 mg/g). It significantly inhibited pancreatic lipase and α-glucosidase, the key enzymes involved in obesity and diabetes. Similarly, incubation of corn with *A. fuscosuccinea* demonstrated high phenolic content and antioxidant activity in DPPH and ABTS assays (Stoffel et al., 2019). Fermentation of lentil with *P. ostreatus* enhances nutritional profile and digestibility of lentil flour. The solid-state fermentation increased quantity of protein content, resistant starch, and polyphenols in a significant manner and also improved in vitro digestion of lentil flour (Asensio-Grau et al., 2020). Fungal mycelia are a good source of antioxidant biomolecules including polyphenols and tocopherols. The mycelia from edible fleshy fungi were studied for their antioxidant properties and *P. ostreatus* strain PQMZ91109 was found to have gallic acid, homogentisic acid, catechin, myricetin, ascorbic acid, and α-tocopherols as a free radical scavenging phytomolecules, while *Coprinus comatus* M8102 strain possessed gallic acid, homogentisic acid, ascorbic acid, β-carotene, lycopene, rutin, apigenin, and α-tocopherol as its chief constituents (Vamanu, 2014). The presence of tocopherol acetate (4.8 ± 0.05 mg/100 g extract) in *P. ostreatus* strain PQMZ91109 has physiological significance in regulating melanin over production and hyperpigmentation (Solano et al., 2006; Vamanu, 2014). The ethanolic extract of earlier harvested fruiting body (after 10 days of fruiting body induction) of *P. eryngii* showed high amount of phenolic acids (7.53 ± 0.39 mg/g dried extract), flavonoid content (1.84 ± 0.10 mg/g dried extract), tocopherol (2.53 ± 0.17 mg/g), and carotenoids (0.73 ± 0.04 mg/g). The ethanolic extract downregulated LPS-induced iNOS (inducible nitric oxide synthase), NO production, COX-2 (cyclooxygenase-2) and PGE2 (prostaglandin E2) and demonstrated anti-inflammatory response in RAW264.7 cells (Lin et al., 2014). The 5% powder of *P eryngii* fruiting bodies has hypocholesterolemic effect on Sprague Dawley female albino rats and promoted excretion of total lipids and cholesterol. Supplementation with 5% powder increased plasma α-lipoprotein and decreased β-lipoprotein and pre-β-lipoprotein, and it had no adverse effect on liver and kidney of model rats (Alam et al., 2011). Nutritional characterization of *A. brasiliensis* revealed high phenolic acid content, polyunsaturated fatty acids (PUFAs), tocopherols, antioxidant, and antibacterial activity. The *A. bisporus* was found to be rich in monounsaturated fatty acids (MUFAs) and ergosterol. The ethanolic extract of *A. brasiliensis* prevented the growth of food-borne pathogen *Listeria monocytogenes* in yogurt and can be exploited as natural food preservatives (Stojković et al., 2014). Tocopherols are natural antioxidant molecules present abundantly in edible fungal mycelium and fruiting bodies. Their incorporation in food items could stabilize the food material and prevent deterioration making food functional and healthier. The mycelia of *G. lucidum*, *P. ostreatus,* and *P. eryngii* have high tocopherol content with 718 ± 12, 687 ± 7, and 473 ± 6 μg/mL, respectively, in extract. The qualitative analysis revealed differences in types of tocopherols in mycelial extracts with *G. lucidum* was rich in δ- and γ-tocopherols, while *P. ostreatus* and *P. eryngii* were rich in β-tocopherols (Bouzgarrou et al., 2018). Fortification of yogurt with tocopherol-rich mycelial extract could enhance the nutritional and therapeutic quality of yogurt. The yogurt supplemented with *G. lucidum* mycelial tocopherol-rich extract demonstrated better nutritional profile (protein and fat content) and antioxidant activity than *P. ostreatus* and *P. eryngii* supplementation without diminishing the original functional value of yogurt (Bouzgarrou et al., 2018). Liquid culture of *Pleurotus djamor* PLO13 mycelium in a medium containing selenium enriched whey powder can be exploited for bioactive compounds production and as a nutritional supplement. The presence of selenium in culture medium enhances the antioxidant activity and β-glucans production. The mycelial biomass can further be used as a dietary supplement for lactose-intolerant customers as it has no lactose (Velez et al., 2019). The ergosterol-rich extract from commercially discarded *A. blazei* fruiting bodies can be used to supplement yogurts to form high value-added food products. Soxhlet extraction demonstrated 58.53 ± 1.72 µg ergosterol content/100 g dry wt. of fruiting bodies. Incorporation of ergosterol-rich extract in yogurts increases its antioxidant activity without altering nutritional composition and fatty acid profiles (Corrêa et al., 2018). Solid-state fermentation of seven cereals by *A. blazei* significantly improved sugars, proteins, phenols, and antioxidant activity. The nutritional improvement was found to be dependent on fermentation time, and different fermentation durations were required for different nutritional parameters and cereals. Millets showed highest enhancement rates in phenols, amino acids, and water-soluble proteins, while wheat showed highest improvement rates in reducing sugars and total protein content after fermentation. Sorghum fermented product revealed highest improvement in DPPH free radical scavenging activity (Zhai et al., 2015).

14.4.2 Vegetable soup

The success of any marketed fortified food product is dependent on its wide acceptance by the consumers. For this, sensory evaluation is done on various parameters like appearance, taste, texture, aroma, and color (Kemp, 2008; Stone, 2018). At the same time, consumers demand for low-energy dietary supplements rich in nutraceuticals to avoid obesity and overweight. A low energy-dense pumpkin-carrot vegetable soup rich in β-glucans from *P. ostreatus* was developed and tested for sensory properties for its wide acceptance. It was found that 2% *P. ostreatus* powder attracted liking of more consumers and liking decreased as the concentration of mushroom powder increased. Thus, β-glucans could be supplemented in the form of vegetable soup (Proserpio et al., 2019).

14.4.3 Noodles

Noodles are consumed worldwide owing to their easy to make recipes and palatability. Cereal flours are the chief ingredient in the formation of noodles, especially dough forming cereals. Rice flours were supplemented with different concentration of β-glucans-rich fractions from widely consumed edible fungus *L. edodes*. Addition of β-glucans-rich fractions to rice flours increases rigidity, extensibility and firmness of rice noodles which are desirable parameters in noodles industry. It significantly reduces cooking loss and swelling index as the concentration of β-glucans-rich fraction increases (Heo et al., 2014).

14.4.4 Biscuit

Biscuits are one of the major cereal-based processed food products consumed in a large scale throughout the world. It can be a vehicle to deliver dietary fibers and other nutraceuticals in a fortified manner. However, due to growing concern about diabetes and other health-related issues, it is necessary to develop or fortify products of low glycemic index with slow digestion and/or absorption of carbohydrates (Zafar et al., 2019; Jenkins et al., 1981, 2008). Inclusion of dietary fiber in food products may provide a way to reduce glycemic index (Jenkins et al., 2008). The edible fungi are rich sources of dietary fibers, and these can be delivered to consumers in the form of biscuits with health benefits. Supplementation of 8% *P. sajor-caju* powder in cereal-based biscuits reduced starch hydrolysis rate and glycemic index of biscuits without compromising its sensory properties. It also increased nutritional quality of biscuits by increasing dietary fibers, protein content and β-glucan. The addition of *P. sajor-caju* powder created disturbances in the morphology of starch granules making it less accessible to starch digestive enzymes and thereby reducing its rate of hydrolysis (Ng et al., 2017). In another study, 10% *P. sajor-caju* powder substitution of wheat flour demonstrated increase in protein content, dietary fiber and ash while decreased carbohydrate in supplemented biscuits. The sensory evaluation demonstrated brown colored, slightly compact, attractive biscuits with pleasant aroma and acceptable taste (Prodhan et al., 2015).

14.4.5 Baked foods

The β-glucan-rich fractions from *L. edodes* can be used as a supplementation in baked foods. The fractions rich in β-glucans were used as a substitute of wheat flour in making cakes. It was found that 1 g of β-glucan/serving did not disturb the air entrapment quality, volume, and texture of baked product while higher concentration of β-glucan-rich fractions caused volume reduction and hardness. The 1 g of β-glucan/serving supplementation may provide health benefits to baked products (Kim et al., 2011b).

14.4.6 Pasta

Pasta is another cereal-based staple food product frequently consumed worldwide owing to its good sensory properties, easy preparation method, low cost and longer shelf life (Bergman et al., 1994). Durum wheat flour (*Triticum durum*) is generally used for making pasta because of its high nutritional content and sensory attributes (Romano et al., 2021; Raiola et al., 2020); however, attempts have been made to develop other cereal flours as an alternative to durum wheat with more bioactive compounds and reduced allergenicity (Romano et al., 2021). Addition of 4% β-glucan-rich fractions from *P. eryngii* to common wheat flour as a substitute produced pasta with enhanced nutritional quality and acceptability score. It significantly increased ordered microstructure with porosity, hardness, adhesiveness and decreased swelling index necessary for developing biofunctional pasta with common wheat matrix (Kim et al., 2016b).

14.4.7 Cheese

Cheese are fermented food product developed from milk and are rich in proteins, bioactive peptides, fatty acids, organic acids, vitamins, and minerals (Zheng et al., 2021; Santiago-López et al., 2018). The aroma, texture and other sensory qualities of cheese are dependent not only on the source of milk but also on the production technologies and microorganisms involved in the cheese development (Zheng et al., 2021; Santiago-López et al., 2018). The common filamentous fungi used in cheese development are species of *Penicillium* viz. *P. camemberti* and *P. roqueforti* (Kumura et al., 2018). The ovine milk-based cheese fortified with β-glucan (0.4% *w*/w) from *P. ostreatus* had 10.30% fat, 8.50% protein, good antioxidant potential, and higher acetic acid concentration. The sensory evaluation was found to be positive for fortified cheese with higher value for flavor characteristics (Kondyli et al., 2022).

14.4.8 Meat products

The oyster mushroom powder (2%) could be used as a substitute for phosphate for stabilizing emulsions in emulsion-type sausages. The polysaccharides present in mushroom powder enhance viscosity and steric hindrance and thereby increase emulsion stability and storage life of meat products (Jung et al., 2022).

14.5 Development of entrepreneurship

Entrepreneurship in EFF is required for the sustainable development of society in terms of nutritious food, medicinal metabolites, and value-added products to cope up demands in day-to-day life.

14.5.1 Fundamental structure of entrepreneurship

Nature works as a sustainable system to produce natural products and these natural products in turn act as a raw material to produce value added products in nature. EFF are natural entrepreneurs, cosmopolitan in occurrence and they work as a natural tool to convert huge biomass into valuable products. Various kinds of interaction are central to their mechanism of action and ecological services. These can be exploited in the development of EFF entrepreneurship.

14.5.2 Functional requirement

EFF entrepreneurship requires many natural, social, and physical assets. However, these assets are easily available to local and marginal population of the world (Marshall and Nair, 2009).

14.5.2.1 Natural assets

The rural and suburban populations generally rely on agriculture for their livelihood. The agriculture system in developing countries are such as it can simultaneously support and flourish EFF production in a large scale. In EFF production, soil and climate play a nominal role, which gives a way to marginal and landless farmers. Agricultural byproducts are used for EFF production like button mushroom, paddy straw mushroom, oyster mushroom, etc.

14.5.2.2 Social assets

Societal connectivity in EFF entrepreneurship provides an opportunity to spread the knowledge and gain to each and every section of the society. It can connect the landless farmers or women to the mainstream agricultural activity. Social, cultural and institutional assets are valuable for reaping the direct and indirect gain of EFF entrepreneurship for different societal units. Social connectivity among the units minimizes the cost and increases the benefits by working together in a societal network. It can generate income for women and make them self-reliant.

14.5.2.3 Human assets

EFF entrepreneurship requires some knowledge and skills for its development. Developing nations are rich in human resources and imparting skills to them would enhance confidence in marginal sections of the society. Men and women including persons with disabilities are able to cope up the need of human resources and related requirements; working together or independently in the society.

14.5.2.4 Physical assets

EFF cultivation and production require some physical assets like clean water, energy, infrastructure, building for storage and transportation facilities. These facilities provide suitable temperature, humidity, ventilation and sanitation necessary for EFF production.

14.5.2.5 Financial assets

Financial assets are basic and important for cultivation of EFF; however, it can be grown by small scale farmer to large scale. Initial requirements to establish a small-scale production unit are raw material and organic waste from locally available resources which are almost free. EFF can be produced and sell in the market within 2–4 months. So, it is beneficial to all those sections which have limited financial assistance.

14.5.3 Analysis of cost and benefits to run entrepreneurship

Feasibility of any mushroom producing unit depends on cost and benefits. Cultivation of EFF plays an important role in helping the local economy by providing sustenance to food security, nutrition, generating additional employment and income through local, regional, and national trade. Generally, the cost of EFF cultivation is lower than small-scale enterprises because cultivation is done on agri-based waste materials. Agri-waste materials are available throughout the world in different localities or parts of the world which can be utilized by the growers in their respective regions. Generally, cultivation of EFF is beneficial to the growers. However, it can be increased by making linkages and collaboration among the farmers as well as marketing areas. It provides valuable food material to grower themselves, to make them healthy as well as sell it to the local and distant markets for making cash benefits. Cash benefits empower them to do or fulfill their other needs of lives.

14.5.4 Success of entrepreneurship

Success of entrepreneurship is basically dependent on zeal and management of inputs and outputs with considering risk factors involved in the process. The EFF cultivation involves infrastructure based on agri-farming with maintenance of agro-climatic conditions required for growth of edible fungi in aseptic environmental conditions. Therefore, rigorous training involvement and sanctity of business trade should be maintained by farmers or growers. These factors will be a milestone for successful cultivation of EFF and income generation as well as maintaining self-esteem in the society. Successful entrepreneurship gives motivation to the growers if they apply innovative technology in an integrated way.

14.6 Conclusion and future prospective

EFF are prominent source of primary and secondary nutrients, and these metabolites are of nutraceutical importance. Also, flavor of EFF is regulated by mycocompounds present in them. The cultivation and production of EFF depend on a few species and majority of EFF are not cultivated on a large scale. However, there is a need to change the cultivation and production trends as there are growing demand for food, nutraceuticals and enzymes important for food security and industrial purposes (Fig. 14.1). EFF can provide an opportunity for income generation for economically marginal population of the world with less input for EFF production. EFF production can be done on agro-industrial wastes and these wastes may further increase the mycoconstituents of industrial significance. There is an opportunity to link agro-industries to mycoceutical industries, and it will also be helpful in maintaining sustainable environment.

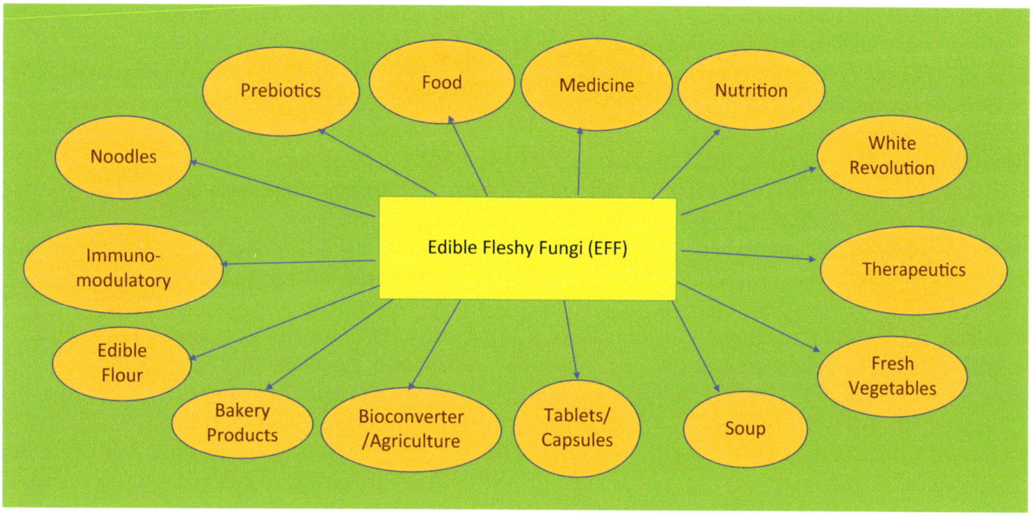

FIG. 14.1

Bioprospection/polyvalent utilization of edible fleshy fungi

References

Afrin, S., Rakib, M.A., Kim, B.H., Kim, J.O., Ha, Y.L., 2016. Eritadenine from edible mushrooms inhibits activity of angiotensin converting enzyme *in vitro*. J. Agric. Food Chem. 64 (11), 2263–2268.

Aisala, H., Sola, J., Hopia, A., Linderborg, K.M., Sandell, M., 2019. Odor-contributing volatile compounds of wild edible Nordic mushrooms analyzed with HS-SPME-GC-MS and HS-SPME-GC-O/FID. Food Chem. 283, 566–578.

Akatin, M.Y., 2013. Characterization of a β-glucosidase from an edible mushroom, *Lycoperdon pyriforme*. Int. J. Food Prop. 16, 1565–1577.

Akiyama, H., Endo, M., Matsui, T., Katsuda, I., Emi, N., Kawamoto, Y., Koike, T., Beppu, H., 2011. Agaritine from *Agaricus blazei* Murrill induces apoptosis in the leukemic cell line U937. Biochim. Biophys. Acta Gen. Subj. 1810 (5), 519–525.

Akpinar, M., Urek, R.O., 2012. Production of ligninolytic enzymes by solid-state fermentation using *Pleurotus eryngii*. Prep. Biochem. Biotechnol. 42 (6), 582–597.

Akpinar, M., Urek, R.O., 2014. Extracellular ligninolytic enzymes production by *Pleurotus eryngii* on agroindustrial wastes. Prep. Biochem. Biotechnol. 44 (8), 772–781.

Alam, N., Yoon, K.N., Lee, J.S., Cho, H.J., Shim, M.J., Lee, T.S., 2011. Dietary effect of *Pleurotus eryngii* on biochemical function and histology in hypercholesterolemic rats. Saudi J. Biol. Sci. 18, 403–409.

Anto, L., Blesso, C.N., 2022. Interplay between diet, the gut microbiome, and atherosclerosis: role of dysbiosis and microbial metabolites on inflammation and disordered lipid metabolism. J. Nutr. Biochem. 105, 108991.

Ao, T., Deb, C.R., 2019. Nutritional and antioxidant potential of some wild edible mushrooms of Nagaland, India. J. Food Sci. Technol. 56 (2), 1084–1089.

Apparoo, Y., Phan, C.W., Kuppusamy, U.R., Sabaratnam, V., 2022. Ergothioneine and its prospects as an anti-ageing compound. Exp. Gerontol. 170, 111982.

Asensio-Grau, A., Calvo-Lerma, J., Heredia, A., Andrés, A., 2020. Enhancing nutritional profile and digestibility of lentil flour by solid state fermentation with *Pleurotus ostreatus*. Food Funct. 11 (9), 7905–7912.

Bach, F., Helm, C.V., Bellettini, M.B., Maciel, G.M., Haminiuk, C.W.I., 2017. Edible mushrooms: a potential source of essential amino acids, glucans and minerals. Int. J. Food Sci. Technol. 52 (11), 2382–2392.

Bach, F., Zielinski, A.A.F., Helm, C.V., Maciel, G.M., Pedro, A.C., Stafussa, A.P., Ávila, S., Haminiuk, C.W.I., 2019. Bio compounds of edible mushrooms: *in vitro* antioxidant and antimicrobial activities. LWT Food Sci. Technol. 107, 214–220.

Backes, E., Kato, C.G., Corrêa, R.C.G., Moreira, R.F.P.M., Peralta, R.A., Barros, L., Ferreira, I.C.F.R., Zanin, G.M., Bracht, A., Peralta, R.M., 2021. Laccases in food processing: current status, bottlenecks and perspectives. Trends Food Sci. Technol. 115, 445–460.

Banerjee, S., Roy, A., Madhusudhan, M.S., Bairagya, H.R., Roy, A., 2019. Structural insights of a cellobiose dehydrogenase enzyme from the basidiomycetes fungus *Termitomyces clypeatus*. Comput. Biol. Chem. 82, 65–73.

Benahmed, A.G., Gasmi, A., Dosa, A., Chirumbolo, S., Mujawdiya, P.K., Aaseth, J., Dadar, M., Bjørklund, G., 2021. Association between the gut and oral microbiome with obesity. Anaerobe 70, 102248.

Berger, B.W., Sallada, N.D., 2019. Hydrophobins: multifunctional biosurfactants for interface engineering. J. Biol. Eng. 13 (10), 1–8.

Bergman, C.J., Gualberto, D.G., Weber, C.W., 1994. Development of a high-temperature-dried soft wheat pasta supplemented with cowpea (*Vigna unguiculata* (L.) Walp.) cooking quality, color, and sensory evaluation. Cereal Chem. 71 (6), 523–527.

Bhanja, S.K., Nandan, C.K., Mandal, S., Bhunia, B., Maiti, T.K., Mondal, S., Islam, S.S., 2012. Isolation and characterization of the immunostimulating β-glucans of an edible mushroom *Termitomyces robustus* var. Carbohydr. Res. 357, 83–89.

Bose, S., Mazumder, S., Mukherjee, M., 2007. Laccase production by the white-rot fungus *Termitomyces clypeatus*. J. Basic Microbiol. 47, 127–131.

Bouzgarrou, C., Amara, K., Reis, F.S., Barreira, J.C.M., Skhiri, F., Chatti, N., Martins, A., Barros, L., Ferreira, I.C.F.R., 2018. Incorporation of tocopherol-rich extracts from mushroom mycelia into yogurt. Food Funct. 9, 3166–3172.

Cao, X., Liu, D., Bi, R., He, Y., He, Y., Liu, J., 2019. The protective effects of a novel polysaccharide from Lentinus edodes mycelia on islet β (INS-1) cells damaged by glucose and its transportation mechanism with human serum albumin. Int. J. Biol. Macromol. 134, 344–353.

Cardwell, G., Bornman, J.F., James, A.P., Black, L.J., 2018. A review of mushrooms as a potential source of dietary vitamin D. Nutrients 10, 1498.

Chachage, M., 2021. The gut-microbiome contribution to HIV-associated cardiovascular disease and metabolic disorders. Curr. Opin. Endocr. Metab. Res. 21, 100287.

Chaurasia, M.K., 2004. Studies on Protein Nutritive Value and Cultural Characteristic of some Fleshy fungi Particularly Mushrooms. Ph. D. Thesis. Department of Botany, DDU Gorakhpur University, Gorakhpur, pp. 1–163.

Cheah, I.K., Halliwell, B., 2012. Ergothioneine; antioxidant potential, physiological function and role in disease. Biochim. Biophys. Acta Mol. basis Dis. 1822 (5), 784–793.

Cheah, I.K., Halliwell, B., 2021. Ergothioneine, recent developments. Redox Biol. 42, 101868.

Chen, Y., Lan, P., 2018. Total syntheses and biological evaluation of the *Ganoderma lucidum* alkaloids Lucidimines B and C. ACS Omega 3, 3471–3481.

Chen, K., Liu, X., Long, L., Ding, S., 2017. Cellobiose dehydrogenase from *Volvariella volvacea* and its effect on the saccharification of cellulose. Process Biochem. 60, 52–58.

Chen, S., Yong, T., Xiao, C., Su, J., Zhang, Y., Jiao, C., Xie, Y., 2018. Pyrrole alkaloids and ergosterols from *Grifola frondosa* exert anti-α-glucosidase and anti-proliferative activities. J. Funct. Foods 43, 196–205.

Chen, G., Wang, M., Zeng, Z., Xie, M., Xu, W., Peng, Y., Zhou, W., Sun, Y., Zeng, X., Liu, Z., 2022. Fuzhuan brick tea polysaccharides serve as a promising candidate for remodeling the gut microbiota from colitis subjects *in vitro*: fermentation characteristic and anti-inflammatory activity. Food Chem. 391, 133203.

Chilanti, G., Da Rosa, L.O., Poleto, L., Branco, C.S., Camassola, M., Fotana, R.C., Dillon, A.J.P., 2022. Effect of different substrates on *Pleurotus* spp. cultivation in Brazil- Ergothioneine and Lovastatin. J. Food Compos. Anal. 107, 104367.

Cho, I.H., Choi, H., Kim, Y., 2006. Difference in the volatile composition of pine-mushrooms (*Tricholoma matsutake* Sing.) according to their grades. J. Agric. Food Chem. 54, 4820–4825.

Corrêa, R.C.G., Brugnari, T., Bracht, A., Peralta, R.M., Ferreira, I.C.F.R., 2016. Biotechnological, nutritional and therapeutic uses of *Pleurotus* spp. (Oyster mushroom) related with its chemical composition: a review on the past decade findings. Trends Food Sci. Technol. 50, 103–117.

Corrêa, R.C.G., Barros, L., Fernandes, A., Sokovic, M., Bracht, A., Peralta, R.M., Ferreira, I.C.F.R., 2018. A natural food ingredient based on ergosterol: optimization of the extraction from *Agaricus blazei*, evaluation of bioactive properties and incorporation in yogurts. Food Funct. 9, 1465–1474.

Cruz-Pereira, J.S., Moloney, G.M., Bastiaanssen, T.F.S., Boscaini, S., Tofani, G., Borras-Bisa, J., Van de Wouw, M., Fitzgerald, P., Dinan, T.G., Clarke, G., Cryan, J.F., 2022. Prebiotic supplementation modulates selective effects of stress on behavior and brain metabolome in aged mice. Neurobiol. Stress 21, 100501.

Daniewski, W.M., Danikiewicz, W., Gołębiewski, W.M., Gucma, M., Łysik, A., Grodner, J., Przybysz, E., 2012. Search for bioactive compounds from *Cantharellus cibarius*. Nat. Prod. Commun. 7 (7), 917–918.

Dare, A., Channa, M.L., Nadar, A., 2022. L-ergothioneine; a potential adjuvant in the management of diabetic nephropathy. Pharmacol. Res.-Mod. Chin. Med. 2, 100033.

Diana, M., Quílez, J., Rafecas, M., 2014. Gamma-aminobutyric acid as a bioactive compound in foods: a review. J. Funct. Foods 10, 407–420.

Dokouhaki, M., Hung, A., Kasapis, S., Gras, S.L., 2021. Hydrophobins and chaplins: novel bio-surfactants for food dispersions a review. Trends Food Sci. Technol. 111, 378–387.

El-Fattah, A.A., Sakr, S., El-Dieb, S., Elkashef, H., 2018. Developing functional yogurt rich in bioactive peptides and gamma-aminobutyric acid related to cardiovascular health. LWT Food Sci. Technol. 98, 390–397.

Ergönül, P.G., Akata, I., Kalyoncu, F., Ergönül, B., 2013. Fatty acids compositions of six wild edible mushroom species. Sci. World J. 2013, 163964.

Falandysz, J., 2008. Selenium in edible mushrooms. J. Environ. Sci. Health C Environ. Carcinog. Ecotoxicol. Rev. 26 (3), 256–299.

Feng, T., Li, Z., Dong, Z., Su, J., Li, Y., Liu, J., 2011. Non-isoprenid botryane sesquiterpenoids from basidiomycete *Boletus edulis* and their cytotoxic activity. Nat. Prod. Bioprospect. 1, 29–32.

Friedman, M., 2015. Chemistry, nutrition, and health-promoting properties of *Hericium erinaceus* (Lion's Mane) mushroom fruiting bodies and mycelia and their bioactive compounds. J. Agric. Food Chem. 63 (32), 7108–7123.

Frishman, W.H., Rapier, R.C., 1989. Lovastatin: an HMG-CoA reductase inhibitor for lowering cholesterol. Med. Clin. North Am. 73 (2), 437–448.

Fukushima-Sakuno, E., 2020. Bioactive small secondary metabolites from the mushrooms *Lentinula edodes* and *Flammulina velutipes*. J. Antibiot. 73 (10), 687–696.

Garavand, F., Daly, D.F.M., Gómez-Mascaraque, L.G., 2022. Biofunctional, structural, and tribological attributes of GABA-enriched probiotic yoghurts containing *Lacticaseibacillus paracasei* alone or in combination with prebiotics. Int. Dairy J. 129, 105348.

Garcia, J., Rodrigues, F., Castro, F., Aires, A., Marques, G., Saavedra, M.J., 2022. Antimicrobial, antibiofilm, and antioxidant properties of *Boletus edulis* and *Neoboletus luridiformis* against multidrug-resistant ESKAPE pathogens. Front. Nutr. 8, 773346.

Geng, X., Tian, G., Zhang, W., Zhao, Y., Zhao, L., Wang, H., Ng, T.B., 2016. A *Tricholoma matsutake* peptide with angiotensin converting enzyme inhibitory and antioxidative activities and antihypertensive effects in spontaneously hypertensive rats. Sci. Rep. 6, 24130.

Ghosh, A.K., Naskar, A.K., Sengupta, S., 1997. Characterisation of a xylanolytic amyloglucosidase of *Termitomyces clypeatus*. Biochim. Biophys. Acta Protein Struct. Mol. Enzymol. 1339 (2), 289–296.

Ghosh, K., Chandra, K., Roy, S.K., Mondal, S., Maity, D., Das, D., Ojha, A.K., Islam, S.S., 2008. Structural investigation of a polysaccharide (Fr. I) isolated from the aqueous extract of an edible mushroom, *Volvariella diplasia*. Carbohydr. Res. 343, 1071–1078.

Ghosh, S., Khatua, S., Dasgupta, A., Acharya, K., 2021. Crude polysaccharide from the milky mushroom, *Calocybe indica*, modulates innate immunity of macrophage cells by triggering MyD88-dependent TLR4/NF-κB pathway. J. Pharm. Pharmacol. 73 (1), 70–81.

Gibson, G.R., Probert, H.M., Loo, J.V., Rastall, R.A., Roberfroid, M.B., 2004. Dietary modulation of the human colonic microbiota: updating the concept of prebiotics. Nutr. Res. Rev. 17, 259–275.

Gil-Ramírez, A., Caz, V., Smiderle, F.R., Martin-Harnandez, R., Largo, C., Tabernero, M., Marín, F.R., Lacomini, M., Reglero, G., Soler-Rivas, C., 2016. Water-soluble compounds from *Lentinula edodes* influencing the HMG-CoA reductase activity and the expression of genes involved in the cholesterol metabolism. J. Agric. Food Chem. 64, 1910–1920.

Green, A.J., Littlejohn, K.A., Hooley, P., Cox, P.W., 2013. Formation and stability of food foams and aerated emulsions: hydrophobins as novel functional ingredients. Curr. Opin. Colloid Interface Sci. 18 (4), 292–301.

Gupta, N., Srivastava, A.K., Pandey, V.N., 2012. Biodiversity and nutraceutical quality of some Indian millets. Proc. Natl. Acad. Sci. India Sect. B Biol. Sci. 82, 265–273.

Hammann, S., Lehnert, K., Vetter, W., 2016. Esterified sterols and their contribution to the total sterols in edible mushrooms. J. Food Compos. Anal. 54, 48–54.

Hanahan, D., Weinberg, R.A., 2011. Hallmarks of cancer: the next generation. Cell 144, 646–674.

He, X., Du, X., Zang, X., Dong, L., Gu, Z., Cao, L., Chen, D., Keyhani, N.O., Yao, L., Qiu, J., Guan, X., 2016. Extraction, identification and antimicrobial activity of a new furanone, grifolaone A, from *Grifola frondosa*. Nat. Prod. Res. 30 (8), 941–947.

Heleno, S.A., Barros, L., Sousa, M.J., Martins, A., Santos-Buelga, C., Ferreira, I.C.F.R., 2011. Targeted metabolites analysis in wild *Boletus* species. LWT Food Sci. Technol. 44, 1343–1348.

Heo, S., Jeon, S., Lee, S., 2014. Utilization of Lentinus edodes mushroom β-glucan to enhance the functional properties of gluten-free rice noodles. LWT Food Sci. Technol. 55, 627–631.

Hirotani, M., Sai, K., Hirotani, S., Yoshikawa, T., 2002. Blazeispirols B, C, E and F, des-*A*-ergostane-type compounds, from the cultured mycelia of the fungus *Agaricus blazei*. Phytochemistry 59 (5), 571–577.

Itoh, H., Ito, H., Hibasami, H., 2008. Blazein of a new steroid isolated from *Agaricus blazei* Murrill (himematsutake) induces cell death and morphological change indicative of apoptotic chromatin condensation in human lung cancer LU99 and stomach cancer KATO III cells. Oncol. Rep. 20 (6), 1359–1361.

Iwashita, K., 2002. Recent studies of protein secretion by filamentous fungi. J. Biosci. Bioeng. 94 (6), 530–535.

Jenkins, D.J.A., Wolever, T.M.S., Taylor, R.H., Barker, H., Fielden, H., Baldwin, J.M., Bowling, A.C., Newman, H.C., Jenkins, A.L., Goff, D.V., 1981. Glycemic index of foods: a physiological basis for carbohydrate exchange. Am. J. Clin. Nutr. 34 (3), 362–366.

Jenkins, A.L., Jenkins, D.J.A., Wolever, T.M.S., Rogovik, A.L., Jovanovski, E., Božikov, V., Rahelić, D., Vuksan, V., 2008. Comparable postprandial glucose reductions with viscous fiber blend enriched biscuits in healthy subjects and patients with diabetes mellitus: acute randomized controlled clinical trial. Croat. Med. J. 49 (6), 772–782.

Johnston, S.R.D., Dowsett, M., 2003. Aromatase inhibitors for breast cancer: lessons from the laboratory. Nat. Rev. Cancer 3 (11), 821–831.

Jonathan, S.G., Adeoyo, O.R., 2011. Evaluation of ten wild Nigerian mushrooms for amylase and cellulase activities. Mycobiology 39 (2), 103–108.

Jung, D.Y., Lee, H.J., Shin, D., Kim, C.H., Jo, C., 2022. Mechanism of improving emulsion stability of emulsion-type sausage with oyster mushroom (*Pleurotus ostreatus*) powder as a phosphate replacement. Meat Sci. 194, 108993.

Kaewnarin, K., Suwannarach, N., Kumla, J., Lumyong, S., 2016. Phenolic profile of various wild dibble mushroom extracts from Thailand and their antioxidant properties, anti-tyrosinase and hyperglycaemic inhibitory activities. J. Funct. Foods 27, 352–364.

Kang, M., Kim, Y., Bolormaa, Z., Kim, M., Seo, G., Lee, J., 2013. Characterization of an antihypertensive angiotensin I- converting enzyme inhibitory peptide from the edible mushroom *Hypsizygus marmoreus*. Biomed. Res. Int. 2013, 283964.

Kansci, G., Mossebo, D.C., Selatsa, A.B., Fotso, M., 2003. Nutrient content of some mushroom species of the genus *Termitomyces* consumed in Cameroon. Nahrung/Food 3, 213–216.

Katsube, M., Watanabe, H., Suzuki, K., Ishimoto, T., Tatebayashi, Y., Kato, Y., Murayama, N., 2022. Food-derived antioxidant ergothioneine improves sleep difficulties in humans. J. Funct. Foods 95, 105165.

Kaya, M., Cam, M., 2022. Eritadenine: pressurized liquid extraction from *Lentinula edodes* and thermal degradation kinetics. Sustain. Chem. Pharm. 29, 100809.

Kemp, S.E., 2008. Application of sensory evaluation in food research. Int. J. Food Sci. Technol. 43, 1507–1511.

Khan, I., Huang, G., Li, X., Leong, W., Xia, W., Hsiao, W.L.W., 2018. Mushroom polysaccharides from *Ganoderma lucidum* and *Poria cocos* reveal prebiotic functions. J. Funct. Foods 41, 191–201.

Khumlianlal, J., Sharma, K.C., Singh, L.M., Mukherjee, P.K., Indira, S., 2022. Nutritional profiling and antioxidant property of three wild edible mushrooms from North East India. Molecules 27, 5423.

Kikuchi, T., Maekawa, Y., Tomio, A., Masumoto, Y., Yamamoto, T., In, Y., Yamada, T., Tanaka, R., 2016. Six new ergostane-type steroids from king trumpet mushroom (*Pleurotus eryngii*) and their inhibitory effects on nitric oxide production. Steroids 115, 9–17.

Kikuchi, T., Horii, Y., Maekawa, Y., Masumoto, Y., In, Y., Tomoo, K., Sato, H., Yamano, A., Yamada, T., Tanaka, R., 2017a. Pleurocins A and B: unusual 11(9 → 7)-abeo-Ergostanes and Eringiacetal B: a 13,14-seco-13,14-epoxyergostane from fruiting bodies of *Pleurotus eryngii* and their inhibitory effects on nitric oxide production. J. Org. Chem. 82 (19), 10611–10616.

Kikuchi, T., Motoyashiki, N., Yamada, T., Shibatani, K., Ninomiya, K., Morikawa, T., Tanaka, R., 2017b. Ergostane-type sterols from king trumpet mushroom (*Pleurotus eryngii*) and their inhibitory effects on aromatase. Int. J. Mol. Sci. 18, 2479.

Kikuchi, T., Kitaura, K., Katsumoto, A., Zhang, J., Yamada, T., Tanaka, R., 2018. Three bisabolane-type sesquiterpenes from edible mushroom *Pleurotus eryngii*. Fitoterapia 129, 108–113.

Kikuchi, T., Isobe, M., Uno, S., In, Y., Zhang, J., Yamada, T., 2019. Strophasterols E and F: rearranged ergostane-type sterols from *Pleurotus eryngii*. Bioorg. Chem. 89, 103011.

Kim, H.Y., Yokozawa, T., Nakagawa, T., Sasaki, S., 2004. Protective effect of γ-aminobutyric acid against glycerol-induced acute renal failure in rats. Food Chem. Toxicol. 42 (12), 2009–2014.

Kim, J., Lau, E., Tay, D., De Blanco, E.J.C., 2011a. Antioxidant and NF-κB inhibitory constituents isolated from *Morchella esculenta*. Nat. Prod. Res. 25 (15), 1412–1417.

Kim, J., Lee, S.M., Bae, I.Y., Park, H., Lee, H.G., Lee, S., 2011b. (1-3) (1-6)-β-glucan-enriched materials from *Lentinus edodes* mushroom as a high-fibre and low-calorie flour substitute for baked foods. J. Sci. Food Agric. 91, 1915–1919.

Kim, H., Lee, C., Park, Y., 2016a. Further characterization of hydrophobin genes in genome of *Flammulina velutipes*. Mycoscience 57 (5), 320–325.

Kim, S.H., Lee, J., Heo, Y., Moon, B., 2016b. Effect of *Pleurotus eryngii* mushroom β-glucan on quality characteristics of common wheat pasta. J. Food Sci. 81 (4), C835–C840.

Kondyli, E., Pappa, E.C., Arapoglou, D., Metafa, M., Eliopoulos, C., Israilides, C., 2022. Effect of fortification with mushroom polysaccharide β-glucan on the quality of ovine soft spreadable cheese. Foods 11 (3), 417.

Koo, M.H., Chae, H., Lee, J.H., Suh, S., Youn, U.J., 2021. Antiinflammatory lanostane triterpenoids from *Ganoderma lucidum*. Nat. Prod. Res. 35 (22), 4295–4302.

Koyama, K., Akiba, M., Imaizumi, T., Kinoshita, K., Takahashi, K., Suzuki, A., Yano, S., Horie, S., Watanabe, K., 2002. Antinociceptive constituents of *Auricularia polytricha*. Planta Med. 68 (3), 284–285.

Krümmel, A., Rodrigues, L.G.G., Vitali, L., Regina, S., Ferreira, S., 2022. Bioactive compounds from *Pleurotus sajor-caju* mushroom recovered by sustainable high-pressure methods. LWT Food Sci. Technol. 160, 113316.

Kulkarni, S., Nene, S.N., Joshi, K.S., 2020. A comparative study of production of hydrophobin like proteins (HYD-LPs) in submerged liquid and solid-state fermentation from white rot fungus *Pleurotus ostreatus*. Biocatal. Agric. Biotechnol. 23, 101440.

Kulkarni, S.S., Nene, S.N., Joshi, K.S., 2022. Identification and characterization of a hydrophobin Vmh3 from *Pleurotus ostreatus*. Protein Expr. Purif. 195-196, 106095.

Kumura, H., Ohtsuyama, T., Matsusaki, Y., Taitoh, M., Koyanagi, H., Kobayashi, K., Hayakawa, T., Wakamatsu, J., 2018. Application of red pigment producing edible fungi for development of a novel type of functional cheese. J. Food Process. Preserv. 2018, e13707.

Lakhundi, S., Siddiqui, R., Khan, N., 2015. Cellulose degradation: a therapeutic strategy in the improved treatment of *Acanthamoeba* infections. Parasit. Vectors 8, 23.

Lau, W.L., Tran, T., Rhee, C.M., Kalantar-Zadeh, K., Vaziri, N.D., 2021. Diabetes and the gut microbiome. Semin. Nephrol. 41 (2), 104–113.

Lee, S.R., Roh, H., Lee, S., Park, H.B., Jang, T.S., Ko, Y., Baek, K., Kim, K.H., 2018. Bioactivity-guided isolation and chemical characterization of antiproliferative constituents from morel mushroom (*Morchella esculenta*) in human lung adenocarcinoma cells. J. Funct. Foods 40, 249–260.

Lee, Y.H., Verma, N.K., Thanabalu, T., 2021. Prebiotics in atopic dermatitis prevention and management. J. Funct. Foods 78, 104352.

Li, P., Deng, Y., Wei, X., Xu, J., 2013. Triterpenoids from *Ganoderma lucidum* and their cytotoxic activities. Nat. Prod. Res. 27 (1), 17–22.

Li, H., Lee, H., Kim, S., Moon, B., Lee, C., 2014. Antioxidant and anti-inflammatory activities of methanol extracts of *Tremella fuciformis* and its major phenolic acids. J. Food Sci. 79 (4), C460–C468.

Li, Y., Elmén, L., Segota, I., Xian, Y., Tinoco, R., Feng, Y., Fujita, Y., Muñoz, R.R.S., Schmaltz, R., Bradley, L.M., Ramer-Tait, A., Zarecki, R., Long, T., Peterson, S.N., Ronai, Z.A., 2020. Prebiotic-induced anti-tumor immunity attenuates tumor growth. Cell Rep. 30 (6), 1753–1766.e6.

Li, H., Tian, Y., Menolli Jr., N., Ye, L., Karunarathna, S.C., Perez-Moreno, J., Rahman, M.M., Rashid, M.H., Phengsintham, P., Rizal, L., Kasuya, T., Lim, Y.W., Dutta, A.K., Khalid, A.N., Huyen, L.T., Balolong, M.P., Baruah, G., Madawala, S., Thongklang, N., Hyde, K.D., Kirk, P.M., Xu, J., Sheng, J., Boa, E., Mortimer, P.E., 2021. Reviewing the world's edible mushroom species: a new evidence-based classification system. Compr. Rev. Food Sci. Food Saf. 20, 1982–2014.

Lin, J., Liu, C., Chen, Y., Hu, C., Juang, L., Shiesh, C., Yang, D., 2014. Chemical composition, antioxidant and anti-inflammatory properties for ethanolic extracts from *Pleurotus eryngii* fruiting bodies harvested at different time. LWT Food Sci. Technol. 55, 374–382.

Liu, F., Luo, K., Yu, Z., Co, N., Wu, S., Wu, P., Fung, K., Kwok, T., 2009. Suillin from the mushroom *Suillus placidus* as potent apoptosis inducer in human hepatoma HepG2 cells. Chem. Biol. Interact. 181, 168–174.

Liu, E., Ji, Y., Zhang, F., Liu, B., Meng, X., 2021. Review on *Auricularia auricula-judae* as a functional food: growth, chemical composition, and biological activities. J. Agric. Food Chem. 69, 1739–1750.

Lo, Y., Lin, S., Ulziijargal, E., Chen, S., Chien, R., Tzou, Y., Mau, J., 2012. Comparative study of contents of several bioactive components in fruiting bodies and mycelia of culinary-medicinal mushrooms. Int. J. Med. Mushrooms 14 (4), 357–363.

Ma, G., Yang, W., Zhao, L., Pei, F., Fang, D., Hu, Q., 2018. A critical review on the health promoting effects of mushrooms. Food Sci. Human Wellness 7, 125–133.

Ma, G., Kimatu, B.M., Yang, W., Pei, F., Zhao, L., Du, H., Su, A., Hu, Q., Xiao, H., 2020. Preparation of newly identified polysaccharide from *Pleurotus eryngii* and its anti-inflammation activities potential. J. Food Sci. 85 (9), 2822–2831.

Maheshwari, G., Gessner, D.K., Neuhaus, K., Most, E., Zorn, H., Eder, K., Ringseis, R., 2021. Influence of a biotechnologically produced oyster mushroom (*Pleurotus sajor-caju*) on the gut microbiota and microbial metabolites in obese zucker rats. J. Agric. Food Chem. 69, 1524–1535.

Maity, P., Nandi, A.K., Pattanayak, M., Manna, D.K., Sen, I.K., Chakraborty, I., Bhanja, S.K., Sahoo, A.K., Gupta, N., Islam, S.S., 2020. Structural characterization of a heteroglycan from an edible mushroom *Termitomyces heimii*. Int. J. Biol. Macromol. 151, 305–311.

Maity, P., Sen, I.K., Chakraborty, I., Mondal, S., Bar, H., Bhanja, S.K., Mandal, S., Maity, G.N., 2021. Biologically active polysaccharide from edible mushrooms: a review. Int. J. Biol. Macromol. 172, 408–417.

Majumder, R., Banik, S.P., Khowala, S., 2016. AkP from mushroom *Termitomyces clypeatus* is a proteoglycan specific protease with apoptotic effect on HepG2. Int. J. Biol. Macromol. 91, 198–207.

Manna, D.K., Nandi, A.K., Pattanayak, M., Maity, P., Tripathy, S., Mandal, A.K., Roy, S., Tripathy, S.S., Gupta, N., Islam, S.S., 2015. A water soluble β-glucan of an edible mushroom *Termitomyces heimii*: structural and biological investigation. Carbohydr. Polym. 134, 375–384.

Mannina, L., Cristinzio, M., Sobolev, A.P., Ragni, P., Segre, A., 2004. High-field nuclear magnetic resonance (NMR) study of truffles (*Tuber aestivum* Vittadini). J. Agric. Food Chem. 52, 7988–7996.

Marathe, S.J., Hamzi, W., Bashein, A.M., Deska, J., Seppänen-Laakso, T., Singhal, R.S., Shamekh, S., 2020. Anti-angiogenic and anti-inflammatory activity of the summer truffle (*Tuber aestivum* Vittad.) extracts and a correlation with the chemical constituents identified therein. Food Res. Int. 137, 109699.

Marshall, E., Nair, N.G., 2009. Make Money by Growing Mushrooms. Rural Infrastructure and Agro-Industries Division. Food and Agricultural Organization of the United Nations, Rome.

Mau, J., Tseng, Y., 1998. Nonvolatile taste components of three strains of *Agrocybe cylindracea*. J. Agric. Food Chem. 46 (6), 2071–2074.

Mau, J., Chyau, C., Li, J., Tseng, Y., 1997. Flavor compounds in straw mushrooms *Volvariella volvacea* harvested at different stages of maturity. J. Agric. Food Chem. 45 (12), 4726–4729.

Meng, M., Huo, R., Wang, Y., Ma, N., Shi, X., Shen, X., Chang, G., 2022. Lentinan inhibits oxidative stress and alleviates LPS-induced inflammation and apoptosis of BMECs by activating the Nrf2 signaling pathway. Int. J. Biol. Macromol. 222 (B), 2375–2391.

Mikolasch, A., Schauer, F., 2009. Fungal laccases as a tool for the synthesis of new hybrid molecules and biomaterials. Appl. Microbiol. Biotechnol. 82, 605–624.

Misgiati, M., Widyawaruyanti, A., Raharjo, S.J., Sukardiman, S., 2021. Ergosterol isolated from *Agaricus blazei* Murill n-hexane extracts as potential anticancer MCF-7 activity. Pharmacogn. J. 13 (2), 418–426.

Möhler, H., 2012. The GABA system in anxiety and depression and its therapeutic potential. Neuropharmacology 62, 42–53.

Mondal, S., Chandra, K., Maiti, D., Ojha, A.K., Das, D., Roy, S.K., Ghosh, K., Chakarborty, I., Islam, S.S., 2008. Chemical analysis of a new fucoglucan isolated from an edible mushroom, *Termitomyces robustus*. Carbohydr. Res. 343, 1062–1070.

Mondal, A., Banerjee, D., Majumder, R., Maity, T.K., Khowala, S., 2016. Evaluation of *in vitro* antioxidant, anticancer and *in vivo* antitumour activity of *Termitomyces clypeatus* MTCC 5091. Pharm. Biol. 54 (11), 2536–2546.

Moumita, S., Das, B., 2022. Assessment of the prebiotic potential and bioactive components of common edible mushrooms in India and formulation of synbiotic microcapsules. LWT Food Sci. Technol. 156, 113050.

Mozaffari, S.A., Salehi, A., Mousavi, E., Zaman, B.A., Nassaj, A.E., Ebrahimzadeh, F., Nasiri, H., Valedkarimi, Z., Adili, A., Asemani, G., Akbari, M., 2022. SARS-CoV-2-associated gut microbiome alteration; a new contributor to colorectal cancer pathogenesis. Pathol. Res. Pract. 239, 154131.

Nakatni, Y., Fukaya, T., Kishino, S., Ogawa, J., 2022. Production of GABA-enriched tomato juice by *Lactiplantibacillus plantarum* KB1253. J. Biosci. Bioeng. 134 (5), 424–431.

Ng, S.H., Robert, S.D., Ahmad, W.A.N.W., Ishak, W.R.W., 2017. Incorporation of dietary fibre-rich oyster mushroom (*Pleurotus sajor-caju*) powder improves postprandial glycaemic response by interfering with starch granule structure and starch digestibility of biscuit. Food Chem. 227, 358–368.

Novak, M., Vetvicka, V., 2008. β-Glucans, history, and the present: immunomodulatory aspects and mechanism of action. J. Immunotoxicol. 5 (1), 47–57.

Nowacka, N., Nowak, R., Drozd, M., Olech, M., Los, R., Malm, A., 2014. Analysis of phenolic constituents, antiradical and antimicrobial activity of edible mushrooms growing wild in Poland. LWT Food Sci. Technol. 59, 689–694.

Nzekoue, F.K., Sun, Y., Caprioli, G., Vittori, S., Sagratini, G., 2022. Effect of the ultrasound-assisted extraction parameters on the determination of ergosterol and vitamin D_2 in *Agaricus bisporus*, *A. bisporus* Portobello, and *Pleurotus ostreatus* mushrooms. J. Food Compos. Anal. 109, 104476.

Ohira, H., Tsutsui, W., Fujioka, Y., 2017. Are short chain fatty acids in gut microbiota defensive players for inflammation and atherosclerosis? J. Atheroscler. Thromb. 24 (7), 660–672.

Pal, S., Banik, S.P., Ghorai, S., Chowdhury, S., Khowala, S., 2010. Purification and characterization of a thermostable intra-cellular b-glucosidase with transglycosylation properties from filamentous fungus *Termitomyces clypeatus*. Bioresour. Technol. 101, 2412–2420.

Pandey, V.N., Srivastava, A.K., 2004. Fleshy fungi of ethnobotanical food use of North Eastern Terai region of U.P. In: Proceedings of national symposium on mushroom, NCMRT, Solan (H.P.).

Pandey, V.N., Kamal, Srivastava, A.K., 1993. Diversity of fungi and its role in biotechnology. In: Proceedings of national conference on biotechnology, Tribhuvan University, Kathmandu, Nepal, April 29-30, 1993.

Phillips, K.M., Ruggio, D.M., Horst, R.L., Minor, B., Simon, R.R., Feeney, M.J., Byrdwell, W.C., Haytowitz, D.B., 2011. Vitamin D and sterol composition of 10 types of mushrooms from retail suppliers in the United States. J. Agric. Food Chem. 59, 7841–7853.

Prodhan, U.K., Linkon, K.M.M.R., Al-Amin, M.F., Alam, M.J., 2015. Development and quality evaluation of mushroom (*Pleurotus sajor-caju*) enriched biscuits. Emir. J. Food Agric. 27 (7), 542–547.

Proserpio, C., Lavelli, V., Laureati, M., Pagliarini, E., 2019. Effect of *Pleurotus ostreatus* powder addition in vegetable soup on β-glucan content, sensory perception, and acceptability. Food Sci. Nutr. 7, 730–737.

Puttaraju, N.G., Venkateshaiah, S.U., Dharmesh, S.M., Urs, S.M.N., Somasundaram, R., 2006. Antioxidant activity of indigenous edible mushrooms. J. Agric. Food Chem. 54, 9764–9772.

Rafeeq, C.M., Vaishnav, A.B., Ali, P.P.M., 2021. Characterisation and comparative analysis of hydrophobin isolated from *Pleurotus floridanus* (PfH). Protein Expr. Purif. 182, 105834.

Rahi, D.K., Malik, D., 2016. Diversity of mushrooms and their metabolites of nutraceutical and therapeutic significance. J. Mycol. 2016, 7654123.

Raiola, A., Romano, A., Shanakhat, H., Masi, P., Cavella, S., 2020. Impact of heat treatments on technological performance of re-milled semolina dough and bread. LWT Food Sci. Technol. 117, 108607.

Ram, R.C., Pandey, V.N., Singh, H.B., 2010. Morphological characterization of edible fleshy fungi from different forest regions. Indian J. Sci. Res. 1 (2), 33–35.

Raman, J., Jang, K., Oh, Y., Oh, M., Im, J., Lakshmanan, H., Sabaratnam, V., 2021. Cultivation and nutritional value of prominent *Pleurotus* spp.: an overview. Mycobiology 49 (1), 1–14.

Ramos, I.M., Poveda, J.M., 2022. Fermented sheep's milk enriched in gamma-amino butyric acid (GABA) by the addition of lactobacilli strains isolated from different food environments. LWT Food Sci. Technol. 163, 113581.

Rao, J.R., Millar, B.C., Moore, J.E., 2009. Antimicrobial properties of shitake mushrooms (*Lentinula edodes*). Int. J. Antimicrob. Agents 33, 591–592.

Ray, R., Saha, S., Paul, S., 2022. Two novel compounds, ergosterol and ergosta-5,8-dien-3-ol, from *Termitomyces heimii* Natarajan demonstrate promising anti-hepatocarcinoma activity. J. Tradit. Chin. Med. Sci. 9, 443–453.

Rizkyana, A.D., Ho, T.C., Roy, V.C., Park, J., Kiddane, A.T., Kim, G., Chun, B., 2022. Sulfation and characterization of polysaccharides from Oyster mushroom (*Pleurotus ostreatus*) extracted using subcritical water. J. Supercrit. Fluids 179, 105412.

Romano, A., Ferranti, P., Gallo, V., Masi, O., 2021. New ingredients and alternatives to durum wheat semolina for a high quality dried pasta. Curr. Opin. Food Sci. 41, 249–259.

Ruthes, A.C., Cantu-Jungles, T.M., Cordeiro, L.M.C., Lacomini, M., 2021. Prebiotic potential of mushroom D-glucans: implications of physicochemical properties and structural features. Carbohydr. Polym. 262, 117940.

Saha, T., Ghosh, D., Mukherjee, S., Bose, S., Mukherjee, M., 2008. Cellobiose dehydrogenase production by the mycelial culture of the mushroom *Termitomyces clypeatus*. Process Biochem. 43, 634–641.

Saha, T., Sasmal, S., Alam, S., Das, N., 2014. Tamarind kernel powder: a novel agro-residue for the production of cellobiose dehydrogenase under submerged fermentation by *Termitomyces clypeatus*. J. Agric. Food Chem. 62, 3438–3445.

Saini, R.K., Rauf, A., Khalil, A.A., Ko, E., Keum, Y., Anwar, S., Alamri, A., Rengasamy, K.R.R., 2021. Edible mushrooms show significant differences in sterols and fatty acid compositions. S. Afr. J. Bot. 141, 344–356.

Salvachúa, D., Prieto, A., Mattinen, M., Tamminen, T., Liitiä, T., Lille, M., Willför, S., Martínez, A.T., Martínez, M.J., Faulds, C.B., 2013. Versatile peroxidase as a valuable tool for generating new biomolecules by homogeneous and heterogeneous cross-linking. Enzym. Microb. Technol. 52 (6–7), 303–311.

Sánchez-Minutti, L., López-Valdez, F., Rosales-Pérez, M., Luna-Suárez, S., 2019. Effect of heat treatments of *Lentinula edodes* mushroom on eritadenine concentration. LWT Food Sci. Technol. 102, 364–371.

Sande, D., De Oliveira, G.P., E Moura, M.A.F., Martins, B.A., Lima, M.T.N.S., Takahashi, J.A., 2019. Edible mushrooms as a ubiquitous source of essential fatty acids. Food Res. Int. 125, 108524.

Santiago-López, L., Aguilar-Toalá, J.E., Hernández-Mendoza, A., Vallejo-Cordoba, B., Liceaga, A.M., González-Cordóva, A.F., 2018. Invited review: bioactive compounds produced during cheese ripening and health effects associated with aged cheese consumption. J. Dairy Sci. 101 (5), 3742–3757.

Sarkar, P., Roy, A., 2014. Molecular cloning, characterization and expression of a gene encoding phosphoketolase from *Termitomyces clypeatus*. Biochem. Biophys. Res. Commun. 447, 621–625.

Sato, M., Miyagi, A., Yoneyama, S., Gisusi, S., Tokuji, Y., Kawai-Yamada, M., 2017. CE-MS-based metabolomics reveals the metabolic profile of maitake mushroom (*Grifola frondosa*) strains with different cultivation characteristics. Biosci. Biotechnol. Biochem. 81 (12), 2314–2322.

Sawangwan, T., Wansanit, W., Pattani, L., Noysang, C., 2018. Study of prebiotic properties from edible mushroom extraction. Agric. Nat. Resour. 52, 519–524.

Scheiblbrandner, S., Ludwig, R., 2020. Cellobiose dehydrogenase: bioelectrochemical insights and applications. Bioelectrochemistry 131, 107345.

Sharma, C., Bhardwaj, N., Sharma, A., Tuli, H.S., Batra, P., Beniwal, V., Gupta, G.K., Sharma, A.K., 2019. Bioactive metabolites of *Ganoderma lucidum*: factors, mechanism and broad-spectrum therapeutic potential. J. Herb. Med. 17-18, 100268.

Shimada, Y., Morita, T., Sugiyama, K., 2003a. Eritadenine-induced alterations of plasma lipoprotein lipid concentrations and phosphatidylcholine molecular species profile in rats fed cholesterol-free and cholesterol enriched diets. Biosci. Biotechnol. Biochem. 67 (5), 996–1006.

Shimada, Y., Morita, T., Sugiyama, K., 2003b. Dietary eritadenine and ethanolamine depress fatty acid desaturase activities by increasing liver microsomal phosphatidylethanolamine in rats. J. Nutr. 133 (3), 758–765.

Shimada, M., Hasegawa, T., Nishimura, C., Kan, H., Kanno, T., Nakamura, T., Matsubayashi, T., 2009. Anti-hypertensive effect of γ-aminobutyric acid (GABA)-rich *Chlorella* on high-normal blood pressure and borderline hypertension in placebo-controlled double-blind study. Clin. Exp. Hypertens. 31 (4), 342–354.

Shimizu, T., Kawai, J., Ouchi, K., Kikuchi, H., Osima, Y., Hidemi, R., 2016. Agarol, an ergosterol derivative from *Agaricus blazei*, induces caspase-independent apoptosis in human cancer cells. Int. J. Oncol. 48 (4), 1670–1678.

Singh, G., Verma, A.K., Kumar, V., 2016. Catalytic properties, functional attributes and industrial applications of β-glucosidases. 3 Biotech 6 (1), 3.

Singh, M., Kamal, S., Sharma, V.P., 2020. Status and trends in world mushroom production-III. World production of different mushroom species in 21st century. Mushroom Res. 29 (2), 75–111.

Sirtori, C.R., 1990. Pharmacology and mechanism of action of the new HMG-CoA reductase inhibitors. Pharmacol. Res. 22 (5), 555–563.

Solano, F., Briganti, S., Picardo, M., Ghanem, G., 2006. Hypopigmenting agents: an updated review on biological, chemical and clinical aspects. Pigment Cell Res. 19 (6), 550–571.

Srivastava, B., Dwivedi, A.K., Pandey, V.N., 2011. Morphological characterization and yield potential of *Termitomyces* spp. mushroom in Gorakhpur forest division. Bull. Environ. Pharmacol. Life Sci. 1 (1), 54–56.

Srivastava, B., Dwivedi, A.K., Pandey, V.N., 2012. Sociobiology and natural adaptation of termite and *Termitomyces* in different forest division of Gorakhpur region. Bull. Environ. Pharmacol. Life Sci. 2 (1), 32–36.

Stajić, M., Vukojević, J., Duletić-Lausević, S., 2009. Biology of *Pleurotus eryngii* and role in biotechnological processes: a review. Crit. Rev. Biotechnol. 29 (1), 55–66.

Stoffel, F., Santana, W.D.O., Fontana, R.C., Gregolon, J.G.N., Kist, T.B.L., Siqueira, F.G.D., Mendoça, S., Camassola, M., 2019. Chemical features and bioactivity of grain flours colonized by macrofungi as a strategy for nutritional enrichment. Food Chem. 297, 124988.

Stojković, D., Reis, F.S., Glamočlija, J., Ćirić, A., Barros, L., Van Griensven, L.J.L.D., Ferreira, I.C.F.R., Soković, M., 2014. Cultivated strains of *Agaricus bisporus* and *A. brasiliensis*: chemical characterization and evaluation of antioxidant and antimicrobial properties for the final healthy product- natural preservatives in yoghurt. Food Funct. 5, 1602–1612.

Stone, H., 2018. Example food: what are its sensory properties and why is that important. NPJ Sci. Food. 2 (11), 1–3.

Su, H., Peng, X., Shi, Q., Huang, Y., Zhou, L., Qiu, M., 2020. Lanostane triterpenoids with anti-inflammatory activities from *Ganoderma lucidum*. Phytochemistry 173, 112256.

Subbiah, K.A., Balan, V., 2015. A comprehensive review of tropical milky white mushroom (*Calocybe indica* P&C). Mycobiology 43 (3), 184–194.

Sunil, C., Xu, B., 2022. Mycochemical profile and health-promoting effects of morel mushroom *Morchella esculenta* (L.)- a review. Food Res. Int. 159, 111571.

Synytsya, A., Míčková, K., Synytsya, A., Jablonsky′, I., Spěváček, J., Erban, V., Kovářiková, E., Čopíková, J., 2009. Glucans from fruit bodies of cultivated mushrooms *Pleurotus ostreatus* and *Pleurotus eryngii*: structure and potential prebiotic activity. Carbohydr. Polym. 76 (4), 548–556.

Takashima, K., Izumi, K., Iwai, H., Takeyama, S., 1973. The hypocholesterolemic action of eritadenine in the rat. Atherosclerosis 17 (3), 491–502.

Tao, Q., Ma, K., Yang, Y., Wang, K., Chen, B., Huang, Y., Han, J., Bao, L., Liu, X., Yang, Z., Yin, W., Liu, H., 2016. Bioactive Sesquiterpenes from the edible mushroom *Flammulina velutipes* and their biosynthetic pathway confirmed by genome analysis and chemical evidence. J. Org. Chem. 81 (20), 9867–9877.

Tian, Y., Nichols, R.G., Roy, P., Gui, W., Smith, P.B., Zhang, J., Lin, Y., Weaver, V., Cai, J., Patterson, A.D., Cantorna, M.T., 2018. Prebiotic effects of white button mushroom (*Agaricus bisporus*) feeding on succinate and intestinal gluconeogenesis in C57BL/6 mice. J. Funct. Foods 45, 223–232.

Tobert, J.A., 2003. Lovastatin and beyond: the history of the HMG-CoA reductase inhibitors. Nat. Rev. Drug Discov. 2 (7), 517–526.

Trivedi, S., Patel, K., Belgamwar, V., Wadher, K., 2022. Functional polysaccharide lentinan: role in anti-cancer therapies and management of carcinomas. Pharmacol. Res. Mod. Chin. Med. 2, 100045.

Tsao, Y., Kuan, Y., Wang, J., Sheu, F., 2013. Characterization of a novel maitake (*Grifola frondosa*) protein that activates natural killer and dendritic cells and enhances antitumor immunity in mice. J. Agric. Food Chem. 61, 9828–9838.

Tsukagoshi, S., Hashimoto, Y., Fujii, G., Kobayashi, H., Nomoto, K., Orita, K., 1984. Krestin (PSK). Cancer Treat. Rev. 11, 131–155.

Vamanu, E., 2014. Antioxidant properties of mushroom mycelia obtained by batch cultivation and tocopherol content affected by extraction procedures. Biomed. Res. Int. 2014, 974804.

Velez, M.E.V., Da Luz, J.M.R., Da Silva, M.D.C.S., Cardoso, W.S., Lopes, L.D.S., Vieira, N.A., Kasuya, M.C.M., 2019. Production of bioactive compounds by the mycelial growth of *Pleurotus djamor* in whey powder enriched with selenium. LWT Food Sci. Technol. 114, 108376.

Venditti, A., Frezza, C., Sciubba, F., Serafini, M., Bianco, A., 2017. Primary and secondary metabolites of an European edible mushroom and its nutraceutical value: *Suillus bellini* (Inzenga) Kuntze. Nat. Prod. Res. 31 (16), 1910–1919.

Wang, M., Zhao, R., 2023. A review on nutritional advantages of edible mushrooms and its industrialization development situation in protein meat analogues. J. Future Foods 3 (1), 1–7.

Wang, Z., Feng, S., Huang, Y., Li, S., Xu, H., Zhang, X., Bai, Y., Qiao, M., 2010. Expression and characterization of a *Grifola frondosa* hydrophobin in *Pichia pastoris*. Protein Expr. Purif. 72 (1), 19–25.

Wang, Y., Bao, L., Yang, X., Li, L., Li, S., Gao, H., Yao, X., Wen, H., Liu, H., 2012. Bioactive sesquiterpenoids from the solid culture of the edible mushroom *Flammulina velutipes* growing on cooked rice. Food Chem. 132, 1346–1353.

Wang, C., Zeng, F., Liu, Y., Pan, Y., Xu, J., Ge, X., Zheng, H., Pang, J., Liu, B., Huang, Y., 2021. Coumarin-rich Grifola *frondosa* ethanol extract alleviate lipid metabolism disorders and modulates intestinal flora compositions of high-fat diet rats. J. Funct. Foods 85, 104649.

Wani, B.A., Bodha, R.H., Wani, A.H., 2010. Nutritional and medicinal importance of mushrooms. J. Med. Plant Res. 4 (24), 2598–2604.

Wu, L., Cheah, I.K., Chong, J.R., Chai, Y.L., Tan, J.Y., Hilal, S., Vrooman, H., Chen, C.P., Halliwell, B., Lai, M.K.P., 2021. Low plasma ergothioneine levels are associated with neurodegeneration and cerebrovascular disease in dementia. Free Radic. Biol. Med. 177, 201–211.

Xu, Y., Xie, L., Zhang, Z., Zhang, W., Tang, J., He, X., Zhou, J., Peng, W., 2021. *Tremella fuciformis* polysaccharides inhibited colonic inflammation in dextran sulphate sodium-treated mice via Foxp3+ T cells, gut microbiota, and bacterial metabolites. Front. Immunol. 12, 648162.

Yang, C., Meng, Q., Zhou, X., Cui, Y., Fu, S., 2019. Separation and identification of chemical constituents of *Morchella conica* isolated from Guizhou Province China. Biochem. Syst. Ecol. 86, 103919.

Yang, L., Kang, X., Dong, W., Wang, L., Liu, S., Zhong, X., Liu, D., 2022. Prebiotic properties of *Ganoderma lucidum* polysaccharides with special enrichment of *Bacteroides ovatus* and *B. uniformis in vitro*. J. Funct. Foods 92, 105069.

Yodthong, T., Kedjarune-Leggat, U., Smythe, C., Sukprasirt, P., Aroonkesorn, A., Wititsuwannakul, R., Pitakpornpreecha, T., 2020. Enhancing activity of *Pleurotus sajor-caju* (Fr.) sing β-1,3-glucanoligosaccharide (*Ps*-GOS) on proliferation, differentiation, and mineralization of MC3T3-E1 cells through the involvement of BMP-2/Runx2/MAPK/Wnt/β-catenin signaling pathway. Biomol. Ther. 10 (2), 190.

Yoshida, M., Ohira, T., Igarashi, K., Nagasawa, H., Samejima, M., 2002. Molecular cloning and characterization of a cDNA encoding cellobiose dehydrogenase from the wood-rotting fungus *Grifola frondosa*. FEMS Microbiol. Lett. 217, 225–230.

Yuan, B., Zhao, L., Rakariyatham, K., Han, Y., Gao, Z., Kimatu, B.M., Hu, Q., Xiao, H., 2017. Isolation of a novel bioactive protein from an edible mushroom *Pleurotus eryngii* and its anti-inflammatory potential. Food Funct. 8, 2175–2183.

Zafar, M.I., Mills, K.E., Zheng, J., Regmi, A., Hu, S.Q., Gou, L., Chen, L., 2019. Low-glycemic index diets as an intervention for diabetes: a systematic review and meta-analysis. Am. J. Clin. Nutr. 110 (4), 891–902.

Zhai, F., Wang, Q., Han, J., 2015. Nutritional components and antioxidant properties of seven kinds of cereals fermented by the basidiomycete *Agaricus blazei*. J. Cereal Sci. 65, 202–208.

Zhang, C., Song, X., Cui, W., Yang, Q., 2021. Antioxidant and anti-ageing effects of enzymatic polysaccharide from *Pleurotus eryngii* residue. Int. J. Biol. Macromol. 173, 341–350.

Zhang, Y., Zhang, Z., Liu, H., Wang, D., Wang, J., Liu, M., Yang, Y., Zhong, S., 2022. A natural selenium polysaccharide from *Pleurotus ostreatus*: structural elucidation, anti-gastric cancer and anti-colon cancer activity *in vitro*. Int. J. Biol. Macromol. 201, 630–640.

Zhang, Z., Wu, D., Li, W., Chen, W., Liu, Y., Zhang, J., Wan, J., Yu, H., Zhou, S., Yang, Y., 2023. Structural elucidation and anti-inflammatory activity of a proteoglycan from spent substrate of *Lentinula edodes*. Int. J. Biol. Macromol. 224, 1509–1523.

Zhao, Y., Zheng, W., 2021. Deciphering the antitumoral potential of the bioactive metabolites from medicinal mushroom *Inonotus obliquus*. J. Ethnopharmacol. 265, 113321.

Zhao, Z., Chen, H., Feng, T., Li, Z., Dong, Z., Liu, J., 2015. Lucidimine A-D, four new alkaloids from the fruiting bodies of *Ganoderma lucidum*. J. Asian Nat. Prod. Res. 17 (12), 1160–1165.

Zhao, Z., Chen, H., Wu, B., Zhang, L., Li, Z., Feng, T., Liu, J., 2017. Matsutakone and matsutoic acid, two (nor) steroids with unusual skeletons from the edible mushroom *Tricholoma matsutake*. J. Org. Chem. 82 (15), 7974–7979.

Zheng, W., Miao, K., Liu, Y., Zhao, Y., Zhang, M., Pan, S., Dai, Y., 2010. Chemical diversity of biologically active metabolites in the sclerotia of *Innonotus obliquus* and submerged culture strategies for up-regulating their production. Appl. Microbiol. Biotechnol. 87, 1237–1254.

Zheng, X., Shi, X., Wang, B., 2021. A review on the general cheese processing technology, flavor biochemical pathways and the influence of yeasts in cheese. Front. Microbiol. 12, 703284.

Zhou, J., Chen, M., Wu, S., Liao, X., Wang, J., Wu, Q., Zhuang, M., Ding, Y., 2020. A review on mushroom-derived bioactive peptides: preparation and biological activities. Food Res. Int. 134, 109230.

Zhu, Z., Huang, R., Huang, A., Wang, J., Liu, W., Wu, S., Chen, M., Chen, M., Xie, Y., Jiao, C., Zhang, J., Wu, Q., Ding, Y., 2022. Polysaccharide from *Agrocybe cylindracea* prevents diet-induced obesity through inhibiting inflammation mediated by gut microbiota and associated metabolites. Int. J. Biol. Macromol. 209 (A), 1430–1438.

Zhuo, R., Yuan, P., Yang, Y., Zhang, S., Ma, F., Zhang, X., 2017. Induction of laccase by metal ions and aromatic compounds in *Pleurotus ostreatus* HAUCC 162 and decolorization of different synthetic dyes by the extracellular laccase. Biochem. Eng. J. 117 (B), 62–72.

Zhuo, R., Yu, H., Yuan, P., Fan, J., Chen, L., Li, Y., Ma, F., Zhang, X., 2018. Heterologous expression and characterization of three laccases obtained from *Pleurotus ostreatus* HAUCC 162 for removal of environmental pollutants. J. Hazard. Mater. 344, 499–510.

Zhuo, R., Zhang, J., Yu, H., Ma, F., Zhang, X., 2019. The roles of *Pleurotus ostreatus* HAUCC 162 laccase isoenzymes in decolorization of synthetic dyes and the transformation pathways. Chemosphere 234, 733–745.

Zuo, T., Wu, X., Wen, W., Lan, P., 2021. Gut microbiome alterations in COVID-19. Genomics Proteomics Bioinformatics 19 (5), 679–688.

Zysset-Burri, D.C., Morandi, S., Herzog, E.L., Berger, L.E., Zinkernagel, M.S., 2023. The role of the gut microbiome in eye diseases. Prog. Retin. Eye Res. 92, 101117.

CHAPTER

15

Fungal endophytes as a potential source in the agricultural industry: An idea for sustainable entrepreneurship

Jentu Giba[a], Tonlong Wangpan[b], and Sumpam Tangjang[b]

[a]Department of Biotechnology, APSCS & T CoE for Bioresources and Sustainable Development, Kimin, Arunachal Pradesh, India, [b]Department of Botany, Rajiv Gandhi University, Doimukh, Arunachal Pradesh, India

15.1 Introduction

Endophytes which colonize the internal tissues of the host plant species are currently gaining attention as a critical target for bioprospection in the exploration of novel biochemical entities (Staniek et al., 2008; Tiwari et al., 2021). Unlike other microorganisms, they are nonpathogenic and are believed to enhance the fitness of their host (Mendes et al., 2013; Philippot et al., 2013). They can also make the chemical-intensive crop production system more sustainable by enhancing plant growth, fitness, and yield by furnishing biotic and abiotic stress tolerance (Castillo et al., 2000; Tanaka et al., 2005). Polyphyletic and highly diverse endophytic fungi mainly belong to the phylum ascomycetes, an anamorphic fungi group (Arnold, 2007). The hyperdiversity of endophytic fungi is based on the assumption that each plant species can be colonized with one or more fungal species (Huang et al., 2007).

15.2 Classification and mode of transmission

Based on the differences in the evolutionary relationship, taxonomy, ecological functions, and host plants, endophytic fungi are classified into clavicipitaceous (C-endophytes) and nonclavicipitaceous (NC-endophytes) (Clay and Schardl, 2002). These two major groups are further divided into four classes based on their symbiotic criteria (Rodriguez et al., 2009), where Class 1 endophytes belong to C-endophytes and Class 2, 3, and 4 belong to NC-endophytes (Table 15.1). Class 1 endophytes can survive in some cool-season grasses. Their transmission mode is primarily vertical, where the offspring are passed on to the next generation with the seed infection. This transmission mode ensures systemic colonization in the floral organs of host plants (Saikkonen et al., 2002). The plants colonized by Class 1 endophytes benefitted from increased biomass, tolerance to drought, and protection against herbivores with the production of toxic chemicals (Cheplick and Clay, 1988). However, acquiring benefits through Class 1 endophytic fungi varies or depends mainly on the host species, its genotype, and environmental conditions (Faeth et al., 2006). On the other hand, Class 2 endophytes transmit vertically

Entrepreneurship with Microorganisms. https://doi.org/10.1016/B978-0-443-19049-0.00018-9

Table 15.1 Classes of endophytes and their mode of transmission (Rodriguez et al., 2009).

Class of endophytes	Endophytic fungi	Host	Host range	Tissue colonized	Transmission	References
Class 1	Clavicipitaceous fungi; *Epichloë*, *Metarhizium*, and others	Grasses	Narrow	Shoots and rhizome s	Vertical and horizontal	Faeth and Saari (2012); Behie and Bidochka (2014)
Class 2	NC-endophyte; *Penicillium*, *Colletotrichum*, *Fusarium*, *Trichoderma*, *Aspergillus*, *Xylaria* sp., and others	NS	Board	Roots, stem, leaves, and rhizome s	Vertical and horizontal	Rodriguez et al. (2009); Hiruma et al. (2018); Waqas et al. (2012)
Class 3	NC-endophytes; Depends on host, local infection and mode of transmission	Mainly tropical tress	Board	Shoot	Horizontal	Rodriguez et al. (2009)
Class 4	NC-endophytes; *Curvularia*, *Cladosporium*, *Aletrnaria*, *Ophiosphaerella*, *Phialocephala*, and others	NS	Board	Roots	Horizontal	Rodriguez et al. (2009); Hamayun et al. (2010); Spagnoletti et al. (2017)

Note: NC – Nonclavicipitaceous; NS – Nonspecific.

or horizontally, colonizing the shoots, roots, and rhizomes of plants. It comprises members of Dikarya and assists the host plant in habitat-specific stress tolerance (Rodriguez et al., 2008). Class 3 endophyte transmits horizontally and has a broad range of hosts exhibiting restricted colonization of shoots. The host range includes vascular, nonvascular, woody, and herbaceous angiosperm of tropical forests (Higgins et al., 2007; Murali et al., 2007; Davis and Shaw, 2008). They infect the plant's leaf tissue in a highly localized manner and exhibit an exceedingly diverse level of species richness. The level of species richness and diversity of endophytic fungi suggest that an individual plant can harbor hundreds of various species of endophytes. This class of endophytes can effectively defend their host plants against several abiotic stresses, such as salinity, nutrient deficiency, drought, and metal toxicity, as well as biotic stresses caused by insect pests and pathogens (Hardoim et al., 2008; Rho et al., 2018; Waller et al., 2005; Manasa et al., 2020; Sampangi-Ramaiah et al., 2020). They also produce pharmaceutically necessary enzymes, secondary metabolites, and phytohormones (Nagarajan et al., 2014; Uzma et al., 2019; Bilal et al., 2018). Class 4 endophytes are known as dark septate endophytes (DSE) and transmit horizontally. They are mostly restricted to the host plant's roots and belong mainly to Ascomycota. They infect the host cell benefitted from increased cellular hyphae with melanized structure and form microsclerotia in the root tissues of the host plant. Class 4 endophytes are not host-specific but are distributed randomly among different plant species and were identified from various ecosystems in more than 600 plant species. Their distribution is quite versatile, but they mainly restrict to host plants in highly stressed environments (Jumpponen, 2001).

15.3 Isolation of fungal isolates

The endophytic fungi may reside in innumerable tissues of the host plant, including roots, branches, leaves, stems, petioles, seeds, flowers, twigs, bark, and fruit, and may also include the xylem of all available plant organs (Sieber, 1989; Clay, 1990; Bills and Polishook, 1991; Kowalski and Kehr, 1992; Guo et al., 2000; Evans et al., 2003; Holmes et al., 2004; Nishijima et al., 2004; Stone et al., 2004; Arnold and Lewis, 2005; Rubini et al., 2005; Crozier et al., 2006). The main type of host plant tissues analyzed for endophyte association are leaves, twigs, branches, and roots (Petrini and Fisher, 1990; Vujanovic and Brisson, 2002) (Fig. 15.1). Thus, for the isolation of endophytes, plant samples are usually collected based on the target tissue of the host plant from the collection site and processed within 36h of collection. Healthy and matured samples are preferred and collected in a collection bag. The samples will be washed under running water to remove debris (Sardul et al., 2014). Furthermore, targeted tissue will be cut into bits of 0.5 cm^2 in the case of leaf tissue (Suryanarayanan et al., 1998), 2.0–3.0 cm for root (Latiffah et al., 2016), and 1.0×0.1 cm for twigs and bark (Suryanarayanan, 1992). The size of tissue fragments is negatively correlated with the estimation of endophytic species richness (Gamboa et al., 2003). Therefore, excising tissue into smaller segment units assures the recovery of greater species genotype diversity. On the other hand, coarsely divided tissue samples can potentially miss out on rare and slow-growing species and need help recovering mixed genotypes of the same species. The isolation of fungal endophytes mainly follows the protocol of Suryanarayanan et al. (1998), where the tissue segment is treated with 70% ethanol for 5 s, followed by treatment with 4% sodium hypochlorite for 90 s and finally rinsed with sterile distilled water for 10 s. The bits are inoculated carefully in freshly prepared PDA (potato dextrose agar) media amended with antibiotics. The plates are sealed and incubated for 12 h in a light chamber and 12 h in a dark period for 28 days at 26 ± 2°C

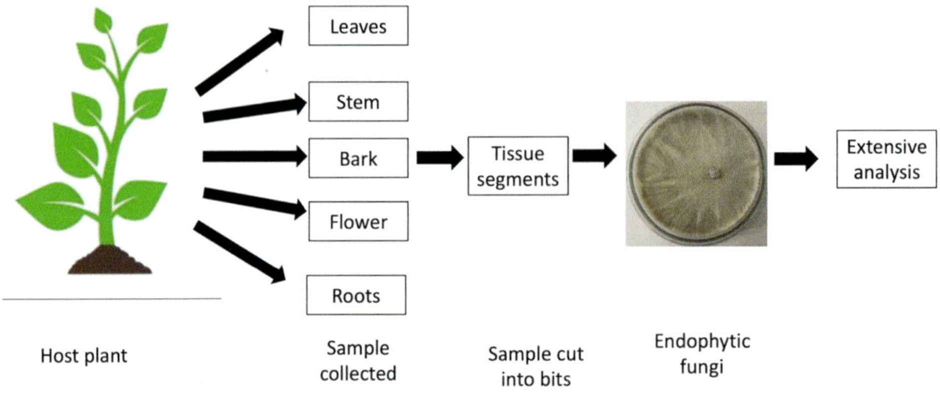

FIG. 15.1

Isolation of endophytic fungi from plant samples.

(Suryanarayanan, 1992). Fungal hyphal emerging from the tissue segments is isolated to obtain axenic culture. The pure culture is then stored in a slant at 4°C for further analysis or storage. Fig. 15.1 explains the overall process of isolation of endophytic fungi from plant samples.

15.4 Morphological identification of endophytic fungi

The endophytic fungi are identified based on the colony and spores' cultural characteristics and morphology (Barnett and Hunter, 1998; Hanlin, 1998; Kiffer and Morelet, 2000). For the identification of fungi, slides are prepared and observed under the microscope. The observation was based on the presence of conidial mycelium, spore attachment, spore characteristics, distinct reverse colony color, and production of diffusible pigments (Barnett and Hunter, 1998). The fungal endophytes are thus characterized under suitable genera and species of fungi (Bhat, 2010). Cultures that repetitively fail to show sporulation on different media are recorded as sterile. Such rare and sterile fungi can further identify based on molecular characterization. Fig. 15.2 depicts the brief strategy for the isolation, identification of endophytic fungi, and the application of its metabolites in agriculture.

15.5 Endophytic fungi in agricultural industries

Extensive use of chemical fertilizers has shown an adverse effect on the environment and high expenses for their production. Thus, in search of a better alternative, the focus has shifted to microbial inoculants, particularly endophytic microorganisms. Endophytes are believed to spend the whole or a part of their life cycle by colonizing the host plant's inter- or healthy intracellular tissues without causing any apparent visible symptoms (Tan and Zou, 2001; Hartley and Gange, 2009). The association between the host plant and fungal endophytes is owing to their unique adaptations that help endophytes grow harmoniously with and within their host (Varma et al., 2012). Along with harmonious interaction with their host, they are also known to produce secondary metabolites, enzymes, and phytohormones, which

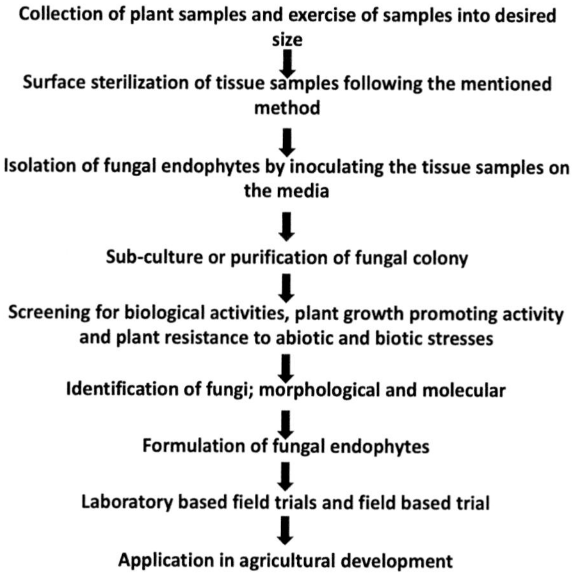

FIG. 15.2

Brief strategy for the isolation, identification of endophytic fungi, and the application of its metabolites in agriculture (Abo Nouh, 2019).

are pharmaceutically and agriculturally important (Shweta et al., 2010; Uzma et al., 2019; Bilal et al., 2018). It was reported that endophytic fungi species isolated from *Cardiospermum halicacabum*, a medicinal plant, confer plant growth in adverse conditions by producing different extracellular enzymes (Chathurdevi and Gowrie, 2016). Also, *Acremonium zeae*, an endophyte isolated from maize, was reported to produce the extracellular enzyme hemicellulose, which could be used for the bioconversion of lignocellulosic biomass to fermentable sugars (Bischoff et al., 2009). Endophytic fungi such as *Sebacina vermifera*, *Colletotrichum*, *Piriformospora indica*, and *Penicillium* were able to display better growth-promoting effects on plants under unfavorable conditions due to their ability to synthesize bioactive metabolites and enzymes (Waller et al., 2005; Redman et al., 2011; Hamilton and Bauerle, 2012). In addition to the production of enzymes, some fungal endophytes have also been reported to produce phytohormones. For instance, one such report mentioned that *Cladosporium sphaerospermum*, isolated from *Glycine* max, induced plant growth in soybean with gibberellic acid production (Hamayun et al., 2009). Plant hormones like indole acetic acid enhancing plant growth were reported to be produced by *Fusarium tricintum* and *Alternaria alternata* (Khan et al., 2015). IAA (Indole acetic acid) acts as a plant growth promoter by enhancing the process of cell elongation and cell division, an essential step for tissue differentiation (Taghavi et al., 2009). The endophytes could be isolated from every plant and part of plants studied to date (Giauque et al., 2019).

It is evident that endophytes could be isolated from every plant and part of plants studied to date (Giauque et al., 2019). Their universal occurrence, ability to sustain in plants as nonpathogenic organism, their ability to enhance the biotic and abiotic stress tolerance of their host plants (Rodriguez et al., 2009), and their attribute in increasing the plant yield and access to soil nutrients (Xia et al., 2019;

White et al., 2019) play a major role as an indicator of a candidate which could serve as a potential tool in crop improvement. *Neotyphodium* sp. (an endophyte) was reported to enhance the tolerance of grass plants to drought and nitrogen starvation by stomatal and osmoregulation (Chhipa and Deshmukh, 2019). When compared, the plants with endophytes significantly consumed less water and enhanced plant biomass (Porcel et al., 2006; Marulanda et al., 2009). Moreover, endophytes, *Chaetomium globosum*, and *Penicillium resedanum* isolated from *Capsicum annuum* enhanced the shoot length and biomass of the plant in drought-stressed conditions (Khan et al., 2012a, 2012b, 2014). *Curvularia* sp., *Curvularia* sp., were not only able to confer thermal tolerance to crops like wheat, watermelon, and tomato (Redman et al., 2002) but also protects *Dichanthelium lanuginosum* from thermal stress, residing in areas where soil temperature reaching up to 57°C (Stierle et al., 1993). Owing to their great potential as a source of secondary metabolites with unique structures, including alkaloids, flavonoids, tetralones, quinones, terpenoids, chinos, phenolic acids, and benzo pyranoses, they have gained significant interest in sustainable agriculture (Jalgaonwala et al., 2011; Pimentel et al., 2011). These compounds can be produced directly by the fungi or the plant due to their interaction with the fungi. Among all the secondary metabolites, alkaloids are the first reported fungal metabolites showing insecticidal activity and resistance against herbivores (Azevedo et al., 2000; Powell and Petroski, 1992). They are categorized into three groups; amines and amides, indole derivatives, and pyrrolizidines in fungal endophytes. Peramine among amine and amides is toxic to insects but shows no such activity against mammals (Dew et al., 1990; Rowan and Latch, 1994). Indole derivatives such as elymoclavine, chanoclavine, and agroclavin isolated from endophyte *Neotyphodium* sp. were also reported to be toxic against some insects and mammals (Ball et al., 1997; Strobel et al., 2004). Sesquiterpenes, one of the terpenoids isolated from an endophytic fungus of *Abies balsamea*, as heptelidic and hydroheptelidic acid toxicity against spruce budworm, *Choristoneura fumiferana* (Calhoun et al., 1992). In the recent decade, a phenolic compound was purified from the ethyl acetate extract of endophytic fungi, *Cladosporium* sp., which significantly increased mortality and affected the development, and survival of polyphagous pest *Spodoptera litura*, otherwise known as tobacco cutworm (Singh et al., 2015).

The mutualistic relationship between plants and fungal endophytes confers that incorporating fungal symbionts could be a valuable strategy for dealing with the impact of climate change on major crops and the expansion of agricultural production onto marginal lands (Redman et al., 2011). They not only increase the availability of nutrients for their host but also increase the susceptibility of plant against plant pathogens, nematodes, and insects, and help their host to take up adaptation based on the new habitat, inducing genetic and physiological modifications of the plant (Lugtenberg et al., 2016; Ali et al., 2018). Further, they correspondingly produce phytohormones and confer rhizoremediation and phytoremediation for resistance against stressed conditions. Different endophytes can solubilize, fix, and mobilize both micro and macro elements for their host plants (Khan et al., 2012a, 2012b; Anitha et al., 2013; Zhou et al., 2014; Jain and Pundir, 2017; Arora and Ramawat, 2017; Lata et al., 2018; Khare et al., 2018; Kumar et al., 2019). The association of endophytic fungi with crops could be a solution for an increase in crop yield and growth without the need to provide extensive fertilizers (Rodriguez et al., 2008). Table 15.2 summarizes the activity of endophytic fungi associated with their host and explains their agricultural potential.

Table 15.2 List of endophytic fungi and their potential in agriculture.

SL. No	Fungal endophytes	Host plant	Benefits to its host	Agricultural potentials	References
1.	Epichloë species	I. *Festuca rubra*	They function as plant growth promoter, increasing the plant's growth with high uptake of nutrients	The relationship between the fungi and grass species is highly integrated involving symbiotic relationship, and alteration, and manipulation of physiology, morphology, and life cycle of the host with increase in its symbiont fitness. They are important in pasture system with delivering economical and sustainable agricultural solutions	Saikkonen et al. (2016); Lugtenberg et al. (2016); Jain and Pundir (2017); Chhipa and Deshmukh (2019)
		II. Temperate grass	Owing to its presence, they produce bioactive compounds in the host which work as deterrent to herbivores and pests		
2.	*Piriformospora indica*	Barley	Enhance the grain yield, and tolerance against mild salt stresses. They also provide resistance against necrotrophic fungi (*Fusarium culmorum*), a root pathogen that causes root rot and biotrophic fungi *Elumeria graminis*, a leaf pathogen	The beneficial effects such as enhancement of nutrient uptake, promotion of plant growth, increase resistance to pathogens and stimulation of growth and production of seeds are also observed in other crop such as wheat, rice and, maize. Further, they control various plant diseases, including powdery mildew, Rhizoctonia root rot, verticillium wilt, eyespot, cyst nematode, black root rot, Fusarium wilt, leaf blight, and yellow leaf mosaic in wheat, tomato, Arabidopsis, barley, and maize	Oelmüller et al. (2009); Lugtenberg et al. (2016); Ali et al. (2018); Chhipa and Deshmukh (2019)
3.	*Epicoccum nigrum*	Isolated from apple tree and included to *Catharanthus roseus*	This mechanism prompted the defense responses of *C. roseus* against "*Candidatus Phytoplasma mali*" and delimited the severity of symptoms	They are known for their biocontrol potential against bacterial pathogen, *pseudomonas savastanoi pv. Savastanoi* (Psv) causing olive knot by reducing its biomass up to 96% and also resistance against fungal pathogen, *Sclerotinia sclerotiorum* in sunflower and *Pythium* in cotton crop. They also promote antibacterial activity against *Phytoplasma* in peach fruit, apple, nectarines, and *Monilinia* sp.	Fávaro et al. (2012); Lugtenberg et al. (2016); Berardo et al. (2018); Chhipa and Deshmukh (2019)
4.	*Trichoderma* sp.	Ubiquitous endophyte	Potential to be isolated from multiple host plant	Some of the strains of fungi have antagonistic and mycoparasitic potentials which inhibit the plant pathogens present in the soil or on plant roots and reduce the severity of plant diseases and promote plant growth and defense. Thus, they are most successful biofungicides used in agriculture.	Hermosa et al. (2012); Mukherjee et al. (2012); Talapatra et al. (2017)

15.6 Conclusion

Recently, pathogenic organisms have been gaining resistance against the existing pesticides, fungicides, and insecticides, while the incorporation of Chemical fertilizers is declining crop quality. The fungal endophytes demonstrate a potential escape route for such problems. Endophytic fungi can influence plant/crop quality by producing several significant bioactive compounds, thereby increasing the host plant's growth, development, diversity, and fitness. Endophytes also influence the plant community diversity, its population dynamic, and the functioning of an ecosystem. It could be used as commercial bioinoculants and biocontrol agents for crops growing under abiotic and biotic stresses, as they confer to surge the tolerance of its host in stressed environments, ensuring sustainable agriculture. The endophytic fungi could be a potential source for developing entrepreneurship in agricultural industries due to the manifold utility of endophytes in modifying their host. Moreover, exploring the novel bioactive compounds of endophytic fungi may serve as the future of the agricultural industry.

References

Abo Nouh, F.A., 2019. Endophytic fungi for sustainable agriculture. Microbial Biosyst. 4 (1), 31–44.

Ali, A.H., Radwan, U., El-Zayat, S., El-Sayed, M.A., 2018. Desert plant-fungal endophytic association: the beneficial aspects to their hosts. Biol. Forum Int. J. 10 (1), 138–145.

Anitha, D., Vijaya, T., Pragathi, D., Reddy, N.V., Mouli, K.C., Venkateswarulu, N., Bhargav, D.S., 2013. Isolation and characterization of endophytic fungi from endemic medicinal plants of Tirumala hills. Int. J. Life Sci. Biotechnol. Pharma Res. 2 (3), 367–373.

Arnold, A.E., 2007. Understanding the diversity of foliar endophytic fungi: progress, challenges, and frontiers. Fungal Biol. Rev. 21, 51–66.

Arnold, A.E., Lewis, L.C., 2005. Ecology and evolution of fungal endophytes and their roles against insects. In: Vega, F.E., Blackwell, M. (Eds.), Insect-fungal Association: Ecology and Evolution. Oxford University Press, New York, USA, pp. 74–96.

Arora, J., Ramawat, K.G., 2017. An Introduction to Endophytes. In: Maheshwari, D.K. (Ed.), Endophytes: Biology and Biotechnology, Sustainable Development and Biodiversity 15. Springer International Publishing AG, Swizterland, pp. 1–16, https://doi.org/10.1007/978-3-319-66541-2_1.

Azevedo, J.L., Maccheroni Jr., W., Periera, J.O., Araujo, W.L., 2000. Endophytic microorganisms: a review on insect control and recent advances on tropical plants. Electron. J. Biotechnol. 3, 40–65. https://doi.org/10.2225/vol3-issue1- fulltext-4.

Ball, O.J.P., Barker, G.M., Prestidge, R.A., Sprosen, J.M., 1997. Distribution and accumulation of the Mycotoxin Lolitrem B in Neotyphodium lolii-infected perennial ryegrass. J. Chem. Ecol. 23, 1435–1449. https://doi.org/10.1023/B:JOEC.0000006474.44100.17.

Barnett, H., Hunter, B., 1998. Illustrated Genera of Imperfect Fungi, fourth ed. APS Press, Minnesota, pp. 8–34.

Behie, S.W., Bidochka, M.J., 2014. Ubiquity of insect-derived nitrogen transfer to plants by endophytic insect-pathogenic fungi: an additional branch of the soil nitrogen cycle. Appl. Environ. Microbiol. 80 (5), 1553–1560. https://doi.org/10.1128/AEM.03338-13.

Berardo, C., Bulai, I.M., Baptista, P., Gomes, T., Venturino, E., 2018. Modeling the endophytic fungus Epicoccum nigrum action to fight the "olive knot" disease caused by Pseudomonas savastanoi pv. savastanoi (Psv) bacteriain Olea europea L. trees. In: Mondaini, R.P. (Ed.), Trends in Biomathematics: Modeling, Optimization and Computational Problems. Switzerland: Springer International Publishing, AG, pp. 189–207.

Bhat, J., 2010. Fascinating Microfungi (Hyphomycetes) of Western Ghats. Broadway Book Centre, p. 221. ISBN:978-3-642-23341-8.

Bilal, L.A.S., Hamayun, M.G.H., Iqbal, A., Ullah, I., Lee, I.J., Hussain, A., 2018. Plant growth promoting endophytic fungi Aspergillus fumigatus TS1 and Fusarium proliferatum BRL1 produce gibberellins and regulates plant endogenous hormones. Symbiosis 97, 1–11. https://doi.org/10.1007/s13199-018-0545-4.

Bills, G.F., Polishook, J.D., 1991. Microfungi Carpinus caroliniana. Can. J. Bot. 9, 1477–1482.

Bischoff, K.M., Wicklow, D.T., Jordan, D.B., de Rezende, S.T., Liu, S., Hughes, S.R., et al., 2009. Extracellular hemicellulolytic enzymes from the maize endophyte Acremonium zeae. Curr. Microbiol. 58, 499–503. https://doi.org/10.1007/s00284-008-9353-z.

Calhoun, L.A., Findlay, J.A., Miller, J.D., Whitney, N.J., 1992. Metabolites toxic to spruce budworm from balsam fir needle endophytes. Mycol. Res. 96, 281–282. https://doi.org/10.1016/S0953-7562(09)80939-8.

Castillo, M.A., Moya, P., Hernandez, E., Primo-Yufera, E., 2000. Susceptibility of Ceratitis capitata Wiedemann (Diptera: Tephritidae) to entomopathogenic fungi and their extracts. Biol. Control 19, 274–282. https://doi.org/10.1006/bcon.2000.0867.

Chathurdevi, G., Gowrie, S.U., 2016. Endophytic fungi isolated from medicinal plant—a source of potential bioactive metabolites. Int. J. Curr. Pharm. Res. 8, 50–56.

Cheplick, G.P., Clay, K., 1988. Acquired chemical defenses of grasses: the role of fungal endophytes. Oikos, 52309–52318.

Chhipa, H., Deshmukh, S.K., 2019. Fungal endophytes: rising tools in sustainable agriculture production. In: Jha, S. (Ed.), Endophytes and Secondary Metabolites. Springer International Publishing, Swizterland, pp. 1–24, https://doi.org/10.1007/978-3-319-76900-4_26-1.

Clay, K., 1990. Fungal endophytes of grasses. Annu. Rev. Ecol. Evol. Syst. 21, 275–297.

Clay, K., Schardl, C.L., 2002. Evolutionary origins and ecological consequences of endophyte symbiosis with grasses. Am. Nat. 160, S99–S127.

Crozier, J., Thomas, S.E., Aime, M.C., Evans, H.C., Holmes, K.A., 2006. Molecular characterization of fungal endophytic morphospecies isolated from stems and pods of Theobroma cacao. Plant Pathol. 55, 783–791.

Davis, E.C., Shaw, A.J., 2008. Biogeographic and phylogenetic patterns in diversity of liverwort-associated endophytes. Am. J. Bot. 95, 914–924.

Dew, R.K., Boissonneault, G.A., Gay, N., Boling, J.A., Cross, R.J., Cohen, D.A., 1990. The effect of the endophyte (*Acremonium coenophialum*) and associated toxin of tall fescue on serum titer response to immunization and spleen cell flow cytometry analysis and response to mitogens. Vet. Immunol. Immunopathol. 26, 285–295. https://doi.org/10.1016/0165-2427(90)90097-c.

Evans, H.C., Holmes, K.A., Thomas, S.E., 2003. Endophytes and mycoparasites associated with an indigenous forest tree, Theobroma gileri, in Ecuador and a preliminary assessment of their potential as biocontrol agents of cocoa diseases. Mycol. Prog. 2, 149–160.

Faeth, S.H., Saari, S., 2012. Fungal grass endophytes and arthropod communities: lessons from plant defence theory and multitrophic interactions. Fungal Ecol. 5 (3), 364–371. https://doi.org/10.1016/j.funeco.2011.09.003.

Faeth, S.H., Gardner, D., Hayes, C.J., Jani, A., Wittlinger, S.K., 2006. Temporal and spatial variation in alkaloid levels in *Achnatherum robustum*, a native grass infected with the endophyte *Neotyphodium*. J. Chem. Ecol. 32 (2), 307–324.

Fávaro, L.C.D.L., Sebastianes, F.L.D.S., Araujo, W.L., 2012. *Epicoccum nigrum* P16, a sugarcane endophyte, produces antifungal compounds and induces root growth. PLoS One 7 (6), 1–10. https://doi.org/10.1371/journal.pone.0036826.

Gamboa, M.A., Laureano, S., Bayman, P., 2003. Measuring diversity of endophytic fungi in leaf fragments: does size matter? Mycopathologia 156, 41–45.

Giauque, H., Connor, E.W., Hawkes, C.V., 2019. Endophyte traits relevant to stress tolerance, resource use and habitat of origin predict effects on host plants. New Phytol. 221, 2239–2249. https://doi.org/10.1111/nph.15504.

Guo, L.D., Hyde, K.D., Liew, E.C.Y., 2000. Identification of endophytic fungi from *Livistona chinensis* based on morphology and rDNA sequences. New Phytol. 147, 617–630.

Hamayun, M., Khan, S.A., Iqbal, I., Na, C.I., Khan, A.L., Hwang, Y.H., et al., 2009. *Chrysosporium pseudomerdarium* produces gibberellins and promotes plant growth. J. Microbiol. 47, 425–430. https://doi.org/10.1007/s12275-009-0268-6.

Hamayun, M., Khan, S.A., Khan, A.L., Rehman, G., Kim, Y.H., Iqbal, I., Lee, I.J., 2010. Gibberellin production and plant growth promotion from pure cultures of *Cladosporium* sp. MH-6 isolated from cucumber (*cucumis sativus* L.). Mycologia 102 (5), 989–995. https://doi.org/10.3852/09-261.

Hamilton, C.E., Bauerle, T.L., 2012. A new currency for mutualism? Fungal endophytes alter antioxidant activity in hosts responding to drought. Fungal Divers. 54, 39–49. https://doi.org/10.1007/s13225-012-0156-y.

Hanlin, R.T., 1998. Combined Keys to Illustrated Genera of Ascomycetes. vols. 1 and 2 APS Press, Minnesota.

Hardoim, P.R., van Overbeek, L.S., van Elsas, J.D., 2008. Properties of bacterial endophytes and their proposed role in plant growth. Trends Microbiol. 16, 463–471. https://doi.org/10.1016/j.tim.2008.07.008.

Hartley, S.E., Gange, A.C., 2009. Impacts of plant symbiotic fungi on insect herbivores: mutualism in a multitrophic context. Ann. Rev. Entomol. 54 (1), 323–342. https://doi.org/10.1146/annurev.ento.54.110807.090614.

Hermosa, R., Viterbo, A., Chet, I., Monte, E., Monte, E., 2012. Plant-beneficial effects of trichoderma and of its genes. Microbiology 158, 17–25.

Higgins, K.L., Arnold, A.E., Miadlikowska, J., Sarvate, S.D., Lutzoni, F., 2007. Phylogenetic relationships, host affinity, and geographic structure of boreal and arctic endophytes from three major plant lineages. Mol. Phylogenet. Evol. 42, 543–555.

Hiruma, K., Kobae, Y., Toju, H., 2018. Beneficial associations between brassicaceae plants and fungal endophytes under nutrient-limiting conditions: evolutionary origins and host-symbiont molecular mechanisms. Curr. Opin. Plant Biol. 44, 145–154. https://doi.org/10.1016/j.pbi.2018.04.009.

Holmes, K.A., Schroers, H.-J., Thomas, S.E., Evans, H.C., Samuels, G.J., 2004. Taxonomy and biocontrol potential of a new species of Trichoderma from the Amazon basin of South America. Mycol. Prog. 3, 199–210.

Huang, W.Y., Cai, Y.Z., Xing, J., Corke, H., Sun, M., 2007. A potential antioxidant resource: endophytic fungi isolated from traditional Chinese medicinal plants. Econ. Bot. 61, 14–30.

Jain, P., Pundir, R.K., 2017. Potential role of endophytes in sustainable agriculture-recent developments and future prospects. In: Maheshwari, D.K. (Ed.), Endophytes: Biology and Biotechnology, Sustainable Development and Biodiversity. vol. 15. Springer International Publishing AG, Swizterland, pp. 145–160, https://doi.org/10.1007/978-3-319-66541-2_1.

Jalgaonwala, R.E., Mohite, B.V., Mahajan, R.T., 2011. Natural products from plant associated endophytic fungi. J. Microbiol. Biotechnol. Res. 1, 21–32.

Jumpponen, A., 2001. Dark septate endophytes are they mycorrhizal? Mycorrhiza 11, 207–211.

Khan, A.L., Hamayun, M., Hussain, J., Kang, S.M., Lee, I.J., 2012a. The newly isolated endophytic fungus *Paraconiothyrium* sp. LK1 produces ascotoxin. Molecules 17, 1103–1112. https://doi.org/10.3390/molecules17011103.

Khan, S.A., Hamayun, M., Khan, A.L., Shinwari, Z.K., 2012b. Isolation of plant growth promoting endophytic fungi from dicots inhabiting coastal sand dunes of Korea. Pak. J. Bot. 44 (4), 1453–1460.

Khan, A.L., Waqas, M., Hussain, J., Al-Harrasi, A., Lee, I.J., 2014. Fungal endophyte *Penicillium janthinellum* LK5 can reduce cadmium toxicity in *Solanum lycopersicum* (Sitiens and Rhe). Biol. Fertil. Soils 50, 75–85. https://doi.org/10.1007/s11274-013-1378-1.

Khan, A.L., Hussain, J., Al-Harrasi, A., Al-Rawahi, A., Lee, I.J., 2015. Endophytic fungi: Resource for gibberellins and crop abiotic stress resistance. Crit. Rev. Biotechnol. 35, 62–74. https://doi.org/10.3109/07388551.2013.800018.

Khare, E., Mishra, J., Arora, N.K., 2018. Multifaceted interactions between endophytes and plant: developments and prospects. Front. Microbiol. 9 (2732), 1–12.

Kiffer, E., Morelet, M., 2000. The Deuteromycetes, Mitosporic Fungi: Classification and Generic Keys. Science Publishers, Enfield, NH.

Kowalski, T., Kehr, R.D., 1992. Endophytic colonization of branch bases in several forest tree species. Sydowia 44, 137–168.

Kumar, V., Soni, R., Jain, L., Dash, B., Goel, R., 2019. Endophytic fungi: recent advances in identification and explorations. In: Singh, B.P. (Ed.), Advances in Endophytic Fungal Research, Fungal Biology. Springer International Publishing AG, Switzerland, pp. 267–281, https://doi.org/10.1007/978-3-030-03589-1_13.

Lata, R., Chowdhury, S., Gond, S., White, J.F., 2018. Induction of abiotic stress tolerance in plants by endophytic microbes. Appl. Microbiol. 66 (4), 268–276.

Latiffah, Z., Muhamad, I., Muhamad, J., Intan, S.M.A., 2016. Molecular characterisation of endophytic fungi from roots of wild banana (Musa acuminata). Trop. Life Sci. Res. 27 (1), 153–162.

Lugtenberg, B.J.J., Caradus, J.R., Johnson, L.J., 2016. Fungal endophytes for sustainable crop production. FEMS Microbiol. Ecol. 92 (12), 1–17.

Manasa, K.M., Vasanthakumari, M.M., Nataraja, K.N., Uma Shaanker, R., 2020. Endophytic fungi of salt adapted Ipomeapes-caprae L. R. Br: their possible role in inducing salinity tolerance in paddy (Oryza sativa L.). Curr. Sci. 118, 1448–1453. https://doi.org/10.18520/cs/v118/i9/1448-1453.

Marulanda, A., Barea, J.M., Azcon, R., 2009. Stimulation of plant growth and drought tolerance by native microorganisms (AM fungi and bacteria) from dry environments: mechanisms related to bacterial effectiveness. J. Plant Growth Regul. 28, 115–124. https://doi.org/10.1007/s00344-009-9079-6.

Mendes, R., Garbeva, P., Raaijmakers, J.M., 2013. The rhizosphere microbiome: significance of plant beneficial, plant pathogenic, and human pathogenic microorganisms. FEMS Microbiol. Rev. 37 (5), 634–663. https://doi.org/10.1111/1574-6976.12028.

Mukherjee, M., Mukherjee, P.K., Horwitz, B.A., Zachow, C., Berg, G., Zeilinger, S., 2012. Trichoderma-plant-pathogen interactions: advances in genetics of biological control. Indian J. Microbiol. 52 (4), 522–529.

Murali, T.S., Suryanarayanan, T.S., Venkatesan, G., 2007. Fungal endophyte communities in two tropical forests of Southern India: diversity and host affiliation. Mycol. Prog. 6, 191–199.

Nagarajan, A., Thirunavukkarasu, N., Suryanarayanan, T.S., Gummadi, S.N., 2014. Screening and isolation of novel glutaminase free L-asparaginase from fungal endophytes. Res. J. Microbiol. 9, 163–176. https://doi.org/10.3923/jm.2014.163.176.

Nishijima, K.A., Chan Jr., H.T., Nishijima, W.T., 2004. A new strain of Mucor isolated from guava fruit in Hawaii. Phytopathology 94, S153.

Oelmüller, R., Sherameti, I., Tripathi, S., Varma, A., Jena, F., Botanik, A., Str, D., 2009. Piriformospora indica, a cultivable root endophyte with multiple biotechnological applications. Symbiosis 49, 1–17.

Petrini, O., Fisher, P.J., 1990. Occurrence of fungal endophytes in twigs of Salix fragilis and Quercus robur. Mycol. Res. 94, 1077–1080.

Philippot, L., Raaijmakers, J.M., Lemanceau, P., van der Putten, W.H., 2013. Going back to the roots: the microbial ecology of the rhizosphere. Nat. Rev. Microbiol. 11, 789–799. https://doi.org/10.1038/nrmicro3109.

Pimentel, M.R., Molina, G., Dionisio, A.P., Marostica, M.R., Pastore, G.M., 2011. Use of endophytes to obtain bioactive compounds and their application in biotransformation process. Biotechnol. Res. Int., 1–11. Article ID: 566286 https://doi.org/10.4061/2011/576286.

Porcel, R., Aroca, R., Cano, C., Bago, A., Ruiz-Lozano, J.M., 2006. Identification of a gene from the arbuscular mycorrhizal fungus Glomus intraradices encoding for a 14-3-3 protein that is upregulated by drought stress during the AM symbiosis. Microb. Ecol. 52, 575–582. https://doi.org/10.1007/s00248-006-9015-2.

Powell, R.G., Petroski, R.J., 1992. Alkaloid toxins in endophyte infected grasses. Nat. Toxins 1, 163–170. https://doi.org/10.1002/nt.2620010304.

Redman, R.S., Sheehan, K.B., Stout, R.G., Rodriguez, R.J., Henson, J.M., 2002. Thermotolerance generated by plant/fungal symbiosis. Science 298, 1581.

Redman, R.S., Kim, Y.O., Woodward, C.J.D.A., Greer, C., Espino, L., Doty, S.L., Rodriguez, R.J., 2011. Increased fitness of rice plants to abiotic stress via habitat adapted symbiosis: a strategy for mitigating impacts of climate change. PLoS One 6 (7), 1–10. https://doi.org/10.1371/journal.pone.0014823.

Rho, H., Hsieh, M., Kandel, S.L., Cantillo, J., Doty, S.L., Kim, S.H., 2018. Do endophytes promote growth of host plants under stress? A meta-analysis on plant stress mitigation by endophytes. Microb. Ecol. 75, 407–418. https://doi.org/10.1007/s00248-017-1054-3.

Rodriguez, R.J., Henson, J., Volkenburgh, E.V., Hoy, M., 2008. Stress tolerance in plants via habitat-adapted symbiosis. Int. Soc. Microbial Ecol. J. 2, 404–416.

Rodriguez, R.J., White Jr., J.F., Arnold, A.E., Redman, R.S., 2009. Fungal endophytes: diversity and functional roles. New Phytol. 182, 314–330.

Rowan, D.D., Latch, G.C.M., 1994. Utilization of endophyte-infected perennial ryegrasses for increased insect resistance. In: Bacon, C.W., White Jr., J.F. (Eds.), Biotechnology of Endophytic Fungi of Grasses. CRC Press, Boca Raton, pp. 169–183.

Rubini, M.R., Silva-Ribeiro, R.T., Pomella, A.W.V., Maki, C.S., Araujo, W.L., dos Santos, D.R., Azevedo, J.L., 2005. Diversity of endophytic fungal community of cacao (*Theobroma cacao* L.) and biological control of *Crinipellis perniciosa*, casual agent of Witches' broom disease. Int. J. Biol. Sci. 1, 24–33.

Saikkonen, K., Ion, D., Gyllenberg, M., 2002. The persistence of vertically transmitted fungi in grass metapopulations. Proc. R. Soc. B Biol. Sci. 269, 1397–1403.

Saikkonen, K., Young, C.A., Helander, M., Schardl, C.L., 2016. Endophytic Epichloë species and their grass hosts: from evolution to applications. Plant Mol. Biol. 90 (6), 665–675.

Sampangi-Ramaiah, M.H., Jagadheesh, D.P., Jambagi, S., Vasantha Kumari, M.M., Oelmüller, R., et al., 2020. An endophyte from salt-adapted Pokkali rice confers salt-tolerance to a salt-sensitive rice variety and targets a unique pattern of genes in its new host. Sci. Rep. 10, 3237. https://doi.org/10.1038/s41598-020-59998-x.

Sardul, S.S., Suneel, K., Ravindra, P.A., 2014. Isolation and identification of endophytic fungi from *Ricinus Communis Linn.* and their antibacterial activity. Int. J. Res. Pharm. Chem. 4 (3), 611–618.

Shweta, S., Zuehlke, S., Ramesha, B.T., Priti, V., Mohana Kumar, P., Ravikanth, G., et al., 2010. Endophytic fungal strains of Fusarium solani, from Apodytes dimidiata E. Mey.exArn (Icacinaceae) produce camptothecin, 10-hydroxycamptothecin and 9 methoxycamptothecin. Phytochemistry 71, 117–122. https://doi.org/10.1016/j.phytochem.2009.09.030.

Sieber, T.N., 1989. Endophytic fungi in twigs of healthy and diseased Norway spruce and white. Mycol. Res. 92, 322–326.

Singh, B., Kaur, T., Kaur, S., Manhas, R.K., Kaur, A., 2015. An alpha-glucosidase inhibitor from an endophytic *Cladosporium* sp. with potential as a biocontrol agent. Appl. Biochem. Biotechnol. 175, 2020–2034. https://doi.org/10.1007/s12010-014-1325-0.

Spagnoletti, F.N., Tobar, N.E., Di Pardo, A.F., Chiocchio, V.M., Lavado, R.S., 2017. Dark septate endophytes present different potential to solubilize calcium, iron and aluminium phosphates. Appl. Soil Ecol. 111, 25–32. https://doi.org/10.1016/j.apsoil.2016.11.010.

Staniek, A., Woerdenbag, J., Kayser, O., 2008. Endophytes exploiting biodiversity for the improvement of natural product-based drug discovery. J. Plant Interact. 3, 75–98. https://doi.org/10.1080/17429140801886293.

Stierle, A., Strobel, G., Stierle, D., 1993. Taxol and taxane production by *Taxomyces andreanae*, an endophytic fungus of pacific yew. Science 260, 214–216. https://doi.org/10.1126/science.8097061.

Stone, J.K., Polishook, J.D., White Jr., J.F., 2004. Endophytic fungi. In: Mueller, G.M., Bills, G.F., Foster, M.S. (Eds.), Biodiversity of Fungi: Inventory and Monitoring Methods. Elsevier Academic Press, New York, USA, pp. 241–270.

Strobel, G., Daisy, B., Castillo, U., Harper, J., 2004. Natural products from endophytic microorganisms. J. Nat. Prod. 67, 257–268. https://doi.org/10.1021/np030397v.

Suryanarayanan, T.S., 1992. Light–incubation: a neglected procedure in mycology. Mycologist 6, 144.

Suryanarayanan, T.S., Kumaresan, V., Johnson, J.A., 1998. Foliar fungal endophytes from two species of the mangrove Rhizophora. Can. J. Microbiol. 44, 1003–1006.

Taghavi, S., Garafola, C., Monchy, S., Newman, L., Hoffman, A., Weyens, N., 2009. Genome survey and characterization of endophytic bacteria exhibiting a beneficial effect on growth and development of poplar. Appl. Environ. Biol. 75, 748–757.

Talapatra, K., Das, A.R., Saha, A.K., Das, P., 2017. In vitro antagonistic activity of a root endophytic fungus towards plant pathogenic fungi. J. Appl. Biol. Biotechnol. 5 (2), 68–71.

Tan, R.X., Zou, W.X., 2001. Endophytes: a rich source of functional metabolites. Nat. Prod. Rep. 18, 448–459. https://doi.org/10.1039/B100918O.

Tanaka, A., Tapper, B.A., Popay, A., Parker, E.J., Scott, B.A., 2005. Symbiosis expressed non-ribosomal peptide synthetase from a mutualistic fungal endophyte of perennial ryegrass confers protection to the symbiotum from insect herbivory. Mol. Microbiol. 57, 1036–1050. https://doi.org/10.1111/j.1365-2958.2005.04747.x.

Tiwari, P., Srivastava, Y., Bae, H., 2021. Endophytes: trend of pharmaceutical design of Endophytes as anti-infective. Curr. Topic Med. Chem. 21, 1572–1586.

Uzma, F., Mohan, C.D., Siddaiah, C.N., Chowdappa, S., 2019. Endophytic fungi: promising source of novel bioactive compounds. In: Singh, B.P. (Ed.), Advances in Endophytic Fungal Research. Springer, Cham, pp. 243–265.

Varma, A., Bakshi, B., Lou, B., Hartmann, A., Oelmueller, R., 2012. *Piriformospora indica*: a novel plant growth-promoting mycorrhizal fungus. Agric. Res. 1, 117–131.

Vujanovic, V., Brisson, J., 2002. A comparative study of endophytic mycobiota in leaves of Acer saccharum in Eastern North America. Mycol. Prog. 1 (147), 154.

Waller, F., Achatz, B., Baltruschat, H., Fodor, J., Becker, K., Fischer, M., et al., 2005. The endophytic fungus Piriformospora indica reprograms barley to salt stress tolerance, disease resistance, and higher yield. Proc. Natl. Acad. Sci. U. S. A. 102, 13386–13391.

Waqas, M., Khan, A.L., Kamran, M., Hamayun, M., Kang, S.M., Kim, Y.H., Lee, I.J., 2012. Endophytic fungi produce gibberellins and Indoleacetic Acid and promotes host-plant growth during stress. Molecules 17 (9), 10754–10773. https://doi.org/10.3390/molecules170910754.

White, J.F., Kingsley, K.L., Zhang, Q., Verma, R., Obi, N., Dvinskikh, S., et al., 2019. Review: endophytic microbes and their potential applications in crop management. Pest Manag. Sci. 75, 2558–2565. https://doi.org/10.1002/ps.5527.

Xia, Y., Sahib, M.R., Amna, A., Opiyo, S.O., Zhao, Z., Gao, Y.G., 2019. Culturable endophytic fungal communities associated with plants in organic and conventional farming systems and their effects on plant growth. Sci. Rep. 9, 1669. https://doi.org/10.1038/s41598-018-38230-x.

Zhou, Z., Zhang, C., Zhou, W., Li, W., Chu, L., Yan, J., 2014. Diversity and plant growth- promoting ability of endophytic fungi from the five flower plant species collected from Yunnan, Southwest China. J. Plant Interact. 9 (1), 585–591.

CHAPTER

Trichoderma bioinoculant: Scope in entrepreneurship and employment generation

16

Raj K. Mishra, Sonika Pandey, Monika Mishra, Utkarsh Singh Rathore, and Krishna Kumar
Division of Crop Protection, ICAR-Indian Institute of Pulses Research, Kanpur, India

16.1 Introduction

In India, biopesticides market is dominated by various technically nonexperienced companies due to the costly registration procedure. However, the market is also differentiated by the presence of a number of large, organized companies such as Pest Control India and International Panaacea Ltd. Major requirements for biopesticides come from cotton, paddy crops, cereals, pulses, and oilseeds as India is the largest consumer of pulses in the world and the second largest producer of cereals which include wheat, millets, maize, paddy etc. As next-generation pesticides, biopesticides are advancing reputation in India.

Need for biopesticides in India is estimated to show signs of high growth in requisites of both volume and value, estimated at particular compounded annual rates of 18.3% and 19% over the period of 2015–20. Biopesticides Market is projected to reach USD million by 2021 from USD 23.92 million in 2015 (Nakkeeran et al., 2016).

16.2 Scope for commercial production of *Trichoderma*

As of today, requirements of less than 1% cropped area are fulfilled by about 140 biopesticides production units existing in the country. There exists an ample gap, which can only be bridged by setting up of more and more units for producing biopesticides. This involves large-scale venture and private-sector involvement.

Various confined small-scale industries producing and marketing *Trichoderma* and *Trichogramma*. There is a scope to improve production and use of biological control agents in the days to come as the demand is on the augment every year.

16.3 Scale of production

Trichoderma biopesticides can be produced either on a small or large scale. Small-scale production is mainly appropriate to village or community-level cooperatives, which can produce and allocate these for restricted exploitation. As the production technology of some of these bioagents is relatively simple,

Entrepreneurship with Microorganisms. https://doi.org/10.1016/B978-0-443-19049-0.00014-1

the local farmers/SHGs can be trained to commence the production. Medium- and large-scale production can be carried out by firms, sugar mills cooperatives engaged in the manufacture and distribution of agro-chemicals. Fertilizer companies, which already own adequate internal hi-tech expertise and marketing resources, are preferably appropriate for producing biopesticides on a large scale. In the same way, seed companies are predominantly well positioned for undertaking the production and marketing of *Trichoderma* responsibly.

16.4 Market potential of *Trichoderma* bioinoculant

In view of the downbeat effects of haphazard case of pesticides, the significance of organic farming and advertising sustainable farming practices is expected that there will be additional possibilities for new units, particularly in the states of Uttar Pradesh, Maharashtra, Gujarat, Rajasthan, Madhya Pradesh, Tamil Nadu, AP, West Bengal and Karnataka, where crops such as sugarcane, pulses, cereals, and vegetable crops are grown in large scale.

16.5 Entrepreneurship and employment generation through *Trichoderma* bioinoculant

Besides their use as biocontrol agents, *Trichoderma* have wider applications in many other sectors for its applications:

16.5.1 Bioremediation

Several pollutants like phenols, cyanides, nitrates, and phosphates are commonly present in the environment. *Trichoderma* species like *T. harzianum* are reported to degrade these pollutants (Huang et al., 2018). Degradation of artificial dyes like pentachlorophenol, endosulfan, and dichlorodiphenyl trichloroethane (DDT) has been reported by several *Trichoderma* species Katayama and Matsumura (1991). *Trichoderma inhamatum* has the ability to degrade Chromium, thus it can play a significant role in bioremediation (Morales-Barrera and Cristiani-Urbina, 2008). Similarly, *Trichoderma harzianum* has the ability to degrade cadmium (Faedda et al., 2012).

16.5.2 Animal feed

Lytic enzymes like cellulase, hemicelluloses, and pettiness produced by *Trichoderma* can be used for the partial hydrolysis of plant cells. By this process, the nutritive value and digestive value of animal feed increases (Ying et al., 2018).

16.5.3 Industrial application

Cellulase is used extensively for the softening of textiles. *Trichoderma* species are well-known producers of cellulase and are used to reduce the lignin content 197. Mutanase obtained from *T. harzianum* is used in the toothpaste industry to degrade plaque. 198. Nut aroma compounds obtained from *T. viride* and *T. atroviride* are used in food industry for their antibiotic properties. 199. In brewery industry, *Trichoderma*-based enzymes are also used. In fruits and vegetable juice industry, *Trichoderma*-based enzymes are used as additives. *Trichoderma*-based enzymes are also used to improve wine tang and increase the fermentation, filtration, and for the excellence of beer. *Trichoderma* have several curative properties so that it can also be employed in pharmaceutical industry (Marra et al., 2019; Han et al., 2019).

16.5.4 Second-generation biofuels

Cellulase and hemicelluloses produced by *Trichoderma reesei* are used in the production bioethanol from farm wastes. These enzymes catalyze the biodegradation of complex substrates into simple sugars, and then, these substrates are exposed to yeast-catalyzed fermentation (Arthe et al., 2008, Schuster and Schmoll, 2010).

16.5.5 Wood preservation

Chemical wood preservation is relatively cheap and effectively prolongs the life of wood but it has many hazardous health and environmental effects (Rai et al., 2019). This needs an innovation in this field. As the antagonistic nature of *Trichoderma* is well known with wood rotting and saprophytic fungi or other molds, it can be used as a wood preservative. Ejechi in 1921 showed the capability of *Trichoderma viride* to prevent *Gloeophyllum* sp. and *G. sepiarium* decay of *Trichophyton sceleroxylon* wood. Similarly, Tucker et al demonstrated the protective role of *Trichoderma* isolates against the wood degradation by Basidiomycetes.

16.5.6 Agriculture and horticulture applications

Several *Trichoderma* species are currently using to protect fruits and vegetables in postharvest storage, for example, *T. viride* and *T. harzianum* are been using to prevent apple blue mold infection. For the management of *Colletotrichum truncatum*, causative agent of brown blotch of cowpea application of *Trichoderma viride* spore suspension has found effective.

16.5.7 *Trichoderma* research in pulse crops

Pulses play a significant role in the Indian agriculture economy. Pulses are the main source of protein in human diets. In pulse production, India accounts for 20% of total pulse production in world. The major fungal disease of pulses are wilt, Dry root rot, collar rot, wet root rot, Ascochyta blight, Botrytis gray mold, black root rot, seed rot, stem rot, and crown rot. *Trichoderma*-based biopesticides are widely used for the management of pulse pathogens (Mishra and Gupta, 2012; Mishra et al., 2016, 2018a,b, 2020b, 2021b). Moreover, 50% of market is covered by the *Trichoderma* based biofertilizers (Whipps, 2001). Potential of *Trichoderma* in several foliar diseases has been explored by seed or soil applications. *Trichoderma* species have also been used as growth-promoting agent. Thus, they play an important role in the integrated disease management.

Chickpea (*Cicer arietinum*) is one of the most popular pulse crops in India. In India, it accounts for the 50% of pulse production. Main pathogen which causes yield loss in chickpea is *Fusarium oxysporum ciceri* apart from this stem rot by *Sclerotinia* and damping off caused by *R. soleni*. Kaur and Mukhopaday (1992) done the preliminary investigations on the usefulness of *Trichoderma harzianum* with fungicides like Vitavax and Ziram for the management of *Fusarium* wilt. Later, Mukherjee et al. (1995) presented the results of Vinclozolin resistant *Trichoderma* against *Botrytis gray* mold disease of chickpea.

Black gram (*Vigna mungo* L.) and green gram (*Vigna radiate)* are also gaining importance all over the world. First report of antagonism against *M. phaseolins* by *Trichoderma* in mungbean under greenhouse conditions was reported by Kehri and Chandra (1991). Raguchander et al. (1993, 1997) also found the same results. In 2001, Dubey and Patel conducted the experiment with six *Trichoderma* isolates (*T. viride, T. harzianum)* and *Gliocladium* against *Thanatephorus cucumeris* and *R. soleni* and found that all the bioagents were effective in mitigating the effect of pathogens. Choudhary et al. (2011) conducted an experiment and found that the application of *Trichoderma viride* increase the plant growth and *M. phaseolina* disease incidence in mungbean. Mishra et al. in 2011 conducted an experiment and found that *Trichoderma* isolate, *T. viride* is effective in controlling different phytopathogens of mungbean like *R. solani, S. rolfsii, M. phaseolina, Alternaria alternata, Fusarium solani*, and *Colletotrichum capsici*. In present time several species of *Trichoderma* have been used commercially. Mostly isolates of *T. viride* and *T. harzianum* are being used for formulation. Dlahanderma and Dalhanderma-1 developed by ICAR-IIPR are found to be the best in respect of their efficacy and shelf life (Kumar et al., 2021; Mishra et al., 2020a,c,d,e,f, 2021a, 2022). *Glomus mosseae*, *Trichoderma harzianum*, and *Verticillium chlamydosporium* were used alone and in combination for the management of wilt disease complex of pigeon pea caused by the nematode *Heterodera cajani* and the fungus.

In a study conducted by Chaudhary and Reena, they found that three species of *Trichoderma* (*T. harzianum, T. konningi*, and *T. viride*) are effective in controlling the wilt disease of lentil (Fig. 16.1).

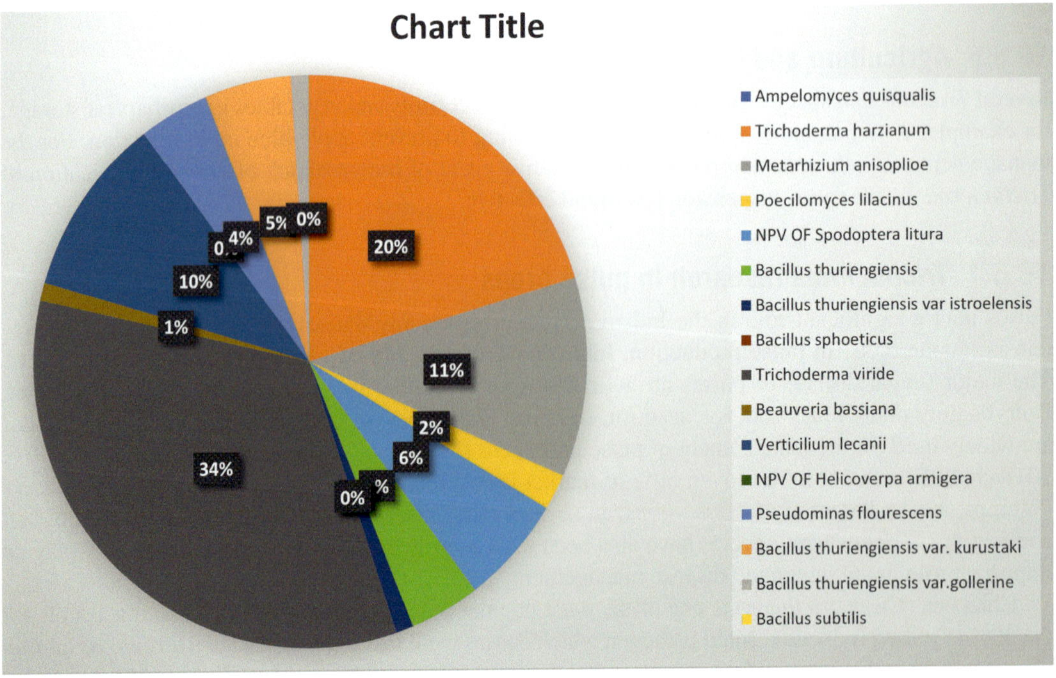

FIG. 16.1

Trichoderma share in biopesticides market (Keswani et al., 2019).

16.6 Factors affecting entrepreneurship

Various research analyses have shown that how important is high tech investment for country growth. Biotechnology is the important and stimulating field for entrepreneurs. Entrepreneurship for the country growth and economic importance is very important. It is very important to think about the entrepreneurship and economic process. Analysts have shown that there is a U-shaped relation between the entrepreneurship and economy. Entrepreneurship analysis is easy to achieve when the homogeneity of the process is high like in the USA, Canada, etc. In countries where heterogeneity is present, it becomes difficult. However, in such countries some models have been created like Anglo-Saxon model, which creates technology parks, nonprofit model which is focused on the stimulation and supporting regional development, promoting technology transfer, BIRAC. This provides incubation facilities for budding entrepreneurs. For biotech industries entrepreneurship is the best option. Since 1980, entrepreneurships have taken into account for economic growth and development. It is very easy to open a biotech company but the main hurdle is to make to survive the company. Because industries are facing global competition and struggle. Factors which affect the entrepreneurship majorly are funding. Funding can be obtained from a variety of sources like (Shimasaki, 2014) funds can be obtained from family, friends, funds can be borrowed from various entrepreneur organist ions etc.

The major factors which affect *Trichoderma* production and entrepreneurship are-

- ✓ Low reliability
- ✓ Target specificity
- ✓ Slow action compared to synthetic pesticides
- ✓ Favorable regulatory system for chemical pesticides
- ✓ Limited biopesticides market

16.7 Economic benefits of entrepreneurship

Entrepreneurship has a positive impact on country growth and development. The various advantages associated with *Trichoderma*-based entrepreneurship are More innovative products at individual level with higher capacity to meet the needs of consumer. Secondly, it will increase the level of performance and competitiveness among industry. Third, it increases the development and national economy. It increases the innovations.

The various aspects which make *Trichoderma* suitable for entrepreneurship are-

- ❖ It has very high level of diversity
- ❖ It offers many challenges
- ❖ It is the best agent to deliver innovative biopesticides products to the market.
- ❖ It has a rising growth rate because of increasing awareness of using chemical pesticides.

The creation of new employment opportunities is dependent on various parameters like expenditure, distribution of productivity growth, and creation of new and innovative business fundamentals. The biobased economy is the key sector for profit investment. Globally, the biobased economy is projected to grow by at least 50% by 2030 (Bio, 2017). For low- and middle-income countries, biobased economy is the best option to invest. Currently, biotechnology is the key drivers for medical innovation.

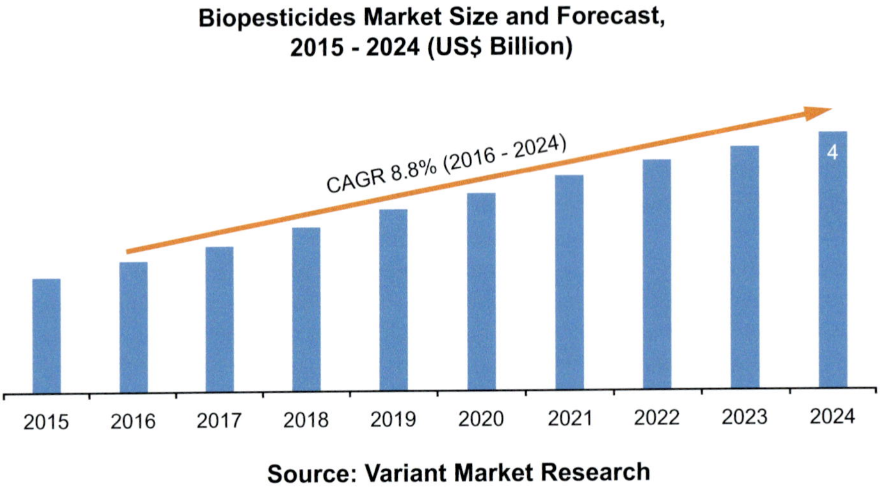

FIG. 16.2

Market potential of biopesticides.

But for agriculture also biotech-based industries are boon. As agriculture is the main source of living for any country, it is the area where invested money will always be benefited. It is ideally suited not only for R&D of new and more efficient products, processes, and services, but also for diversification that increase productivity and scale down waste: less space, less matter, less waste, for more; enhancing bioeconomies and transnational value creation; retaining the return on investment in education and recruiting new talent; and attracting young professionals to the most dynamic sectors of biobusiness (Fig. 16.2).

In 1985, government of India had included IPM in the national policy statement but a major initiative was taken by the department of agriculture and cooperation (DAC), ministry of agriculture by launching a scheme on and modernization of pest management approach in India 1991–92. In 1993, the 8th 5-year plan of India AICRP-BC&W project was further upgraded as an independent Project Directorate of biological control (PDBC) with its headquarter in Bangalore. In India, Bio-Control Research Laboratories (BCRL), which is governed by plant protection research institute, is the major player in the commercial production of biocontrol agents, presently BCRL selling formulations of *T. viride*, *T. harzianum*, and *Beauveria bassiana*. DBT is also providing support for the development of bioformulations. Apart from DBT, DST, and ICAR have made several schemes with the major emphasis on biopesticides.

A total of 361 plant protection quarantine and storage (DPPQS) centers are present. Biopesticide consumption in India has increased since the last decades. Biopesticide production has increased from 83 metric ton to 686 in 1999–2000. A standing committee during 2012–13 has submitted a report in the 15 Lok Sabha on the production and availability of pesticides in India. It was mentioned in the report that use of biopesticides has increased from 123 metric ton (1994–95) to 8110 MT in 2011–12. Data from DPQRS have clearly mentioned that 40% increase in the biopesticide market has been observed from 2014 to 2015 to 2018–19. In present time, there are around 970 biopesticides, which are registered with the central insecticides board and registration committee the percentage of bacteria fungi virus and other agents in biopesticide market is 29, 66, 4, and 1. Various central and state agriculture universities are involved in the commercial production of different biopesticides. Tamil Nadu Agricultural University (TNAU), Coimbatore, Central Plantation Crops Research Institute (CPCRI), Indian Institute of Horticultural Research, Bangalore, Central Research Institute for dry land Agriculture, Hyderabad, Directorate of Oilseed Research (ICAR), Hyderabad and Kerala agricultural University (KAU), Kerala, are known to have dedicated biopesticides production units. Assam Agriculture University and Central Agricultural University, Manipur are producing biopesticides against invasive pests. In north, Indian Agricultural Research Institute (IARI), New Delhi, Punjab Agricultural University (PAU), Punjab, G.B. Pant University of Agriculture & Technology (GBPUA & T), Uttarakhand, are involved in the production of biopesticides, whereas in Central Uttar Pradesh, Indian Institute of Sugarcane Research (IISR) and Central Institute for Subtropical Horticulture, and Directorate of Plant Protection Quarantine & Storage in Lucknow, which works under The Central Integrated Pest Management Centre are the major government agencies involved in manufacturing of biopesticides. Apart from these, with governmental support, several Krishi Vigyan Kendras (KVK) and State biocontrol labs have also been developed.

In India, 70% of biopesticides are produced by private firms. Major biopesticides producing companies are as follows: Ajay Biotech (India) Ltd., Pune, Bharat Biocon Pvt. Ltd. (Chhattisgarh), Microplex Biotech & Agrochem Pvt. (Mumbai), Excel Crop Care Ltd., Mumbai and Govinda Agro Tech Ltd. (Nagpur), Jai Biotech Industries, Satpur, Nasik, Ganesh Biocontrol System, Rajkot, Gujarat Chemicals and Fertilizers Trading Company, Baroda, Gujarat Eco Microbial Technologies Pvt. Ltd., Vadodara, ChaitraAgri Organics, Mysore, Deep Farm Inputs (P) Ltd., KanBiosys Pvt. Ltd., Pune Indore Biotech Inputs & Research Pvt. Ltd., Indore, Romvijay Biotech Pvt. Ltd., Pondicherry Neyyattinkara, Kerala, Devi Biotech (P) Ltd., Madurai, T. Stanes & Company Ltd., Coimbatore, Harit Bio ControlLab., Yavatmal and Hindustan Bioenergy Ltd., Lucknow. Eid Parry, T Stanes Fortune Biotech, Excel Crop care, International Panaacea limited, Biotech International Limited (BIL), Kan biosys, Pest control India (PCI), P J Margo, Prathibha Biotech, Camson Biotech, Zytex. Some foreign companies like Abott have also start working in collaboration with India for biopesticides production. These companies mainly produce biopesticides based on *Trichoderma* and *Pseudomonas*.

16.8 Formulation technology

For *Trichoderma* bioformulation production, the most important thing which should always be kept in mind is that it should be safe, effective, should have longer shelf life and it must be easy to use. In present time for *Trichoderma,* wettable powder and wettable granules are most common to applying in different crops including pulses (Mishra et al., 2020b). A liquid-based formulation of *Trichoderma* is not very much popular and viable for longer time. But from the last few years, researches are trying to develop secondary metabolite-based liquid formulation of *Trichoderma*. Various carrier materials like talc, corn cob, fly ash, sawdust, charcoal, Plaster of Paris are being used for the development of solid formulations. The main problem associated with formulation development is how to maintain the shelf life and microbial activity in carrier materials (Fig. 16.3).

FIG. 16.3

Process of *Trichoderma* formulation development through solid state fermentation developed by ICAR-IIPR, Kanpur.

16.9 Scope and opportunities

If you think about global use of biopesticides (1 kg per hectare), total demand for India should be 100000MT instead of 2890MT. This clearly signifies a huge growth of *Trichoderma*-based biopesticides in India. Due to increasing awareness about environment and human safety and high cost of biopesticides, there is great need to develop biopesticides market at the larger area. India has a rich biodiversity which offers a large scope for natural biological control organisms as well as plant materials. Our ancient culture is also a very important source which can provide valuable information for developing *Trichoderma* biopesticides.

16.10 Conclusion

The biopesticides market is growing continuously due to the increased pest resistance. In Indian perspective, there is a great demand to increase the biopesticides production. *Trichoderma*-based biopesticides have drawn attention as a safer alternative for managing crop disease. In developed countries like USA, there are monitoring agencies, which monitor the release, efficacy and potential hazardous risks associated with the biopesticides. But in developing countries like India, there is a need to develop public-private sector approach for the development, marketing, and risk assessment.

Production of *Trichoderma*-based biopesticides is knowledge-based business in present times, and this business has a wide application in agriculture and that's why in present time it is driving the attention of many enterprises. This business will create numerous jobs in future. This business has significantly contributed to the economy of the nation. This is promoting sustainable and constant economic growth. *Trichoderma*-based biopesticides presents opportunities for the creation and exploitation of different inter- and intranational partnerships for commercial and economical benefits. However, this area needs support from policymakers, businessmen, educators, and government bodies for strengthening and establishment. Collaboration between academic and industry is the best way to accelerate this business.

References

Arthe, R., Rajesh, R., Rajesh, E.M., Rajendran, R., Jeyachandran, S., 2008. Production of bio-ethanol from cellulosic cotton waste through microbial extracellular enzymatic hydrolysis and fermentation. Elec. J. Environ. Agric. Food Chem. 7, 2948–2958.

Bio, 2017. The Biobased Economy: Measuring Growth and Impacts. Biotechnology Innovation Organization. https://www.bio.org/sites/default/files/Biobased_Economy_.

Choudhary, S., Pareek, S., Saxena, J., 2011. Efficacy of *Trichoderma viride* amended composts on growth and dry root rot incidence in mungbean caused by *Macrophomina phaseolina*. Indian Phytopathol. 64 (1), 102–104.

Faedda, R., Puglisi, I., Sanzaro, V., Petrone, G., Cacciola, S.O., 2012. Expression of genes of Trichoderma harzianum in response to the presence of cadmium in the substrate. Nat. Prod. Res. 26, 2301–2308.

Han, M., Qin, D., Ye, T., Yan, X., Wang, J., Duan, X., Dong, J., 2019. An endophytic fungus from *Trichoderma harzianum* SWUKD3. 1610 that produces nigranoic acid and its analogues. Nat. Prod. Res. 33, 2079–2087.

Huang, Y., Xiao, L., Li, F., Xiao, M., Lin, D., Long, X., Wu, Z., 2018. Microbial degradation of pesticide residues and an emphasis on the degradation of cypermethrin and 3-phenoxy benzoic acid: a review. Molecules 23, 2313.

Katayama, A., Matsumura, F., 1991. Photochemically enhanced microbial degradation of environmental pollutants. Environ. Sci. Technol. 25, 1329–1333.

Kaur, N.P., Mukhopadhyay, A.N., 1992. Integrated control of chickpea wilt complex by Trichoderma and chemical methods in India. Int. J. Pest Manag. 38 (4), 372–375.

Kehri, K.H., Chandra, S., 1991. Antagonism of *Trichoderma viride* to *Macrophomina phaseolina* and its application in the control of dry root-rot of mung. Indian Phytopath. 44 (1), 60–63.

Keswani, C., Dilnashin, H., Birla, H., Singh, S., 2019. Regulatory barriers to agricultural research commercialization: a case study of biopesticides in India. Rhizosphere 11. https://doi.org/10.1016/j.rhisph.2019.100155.

Kumar, K., Thakur, P., Rathore, U.S., Kumar, S., Amaresan, N., Mishra, R.K., 2021. Application of potential multitrait *Trichoderma* spp. for management of wilt and enhancing growth in tomato. Vegetos.

Marra, R., Nicoletti, R., Pagano, E., DellaGreca, M., Salvatore, M.M., Borrelli, F., Lombardi, N., Vinale, F., Woo, S.L., Andolfi, A., 2019. Inhibitory e ect of trichodermanone C, a sorbicillinoid produced by *Trichoderma citrinoviride* associated to the green alga Cladophora sp., on nitrite production in LPS-stimulated macrophages. Nat. Prod. Res. 33, 3389–3397.

Mishra, R.K., Gupta, R.P., 2012. *In vitro* evaluation of plant extracts, bio-agents and fungicides against purple blotch and Stemphylium blight of onion. J. Med. Plant Res. 6 (48), 5840–5843.

Mishra, R.K., Naimuddin, M.M., Sharma, S., 2016. *Trichoderma*: potential biocontrol agents for pulses. Dalhan Alok 14, 80–86.

Mishra, R.K., Monika, M., Naimuddin, Krishna, K., 2018a. *Trichoderma asperellum*: a potential biocontrol agents against wilt of pigeonpea caused by *Fusarium udum* Butler. J. Food Legum. 31 (1), 50–53.

Mishra, R.K., Bohra, A., Naimuddin, Kumar, K., Sujayanand, G.K., Saabale, P.R., Naik, S.J., Sarma, B.K., Kumar, D., Mishra, M., Srivastava, D.K., Singh, N.P., 2018b. Utilization of biopesticides as sustainable solutions for management of pests in legume crops: achievements and prospects. Egypt. J. Biol. Pest Control. https://doi.org/10.1186/s41938-017-0004-1.

Mishra, R.K., Pandey, S., Mishra, M., Rathore, U.S., Naimuddin, Krishna, K., Singh, B., 2020a. Assessment of biocontrol potential of *Trichoderma* isolates against wilt in pulses. J. Food Legum. 33, 50–54.

Mishra, R.K., Pandey, S., Mishra, M., Rathore, U.S., Naimuddin, Mohammad, A., Krishna, K., Singh, B., 2020b. Molecular identification, enzyme profiling and biocontrol potential of *Trichoderma* isolates against wilt disease of major pulse crops. J. Food Legum. 33 (1), 50–54.

Mishra, R.K., Monika, M., Sonika, P., Naimuddin, Saabale, P.R., Bansa, S., 2020c. DALHANDERMA (IIPRTh-31): multi-trait *Trichoderma* based formulation for management of wilt diseases of pulse crops. J. Food Legum. 33, 123–126.

Mishra, R.K., Sonika, P., Monika, M., Arvind, K.K., Rathore, U.S., Naimuddin, 2020d. Characterization of biocontrol genes (ech-42 and Xyn-2) from *Trichoderma* spp. ICAR-IIPR Newsletter 31, 3–4.

Mishra, R.K., Sonika, P., Monika, M., Rathore, U.S., Naimuddin, Mohd, A., Bansa, S., 2020e. Potential indigenous *Trichoderma* spp. identified from pulses rhizosphere. ICAR-IIPR Newsletter 31, 5.

Mishra, R.K., Sonika, P., Monika, M., Rathore, U.S., Saloni, R., 2020f. *Trichoderma* for bioremediation and phytoremediation in sustainable soil and plant health management. ENVIS NBRI Newsletter 16.

Mishra, R.K., Sonika, P., Monika, M., Rathore, U.S., Kulbhushan, M.T., Krishna, K., 2021a. *Trichoderma*: an effective and potential biocontrol agent for sustainable management of pulses pathogens. In: Trichoderma: Agricultural Applications and Beyond. Springer, pp. 159–180.

Mishra, R.K., Mishra, M., Pandey, S., Saabale, P.R., Naimuddin, 2021b. Exploring the indigenous Trichoderma strains from pulses rhizosphere and their biocontrol potential against *Fusarium oxysporum* f.sp. *ciceri* in chickpea. Indian Phytopathol. https://doi.org/10.1007/s42360-021-00416-1.

Mishra, R.K., Pandey, S., Mishra, M., Rathore, U.S., Nimuddin, Singh, B., 2022. Trichoderma dalhani phaslo evam mrida swasthya ke liye Vardaan.

Morales-Barrera, L., Cristiani-Urbina, E., 2008. Hexavalent chromium removal by a *Trichoderma inhamatum* fungal strain isolated from tannery effluent. Water Air Soil Pollut. 187, 327–336.

Mukherjee, P.K., Haware, M.P., Jayanthi, S., 1995. Preliminary investigations in integrated biocontrol of Botrytis gray mold of chickpea. Indian Phytopath. 48 (2), 141–149.

Nakkeeran, S., Renukadevi, P., Aiyanathan, K.E.A., 2016. Exploring the potential of *Trichoderma* for the management of seed and soil-borne diseases of crops. In: Muniappan, R., Heinrichs, E. (Eds.), Integrated Pest Management of Tropical Vegetable Crops. Springer, Dordrecht, https://doi.org/10.1007/978-94-024-0924-6_4.

Raguchander, T., Samiappan, R., Arjunan, G., 1993. Biocontrol of Macrophomina root rot of mungbean. Indian Phytopath. 46 (4), 379–382.

Raguchander, T., Rajappan, K., Samiappan, R., 1997. Evaluating methods of application of biocontrol agent in the control of mungbean root rot. Indian Phytopath. 50 (2), 229–234.

Rai, P.K., Lee, S.S., Zhang, M., Tsang, Y.F., Kim, K.-H., 2019. Heavy metals in food crops: health risks, fate, mechanisms, and management. Environ. Int. 125, 365–385.

Schuster, A., Schmoll, M., 2010. Biology and biotechnology of trichoderma. Appl. Microbiol. Biotechnol. 87, 787–799.

Shimasaki, C.D. (Ed.), 2014. Biotechnology Entrepreneurship. Starting, Managing, and Leading Biotech Companies. Elsevier, Waltham, MA.

Whipps, J.M., 2001. Microbial interactions and biocontrol in the rhizosphere. J. Exp. Bot. 52, 487–511. https://doi.org/10.1093/jexbot/52.suppl_1.487.

Ying, W., Shi, Z., Yang, H., Xu, G., Zheng, Z., Yang, J., 2018. Effect of alkaline lignin modification on cellulase–lignin interactions and enzymatic saccharification yield. Biotechnol. Biofuels 11, 214.

CHAPTER

17 Arbuscular mycorrhizal fungi in alleviation of biotic stress tolerance in plants: A new direction in sustainable agriculture

Ashish Kumar[a], Joystu Dutta[b], Nagendra Kumar Chandrawanshi[c], Alka Ekka[a], and Santosh Kumar Sethi[d]

[a]Department of Biotechnology, Guru Ghasidas Vishwavidyalaya, Bilaspurr, Chhattisgarh, India, [b]Department of Environmental Science, Sant Gahira Guru Vishwavidyalaya, Sarguja Ambikapur, Chhattisgarh, India, [c]School of Studies in Biotechnology, Pt. Ravishankar Shukla University, Raipur, Chhattisgarh, India, [d]School of Biotechnology, Gangadhar Meher University, Sambalpur, Odisha, India

17.1 Introduction

Paleontological and molecular genetic evidences point out to the fact that the mycorrhizal relationship between various soil fungi and plant roots has existed for a billions of years (Taylor et al., 1995). The mycorrhizal relationship is thought to have helped plant colonization of soil, and this coevolution may have involved both physiological and ecological interactions. Interactions with other functional classes of microbes, both harmful and beneficial, are among the ecological interactions. In natural environments where microbial equilibrium (including mycorrhizas) has not been affected, we currently find that plants have very less or no disease symptoms, or at least no strong effect of disease on their development and survival. Plants grown in disrupted agroecosystems, on the other hand, where mycorrhizas may be absent or less functional, may experience serious disease occurrence and extreme plant production loss. Since most vascular plants can develop mycorrhizas, it makes sense to try to link the mycorrhizal relationship to the occurrence and seriousness of plant diseases, and considerable research has been done in this area in recent decades. In order to promote most ecofriendly and sustainable agriculture production, such study is further provoked by the need to discover alternative ways of combating pathogens without the use of costly and potentially dangerous chemicals (Bethlenfalvay and Linderman, 1992). There have been several attempts to summarize the literature on mycorrhiza-disease relationships (Hooker et al., 1994; Jalali and Jalali, 1991; Linderman and Paulitz, 1990; St-Arnaud et al., 1995). Furthermore, the responses of plants with and without mycorrhizas to numerous pathogens (bacteria, fungi, virus, or nematode) can be especially different, making comparisons difficult. The intrinsic variations between the various types of mycorrhizas produced on different plant groups

Entrepreneurship with Microorganisms. https://doi.org/10.1016/B978-0-443-19049-0.00006-2

further complicate general interpretations. Furthermore, one could contemplate the evolutionary progress of plants that do not form mycorrhizas. Despite this, a number of researchers have addressed relations and the processes that seem to be involved (van der Heijden et al., 2015).

Arbuscular mycorrhizal fungi (AMF) colonize plant roots thriving in the rhizosphere and are beneficial to overall development of plants and smooth functioning of metabolic activities. AMF improves the rhizosphere soil distinctiveness and health of host plants by supporting with mineral nutrients, especially phosphorus, and inhibiting the translocation of toxic ions including sodium and other metals. Plants have innate resistance mechanisms to avoid the harmful impacts of various environmental stresses. The antioxidant mechanisms aided by osmolytes aggregation and the selective ion absorption are some of the most important resistance mechanisms observed in plants (Sharma et al., 2019). The potential of AMF in phytological growth functions and biotic/abiotic stress tolerance is well studied. This chapter discusses a comprehensive review updates on AMF with emphasis on AMF-induced improvements in plant antioxidant and osmolyte metabolisms that help plants cope with natural as well as anthropogenic stresses that include extreme climate change, natural disasters, deforestation, atmospheric pollution, increasing concentration of greenhouse gases (GHG), ozone depletion, radioactive fallout, and chemical leakages. Furthermore, several key possible future goals have been defined to better our understanding of beneficial impacts of AMF on plant growth enhancement under both natural and under the impact of several stress factors. Environmental factors, such as soil quality, salinity, soil organic carbon, water retention capacity, altered environmental temperature, erratic precipitation pattern, and the occurrence or deficiency of macro- and micronutrients, presence or absence of xenobiotic compounds, and heavy metals, are the major determinants of altered plant development. The net impact of these factors can be seen in crop growth and productivity. Modifications in the morphology and biochemistry of plant species caused by these environmental conditions result in altered growth and yield, posing a significant threat to global food crisis (Hashem et al., 2015). Biotic and abiotic stress factors play a crucial role in determining physiological activities of different plant species with variations in the defense mechanisms (Hashem et al., 2014; Alqarawi et al., 2014a; Sharma et al., 2019). Environmental stressors such as aridity and extreme rise in temperature associated with low water holding capacity off soil alters plant growth and metabolic activities (Barnawal et al., 2014). The situation has been made worse by global climate change during the last decade (Ainsworth et al., 2012). The continued use of metal and salt rich water for irrigation results in the reduction of agriculturally productivity. Stress causes toxic effects, which adversely affects essential physiological process and biochemical pathways processes such as photosynthesis and ion homeostasis, eventually resulting in a decrease in plant growth and yield (Porcel et al., 2012; Khan et al., 2015). Stress caused alterations in photosynthesis are associated with impeded carbon and nitrogen metabolism, and in legumes, salinity alters the nitrogen fixation and hence the growth, as well as the yield (Tejera et al., 2004). Stressed plants produce and accumulate more toxic reactive oxygen species (ROS), which have the ability to destroy membrane lipids, degrade nucleic acids, and oxidize proteins (Mittler, 2002; Hashem et al., 2016). Under stress, superoxide, hydroxyl, and peroxide radicals are the most common ROS generated (Ahmad et al., 2010). ROS build up in sensitive tissues like leaves, causing oxidation of the above macromolecular structures and thereby disrupting the plant's cellular physiology. Mitochondria, chloroplasts, and peroxisomes are the primary sources of reactive oxygen species (ROS). ROS are mainly generated in chloroplasts, mitochondria, and peroxisomes (Mittler, 2002; Ahmad et al., 2010).

17.2 Characteristics of AMF symbiosis

The most common beneficial relationship between plants and microorganisms is the arbuscular mycorrhizal (AM) symbiosis (Parniske, 2008). Several studies (Siddiqui and Mahmood, 1998; Nakmee et al., 2016) have shown that they aid plant nutrition and growth in stressed habitats, as well as improving a number of essential ecosystem processes. The most frequent form of mycorrhizal involvement is arbuscular mycorrhizas. They are produced by a diverse range of host plants (approximately 65% of all recognized land plant species) (Wang and Qiu, 2006), including several important agricultural crop species like soybean, corn, rice, and wheat. AMF are obligate biotrophs, meaning they must complete their life cycle and develop the next generation of spores on their autotrophic host. The spores can germinate in the absence of a host, but root exudates cause an increase in hyphal branching and metabolic activity in the spores (Tamasloukht et al., 2003; Gachomo et al., 2009). Plant roots, for example, produce strigolactones, which may cause AM fungal spores to emerge in a presymbiotic state (Akiyama and Hayashi, 2006). AM fungi form an unique appressorium—the hyphopodium—on the surface of host plant root. The prepenetration apparatus, which directs the fungal hyphae through the root cells into the cortex, allows fungal hyphae expanding from this hyphopodium to penetrate further into root. Hyphae penetrate the apoplast in the cortex, expand laterally along the roots axis, and invade internal root cortical cells. Tiny hyphal branches reach the cell of "typical" AM connections of the "Arum type," which grow into distinctive highly branched arbuscules by persistent dichotomous branching. Mycorrhizas of the "Paris type," on the other hand, disperse the fungus mainly from cell to cell and grow substantial intracellular hyphal bundles with arbuscular branching (Smith and Read, 2008). The fungus cannot penetrate the plant symplast and is kept out of the cytoplasm of the host by the host's expanded periarbuscular membrane (PAM). In the root apoplast, certain fungi form vesicles are fungal storage organs.

Besides its coenocytic origin, the intraradical mycelium (IRM), which forms inside the root, varies phenotypically and functionally from the extraradical mycelium (ERM), which develops into the land. The ERM receives nutrients and water from the soil and passes them on to the host plant root. On the other side, the IRM releases nutrients into the interfacial apoplast and exchanges them for carbon out from the host. These carbon products are used by the fungus to sustain and expand the ERM, for cell metabolism (e.g., successful absorption processes, nitrogen assimilation), as well as for the growth of spores that can colonize a new generation of host plants.

Continuous signal communication between both the two symbionts is needed for the development of a mycorrhizal symbiosis, which activates synchronized differentiation of the both partners. Furthermore, the close association of arbuscules within root cells usually requires a partial activation of plant defensive responses (Liao et al., 2018). Mycorrhizal hyphae can colonize areas of the soil that plant roots never can enter, and they can also obtain nutrients due to active transporters. Mostly, the fungus can assist the plant to obtain phosphorus, but it can also help with other low-mobility nutrients including ammonium, copper, potassium, iron, molybdenum, sulfur, and zinc. As a result, the plant should provide carbohydrates to the fungus to satisfy their needs, though this has no harmful consequences for the plant due to photosynthetic reinforcement and decreased root growth (Berruti et al., 2016; Chen et al., 2018). Furthermore, it is commonly assumed that inoculating host plants with mycorrhizal fungi confers resistance to a variety of fungal and bacterial pathogens (Atera et al., 2012; Kumari and Prabina, 2019).

17.3 AMF-Plant defense and disease resistance mechanisms

Fungal colonization of mycorrhizal roots can be very intense, both inter and intracellularly, and can cover more than 90% of the total root length. Many researchers have reported that AMF colonization benefit their host plants, however, provided that AMF possess and release several molecular signals (e.g., chitin oligomers) that can be sensed by plants, which have been shown to cause defensive responses in different plant species, it remains a mystery how plants can survive such high levels of colonization without staging a defense response (Wan et al., 2008; Boller and Felix, 2009). As a result, it has been argued that AMF is associated with the repression of protection. Indeed, plant mutants deficient in genes involved in mycorrhizal signaling and AMF establishment frequently exhibit typical defensive responses when infected with AMF, implying that such fungi have powerful signaling molecules that cause defense, which are inhibited during normal AMF growth. Pathogens typically develop defense inhibitors (widely recognized as effectors), and some effectors have recently been discovered in the genomes of AMF9 (Kamel et al., 2017). However, only a small percentage of them have undergone functional analysis (Kloppholz et al., 2011). Although the host's deference mechanisms must be weakened to allow AMF invasion and root colonization, general defense must stay operational to deal with rhizospheric pathogens.

Mycorrhizal plants, on the other hand, are often found to be more disease resistant (Pozo, 2007; Jung et al., 2012; Cameron et al., 2013). Experiments with split root systems showed that the whole plant is protected from pathogenic attack and interactions (Rodriguez et al., 2019). This may be a general improvement in plant health due to better nutrition, or a systematic induction of protection status, referred to as systemic acquired resistance (SAR). Induced systemic resistance (ISR) may refer to increased efficacy of pathogenic resistance mechanism displayed by mycorrhizal plant associations (Conrath et al., 2006). These plant-protective effects of AMF are of particular concern for long-term plant protection strategies (Solaiman et al., 2014). AMF are mainly used to protect plants from soil-borne bacterial pathogens (Jung et al., 2012; Cameron et al., 2013). AMF and other microbes interacting with their mycelium may directly interact with rhizospheric pathogens by releasing antimicrobial compounds.

Mycorrhizal symbiosis regulation is a specialized mechanism involving several regulatory elements operating at various stages. Defense phytohormones have been shown to play a crucial role in regulating mycorrhizal interactions over recent years, from early recognition and colonization events to final arbuscular development and degradation (Liao et al., 2018). Numerous scholars such as (Hohmann and Messmer, 2017; Jacott et al., 2017) have examined the capacity of mycorrhizal fungi to enable ISR in plants to protect them from pathogens. Mycorrhizal fungi activate plant defense mechanisms similar to biotrophic pathogen and regulate plant responses for colonization success (Hohmann and Messmer, 2017). Plant tolerance to phytopathogenic nematodes by mycorrhizal fungi has been documented (Poveda et al., 2020). Observations made by Vos et al. (2012) demonstrated how inoculating tomato roots with *F. mosseae* in one compartment decreased the infection rates of *M. incognita* and *Pratylenchus penetrans* in the compartment without fungus by altering root exudation. PR protein encoding gene activation that facilitates ROS detoxification has been linked to the root defense mechanism. Enzymes such as superoxide dismutase (SOD) or glutathione S-transferase that participates in shikimate pathway and lignin biosynthesis creates precursors of different aromatic secondary metabolites against nematodes (Schouteden et al., 2015; Sharma and Sharma, 2017; Balestrini et al., 2019). Strigolactone, an emerging phytohormone, plays a crucial role while displaying a bouquet of functions such as growth stimulant of parasitic plants, in determination of plant architecture, in promotion of symbiotic processes by arbuscular mycorrhiza (Mishra et al., 2017). Despite of their immense developmental potential, the strigolactone research in the last few years has also established their significance in adverse environmental conditions (Rochange et al., 2019). Strigolactone-mediated signaling suppresses jasmonate aggregation and facilitates RKN infection in rice (Lahari et al., 2019). Strigolactones facilitate in enhancing defense mechanism of certain plant species such as tomatoes against nematode attacks (Le-Xu et al., 2018). Although further research is needed, another approach to modulate response of plants by mycorrhization may be to change strigolactones-plant development, which could affect plant-nematode interactions. Colonization of pines (*Pinus thunbergii*) resulted in a long-term SAR against the nematode *Bursaphelenchus xylophilus*, which is carried by Monochamus feeds and beetles by colonization of the vascular bundles. Interestingly, such fungi increase plant defenses against nematode infections, such as the grapevine fanleaf virus (GFLV) transmitted by the nematode vector *Xiphinema index*, which AMF *R. intraradices* confers systemic resistance (Hao et al., 2018). Other instances of systemic resistance offered by AMF include decreased nematode infection due to the activity of phenolics and defensive plant enzymes including peroxidase, superoxide dismutase, and polyphenol oxidase, as well as a significant reduction in hydrogen peroxide and malondialdehyde concentration in tomato roots infected with *R. irregularis*, which increased plant development (Sharma and Sharma, 2017a). In tomato, split-root treatment of *G. mosseae* against migratory and stationary nematodes *P. penetrans* and *M. incognita* indicated that only systemically AMF can suppress infection caused by both types of nematodes (Vos et al., 2012). Using AMF *G. intraradices*, the migratory nematodes *Pratylenchus coffeae* and *Radopholus similis* were successfully controlled in banana (*Musa* spp.) split-roots (Elsen et al., 2008). In some cases, it is unclear if a direct mechanism or ISR is combating nematodes, or whether both are at work at that time. One example of this kind is the decrease of *P. penetrans* infestation in apple seedlings by AMF (Ceustermans et al., 2018), as well as the decrease of *Meloidogyne arenaria* in red ginger (*Alpinia purpurata*) by *Gigaspora albida*, *Acaulosporalongula*, and *Claroideoglomus etunicatum* (Da Silva-Campos et al., 2017), or *G. dussii* and *F. mosseae* regulate *Scutellonema bradys*, a migratory endoparasitic nematode, in yam (*Dioscorea* spp.) (Tchabi et al., 2016). In wheat, however, root treatment of *Funneliformis coronatum*, *Claroideoglomus entunicatum, F. mosseae,* and *R. irregularis* led in an increase in populations of the plant-parasitic nematode *Pratylenchus neglectus*, according to Frew et al. (2018). This might be owing to a decrease in root benzoxazinoid glucoside aggregation, a key nematode defense chemical required for the fungus' root colonization.

Finally, it may be suggested that AMF is capable to induce plant defense mechanisms in their host plants. Few studies have explained that simply plant-AMF interaction known as mycorrhization (Myc; root colonization and arbuscular formation) and mycorrhizal responsiveness (MR; ability of the plant to respond to AMF have been proposed as two general ways to evaluate the plant-arbuscular mycorrhizal fungi interaction. Mycorrhization has an effect on phytohormones such as jasmonic acid (JA), as well as defense-related compounds. The plant in turn regulates the symbiosis via various phytohormone pathways. It is proposed to include AMF-mediated disease resistance as an additional measure of mycorrhizal responsiveness. Genotype selection must take place in conditions that do not inhibit the plant-microbe relationship in order to identify variations in efficiency. As a result, the importance of organic breeding programs is emphasized (Hohmann and Messmer, 2017).

17.4 AMF and plant biotic stress tolerance

Microsymbionts are becoming more important in the control of plant biotic stresses (Tariq et al., 2017). Farmers in India and worldwide have used large amounts of harmful pesticides to control their crops from pathogenic fungi, bacteria, viruses, and nematodes that cause significant yield losses. The symbiotic relationship of AMF and plant is a typical example of a mutualistic relationship that can control plant growth and development. The mycelial network of the fungi spreads under the roots of the plant facilitating nutrient uptake that would otherwise be unavailable. The mycelium of fungi inhabits the roots of several plants, even though they are from different species, form a common mycorrhizal network (CMN). This CMN is regarded as a key component of the terrestrial environment, with major effects on various plant populations, especially invasive plants and fungal-mediated nitrogen (N), and phosphorus (P) transport to plants (Pringle et al., 2009). Communal nutrients, as well as other associated effects, are transferred from fungi to plants, which is likely why AMF improves plant immunity to biotic and abiotic stresses (Plassard and Dell, 2010). They have the potential to change soil characteristics and thereby promote plant growth in natural as well as in stressful conditions (Navarro et al., 2014; Alqarawi et al., 2014b,a). Plants tolerance to environmental biotic and abiotic stresses increases as result of AMF colonization, which causes many improvements in their morpho-physiological characteristics (Alqarawi et al., 2014a,b; Hashem et al., 2015). AMF are natural growth promoters in a wide range of terrestrial flora and are being used as bio-inoculants, and experts recommend that they can be used as popular biofertilizers in the production of sustainable crops (Barrow, 2012). Several studies have shown that mycorrhizal fungi aid plant immunity to biotic stress caused by soil-borne potential pathogens that interact with a wide range of plant organisms. This has been repeatedly demonstrated among pathogenic fungi or Oomycetes such as Fusarium, Rhizoctonia, Thievalopsis, Verticillium, Phytophthora, Aphanomyces, and Pythium, as well as nematodes from the genera Meloidogyne, Heterodera, Radophol, and Pratylenchus (Whipps, 2004; Harrier and Watson, 2004). The majority of the study has been done under very controlled and scheduled conditions during the early stages of plant growth, but a few experiments performed in the field or in a greenhouse under real-world cultivation conditions back up these findings (Bodker et al., 2002; Utkhede, 2006). Major findings have been obtained for the tomato crop which is one of the most widely grown vegetables crop worldwide. According to current cultivation methods, this crop is vulnerable to multiple bacteria, fungi, and nematodes, resulting in a significant decrease in fruit yield (Engindeniz, 2006; Kumar et al., 2014, 2018). Though the tomato plant is not very friendly to mycorrhizal fungi in terms of plant development (Smith et al., 2009), it obviously benefits from mycorrhizal inoculation when faced with plant root pathogens such as *Rhizoctonia solani*, *Fusarium oxysporum f.* sp. *radicis-lycopersici*, *Meloidogyne incognita*, and *Phythophthora parasitica*. When compared to pathogen-infected mycorrhizal noninoculated plants, root colonization by mycorrhizal fungi will reduce root infection and disease severity caused by potential pathogens, resulting in increased fruit yield (14.3%) and fresh plant weight (up to 198%). Besides numerous studies have shown that AMF decreases the total yield losses by protecting the plants from various plant pathogens (Nguvo and Gao, 2019; Vos et al., 2012; Kumari and Prabina, 2019) (Table 17.1). Various studied reported that AMF inoculation can reduce the incidence of charcoal root-rot disease in soybean. Arbuscular colonization raised shoot dry weight in the presence of Fusarium (Vos et al., 2012). The defensive

Table 17.1 Detail of several AMF assisting plants in dealing with a variety of biotic stresses.

Name of Host Plants	Name of Pathogen	AMF Strains	AMF Inoculation-Related response	References
Solanum lycopersicum L.	*M. incognita*	*F. mosseae*	Developed systemic resistance against migratory nematode *P. penetrans* and sedentary nematode *M. incognita*	Vos et al. (2012)
L. esculentum	*Fusarium oxysporum* f. sp. Lycopersici	*Glomus* sp.	Antimicrobial compounds produced by the rhizospheric root prevented the fungal pathogen's mycelial growth. Plant growth, yield, dry weight, chlorophyll, N, P, K content, were all improved	Kumari and Prabina (2019)
Glycine max L.	*Macrophomina phaseolina*	*R. irregularis*	Increase number of functional leaves and plant height	Spagnoletti et al. (2020)
Solanum lycopersicum L.	*M. javanica*	*F. mosseae*	Galling, nematode replication, and female morphometric parameters in tomato plants were also affected	Siddiqui and Mahmood (1998)
Solanum lycopersicum L.	*Alternaria solani Fusarium oxysporum*	*F. mosseae*	Eliminated tomato diseases	Song et al. (2015)
Saccharum officinarum L.	*Striga hermonthica*	*G. etunicatum, G. margarita S. fulgida,*	Plant growth, biomass, and environmental responses were stimulated	Manjunatha et al. (2018)
Solanum lycopersicum L.	*Cladosporium fulvum*	*F. mosseae*	Increased plant growth, dry weight, net photosynthesis rate, and total chlorophyll content and disease resistance	Wang (2017)
Astragalus adsurgens	*Erysiphe pisi*	*C. etunicatum, F. mosseae G. versiforme*	Increased the root and shoot development and disease resistance to powdery mildew	Liu et al. (2018)
Capsicum annum	*Pythium Aphanidermatum*	*Glomus* sp.	The fungal pathogen's mycelial growth decreased disease occurrence, and improved crop growth and yield	Kumari and Srimeena (2019)

Continued

Table 17.1 Detail of several AMF assisting plants in dealing with a variety of biotic stresses—cont'd

Name of Host Plants	Name of Pathogen	AMF Strains	AMF Inoculation-Related response	References
Cicer arietinum L.	*Fusarium* wilt	*R. Fasciculatum Glomushoi*	Increased N and P contents in treated plants	Singh et al. (2010)
Cucumis melo L.	*Fusarium* wilt	*F. mosseae*	The greatest ability to reduce disease occurrence	Martínez-Medina et al. (2011)
Arachis hypogaea L.	*Sclerotium rolfsii*	*R. fasciculatum, A. laevis, Sclerocystisdussii, and Gigaspora margarita*	Eliminated the harmful special effects of *S. rolfsii*	Kulkarni et al. (2011)
Cucumis sativus L. *Solanum melongena* L.	*Pseudomonas lacrymans, Verticillium dahlia*	*G. versiforme*	Symptoms of wilt caused by *V. dahliae* were reduced	Jones et al. (2014)
S. tuberosum	*Potato virus* Y	*R. irregularis*	PVY-infected plants inoculated with the AMF showed mild symptoms and considerable stimulation in shoot growth	Thiem et al. (2014)
Nicotiana tabacum L.	Cucumber green mottle mosaic virus Tobacco mosaic virus	*R. irregularis*	As compared to nonmycorrhizal plants, mycorrhizal plants had lower disease symptoms and virus titers	Stolyarchuk et al. (2009)
Zea mays L.	*Striga hermonthica*	*G. margarita Utellosporafulgida*	Decreased Striga plant occurrence, phosphate content and plant biomass	Othira (2012)
Sorghum bicolor (L.) Moench	*S. hermonthica*	*F. mosseae*	Improved the sorghum yield and performance	Isah et al. (2013)

effect of AMF colonization is known as mycorrhiza-induced resistance (Pozo, 2007; Jung et al., 2012; Nguvo and Gao, 2019). This offers systemic defense against a broad spectrum of pathogens and shares distinctiveness with systemic acquired resistance (SAR) after pathogen attack and induced systemic resistance (ISR) after root colonization of nonpathogenic rhizobacteria Mycorrhiza-induced disease resistance was described by (Cameron et al., 2013). AMF enhances the synthesis of antioxidant enzymes in plants, which can function as a defense agent against pathogens and other stresses. Furthermore, the activation of plant defense mechanisms, increased nutritional status of the host plant, improvements in root growth and morphology, competition for colonization sites and host photosynthates, and microbial changes in the mycorrhizosphere have all been linked to reduced pathogen damage by AMF (Vos et al., 2012; Siddiqui and Mahmood, 1998; Akhtar and Siddiqui, 2007). Since mycorrhizae may help tissues regrowth after attacks, improving plant growth can have a beneficial impact. However, since plant nutrition increases, it becomes more nutritive or appealing to herbivorous insects (Hoffmann et al., 2010).

The role of AMF in plant defense against pathogenic fungi and nematodes has been well established (Schouteden et al., 2015). The success of the interactions varies depending on the host plant (Schubler et al., 2001). The colonization of Arbuscular mycorrhizal fungi *F. mosseae* on tomato roots resulted the systematic resistance against the migratory nematode *Pratylenchus penetrans* and the sedentary nematode *Meloidogyne incognita* (Vos et al., 2012). In AMF-colonized plants, nematode contamination was decreased by 45% and 87%, respectively, for *M. incognita* and *P. penetrans* relative to controls. Further research on root exudates has shown that the reduction of nematode infection in mycorrhizal plants is most likely due to AMF altering their root exudates (Vos et al., 2011). In reality, mycorrhizal root exudates decreased nematode penetration and temporarily paralyzed nematodes when applied to mycorrhizal plants, when compared to the use of water or nonmycorrhizal root exudates. In tomato plants inoculated with *Meloidogyne javanica*, root colonization by *F. mosseae* reduced galling, nematode reproduction, and female morphometric parameters (Siddiqui and Mahmood, 1998). In terms of pathogenic fungi, *F. mosseae* mycorrhizal inoculation greatly reduced tomato diseases caused by *Alternaria solani* and *Fusarium oxysporum,* respectively. When plants were inoculated with AMF and sprayed with hormonal inducers (Salicylic acid and Jasmonic acid), the beneficial impact was much stronger, implying a synergistic and cooperative effect between the two, contributing to improved disease tolerance induction and control (El-Khallal, 2007). These beneficial impacts of AMF have also been observed in pathogen-infected chickpea (Gallou et al., 2011; Akhtar and Siddiqui, 2007). In contrast to its well-known effect on pathogenic fungi and nematodes, few studies on the effects of AMF on herbivore insects have been conducted (Koricheva et al., 2009; Gange et al., 2003). The effects of mycorrhizal fungi on herbivore insects varied depending on the parameter calculated and the degree of herbivore feeding specialization, according to a metaanalysis conducted by Koricheva et al. (2009). In *Plantago lanceolate* L., mycorrhizal association increased leaves tolerance to the chewing insect *Arctia caja*, while mycorrhizal plants performed better for the sucking insect *Myzus persicae* (Sulzer) (Sharma and Sharma, 2017). It is recorded that, the effects of three species of AMF on parasitism rates were depending on the species of AMF, while parasitism of *Chromatomyia syngenesiae* by *Diglyphus isaea* was lower on mycorrhizal plants (Gange et al., 2003). The African witch weed *Striga* sp. found primarily in sub-Saharan Africa is one of the biotic constraints that has a significant impact on production in developing countries. Owing to high infestation rates, this parasitic plant is a socioeconomic crisis that has destroyed farms in tropical countries (Atera et al., 2012). Striga germination can be inhibited or suppressed by soil microorganisms, including AMF (Jones et al., 2014). Striga and cereal interactions may be influenced by AMF (Lendzemo et al., 2006). These researchers discovered that AMF harmed *Striga* seed germination, decreased the number of *Striga* seedlings attaching and emerging, and slowed *Striga* emergence. Studies carried out by Manjunatha et al. (2018) and confirmed the effectiveness of AMF in protecting sugarcane against *Striga* infestation as well as promoting crop growth and reducing the soil *Striga* seed bank. AMF enhanced the performance of the plant host, allowing it to better withstand *Striga* damage. AMF improved the plant host's efficiency, helping it to better withstand Striga damage (Lendzemo et al., 2006).

17.5 Conclusion

The mycorrhizal relationship with land plant roots has existed for hundreds of millions of years, and it logically entails interactions with other functional classes of soil microbes. It is clear that the plant-AMF association has enormous potential in terms of innovative and more effective crop management activities, but it is also clear that we have just scratched the surface of this potential. It is reasonable to assume that mycorrhiza played a role in protecting plants from pathogenic invasion. However, experimental demonstration of the phenomenon is challenging owing to variations in experimental methods, variety of pathogens, and plethora of diseases. Many studies have concentrated on interaction processes such as (a) improved nutrition, (b) competition for nutrition and contamination sites, (c) morphological improvements, (d) changes in phytochemical compounds in host plants, (e) diminishment of abiotic stress, and (f) microbial variations in the mycorrhizosphere. Any of more processes may be involved, depending on the disease and the climate, but variations in microbial communities in the mycorrhizosphere seem to be the better cause, despite being the least documented. The mycorrhizosphere framework, in which soil microbes are affected by both rhizodeposition of substances from roots and secretion from arbuscular mycorrhizal (AM) fungal hyphae, requires further investigations. The microbial communities in the mycorrhizosphere can vary over time and are affected by microbes present in soil or substrate, the development of the AMF relationship intraradially along with extraradially, and the processes of selective enrichment of different functional groups of microbes, including those which can contribute to root pathogen antagonistic capacity. Documentation is presented to demonstrate how the AMF colonization works to alleviate from fungal, bacterial and plant diseases. According to the findings, early establishment of plant-AMF relationship enhances the level of various biotic stress tolerance in host plants that may be a disease management strategy and a new direction in sustainable agriculture.

References

Ahmad, P., Jaleel, C.A., Salem, M.A., Nabi, G., Sharma, S., 2010. Roles of enzymatic and non-enzymatic antioxidants in plants during abiotic stress. Crit. Rev. Biotechnol. 30, 161–175.

Ainsworth, E.A., Yendrek, C.R., Sitch, S., Collins, W.J., Emberson, L.D., 2012. The effects of tropospheric ozone on net primary productivity and implications for climate change. Annu. Rev. Plant Biol. 63, 637–661.

Akhtar, M.S., Siddiqui, Z.A., 2007. *Glomus intraradices*, *Pseudomonas alcaligenes*, and *Bacillus pumilus*: effective agents for the control of root-rot disease complex of chickpea (*Cicer arietinum* L.). J. Gen. Plant Pathol. 74, 53–60.

Akiyama, K., Hayashi, H., 2006. Strigolactones: chemical signals for fungal symbionts and parasitic weeds in plant roots. Ann. Bot. 97, 925–931.

Alqarawi, A.A., Abd-Allah, E.F., Hashem, A., 2014a. Alleviation of salt-induced adverse impact *via* mycorrhizal fungi in *Ephedra aphylla* Forssk. J. Plant Interact. 9 (1), 802–810.

Alqarawi, A.A., Hashem, A., Abd Allah, E.F., Alshahrani, T.S., Huqail, A.A., 2014b. Effect of salinity on moisture content, pigment system, and lipid composition in Ephedra alata Decne. Acta Biol. Hung. 65 (1), 61–71.

Atera, E.A., Itoh, K., Azuma, T., Ishii, T., 2012. Farmers' perspectives on the biotic constraint of *Striga hermonthica* and its control in western Kenya. Weed Biol. Manag. 12, 53–62.

Balestrini, R., Rosso, L.C., Veronico, P., Melillo, M.T., De Luca, F., Fanelli, E., et al., 2019. Transcriptomic responses to water deficit and nematode infection in mycorrhizal tomato roots. Front. Microbiol. 10, 1807.

Barnawal, D., Bharti, N., Maji, D., Chanotiya, C.S., Kalra, A., 2014. ACC deaminase-containing Arthrobacter protophormiae induces NaCl stress tolerance through reduced ACC oxidase activity and ethylene production resulting in improved nodulation and mycorrhization in *Pisum sativum*. J. Plant Physiol. 171, 884–894.

Barrow, C.J., 2012. Biochar potential for countering land degradation and for improving agriculture. Appl. Geogr. 34, 21–28.

Berruti, A., Lumini, E., Balestrini, R., Bianciotto, V., 2016. Arbuscular mycorrhizal fungi as natural biofertilizers: let's benefit from past successes. Front. Microbiol. 6, 1559.

Bethlenfalvay, G.J., Linderman, R.G. (Eds.), 1992. Mycorrhizae in Sustainable Agriculture, ASA Spec. Publ. No. 54. Amer. Soc. Agronomy Press, Madison, WI.

Bodker, L., Kjoller, R., Kristensen, K., Rosendahl, S., 2002. Interactions between indigenous arbuscular mycorrhizal fungi and Aphanomyces euteiches in field grown pea. Mycorrhiza 12, 7–12.

Boller, T., Felix, G., 2009. A renaissance of elicitors: perception of microbeassociated molecular patterns and danger signals by pattern-recognition receptors. Annu. Rev. Plant Biol. 60, 379–406.

Cameron, D.D., Neal, A.L., van Wees, S.C.M., Ton, J., 2013. Mycorrhizainduced resistance: more than the sum of its parts? Trends Plant Sci. 18, 539–545.

Ceustermans, A., Van Hemelrijck, W., Van Campenhout, J., Bylemans, D., 2018. Effect of arbuscular mycorrhizal fungi on *Pratylenchus penetrans* infestation in apple seedlings under greenhouse conditions. Pathogens 7 (4), 76.

Chen, M., Arato, M., Borghi, L., Nouri, E., Reinhardt, D., 2018. Beneficial services of arbuscular mycorrhizal fungi—from ecology to application. Front. Plant Sci. 9, 1270.

Conrath, U., Beckers, G.J., Flors, V., García-Agustín, P., Jakab, G., Mauch, F., Newman, M.A., Pieterse, C.M., Poinssot, B., Pozo, M.J., Pugin, A., Schaffrath, U., Ton, J., Wendehenne, D., Zimmerli, L., Mauch-Mani, B., 2006. Priming: getting ready for battle. Mol. Plant-Microbe Interact. 19, 1062–1071.

Da Silva-Campos, M.A., Da Silva, F.S.B., Yano-Melo, A.M., De Melo, N.F., Maia, L.C., 2017. Application of arbuscular mycorrhizal fungi during the acclimatization of *Alpinia purpurata* to induce tolerance to *Meloidogyne arenaria*. Plant Pathol. J33, 329–336.

El-Khallal, S.M., 2007. Induction and modulation of resistance in tomato plants against fusarium wilt disease by bioagent fungi (arbuscular mycorrhiza) and/or hormonal elicitors (jasmonic acid & salicylic acid): 2-changes in the antioxidant enzymes, phenolic compounds and pathogen related- proteins. Aust. J. Basic Appl. Sci. 1, 717–732.

Elsen, A., Gervacio, D., Swennen, R., De Waele, D., 2008. AMF-induced biocontrol against plant parasitic nematodes in *Musa* sp.: a systemic effect. Mycorrhiza 18, 251–256.

Engindeniz, S., 2006. Economic analysis of pesticide use on processing tomato growing: a case study for Turkey. Crop Prot. 25, 534–541.

Frew, A., Powell, J.R., Glauser, G., Bennett, A.E., Johnson, S.N., 2018. Mycorrhizal fungi enhance nutrient uptake but disarm defences in plant roots, promoting plant-parasitic nematode populations. Soil Biol. Biochem. 126, 123–132.

Gachomo, E., Allen, J.W., Pfeffer, P.E., Govindarajulu, M., Douds, D.D., Jin, H.R., Nagahashi, G., Lammers, P.J., Shachar-Hill, Y., Bucking, H., 2009. Germinating spores of *Glomus intraradices*can use internal and exogenous nitrogen sources for *de novo* biosynthesis of amino acids. New Phytol. 184 (2), 399–411.

Gallou, A., Mosquera, H.P.L., Cranenbrouck, S., Suarez, J.P., Declerck, S., 2011. Mycorrhiza induced resistance in potato plantlets challenged by Phytophthora infestans. Physiol. Mol. Plant Pathol. 76, 20–26.

Gange, A.C., Brown, V.K., Aplin, D.M., 2003. Multitrophic links between arbuscular mycorrhizal fungi and insect parasitoids. Ecol. Lett. 6, 1051–1055.

Hao, Z., van Tuinen, D., Fayolle, L., Chatagnier, O., Li, X., Chen, B., Gianinazzi, S., Gianinazzi-Pearson, V., 2018. Arbuscular mycorrhiza affects grapevine fanleaf virus transmission by the nematode vector *Xiphinema index*. Appl. Soil Ecol. 129, 107–111.

Harrier, L.A., Watson, C.A., 2004. The potential role of arbuscular mycorrhizal (AM) fungi in the bioprotection of plants against soil-borne pathogens in organic and/or other sustainable farming systems. Pest Manag. Sci. 60, 149–157.

Hashem, A., Abd Allah, E.F., Alqarawi, A.A., Aldubise, A., Egamberdieva, D., 2015. Arbuscular mycorrhizal fungi enhances salinity tolerance of *Panicum turgidum* Forssk by altering photosynthetic and antioxidant pathways. J. Plant Interact. 10 (1), 230–242.

Hashem, A., Abd Allah, E.F., Alqarawi, A.A., Egamberdieva, D., 2016. Bioremediation of adverse impact of cadmium toxicity on Cassia italica Mill by arbuscular mycorrhizal fungi. Saudi J. Biol. Sci. 23, 39–47.

Hashem, A., Abd Allah, E.F., Alqarawi, A.A., El-Didamony, G., Alwhibi Mona, S., Egamberdieva, D., Ahmad, P., 2014. Alleviation of adverse impact of salinity on faba bean (*Vicia faba* L.) by arbuscular mycorrhizal fungi. Pak. J. Bot. 46, 2003–2020.

Hoffmann, D., Vierheilig, H., Schausberger, P., 2010. Arbuscular mycorrhiza enhances preference of ovipositing predatory mites for direct prey-related cues. Physiol. Entomol. 36, 90–95.

Hohmann, P., Messmer, M., 2017. Breeding for mycorrhizal symbiosis: focus on disease resistance. Euphytica 213, 1–11.

Hooker, J.E., Jaizme-Vega, M., Atkinson, D., 1994. Biocontrol of plant pathogens using arbuscular mycorrhizal fungi. In: Gianinazzi, S., Schuepp, H. (Eds.), Impact of Arbuscular Mycorrhizas on Sustainable Agriculture and Natural Ecosystems. Birkhduser Verlag, Basel, pp. 191–200.

Isah, K., Kumar, N., Lagoke, S.O., Atayese, M., 2013. Management of Striga hermonthica on sorghum (sorghum bicolor) using arbuscular mycorrhizal fungi (Glomus mosae) and NPK fertilizer levels. Pak. J. Biol. Sci. 16, 1563–1568.

Jacott, C.N., Murray, J.D., Ridout, C.J., 2017. Trade-offs in arbuscular mycorrhizal symbiosis: disease resistance, growth responses and perspectives for crop breeding. Agronormy 7, 75.

Jalali, B.L., Jalali, I., 1991. Mycorrhiza in plant disease control. In: Arora, D.K., Rai, B., Mukerji, K.G., Knudsen, G.R. (Eds.), Handbook of Applied Mycology. Soil and Plants. vol. 1. Marcel Dekker, New York, NY, pp. 131–154.

Jones, N., Madhura, A., Prashant, S., Ramesh, B., Jagadeesh, K., Asha, A., 2014. Evaluation of arbuscular mycorrhizal fungi for suppression of *Striga hermonthica*, a parasitic weed in sorghum. In: Proceedings of the Biennial Conference on Emerging Challenges in Weed Management, Jabalpur, India, 15–17 February, p. 227.

Jung, S.C., Martinez-Medina, A., Lopez-Raez, J.A., Pozo, M.J., 2012. Mycorrhiza-induced resistance and priming of plant defenses. J. Chem. Ecol. 38, 651–664.

Kamel, L., Tang, N.W., Malbreil, M., San Clemente, H., Le Marquer, M., Roux, C., Frei Dit, F.N., 2017. The comparison of expressed candidate secreted proteins from two arbuscular mycorrhizal fungi unravels common and specific molecular tools to invade different host plants. Front. Plant Sci. 8, 124.

Khan, M.I.R., Nazir, F., Asgher, M., Per, T.S., Khan, N.A., 2015. Selenium and sulfur influence ethylene formation and alleviate cadmium-induced oxidative stress by improving proline and glutathione production in wheat. J. Plant Physiol. 173, 9–18.

Kloppholz, S., Kuhn, H., Requena, N., 2011. A secreted fungal effector of *Glomus intraradices* promotes symbiotic biotrophy. Curr. Biol. 21, 1204–1209.

Koricheva, J., Gange, A.C., Jones, T., 2009. Effects of mycorrhizal fungi on insect herbivores: a meta-analysis. Ecology 90, 2088–2097.

Kulkarni, S.A., Kulkarni, S., Sreenivas, M.N., 2011. Interaction between vesicular-arbuscular (V-A) mycorrhizae and *Sclerotium rolfsii* Sacc. in groundnut. J. Farm Sci. 10, 919–921.

Kumar, A., Kl, T., Datta, D., Singh, M., 2014. Marker assisted gene pyramiding for enhanced tomato leaf curl virus disease resistance in tomato cultivars. Biol. Plant. 58 (4), 792–797.

Kumar, A., Datta, D., Pandey, S., Singh, S., Das, S.P., 2018. Inheritance and tagging of tomato leaf curl virus (ToLCV) disease resistance genes in exotic tomato genotypes. Indian J. Biotechnol. 17, 357–363.

Kumari, S.M.P., Prabina, B.J., 2019. Protection of tomato, *Lycopersicon esculentum* from wilt pathogen, *fusarium oxysporum*f. sp. lycopersici by arbuscular mycorrhizal Fungi, *Glomus* sp. Int. J. Curr. Microbiol. App. Sci. 8, 1368–1378.

Kumari, S.M.P., Srimeena, N., 2019. Arbuscular mycorrhizal Fungi (AMF) induced defense factors against the damping-off disease pathogen, pythium aphanidermatum in Chilli (*Capsicum annum*). Int. J. Curr. Microbiol. App. Sci. 8, 2243–2248.

Lahari, Z., Ullah, C., Kyndt, T., Gershenzon, J., Gheysen, G., 2019. Strigolactones enhance root-knot nematode (*Meloidogyne graminicola*) infection in rice by antagonizing the jasmonate pathway. New Phytol. 224, 454–465.

Lendzemo, V.W., Van Ast, A., Kuyper, T.W., 2006. Can arbuscular mycorrhizal Fungi contribute to Striga management on cereals in Africa? Outlook Agric. 35, 307–311.

Le-Xu, C.W., Oelmüller, R., Zhang, W., 2018. Role of phytohormones in *Piriformospora indica*-induced growth promotion and stress tolerance in plants: more questions than answers. Front. Microbiol. 9, 1646.

Liao, D., Wang, S., Cui, M., Liu, J., Chen, A., Xu, G., 2018. Phytohormones regulate the development of arbuscular mycorrhizal symbiosis. Int. J. Mol. Sci. 19, 3146.

Linderman, R.G., Paulitz, T.C., 1990. Mycorrhizal-rhizobacterial interactions. In: Hornby, D. (Ed.), Biological Control of Soil-Borne Plant Pathogens. CAB International, Wallingford, pp. 261–283.

Liu, Y., Feng, X., Gao, P., Li, Y., Christensen, M.J., Duan, T., 2018. Arbuscular mycorrhiza fungi increased the susceptibility of Astragalus adsurgens to powdery mildew caused by Erysiphe pisi. Mycology 9, 223–232.

Manjunatha, H.P., Jones Nirmalnath, P., Chandranath, H.T., Shiney, A., Jagadeesh, K.S., 2018. Field evalualtion of native arbuscular mycorrhizal fungi in the management of Striga in sugarcane (*Saccharum officinarum* L.). J. Pharm. Phytochem. 7, 2496–2500.

Martínez-Medina, A., Roldan, A., Pascual, J.A., 2011. Interaction between arbuscular mycorrhizal fungi and *Trichoderma harzianum*under conventional and low input fertilization field condition in melon crops: growth response and fusarium wilt biocontrol. Appl. Soil Ecol. 47, 98–105.

Mishra, S., Upadhyay, S., Shukla, R., 2017. The role of Strigolactones and their potential cross-talk under hostile ecological conditions in plants. Front. Physiol. 10, 2017.

Mittler, R., 2002. Oxidative stress, antioxidants and stress tolerance. Trends Plant Sci. 7, 405–410.

Nakmee, P.S., Techapinyawat, S., Ngamprasit, S., 2016. Comparative potentials of native arbuscular mycorrhizal fungi to improve nutrient uptake and biomass of *Sorghum bicolor* Linn. Agric. Nat. Resour. 50, 173–178.

Navarro, J.M., Perez-Tornero, O., Morte, A., 2014. Alleviation of salt stress in citrus seedlings inoculated with arbuscular mycorrhizal fungi depends on the root stock salt tolerance. J. Plant Physiol. 171 (1), 76–85.

Nguvo, K.J., Gao, X., 2019. Weapons hidden underneath: bio-control agents and their potentials to activate plant induced systemic resistance in controlling crop Fusarium diseases. J. Plant Dis. Prot. 126, 177–190.

Othira, J.O., 2012. Effectiveness of arbuscular mycorrhizal fungi in protection of maize (*Zea mays* L.) against witchweed (*Striga hermonthica* Del Benth) infestation. J. Agric. Biotechnol. Sustain. Dev. 4, 37–44.

Parniske, M., 2008. Arbuscular mycorrhiza: the mother of plant root endosymbioses. Nat. Rev. Microbiol. 6, 763–775.

Plassard, C., Dell, B., 2010. Phosphorus nutrition of mycorrhizal trees. Tree Physiol. 30, 1129–1139.

Porcel, R., Aroca, R., Ruiz-Lozano, J.M., 2012. Salinity stress alleviation using arbuscular mycorrhizal fungi. A review. Agron. Sustain. Dev. 32, 181–200.

Poveda, J., Abril-Urias, P., Escobar, C., 2020. Biological control of plant-parasitic nematodes by filamentous Fungi inducers of resistance: trichoderma, mycorrhizal and endophytic Fungi. Front. Microbiol. 25, 2020.

Pozo, M.J., 2007. Azcon-Aguilar C. Unraveling mycorrhizainduced resistance. Curr. Opin. Plant Biol. 10, 393–398.

Pringle, A., Bever, J.D., Gardes, M., Parrent, J.L., Rillig, M.C., Klironomos, J.N., 2009. Mycorrhizal symbioses and plant invasions. Annu. Rev. Ecol. Evol. Syst. 40, 699–715.

Rochange, S., Goormachtig, S., Lopez-Raez, J.A., Gutjahr, C., 2019. The role of strigolactones in plant–microbe interactions. In: Koltai, H., Prandi, C. (Eds.), Strigolactones-Biology and Applications. Springer, Cham, pp. 121–142.

Rodriguez, P.A., Rothballer, M., Chowdhury, S.P., Nussbaumer, T., Gutjahr, C., Falter-Braun, P., 2019. Systems biology of plant-*microbiome* interactions. Mol. Plant 12 (6), 804–821 (ISSN 1674-2052).

Schouteden, N., De Waele, D., Panis, B., Vos, C.M., 2015. Arbuscular mycorrhizal Fungi for the biocontrol of plant-parasitic nematodes: a review of the mechanisms involved. Front. Microbiol. 6, 1280.

Schubler, A., Schwarzott, D., Walker, C., 2001. A new fungal phylum, the Glomeromycota: phylogeny and evolution. Mycol. Res. 105, 1413–1421.

Sharma, A., Shahzad, B., Kumar, V., Kohli, S.K., Sidhu, G., Bali, A.S., Handa, N., Kapoor, D., Bhardwaj, R., Zheng, B., 2019. Phytohormones regulate accumulation of Osmolytes under abiotic stress. Biomol. Ther. 9 (7), 285.

Sharma, I.P., Sharma, A.K., 2017. Co-inoculation of tomato with an arbuscular mycorrhizal fungus improves plant immunity and reduces root-knot nematode infection. Rhizosphere 4, 25–28.

Siddiqui, Z.A., Mahmood, I., 1998. Effect of a plant growth promoting bacterium, an AM fungus and soil types on the morphometrics and reproduction of *Meloidogyne javanica*on tomato. Appl. Soil Ecol. 8, 77–84.

Singh, P.K., Singh, M., Vyas, D., 2010. Biocontrol of fusarium wilt of chickpea using arbuscular mycorrhizal Fungi and *Rhizobium leguminosorum* Biovar. Caryologia 63, 349–353.

Smith, F.A., Grace, E.J., Smith, S.E., 2009. More than a carbon economy: nutrient trade and ecological sustainability in facultative arbuscular mycorrhizal symbioses. New Phytol. 182, 347–358.

Smith, S.E., Read, D.J., 2008. Mycorrhizal Symbiosis, third ed. Academic Press, London.

Solaiman, Z.M., Abbott, L.K., Varma, A., 2014. Mycorrhizal Fungi: Use in Sustainable Agriculture and Land Restauration. Springer, Berlin.

Song, Y., Chen, D., Lu, K., Sun, Z., Zeng, R., 2015. Enhanced tomato disease resistance primed by arbuscular mycorrhizal fungus. Front. Plant Sci. 6, 6.

Spagnoletti, F.N., Cornero, M., Chiocchio, V., Lavado, R.S., Roberts, I.N., 2020. Arbuscular mycorrhiza protects soybean plants against *Macrophominaphaseolina* even under nitrogen fertilization. Eur. J. Plant Pathol. 156, 839–849.

St-Arnaud, M., Hamel, C., Caron, M., Fortin, J.A., 1995. Endomycorhizes VA et sensibility des plantes aux maladies: synthase de la liturature et mecanismesd'interaction potentials. In: Fortin, J.A., Charest, C., Piche, Y. (Eds.), La Symbiose mycorhizienne—Etat des connaissances. Orbis Publishing, Frelighsburg, pp. 51–87.

Stolyarchuk, I.M., Shevchenko, T.P., Polischuk, V.P., Kripka, A.V., 2009. Virus infection course in different plant species under influence of arbuscular mycorrhiza. Microbiology 3, 70–75.

Tamasloukht, M., Séjalon-Delmas, N., Kluever, A., Jauneau, A., Roux, C., Bécard, G., Franken, P., 2003. Root factors induce mitochondrial related gene expression and fungal respiration during developmental switch from Asymbiosis to Presymbiosis in the arbuscular mycorrhizal fungus *Gigaspora rosea*. Plant Physiol. 131, 1468–1478.

Tariq, M., Hameed, S., Khan, H.U., 2017. Role of microsymbionts in plant microbe Symbiosis. J. Appl. Microbiol. Biochem. 1 (2), 6.

Taylor, T.N., Ramy, W., Hass, H., Kerp, H., 1995. Fossil arbuscular mycorrhizae from the early Devo. Mycologia 87, 560–573.

Tchabi, A., Hountondji, C.C., Ogunsola, B., Lawouin, L., Coyne, D.L., Wiemken, A., Oehl, F., 2016. Effect of two species of arbuscular mycorrhizal fungi inoculation on development of micro-propagated yam plantlets and suppression of *Scutellonemabradys* (Tylenchideae). J. Entomol. Nematol. 8, 1–10.

Tejera, N.A., Campos, R., Sanjuan, J., Lluch, C., 2004. Nitrogenase and antioxidant enzyme activities in *Phaseolus vulgaris* nodules formed by *Rhizobium tropici* isogenic strains with varying tolerance to salt stress. J. Plant Physiol. 161, 329–338.

Thiem, D., Szmidt-Jaworska, A., Baum, C., Muders, K., Niedojadło, K., Hrynkiewicz, K., 2014. Interactive physiological response of potato (*Solanum tuberosum* L.) plants to fungal colonization and potato virus Y (PVY) infection. Acta Mycol 1, 291–303.

Utkhede, R., 2006. Increased growth and yield of hydroponically grown greenhouse tomato plants inoculated with arbuscular mycorrhizal fungi and *fusarium oxysporum* f. sp. radicislycopersici. BioControl 51, 393–400.

van der Heijden, M.G.A., Martin, F.M., Selosse, M.A., Sanders, I.R., 2015. Mycorrhizal ecology and evolution: the past, the present, and the future. New Phytol. 205, 1406–1423.

Vos, C., Claerhout, S., Mkandawire, R., Panis, B., De Waele, D., Elsen, A., 2011. Arbuscular mycorrhizal fungi reduce root-knot nematode penetration through altered root exudation of their host. Plant Soil 354, 335–345.

Vos, C., Tesfahun, A., Panis, B., De Waele, D., Elsen, A., 2012. Arbuscular mycorrhizal fungi induce systemic resistance in tomato against the sedentary nematode *Meloidogyne incognita* and the migratory nematode *Pratylenchus penetrans*. Appl. Soil Ecol. 61, 1–6.

Wan, J., Zhang, X.C., Stacey, G., 2008. Chitin signaling and plant disease resistance. Plant Signal. Behav. 3, 831–833.

Wang, B., Qiu, Y.L., 2006. Phylogenetic distribution and evolution of mycorrhizas in land plants. Mycorrhiza 16 (5), 299–363.

Wang, F., 2017. Occurrence of arbuscular mycorrhizal fungi in mining-impacted sites and their contribution to ecological restoration: mechanisms and applications. Crit. Rev. Environ. Sci. Technol. 47, 1–57.

Whipps, J.M., 2004. Prospects and limitations for mycorrhizas in biocontrol of root pathogens. Can. J. Bot. 82, 1198–1227.

CHAPTER

18 Biological synthesis of gold nanoparticles by microbes: Mechanistic aspects, biomedical applications, and future prospects

Gagan Kumar Panigrahi and Kunja Bihari Satapathy

School of Applied Sciences, Centurion University of Technology and Management, Khurda, Odisha, India

18.1 Introduction

Nanomaterials have widespread applications and are considered to be an illustrious component of several manufacturing sectors (Salata, 2004; Narkeviciute et al., 2016; Fig. 18.1). Though the importance of nanomaterials is numerous, their synthesis has always been the key factor. The nanoparticles (NPs) such as zinc, gold, silver, titanium, and many more are always been of profound interest (Reddy et al., 2007, 2011). Present scenario demands effective and eco-friendly synthesis of NPs, and this is barely possible only when biological methodologies including green synthesis routes would be prioritized (Duran and Seabra, 2009). Among many, gold nanoparticles (AuNPs) have gained substantial attention owing to its biocompatibility nature, stability, and resistance properties (Alkilany and Murphy, 2010). AuNPs have been effectively used in a wide range of applications including cancer diagnostics, gene delivery approaches, genetic engineering, drug delivery processes, and many more (Arvizo et al., 2010; Yang et al., 2015; Daraee et al., 2016; Panigrahi and Satapathy, 2021, Sahoo et al., 2023b; Sahoo and Satapathy, 2021). Synthesis of metal NPs has been tricky and mostly chemical methods are employed, which are not eco-friendly (Zhang et al., 2007; Reddy et al., 2008, 2014; Hassan et al., 2014). Usage of expensive and hazardous reducing agents such as amino acid derivatives, citrate, PEG 4000 during synthesis of NPs through chemical route leads to increase of toxic components in the environment and may severely affect the sustenance of life forms (Kimling et al., 2006; Roy and Lahiri, 2006; Sugunan and Dutta, 2006; Akbarzadeh et al., 2009; Ravindra, 2009; Panigrahi et al., 2021c; Panigrahi and Satapathy, 2020b). This is an alarming state to remodify our approach of synthesizing NPs by prioritizing environmentally friendly processes such as use of biological systems like bacteria, yeasts, fungi, and plants (Fig. 18.2). Microbes can be used economically and effectively for the synthesis of AuNPs (He et al., 2007; Dhillon et al., 2012). Most importantly, synthesis of microbe-mediated NPs is concurrent with chemical synthesis approach as far as features of NPs are concerned. Microbial synthesis of AuNPs includes intracellular and extracellular synthesis routes. For instance, *Acidithiobacillus thiooxidans*,

Entrepreneurship with Microorganisms. https://doi.org/10.1016/B978-0-443-19049-0.00012-8

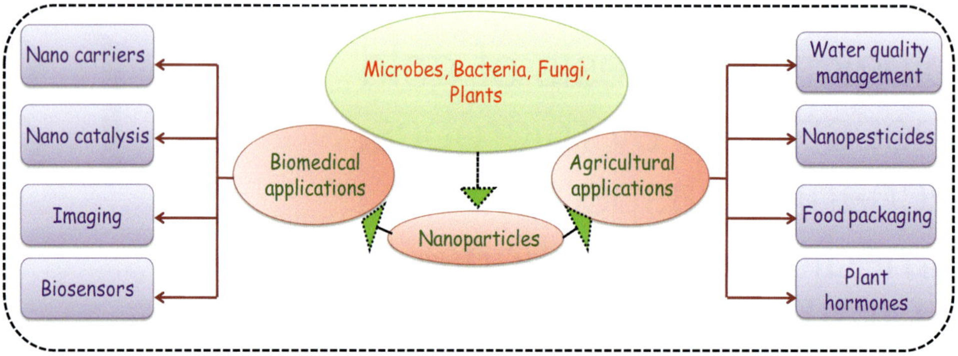

FIG. 18.1

Wide ranges of applications of nanoparticles.

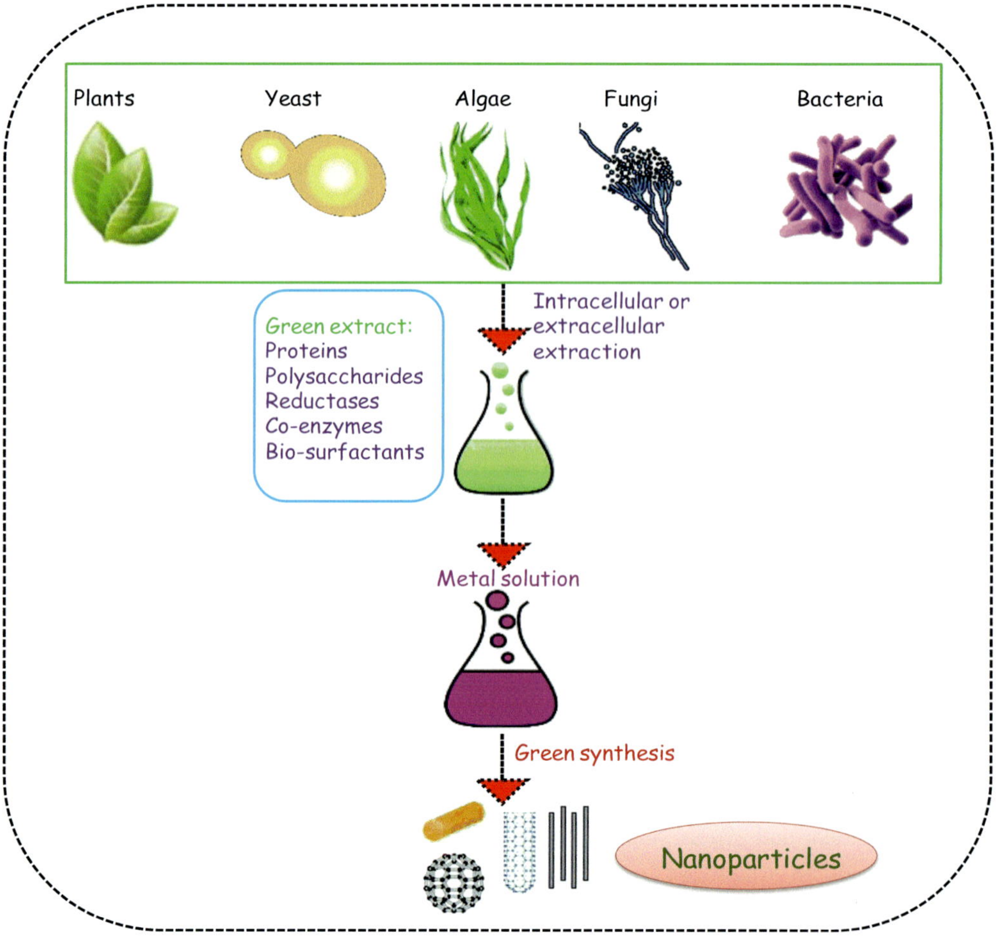

FIG. 18.2

A schematic flow diagram representing the green synthesis of nanoparticles.

Hormoconis resinae, and *Rhodococcus* sp. have been effectively used for intracellular synthesis of NPs (Ahmad et al., 2003a; Varshney et al., 2009; Lengke et al., 2005). Similarly, *Thermomonospora* sp. and *Fusarium oxysporum* have been involved in extracellular synthesis (Ahmad et al., 2003b, 2003c; Panigrahi et al., 2021b; Sahoo et al., 2023a). The cyanobacterium *Plectonema boryanum* has been implicated in both intracellular and extracellular synthesis of NPs (Lengke et al., 2006a, 2006b; Panigrahi et al., 2021c; Panigrahi and Satapathy, 2020c). Growth parameters including growth stage of cells, pH, and temperature significantly affects the size and shape of AuNPs (Gericke and Pinches, 2006). Extracellular synthesis of NPs is relatively preferred as the isolation process is compatible to that of intracellular production. As far as biogenic synthesis of AuNPs is concerned, many microbes such as *Alternaria alternata*, *Helminthosporium solani*, *Penicillium brevicompactum*, *Hansenula anomala*, and *Rhizopus oryzae* have been successfully used (Siddiqi and Husen, 2016). Even though the bacterial-mediated NPs synthesis has been successful, fungus can be a better alternative as fungus are considered to secrete more enzymes that would boost the biosynthesis of NPs and consequently would lead to large-scale production. Even the presence of mycelia which enhances the surface area plays a significant role in scaling up of biosynthesis process (Kitching et al., 2015). Across the globe, hundreds of plant species are constantly infected by an extremely fierce fungal pathogen, *Macrophomina phaseolina* (Mihail and Taylor, 1995; Panigrahi et al., 2021a, 2021c; Panigrahi and Satapathy, 2021). Essentially, it is anamorphic Basidiomycetes. The common symptoms of this fungal infected disease are charcoal rot, stem blight, stem rot, and damping off (Jung et al., 2020; Khan, 2007). Excitingly, *M. phaseolina* secretes a range of hydrolytic enzymes resulting in the degradation of the plant cell wall and thus allowing it to enter into the host tissue. Reportedly, *M. phaseolina* harbors a lot of paralogs for the oxidoreductase class of enzymes as confirmed from genomic studies (Islam et al., 2012; Sreedharan et al., 2019; Panigrahi and Satapathy, 2020a). Primarily, these oxidoreductases facilitate the fungus to sustain in unfavorable environmental conditions and also mediate the infection event. Remarkably, this group of enzymes catalyzes the synthesis of NPs (Ramezani et al., 2010). Relative to the activity of these oxidoreductase enzymes in other fungal species, *M. phaseolina* displays higher rate of activity and thus can be substantially exploited which would be economical as well as eco friendly. Importantly, since this fungal pathogen is known for its infective approach, instead they could be channelized in biogenic synthesis of NPs, which would ultimately serve for growth of plants.

18.2 Microorganisms-mediated synthesis of nanoparticles

A variety of approaches have been extensively employed for the synthesis of metal NPs (MtNPs). These methods include physical, chemical, and biological strategies. Moreover, both physical and chemical methods possess disadvantages such as high heat generation, expensive process, environmentally toxic, and low production yield (Gahlawat and Choudhury, 2019; Soni et al., 2018). The major drawback of these synthetic procedures is that they make use of toxic chemicals leading to environmental exertion (Pal et al., 2019). Thus, there is a need of eco-friendly method for synthesis of MtNPs. Green synthesis of NPs is treated as a suitable alternative approach and being widely used nowadays. Green synthesis preferentially makes use of biological routes such as degradable polymers, polysaccharides, microbial enzymes, microorganisms, and plants (Roychoudhury, 2020). Green synthesis approaches are advantageous than the physical and chemical methods owing to their cost-effective, eco-friendly, and simple attributes and thus have gained substantial importance (Roychoudhury, 2020). Recently, synthesis and research on MtNPs have been significantly enhanced owing to their inventive and varied applications in diverse fields including agriculture, environmental sciences, and medical sciences (Gahlawat and Choudhury, 2019; Kato and Suzuki, 2020). Broadly, top-down and bottom-up approaches are considered for the synthesis of MtNPs. In case of top-down approach, bulk materials are dissociated into nano-sized materials, using several physical and chemical techniques (Gahlawat and Choudhury, 2019; Prasad, 2019). But, the major setback of this approach is that the synthesized MtNPs bears flawed surface structures. This approach is also expensive and time-consuming, making it difficult for large-scale synthesis (Prasad, 2019), whereas, in bottom-up strategy, self-assembly at atomic scales leads to the synthesis of NPs (Su and Chang, 2018; Fig. 18.3).

Bottom-up approaches include chemical and biological methods (Gahlawat and Choudhury, 2019). Among various green synthesis routes, microorganisms-mediated nanosynthesis is treated to be superior as the growth rate of microorganisms is too high, and also, they are easy to grow in ambient environmental conditions (Prasad, 2019). Several microorganisms can act as biofactories, primarily for the environmental-friendly synthesis of MtNPs including gold, silver, nickel, copper, zinc, palladium, and titanium (Tables 18.1 and 18.2). MtNPs of defined composition, monodispersity, size, and shape can also be achieved (Gahlawat and Choudhury, 2019; Kato and Suzuki, 2020; Khan et al., 2018). Enzymatic conversion of metal ions into corresponding elemental form is possible through microorganisms as they

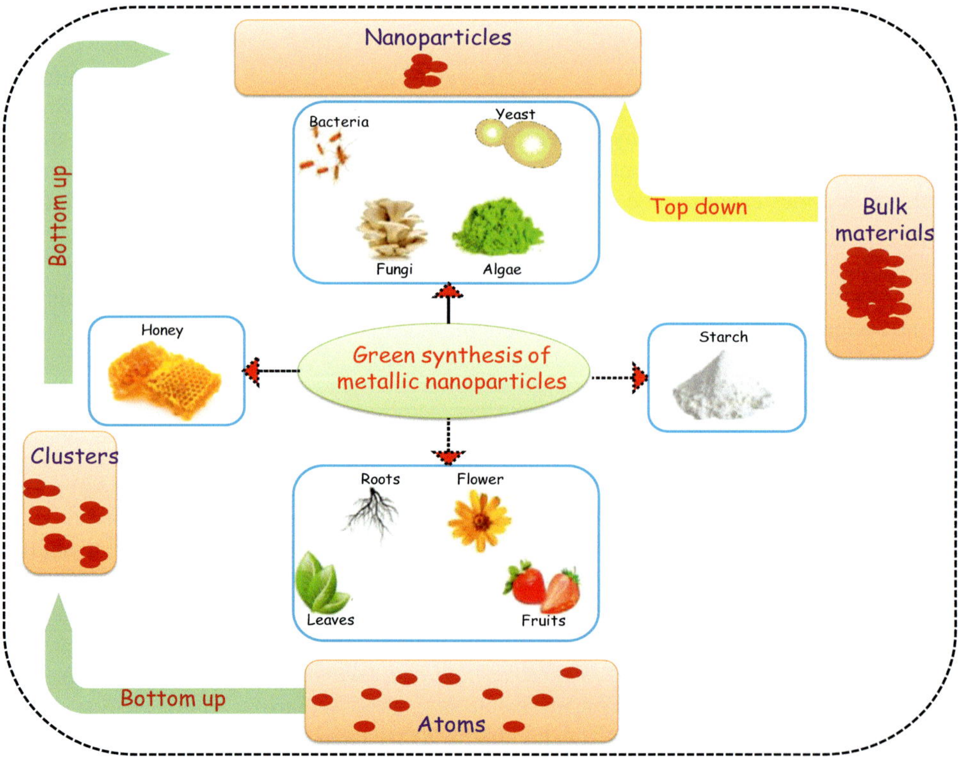

FIG. 18.3

Top-down and bottom-up approaches for the green synthesis of nanoparticles.

Table 18.1 Synthesis of MtNPs using various microorganisms.

Microorganisms		Size of NPs (nm)	MtNPs	References
Bacteria	*Klebsiella pneumoniae*	5–32, 2–100, 1–5	Ag(0)	Ahmad et al. (2007a), Minaeian et al. (2008), Narges et al. (2009)
	Enterobacteria	28.2–122	Ag(0)	Ahmad et al. (2007b)
	Escherichia coli	20	Au(0)	Du et al. (2007)
	Plectonema boryanum	1–20	Pd(0)	Maggy et al. (2007)
	Shewanella algae	Around 5	Pt(0)	Yasuhiro et al. (2007)
	Lactobacillus sp.	15–35	TiO_2	Jha et al. (2009)
	Staphylococcus aureus	160–180	Ag(0)	Nanda and Saravanan (2009)
	Bacillus subtilis	5–60	Ag(0)	Saifuddin et al. (2009)
Fungi				
	Helminthosporium solani	2–70	Au(0)	Kumar et al. (2008)
	Penicillium	<135	Au(0)	Sadowski et al. (2008)
	Penicillium fellutanum	5–25	Ag(0)	Kathiresan et al. (2009)
	Hormoconis resinae	20–80	Ag(0)	Varshney et al. (2009)
	Trichoderma reesei	5–50	Ag(0)	Mansoori et al. (2010)
	Fusarium oxysporum	5–15	Ag	Ahmad et al. (2003a)
	Fusarium oxysporum	4–5	Barium titanate	Bansal et al. (2006)
Yeast	*Yarrowia lipolytica*	15	Au(0)	Mithila et al. (2009)

Table 18.2 Synthesis of AuNPs using various microorganisms.

Microorganisms		Size of AuNPs (nm)	References
Bacteria	*Brevibacterium casei*	10–50	Kalishwaralal et al. (2010)
	Escherichia coli	20–30	Brown et al. (2000)
	Plectonema boryanum	10–20	Lengke et al. (2006a)
	Plectonema boryanum	10	Lengke et al. (2006b)
	Rhodococcus sp.	5–15	Ahmad et al. (2003c)
	Rhodopseudomonas capsulata	10–20	He et al. (2007)
	Sargassum wightii	8–12	Singaravelu et al. (2007)
	Shewanella algae	10–20	Konishi et al. (2007)
	Shewanella oneidensis	9–20	Suresh et al. (2011)
	Ureibacillus thermosphaericus	35–75	Juibari et al. (2015)
Fungi	*Macrophomina phaseolina*	14–16	Sreedharan et al. (2019)
	Alternaria alternata	10–20	Sarkar et al. (2012)
	Aspergillus niger	12–20	Bhambure et al. (2009)

Continued

Table 18.2 Synthesis of AuNPs using various microorganisms—cont'd

Microorganisms		Size of AuNPs (nm)	References
	Aspergillus niger	10–20	Xie et al. (2007)
	Aspergillus oryzae var. *viridis*	10–60	Binupriya et al. (2010)
	Aspergillus sydowii	9–16	Vala (2015)
	Aureobasidium pullulans	29–36	Zhang et al. (2011)
	Candida albicans	5	Ahmad et al. (2013)
	Candida albicans	20–40	Chauhan et al. (2011)
	Candida albicans	60–80	Sathish Kumar et al. (2011)
	Colletotrichum sp.	8–40	Shankar et al. (2003)
	Coriolus versicolor	20–100	Sanghi and Verma (2010)
	Cylindrocladium floridanum	19.05	Narayanan and Sakthivel (2013)
	Cylindrocladium floridanum	5–35	Narayanan and Sakthivel (2011a, 2011b)
	Fusarium oxysporum	2–50	Mandal et al. (2006)
	Fusarium semitectum	10–80	Sawle et al. (2008)
	Hormoconis resinae	3–20	Mishra et al. (2010)
	Helminthosporium solani	2–70	Kumar et al. (2008)
	Neurospora crassa	32	Castro et al. (2011)
	Penicillium brevicompactum	10–50	Mishra et al. (2011)
	Penicillium rugulosum	20–40	Mishra et al. (2012)
	Penicillium sp.	30–50	Du et al. (2011)
	Phanerochaete chrysosporium	10–100	Sanghi et al. (2011)
	Pichia jadinii	30–100	Gericke and Pinches (2006)
	Rhizopus oryzae	16–25	Das et al. (2012)
	Saccharomyces cerevisiae	100	Lin et al. (2005)
	Sclerotium rolfsii	25–50	Narayanan and Sakthivel (2011a, 2011b)
	Trichoderma koningii	30–40	Maliszewska et al. (2009); Maliszewska (2013)
	Verticillium sp.	20–30	Mukherjee et al. (2001)
	Verticillium volvacea	20–150	Philip (2009)
	Yarrowia lipolytica	15	Agnihotri et al. (2009)

would provide the suitable reducing environment (Khan et al., 2018). Microorganisms can tolerate heavy metals to certain extent and this factor modulates the capacity of microorganisms to synthesize MtNPs. Enhanced metal stresses affect microbial activities to a greater extent and enables them to reduce metal ions (Fig. 18.4). Mostly, microorganisms which live in habitats rich in metal are particularly resistant to those metals and subsequently this approach mimics the biomineralization process (Yusof et al., 2019). In case of intracellular biosynthesis of MtNPs, transport systems in microorganisms involving the cell wall play a critical role as it attracts positively charges metal ions toward it. After being transported inside the cells, by metabolic reactions the ions are reduced to form NPs (Tiquia-Arashiro and Rodrigues, 2016; Hulkoti and Taranath, 2014). Similarly, the extracellular biosynthesis of MtNPs

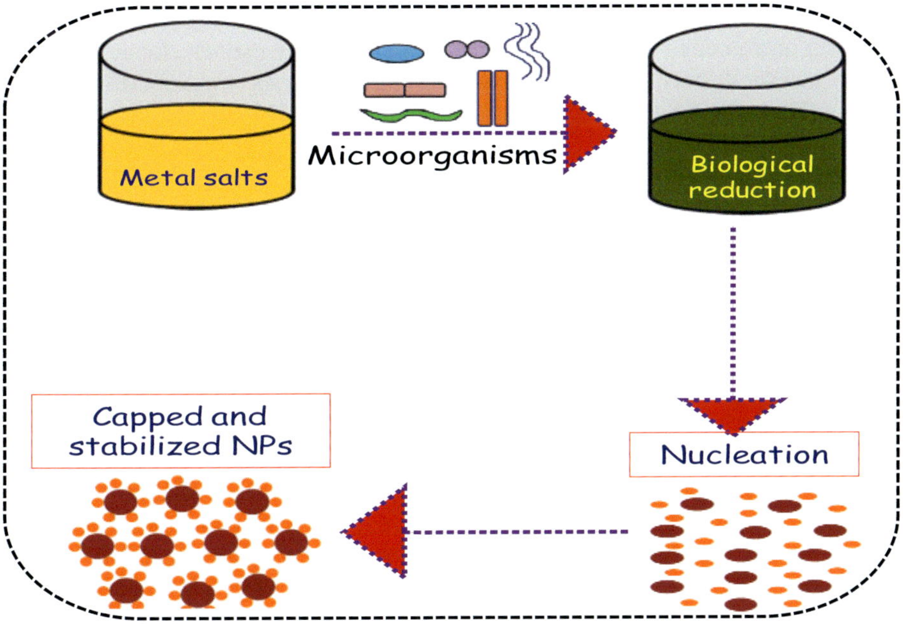

FIG. 18.4

Microorganisms-mediated syntheses of nanoparticles.

is favored by several reductase enzymes, which are primarily localized in the cell wall. Reductases mediate the synthesis of metallic ions and also the presence of several biological molecules such as proteins and polysaccharides favor the synthesis process (Yusof et al., 2019; Hulkoti and Taranath, 2014). Strikingly, proteins secreted from microorganisms stabilize the entire event and prevents any agglomeration (Yusof et al., 2019). Microbes are preferably cultured in appropriate growth medium maintained at optimum temperature and pH (Prasad, 2019). The cultured microbes are harvested post incubation and after being processed thoroughly with sterile water are exposed to metal salt solution. Cell-free (CF) approaches are also been widely used were the cell-free extracts (CFE) or culture supernatant are used to serve the purpose of synthesizing MtNPs (Kato and Suzuki, 2020). In the CF approach, suitable enzymes are employed which primarily act as capping and reducing agents. Here, the microbes are removed and suspended in sterile water and subsequently the CFE is collected and finally is exposed to metal salt solutions (Soni et al., 2018; Kato and Suzuki, 2020; Ali et al., 2020; Singh et al., 2015). Usually, a change in color during the cell-free process indicates the synthesis of NPs. For instance, upon the synthesis of AuNPs, the color of the reaction mixture changes from pale yellow to dark purple and similarly the color changes from pale yellow to deep brown during the synthesis of silver NPs (Soni et al., 2018; Ali et al., 2020; Kalimuthu et al., 2008). There are numerous critical physiological factors such as pH, metal salt concentration, pressure, incubation time significantly influence the synthesis of MtNPs, thus optimizing these factors is highly important, which would result in synthesis of NPs of desired morphology and composition (Jeyaraj et al., 2019; Singh et al., 2020). Postsynthesis of MtNPs, purification is essential to eliminate any unreacted biomolecule (Singh et al., 2020). Consequently, MtNPs derived from microorganisms are characterized by numerous analytical techniques including UV–visible spectroscopy, Fourier transform infrared (FTIR) spectroscopy, transmission electron microscopy (TEM), scanning electron microscopy (SEM), atomic force microscopy (AFM), X-ray powder diffraction (XRD), energy dispersive x-ray spectroscopy (EDS), and dynamic light scattering (DLS) (Ingale and Chaudhari, 2013; Jeyaraj et al., 2019; Strasser et al., 2010).

18.3 Synthesis of gold nanoparticles: Synthetic approach

Numerous solution-based methods have been developed and been practiced since last decade for the synthesis of AuNPs. AuNPs (AuNPs) of desired shape, size, and functionality is synthesized by following a wide array of methods (Soni et al., 2018; Pal et al., 2019). Synthetic method involving the use of citric acid onto the hydrogen tetrachloroaurate ($HAuCl_4$) results in synthesis of AuNPs, whereby the citrate acts as both stabilizing and reducing agent (Fig. 18.5). Furthermore, refinements to this approach were made to control the AuNPs size (Gahlawat and Choudhury, 2019; Jeyaraj et al., 2019). Following this approach, spherical AuNPs with diameters of 10–30 nm is mostly synthesized (Khan et al., 2018; Yusof et al., 2019; Singh et al., 2020). The major drawback during this synthesis process is the irreversible aggregation of the AuNPs, essentially during the functionalization process. Various approaches have been formularized to overcome this issue, primarily by use of several chemical reagents such as thioctic acid and Tween-20 (Jeyaraj et al., 2019). Again, large-scale production remains challenging after the use of these chemicals (Jeyaraj et al., 2019). Interestingly, alkanethiol-stabilized AuNPs were synthesized following a biphasic reduction protocol (Yeh et al., 2012). This method has been very productive as it produces low-dispersity AuNPs ranging from 1.5 to 5 nm, preferably by altering the reaction conditions such as reaction temperature, gold-to-thiol ratio, and reduction rate (Su and Chang, 2018; Kato and Suzuki, 2020; Roychoudhury, 2020). As compared to other synthetic processes, alkanethiol-protected AuNPs display higher stability, preferably due to the synergic effect of van der Waals and thiol-gold attractions (Yusof et al., 2019). Consequently, AuNPs thus prepared can be extensively dried and can be applied as precursors for further processes (Kalimuthu et al., 2008).

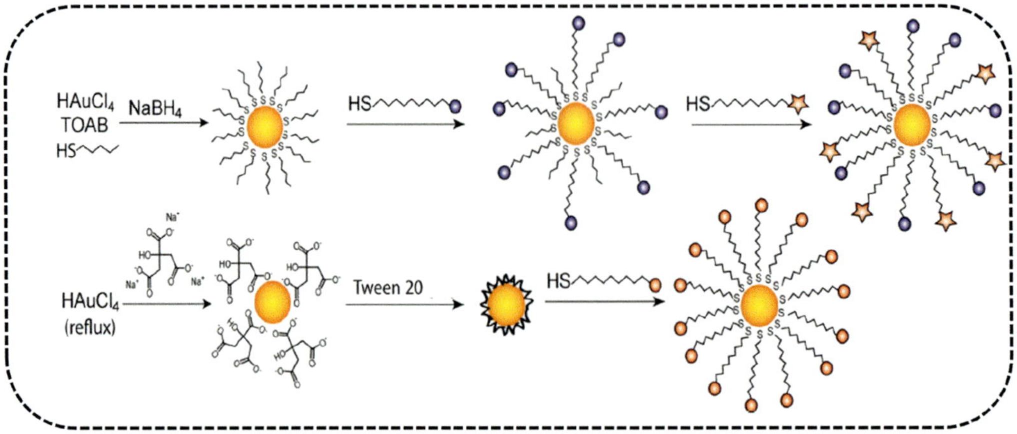

FIG. 18.5

Synthesis of AuNPs by reduction of $HAuCl_4$.

18.4 Plant and microorganisms as source for synthesis of gold nanoparticles

Chemical methods are extensively used for the synthesis of MtNPs, but the hazardous nature of the chemicals used make the entire process disadvantageous. Importantly, the NPs synthesized through chemical routes may also have adverse effects in biomedical applications (Shankar et al., 2004; Noruzi et al., 2011). Thus, it would be great enough to make use of greener ways for synthesizing MtNPs including AuNPs, which would be eco-friendly and cost-effective process. AuNPs could be synthesized by various biological sources including plants, enzymes and microorganisms (Mohanpuria et al., 2008; Singh et al., 2013). Recently, the biosynthesis of AuNPs from various plant sources such as *Aloe vera, Medicago sativa, Azadirachta indica, Pelargonium graveolens, Coriandrum sativum, Cymbopogon citratus* (lemongrass), *Terminalia catappa,* and *Cinnamomum camphora* have been possible (Herizchi et al., 2016). Extracts from plants such as *Memecylon umbellatum, Cinnamomum zeylanicum, Cochlospermum gossypium, Terminalia chebula, Brevibacterium casei, Citrus sinensis, Macrotyloma uniflorum, Citrus limon, Citrus reticulata, Memecy lonedule, Mangifera indica, Murraya koenigii,* and *Piper pedicellatum* also been used for the synthesis of AuNPs (Herizchi et al., 2016). *Zingiber officinale* extracts have been successfully used to synthesize AuNPs of size ranging from 5 to 15 nm as the extracts of the plant act as both reducing and stabilizing agent (Kumar et al., 2011; Fig. 18.6). Similarly, the presence of vitamin C in *Allium cepa* (onion) extracts makes it a suitable source for the synthesis of AuNPs (Parida et al., 2011). Numerous microorganisms including bacterial and fungal species have been reported for been able to mediate the synthesis of AuNPs either through intracellular or extracellular mode (Fig. 18.7, Table 18.2). *M. phaseolina* has also been effective in the synthesis of AuNPs (Sreedharan et al., 2019). The mycosynthesis of AuNPs from *M. phaseolina* is confirmed using UV–vis spectroscopy. It has been reported that color change from light yellow to pink/purple corresponds to the formation of AuNPs (Sawle et al., 2008; Bhambure et al., 2009). AuNPs shows absorbance peak at 540 nm. Other fungal species such as *Cylindrocladium floridanum* and *Coriolus versicolor* shows similar absorbance peaks (Narayanan and Sakthivel, 2013). The final yield of AuNPs was found to be

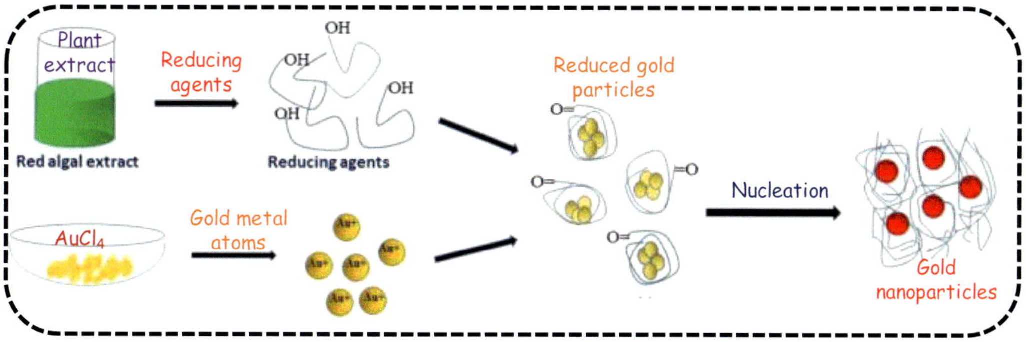

FIG. 18.6

Synthesis of AuNPs mediated by plant extract.

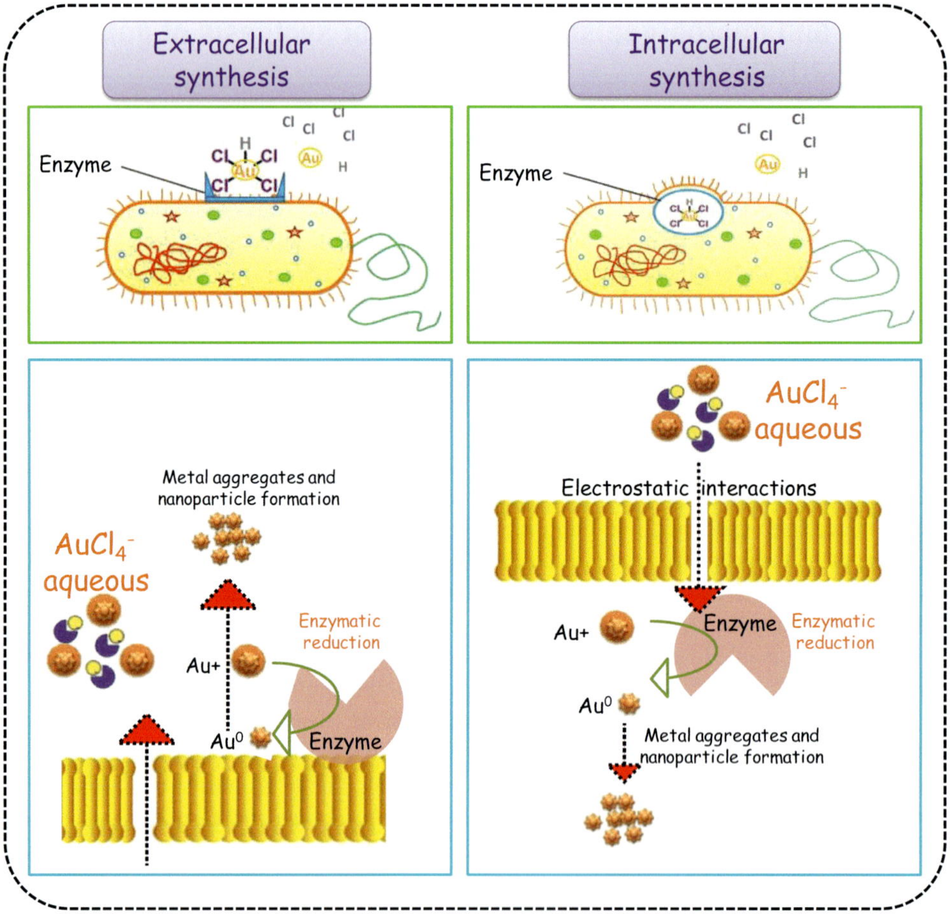

FIG. 18.7

Extracellular and intracellular mode of synthesis of AuNPs mediated by microorganisms.

10 mg/mL of the reaction mixture used (Sreedharan et al., 2019). It is of the view that the biofabrication of AuNPs primarily involves the synthesis of elemental metal from metal salts by the bio-reduction process. Consequently, the organic molecules secreted from the fungus mediate the stabilization of the synthesized NPs.

18.5 Gold nanoparticles: Its application in bionanotechnology

AuNPs have a wide range of applications (Fig. 18.8). They are imperative components for biomedical applications. AuNPs are extensively used for diagnostics and in the field of therapeutics. AuNPs easily conjugate with biomolecules such as oligonucleotides and antibodies which enables them to be used for the detection of target molecules. This allows the use of AuNPs for diagnostics applications, primarily

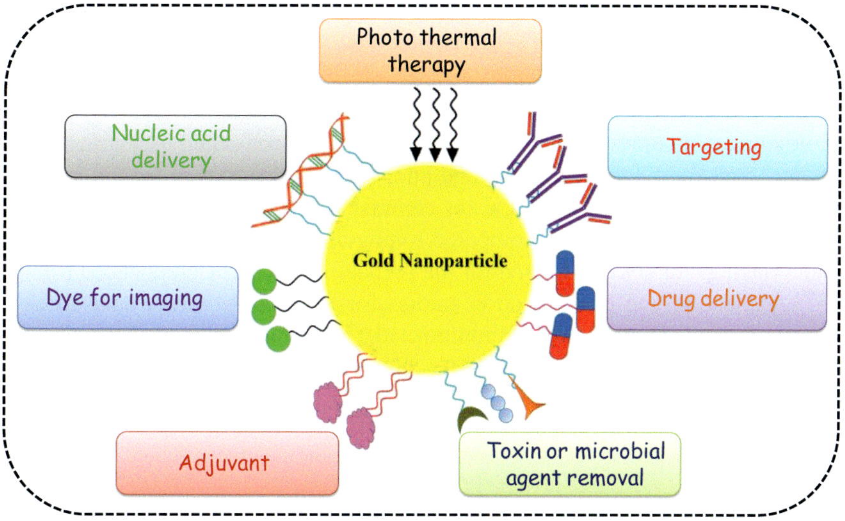

FIG. 18.8

Multifaceted applications of AuNPs in bio nanotechnology.

in cancer (Soni et al., 2018; Pal et al., 2019). An ultrasensitive method, referred to as bio-barcode assay largely modulated by AuNPs principally detects target various biomolecules. This assay makes use of conjugated AuNPs against specific target molecules for detection and quantification of nucleic acids and proteins. For instance, prostate-specific antigen (PSA) present at very low level, i.e., 330 fg/mL, was successfully detected (Yeh et al., 2012). Cancer cells and small molecules are also easily detected by the use of specific aptamer-conjugated AuNPs (Prasad, 2019; Kato and Suzuki, 2020). Lymphoma cells were detected by an aptamer-nanoparticle strip biosensor (Tiquia-Arashiro and Rodrigues, 2016; Yusof et al., 2019). Recently, developed chemical nose method that makes use of AuNP and fluorophore has been demonstrated to be highly sensitive against several biomolecular targets (Yeh et al., 2012). Based on specific lock and key approach, AuNP-fluorophore conjugates function in recognizing their cognate molecules (Singh et al., 2015). AuNPs are also proven to recognize and differentiate normal, cancerous, and metastatic cells (Kalimuthu et al., 2008; Strasser et al., 2010). An enzyme-amplified array sensing (EAAS) approach has been developed which enhances the sensitivity through enzymatic catalysis, where AuNPs play a significant role (Singh et al., 2020). In biomedical treatment, AuNPs plays a significant role as they have the ability to get penetrated into the cell membrane (Jeyaraj et al., 2019; Singh et al., 2020). Functionalized AuNPs with an orderly arrangement of amphiphilic molecules can penetrate into the cell and thus can mediate delivery of drugs into the destined locus (Su and Chang, 2018). Effective drug delivery and targeting strategies have been employed to strategically act on deformed cells including cancerous cells (Khan et al., 2018). RNA-AuNP conjugates successfully down-regulates the expression of luciferase activity, demonstrating gene knock down ability (Yeh et al., 2012). AuNPs are also been widely used for gene transfection assays (Ingale and Chaudhari, 2013; Pal et al., 2019). AuNPs can associate with specific drugs by both covalent and non-covalent interactions (Su and Chang, 2018). The AuNPs possess cationic charge on its surface which facilitates the NPs to enter into the cell (Prasad, 2019). Functionalized AuNPs has been reported to elevate the anti-proliferative effect against leukemia cells, plasma cell disorder (Strasser et al., 2010; Ingale and Chaudhari, 2013; Singh et al., 2015; Ali et al., 2020). The protein level of cell-cycle proteins p21 and p27 is upregulated, which in turn inhibits the growth of multiple myeloma cells (Kato and Suzuki, 2020). Remarkably, due to flexible electronic and optical properties of AuNPs, they have been widely used for cell imaging techniques, such as Raman spectroscopy, computed tomography, optical coherence tomography, dark-field light scattering, and photothermal heterodyne imaging technique (Singh et al., 2015; Ali et al., 2020). Also, AuNPs act as contrast agents as reports suggest that these NPs are able to elevate the vascular contrast in computed tomography imaging (Pal et al., 2019). For small-animal Raman imaging, AuNPs are applied for the preparation of surface-enhanced Raman scattering (SERS) NPs. Localization and colocalization studies for SERS NPs within deep tissues have been studied using AuNPs coated with silica (Kalimuthu et al., 2008; Strasser et al., 2010; Singh et al., 2015; Kato and Suzuki, 2020). AuNPs have wide range of applications and engineering these NPs would mediate multifaceted processes in the near future.

18.6 Conclusion and future prospects

Recently, green syntheses of NPs have been prioritized owing to their eco-friendly attribute. Over the last decade, microorganisms have been used to synthesize NPs. The synthesis of NPs using microorganisms is considered to be a slow process relative to physical and chemical processes. Lowering the synthesis time would enhance the efficiency of the entire approach. Specific size of the NPs and monodispersity is also critical and must be further explored. The stability of NPs synthesized from microorganisms needs to be evaluated and must be enhanced for their effective use in different application fields. Intracellular and extracellular synthesis of NPs has been extensively studied and is found that the later has more advantages since additional costly process such as ultrasound and chemical reactions are not required unlike the intracellular approach. Detailed processes and mechanism(s) related to the synthesis of NPs from microorganisms are yet to be deciphered, as the synthesis is still not in larger scale. The field of nanotechnology will be greatly improved if the biosynthetic approaches toward synthesis of NPs can be understood. There are ample aspects, which need to be revealed, particularly the detailed role of microorganisms in the synthesis of NPs. Selection of microorganisms is equally challenging. Optimum conditions needs to be controlled so as to effectively obtain the synthesized NPs. The combination pattern independently dispersed with microorganisms in the solution or coated on the microorganism or is one of the critical factors which determine the reaction rate. Interestingly, longer time is required for microbial processes to occur. For instance, due to slow microbial reaction rate, long retention time is observed for biological waste water treatments. Similarly, extreme cold conditions would not favor microbial processes to occur.

Even though the detailed molecular mechanisms are not known, it is evident that microorganisms take counter measures to defend themselves from toxic metallic ions, and thus, their cell wall which is negatively charged secretes sticky secretion to attract the ions. The enzymes released by microbes mediate the conversion of the detrimental ions into non-harmful and thus have an explicit role in the synthesis of NPs. Still, approaches are sought to better control the monodispersity of NPs and particle size and it is quite revealed that the microbial species and the associated environmental factors greatly affect the synthesis of MtNPs. The biological methods to fabricate NPs are still in the embryonic stage. Efforts should be made to develop the biosynthesis process, moreover, to realize the application of nanoparticle biosynthesis in practice.

References

Agnihotri, M., Joshi, S., Ravikumar, A., Zinjarde, S., Kulkarni, S., 2009. Biosynthesis of gold nanoparticles by the tropical marine yeast *Yarrowia lipolytica* NCIM 3589. Mater. Lett. 63, 1231–1234.

Ahmad, A., Mukherjee, P., Senapati, S., Mandal, D., Khan, M.I., Kumar, R., Sastry, M., 2003a. Extracellular biosynthesis of silver nanoparticles using the fungus *Fusarium oxysporum*. Colloids Surf. B Biointerfaces 28, 313–318.

Ahmad, A., Senapati, S., Khan, M.I., Kumar, R., Sastry, M., 2003b. Extracellular biosynthesis of monodisperse gold nanoparticles by a novel extremophilic actinomycete, *Thermomonospora* sp. Langmuir 19, 3550–3553.

Ahmad, A., Senapati, S., Khan, M.I., Kumar, R., Ramani, R., Srinivas, V., Sastry, M., 2003c. Intracellular synthesis of gold nanoparticles by a novel alkalotolerant actinomycete, Rhodococcus species. Nanotechnology 14, 824–828.

Ahmad, R.S., Ali, F., Hamid, R.S., Sara, M., 2007a. Synthesis and effect of silver nanoparticles on the antibacterial activity of different antibiotics against *Staphylococcus aureus* and *Escherichia coli*. Nanomed. Nanotechnol. 3, 168–171.

Ahmad, R.S., Sara, M., Hamid, R.S., Hossein, J., Ashraf-Asadat, N., 2007b. Rapid synthesis of silver nanoparticles using culture supernatants of Enterobacteria: a novel biological approach. Process Biochem. 42, 919–923.

Ahmad, T., Wani, I.A., Manzoor, N., Ahmed, J., Asiri, A.M., 2013. Biosynthesis, structural characterization and antimicrobial activity of gold and silver nanoparticles. Colloids Surf. B Biointerfaces 107, 227–234.

Akbarzadeh, A., Davood, Z.A.F., Mohammad, R.M., Norouzian, D., Tangestaninejad, S., Moghadam, M., Bararpour, N., 2009. Synthesis and characterization of gold nanoparticles by tryptophane. Am. J. Appl. Sci. 6, 691–695.

Ali, M., Ahmed, T., Wu, W., Hossain, A., Hafeez, R., Islam Masum, M., et al., 2020. Advancements in plant and microbe-based synthesis of metallic nanoparticles and their antimicrobial activity against plant pathogens. Nanomaterials 10 (6), 1146.

Alkilany, A.M., Murphy, C.J., 2010. Toxicity and cellular uptake of gold nanoparticles: what we have learned so far? J. Nanopart. Res. 12, 2313–2333.

Arvizo, R., Bhattacharya, R., Mukherjee, P., 2010. Gold nanoparticles: opportunities and challenges in nanomedicine. Expert Opin. Drug Deliv. 7, 753–763.

Bansal, V., Poddar, P., Ahmad, A., Sastry, M., 2006. Room-temperature biosynthesis of ferroelectric barium titanate nanoparticles. J. Am. Chem. Soc. 128, 11958–11963.

Bhambure, R., Bule, M., Shaligram, N., Kamat, M., Singhal, R., 2009. Extracellular biosynthesis of gold nanoparticles using *Aspergillus niger*—its characterization and stability. Chem. Eng. Technol. 32, 1036–1041.

Binupriya, A.R., Sathishkumar, M., Vijayaraghavan, K., Yun, S.I., 2010. Bioreduction of trivalent aurum to nanocrystalline gold particles by active and inactive cells and cell free extract of *Aspergillus oryzae* var. viridis. J. Hazard. Mater. 177, 539–545.

Brown, S., Sarikaya, M., Johnson, E.A., 2000. A genetic analysis of crystal growth. J. Mol. Biol. 299, 725–735.

Castro, L.E., Vilchis, N.A.R., Avalos, B.M., 2011. Biosynthesis of silver, gold and bimetallic nanoparticles using the filamentous fungus *Neurospora crassa*. Colloids Surf. B Biointerfaces 83, 42–48.

Chauhan, A., Zubair, S., Tufail, S., Sherwani, A., Sajid, M., Raman, S.C., Azam, A., Owais, M., 2011. Fungus-mediated biological synthesis of gold nanoparticles: potential in detection of liver cancer. Int. J. Nanomedicine 6, 2305–2319.

Daraee, H., Eatemadi, A., Abbasi, E., Aval, S.F., Kouhi, M., Akbarzadeh, A., 2016. Application of gold nanoparticles in biomedical and drug delivery. Artif. Cells Nanomed. Biotechnol. 44, 410–422.

Das, S.K., Dickinson, C., Laffir, F., Brougham, D.F., Marsili, E., 2012. Synthesis, characterization and catalytic activity of gold nanoparticles biosynthesized with *Rhizopus oryzae* protein extract. Green Chem. 14, 1322–1344.

Dhillon, G.S., Brar, S.K., Kaur, S., Verma, M., 2012. Green approach for nanoparticle biosynthesis by fungi: current trends and applications. Crit. Rev. Biotechnol. 32, 49–73.

Du, L.W., Jiang, H., Liu, X.H., Wang, E.K., 2007. Biosynthesis of gold nanoparticles assisted by *Escherichia coli* DH5α and its application on direct electrochemistry of hemoglobin. Electrochem. Commun. 9, 1165–1170.

Du, L., Xian, L., Feng, J.X., 2011. Rapid extra–/intracellular biosynthesis of gold nanoparticles by the fungus *Penicillium* sp. J. Nanopart. Res. 13, 921–930.

Duran, N., Seabra, A.B., 2009. Metallic oxide nanoparticles: state of the art in biogenic syntheses and their mechanisms. Appl. Microbiol. Biotechnol. 95, 275–288.

Gahlawat, G., Choudhury, A.R., 2019. A review on the biosynthesis of metal and metal salt nanoparticles by microbes. RSC Adv. 9 (23), 944–967.

Gericke, M., Pinches, A., 2006. Microbial production of gold nanoparticles. Gold Bull. 39, 22–28.

Hassan, M., Haque, E., Reddy, K.R., Minett, A.I., Chen, J., Gomes, V.G., 2014. Edge-enriched graphene quantum dots for enhanced photoluminescence and supercapacitance. Nanoscale 6, 11988–11994.

He, S., Guo, Z., Zhang, Y., Zhang, S., Wang, J., Gu, N., 2007. Biosynthesis of gold nanoparticles using the bacteria *Rhodopseudomonas capsulata*. Mater. Lett. 61, 3984–3987.

Herizchi, R., Abbasi, E., Milani, M., Akbarzadeh, A., 2016. Current methods for synthesis of gold nanoparticles. Artif. Cells Nanomed. Biotechnol. 44 (2), 596–602.

Hulkoti, N.I., Taranath, T., 2014. Biosynthesis of nanoparticles using microbes—a review. Colloids Surf. B 121, 474–483.

Ingale, A.G., Chaudhari, A., 2013. Biogenic synthesis of nanoparticles and potential applications: an eco-friendly approach. J. Nanomed. Nanotechol. 4 (165), 1–7.

Islam, M.S., Haque, M.S., Islam, M.M., 2012. Tools to kill: genome of one of the most destructive plant pathogenic fungi *Macrophomina phaseolina*. BMC Genomics 13, 493.

Jeyaraj, M., Gurunathan, S., Qasim, M., Kang, M.-H., Kim, J.-H., 2019. A comprehensive review on the synthesis, characterization, and biomedical application of platinum nanoparticles. Nanomaterials 9 (12), 1719.

Jha, A.K., Prasad, K., Kulkarni, A.R., 2009. Synthesis of TiO_2 nanoparticles using microorganisms. Colloids Surf. B 71, 226–229.

Juibari, M.M., Yeganeh, L.P., Abbasalizadeh, S., Azarbaijani, R., Mousavi, S.H., Tabatabaei, M., Jouzani, G.S., Salekdeh, G.H., 2015. Investigation of a hot spring extremophilic *Ureibacillus thermosphaericus* strain thermos-BF for extracellular biosynthesis of functionalized gold nanoparticles. Bionanoscience 5, 233–241.

Jung, H.W., Panigrahi, G.K., Jung, G-Y., Lee, Y.J., Shin, K.H., Sahoo, A., Choi, E.S., Lee, E., Kim, K.M., Yang, S.H., Jeon, J.S., Lee, S.C., Kim, S.H., 2020. PAMP-triggered immunity involves proteolytic degradation of core nonsense-mediated mRNA decay factors during early defense response. Plant Cell 32 (4), 1081–1101. https://doi.org/10.1105/tpc.19.00631.

Kalimuthu, K., Babu, R.S., Venkataraman, D., Bilal, M., Gurunathan, S., 2008. Biosynthesis of silver nanocrystals by *Bacillus licheniformis*. Colloids Surf. B 65 (1), 150–153.

Kalishwaralal, K., Deepak, V., Pandian, S.R.K., 2010. Biosynthesis of silver and gold nanoparticles using *Brevibacterium casei*. Colloids Surf. B. Biointerfaces 77, 257–262.

Kathiresan, K., Manivannan, S., Nabeel, M.A., Dhivya, B., 2009. Studies on silver nanoparticles synthesized by a marine fungus, *Penicillium fellutanum* isolated from coastal mangrove sediment. Colloids Surf. B Biointerfaces 71, 133–137.

Kato, Y., Suzuki, M., 2020. Synthesis of metal nanoparticles by microorganisms. Crystals 10 (7), 589.

Khan, S.K., 2007. *M. phaseolina* as causal agent for charcoal rot of sunflower. Mycopathologia 5, 111–118.

Khan, T., Abbas, S., Fariq, A., Yasmin, A., 2018. Microbes: nature's cell factories of nanoparticles synthesis. In: Prasad, R., Jha, A.K., Prasad, K. (Eds.), Exploring the Realms of Nature for Nanosynthesis. Springer, Cham, pp. 25–50.

Kimling, J., Maier, M., Okenve, B., Kotaidis, V., Ballot, H., Plech, A., 2006. Turkevich method for gold nanoparticle synthesis revisited. J. Phys. Chem. B 110, 15700–15707.

Kitching, M., Ramani, M., Marsili, E., 2015. Fungal biosynthesis of gold nanoparticles: mechanism and scale up. Microb. Biotechnol. 8, 904–917.

Konishi, Y., Tsukiyama, T., Tachimi, T., Saitoh, N., Nomura, T., Nagamine, S., 2007. Microbial deposition of gold nanoparticles by the metalreducing bacterium *Shewanella algae*. Electrochim. Acta 53, 186–192.

Kumar, S.A., Peter, Y.A., Nadeau, J.L., 2008. Facile biosynthesis, separation and conjugation of gold nanoparticles to doxorubicin. Nanotechnology, 495101. 1-10.

Kumar, K.P., Paul, W., Sharma, C.P., 2011. Green synthesis of gold nanoparticles with *Zingiberofficinale* extract: characterization and blood compatibility. Process Biochem. 46, 2007–2013.

Lengke, M.F., Southam, G., Cosmochim, G., 2005. The effect of thiosulfate oxidizing bacteria on the stability of the gold-thiosulfate complex. Geochim. Cosmochim. Acta 69, 3759–3772.

Lengke, M.F., Fleet, M.E., Southam, G., 2006a. Morphology of gold nanoparticles synthesized by filamentous cyanobacteria from gold (I)-thiosulfate and gold(III)–chloride complexes. Langmuir 22, 2780–2787.

Lengke, M.F., Ravel, B., Fleet, M.E., Wanger, G., Gordon, R.A., Southam, G., 2006b. Mechanisms of gold bioaccumulation by filamentous cyanobacteria from gold (III)-chloride complex. Environ. Sci. Technol. 40, 6304–6309.

Lin, Z., Wu, J., Xue, R., Yang, Y., 2005. Spectroscopic characterization of Au^{3+} biosorption by waste biomass of *Saccharomyces cerevisiae*. Spectrochim. Acta A Mol. Biomol. Spectrosc. 61, 761–765.

Maggy, F.L., Michael, E.F., Gordon, S., 2007. Synthesis of palladium nanoparticles by reaction of filamentous cyanobacterial biomass with a palladium(II) chloride complex. Langmuir 23, 8982–8987.

Maliszewska, I., 2013. Microbial mediated synthesis of gold nanoparticles: preparation, characterization and cytotoxicity studies. Dig. J. Nanomater. Biostruct. 8, 1123–1131.

Maliszewska, I., Aniszkiewicz, Ł., Sadowski, Z., 2009. Biological synthesis of gold nanostructures using the extract of *Trichoderma koningii*. Acta Phys. Pol. A 116, 163–165.

Mandal, D., Bolander, M.E., Mukhopadhyay, D., Sarkar, G., Mukherjee, P., 2006. The use of microorganisms for the formation of metal nanoparticles and their application. Appl. Microbiol. Biotechnol. 69, 485–492.

Mansoori, G.A., Brandenburg, K.S., Shakeri-Zadeh, A., 2010. A comparative study of two folate-conjugated gold nanoparticles for cancer nanotechnology applications. Cancers (Basel) 2 (4), 1911–1928. https://doi.org/10.3390/cancers2041911.

Mihail, J.D., Taylor, S.J., 1995. Interpreting variability among isolates for *Macrophomina phaseolina* in pathogenicity, pycnidium production and chlorate utilization. Can. J. Bot. 10, 1596–1603.

Minaeian, S., Shahverdi, A.R., Nohi, A.S., Shahverdi, H.R., 2008. Extracellular biosynthesis of silver nanoparticles by some bacteria. J. Sci. IAU 17, 1–4.

Mishra, A.N., Bhadaurla, S., Singh Gaur, M., Pasricha, R., 2010. Extracellular microbial synthesis of gold nanoparticles using fungus *Hormoconis resinae*. JOM 62, 45–48.

Mishra, A., Tripathy, S.K., Wahab, R., Jeong, S.-H., Hwang, I., Yang, Y.B., Kim, Y.S., Shin, H.S., Yun, S.I., 2011. Microbial synthesis of gold nanoparticles using the fungus *Penicillium brevicompactum* and their cytotoxic effects against mouse mayo blast cancer C2C12 cells. Appl. Microbiol. Biotechnol. 92, 617–630.

Mishra, A., Tripathy, S.K., Yuna, S.I., 2012. Fungus mediated synthesis of gold nanoparticles and their conjugation with genomic DNA isolated from *Escherichia coli* and *Staphylococcus aureus*. Process Biochem. 47, 701–711.

Mithila, A., Swanand, J., Ameeta, R.K., Smita, Z., Sulabha, K., 2009. Biosynthesis of gold nanoparticles by the tropical marine yeast *Yarrowia lipolytica* NCIM 3589. Mater. Lett. 63, 1231–1234.

Mohanpuria, P., Rana, N.K., Yadav, S.K., 2008. Biosynthesis of nanoparticles: technological concepts and future applications. J. Nanopart. Res. 10, 507–517.

Mukherjee, P., Ahmad, A., Mandal, D., Senapati, S., Sainkar, S.R., Khan, M.I., Ramani, R., Parischa, R., Ajaykumar, P.V., Alam, M., Sastry, M., Kumar, R., 2001. Bioreduction of $AuCl^{-4}$ ions by the fungus, *Verticillium sp.* and surface trapping of the gold nanoparticles formed. Angew. Chem. Int. Ed. 40, 3585–3588.

Nanda, A., Saravanan, M., 2009. Biosynthesis of silver nanoparticles from *Staphylococcus aureus* and its antimicrobial activity against MRSA and MRSE. Nanomed. Nanotechnol. 5, 452–456.

Narayanan, K.B., Sakthivel, N., 2011a. Facile green synthesis of gold nanostructures by NADPH-dependent enzyme from the extract of *Sclerotium rolfsii*. Colloids Surf. A Physicochem. Eng. Asp. 380, 156–161.

Narayanan, K.B., Sakthivel, N., 2011b. Synthesis and characterization of nanogold composite using *Cylindrocladium floridanum* and its heterogeneous catalysis in the degradation of 4-nitrophenol. J. Hazard. Mater. 189, 519–525.

Narayanan, K.B., Sakthivel, N., 2013. Mycocrystallization of gold ions by the fungus *Cylindrocladium floridanum*. World J. Microbiol. Biotechnol. 29, 2207–2211.

Narges, M., Shahram, D., Seyedali, S., Reza, A., Khosro, A., Saeed, S., Sara, M., Hamid, R.S., Ahmad, R.S., 2009. Biological synthesis of very small silver nanoparticles by culture supernatant of *Klebsiella pneumonia*: the effects of visible-light irradiation and the liquid mixing process. Mater. Res. Bull. 44, 1415–1421.

Narkeviciute, I., Chakthranont, P., Mackus, A.J.M., Hahn, C., Pinaud, B.A., Bent, S.F., Jaramillo, T.F., 2016. Tandem core-shell Si-Ta3N5 photoanodes for photoelectrochemical water splitting. Nano Lett. 16, 7565–7572.

Noruzi, M., Zare, D., Khoshnevisan, K., Davoodi, D., 2011. Rapid green synthesis of gold nanoparticles using *Rosa hybrida* petal extract at room temperature. Spectrochim. Acta A Mol. Biomol. Spectrosc. 79, 1461–1465.

Pal, G., Rai, P., Pandey, A., 2019. Green synthesis of nanoparticles: a greener approach for a cleaner future. In: Shukla, A.K., Iravani, S. (Eds.), Green Synthesis, Characterization and Applications of Nanoparticles. Elsevier, pp. 1–26.

Panigrahi, G.K., Sahoo, S.K., Sahoo, A., Behera, S., Sahu, S.R., Dash, A., Satapathy, K.B., 2021a. Bioactive molecules from plants: a prospective approach to combat SARS-Cov-2. Adv. Trad. Med., 1–14. https://doi.org/10.1007/s13596-021-00599-y.

Panigrahi, G.K., Sahoo, A., Satapathy, K.B., 2021b. Insights to plant immunity: defense signaling to epigenetics. Physiol. Mol. Plant Pathol. 113, 1–7. https://doi.org/10.1016/j.pmpp.2020.101568.

Panigrahi, G.K., Sahoo, A., Satapathy, K.B., 2021c. Differential expression of selected Arabidopsis resistant genes under abiotic stress conditions. Plant Sci. Today 8 (4), 859–864. https://doi.org/10.14719/pst.2021.8.4.1213.

Panigrahi, G.K., Satapathy, K.B., 2020a. Sacrificed surveillance process favours plant defense: a review. Plant Arch. 20 (1), 2551–2559.

Panigrahi, G.K., Satapathy, K.B., 2020b. Formation of Arabidopsis poly(A)-specific ribonuclease associated processing bodies in response to pathogenic infection. Plant Arch. 20 (2), 4907–4912.

Panigrahi, G.K., Satapathy, K.B., 2020c. Arabidopsis DCP5, a decapping complex protein interacts with Ubiquitin-5 in the processing bodies. Plant Arch. 20 (1), 2243–2247.

Panigrahi, G.K., Satapathy, K.B., 2021. Pseudomonas syringae pv. syringae infection orchestrates the fate of the Arabidopsis J domain containing cochaperone and decapping protein factor 5. Physiol. Mol. Plant Pathol. 113 (101598), 1–9. https://doi.org/10.1016/j.pmpp.2020.101598.

Parida, U.K., Bindhani, B.K., Nayak, P., 2011. Green synthesis and characterization of gold nanoparticles using onion (*Allium cepa*) extract. World J. Nano Sci. Eng. 1, 93–98.

Philip, D., 2009. Biosynthesis of au, ag and au-ag nanoparticles using edible mushroom extract. Spectrochim. Acta Mol. Biomol. Spectrosc. 73, 374–381.

Prasad, R., 2019. Microbial Nanobionics. vol. 1 Springer, p. 321. (1): XVII.

Ramezani, F., Ramezani, M., Talebi, S., 2010. Mechanistic aspects of biosynthesis of nanoparticles by several microbes. Nanocon 10, 1. 12.

Ravindra, P., 2009. Protein-mediated synthesis of gold nanoparticles. Mater. Sci. Eng. B 163, 93–98.

Reddy, K.R., Lee, K.-S., Iyenger, A.G., 2007. Self-assembly directed synthesis of poly (ortho-toluidine)-metal (gold and palladium) composite nanospheres. J. Nanosci. Nanotechnol. 7, 3117–3125.

Reddy, K.R., Sin, B.C., Yoo, C.H., Park, W., Ryu, K.S., Lee, J.-S., Sohn, D., Lee, Y., 2008. A new one-step synthesis method for coating multi-walled carbon nanotubes with cuprous oxide nanoparticles. Scr. Mater. 58, 1010–1013.

Reddy, K.R., Nakata, K., Ochiai, T., Murakami, T., Tryk, D.A., Fujishima, A., 2011. Facile fabrication and photocatalytic application of Ag nanoparticles-TiO_2 nanofiber composites. J. Nanosci. Nanotechnol. 11, 3692–3695.

Reddy, K.R., Gomes, V., Hassan, M., 2014. Carbon functionalized TiO_2 nanofibers for high efficiency photocatalysis. Mater. Res. Express 1, 1–15.

Roy, K., Lahiri, S., 2006. A green method for synthesis of radioactive gold nanoparticles. Green Chem. 8, 1063–1066.

Roychoudhury, A., 2020. Yeast-mediated green synthesis of nanoparticles for biological applications. Indian J. Pharm. Biol. Res. 8 (3), 26–31.

Sadowski, Z., Maliszewska, I.H., Grochowalska, B., Polowczyk, I., Kozlecki, T., 2008. Synthesis of silver nanoparticles using microorganisms. Mater. Sci. Pol. 26, 419–424.

Sahoo, A., Satapathy, K.B., 2021. Differential expression of Arabidopsis EJC core proteins under short-day and long-day growth conditions. Plant Sci. Today 8 (4), 815–819. https://doi.org/10.14719/pst.2021.8.4.1214.

Sahoo, A., Satapathy, K.B, Panigrahi, G.K., 2023a. Ectopic expression of disease resistance protein promotes resistance against pathogen infection and drought stress in Arabidopsis. Physiol. Mol. Plant Pathol. 124 (101949), 1–7. https://doi.org/10.1016/j.pmpp.2023.101949.

Sahoo, A., Satapathy, K.B., Panigrahi, G.K., 2023b. Security check: plant immunity under temperature surveillance. J. Plant Biochem. Biotechnol. https://doi.org/10.1007/s13562-023-00846-0.

Saifuddin, N., Wong, C.W., Yasumira, A.A.N., 2009. Rapid biosynthesis of silver nanoparticles using culture supernatant of bacteria with microwave irradiation. E-J. Chem. 6, 61–70.

Salata, O., 2004. Applications of nanoparticles in biology and medicine. J. Nanobiotechnol. 2 (3), 1–5.

Sanghi, R., Verma, P., 2010. pH dependent fungal proteins in the "green" synthesis of gold nanoparticles. Adv. Mater. Lett. 1, 193–199.

Sanghi, R., Verma, P., Puri, S., 2011. Enzymatic formation of gold nanoparticles using *Phanero chaetechryso sporium*. Adv. Chem. Eng. Sci. 1, 154–162.

Sarkar, J., Ray, S., Chattopadhyay, D., Laskar, A., Acharya, K., 2012. Mycogenesis of gold nanoparticles using a phytopathogen *Alternaria alternata*. Bioprocess Biosyst. Eng. 35, 637–643.

Sathish Kumar, K., Amutha, R., Arumugam, P., Berchmans, S., 2011. Synthesis of gold nanoparticles: an ecofriendly approach using *Hansenula anomala*. ACS Appl. Mater. Interfaces 3, 1418–1425.

Sawle, B.D., Salimath, B., Deshpande, R., Bedre, M.R., Prabhakar, B.K., Venkataraman, A., 2008. Biosynthesis and stabilization of Au and Au-Ag alloy nanoparticles by fungus, *Fusarium semitectum*. Sci. Technol. Adv. Mater. 9, 1–10.

Shankar, S.S., Ahmad, A., Pasricha, R., Sastry, M., 2003. Bioreduction of chloroaurate ions by *Geranium* leaves and its endophytic fungus yields gold nanoparticles of different shapes. J. Mater. Chem. 13, 1822–1826.

Shankar, S.S., Rai, A., Ankamwar, B., Singh, A., Ahmad, A., Sastry, M., 2004. Biological synthesis of triangular gold nanoprisms. Nat. Mater. 3, 482–488.

Siddiqi, K.S., Husen, A., 2016. Fabrication of metal nanoparticles from fungi and metal salts: scope and application. Nanoscale Res. Lett. 11, 98–108.

Singaravelu, G., Arockiamary, J.S., Kumar, V.G., Govindaraju, K., 2007. A novel extracellular synthesis of monodisperse gold nanoparticles using marine alga, *Sargassum wightii* Greville. Colloids Surf. B Biointerfaces 57, 97–101.

Singh, M., Kalaivani, R., Manikandan, S., Sangeetha, N., Kumaraguru, A.K., 2013. Facile green synthesis of variable metallic gold nanoparticle using *Padina gymnospora*, a brown marine macroalga. Appl. Nanosci. 3, 145–151.

Singh, R., Shedbalkar, U.U., Wadhwani, S.A., Chopade, B.A., 2015. Bacteriagenic silver nanoparticles: synthesis, mechanism, and applications. Appl. Microbiol. Biotechnol. 99 (11), 4579–4593.

Singh, A., Gautam, P.K., Verma, A., Singh, V., Shivapriya, P.M., Shivalkar, S., et al., 2020. Green synthesis of metallic nanoparticles as effective alternatives to treat antibiotics resistant bacterial infections: a review. Biotechnol. Rep. 25, 1–14.

Soni, M., Mehta, P., Soni, A., Goswami, G.K., 2018. Green nanoparticles: synthesis and applications. IOSR J. Biotechnol. Biochem. 4, 78–83.

Sreedharan, S.M., Gupta, S., Saxena, A.K., Singh, R., 2019. *Macrophomina phaseolina*: microbased biorefinery for gold nanoparticle production. Ann. Microbiol. 3, 1–11.

Strasser, P., Koh, S., Anniyev, T., Greeley, J., More, K., Yu, C., et al., 2010. Lattice strain control of the activity in dealloyed core-shell fuel cell catalysts. Nat. Chem. 2 (6), 454–460.

Su, S.S., Chang, I., 2018. Review of production routes of nanomaterials. In: Brabazon, D., Pellicer, E., Zivic, F., Sort, J., Baró, M.D., Grujovic, N., Choy, K.-L. (Eds.), Commercialization of Nanotechnologies—A Case Study Approach. Springer, pp. 15–29.

Sugunan, A., Dutta, J., 2006. Novel synthesis of gold nanoparticles in aqueous media. Mater. Res. Soc. Symp. Proc., 55.

Suresh, A.K., et al., 2011. Biofabrication of discrete spherical gold nanoparticles using the metal-reducing bacterium *Shewanella oneidensis*. Acta Biomater. 7, 2148–2152.

Tiquia-Arashiro, S., Rodrigues, D.F., 2016. Nanoparticles synthesized by microorganisms. In: Extremophiles: Applications in Nanotechnology. Springer, pp. 1–51.

Vala, A.K., 2015. Exploration on green synthesis of gold nanoparticles by a marine-derived fungus *Aspergillus sydowii*. Environ. Prog. Sustain. Energy 34, 194–197.

Varshney, R., Mishra, A.N., Bhadauria, S., Gaur, M.S., 2009. A novel microbial route to synthesize silver nanoparticles using fungus *Hormoconis resinae*. Dig. J. Nanomater. Biostruct. 4, 349–355.

Xie, J., Lee, J.Y., Wang, D.I.C., Ting, Y.P., 2007. High-yield synthesis of complex gold nanostructures in a fungal system. J. Phys. Chem. C 111, 16858–16865.

Yang, X., Yang, M., Bo, P., Vara, M., Xia, Y., 2015. Gold nanomaterials at work in biomedicine. Chem. Rev. 115, 10410–10488.

Yasuhiro, K., Kaori, O., Norizoh, S., Toshiyuki, N., Shinsuke, N., Hajime, H., Yoshio, T., Tomoya, U., 2007. Bioreductive deposition of platinum nanoparticles on the bacterium *Shewanella algae*. J. Biotechnol. 128, 648–653.

Yeh, Y.C., Creran, B., Rotello, V.M., 2012. Gold nanoparticles: preparation, properties, and applications in bionanotechnology. Nanoscale 4 (6), 1871–1880.

Yusof, H.M., Mohamad, R., Zaidan, U.H., 2019. Microbial synthesis of zinc oxide nanoparticles and their potential application as an antimicrobial agent and a feed supplement in animal industry: a review. J. Anim. Sci. Biotechnol. 10 (1), 1–22.

Zhang, Y.P., Lee, S.H., Reddy, K.R., Gopalan, A.I., Lee, K.P., 2007. Synthesis and characterization of core-shell SiO_2 nanoparticles/poly(3-aminophenylboronic acid) composites. J. Appl. Polym. Sci. 104, 2743–2750.

Zhang, X., He, X., Wang, K., Yang, X., 2011. Different active biomolecules involved in biosynthesis of gold nanoparticles by three fungus species. J. Biomed. Nanotechnol. 7, 245–254.

CHAPTER

19

Biofilm and its impact on microbial-induced corrosion: An entrepreneurship and industrial perspective

Neha Sharma[a], Devinder Toor[b], and Udita Tiwari[c]

[a]Clinical Research Division, Department of Biosciences, School of Basic and Applied Sciences, Greater Noida, Uttar Pradesh, India, [b]Amity Institute of Virology and Immunology, Amity University Uttar Pradesh, Noida, India, [c]Department of Biochemistry, School of Life Sciences, Dr. Bhimrao Ambedkar University, Agra, Uttar Pradesh, India

19.1 Introduction

Biofilm refers to a complex matrix formed by a consortium of microorganisms or by a single type of microorganism. The complex matrix is formed by self-generated extracellular polymeric substances (EPS), which inculcate a certain set of characteristics such as resistance toward antimicrobials or long-term persistence to evolve substantially as compared to free-living microorganisms. There are multiple aspects of biofilms in the context of their pros and cons, which are well-studied and documented in the context of an industrial perspective. Biofilms in some cases offer protection against corrosion, whereas, in a majority of cases, the biofilms intensify the corrosion process due to the diversity of microbial consortia which can promote a pH gradient and hence provide electrochemical conditions at the metal-solution interface to induce corrosion.

19.2 Microbiologically induced corrosion (MIC)

Microbiologically induced corrosion (MIC) is well-documented long-term concern in various industries across the globe. Despite the substantial efforts that have been made, MIC remains a threat to industrial infrastructure, which accounts for a huge economic loss. MIC has been well studied and reported to be accountable for damage to surfaces such as metals, concrete, and other engineering materials, but there is a dearth of studies to clarify the underlying mechanisms and to distinguish between the corrosion caused by microbes or the secondary factors such as environmental factors.

To further acknowledge the need to investigate MIC, there is a prerequisite to understand the serious implications of MIC and to explore the treatment and management strategies. With an advanced understanding of the interaction of biofilms with surfaces (such as metal and concrete), on-site diagnosis of MIC, and laboratory investigations, there is a scope for appropriate and timely diagnosis and further exploring newer methods to manage and mitigate MIC.

Entrepreneurship with Microorganisms. https://doi.org/10.1016/B978-0-443-19049-0.00002-5

19.2.1 Economic loss

Economic loss attributed to MIC is underreported due to the secondary factors such as environmental damage and lack of appropriate and timely diagnosis. There are no reports that have documented the direct economic loss due to microbial-induced corrosion across the globe. Cumulative damage due to MIC can be estimated through various research reports and case studies, which have documented the damage and associated cost due to MIC in oil, gas, petroleum, and various other industries. MIC has been reported to be accountable for 20% of the annual losses, and it has been reported that up to 20% of the overall corrosion in the marine system is attributed to MIC (El-Bassi et al., 2020; Sachan and Singh, 2020). Other studies have suggested that MIC has been the root cause for approximately 40% of pipe corrosion in the oil industry, which leads to a massive maintenance cost of US$100 million in various processes such as production, transportation, and storage of oil every year in many countries. Major incidences of pipeline failure and leakage due to MIC have been reported in New Mexico and Alaska in 2000 and 2006, respectively, which lead to havoc losses and a spike in world oil prices (Xu and Tingyue, 2014). Collectively, billions of US$/year have been spent to overcome the losses due to MIC.

19.3 Biofilm interaction on metal surfaces: An overview

Biofilm is a consortium of microorganisms that produces corrosive metabolites during the process of metabolism, which leads to the damage of engineering materials used for construction such as metals, concrete, plastics, scaffolds, and other engineered materials. This is supported by a range of environmental conditions available in various systems of industries such as cooling towers, pipelines, heat exchangers, and condensers in oil and gas industries, power plants, manufacturing industries, etc. Major materials used in industrial infrastructure are mild steel, carbon steel, stainless steel, aluminum, copper, titanium, and concrete. Although steel and copper offer a robust resistance towards corrosion, the corrosive metabolites evade the protective layer and interact with the surfaces through electrochemical reactions and initiate a point of corrosion, which further leads to a destructive environment and damages the surface to an extent. Multiple aspects of biofilm formation and sustainability are required to be investigated to understand the mechanism so that early diagnosis can be performed. Lack of timely diagnosis and inappropriate intervention contributes to the corrosion burden.

19.3.1 Sulfate reducing bacteria

Sulfate-reducing bacteria (SRB) are well-studied and known for their corrosive capacity on metallic structures. SRBs are anaerobic microorganisms and have been documented to corrode carbon steel, stainless steel, aluminum alloy, copper, and concrete in various industrial infrastructures. There are several reports that have shown the direct correlation of SRBs with metallic and concrete corrosion and studies simulated in laboratories have confirmed the massive magnitude of corrosion due to SRBs' interaction with the metal and other engineered scaffolds in the industrial infrastructure. SRBs produce an enormous amount of hydrogen sulfide which facilitates anaerobiosis. Some of the strains of SRB are strict anaerobes and some are facultative anaerobes that can survive in the presence of oxygen, which suggests the long-term impact of SRB biofilm on metal surfaces.

19.3.1.1 Carbon steel

Sulfate-reducing bacterium *Desulfotomaculum nigrificans* present under the deposits have been reported to synergistically promote weight loss and deep pitting patterns on carbon steel as compared to the control conditions (Liu et al., 2019). *Desulfovibrio vulgaris* were isolated from a gas transmission pipeline in Malaysia to perform the corrosion study on carbon steel through the weight loss method, and it was observed that the isolated SRB consortium imposed substantial damage to carbon steel as compared to the single strain obtained from ATCC (Abdullah et al., 2014). In a recent study, it has been reported that both localized and overall corrosion has been observed to be enhanced in presence of SRBs accumulated under the deposits (Yang et al., 2022). Corrosion due to the direct interaction of the organism with the metal surface is different from the damage caused by deposits formed through the biofilm formation which provides a support system for the sustainable growth and prolonged exposure of the metal surface with organisms (Batmanghelich et al., 2017). Additionally, it has been reported that a lack of nutrients might promote a situation in which SRBs avail nutrients through metal ions (Liu et al., 2019). Another mechanism has also been observed that the rate of corrosion in carbon steel in soil extract solutions was comparatively slower during the growth phase of SRBs than the death phase (Chen et al., 2021). Although vigorous research has been done to establish the significant role of SRB in metallic corrosion, the species-specific interactions and the contribution of the secondary environmental factors are still unexplored. It is therefore required to understand the role of these factors so that differential diagnosis can be done for the mitigation of the MIC.

19.3.1.2 Stainless steel

Ferrous sulfide is the main corrosion product of SRBs, which intensifies the corrosion process on the metal surfaces and the formation of crystalline biofilm of *D. desulfuricans* on 2205 duplex stainless steel impeded the corrosion resistance (Dec et al., 2016). The passive film formed by sulfide impedes the corrosion resistance potential of stainless steel, and alloys in heat exchangers, cooling towers, pipelines, and other engineering materials in power plants, oil and gas industries(Dec et al., 2018; Prithiraj et al., 2019). Similarly, corrosion pitting on the stainless steel surface in seawater and soil due to the SRB biofilm has been also been reported (Tran et al., 2021; Yoffe, 1997).

19.3.1.3 Aluminum alloys

Aluminum alloys are widely used for thermal desalination due to their excellent corrosion resistance but due to the consortium of microorganisms present in seawater and sea mud, there is a favorable environment for the induction of microbial corrosion. The anaerobic corrosion of aluminum alloys by SRB was reported way back in 1968, and the magnitude of corrosion in presence of SRB was huge as compared to the control experiments (Tiller and Booth, 1968). In the past decade, it has been documented that the hydrogenase activity of SRBs promotes microbial corrosion on aluminum surfaces. Earlier, the widely accepted mechanism behind this phenomenon was observed to be scavenging 'cathodic hydrogen', but more recently, it has been reported that the bi-cathodic nature of SRB might be the plausible cause of pitting corrosion. In this mechanism, SRBs obtain electrons from metals through extracellular electron transfer (EET) in absence of carbon sources, this mechanism leads to more severe damage than the bi-cathodic or metabolic mechanism of corrosion (Guan et al., 2020). Similarly, a higher corrosion rate of aluminum alloys has been reported in SRB-containing sea mud, which increased the size of the pits due to cathodic depolarization (Liu et al., 2014). Collectively, SRBs have the potential to initiate and accelerate the corrosion rate in aluminum alloys.

19.3.1.4 Copper

The impact of SRBs on copper corrosion can be investigated through surface analysis and electrochemical measurement studies. Although copper exhibits better corrosion resistance than carbon steel, it is too vulnerable to microbial corrosion and the accepted mechanism is the production of hydrogen sulfide ions, which diffuse through the copper surface and form copper sulfide. Thus, the magnitude of production of hydrogen sulfide (H_2S) by SRBs provides the conditioning of copper surface with a sulfide layer and leads to pitting and intergranular corrosion (Tiller and Booth, 1968). This mechanism has also been confirmed by thermodynamic studies to understand the phenomenon of microbial corrosion on copper with *D. vulgaris* as a model organism (Dou et al., 2020; Huang et al., 2004).

19.3.1.5 Concrete

Concrete has been the material of primary choice for the construction of industrial infrastructure like sewerage systems and wastewater treatment plants due to its sustainability, easy access, and low cost. The chemical environment of the pipelines (organic and inorganic salts and gases) facilities promotes the growth of bacteria with corrosive properties. The anaerobic conditions of the sewerage systems provide the ideal condition for the growth of the MIC-related organisms, *Desulfovibrio* and *Desulfomaculum,* which produces an enormous amount of hydrogen sulfide(Chaudhari et al., 2022). Although the primary cause of MIC in concrete is not SRB, the H_2S produced by SRBs forms a layer over the concrete surface, which is the initiation step of the corrosion process(Chaudhari et al., 2022). The major contributors of MIC in concrete are sulfur-oxidizing bacteria (SOBs), and the mechanism and the potential role of SOBs in promoting MIC will be discussed in later sections.

19.3.2 Sulphur oxidizing bacteria

Sulfur oxidizing bacteria (SOB) facilitates the production of sulfuric acid via the utilization of energy obtained from reduced sulfur compounds which further promotes biogenic sulfuric acid (BSA) corrosion of metallic and concrete surfaces. Few strains of SOB viz. *Acidithiobacillus thiooxidans*, *Thiomonas intermedia*, and *Thiomonas perometabolis* have been investigated to assess their role in promoting microbial corrosion of metals and concrete. *Acidithiobacillus thiooxidans* has been reported to produce an enormous amount of sulfuric acid, which interacts with the components of the concrete surface and forms a layer of gypsum and ettringite (Huber et al., 2016; Wei et al., 2013). The formation of salts due to *A. thiooxidans* leads to the compromised structural integrity of concrete surface due to the formation of internal cracks (Wells and Melchers, 2015). This mechanism has been observed in SOBs isolated from the pipelines surfaces of waste-water treatment plants and marine sediments (Beech and Campbell, 2008), and it was reported that both SRB and SOB might be the responsible organisms for accelerated low water corrosion (ALWC) and can survive in an acidic environment. SOBs are also capable of forming biofilms on titanium alloys, which are widely used in medical equipment, implants, heat exchangers in power plants, aircraft, and ships. Aerobic strains of SOB viz. *Acidithiobacillus thiooxidans* and *Acidithiobacillus ferrooxidans* have been investigated (Beech and Campbell, 2008) and reported to impose a serious threat to human health due to their capability to colonize the titanium surface and have better sustainability with the changing environment as compared to SRBs (Mattes et al., 2013). Biogenic sulfides also promote pitting corrosion in stainless steel structures and lead to long-term damage of the surface.

19.3.3 Manganese oxidizing bacteria (MOB)

MOBs oxidize divalent, soluble Mn (II) to insoluble manganese oxides and the deposition of manganese oxide on the surface induces corrosion. Various reports have analyzed the plausible role of MOBs in the corrosion of steel and the metabolism of *Leptothrix discophora* forms a layer of manganese oxide (MnO_2) and further creates an anaerobic zone which further supports the growth of SRBs(Rajasekar et al., 2007). This mechanism has been widely observed in metallic pipelines of sewage treatment plants. Other strains of MOB viz. *Bacillus cereus* ACE4 and *Pseudoxanthomonas* sp. have also been reported to initiate and enhance localized corrosion on steel surfaces in oil and gas industries (Anandkumar et al., 2011).

19.3.4 Iron-oxidizing bacteria

Among all MIC-causing microorganisms, iron-reducing bacteria (IRB) has been least explored for its role in microbial corrosion. *Shewanella* genera of IRB are known to reduce insoluble iron oxides into soluble ferrous complexes and promote corrosion of IRB (Welikala et al., 2022). *S. oneidensis* strain MR-1 has been reported to induce and enhance steel corrosion (Herrera and Videla, 2009), whereas *Geobacter sulfurreducens* an iron-reducing bacterium has been reported to form a protective phosphate layer that induces resistance against corrosion even in the presence of air and moisture (Cote et al., 2015). There are reports that suggest that IRBs can enhance corrosion or might form a protective environment against corrosion (Esnault et al., 2011; Herrera and Videla, 2009), but no conclusion has been made, and thus, further investigations are required to establish the role of IRBs in microbial corrosion of metals. Biofilm formation by iron-oxidizing bacteria *Sphaerotilus* spp. has been investigated for its role in pitting corrosion on steel 316L and 304L, and it was observed that 316L exhibits more resistance to microbial corrosion rather than 304L (Starosvetsky et al., 2008).

19.3.5 Fungal biofilms

During the 1970s, the mechanism of adhesion of fungi biofilms on various surfaces such as paint coatings, adhesives, and glass was investigated in the context of biodeterioration, and the studies were more focused on the toxins and metabolites produced by fungi and their impact, which provides a biological milieu for the destruction of surfaces (Siegel et al., 1983). Later in the 1990s, *Cladosporium resinae* known as the kerosene fungus was investigated primarily in the context of food contamination, but it was found to be more significantly associated with corrosion of metal and metal alloys. Later, it was reported that *Penicillium* biofilm play a role in the microbial corrosion of stainless steel in 1983. It has been observed that *Penicillium* led to the dissolution of iron from the stainless steel and initiated the point of corrosion but the attachment and interaction between fungi and the metal surface were not considered to play a significant role in the dissolution of the metal (Siegel et al., 1983). Additionally, fungi biofilm on metal wires in humid conditions suggested the role of fungi in the initiation of localized corrosion. Fungal-mediated corrosion has also been reported in hydrocarbon storage tanks and transport pipelines since they provide the appropriate environment with the carbon and hydrogen sources and the presence of water facilitates the humid conditions required for the growth of the fungal species. *Cladosporium*, *Aspergillus niger*, and *Penicillium* sp. impose additional burden by biofouling of hydrocarbons and generating a biogenic acidic environment, which promotes the degradation of metallic surfaces, paint coatings, and rapidly increases corrosion. The mechanism disbursed by fungal sp. involves the production of water by utilization of hydrocarbons, adherence of fungal biofilms through mycelia on metal surfaces, and the differential aeration are ideal conditions for the crevice corrosion of mild and carbon steel. In another study, *Penicillium* sp., *Candida* sp., and *Aspergillus* sp. were isolated from the hydrocarbon storage tank, and weight loss corrosion study of Mild steel (API 5LX grade) was performed (Mohanan et al., 2005). The results confirmed the potential role of fungal species in degrading diesel oil and endorsing the corrosion of mild steel by converting the ferrous ions into ferric oxide. The plausible role of fungal strains in corrosion of metals is not restricted to steel surfaces but aluminum and aluminum-based alloys are also likely to be damaged by microbial corrosion. Micromycetes of eight strains viz. *Aspergillus niger*, *Aspergillus oryzae*, *Aspergillus terreus*, *Alternaria alternate*, *Penicillium chrysogenum*, *Penicillium funiculosum*, *Penicillium cyclopium,* and *Trichoderma viridae* were cultured and the biofilm formed on aluminum and aluminum-based alloys (Belov et al., 2008). This study was performed for 3 months and emphasized the distinctiveness of the corrosion mechanism at the initial stages as well as the later stages of the fungal growth when the nutrients are exhausted, and the growth phase of organisms is expected to be declined. The corrosion pattern observed in this study confirmed the formation of aluminum compounds at the later stages and stated the significant role of fungal attachment in the initiation of corrosion. Oxalic acid is the main metabolite produced by *Aspergillus niger,* which is reported to be responsible for the corrosion of aluminum, and it enhanced the corrosion rate by four times as compared to the control conditions and the biofilm formation lead to micro pitting of the metal surface(Dai et al., 2016). Similar pattern has been observed through *Acremonium kiliense* in aluminum corrosion(Imo and Chidiebere, 2019).More recently fungal species have been investigated in B20 Biodiesel Storage Tanks and the pitting corrosion has been reported under the fungal biofilm (Stamps et al., 2020). Filamentous fungal species of *Trichocomaceae* is not a well-known contaminant of hydrocarbon fuels, but it has been detected in storage tanks and is also a potent corroding agent as observed in hydrocarbon fuel tank. Collectively, many fungal strains can form biofilm on the metal surfaces which induces the corroding environment and also deteriorates the hydrocarbons in storage and transport pipelines in various industries. Still, there are many fungal strains that are potent corroding agents but are understudied and need to be investigated for a better preventive measure to protect hydrocarbon biofouling and corrosion of metal surfaces.

19.3.6 Algae

Algae are oxygen-producing photosynthetic organisms and form a thick and dense biofilm over the surfaces in presence of adequate light and humid conditions. Marine environments exhibit the best-suited conditions for the development of algae. The initial stages of development of algal biofilm do not promote corrosion but suffice aerobic bacterial growth and initiate corrosion mechanism at the later stages by creating an anoxic environment under the dense layers of the biofilm formed on metal or concrete surfaces. This formation of differential zones of oxygen inculcates various aspects of corrosion through the production of organic acids, and acceleration of growth of aerobic and anaerobic bacteria involved in the corrosion process. Involvement of algae in MIC has been proposed in the 1980s, but the majority of studies then focused on the role of microalgae in biofouling and the underlying mechanism was not clear (Iverson, 1987; Terry and Edyvean, 1986). Later, studies emphasized the probable role of algae in microbial corrosion of metal surfaces in seawater and other marine structure, and it was observed that dense algal biofilms create cathodes and anodes and initiate localized corrosion (George et al., 2000; Jahan et al., 2004).

Javaherdashti et al. (2009) reviewed the mechanism of corrosion of reinforced concrete structures due to algal biofilm formation (Javaherdashti et al., 2009). It has been proposed that there can be four distinct perspectives to understand the underlying mechanisms of algal corrosion. This includes the uptake of metals by algal biofilms, creating a favorable environment for the growth of other microorganisms such as sulfur-oxidizing bacteria, development of differential oxygen zones, and production of corrosive metabolites such as oxalic acid. Other studies have also confirmed the role of algae in concrete deterioration and corrosion (Olivia et al., 2012). As it is well known that algae and bacteria coexist and influence the metabolic activities of one another, so it is not possible to distinguish between the contribution of both algae and bacteria in MIC in such a scenario. Thus, the challenge is to identify the real culprit so that early intervention or preventive measures can be designed based on the causative agent of MIC, whereas this symbiotic effect can also provide resistance toward microgalvanic corrosion by maintaining an oxygen barrier inside and outside the biofilm (Dong et al., 2022) as observed by Dong et al. that the addition of *Phaeodactylum tricornutum* to the culture medium of *Bacillus altitudinis* inhibited the pitting of 304 stainless steel.

More recently, the role of *Shewanella* algae in marine biofilms has been investigated to understand its role on the MIC of 316L steel surface (Kalnaowakul et al., 2020). The study concluded that the deterioration of metal surface through pitting corrosion was more severe (pit depth: 9.8 μm) when compared to the pits (pit depth: 1.4 μm) formed in the abiotic medium conditions which clarify the role of algal biofilm in accelerating the corrosion of steel surfaces. Thus, algal biofilms play a dual role in MIC and the synergistic effect of the biofilm consortium can be damaging or protective provided the favorable environment so that microorganisms can thrive and survive.

19.4 MIC of titanium: A metal of primary choice in industrial equipment and implants

Titanium displays key characteristics like resistance to corrosion, high strength, and high-temperature resistance, which makes it a popular choice of material in oil and gas, power plants, and biomedical industries. On contrary, the cost associated with titanium is very high and the damage due to biofilm accumulation on the titanium-based structure such as heat exchangers and dental implants can levy both the financial constraint and health concerns, respectively. Studies have shown the formation of titanium sulfide on titanium surfaces in the presence of SRBs (*Desulfovibrio vulgaris*), which further promotes pitting corrosion (Costa et al., 2021; Rao et al., 2005). Furthermore, titanium is frequently used in dental implants and the risk associated is the induction of infection by pathogenic microorganisms, viz. *Streptococcus mutans*, *Porphyromonas gingivalis,* and *Prevotela intermedia.* These microbial strains are capable of releasing corrosive metabolites such as lactic acid and the corrosion study of *S. mutans* confirmed the correlation of MIC with titanium due to the formation of biofilm on the implant in the presence of saliva(Souza et al., 2013). The aerobic marine bacterium *Pseudomonas aeruginosa,* which stimulates an acidic environment and is known for its role in MIC penetrates the passive biofilm layer and induces corrosion on the titanium surface (Saleem Khan et al., 2019). Although there are reports to establish the impact MIC causing organisms on titanium still, it offers better and long-term corrosion resistance as compared to other engineered materials.

19.5 MIC Diagnostics standards and protocols: An entrepreneurial perspective

In the previous sections, the competence of biofilm that is a consortium of microorganisms in inducing and enhancing corrosion has been discussed, which emphasizes the magnitude of economic loss due to MIC in industrial equipment. The information mentioned in the literature and the knowledge about the characteristics of biofilm, the pattern of corrosion initiation, and progression towards a severe stage can be crucial in the apt and appropriate detection and diagnosis of MIC. Visual inspections, chemical testing, corrosion monitoring, and evaluation through state-of-the-art technologies are the major tools to regulate and control corrosion in industries. Despite ample information on the subject and the newest technologies and testing methods offered by R&D, consultancies, and firms, it is not feasible to mark a clear distinction between chemical corrosion and microbial corrosion. Moreover, there are multiple challenges such as preliminary tests to confirm the presence of organisms, and since microorganisms are ubiquitous, thus confirmatory tests are required to identify the MIC causing organisms to establish their role in corrosion and further necessary action can be taken to mitigate MIC. The key characteristics of microbial corrosion are pitting, crevice, under deposits, galvanic corrosion, and dealloying, but these features of corrosion morphology can be overlapped, and there can be other factors responsible for the corrosion process. Therefore, the precise identification of MIC is one of the major challenges, which leads to underreporting or underestimation of MIC in various industries.

The National Association of Corrosion Engineers (NACE) has developed and published a standard method (TM0212-2018-SG) for the detection of MIC applicable to internal surfaces of pipelines which includes sampling methods, and testing for the presence of MIC. According to the TM0212-2012 standard, there are three prerequisites for the confirmation of internal corrosion, assuming the exclusion of any other source of contamination there should be a significant difference in the microbial profile of the internal and external surface, presence of microbial metabolites such as sulfur deposits and organic acids and the conditions on site of corrosion should correspond to the conditions required for the growth of organism isolated and identified(Larsen, 2020). Furthermore, the American society for testing and materials (ASTM) has also provided standardized methods and techniques, e.g., ASTM C1894-22 and ASTM D4412-84 for the detection and testing of MIC in industries. Although these standards and tests are beneficial and informative in the detection of MIC, most of these guidelines focus on the microbiological aspect and are not able to provide a better insight into the engineering and metallurgical specifics of MIC. Therefore, there is a need for a guiding principle that specifies microbiological, metallurgical, chemical, and microbe-metal interactions in MIC to attain a better approach for early and efficient detection of the same. In the following sections, we have discussed the current scenario and prospects of various approaches available in the market for the detection of MIC.

19.5.1 On-site diagnosis

Corrosion monitoring is crucial to early and on-site diagnosis to confirm the role of microbes in the initiation and progression of corrosion. Early signs and the characteristics features as suggested by NACE and ASTM standards discussed in previous sections can be an initial trigger for intervention so that the problem can be handled before the alarming situation. As the preliminary screening of the corrosion site confirms the characteristics of MIC, qualitative and quantitative screening of microbes can be done using preferred methods and techniques. However, to carry out the precise diagnosis of microbes sampling should be done carefully. The sampling procedure includes details of operating conditions, date, timing, biotic and abiotic factors, and microbiological characteristics to obtain a trend in the development of the process of corrosion. Additionally, the severity of corrosion should be noted based on the morphological features and both planktonic and sessile organisms should be considered for collection and evaluation in the laboratory settings. Collection of liquid samples such as hydrocarbons, water, and sludge from the oil and gas industry, power plants, and wastewater treatment plants should be done in sterile containers. Also, the composition of the collected sample is certainly influenced by the location of the sampling, and physical factors such as temperature, humidity, and contaminants. For efficient sampling and the true simulation of the source of sampling in the laboratory, sampling should be done with proper guidelines.

19.5.2 MIC detection

Samples collected from the site of corrosion can be screened for several chemical, microbiological, molecular, microscopic, and metallurgical tests and the accurate identification of the organisms and their impact on the metal and nonmetal surfaces can be done in the laboratories (Fig. 19.1). However, the corrosion potential of a specific microorganism can be evaluated in a simulated manner, but it is not the true representation of the on-site industrial settings(Canales et al., 2021). In an industrial field set-up, MIC is a complex process, and no critical point or threshold value has been assigned for the number of microorganisms present at the site of corrosion whether internal or external surfaces. The direct correlation between the number of organisms (planktonic or sessile) detected on the site of corrosion and metallurgical damage of the metal surface cannot be established without taking account of other corrosion kinetics such as the synergistic effect of consortia of microorganisms, favorable chemical conditions such as under deposits, biogenic acids production and preexisting damage of the surfaces due to any other physical or chemical factors.

19.5.2.1 Microbiological and biochemical methods

Enzyme catalysis persuades anodic/cathodic shifts process in the mechanism of microbial corrosion but the direct impact of enzymes in MIC is understudied. However, an enzymatic profile can be a key indicator of the presence of metabolically active microbes at the site of corrosion (Landoulsi et al., 2008). Biochemical methods are majorly performed to check the metabolic profile of microorganisms

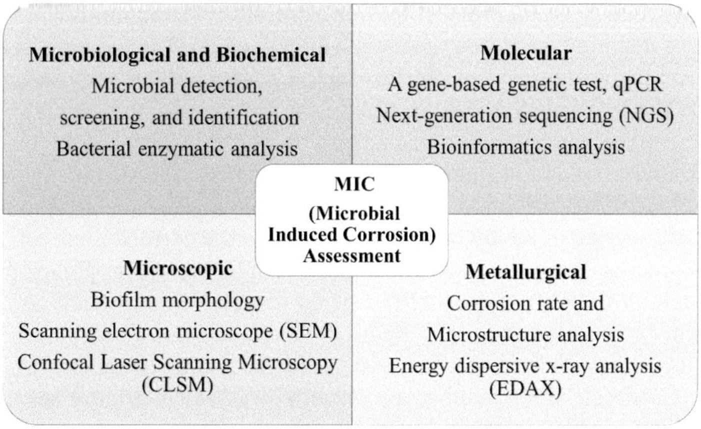

FIG. 19.1

Assessment of microbial-induced corrosion (MIC) through multidisciplinary approaches.

through the detection of key enzymes such as hydrogenase which is produced by bacteria and utilizes hydrogen as an energy source, and adenosine phosphosulfate (APS) reductase which is reportedly associated with SRB. Also, peroxidases and catalases play a significant role in the aerobic corrosion of steel and concrete surfaces. The biochemical tests are cost-effective, easily accessible, and easy to perform with the kits available in the market. The results of biochemical tests are reliable and promising and thus offer a good and simple detection method to check for the presence of microbes at the site of corrosion.

The consortium of microorganisms present in the biofilm is diverse and thus imposes the challenge to perform an accurate identification of the causative agent for MIC. Microbiological methods, viz., most probable number (MPN) method and other culture-dependent methods, are the primary choice of testing methods due to their potential to identify a broad spectrum of microbes with simple techniques and the standard methodology using synthetic culture media to assess microbes responsible for MIC has been developed and published by ASTM and NACE. These standards and methodologies offer a common platform for testing and ruling out the possibility of differential diagnosis by various methods practiced for MIC detection. Although microbiological methods are a popular choice for MIC detection, it is a semi-quantitative method and since there is no critical point or cut-off set for the number of microbes to be detected on the site of corrosion, it cannot be a true indicator or the final verdict for the diagnosis of MIC. One of the studies has reported that a range between 1 and 10,000 bacteria per milliliter or cm^2 can be the typical number of bacteria associated with MIC in the oil and gas industry (Al-Shamari et al., 2013), but this number cannot be assigned and considered in all the situation of MIC and might vary from case to case due to the factors like presence or absence of oxygen, temperature, pH, and chemical contaminants. There are a few drawbacks to this method since there can be cross-contamination and one must perform separate assessments for different organisms which are tedious and time-consuming. So, there are fewer methodologies but since they are validated and directed by international standards, they most frequently used methods for the detection of MIC.

19.5.2.2 Microscopic and metallurgical methods

Applications of microscopy in microbial corrosion are truly remarkable, it can be utilized to confirm the presence of microorganisms on the metal surface, enumerate bacterial species irrespective of their species, and the progression of biofilm can be observed. Direct observation can be made about the amorphous and crystalline pattern of biofilm and the initial stages of biofilm and its interaction with the surface can be studied using microscopy techniques. Imaging systems such as epifluorescence microscopy and scanning electron microscopy (SEM) are the most widely used techniques to study microbial corrosion. SEM can be used to analyze the attachment of biofilm and the magnitude of destruction of the metallic surfaces due to the corrosion process. Another significant application of microscopy techniques is the analysis of the detection of pits, the morphology of pitting, and the depth of the pit(Larsen, 2020). Energy dispersive x-ray analysis (EDAX)is a component of SEM that can be applied to investigate the elemental composition of deposits collected from the site of corrosion, which further helps in the identification of the probable microorganism involved in MIC. Confocal laser scanning microscopy (CLSM) is a more sensitive technique to investigate the detailed morphological features of biofilm via the generation of 3D structure of biofilm. Imaging techniques are widely used due to their specificity and noninvasiveness procedures.

Moreover, a more compact, and table-top version of SEM is introduced which might be easier to set up in fields for early diagnosis of MIC. An approach to quantification assessment of MIC through biofilm attachment and estimation of bacterial clusters has been proposed by a group of researchers in Italy and the results suggested better identification of organisms as compared to other techniques (Iannucci et al., 2019).

19.6 Molecular biology methods

Molecular methods such as polymerase chain reaction (PCR), PCR variants, 16S sequencing, and nucleotide sequencing technology are the gold standard for the detection and analysis of biological markers. These methods provide better efficiency, apt and early diagnosis, and identification of microbes isolated from the site of microbial corrosion. These techniques also facilitate the prescreening of microbes in the zones which are more prone to MIC. Furthermore, with the advancement of molecular biology techniques, the diagnosis of MIC and identification of microbes associated with MIC has been improvised which further enhances the selection and efficiency of treatment strategies. In the recent past, these techniques have been utilized for the grouping of microbes isolated from a MIC such as sulfate reducers and methanogen. Additionally, omics technologies have helped in the identification and establishment of a particular type of microbiome present in a specific MIC zone. This information provides the basis for a customized treatment strategy and appropriate mitigation of MIC. Although conventional microbiological methods are imperative to screen a wide spectrum of microorganisms, there might be an overlap in the detection of biofilm-forming bacteria, which may or may not promote MIC. To overcome such issues molecular methods can be applied which are well-designed and defined. Recently, it has been observed that molecular methods such as most probable number-PCR(MPN-PCR), qPCR(functional gene-based quantitative PCR), and next-generation sequencing (NGS) can be implemented for the assessment of viable, and nonviable organisms and identification of the desired species which is documented to be associated with MIC (Hutcherson and Reed, 2020; Martin-Sanchez et al., 2016). Utilization of molecular methods is not restricted to screening and identification of microbes; it also facilitates understanding of the interaction of microbial communities and the differentiation between the impact of microbes that are present in the inner surface of the biofilm as compared to the superficial outer surface.

MIC in industries such as oil refineries, the gas sector, and power plants is well documented and is the biggest concern for the infrastructure and its components. This scenario has emphasized the increased demand for MIC testing, which has led to various significant initiatives in this direction. Many research groups, organizations, and multidisciplinary analytical labs have established their microbial corrosion testing setups and started outsourcing services for microbial detection, on-site diagnosis, and failure analysis due to microbial corrosion (Table 19.1). Although other multidisciplinary laboratories are also providing the services such as microbiological tests, chemical analysis of deposits, and microscopic and metallurgical analysis of corroded samples still there is a dearth of services that can provide comprehensive testing of MIC. Conclusively, the field of MIC research and testing has been in high demand in the past few years and offers plenty of opportunities in the context of entrepreneurial development.

Table 19.1 Global distribution of services and products for MIC testing.

MIC testing	Service provider
Microbial-induced corrosion testing	Situ Biosciences, USA
Microbial-induced corrosion testing	Estech Lab, USA
DNA-based monitoring MIC	LuminUltra, Canada
BAC-PAK bacterium test kit for MIC causing bacterium	Potter Electric Signal Company, LLC, USA
Hydrogenase test kit for MIC	CAPRROCO, internal corrosion monitoring specialists, Canada
Consultancy services for corrosion management	PENSPEN, UK
Group test of MIC-associated bacteria: SRB, APB, HAB, SLYME, IRB	EMSL Analytical, Inc., USA and Canada
Genetic MIC test services	GTI Labs, USA
MIC threat assessment	DNV lab Ohio, USA
Failure investigations, MIC risk assessment, consultancy advice, and peer-review of technical reports.	Curtin Corrosion Center, Curtin University, Australia
Corrosion testing and metallurgical testing	Lucid Labs, Hyderabad, India
Census (qPCR) MIC	Microbial Insights, USA
MIC testing	Eurofins, Global international network

Acknowledgment

We would like to thank Dr. Anand Varma, Senior Manager, NETRA-NTPC Limited for his valuable suggestions on the metallurgical aspects of MIC in this chapter.

References

Abdullah, A., Yahaya, N., Noor, N.M., Rasol, R.M., 2014. Microbial corrosion of API 5L X-70 carbon steel by ATCC 7757 and consortium of sulfate-reducing bacteria. J. Chem. 2014, 1–7. http://www.hindawi.com/journals/jchem/2014/130345/.

Al-Shamari, A.R., Al-Mithin, A.W., Olabisi, O., Mathew, A., 2013. Developing a metric for microbilogically influenced corrosion (MIC) in oilfield water handling systems. In: CORROSION 2013: NACE-2013-2299.

Anandkumar, B., et al., 2011. Studies on microbiologically influenced corrosion of SS304 by a novel manganese oxidizer, Bacillus flexus. Biofouling 27 (6), 675–683. http://www.tandfonline.com/doi/abs/10.1080/08927014.2011.597001.

Batmanghelich, F., Li, L., Seo, Y., 2017. Influence of multispecies biofilms of pseudomonas aeruginosa and Desulfovibrio vulgaris on the corrosion of cast iron. Corros. Sci. 121, 94–104. https://linkinghub.elsevier.com/retrieve/pii/S0010938X16306771.

Beech, I.B., Campbell, S.A., 2008. Accelerated low water corrosion of carbon steel in the presence of a biofilm harbouring sulphate-reducing and sulphur-oxidising bacteria recovered from a marine sediment. Electrochim. Acta 54 (1), 14–21. https://linkinghub.elsevier.com/retrieve/pii/S0013468608007330.

Belov, D.V., et al., 2008. Corrosion of aluminum and its alloys under the effect of microscopic fungi. Prot. Met. 44 (7), 737–742. http://link.springer.com/10.1134/S0033173208070151.

Canales, C., et al., 2021. Testing the test: a comparative study of marine microbial corrosion under laboratory and field conditions. ACS Omega 6 (20), 13496–13507. https://pubs.acs.org/doi/10.1021/acsomega.1c01762.

Chaudhari, B., Panda, B., Šavija, B., Paul, S.C., 2022. Microbiologically induced concrete corrosion: a concise review of assessment methods, effects, and corrosion-resistant coating materials. Materials 15 (12), 4279. https://www.mdpi.com/1996-1944/15/12/4279.

Chen, L., Wei, B., Xianghong, X., 2021. Effect of sulfate-reducing bacteria (SRB) on the corrosion of buried pipe steel in acidic soil solution. Coatings 11 (6), 625. https://www.mdpi.com/2079-6412/11/6/625.

Costa, R.C., et al., 2021. Microbial corrosion in titanium-based dental implants: how tiny bacteria can create a big problem? J. Bio- Tribo-Corrosion 7 (4), 136. https://link.springer.com/10.1007/s40735-021-00575-8.

Cote, C., Rosas, O., Basseguy, R., 2015. Geobacter sulfurreducens: an iron reducing bacterium that can protect carbon steel against corrosion? Corros. Sci. 94, 104–113. https://linkinghub.elsevier.com/retrieve/pii/S0010938X15000554.

Dai, X., et al., 2016. Corrosion of aluminum alloy 2024 caused by Aspergillus niger. Int. Biodeterior. Biodegradation 115, 1–10. https://linkinghub.elsevier.com/retrieve/pii/S0964830516302323.

Dec, W., Jaworska-Kik, M., Simka, W., Michalska, J., 2018. Corrosion behaviour of 2205 duplex stainless steel in pure cultures of sulphate reducing bacteria: SEM studies, electrochemical characterisation and biochemical analyses. Mater. Corros. 69 (1), 53–62. https://onlinelibrary.wiley.com/doi/10.1002/maco.201709649.

Dec, W., et al., 2016. The effect of sulphate-reducing bacteria biofilm on passivity and development of pitting on 2205 duplex stainless steel. Electrochim. Acta 212, 225–236. https://linkinghub.elsevier.com/retrieve/pii/S0013468616315481.

Dong, Y., Song, G.-L., Zheng, D., 2022. Naturally effective inhibition of microbial corrosion by bacterium-alga symbiosis on 304 stainless steel. J. Clean. Prod. 356, 131823. https://linkinghub.elsevier.com/retrieve/pii/S0959652622014330.

Dou, W., et al., 2020. Corrosion of Cu by a sulfate reducing bacterium in anaerobic vials with different headspace volumes. Bioelectrochemistry 133, 107478. https://linkinghub.elsevier.com/retrieve/pii/S1567539419306929.

El-Bassi, L., et al., 2020. Investigations on biofilm forming bacteria involved in biocorrosion of carbon steel immerged in real wastewaters. Int. Biodeterior. Biodegradation 150, 104960. https://linkinghub.elsevier.com/retrieve/pii/S0964830520301074.

Esnault, L., et al., 2011. Metallic corrosion processes reactivation sustained by iron-reducing bacteria: implication on long-term stability of protective layers. Phys. Chem. Earth, Parts A/B/C 36 (17–18), 1624–1629. https://linkinghub.elsevier.com/retrieve/pii/S1474706511002956.

George, R.P., et al., 2000. Microbiologically influenced corrosion of AISI Type 304 stainless steels under fresh water biofilms. Mater. Corros. 51 (4), 213–218. https://onlinelibrary.wiley.com/doi/10.1002/(SICI)1521-4176(200004)51:4%3C213::AID-MACO213%3E3.0.CO;2-J.

Guan, F., et al., 2020. Interaction between sulfate-reducing bacteria and aluminum alloys—corrosion mechanisms of 5052 and Al-Zn-In-Cd aluminum alloys. J. Mater. Sci. Technol. 36, 55–64. https://linkinghub.elsevier.com/retrieve/pii/S1005030219302233.

Herrera, L.K., Videla, H.A., 2009. Role of iron-reducing bacteria in corrosion and protection of carbon steel. Int. Biodeterior. Biodegradation 63 (7), 891–895. https://linkinghub.elsevier.com/retrieve/pii/S0964830509001139.

Huang, G., Chan, K.-Y., Fang, H.H.P., 2004. Microbiologically induced corrosion of 70Cu-30Ni alloy in anaerobic seawater. J. Electrochem. Soc. 151 (7), B434. https://iopscience.iop.org/article/10.1149/1.1756153.

Huber, B., et al., 2016. Characterization of sulfur oxidizing bacteria related to biogenic sulfuric acid corrosion in sludge digesters. BMC Microbiol. 16 (1), 153. http://bmcmicrobiol.biomedcentral.com/articles/10.1186/s12866-016-0767-7.

Hutcherson, J., Reed, M., 2020. On Site QPCR Application for Rapid Microbial Detection and Quantification. NACE Corrosion.

Iannucci, L., et al., 2019. An imaging system for microbial corrosion analysis. In: 2019 IEEE International Instrumentation and Measurement Technology Conference (I2MTC), IEEE, pp. 1–6. https://ieeexplore.ieee.org/document/8826965/.

Imo, E., Chidiebere, A., 2019. fungal influenced corrosion of aluminium in the presence of Acremonium kiliense. Int. J. Appl. Microbiol. Biotechnol. Res., 1–6.

Iverson, W.P., 1987. Microbial Corrosion of Metals. pp. 1–36. https://linkinghub.elsevier.com/retrieve/pii/S0065216408700777.

Jahan, K., et al., 2004. Metal Uptake by Algae. pp. 223–232.

Javaherdashti, R., et al., 2009. On the impact of algae on accelerating the biodeterioration/biocorrosion of reinforced concrete: a mechanistic review. Eur. J. Sci. Res. 36, 394–406.

Kalnaowakul, P., Dake, X., Rodchanarowan, A., 2020. Accelerated corrosion of 316L stainless steel caused by Shewanella algae biofilms. ACS Appl. Bio Mater. 3 (4), 2185–2192. https://pubs.acs.org/doi/10.1021/acsabm.0c00037.

Landoulsi, J., et al., 2008. Enzymatic approach in microbial-influenced corrosion: a review based on stainless steels in natural waters. Environ. Sci. Technol. 42 (7), 2233–2242. https://pubs.acs.org/doi/10.1021/es071830g.

Larsen, K.R., 2020. Diagnosing microbiologically influenced corrosion in a pipeline. In: Materials Performance.

Liu, F., et al., 2014. The corrosion of two aluminium sacrificial anode alloys in SRB-containing sea mud. Corros. Sci. 83, 375–381. https://linkinghub.elsevier.com/retrieve/pii/S0010938X14001073.

Liu, H., et al., 2019. Microbiologically influenced corrosion of carbon steel beneath a deposit in CO2-saturated formation water containing Desulfotomaculum nigrificans. Front. Microbiol. 10. https://www.frontiersin.org/article/10.3389/fmicb.2019.01298/full.

Martin-Sanchez, P.M., Gorbushina, A.A., Kunte, H.-J., Toepel, J., 2016. A novel QPCR protocol for the specific detection and quantification of the fuel-deteriorating fungus Hormoconis resinae. Biofouling 32 (6), 635–644.

Mattes, T.E., et al., 2013. Sulfur oxidizers dominate carbon fixation at a biogeochemical hot spot in the dark ocean. ISME J. 7 (12), 2349–2360. http://www.nature.com/articles/ismej2013113.

Mohanan, S., et al., 2005. The role of fungi on diesel degradation, and their influence on corrosion of API 5LX steel. Corrosion Prevent. Control 52, 123–130.

Olivia, M., et al., 2012. The influence of micro algae on corrosion of steel in fly ash geopolymer concrete: a preliminary study. Adv. Mater. Res. 626, 861–866. https://www.scientific.net/AMR.626.861.

Prithiraj, A., Otunniyi, I.O., Osifo, P., van der Merwe, J., 2019. Corrosion behaviour of stainless and carbon steels exposed to sulphate—reducing bacteria from industrial heat exchangers. Eng. Fail. Anal. 104, 977–986. https://linkinghub.elsevier.com/retrieve/pii/S1350630718310756.

Rajasekar, A., et al., 2007. Biodegradation and corrosion behavior of manganese oxidizer Bacillus cereus ACE4 in diesel transporting pipeline. Corros. Sci. 49 (6), 2694–2710. https://linkinghub.elsevier.com/retrieve/pii/S0010938X06004136.

Rao, T.S., et al., 2005. Pitting corrosion of titanium by a freshwater strain of sulphate reducing bacteria (Desulfovibrio vulgaris). Corros. Sci. 47 (5), 1071–1084. https://linkinghub.elsevier.com/retrieve/pii/S0010938X04002161.

Sachan, R., Singh, A.K., 2020. Comparison of microbial influenced corrosion in presence of iron oxidizing bacteria (strains DASEWM1 and DASEWM2). Constr. Build. Mater. 256, 119438. https://linkinghub.elsevier.com/retrieve/pii/S0950061820314434.

Saleem Khan, M., et al., 2019. Microbiologically influenced corrosion of titanium caused by aerobic marine bacterium pseudomonas aeruginosa. J. Mater. Sci. Technol. 35 (1), 216–222. https://linkinghub.elsevier.com/retrieve/pii/S1005030218302093.

Siegel, S.M., Siegel, B.Z., Clark, K.E., 1983. Bio-corrosion: solubilization and accumulation of metals by fungi. Water Air Soil Pollut. 19 (3), 229–236. http://link.springer.com/10.1007/BF00599050.

Souza, J.C.M., et al., 2013. Corrosion behaviour of titanium in the presence of streptococcus mutans. J. Dent. 41 (6), 528–534. https://linkinghub.elsevier.com/retrieve/pii/S0300571213000882.

Stamps, B.W., et al., 2020. In situ linkage of fungal and bacterial proliferation to microbiologically influenced corrosion in B20 biodiesel storage tanks. Front. Microbiol. 11. https://www.frontiersin.org/article/10.3389/fmicb.2020.00167/full.

Starosvetsky, J., et al., 2008. Electrochemical behaviour of stainless steels in media containing iron-oxidizing bacteria (IOB) by corrosion process modeling. Corros. Sci. 50 (2), 540–547. https://linkinghub.elsevier.com/retrieve/pii/S0010938X07002296.

Terry, L.A., Edyvean, R.G.J., 1986. Chapter 15. Recent investigations into the effects of algae on corrosion. In: Studies in Environmental Science, pp. 211–229. https://linkinghub.elsevier.com/retrieve/pii/S0166111608721812.

Tiller, A.K., Booth, G.H., 1968. Anaerobic corrosion of aluminium by sulphate-reducing bacteria. Corros. Sci. 8 (7), 549–555. https://linkinghub.elsevier.com/retrieve/pii/S0010938X68800095.

Tran, T.T.T., Kannoorpatti, K., Padovan, A., Thennadil, S., 2021. Effect of PH regulation by sulfate-reducing bacteria on corrosion behaviour of duplex stainless steel 2205 in acidic artificial seawater. R. Soc. Open Sci. 8 (1), 200639. https://royalsocietypublishing.org/doi/10.1098/rsos.200639.

Wei, S., et al., 2013. Microbiologically induced deterioration of concrete: a review. Braz. J. Microbiol. 44 (4), 1001–1007. http://www.scielo.br/scielo.php?script=sci_arttext&pid=S1517-83822013000400001&lng=en&nrm=iso&tlng=en.

Welikala, S., et al., 2022. Biofilm development on carbon steel by iron reducing bacterium Shewanella putrefaciens and their role in corrosion. Metals 12 (6), 1005. https://www.mdpi.com/2075-4701/12/6/1005.

Wells, T., Melchers, R.E., 2015. Modelling concrete deterioration in sewers using theory and field observations. Cem. Concr. Res. 77, 82–96. https://linkinghub.elsevier.com/retrieve/pii/S000888461500191X.

Xu, D., Tingyue, G., 2014. Carbon source starvation triggered more aggressive corrosion against carbon steel by the Desulfovibrio vulgaris biofilm. Int. Biodeterior. Biodegradation 91, 74–81. https://linkinghub.elsevier.com/retrieve/pii/S0964830514000730.

Yang, J., Wang, Z.B., Qiao, Y.X., Zheng, Y.G., 2022. Synergistic effects of deposits and sulfate reducing bacteria on the corrosion of carbon steel. Corros. Sci. 199, 110210. https://linkinghub.elsevier.com/retrieve/pii/S0010938X22001287.

Yoffe, P., 1997. 3161 Stainless steel tubes corrosion influenced by SRB in sea water. In: Corrosion97: NACE-97219.

CHAPTER

20 Recombinant fungal pectinase and their role towards fostering modern agriculture

Subhadeep Mondal[a], Suman Kumar Halder[b], and Keshab Chandra Mondal[b]

[a]*Centre for Life Sciences, Vidyasagar University, Midnapore, West Bengal, India,* [b]*Department of Microbiology, Vidyasagar University, Midnapore, West Bengal, India*

20.1 Introduction

Pectin, a hetero polysaccharide, constitutes one of the major structural units of the middle lamella and primary cell wall of plants, where it is interwoven with the other structural polymers of the plant: celluloses and hemicelluloses. As the significant constituents of the middle lamella of plant cell walls, pectin serves as cementing and lubricating agents and occupies one-third of the primary cell wall's dry weight. Pectin substances are responsible for providing the characteristic texture of fruits and vegetables during growth and maturation. Additionally, they are responsible for several cellular functions of plants like growth and development, defence, seed hydration, ionic bonding, pH balance, lignification, accumulation of protease inhibitors, cellular expansion, abscission of leaf, and fruit development (Lara-Marquez et al., 2011). The α-1,4-D-galacturonan partially esterified with a methyl group is present in the main chain of pectin. De-methylated pectin is designated as polygalacturonic acid or pectic acid (pectate). Pectin was first isolated and roughly characterized by Henri Braconnot in 1825 (Braconnot, 1825).

The pectin's degradation or depolymerization is mainly based on the diverse group of pectinolytic enzymes broadly classified as hydrolase class, lyase class, de-esterifying class, and rhamnogalacturonan-degrading classes (Samanta, 2021). Microbial pectinases account for 25% of the global food and industrial enzymes (Murad and Azzaz, 2011). The commercial application of pectinases was started in the 1930s for the manufacture of wines and fruit juices. Pectinases are showing accelerating improvement in its market value (Kavuthodi and Sebastian, 2018; Garg et al., 2016), maintaining the average annual growth rate of 2.86% from 27.6 million $ in 2013 to 30.0 million $ in 2016, and estimated to boost up to 35.5 million $ by 2021 (Global Pectinase Market Research Report, 2017).

The industrial-level production of pectinases was achieved through solid-state fermentation (SSF) or submerged fermentation (SmF). The production cost, recovery of the final product with better quality carried out through SSF compared to the SmF. The environmental conditions of SSF support the fungal growth (Amin et al., 2019). The production of fungal pectinase through recombinant DNA

Entrepreneurship with Microorganisms. https://doi.org/10.1016/B978-0-443-19049-0.00003-7

technologies (RDT) manages industrial sectors' requirements such as superior substrate specificity, enzymatic activity, stability, and tolerance to the stress conditions. In the contemporary era, innovative technologies at the molecular level have been constantly developed, which accelerate pectinase's efficiency produced from the filamentous fungi (Singh et al., 2019). In this present review, we shall provide a brief account of the various pectic substances with their respective degrading pectin lytic enzymes, fungal production of pectinases, different recombinant strategies for fungal pectinase production with their corresponding application in agriculture.

20.2 Pectic substances

Pectin is defined as a heteropolysaccharide (exist in the form of calcium pectate or magnesium pectate) predominantly containing galacturonic acid (GalA) residues, in which variable fractions of the acid groups exist as methoxyl esters, whereas a certain quantity of neutral sugars perhaps exists as side chains. The heteropolysaccharide structure of pectin consists of 17 varieties of monosaccharides associated in more than 20 different linkages, making it the most complex macromolecule in nature (Voragen et al., 2009). The main chain of pectin comprises α-1,4-D-galacturonan, partially esterified with a methyl group. De Vries et al. (1983) identified an arrangement of "smooth" homogalacturonic regions and branched "hairy" regions containing most of the neutral sugars. All the discovered pectins are consisted of the same repetitive units, although the quantity and fine chemical structure of these units differ (Voragen et al., 2009).

20.2.1 Homogalacturonan

Homogalacturonan (HG) constitutes the principal pectin in plant cell walls, considering up to 60% of the total plant pectin. The HG polymer comprises of a backbone ofα-1,4-linked GalA residues, methyl esterified at C-6 and O-acetylated at O-2 and O-3. Methylesterification determines pectin's physical properties; therefore, it attains special attention to the researcher (Jayani et al., 2005).

20.2.2 Xylogalacturonan

Xylogalacturonan (XGA) is a substituted form of HG, containing β-(1→3) linked D-xylose residue as a single unit side chain. A portion of the Gal A residues is methyl-esterified in XGA, and such methyl esters are equally allotted within the substituted and unsubstituted GalA residues (Le Goff et al., 2001).

20.2.3 Rhamnogalacturonan I

The rhamnogalacturonan I (RG I) backbone is consists of [→2)-α-L-Rhap-(1→4)-αD-GalpA-(1→] repetitive units. The neutral sugars, mainly galactose and arabinose, are linked as a side chain by replacing rhamnosyl residues of RG I at the O-4 (Caffall and Mohnen, 2009).

20.2.4 Rhamnogalacturonan II

Rhamnogalacturonan II (RG II) is an individual region within HG. In the plant kingdom, the structure of RGII is tremendously conserved, and its breakdown is dependent on the action of endo polygalacturonase.

Unusual moieties such aceric acid, 3-deoxy-lyxo-2-heptulosaric acid (DHA), and 3-deoxy-manno-2-octulosonic acid (KDO) enrich its group of four different side chains, which are linked to around nine GalA (some of them methyl-esterified) residues long HG fragment. RGII and boron produce a complex borate-diol ester that crosses linked two HG molecules (O'Neill et al., 2001; Ishii and Matsunaga, 2001).

20.2.5 Pectinolytic enzymes

Nature is the rich source of pectinolytic enzymes, synthesized by various organisms such as plants, protozoa, nematodes, insects, bacteria, and fungi. Among them, microbial pectinase gains special attention due to plant-microbe interaction and their plant biomass degradation ability. Different types of pectinolytic enzymes are enlisted in Table 20.1.

20.3 Production of fungal pectinases

The commercialized production of microbial pectinase conducted through submerged fermentation (SmF) or solid-state fermentation (SSF). In the case of SSF, where numerous types of plant biomass are used as a substrate, fungi are more efficient than bacteria for pectinase production due to the low moisture conditions of SSF. The benefits of SSF over SmF are economically feasible fermentative production, negligible risk of contamination, simple recovery of the protein-enriched end-product with intensified enzymatic activities, production of enzymes with intensified enzyme activities (Mondal et al., 2019). The production cost of industrial enzymes depends on the price of the growth medium of up to 30–40. Therefore, a higher level of interest is paid to optimize economic enzyme production conditions (Palaniyappan et al., 2009). However, some conditions such as pH, incubation period, temperature, carbon source, and nitrogen source control pectinase's fermentative production. The medium's pH regulates the growth of the fungal strain used, membrane permeability, and enzymes' stability; typically, acidic pH supports the optimum production of fungal pectic enzymes (Patidar et al., 2018). However, Khatri et al. (2015), and Yadav et al. (2017) reported the alkaline pectinase production from the filamentous fungi (*Aspergillus niger, Fusarium lateritum*). Darah et al. (2015) reported that the highest amount of pectinase production from *Aspergillus niger* achieved at pH 5 utilizing pomelo as SSF substrate; similarly, Banu et al. (2010) found that maximum production of pectinase from *Penicillium chrysogenum* was attained at pH 6.5. The duration of pectinolytic enzyme production depends on the nature of the solid substrate and fungal strain used. Typically, an incubation period of 1 to 7 days is required for different fungi to reach the maximum yield of pectinolytic enzymes. During fermentation, the optimum level of polygalacturonase activity reached the 25th hour using *Aspergillus awamori* mediated SSF of grape pomace (Botella et al., 2005). Wong et al. (2017) documented that a cultivation period of 129 h required for maximum production of pectinase in SSF by *A. fumigatus*. Mondal et al. (2020) reported that pectinase's maximum production was done at 30 °C using *A. fumigatus*. When the medium's temperature reached to 60 °C, above 50% of the initial pectinase activity loosed. For low-cost media design, especially in SSF, researchers usually emphasize the factors like carbon and nitrogen sources and their concentrations. Banu et al. (2010) found that ammonium sulfate has improved the pectinase production from *P. chrysogenum*. Rangarajan et al. (2010) reported that organic nitrogen sources supported higher *endo-* and *exo-*pectinases activities compared to the inorganic nitrogen sources. When they utilized soybean meal (4%) as a nitrogen source, they observed that the

Table 20.1 Comparative analysis of functional parameters of the various types of pectinases (Samanta, 2021; Garg et al., 2016; Voragen et al., 2009; Jayani et al., 2005).

Enzymes	Mode of action	End products
Hydrolase class		
Endopolygalacturonase (EC 3.2.1.15)	Randomly attacks internal portion of the pectic acid 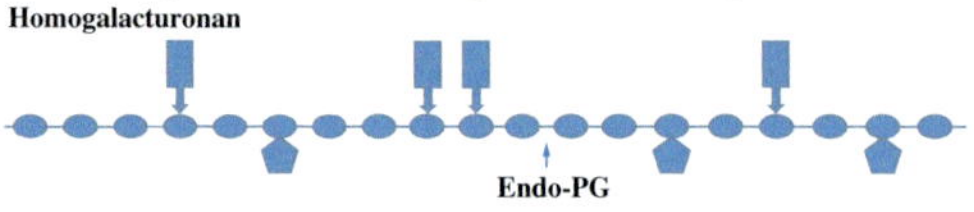	Oligogalacturonates
ExopolyGalacturonase (EC 3.2.1.67)	Attacks terminal site of the nonreducing end of polygalacturonic acid 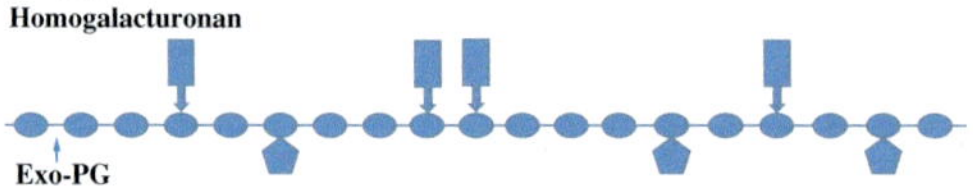	Monogalacturonates
Lyases class		
Pectin lyase (EC 4.2.2.10)	Random attack through trans elimination on pectin 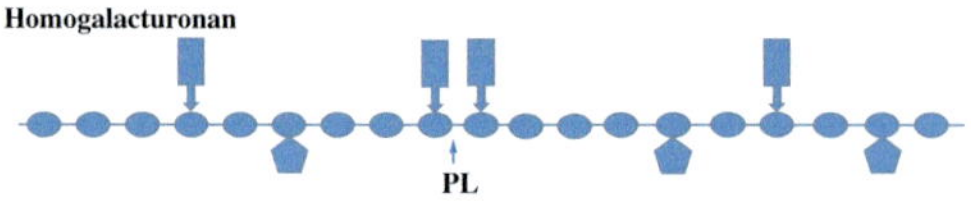	Unsaturated methyl oligo-galacturonates
Pectate lyase (EC 4.2.2.2)	Random attack through trans elimination on pectic acid 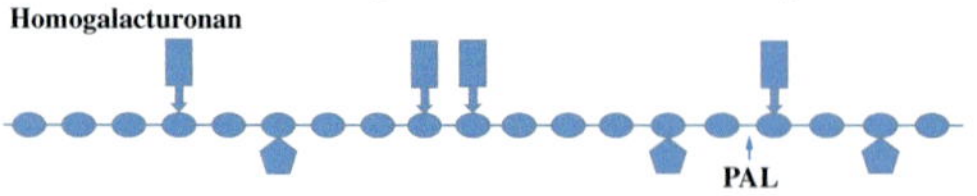	Unsaturated oligo-galacturonates
De-esterifying class		
Pectin methyl esterase (EC 3.1.1.11)	Random attack to break the methyl ester group of galacturonate unit of pectin. 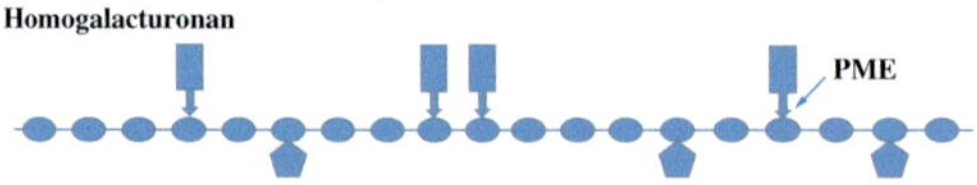	Pectic acid plus methanol
Pectin acetyl esterase (EC 3.1.1.6)	Random attack to break the acetyl ester group of galacturonate unit of pectin. 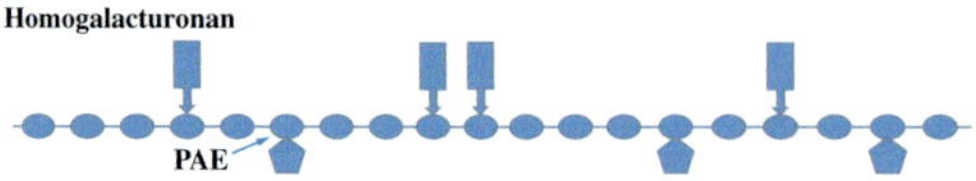	Pectic acid plus ethanol
Act on rhamnogalacturonan		
Rhamnogalacturonan hydrolases (EC 3.2.1.171)	Hydrolysis of rhamnogalacturonan chain 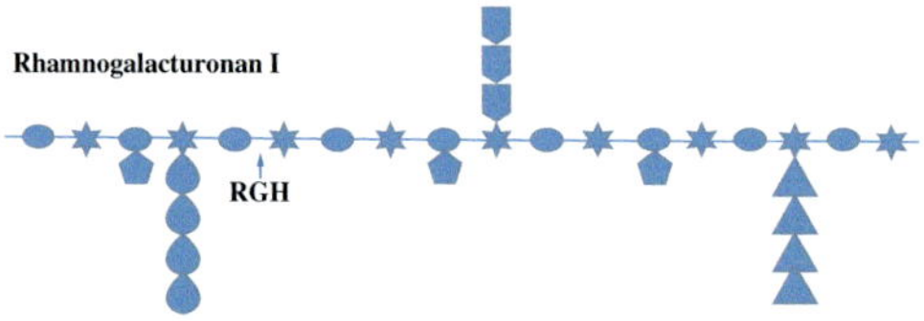	Oligogalacturonates

Table 20.1 Comparative analysis of functional parameters of the various types of pectinases (Samanta, 2021; Garg et al., 2016; Voragen et al., 2009; Jayani et al., 2005)—cont'd

Enzymes	Mode of action	End products
Rhamnogalacturonan lyases (EC 4.2.2.23)	Random trans elimination of rhamnogalacturonan chain 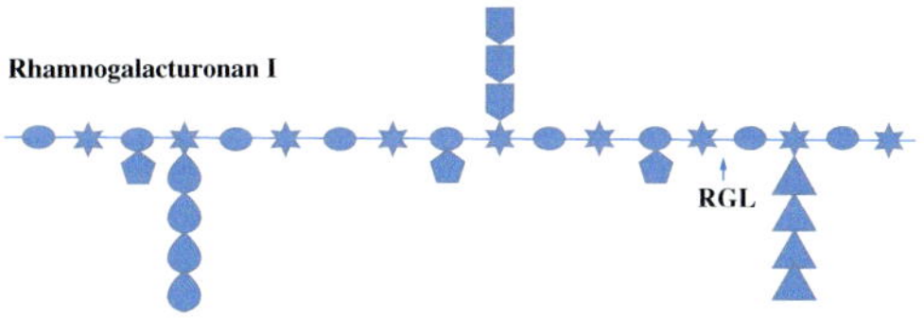	Unsaturated galacturonosyluronic at non-reducing end and rhamnopyranose at reducing end
Rhamnogalacturonan acetylesterase (EC 3.1.1.86)	Hydrolysis of acetyl groups of galacturonic acid in rhamnogalacturonan I 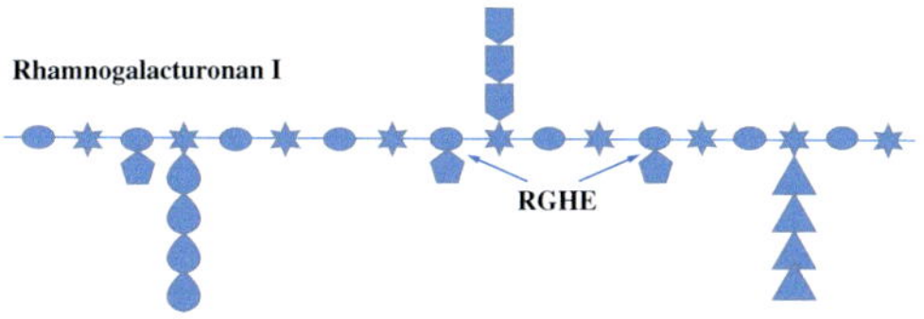	
Xylogalacturonan hydrolase (EC: 3.2.1.-)	Hydrolysis of glycosidic linkages of xylogalacturonan through *endo*-acting mode 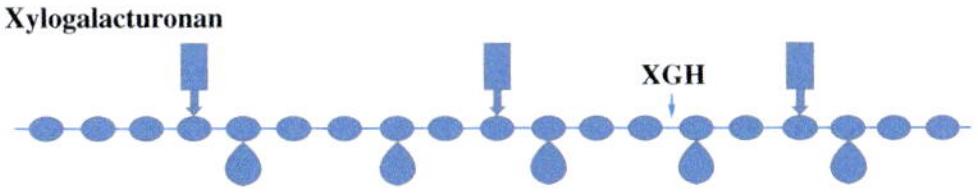	Xylose-galacturonate dimers
Rhamnogalacturonan galacturono hydrolase (EC 3.2.1.173)	Hydrolysis of glycosidic linkage at the non-reducing end of rhamnogalacturonan I 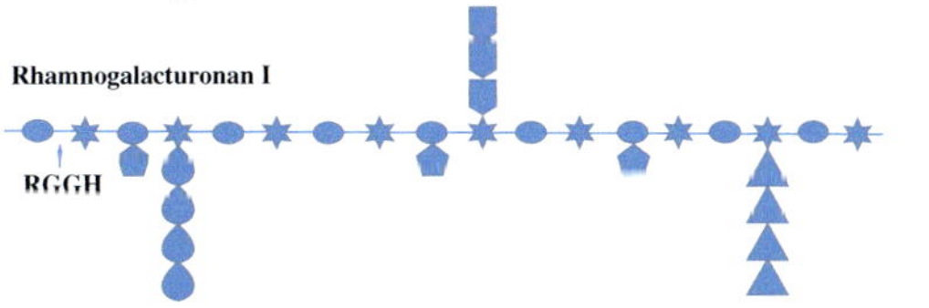	Galacturonic acid
Rhamnogalacturonan rhamnohydrolase (EC 3.2.1.174)	Hydrolysis of glycosidic linkage at the terminal end of rhamnogalacturonan I 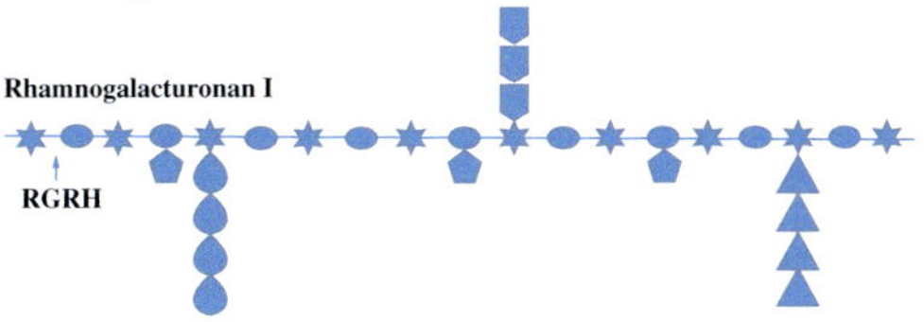	Rhamnose

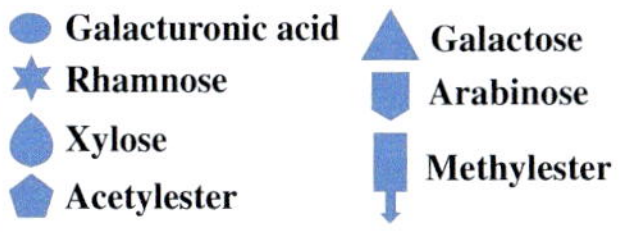

exo-pectinase activity of 5128 IUg^{-1} and endo-pectinase activity of 793 $IU g^{-1}$. Patil and Dayanand (2006) found that a 4–6% increment in the glucose level led to improved pectinase production in the case of SmF, whereas the same result obtained in SSF by the addition of 6–8% sucrose. Ortiz et al. (2017) reported that combination wheat bran mixed with orange peel and lemon peels supported the maximum pectinase production using *Aspergillus giganteus* NRRL10. Some well-known fungal pectinase producers with their mode of production and respective substrate are mentioned in Table 20.2.

20.3.1 Recombinant pectinase

Various strategies of RDT have drawn substantial research attention in the modern biotechnological period for the effectual production of industrially applicable enzymes. Naturally, synthesized fungal pectinases do not have all essential features such as enhanced stability at high pH and temperature, salt tolerance, higher yield, and higher specific rate, which are fundamental requirements for their industrial exploitation.

Mutagenesis is one of the primitive strategies for recombinant strain development to achieve the overproduction of enzymes. An important mutagenesis approach is a site-directed mutagenesis (SDM), based on introducing specific, targeted alterations in double-stranded DNA (Gai et al., 2018). The oldest mutagenesis approach is the controlled treatment with physical or chemical mutagens causes an alteration in the genomic DNA structure of an organism led to the production of mutant strains; some of them showed overexpression of protein secretion (enzymes) (Huang et al., 2019). CRISPR/Cas9 (clustered regularly interspaced short palindromic repeats/CRISPR-associated protein 9) is an innovative genome-editing strategy involving two principal components: a Cas9 endonuclease and a single chimeric guide RNA (sgRNA). The sgRNA is a 20 bp proto spacer sequence, which grabs the site of the target. The developed gRNA escorts Cas9 to the target site, and Cas9 then induces a double-strand

Table 20.2 List of different types of pectinase produced by fungi with their respective production modes and utilized substrates.

Enzymes	Fungi	Mode of fermentation	Substrate	Reference
Pectin methylesterase	*Aspergillus tubingensis*	SSF	Papaya peel	Patidar et al. (2016)
Exo polygalacturonase	*Penicillium viridicatum*	SmF	1.5% ground orange bagasse, 1.5% wheat bran, $(NH_4)_2SO_4$ and $MgSO_4.7H_2O$.	Gomes et al. (2009)
Endo polygalacturonase	*Aspergillus niger*	SSF	Wheat bran, citric pectin	Hendges et al. (2011)
Pectin lyase	*Penicillium expansum*	SSF	Agricultural waste	Atalla et al. (2019)
Pectate lyase	*Penicillium viridicatum*	SSF	Wheat bran and orange bagasse	Ferreira et al. (2010)
Protopectinase	*Aspergillus awamori*	SSF	Wheat bran	Nagai et al. (2000)
Pectinase	*Piriformospora indica*	SmF	Kafer medium containing pectin	Heidarizadeh et al. (2018)

break (DSB) in the target DNA, followed by repairing this DSB by the inbuilt repair system of the host cell. This system permits random insertions and deletions within the target sequence, make it suitable to insert and delete genes (Kun et al., 2020). Promoters take part in avital role in regulating pectinase gene expression. Introducing a strong promoter upstream of the suitable gene permits the preferred level of gene expression (Teixeira et al., 2011). Codon optimization is typically dependent on the phenomenon codon bias, i.e., explained as one codon primarily used among the other codons of an amino acid in a particular organism. Altering the preferable host codon with the infrequent codons led to improved heterologous expression (Karaoğlan and Erden-Karaoğlan, 2020). Multiple gene expression depends on an evolutionary approach "tandem gene duplication," i.e., really mirroring in synthetic biology for boosting protein expression (Ting-Ting et al., 2006). Different types of recombinant approaches for the proficient production of fungal pectinase are summarized in Table 20.3.

However, in the case of pectinolytic enzymes along with the aforementioned RDT strategies, the maximum efforts have been applied for the homologous or heterologous expression of a specific gene of interest, i.e., the gene responsible for synthesizing a particular pectinolytic enzyme.

Table 20.3 Recombinant strategies for proficient fungal pectinase production.

Pectinase	*Promoter engineering* A recombinant strain *of Penicillum griseoroseum* having the *plg1* and *pgg2* gene under the regulation of the *gpd* promoter gene of *Aspergillus nidulans*, resulting in the enhanced activities of pectin lyase (PL) and polygalacturonase (PG), respectively	Teixeira et al. (2011)
	Multiple copy gene expression: A recombinant strain of *Penicillium griseoroseum* having two copy the polygalacturonase gene *pgg2* under the control of *the gpd* promoter of *the Aspergillus nidulans* led to the 1100-fold increment in the enzyme activity	Teixeira et al. (2014)
	Mutagen mediated: Ultraviolet (UV) and nitroso guanidine (NTG)-treated *Aspergillus tubingensis* showed a 161.44% increment in pectinase activity compared to the parental strain Rn14-88A	Huang et al. (2019)
	Site-directed mutagenesis: The polygalacturonase gene from *Saccharomyces cerevisiae* expressed in the methylotrophic yeast *Pichiapastoris* resulting in significant increase of polygalacturonase activity in *P. pastoris*. The SDM of polygalacturonase proteins in *P. pastoris* showed that aspartic acid residues at 179, 200, and 201 and histidine at 222 were important for polygalacturonase activity	Blanco et al. (2002)
	CRISPR-CAS9: CRISPR-CAS9 mediated quick constitutive production of the GaaR transcription factor in *Aspergillus niger*, leading to the enhanced *endo-* and *exo-*polygalacturonases, and pectin lyases production	Kun et al. (2020)
	Codon optimization: Codon-optimized and native endo-polygalacturonase genes from *Aspergillus niger* were transformed into the *Pichiapastoris* X33 strain; the newly produced recombinant strain under the control of ethanol-inducible ADH2 promoter led to the 1.24-fold increment enzyme production	Karaoğlan and Erden-Karaoğlan (2020)

For example, Gainvors et al. (2000) reported that the cloning of open reading frame (ORF) of endo-polygalacturonase gene (PGL1) behind the ADH1 promoter led to the hyperactive expression of endo-polygalacturonase in the yeast *Saccharomyces cerevisiae*. Zheng et al. (2020) reported the production of a recombinant *Pichia pastoris* containing a thermo stable alkaline pectate lyase gene (BspPel) from *Bacillus sp*. RN1 and the pectate lyase synthesized by the recombinant were efficient in degumming ramie fiber. Similarly, a recombinant *Penicillium griseoroseum* 105 overexpress an extracellular pectin lyase (PL) when PL gene is cloned near the strong promoter gpdA from *Aspergillus nidulans* (Cardoso et al., 2010). Jiang et al. (2014) showed the recombinant expression of pectin methylesterase resulting from the cloning of (AFPME) gene from *Aspergillus flavus* in *P. pastoris*. The rha1 [rhamnogalacturonan acetylesterase (RGAE) from *Aspergillus aculeatus*] cDNA, including the N-terminal signal peptide was cloned and overexpressed in *A. oryzae*, which could not have RGHE activity (Kauppinen et al., 1995).

20.4 Pectinase in the agricultural field

From the last century, pectinases are thoroughly utilized to smoothly accelerate several agricultural practices in an eco-friendly way and find a strong opponent to replace the perilous chemicals utilized in agriculture. Pectinases have a wide array of application in the agricultural sector directly (bio scouring of cotton fibers, oil extraction, plant fiber retting, and degumming, coffee, cocoa, and tea fermentation) or indirectly (fruit and juice preparation, jams and jellies preparation, animal feed preparation, dietary fiber production, waste management). Here, we briefly discuss only the direct roles of fungal pectinases in agricultural sectors.

20.5 Bioscouring

Bioscouring involves processing natural textile fibers such as cotton to eliminate waxes and pectins, together with spinning oils and other impurities of the plant cell cuticle. Conventional chemical bleaching with hot NaOH solution led to abolishing the entire cuticle waxy layer of fabric containing an extra burden of alkaline waste (Degani, 2021). Ahlawat et al. (2009) reported that various physical properties such as whiteness, tensile strength, and tearness of fabric were ameliorated upon with pectinase treatment in contrast to the regular alkaline-scoured fabrics. It was noticed that when pectinases in combination with lipases are applied, the time requirement for bioscouring is considerably reduced and significantly improved the nature of dyeing as well as the cotton fibers (Kalantzi et al., 2010). Abdulrachman et al. (2017) constructed a recombinant strain of *Pichia pastoris* carrying the *endo*-polygalacturonase gene from *Aspergillus aculeatus*; the recombinant enzyme was suitable for bioscouring of cotton.

20.6 Oil extraction

Pectinase with the other cell wall-degrading enzymes (CWDE) such as cellulases and hemicellulases were intensively studied for oil extraction from numerous sources such as flaxseed, olives, palm, etc. (Mehanni et al., 2017; Ortiz et al., 2017; Anand et al., 2020). During oil processing, alkaline pectinases allow extraction in the aqueous phase; led to overproduction of oil with improved stability and organoleptic features (polyphenolic and vitamin E) (Kashyap et al., 2001). It has been found that oil production is affected by the enzyme concentration as well as extraction conditions like temperature and pH (Garg et al., 2016). Oumer and Abate in 2017 reported that citrus oil extraction with the help of pectinase eradicates the emulsifying characters of pectin, which influences the oil extraction from the extracts of citrus peel (Oumer and Abate, 2017). During oil extraction, pectinase treatment enriches the oil quality through color intensity, fatty acids content, and value of peroxide, in contrast, to organically processed oil and additionally, provides the refining process relatively inexpensive by the withholding of phospholipid contents in the solid phase (Mehanni et al., 2017)). Olivex produced from *A. aculeatus*, an enzymatic cocktail of pectinase with a reduced cellulase and hemicellulose concentration, led to improving oil production and its stability (Chiacchierini et al., 2007).

20.7 Coffee, tea, and tobacco fermentation

Coffee beans are enclosed with a viscous, mucilaginous (17% by mass) viscous coat, having 84% moisture with 8.9% protein, 4% sugars, 2.8% pectin, and 0.9% ash contents. Microbial pectinases can be utilized to eliminate the mucilage coat from coffee beans. A crude pectinase obtained from *A. niger* CFR quickens the breakdown of the mucilaginous coat of coffee beans of up to 54% and 71% after 60 and 120 min duration of the fermentation process, respectively. A fermentative duration of 3.5 h was required for the complete eradication of pectin (Murthy and Naidu, 2011; Anand et al., 2020).

In the middle lamella of the tealeaves cell wall, enrich with 5–6% pectic substances, which are degraded by pectinases to accelerate tea processing. Therefore, reducing water-soluble pectin content led to the decline in the foaming nature of instant tea powders. Administration of fungal pectinases along with cellulase and xylanase enhance the components of black tea such as thea flavin (TF 24.77%), thea rubigen (TR 21.52%), caffeine (CAF), high polymerized substances (HPS 21.54%), total liquor color (TLC), total soluble solids (TSS 17.49%), and dry matter content (DMC) (Murugesan et al., 2002, Marimuthu, 2000). Regarding this respect, it has been found that crude extract of microbial enzymes such as cellulase, hemicellulase, pectinase, proteinase obtained from *A. oryzae, A. wentii, A. tamari, A. japonicus, A. awamori, and Trichoderma koningii*; inclined the fermentation conditions as well as improved the concentration of TF, TR, HPS, TLC, and TSS of up to 45%, 48%, 33%, 19%, and 3%, respectively (Senthilkumar, 2000). However, enzyme treatment was effective at the working temperature and pH range of 15–50 °C and 2.5–6.0 to accelerate the tea fermentation. Siddiqui et al., 2012 reported that the enzyme concentration should be properly maintained to circumvent the tea leaves' mutilation. Excessive fermentation reduces the quality (color, flavor, and aroma) and glazeness of the tea leaves while raising the tealeaves' spoilage rate.

Pectinase-producing microorganisms are employed in tobacco fermentation. The pectin-degrading fungus *A. tubingensis* GYC 501 was applied to improve the quality of tobacco cut-stem and optimize its fermentation state (Long et al., 2017). During the optimum fermentative conditions of pH 4.8, 0.3 mol/L buffer solution, 50% cut-stem moisture content at 42 °C for 12 h, the pectin degradation rate in tobacco cut-stem reached up to 34%.

20.8 Plant fiber retting and degumming

Pectinases are liable for the retting and exclusion of pectin enrich gumming agents throughout the processing of jute, linen from flax, fibers from hemp (*Cannabis sativa*), ramie, kenaf (*Hibiscus cannabinus*), and coir from coconut husks (Brühlmann et al., 1994). Pectinases treatment of fibers, including jute, flax, and ramie, magnifies the mechanical properties through enhancing tensile strength and brightness. An enzymatic mixture of pectinases along with xylanase was found to be efficient for eco-friendly degumming of fibers (Jayani et al., 2005). The application of enzymes further lowers the expenditure related to the chemicals and energy. Within pectinases, pectate lyase synthesized by actinomycetes was most proficient for detaching bast fiber by extracting gums (Brühlmann et al., 1994). The microbial fermentation of degumming and retting of fiber crops perform in an open alkaline system (pH 10) by numerous microbial genera such as *Cladosporium, Penicillium, Aspergillus,* and *Bacillus* spp. High fermentative pH additionally lowers the probabilities of contamination. For the first time, Yadav et al. (2009) were documented the enzymatic retting through the pectin lyase from *A. terricola* MTCC7588 of natural fibers, viz., *Cannabis sativa and Linum usitatissimum.* Sharma et al. (2011) reported that the pectinase obtained from a novel yeast strain *Pseudozyma* sp. SPJ has been efficiently utilized in the degumming of flax fibers that enhanced fineness and lowers energy consumption compared to the traditional approaches. Molina et al. (2001) constructed a recombinant *Penicillium chrysogenum* through UV mutagenesis, and it synthesizes recombinant pectin lyase that was efficient in degumming textile fiber. Pectin lyase derived from a novel fungal strain *Oidiodendron echinulatum* MTCC 1356 and *A. flavus* MTCC 10938 exhibited proficient retting of sunn hemp fibers (Yadav et al., 2012).

20.9 Conclusion

Fungal pectinases are comprehensively utilized in the agricultural sector and other biotechnological industries because of their production on low-cost substrates. With the modern molecular engineering techniques and continuous searching for novel fungal strains, the drawbacks in the production and potentiality of fungal pectinase should be overcome. Searching should be continued to produce more potent fungal pectinase for the exploitation in the innovative purposes of agriculture. It is also our paramount liability to preserve our mother nature by sustainably using its resources with minimum distortion to deliver a safe environment to our future generations. There is no doubt that pectinolytic enzymes should play an indispensable role in modernizing agriculture to accomplish our future generations' demand.

References

Abdulrachman, D., Thongkred, P., Kocharin, K., Nakpathom, M., Somboon, B., Narumol, N., Chantasingh, D., 2017. Heterologous expression of *Aspergillus aculeatus* endo-polygalacturonase in *Pichia pastoris* by high cell density fermentation and its application in textile scouring. BMC Biotechnol. 17 (1), 1–9.

Ahlawat, S., Dhiman, S.S., Battan, B., Mandhan, R.P., Sharma, J., 2009. Pectinase production by *Bacillus subtilis* and its potential application in biopreparation of cotton and micropoly fabric. Process Biochem. 44 (5), 521–526.

Amin, F., Bhatti, H.N., Bilal, M., 2019. Recent advances in the production strategies of microbial pectinases—a review. Int. J. Biol. Macromol. 122, 1017–1026.

Anand, G., Yadav, S., Gupta, R., Yadav, D., 2020. Pectinases: from microbes to industries. In: Chowdhary, P., Raj, A., Verma, D., Akhter, Y. (Eds.), Microorganisms for Sustainable Environment and Health. Elsevier, p. 287.

Atalla, S.M., Gamal, N.G.E., Awad, H.M., Ali, N.F., 2019. Production of pectin lyase from agricultural wastes by isolated marine *Penicillium expansum* RSW_SEP1 as dye wool fiber. Heliyon 5 (8), e02302.

Banu, A.R., Devi, M.K., Gnanaprabhal, G.R., Pradeep, B.V., Palaniswamy, M., 2010. Production and characterization of pectinase enzyme from *Penicillium chrysogenum*. Indian J. Sci. Technol. 3 (4), 377–381.

Blanco, P., Thow, G., Simpson, C.G., Villa, T.G., Williamson, B., 2002. Mutagenesis of key amino acids alters activity of a *Saccharomyces cerevisiae* endo-polygalacturonase expressed in *Pichia pastoris*. FEMS Microbiol. Lett. 210 (2), 187–191.

Botella, C., De Ory, I., Webb, C., Cantero, D., Blandino, A., 2005. Hydrolytic enzyme production by *Aspergillus awamori* on grape pomace. Biochem. Eng. J. 26 (2–3), 100–106.

Braconnot, H., 1825. Recherches sur un nouvel acide universellement répandu dans tous les végétaux. In: Abago GE (Eds.), Ann. Chim. Phys. 28 (2), 173–178.

Brühlmann, F., Kim, K.S., Zimmerman, W., Fiechter, A., 1994. Pectinolytic enzymes from actinomycetes for the degumming of ramie bast fibers. Appl. Environ. Microbiol. 60 (6), 2107–2112.

Caffall, K.H., Mohnen, D., 2009. The structure, function, and biosynthesis of plant cell wall pectic polysaccharides. Carbohydr. Res. 344 (14), 1879–1900.

Cardoso, P.G., Teixeira, J.A., de Queiroz, M.V., de Araújo, E.F., 2010. Pectin lyase production by recombinant *Penicillium griseoroseum* strain 105. Can. J. Microbiol. 56 (10), 831–837.

Chiacchierini, E., Mele, G., Restuccia, D., Vinci, G., 2007. Impact evaluation of innovative and sustainable extraction technologies on olive oil quality. Trends Food Sci. Technol. 18 (6), 299–305.

Darah, I., Taufiq, M.M.J., Lim, S.H., 2015. Pectinase production by *Aspergillus niger* LFP-1 using pomelo peels as substrate: an optimization study using shallow tray system. Indian J. Biotechnol. 14 (4), 552–558.

De Vries, J.A., Den Uijl, C.H., Voragen, A.G.J., Rombouts, F.M., Pilnik, W., 1983. Structural features of the neutral sugar side chains of apple pectic substances. Carbohydr. Polym. 3 (3), 193–205.

Degani, O., 2021. Synergism between cutinase and pectinase in the hydrolysis of cotton fibers' cuticle. Catalysts 11 (1), 84.

Ferreira, V., Da Silva, R., Silva, D., Gomes, E., 2010. Production of pectate lyase by *Penicillium viridicatum* RFC3 in solid-state and submerged fermentation. Int. J. Microbiol. 2010. https://doi.org/10.1155/2010/276590.

Gai, Y., Chen, J., Zhang, S., Zhu, B., Zhang, D., 2018. Property improvement of α-amylase from *Bacillus stearothermophilus* by deletion of amino acid residues arginine 179 and glycine 180. Food Technol. Biotechnol. 56 (1), 58–64.

Gainvors, A., Nedjaoum, N., Gognies, S., Muzart, M., Nedjma, M., Belarbi, A., 2000. Purification and characterization of acidic endo-polygalacturonase encoded by the PGL1-1 gene from *Saccharomyces cerevisiae*. FEMS Microbiol. Lett. 183 (1), 131–135.

Garg, G., Singh, A., Kaur, A., Singh, R., Kaur, J., Mahajan, R., 2016. Microbial pectinases: an ecofriendly tool of nature for industries. 3 Biotech 6 (1), 1–13.

Global Pectinase Market Research Report, 2017. http://www.marketresearchstore.com/report/global-pectinase-marketresearch-report2017-190713.

Gomes, E., Leite, R.S.R., Da Silva, R., Silva, D., 2009. Purification of an exopolygalacturonase from *Penicillium viridicatum* RFC3 produced in submerged fermentation. Int. J. Microbiol. https://doi.org/10.1155/2009/631942.

Heidarizadeh, M., Rezaei, P.F., Shahabivand, S., 2018. Novel pectinase from *Piriformospora indica*, optimization of growth parameters and enzyme production in submerged culture condition. Turk. J. Biochem. 43 (3), 289–295.

Hendges, D.H., Montanari, Q., Malvessi, E., Silveira, M.M.D., 2011. Production and characterization of endo-polygalacturonase from *Aspergillus niger* in solid-state fermentation in double-surface bioreactor. Braz. Arch. Biol. Technol. 54 (2), 253–258.

Huang, D., Song, Y., Liu, Y., Qin, Y., 2019. A new strain of *Aspergillus tubingensis* for high-activity pectinase production. Braz. J. Microbiol. 50 (1), 53–65.

Ishii, T., Matsunaga, T., 2001. Pectic polysaccharide rhamnogalacturonan II is covalently linked to homogalacturonan. Phytochemistry 57 (6), 969–974.

Jayani, R.S., Saxena, S., Gupta, R., 2005. Microbial pectinolytic enzymes: a review. Process Biochem. 40 (9), 2931–2944.

Jiang, X., Jia, Q., Chen, L., Chen, Q., Yang, Q., 2014. Recombinant expression and inhibition mechanism analysis of pectin methylesterase from *Aspergillus flavus*. FEMS Microbiol. Lett. 355 (1), 12–19.

Kalantzi, S., Mamma, D., Kalogeris, E., Kekos, D., 2010. Improved properties of cotton fabrics treated with lipase and its combination with pectinase. Fibres Text. East. Eur. 18 (5), 86–92.

Karaoğlan, M., Erden-Karaoğlan, F., 2020. Effect of codon optimization and promoter choice on recombinant endo-polygalacturonase production in *Pichia pastoris*. Enzyme Microb. Technol., 109589.

Kashyap, D.R., Vohra, P.K., Chopra, S., Tewari, R., 2001. Applications of pectinases in the commercial sector: a review. Bioresour. Technol. 77 (3), 215–227.

Kauppinen, S., Christgau, S., Kofod, L.V., Halkier, T., Dörreich, K., Dalb, H., 1995. Molecular cloning and characterization of a rhamnogalacturonan acetylesterase from *Aspergillus aculeatus*: synergism between rhamnogalacturonan degrading enzymes. J. Biol. Chem. 270 (45), 27172–27178.

Kavuthodi, B., Sebastian, D., 2018. Review on bacterial production of alkaline pectinase with special emphasis on Bacillus species. Biosci. Biotechnol. Res. Commun. 11, 18–30.

Khatri, B.P., Bhattarai, T., Shrestha, S., Maharjan, J., 2015. Alkaline thermostable pectinase enzyme from *Aspergillus niger* strain MCAS2 isolated from Manaslu conservation area, Gorkha, Nepal. SpringerPlus 4 (1), 1–8.

Kun, R.S., Meng, J., Salazar-Cerezo, S., Mäkelä, M.R., de Vries, R.P., Garrigues, S., 2020. CRISPR/Cas9 facilitates rapid generation of constitutive forms of transcription factors in *Aspergillus niger* through specific on-site genomic mutations resulting in increased saccharification of plant biomass. Enzyme Microb. Technol. 136, 109508.

Lara-Marquez, A., Zavala-Paramo, M.G., Lopez-Romero, E., Calderon-Cortes, N., Lopez-Gomez, R., Conejo-Saucedo, U., Cano-Camacho, H., 2011. Cloning and characterization of a pectin lyase gene from *Colletotrichum lindemuthianum* and comparative phylogenetic/structural analyses with genes from phytopathogenic and saprophytic/opportunistic microorganisms. BMC Microbiol. 11 (1), 1–15.

Le Goff, A., Renard, C.M.G.C., Bonnin, E., Thibault, J.F., 2001. Extraction, purification and chemical characterisation of xylogalacturonans from pea hulls. Carbohydr. Polym. 45 (4), 325–334.

Long, Z., Chen, H., Zou, K., Sun, J., Li, J., Shen, P., 2017. Application of pectin-degrading fungus *Aspergillus tubingensis* GYC 501 in improvement of tobacco cut-stem quality and optimization of its fermentation conditions. Acta Agric. Jiangxi 29 (2), 95–98.

Marimuthu, S., 2000. Effect of addition of biopectinase on biochemical composition of CTC black tea. Rec. Adv. Plant Crop. Res., 265–269.

Mehanni, A.E.S., WHM, E.-.R., Melo, A., Casal, S., Ferreira, I.M., 2017. Enzymatic extraction of oil from *Balanites Aegyptiaca* (desert date) kernel and comparison with solvent extracted oil. J. Food Biochem. 41 (2), e12270.

Molina, S.M., Pelissari, F.A., Vitorello, C., 2001. Screening and genetic improvement of pectinolytic fungi for degumming of textile fibers. Braz. J. Microbiol. 32 (4), 320–326.

Mondal, S., Halder, S.K., Mondal, K.C., 2019. Fungal enzymes for bioconversion of lignocellulosic biomass. In: Jadav, A.N., Singh, S., Mishra, S., Gupta, A. (Eds.), Recent Advancement in White Biotechnology Through Fungi. Springer, Cham, pp. 349–380.

Mondal, S., Soren, J.P., Mondal, J., Rakshit, S., Halder, S.K., Mondal, K.C., 2020. Contemporaneous synthesis of multiple carbohydrate debranching enzymes from newly isolated Aspergillus fumigatus SKF-2 under solid state fermntation: a unique enzyme mixture for proficient saccharification of plant bioresources. Ind. Crop Prod. 150, 112409.

Murad, H.A., Azzaz, H.H., 2011. Microbial pectinases and ruminant nutrition. Res. J. Microbiol. 6 (3), 246–269.

Murthy, P.S., Naidu, M.M., 2011. Improvement of robusta coffee fermentation with microbial enzymes. Eur. J. Appl. Sci. 3 (4), 130–139.

Murugesan, G.S., Angayarkanni, J., Swaminathan, K., 2002. Effect of tea fungal enzymes on the quality of black tea. Food Chem. 79 (4), 411–417.

Nagai, M., Ozawa, A., Katsuragi, T., Sakai, T., 2000. Purification and characterization of acid-stable protopectinase produced by *Aspergillus awamori* in solid-state fermentation. Biosci. Biotechnol. Biochem. 64 (7), 1337–1344.

O'Neill, M.A., Eberhard, S., Albersheim, P., Darvill, A.G., 2001. Requirement of borate cross-linking of cell wall rhamnogalacturonan II for Arabidopsis growth. Science 294 (5543), 846–849.

Ortiz, G.E., Ponce-Mora, M.C., Noseda, D.G., Cazabat, G., Saravalli, C., López, M.C., Albertó, E.O., 2017. Pectinase production by *Aspergillus giganteus* in solid-state fermentation: optimization, scale-up, biochemical characterization and its application in olive-oil extraction. J. Ind. Microbiol. Biotechnol. 44 (2), 197–211.

Oumer, O.J., Abate, D., 2017. Characterization of pectinase from *Bacillus subtilis* strain Btk 27 and its potential application in removal of mucilage from coffee beans. Enzyme Res. https://doi.org/10.1155/2017/7686904.

Palaniyappan, M., Vijayagopal, V., Viswanathan, R., Viruthagiri, T., 2009. Screening of natural substrates and optimization of operating variables on the production of pectinase by submerged fermentation using *Aspergillus niger* MTCC 281. Afr. J. Biotechnol. 8 (4).

Patidar, M.K., Nighojkar, S., Kumar, A., Nighojkar, A., 2018. Pectinolytic enzymes-solid state fermentation, assay methods and applications in fruit juice industries: a review. 3 Biotech 8 (4), 1–24.

Patidar, M.K., Nighojkar, A., Nighojkar, S., Kumar, A., 2016. Purification and characterization of pectin methylesterase produced in solid state fermentation by *Aspergillus tubingensis*. Biotechnol. J., 1–10.

Patil, S.R., Dayanand, A., 2006. Production of pectinase from deseeded sunflower head by *Aspergillus niger* in submerged and solid-state conditions. Bioresour. Technol. 97 (16), 2054–2058.

Rangarajan, V., Rajasekharan, M., Ravichandran, R., Sriganesh, K., Vaitheeswaran, V., 2010. Pectinase production from orange peel extract and dried orange peel solid as substrates using *Aspergillus niger*. Int. J. Biotechnol. Biochem. 6 (3), 445–453.

Samanta, S., 2021. Microbial pectinases: a review on molecular and biotechnological perspectives. J. Microbiol. Biotechnol. Food Sci., 248–266.

Senthilkumar, R.S., 2000. Microbial enzymes for processing of tealeaf. Rec. Adv. Plant Crop Res., 273–276.

Sharma, S., Mandhan, R.P., Sharma, J., 2011. *Pseudozyma sp.* SPJ: an economic and eco-friendly approach fordegumming of flax fibers. World J. Microbiol. Biotechnol. 27 (11), 2697–2701.

Siddiqui, M., Pande, V., Arif, M., 2012. Production, purification, and characterization of polygalacturonase from *Rhizomucor pusillus* isolated from decomposting orange peels. Enzyme Res. https://doi.org/10.1155/2012/138634.

Singh, R.S., Singh, T., Pandey, A., 2019. Microbial enzymes—an overview. Adv. Enzyme Technol., 1–40.

Teixeira, J.A., Gonçalves, D.B., De Queiroz, M.V., De Araujo, E.F., 2011. Improved pectinase production in *Penicillium griseoroseum* recombinant strains. J. Appl. Microbiol. 111 (4), 818–825.

Teixeira, J.A., Nogueira, G.B., de Queiroz, M.V., de Araújo, E.F., 2014. Genome organization and assessment of high copy number and increased expression of pectinolytic genes from *Penicillium griseoroseum*: a potential heterologous system for protein production. J. Ind. Microbiol. Biotechnol. 41 (10), 1571–1580.

Ting-Ting, Y.A.O., Yan-Min, W.A.N.G., Jian-Long, G.U., Zheng-Xiang, W.A.N.G., 2006. Overproduction of glucoamylase by recombinant *Aspergillus niger* harboring multiple copies of glaA. Chin. J. Biotechnol. 22 (4), 567–571.

Voragen, A.G., Coenen, G.J., Verhoef, R.P., Schols, H.A., 2009. Pectin, a versatile polysaccharide present in plant cell walls. Struct. Chem. 20 (2), 263–275.

Wong, L.Y., Saad, W.Z., Mohamad, R., Tahir, P.M., 2017. Optimization of cultural conditions for polygalacturonase production by a newly isolated *Aspergillus fumigatus* R6 capable of retting kenaf. Ind. Crop Prod. 97, 175–183.

Yadav, S., Yadav, P.K., Yadav, D., Yadav, K.D.S., 2009. Purification and characterization of pectin lyase produced by *Aspergillus terricola* and its application in retting of natural fibers. Appl. Biochem. Biotechnol. 159 (1), 270–283.

Yadav, S., Dubey, A.K., Anand, G., Yadav, D., 2012. Characterization of a neutral pectin lyase produced by *Oidiodendron echinulatum* MTCC 1356 in solid state fermentation. J. Basic Microbiol. 52 (6), 713–720.

Yadav, S., Maurya, S.K., Anand, G., Dwivedi, R., Yadav, D., 2017. Purification and characterization of a highly alkaline pectin lyase from *Fusarium lateritum* MTCC 8794. Biologia 72 (3), 245–251.

Zheng, X., Zhang, Y., Liu, X., Li, C., Lin, Y., Liang, S., 2020. High-level expression and biochemical properties of a thermo-alkaline pectate lyase from *Bacillus sp.* RN1 in *Pichia pastoris* with potential in ramie degumming. Front. Bioeng. Biotechnol. 8, 850.

CHAPTER

Gold nanoparticles: A potential tool to enhance the immune response against viral infection

21

Gayathri A. Kanu[a], Raed O. AbuOdeh[a], and Ahmed A. Mohamed[b]

[a]Department of Medical Laboratory Sciences, College of Health Sciences, University of Sharjah, Sharjah, United Arab Emirates, [b]Department of Chemistry, University of Sharjah, Sharjah, United Arab Emirates

21.1 Introduction

The virus that causes coronavirus disease 2019 (COVID-19), a global pandemic that resulted in approximately 6 million fatalities and over 414 million infected persons improving, is known as severe acute respiratory syndrome coronavirus 2 (SARS-CoV-2) (Aghamirza Moghim Aliabadi et al., 2022; Yasamineh et al., 2022a). Patients with COVID-19 infections have supposedly experienced mild to severe respiratory tract infections and symptoms like fever, coughing, and dyspnea that could appear from the first week up to 14 days after infection exposure (Ahmadi et al., 2022; Hashemi et al., 2021). World Health Organization (WHO) proclaimed a COVID pandemic as a result of the SARS-CoV-2 infection's quick global spread and the harm it caused (Park, 2020). Through Emergency Use Authorization (EUA) approvals, numerous healthcare innovations, including rapid testing kits, vaccinations, and antiviral medicines for COVID, have been made available to use immediately. Most governments have launched widespread immunization campaigns to avoid infection and minimize financial drop and human harm. However, rare sharply increasing viral variant infections are to blame for the variable infection rates. A strong warning that the emergence or reemergence of viral illnesses is uncertain but unavoidable is the current pandemic situation.

To control potential viral infections in the future, essential to generate technology-based countermeasures for the detection, hindrance, and treatment in terms of the application of entrepreneurship. Gold nanomaterials are advantageous in several ways when used in the creation of technological countermeasures, including their ease of functionalization, control over surface chemistry, and availability as delivery vehicles (Chintagunta et al., 2021). Because of their extremely small sizes, AuNPs can effectively enter biological cellular organizations. Additionally, nanobiomedical expertise has received a lot of attention for a variety of reasons, including the effective and directed delivery of medications, genes, and healing molecules to specific organs or cells, imaging, and the early-stage correct identification of viruses (Kang et al., 2021; Nabizadeh et al., 2022).

Numerous investigations have demonstrated that AuNPs can operate as an antiviral agent against SARS-CoV-2 (Vijayakumar and Ganesan, 2012). Due to their function in the operation of immune chromatography, AuNPs are primarily used for virus detection. Even while intriguing research on the

Entrepreneurship with Microorganisms. https://doi.org/10.1016/B978-0-443-19049-0.00015-3

COVID-19 virus susceptibility to gold antiviral effects is being done, neither a therapy nor a vaccine has yet been released. Determining the role that AuNPs play in the generation of vaccines and antibodies is an excellent approach to understanding the business opportunities available now and in the future.

21.2 Gold nanoparticles in vaccine synthesis against COVID-19

Alpha-, beta-, gamma-, and delta-CoV are the four major subgroups of coronaviruses (CoVs), while the two main forms are CoVs alpha and beta (Chan et al., 2015; Yousefi et al., 2020, 2022). The pandemic situation, which includes asymptomatic upper and lower respiratory tract infections, is caused by SARS-CoV-2, a member of the CoVs (Hu et al., 2021; Wang et al., 2020). In addition, CoVs are positive-sense, single-stranded RNA (+ss-RNA) (Yousefi and Eslami, 2022). This virus contains many short open reading frames (ORFs). The ORF gene family encodes structural proteins such as the spike (S) glycoprotein, envelope (E), membrane (M), nucleocapsid (N), and nonstructural proteins (NSP) (Islam et al., 2020). Two overlapping polyproteins that are required for viral transcription and reproduction are encoded by the SARS-CoV-2 replicate gene (Khaledi et al., 2021). In this, two-thirds of the RNA in the 5′-UTR section is made up of ORF1a/b (Khailany et al., 2020; Kim et al., 2020). The entire viral genome is produced by holo-RdRp RdRp activities (Comber and Bamezai, 2015). Understanding the structure and replication process of CoVs and identifying the best carrier and delivery method that can affect a robust immune response is a crucial phase that can lead to successful vaccine development.

In several disease models, the utilization of AuNPs and their effects on the immune system has been investigated. The creation of new-age vaccinations can build on gold nanocarriers as an adjuvant. The nanoparticle size, shape, administration method, and antigen tagging mechanism are all crucial in this regard (Salazar-Gonzalez et al., 2015). Characterizing nanoparticle-based on their immunogenicity, immune toxicity, the mechanism of immune suppression, and immunomodulation has made significant strides in recent years (Fadeel, 2019; Frey et al., 2018; Kelly et al., 2019). Antigens can be easily chemically attached to AuNPs and their formulations, making them a good candidate for use in the production of nanovaccines (Fai and Kumar, 2021). Demenev et al. first employed AuNPs to create vaccines in 1996 when they tested the efficacy of an experimental vaccination against tick-borne encephalitis (Demenev et al., 1996). According to the study, a vaccine can be created by conjugating soluble antigens with AuNPs. Since nanoparticles and virus particles are the same in proportions, they might efficiently penetrate a cell and cause the transmitted gene construct (DNA or mRNA) to produce antigens, or they could directly goal immune cells for the transport of antigens. At the cellular level, nanoparticles can target both the innate and the adaptive immune systems which are known for macrophages, monocytes, neutrophils, and T and B cells. However, in the case of COVID, it has been demonstrated to benefit from cellular and humoral immunity targets (Graham et al., 2012; Yang et al., 2004).

AuNPs have great potential in the entrapment of antigens to prevent antigen damage and enhance their reliability, secondly coentrapment of antigen and immune-stimulatory factor(s) the antigen-presenting cells (APCs) can easily phagocytose and produce particles; and finally surface localization of nanoparticles with functional moieties and directing towards ligands that allow organ- and cell-particular binding to lymphoid organs and APCs (Figs. 21.1 and 21.2) (Yasamineh et al., 2022b).

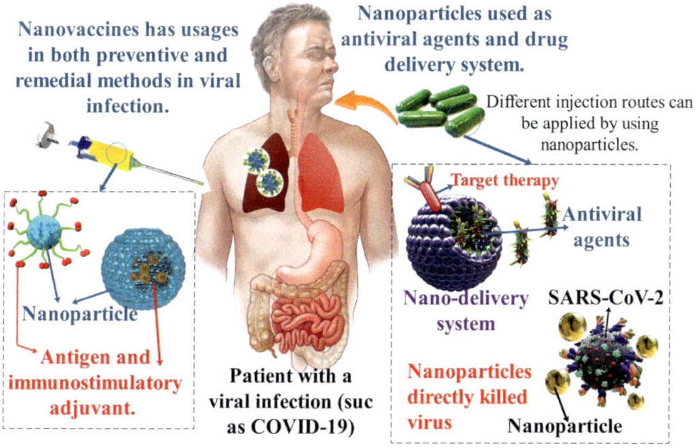

FIG. 21.1

An overview of viral infection and uses of nanomaterial.

Reprinted with permission from Yasamineh, S., Kalajahi, H.G., Yasamineh, P., Yazdani, Y., Gholizadeh, O., Tabatabaie, R., Afkhami, H., Davodabadi, F., Pahlevan, D., Firouzi-Amandi, A., 2022b. An overview on nanoparticle-based strategies to fight viral infections with a focus on COVID-19. J. Nanobiotechnol. 20, 1–26.

21.3 Synthesis of AuNPs

Gold(III) ions are converted to metallic gold in a redox process that produces gold nanoparticles. Various reducing agents are used including ascorbic acid (Andreescu et al., 2006), formaldehyde (Zhang et al., 2009), hydrazine (Pal et al., 2007), and borohydride (Zhou et al., 2006), and most often used gold compound is $HAuCl_4$. The most frequent reducing agents are sodium citrate and $NaBH_4$. As a result, a variety of techniques are available to fabricate AuNPs with different surface ligand compositions, sizes, and shapes. The ligands localized to the surface of AuNPs are crucial because they enable nanoparticle stability and serve as the building blocks for nanoparticle bioconjugation. The procedures typically result in either aqueous or organic solvent-dispersed AuNPs. This review will concentrate on the usage of AuNPs that are dispersible in aqueous solutions for the biological applications.

21.4 Interaction of gold nanoparticles with immune cells

Immune system cells form the first line of defense against AuNP invasion of animal tissues and cells. As a result, there is undeniable interest in studying how AuNPs interact with phagocytes, intracellular uptake mechanisms, and immune cell responses to AuNPs. Shukla et al. used three microscopic techniques to study the uptake of 3 nm AuNPs into RAW264.7, which may be the first comprehensive study of these difficulties. According to Shukla et al., AuNPs inhibit the production of reactive oxygen species and do not stimulate the production of the proinflammatory cytokines TNF-a and IL1-b. They are biocompatible, noncytotoxic, and nonimmunogenic, which contradicts the findings of Yen et al.,

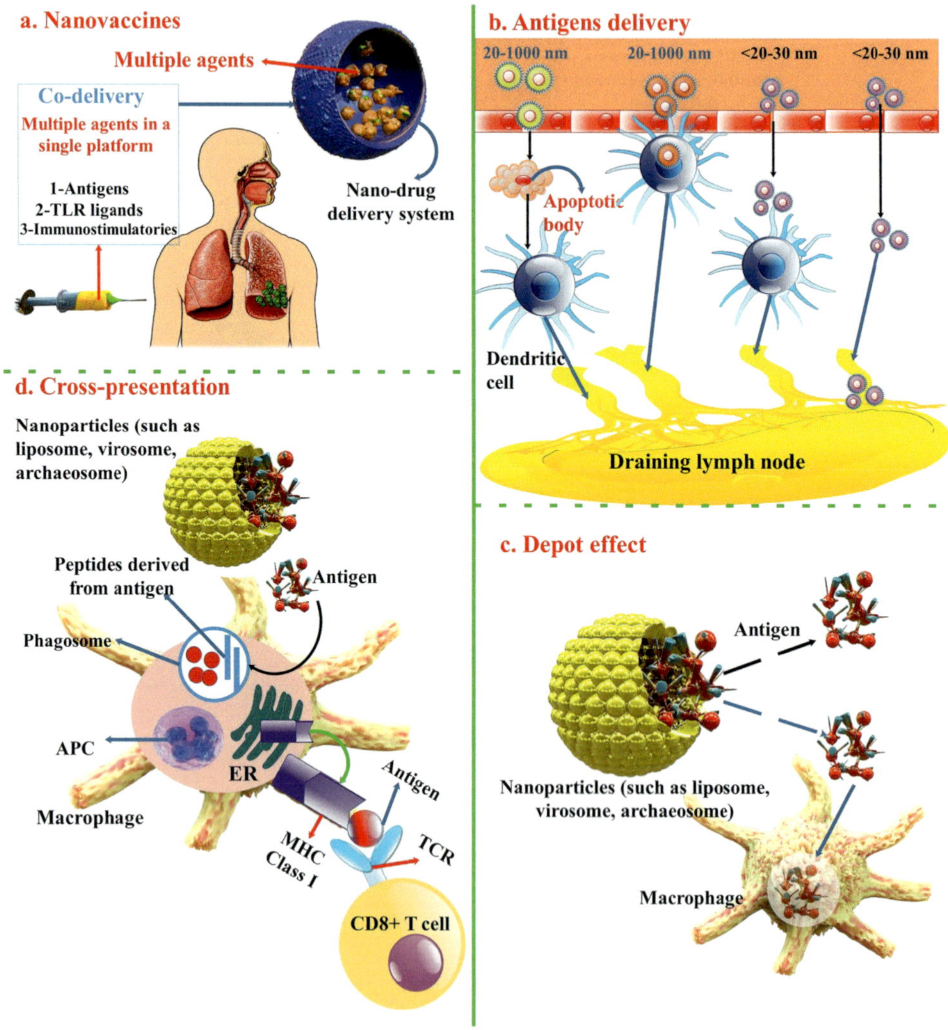

FIG. 21.2

An explanation of immunological responses nanovaccines can trigger. (A) Because NPs may transport antigens and a variety of immune-stimulatory molecules, they can be used as a vaccine complex for several infectious diseases (TLR ligands and adjuvants). Nanovaccines' ability to stimulate the immune system is linked to many mechanisms, including the antigen release effect which considered the release of vaccine antigens gradually to uptake in cells that present antigens. (B) NP-mediated antigen transport (dimensions-related permeation and tissue or organ targeting). (C) A long-lasting and incessant release of steady antigen is provided by the depot effect. (D) Antigen-specific cytotoxic T cells are activated by the cross-presentation of the antigen carried by the NPs (cytosolic transport). T cell receptor, endoplasmic reticulum, and antigen-presenting cell (APC).

Reprinted with permission from Yasamineh, S., Kalajahi, H.G., Yasamineh, P., Yazdani, Y., Gholizadeh, O., Tabatabaie, R., Afkhami, H., Davodabadi, F., Pahlevan, D., Firouzi-Amandi, A., 2022b. An overview on nanoparticle-based strategies to fight viral infections with a focus on COVID-19. J. Nanobiotechnol. 20, 1–26.

who found that injection of AuNPs resulted in a decrease in macrophage numbers and an increase in macrophage size, as well as an increase in IL-1, IL-6, and TNF-a production. (Shukla et al., 2005; Yen et al., 2009).

To overcome the difficulties mentioned above, AuNPs must be decorated on the surface with various ligands, such as monoclonal antibodies, peptides, oligomers, or small molecules like mannose, legumain, and so on, to bind to specific receptors that are overexpressed on the surface of macrophages and perform these functions simultaneously (Poupot et al., 2018). This has been demonstrated by numerous experiments that used significantly larger hollow NSphs capped with dextran AuNPs to produce and match Shukla's results (Lim et al., 2008; Zhang et al., 2011). The effects of nanoparticles on monocytes and macrophages include changes in bone marrow activation, mobilization and recruitment of monocytes, increased microvascular permeability, polarization changes, and so on. Furthermore, nanoparticles can use the inherent properties of monocytes and/or macrophages to improve their ability to target while avoiding immune system clearance. The use of AuNPs as adjuvant aids in the synthesis of high-affinity antibodies to haptens and full antigens, as well as the development of successful vaccinations. Another method for activating immune-competent cells with colloidal gold is to combine AuNPs with CpG oligodeoxynucleotides (ODNs). Colloidal gold is used as a carrier and adjuvant in a technique devised by Dykman et al. to produce antibodies to various antigens (Dykman et al., 2004, 2010; Dykman and Khlebtsov, 2017). Finally, the experimental results support the following claims: Using the "gold immunization" method, one can produce antibodies to haptens that are extremely difficult to produce using conventional methods, such as various antibiotics, vitamins, and nonimmunogenic peptides. Even when conventional procedures can elicit an immune response, the amount of antigen used for immunization, in this case, is far less than that used in conventional methods. An immunological response was obtained in studies with a variety of antigens conjugated with AuNPs without the use of additional adjuvants. When AuNPs are used as antigen carriers, they increase the phagocytic activity of lymphoid cells and cause the production of inflammatory mediators.

21.5 Mechanism of action of AuNPs in the immune response against COVID-19

The primary objective of vaccination is to induce the immune system to eliminate harmful pathogens from the body, thereby providing long-term protection against infections and pathogens. In the case of AuNPs, this can be achieved by the following factors (1) their capacity to repeatedly deliver viral antigens, (2) their capability to mimic the virus in terms of size and structure without requiring an actual infection, (3) their capacity to activate dendritic cells (DCs) and antigen cross-presentation, and (4) their capacity to efficiently activate follicular B cells and trigger a humoral immune response. AuNPs can stimulate various immune cells on their own, enhancing host immunity by growing the production of resistance genes against infection (Smith et al., 2013). Numerous types of NPs such as gold, carbon, dendrimers, polymers, and liposomes can induce cytokine production and antibody responses (Fifis et al., 2004; Mottram et al., 2007; Scholer et al., 2001, 2002; Shvedova, 2005; Vallhov et al., 2006).

21.6 Advantages and entrepreneurship of AuNPs in nanovaccine developments

Due to their high surface area to volume ratio, AuNPs offer a strong binding affinity, unique electronic structures, Plasmon excitation, and enormous surface energies (Fig. 21.3) (Personick et al., 2011). Additionally, AuNPs can engage with many functional groups or ligands with great affinity. Colloidal gold has already been utilized in the treatment of a wide range of disorders with a low level

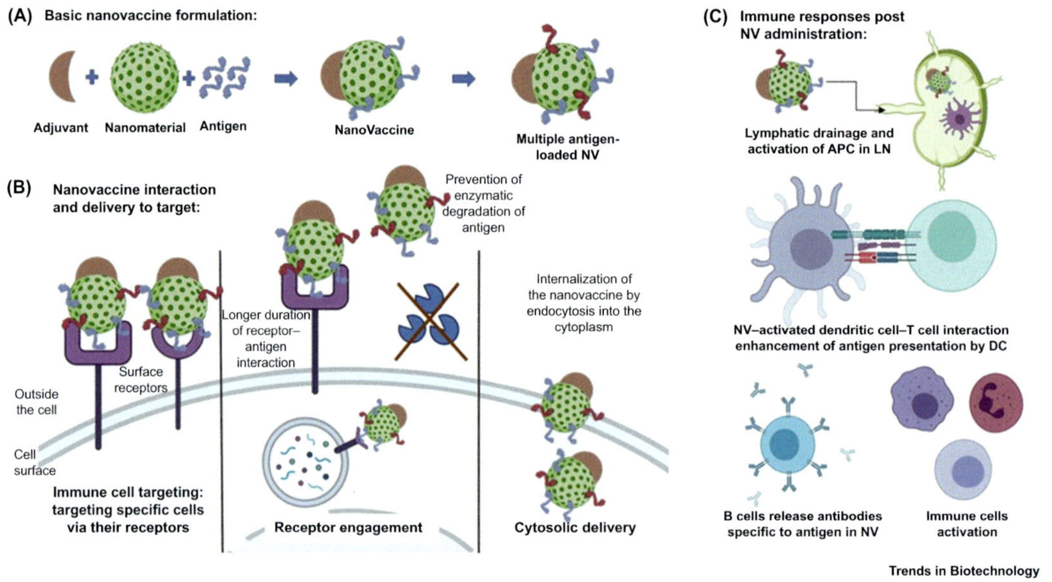

FIG. 21.3

Basic principles and importance of nanovaccines: (A) A chosen antigen that has been coupled to a nanomaterial and an adjuvant to stimulate an immune response make up nanovaccines. The surface of the NPs can be loaded with a variety of antigen epitopes (shown by *red* and *blue* antigens). Depending on the infection, tissue type, and needed immune response, several types of nanomaterial and adjuvants are used. (B) NPs help vaccines more effectively target the desired cell and its receptors, reducing adverse effects. They prolong antigen-receptor interaction, which strengthens the immunological response. The antigen can be delivered into the target cell's cytoplasm using particular kinds of NPs. Antigens are better protected against enzymatic or proteolytic cleavage when they are packaged within NPs. (C) NPs can pass through the lymphatic drainage system and activate APCs within the lymph nodes. (D) NPs support the DC-T cell contact required for enhancing the subsequent immunological response. They control the release of cytokines that are both pro- and antiinflammatory and stimulate dendritic cells. (E) NP-mediated vaccine administration also influences the differentiation, maturation, and activation of lymphocytes and monocytes as well as the generation of antibodies by plasma B cells. *APC*, Antigen-presenting cell; *DC*, dendritic cell; *LN*, lymph node; *NP*, nanoparticle; *NV*, nanovaccine.

Reprinted with permission from Azharuddin, M., Zhu, G.H., Sengupta, A., Hinkula, J., Slater, N.K.H., Patra, H.K., 2022. Nano toolbox in immune modulation and nanovaccines. Trends Biotechnol. 40 1195-1212.

of cytotoxicity due to its innate magnetic and optical capabilities. Additionally, pathogen-specific antibodies have been created using AuNPs coated with antigenic peptides (Barhate et al., 2014; Chen et al., 2010; Dakterzada et al., 2016; Dykman et al., 2015; Gao et al., 2015; Gregory et al., 2012, 2015; Khlebtsov et al., 2013; Kumar et al., 2015; Manea et al., 2008; Parween et al., 2011; Quach et al., 2018; Rodriguez-Del Rio et al., 2015; Safari et al., 2012; Sanchez-Villamil et al., 2019; Torres et al., 2015). For regulated release to the target site, the AuNPs-based drug or antigen delivery system performs better (Lee et al., 2018). We can employ AuNPs nanoformulations by attaching particular molecules or antibodies to their surface. This makes it possible to target certain cell types effectively, which results in a profile of the immune response that is specific to the site and minimal off-target dissemination (Kalishwaralal et al., 2020). The AuNPs possess strong adjuvant qualities that strengthen the host's immune response. Different immune reactions are elicited upon administration by different combinations of size, shape, charge, and surface functionalization (Liu et al., 2018).

21.7 Conclusions

There is a critical need to use AuNPs in vaccine development and clinical settings. If the AuNPs overcome some challenges, it is a big advantage to a setup for large-scale with highly consistent AuNPs production. This will be a successful milestone in entrepreneurship and a major requirement that research is rapidly shifting to in industry and pharmaceuticals.

References

Aghamirza Moghim Aliabadi, H., Eivazzadeh Keihan, R., Beig Parikhani, A., Fattahi Mehraban, S., Maleki, A., Fereshteh, S., Bazaz, M., Zolriasatein, A., Bozorgnia, B., Rahmati, S., 2022. COVID 19: a systematic review and update on prevention, diagnosis, and treatment. MedComm. 3, e115.

Ahmadi, K., Farasat, A., Rostamian, M., Johari, B., Madanchi, H., 2022. Enfuvirtide, an HIV-1 fusion inhibitor peptide, can act as a potent SARS-CoV-2 fusion inhibitor: an in silico drug repurposing study. J. Biomol. Struct. Dyn. 40, 5566–5576.

Andreescu, D., Sau, T.K., Goia, D.V., 2006. Stabilizer-free nanosized gold sols. J. Colloid Interface Sci. 298, 742–751.

Barhate, G.A., Gaikwad, S.M., Jadhav, S.S., Pokharkar, V.B., 2014. Structure function attributes of gold nanoparticle vaccine association: effect of particle size and association temperature. Int. J. Pharm. 471, 439–448.

Chan, J.F.W., Lau, S.K.P., To, K.K.W., Cheng, V.C.C., Woo, P.C.Y., Yuen, K.-Y., 2015. Middle East respiratory syndrome coronavirus: another zoonotic betacoronavirus causing SARS-like disease. Clin. Microbiol. Rev. 28, 465–522.

Chen, Y.-S., Hung, Y.-C., Lin, W.-H., Huang, G.S., 2010. Assessment of gold nanoparticles as a size-dependent vaccine carrier for enhancing the antibody response against synthetic foot-and-mouth disease virus peptide. Nanotechnology 21, 195101.

Chintagunta, A.D., Nalluru, S., Sampath Kumar, N.S., 2021. Nanotechnology: an emerging approach to combat COVID-19. Emergent Mater. 4, 119–130.

Comber, J.D., Bamezai, A., 2015. Gold nanoparticles (AuNPs): a new frontier in vaccine delivery. J. Nanomed. Biother. Discov. 5.

Dakterzada, F., Mobarez, A.M., Roudkenar, M.H., Mohsenifar, A., 2016. Induction of humoral immune response against Pseudomonas aeruginosa flagellin (1-161) using gold nanoparticles as an adjuvant. Vaccine 34, 1472–1479.

Demenev, V.A., Shchinova, M.A., Ivanov, L.I., Vorob'eva, R.N., Zdanovskaia, N.I., Neba kina, N.V., 1996. Perfection of methodical approaches to designing vaccines against tick-borne encephalitis. Vopr. Virusol. 41, 107–110.
Dykman, L.A., Khlebtsov, N.G., 2017. Immunological properties of gold nanoparticles. Chem. Sci. 8, 1719–1735.
Dykman, L.A., Staroverov, S.A., Bogatyrev, V.A., Shchyogolev, S.Y., 2010. Adjuvant properties of gold nanoparticles. Nanotechnol. Russ. 5, 748–761.
Dykman, L.A., Staroverov, S.A., Mezhenny, P.V., Fomin, A.S., Kozlov, S.V., Volkov, A.A., Laskavy, V.N., Shchyogolev, S.Y., 2015. Use of a synthetic foot-and-mouth disease virus peptide conjugated to gold nanoparticles for enhancing immunological response. Gold Bull. 48, 93–101.
Dykman, L.A., Sumaroka, M.V., Staroverov, S.A., Zaitseva, I.S., Bogatyrev, V.A., 2004. Immunogenic properties of colloidal gold. Biol. Bull. Russ. Acad. Sci. 31, 75–79.
Fadeel, B., 2019. Hide and seek: nanomaterial interactions with the immune system. Front. Immunol. 10, 133.
Fai, T.K., Kumar, P.V., 2021. Revolution in the synthesis, physio-chemical and biological characterization of gold nanoplatform. Curr. Pharm. Des. 27, 2482–2504.
Fifis, T., Gamvrellis, A., Crimeen-Irwin, B., Pietersz, G.A., Li, J., Mottram, P.L., McKenzie, I.F.C., Plebanski, M., 2004. Size-dependent immunogenicity: therapeutic and protective properties of nano-vaccines against tumors. J. Immunol. 173, 3148–3154.
Frey, M., Bobbala, S., Karabin, N., Scott, E., 2018. Influences of nanocarrier morphology on therapeutic immunomodulation. Nanomedicine 13, 1795–1811.
Gao, W., Fang, R.H., Thamphiwatana, S., Luk, B.T., Li, J., Angsantikul, P., Zhang, Q., Hu, C.-M.J., Zhang, L., 2015. Modulating antibacterial immunity via bacterial membrane-coated nanoparticles. Nano Lett. 15, 1403–1409.
Graham, R.L., Becker, M.M., Eckerle, L.D., Bolles, M., Denison, M.R., Baric, R.S., 2012. A live, impaired-fidelity coronavirus vaccine protects in an aged, immunocompromised mouse model of lethal disease. Nat. Med. 18, 1820–1826.
Gregory, A.E., Judy, B.M., Qazi, O., Blumentritt, C.A., Brown, K.A., Shaw, A.M., Torres, A.G., Titball, R.W., 2015. A gold nanoparticle-linked glycoconjugate vaccine against Burkholderia mallei. Nanomed. Nanotechnol. Biol. Med. 11, 447–456.
Gregory, A.E., Williamson, E.D., Prior, J.L., Butcher, W.A., Thompson, I.J., Shaw, A.M., Titball, R.W., 2012. Conjugation of Y. pestis F1-antigen to gold nanoparticles improves immunogenicity. Vaccine 30, 6777–6782.
Hashemi, B., Akram, F.-A., Amirazad, H., Dadashpour, M., Sheervalilou, M., Nasrabadi, D., Ahmadi, M., Sheervalilou, R., Reza, M.A.S., Ghazi, F., 2021. Emerging importance of nanotechnology-based approaches to control the COVID-19 pandemic; focus on nanomedicine iterance in diagnosis and treatment of COVID-19 patients. J. Drug Deliv. Sci. Technol., 102967.
Hu, B., Guo, H., Zhou, P., Shi, Z.-L., 2021. Characteristics of SARS-CoV-2 and COVID-19. Nat. Rev. Microbiol. 19, 141–154.
Islam, M.R., Hoque, M.N., Rahman, M.S., Alam, A.S.M., Akther, M., Puspo, J.A., Akter, S., Sultana, M., Crandall, K.A., Hossain, M.A., 2020. Genome-wide analysis of SARS-CoV-2 virus strains circulating worldwide implicates heterogeneity. Sci. Rep. 10, 1–9.
Kalishwaralal, K., Luboshits, G., Firer, M.A., 2020. Synthesis of gold nanoparticle: peptide-drug conjugates for targeted drug delivery. In: Drug Delivery Systems. Springer, pp. 145–154.
Kang, J., Tahir, A., Wang, H., Chang, J., 2021. Applications of nanotechnology in virus detection, tracking, and infection mechanisms. Wiley Interdiscip. Rev. Nanomed. Nanobiotechnol. 13, e1700.
Kelly, H.G., Kent, S.J., Wheatley, A.K., 2019. Immunological basis for enhanced immunity of nanoparticle vaccines. Expert Rev. Vaccines 18, 269–280.
Khailany, R.A., Safdar, M., Ozaslan, M., 2020. Genomic characterization of a novel SARS-CoV-2. Gene Rep. 19, 100682.
Khaledi, M., Yousefi Nojookambari, N., Afkhami, H., Sameni, F., Yazdansetad, S., 2021. A review on phylogenetic assessment and cytopathogenesis of filoviruses, retroviruses, and coronaviruses transmitted from bat to human. Cell. Mol. Res. (Iran. J. Biol.).

Khlebtsov, N., Bogatyrev, V., Dykman, L., Khlebtsov, B., Staroverov, S., Shirokov, A., Matora, L., Khanadeev, V., Pylaev, T., Tsyganova, N., 2013. Analytical and theranostic applications of gold nanoparticles and multifunctional nanocomposites. Theranostics 3, 167.

Kim, D., Lee, J.-Y., Yang, J.-S., Kim, J.W., Kim, V.N., Chang, H., 2020. The architecture of SARS-CoV-2 transcriptome. Cell 181, 914–921.

Kumar, R., Ray, P.C., Datta, D., Bansal, G.P., Angov, E., Kumar, N., 2015. Nanovaccines for malaria using Plasmodium falciparum antigen Pfs25 attached gold nanoparticles. Vaccine 33, 5064–5071.

Lee, J.-H., Cho, H.-Y., Choi, H.K., Lee, J.-Y., Choi, J.-W., 2018. Application of gold nanoparticle to plasmonic biosensors. Int. J. Mol. Sci. 19, 2021.

Lim, Y.T., Cho, M.Y., Choi, B.S., Noh, Y.-W., Chung, B.H., 2008. Diagnosis and therapy of macrophage cells using dextran-coated near-infrared responsive hollow-type gold nanoparticles. Nanotechnology 19, 375105.

Liu, Y., Crawford, B.M., Vo-Dinh, T., 2018. Gold nanoparticles-mediated photothermal therapy and immunotherapy. Immunotherapy 10, 1175–1188.

Manea, F., Bindoli, C., Fallarini, S., Lombardi, G., Polito, L., Lay, L., Bonomi, R., Mancin, F., Scrimin, P., 2008. Multivalent, saccharide functionalized gold nanoparticles as fully synthetic analogs of type A Neisseria meningitidis antigens. Adv. Mater. 20, 4348–4352.

Mottram, P.L., Leong, D., Crimeen-Irwin, B., Gloster, S., Xiang, S.D., Meanger, J., Ghildyal, R., Vardaxis, N., Plebanski, M., 2007. Type 1 and 2 immunity following vaccination is influenced by nanoparticle size: formulation of a model vaccine for respiratory syncytial virus. Mol. Pharm. 4, 73–84.

Nabizadeh, Z., Nasrollahzadeh, M., Daemi, H., Eslaminejad, M.B., Shabani, A.A., Dadashpour, M., Mirmohammadkhani, M., Nasrabadi, D., 2022. Micro-and nanotechnology in biomedical engineering for cartilage tissue regeneration in osteoarthritis. Beilstein J. Nanotechnol. 13, 363–389.

Pal, A., Shah, S., Devi, S., 2007. Synthesis of Au, Ag and Au-Ag alloy nanoparticles in aqueous polymer solution. Colloids Surf. A Physiochem. Eng. Asp. 302, 51–57.

Park, S.E., 2020. Epidemiology, virology, and clinical features of severe acute respiratory syndrome coronavirus 2 (SARS-CoV-2; coronavirus disease-19). Pediatr. Infect. Vaccine 27, 1.

Parween, S., Gupta, P.K., Chauhan, V.S., 2011. Induction of humoral immune response against PfMSP-119 and PvMSP-119 using gold nanoparticles along with alum. Vaccine 29, 2451–2460.

Personick, M.L., Langille, M.R., Zhang, J., Mirkin, C.A., 2011. Shape control of gold nanoparticles by silver underpotential deposition. Nano Lett. 11, 3394–3398.

Poupot, R., Goursat, C., Fruchon, S., 2018. Multivalent nanosystems: targeting monocytes/macrophages. Int. J. Nanomedicine 13, 5511.

Quach, Q.H., Ang, S.K., Chu, J.-H.J., Kah, J.C.Y., 2018. Size-dependent neutralizing activity of gold nanoparticle-based subunit vaccine against dengue virus. Acta Biomater. 78, 224–235.

Rodriguez-Del Rio, E., Marradi, M., Calderon-Gonzalez, R., Frande-Cabanes, E., Penades, S., Petrovsky, N., Alvarez-Dominguez, C., 2015. A gold glyco-nanoparticle carrying a listeriolysin O peptide and formulated with Advax delta inulin adjuvant induces robust T-cell protection against listeria infection. Vaccine 33, 1465–1473.

Safari, D., Marradi, M., Chiodo, F., Th Dekker, H.A., Shan, Y., Adamo, R., Oscarson, S., Rijkers, G.T., Lahmann, M., Kamerling, J.P., 2012. Gold nanoparticles as carriers for a synthetic Streptococcus pneumoniae type 14 conjugate vaccine. Nanomedicine 7, 651–662.

Salazar-Gonzalez, J.A., Gonzalez-Ortega, O., Rosales-Mendoza, S., 2015. Gold nanoparticles and vaccine development. Expert Rev. Vaccines 14, 1197–1211.

Sanchez-Villamil, J.I., Tapia, D., Torres, A.G., 2019. Development of a gold nanoparticle vaccine against enterohemorrhagic Escherichia coli O157: H7. MBio 10, e01869-19.

Scholer, N., Hahn, H., Muller, R.H., Liesenfeld, O., 2002. Effect of lipid matrix and size of solid lipid nanoparticles (SLN) on the viability and cytokine production of macrophages. Int. J. Pharm. 231, 167–176.

Scholer, N., Olbrich, C., Tabatt, K., Muller, R.H., Hahn, H., Liesenfeld, O., 2001. Surfactant, but not the size of solid lipid nanoparticles (SLN) influences viability and cytokine production of macrophages. Int. J. Pharm. 221, 57–67.

Shukla, R., Bansal, V., Chaudhary, M., Basu, A., Bhonde, R.R., Sastry, M., 2005. Biocompatibility of gold nanoparticles and their endocytotic fate inside the cellular compartment: a microscopic overview. Langmuir 21, 10644–10654.

Shvedova, A.A., 2005. Unusual inflammatory and fibrogenic pulmonary responses to single-walled carbon nanotubes in mice. AJP-Lung Cell. Mol. Physiol. 298, L698.

Smith, D.M., Simon, J.K., Baker Jr., J.R., 2013. Applications of nanotechnology for immunology. Nat. Rev. Immunol. 13, 592–605.

Torres, A.G., Gregory, A.E., Hatcher, C.L., Vinet-Oliphant, H., Morici, L.A., Titball, R.W., Roy, C.J., 2015. Protection of non-human primates against glanders with a gold nanoparticle glycoconjugate vaccine. Vaccine 33, 686–692.

Vallhov, H., Qin, J., Johansson, S.M., Ahlborg, N., Muhammed, M.A., Scheynius, A., Gabrielsson, S., 2006. The importance of an endotoxin-free environment during the production of nanoparticles used in medical applications. Nano Lett. 6, 1682–1686.

Vijayakumar, S., Ganesan, S., 2012. Gold nanoparticles as an HIV entry inhibitor. Curr. HIV Res. 10, 643–646.

Wang, Q., Zhang, Y., Wu, L., Niu, S., Song, C., Zhang, Z., Lu, G., Qiao, C., Hu, Y., Yuen, K.-Y., 2020. Structural and functional basis of SARS-CoV-2 entry by using human ACE2. Cell 181, 894–904.

Yang, Z., Kong, W., Huang, Y., Roberts, A., Murphy, B.R., Subbarao, K., Nabel, G.J., 2004. A DNA vaccine induces SARS coronavirus neutralization and protective immunity in mice. Nature 428, 561–564.

Yasamineh, S., Kalajahi, H.G., Yasamineh, P., Gholizadeh, O., Youshanlouei, H.R., Matloub, S.K., Mozafari, M., Jokar, E., Yazdani, Y., Dadashpour, M., 2022a. Spotlight on therapeutic efficiency of mesenchymal stem cells in viral infections with a focus on COVID-19. Stem Cell Res Ther 13, 1–23.

Yasamineh, S., Kalajahi, H.G., Yasamineh, P., Yazdani, Y., Gholizadeh, O., Tabatabaie, R., Afkhami, H., Davodabadi, F., Pahlevan, D., Firouzi-Amandi, A., 2022b. An overview on nanoparticle-based strategies to fight viral infections with a focus on COVID-19. J. Nanobiotechnol. 20, 1–26.

Yen, H.J., Hsu, S.H., Tsai, C.L., 2009. Cytotoxicity and immunological response of gold and silver nanoparticles of different sizes. Small 5, 1553–1561.

Yousefi, B., Banihashemian, S.Z., Feyzabadi, Z.K., Hasanpour, S., Kokhaei, P., Abdolshahi, A., Emadi, A., Eslami, M., 2022. Potential therapeutic effect of oxygen-ozone in controlling of COVID-19 disease. Med. Gas Res. 12, 33.

Yousefi, B., Eslami, M., 2022. Genetic and structure of novel coronavirus COVID-19 and molecular mechanisms in the pathogenicity of coronaviruses. Rev. Med. Microbiol. 33, e180–e188.

Yousefi, B., Valizadeh, S., Ghaffari, H., Vahedi, A., Karbalaei, M., Eslami, M., 2020. A global treatments for coronaviruses including COVID 19. J. Cell. Physiol. 235, 9133–9142.

Zhang, Q., Hitchins, V.M., Schrand, A.M., Hussain, S.M., Goering, P.L., 2011. Uptake of gold nanoparticles in murine macrophage cells without cytotoxicity or production of pro-inflammatory mediators. Nanotoxicology 5, 284–295.

Zhang, D., Neumann, O., Wang, H., Yuwono, V.M., Barhoumi, A., Perham, M., Hartgerink, J.D., Wittung-Stafshede, P., Halas, N.J., 2009. Gold nanoparticles can induce the formation of protein-based aggregates at physiological pH. Nano Lett. 9, 666–671.

Zhou, N., Wang, J., Chen, T., Yu, Z., Li, G., 2006. Enlargement of gold nanoparticles on the surface of a self-assembled monolayer modified electrode: a mode in biosensor design. Anal. Chem. 78, 5227–5230.

Further reading

Chauhan, G., Madou, M.J., Kalra, S., Chopra, V., Ghosh, D., Martinez-Chapa, S.O., 2020. Nanotechnology for COVID-19: therapeutics and vaccine research. ACS Nano 14, 7760–7782.

de Souza, G.A.P., Rocha, R.P., Goncalves, R.L., Ferreira, C.S., de Mello Silva, B., de Castro, R.F.G., Rodrigues, J.F.V., Junior, J., Malaquias, L.C.C., Abrahao, J.S., 2021. Nanoparticles as vaccines to prevent arbovirus infection: a long road ahead. Pathogens 10, 36.

CHAPTER

22 Fungal diversity and studies on euthermal hot spring water from Aravali region Maharashtra, India

Sulabha B. Deokar[a] and Girish R. Pathade[b]

[a]*Department of Biotechnology, Nowrosjee Wadia College, Pune, Maharashtra, India,* [b]*Department of Microbiology, Haribhai V. Desai College, Pune, Maharashtra, India*

22.1 Hot springs

A hot spring is described as a spring that discharges hot water at the temperature above normal local groundwater. It is also called thermal spring (Todd, 1980). Hot spring eco-system as lentic community has been studied to lesser extent. Hot spring is any geothermal natural spring with a water temperature above the average ambient ground temperature. Hot spring is a natural spring that is produced by the emergence of geothermally heated groundwater from the earth's crust. The hot springs from the west coast of Maharashtra are the circulation of water through several graben and faults which emerged during the foundering and attenuation of the continental crust before the emergence of the huge number of lavas along the coast (Chandrasekharam and Parthasarathy, 1978).

22.1.1 Aravali hot spring

Several hot springs are situated on the west coast of India which are extended over 300 km all along the west coast. The hot water spring of Aravali is situated in the village of Aravali in Ratnagiri district of Maharashtra, India. The thermal spring is located on the south side of the ridge above the Gad River. The temperature of thermal spring water is around 42°C. The thermal spring illustrates extreme environmental conditions such as alkalinity, metal concentration, high temperature, high sulfur concentration, etc.

The geothermal province of the west coast is supposed to be one of the significant geothermal prospects in India (Gupta et al., 1976). Aravali is a euthermal type of hot spring. It is situated in the Ratnagiri district of Maharashtra, India, and lies between 17° 08′ North latitude and 73° 19′ East longitude (Fig. 22.1).

Geothermal atlas of India was published by the Geological Survey of India in 1991 and in 2002; GSI compiled the geothermal energy resources of India in a special publication (Thussu, 2002).

Entrepreneurship with Microorganisms. https://doi.org/10.1016/B978-0-443-19049-0.00023-2

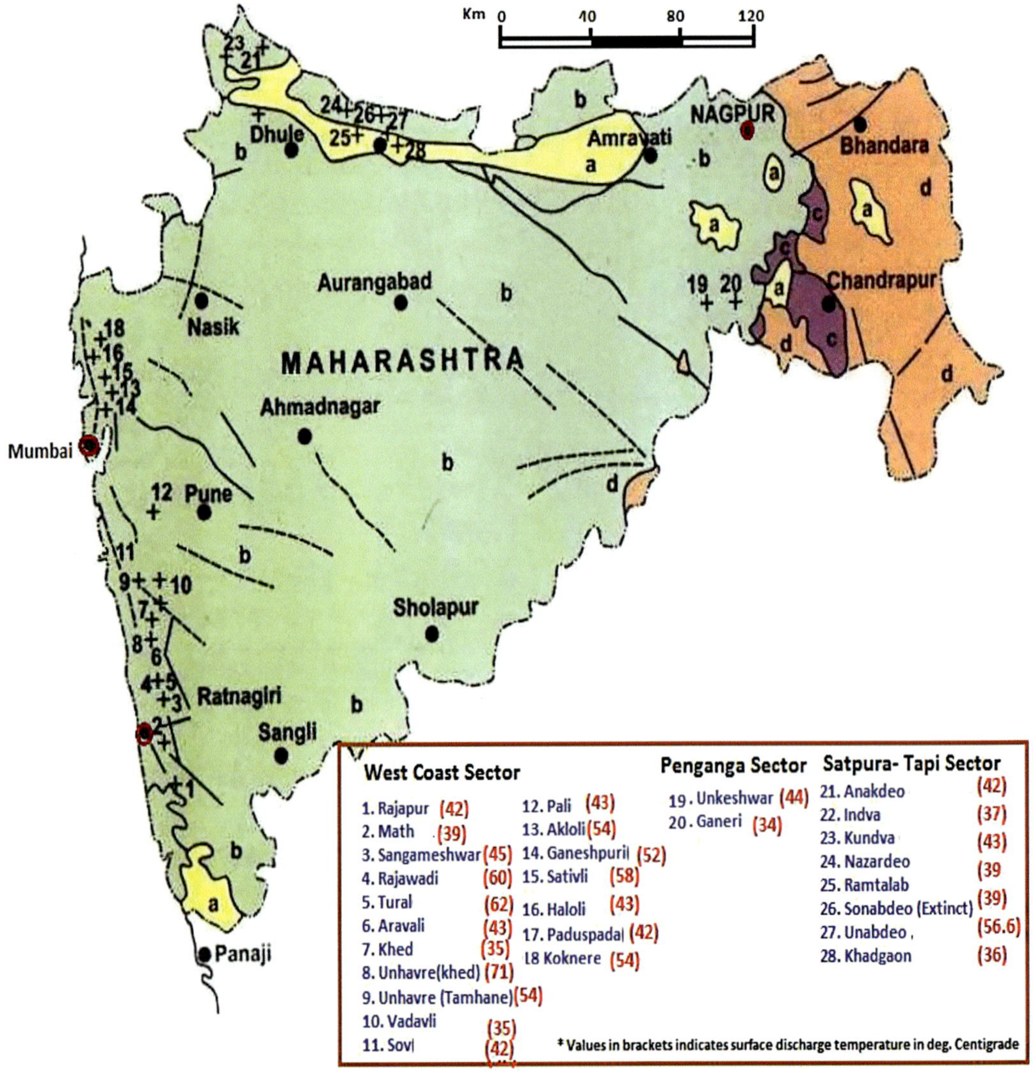

FIG. 22.1

Thermal springs of Maharashtra.

From Thussu, J.L., 2002. Geothermal Energy Resources of India. Geol. Surv. India, Spl. Pub. No. 69, p. 113.

22.1.2 Microorganisms of hot springs and their enzymatic activities

Temperature of thermal spring is one of the most important strand constituents directly affecting metabolism of organisms. Naturally occurring extreme temperatures of thermal springs in which microorganism can live are in a wider range. Some bacterial and fungal species can survive at temperatures approaching the boiling point. Extreme thermophilic microorganisms have an optimum temperature of growth above 70°C, and the ability of microorganisms to survive in extreme environmental condition have been considered the best resource of industrially important thermostable enzymes (Horikoshi and Akiba, 1982).

22.1.3 Fungi

In the greatest ecosystems, fungal species are the most important decomposers, playing a vital role in nutrient cycling as saprotrophs and symbionts that decompose biological and organic matter into inorganic particles. Thermophilic fungal species are those that have a maximum temperature range for growth at or above 50°C and a minimum temperature for growth at or above 20°C (Brock, 1978). Thermophilic fungi are important components in hot spring ecosystems; their ubiquity is mostly due to minor size and easy dispersal, their capacity to grow and reproduce also under extreme conditions. Extensive screening operations for alkaline fungi may lead to the finding of new extremozymes and could be useful in detergents, whereas alkaline tolerant protease is an important part (Horikoshi and Akiba, 1982). The exact fungal diversity and function in extreme environments is not completely understood yet.

Thermophilic fungi are a small group of eukaryotic microorganisms that have an exclusive mechanism of growing at elevated temperature ranges extending up to 60–62°C. They execute a crucial role in the decomposition of plant and animal structural polymers, such as, cellulose, hemicelluloses, and lignin, animal proteins thus contributing to the continuation of the global carbon and nitrogen cycle. Kambura et al. (2016) studied fungal diversity in water and sediment samples from the thermal springs of Lake Magadi and Little Magadi in Kenya. Moderate and weak alkalitolerant fungal species such as *Penicillium* sp., *Cladosporium* sp., *Alternaria*, and *Fusarium* sp. Reported that these species can grow optimally at neutral or below neutral pH values in laboratory conditions.

Ojha et al. (2017) isolated and identified two plastic degrading potential fungal strains, namely *P. chrysogenum* NS10 (KU559907) and *P. oxalicum* NS4 (KU559906) from a soil sample collected from the plastic dumping ground. These two fungal isolates are applied to degrade polyethylene sheets high-density polyethylene and low-density polyethylene in laboratory conditions.

22.1.4 Bioplastic degradation

Microbes such as fungal and bacterial species play an important role in the biodegradation of collectively bioplastics and synthetic plastics in nature. The biodegradation of bioplastics and plastics continues actively under various environments according to their natural properties, because the microorganisms liable for the degradation varied from each other. They have their own optimal ecological conditions in the soil. Polymers predominantly plastics are potential substrates for heterotrophic microorganisms (Glass and Swift, 1989).

22.1.5 Oxo-biodegradable plastic

Oxo-biodegradable plastic made from polymers such as polystyrene, polypropylene, and polyethylene containing extra elements for example, metal salts. Oxo-biodegradable plastic degrade and biodegrades in the open atmosphere in all environmental conditions. The supplementary metal salts catalyze the degradation process to enhance the rate of degradation so that oxo-plastic will degrade abiotically as well as biotically, in the presence of oxygen, much more rapidly than usual plastic material. Oxo-degradable plastics are plastic materials prepared to smash down into smaller fragments, in the presence of sunlight and having minute amounts of metal salts like cobalt, iron, or manganese.

22.2 Materials and methods

22.2.1 Collection of water samples from hot springs of Maharashtra regions

The present investigation was undertaken in the Department of Environmental Science, Fergusson College, Pune, affiliated to the Savitribai Phule Pune University, Maharashtra, India. Euthermal Aravali springs from Maharashtra, India were selected for this study.

22.2.2 Water sampling and locations of thermal springs under study

In the present investigation, euthermal Aravali spring from the Maharashtra region was selected for this study (Table 22.1). Water and sediment samples from hot water spring ware collected in one liter sterile thermos stainless steel container at a depth of 0.6–1.5 m from the water surface with the help of an extendable metal sampler.

22.2.3 Primary screening of the water samples for the study of fungal diversity

Hot spring water-based potato dextrose broth was used for fungal species enrichment purposes. Serial dilutions of enriched samples were performed from 10^{-1} to 10^{-7}. Last four dilutions were plated on Czapek Dox Agar agar plates. Colonies on selective media showing characteristic features were selected and confirmed by colony characters. These isolates were selected for further study.

22.2.4 Genetic characterization of promising isolates by 16-S rDNA sequencing

Advanced molecular biology techniques are excellent tools for complementing the identification of microorganism at species and subspecies level 18S rDNA sequencing at the stable part of genetic code has provided the most evolutionary chronometer because it is universally present in all fungi and over time its functions are unchanged 18S rDNA. Fungal genes (15,000 bp) are enough for bioinformatics studies (Patel, 2001).

22.2.5 Identification of fungal isolates

Identification of fungal isolates from hot springs water was carried out on the basis of morphological, colony characteristics, and microscopic examination with lacto phenol cotton blue stain.

Table 22.1 Euthermal Aravali spring.

Sr. no	Thermal spring	Sample I.D.	Latitude	Longitude
1	Aravali (Ratnagiri)	HWS-4	17.3096°N	73.5270°E

Morphological characteristics studied for identification were colony diameter, color (conidia), exudates and colony texture, the color of mycelia, etc. The characteristics used for microscopic examinations under the 10× and 45× lenses (Olympus-CXZli) were conidial heads, color, and length of vesicles, seriation, the shape of vesicle, metula covering, shapes size, and color of conidia (McClenny, 2005; Samson et al., 2013). Genus level identification was completed by using the key as cited in the text by Samson et al. (2013).

22.2.6 Molecular identification of fungal strain

The identification of isolate was carried out at the sequencing facility of National Centre for Microbial Resource (NCMR), National Centre for Cell Science, Pune. At the facility, genomic DNA was isolated by the standard phenol/chloroform extraction method (Sambrook et al., 1989), followed by PCR amplification of the ITS regions using universal primers ITS1 [5′-TCC GTA GGT GAA CCT GCG G -3′] and ITS4 [5′-TCC TCC GCT TAT TGA TAT GC -3′]. The amplified ITS PCR product was purified by PEG-NaCl precipitation and directly sequenced on an ABI 3730XL automated DNA sequencer (Applied Biosystems, Inc., Foster City, CA) as per the manufacturer's instructions. Essentially, sequencing was carried out from both ends so that each position was read at least twice. Assembly was carried out using a Lasergene package followed by NCBI BLAST against sequences from type material for tentative identification (Boratyn et al., 2013).

22.2.7 Determination of G + C content and phylogenetic tree

G + C content of promising isolates obtained from the secondary screening and having a higher potential for the environment and industrial applications were determined by—Genomics % G + C content calculator. Phylogenetic tree construction—The data obtained after 16S rDNA sequencing was used. Sequences having high similarities aligned using molecular evolutionary genetics analysis version six (MEGA-6) software and the aligned sequences applied to construct phylogenetic tree construction on the basis of maximum like hood (Tamura et al., 2013, https://blast.ncbi.nlm.nih.gov/Blast.cgi).

22.3 Bioplastic biodegradation

22.3.1 Collection of bioplastic sample

Environment friendly oxo-biodegradable plastic (OBP) was selected for this study. The plastic sample was collected from the local market, Vardhaman Packing, Pune-411 057.

22.3.2 Screening of bioplastic degrading organisms by clear-zone test

This method is a semiquantitative method. This is an agar plate test in which the polymer is dispersed as very fine particles within the synthetic agar medium, which results in the agar having an opaque appearance.

The emulsion of oxo-biodegradable plastic was prepared by homogenizing OBP (2 g/L) along with 0.05% (w/v) Triton X-100 in minimal salt medium (MSM) according to the method described by Nishida and Tokiwa (1993). Agar agar (20 gm/L) was added to the emulsion and sterilized by autoclaving. Plates were poured after autoclaving. After inoculation with isolates, the formation of a clear zone around the colony indicates the organisms are able to depolymerize the OBP polymer.

Primary screening of total four fungal isolates was tested for their bioplastic degradation activity. The promising isolates showing positive clear zone test were selected for secondary screening and on the basis of the S/R ratio promising isolates showing higher zone of clearance were selected for further study.

Formula: S/R ratio = Diameter of zone of hydrolyses (mm)/diameter of growth (mm).

22.3.3 Physical analysis and pretreatment of OBP film

OBP films were cut 2×2 cm in size and the thickness of the OBP film was measured using a digital Vernier caliper and micrometer screw gauge. Tensile strength of the OBP film was measured using C-clamps and traveling microscope. Surface sterilized using 70% ethanol exposed to the ultraviolet light for 5 min at 10,771 J/m^2 dose. Ethanol acts as a disinfecting agent for the OBP film and removes organic matter adhered to its surface. Lastly, the OBP films were aseptically transferred to a Petri dish and incubated at 45–50°C overnight.

22.3.4 Fungal degradation of OBP film in laboratory condition

Previously weighed films of OBP were aseptically transferred to the conical flasks containing 200 mL of minimal medium with pH 5.5, inoculated with promising fungal isolate *Aspergillus versicolor* strain GSF-A1. Control was maintained with OBP film in the microbe-free sterile potato-dextrose media for fungi (Ojha et al., 2017).

The experiment was carried out in a set of 4 flasks, each carrying 8 mg of OBP film in 100 mL media and inoculated with standardized fungal spore suspension 10 μL as inoculum. The flasks were incubated in an orbital shaker incubator at 42°C at 150 rpm for 1 month. Results were calculated after 10, 20, 30 days. The experiment was carried out in triplicates and a mean value was taken into consideration. Each set of flask was used for data collection such as physical characteristics of OBP film and measurement of weight loss at 10 days intervals.

22.3.5 Determination of weight loss

The residual OBP films were taken from the broth cultures. Fungal biofilm adhering to the OBP film surface was washed away by a 2% (v/v) aqueous sodium dodecyl sulphate solution for 2–3 h and finally with distilled water (Hadad et al., 2005). The washed OBP films were air-dried and weighed. The dry weights of semidegraded OBP films from the culture media were taken in intervals (i.e., day 0, day 10, day 20, day 30) from the data collected, and weight loss of the OBP films was calculated.

Percentage of decreasing OBP film weight was calculated by the formula:

$$\%\text{Decrease of plastic weight} = R1 - R2/R1 \times 100\%$$

where, R1 = initial weight of plastic film, R2 = final weight of plastic film.

22.3.6 Detection of change in pH

Exoenzymes secreted by the fungal breakdown the complex polymers into the short chains or monomers that are small enough to assimilate through the cell walls to be utilized as carbon and energy sources by a process of depolymerization (Dey et al., 2012). The change in the pH value of the liquid medium containing OBP as a sole carbon source was determined by using pH meter and compared with its respective initial pH value after an interval of 10 days.

22.3.7 Scanning electron microscopy (SEM) of OBP film

The OBP films were scanned for surface changes through scanning electron microscopy, before incubation (control) and after incubation of ten days for fungal isolates (Geed et al., 2016). The OBP films were mounted on the copper stubs along with gold paint after washing with distilled water. Gold coating was done in a vacuum by evaporation to make the OBP film conducting in nature. The images of the test samples were compared with those of abiotic control. energy dispersive spectroscopy (EDS) analysis based on X-rays emitted from sample processing while electron irradiation. SEM analysis was carried out by central facility center, Savitribai Phule Pune University, Pune.

22.3.8 Fourier transform infrared (FT-IR) spectroscopy analysis

FT-IR was used for detection of chemical changes such as appearance or disappearance of new functional groups and chemical bond scission of fungal strains. FT-IR spectrum of biodegraded OBP film with abiotic control was taken and compared with the results of the test sample. Changes in the chemical properties of the polymer in synthetic media, including the formation or disappearance of functional groups were determined by FT-IR.

22.4 Results

According to Vouk (1950), the temperature of the discharged hot water spring ranges from 40°C to 60°C is considered as "Euthermal" spring, so Aravali hot spring is considered as euthermal spring having temperature 42°C.

22.4.1 Identification of fungal diversity

See Tables 22.2 and 22.3 and Plate 22.1.

22.4.2 Molecular identification of fungal isolate strain GSF-A1

16S rDNA sequencing and G + C content of promising isolate GSF-A1 (Fig. 22.2):

Phylogenetic tree and G + C content of promising isolates (Tables 22.4–22.6 and Fig. 22.3)

Table 22.2 Colony characteristics and morphology of fungal isolates.

Sr.no	Characteristics	GSF-A1	GSF-A2	GSF-A3	GSF-A4
1.	Medium used	Czapek Dox Agar	Czapek Dox Agar	Czapek Dox Agar	Czapek Dox Agar
2.	Incubation temperature (°C)	42	42	42	42
3.	Incubation period (days)	7	4	3	4
4.	Colony	Velvety	Velvety		Velvety
5.	Color	Pale green	Yellow to brown	Green	Bluish green with a white border
6.	Size (mm)	3	4	3–7	2–4
7.	Shape	Irregular	Irregular	Irregular	Irregular
8.	Consistency	Rough	Dry	Rough	Dry powdery
9.	Reverse colony	Yellowish	Red brown	Olive green	Yellow
10.	Microscopic observation				
	(i) Hyphae	Septate	Septate	Septate	Branched
	(ii) Conidiophore	Branched	Smooth	Smooth walled	Round
	(iii) Conidia	Slightly rough, round conidia, spiny chain	Yellow numerous	short, columnar, and biseriate	Chain of conidia
	(iv) Phialides/annel	Loosely radiate	Biseriate	Flask shaped	Flask shaped
11.	Species	*Aspergillus versicolor*	*Aspergillus flavus*	*Aspergillus nidulans*	*Penicillium species*

Table 22.3 Summary of output obtained from colony characteristics and microscopic observations of different fungal isolates from the hot water spring HWS-4 Aravali hot spring.

Isolate ID	Species identified
GSF-A1	*Aspergillus versicolor*
GSF-A2	*Aspergillus flavus*
GSF-A3	*Aspergillus nidulans*
GSF-A4	*Penicillium species*

22.5 Bioplastic biodegradation

Fungal degradation of OBP by total of four promising isolates was analyzed by clear-zone test on OBP emulsion agar, then the most promising fungal secondary isolate was selected for this study. Fungal degradation of OBP film in situ was estimated and analyzed by dry weight reduction of OBP film, change in pH of the medium, scanning electron microscopy (SEM) and Fourier transferred infrared (FT-IR) spectroscopy of the OBP film surface. Fungal isolate *Aspergillus versicolor* strain GSF-A1 exhibited depolymerization of OBP film efficiently in aerobic conditions (Table 22.7, Plate 22.2).

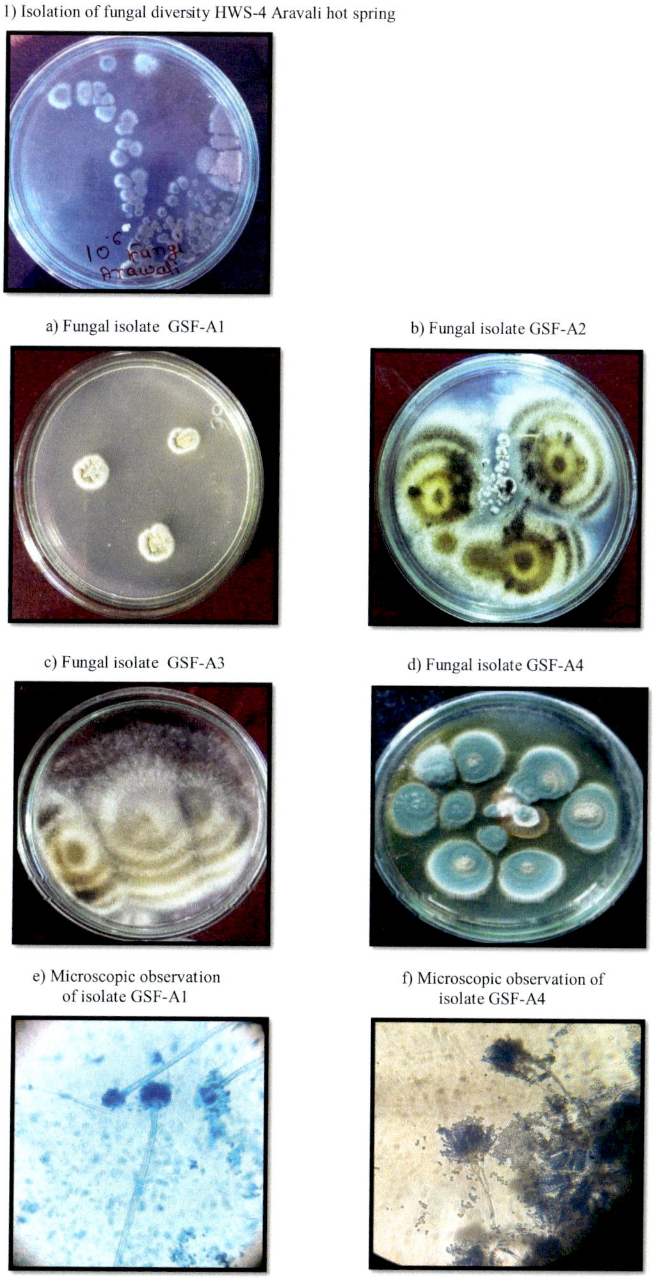

1) Isolation of fungal diversity HWS-4 Aravali hot spring
a) Fungal isolate GSF-A1
b) Fungal isolate GSF-A2
c) Fungal isolate GSF-A3
d) Fungal isolate GSF-A4
e) Microscopic observation of isolate GSF-A1
f) Microscopic observation of isolate GSF-A4

PLATE 22.1

Fungal diversity of HWS-4 Aravali hot spring.

```
GAACCTGCGGAAGGATCATTACTGAGTGCGGGCTGCCTCCGGGCGCCCAACCTC
CCACCCGTGAATACCTAACACTGTTGCTTCGGCGGGGAACCCCCTCGGGGGCGA
GCCGCCGGGGACTACTGAACTTCATGCCTGAGAGTGATGCAGTCTGAGTCTGAA
TATAAAATCAGTCAAAACTTTCAACAATGGATCTCTTGGTTCCGGCATCGATGAA
GAACGCAGCGAACTGCGATAAGTAATGTGAATTGCAGAATTCAGTGAATCATCG
AGTCTTTGAACGCACATTGCGCCCCCTGGCATTCCGGGGGGCATGCCTGTCCGAG
CGTCATTGCTGCCCATCAAGCCCGGCTTGTGTGTTGGGTCGTCGTCCCCCCCGGG
GGACGGGCCCGAAAGGCAGCGGCGGCACCGTGTCCGGTCCTCGAGCGTATGGG
GCTTTGTCACCCGCTCGACTAGGGCCGGCCGGGCGCCAGCCGACGTCTCCAACC
ATTTTTCTTCAGGTTGACCTCGATCAGGTAGGGATACCCGCTGAACTTAAGCATA
TCAATAAGCGG
```

FIG. 22.2

Genomic sequences of isolate GSF-A1, sequence text (in FASTA format).

Table 22.4 Sequence producing significant alignments of GSF-A1.

Sequences producing significant alignments:

Select: All None Selected:6

Alignments Download GenBank Graphics Distance tree of results

	Description	Max score	Total score	Query cover	E value	Ident	Accession
✔	Aspergillus sp. small subunit ribosomal RNA gene, partial sequence; internal transcribed spacer 1, 5.8S riboso	993	993	100%	0.0	99%	MF440628.1
✔	Aspergillus sydowii genomic DNA sequence contains ITS1, 5.8S rRNA gene, ITS2, strain FMR 14440	993	993	100%	0.0	99%	LN898735.1
✔	Aspergillus sp. BAB-5762 18S ribosomal RNA gene, partial sequence; internal transcribed spacer 1, 5.8S ribos	993	993	100%	0.0	99%	KX228408.1
✔	Fungal sp. strain OTU48 18S ribosomal RNA gene, partial sequence; internal transcribed spacer 1, 5.8S ribos(	993	993	100%	0.0	99%	KT923212.1
✔	Aspergillus versicolor genes for ITS1, 5.8S rRNA, ITS2, partial and complete sequence, strain: DY20.1.1	993	993	100%	0.0	99%	LC105698.1

Table 22.5 Summary of the closest neighbor(s) for GSF-A1.

Strain No	Closest neighbor	Accession no. %	Similarity
GSF-A1	*Aspergillus versicolor*	ATCC 9577 NR_131277.1	99%

Table 22.6 Summary of output obtained for molecular identification of hot water spring fungal isolate using BLAST.

Isolate ID	Length of sequence in bp	G+C content (%)	Accession no of closest relative obtained in BLAST	Description	% similarity
GSF-A1	554	58.1	ATCC 9577 NR_131277.1	*Aspergillus versicolor*	99%

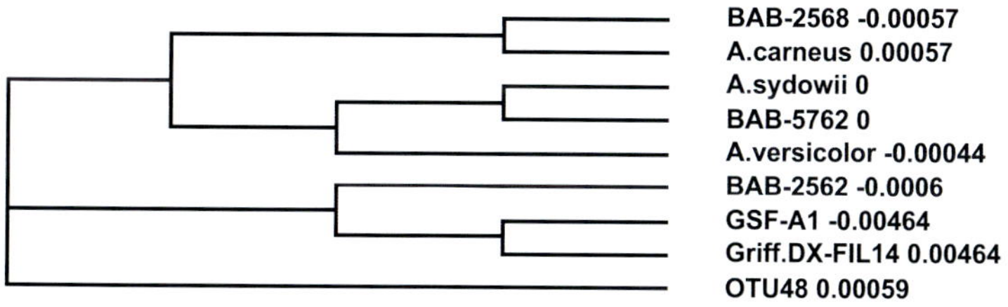

FIG. 22.3

Phylogenetic tree showing relationship with novel strain GSF-A1 with other *Aspergillus* species.

Table 22.7 Visual changes in OBP films after biodegradation after 10 and 20 days by fungal isolate *Aspergillus versicolor* strain GSF-A1.

Sr. no.	Isolate ID	Physical properties		Observation		
				Initial	After 10 days	After 20 days
1.	*Aspergillus versicolor* strain GSF-A1	(i)	Thickness	21 μm	17.5 μm	3.6 μm
		(ii)	Surface	Smooth	Rough hairy fragments	Rough hairy fragments
		(iii)	Color	Green	Faint green	Colorless
		(iv)	Formation of holes	No	Yes	Yes
		(v)	Formation of biofilm on surface	No	Bulgy and fluffy biofilm	Bulgy and fluffy biofilm
		(vi)	Defragmentation	No	Yes	Yes
		(vii)	Tensile strength	1.55 MPa	0.25 MPa	Nil
		(viii)	Percent (%) degradation	0	73	90

a) Bioplastic before degradation

b) Degraded and decolorized OBP film by GSF-A1 after 20 days

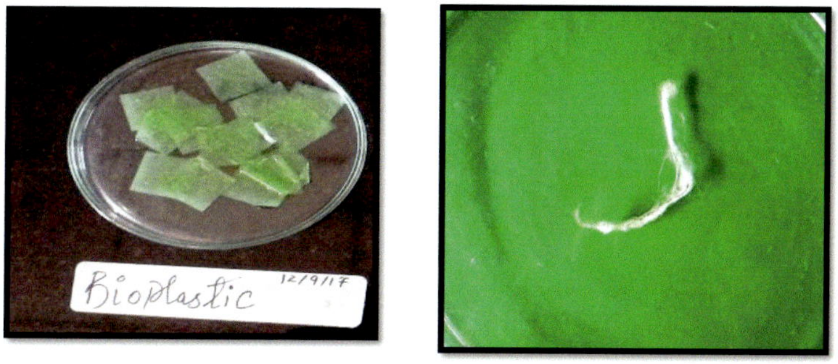

PLATE 22.2

Bioplastic degradation.

22.5.1 Screening of bioplastic degrading fungal species

22.5.1.1 Clear-zone test and S.R. value

Microbial degradation of OBP total of four promising isolates were analyzed by clear-zone test on OBP emulsion agar, Out of four different strains, the most promising fungal strain was selected on the basis of highest S.R. value for this study and designated as *Aspergillus versicolor* strain GSF-A1 (Table 22.8).

22.5.1.2 Determination of weight loss

See Fig. 22.4.

22.5.1.3 Change in pH

Fungal isolate *Aspergillus versicolor* strain GSF-A1 showed the production of acids and some metabolites with the indication of the change in a decrease of pH value supporting the metabolic activity of strains on the OBP film and also exhibits the potential degradation. The pH value of the medium

Table 22.8 S.R. value of selected after secondary screening of promising isolates.

Isolate ID	Diameter of colony (mm)	Diameter of zone (mm)	S.R. value
GSF-A1	10	40	4
GSF-A2	3.5	7	2
GSF-A3	6	18.1	3.01
GSF-A4	3	9	3

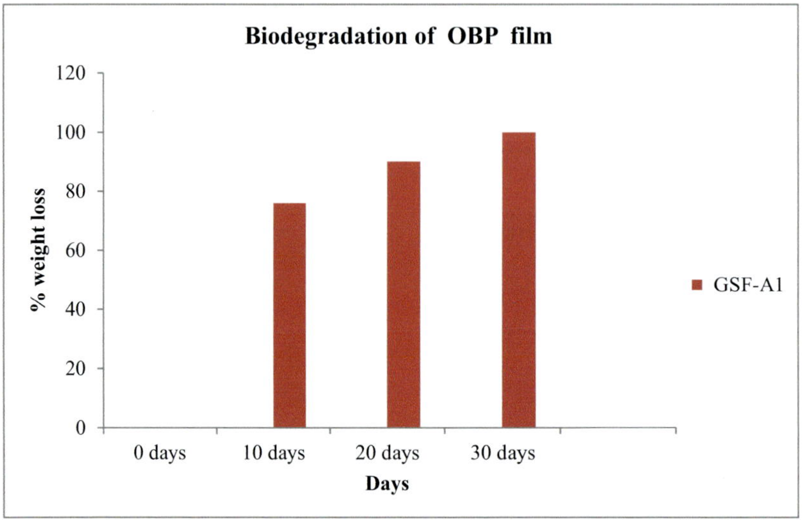

FIG. 22.4

% losses in weight of OBP film by promising fungal isolate *Aspergillus versicolor* strain GSF-A1.

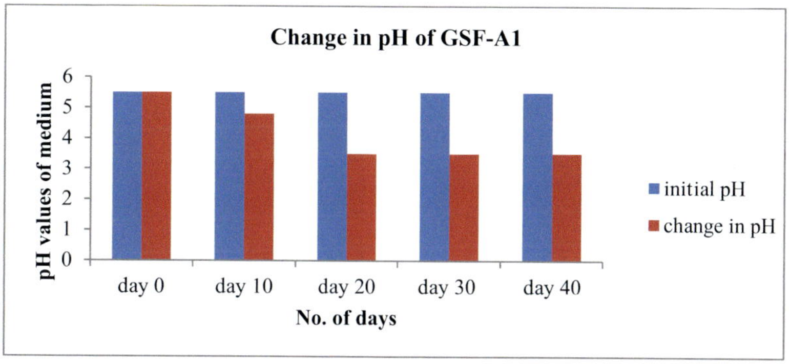

FIG. 22.5

Change in pH value of promising fungal isolate *Aspergillus versicolor* strain GSF-A1 containing medium.

containing OBP film and GSF-A1 was reduced drastically from 5.5 to 3.6 after 30 days. The reduction in pH value indicates culture is still metabolically active and confirms the usage of OBP film as a sole source of carbon (Fig. 22.5).

22.5.1.4 The SEM images of OBP film before and after biodegradation

The SEM images of OBP film before and after biodegradative promising isolate were able to break down the complex polymer of oxo-biodegradable plastic into its fragments and monomers. The cracks and grooves in the image confirmed (Plate 22.3) give further confirmation about the fragility of bioplastic sheet fungal degradation.

22.5.1.5 FT-IR analysis of OBP film

Fourier transform infrared spectra of OBP film after incubation of 20 days with GSF-A1. reports exhibited that the shortening of peaks was due to the degradation of the polymer (Plate 22.4).

22.6 Discussion

In the present investigation, fungal strain was isolated from the Aravali hot springs of Maharashtra. *Aspergillus versicolor* GSF-A1 strain was found to be beneficial for biodegradation of oxo-biodegradable plastic film having applicable evidence. Tambe et al. (2016) studied synthesis of microbial biodegradable starch-based bioplastic and evaluated ex situ biodegradation by *Bacillus amyloliquefaciens and Micrococcus luteus.*

Das and Kumar (2015) studied biodegradation of low density polyethylene by *Bacillus amyloliquefaciens* is reported potential of this organism strain to degrade LDPE film in short span of time. Aamer Ali et al. (2015) studied Degradation of poly (ε-caprolactone) by a thermophilic bacterium *Ralstonia* sp. strain MRL-TL isolated from hot spring Tatta Pani.

FT-IR and SEM reports clearly confirm the significant superficial degradation of oxo-biodegradable plastic film and also the change in structural and functional characteristics. The environmental factors like elevated temperature, sunlight, and moisture content play an important role in the enhancement of biodegradation of OBP film.

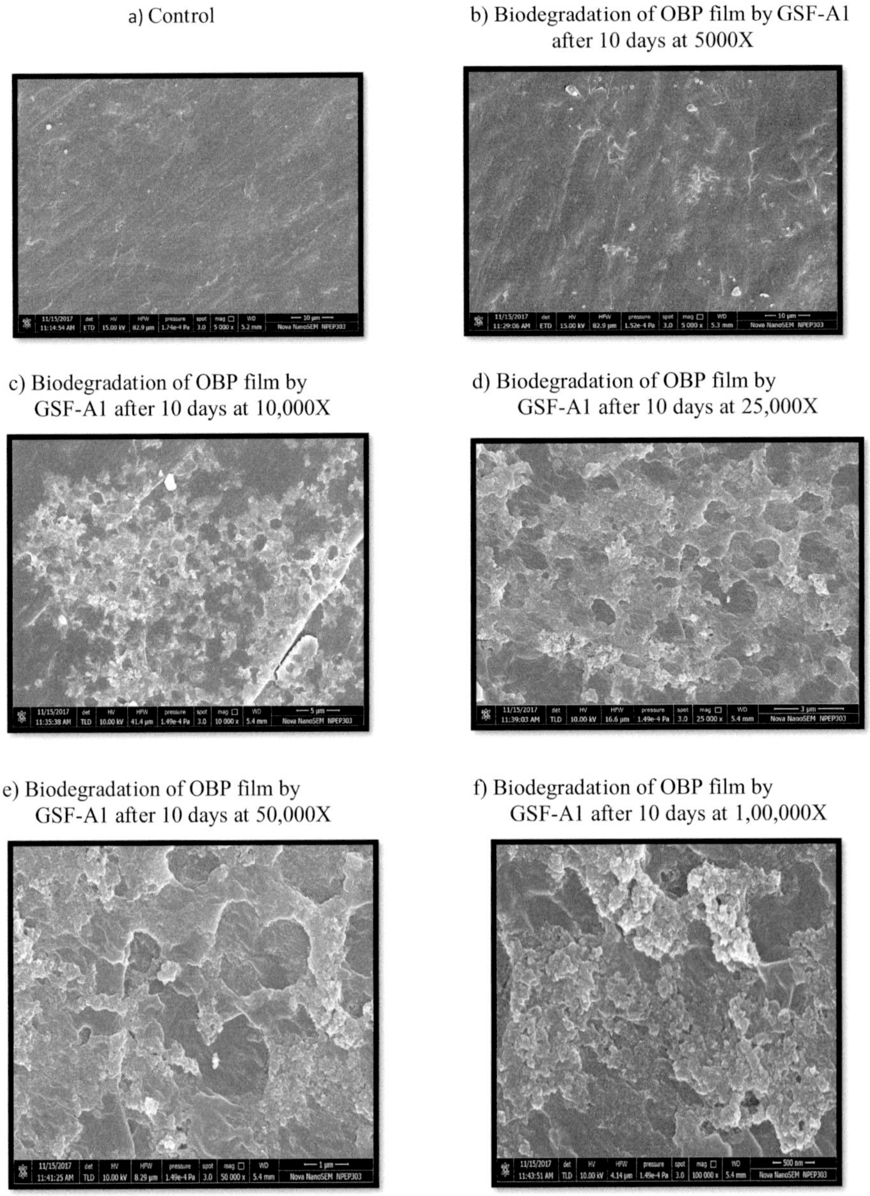

a) Control
b) Biodegradation of OBP film by GSF-A1 after 10 days at 5000X
c) Biodegradation of OBP film by GSF-A1 after 10 days at 10,000X
d) Biodegradation of OBP film by GSF-A1 after 10 days at 25,000X
e) Biodegradation of OBP film by GSF-A1 after 10 days at 50,000X
f) Biodegradation of OBP film by GSF-A1 after 10 days at 1,00,000X

PLATE 22.3

(A and B) SEM report confirmed the filament perforation and decrease of viscosity during degradation of OBP film by *Aspergillus versicolor* strain GSF-A1 indicated hydrolysis and polymer backbone cleavage. (C–F) SEM reports confirmed the filament perforation and decrease of viscosity during degradation of OBP film indicated hydrolysis and polymer backbone cleavage by promising fungal isolate *Aspergillus versicolor* strain GSF-A1.

a) Control sample of OBP film for FT-IR analysis

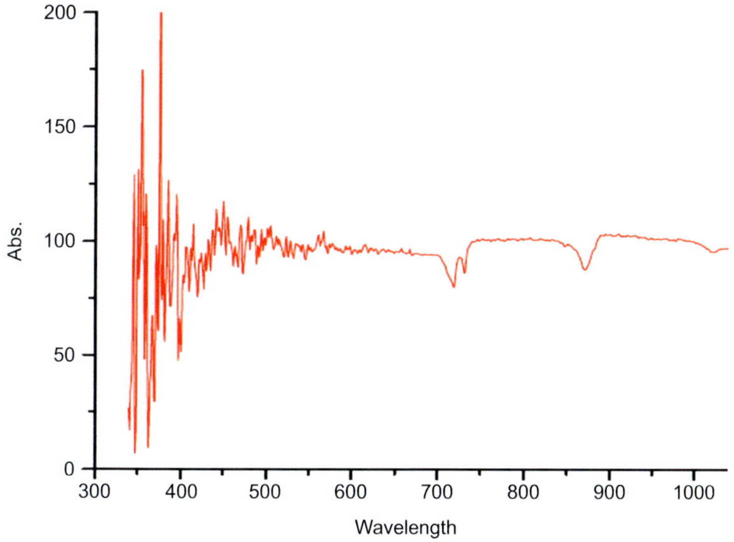

b) FT-IR analysis of OBP film after promising fungal isolate *Aspergillus versicolor* strain GSF-A1 action

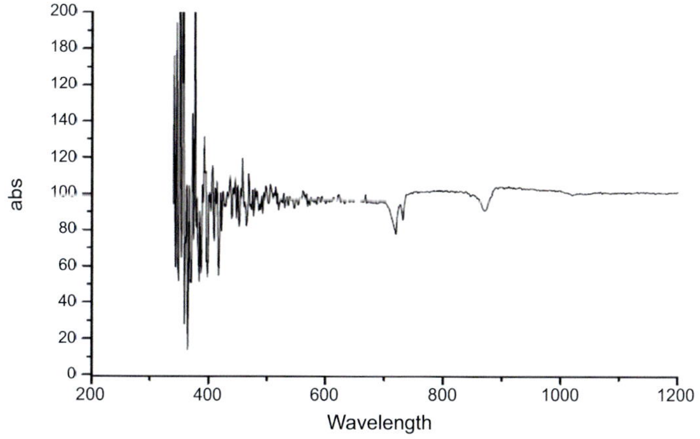

PLATE 22.4

Fourier transform infrared (FT-IR) analysis of OBP film.

All the biotic degradation studies indicated that the isolated fungal strain *Aspergillus versicolor* GSF-A1 exhibited potential degradation which was in a very short span of time that is within 10 days 73% and 20 days 90% and 30 days 100% (Table 22.9).

Gajendiran et al. (2016) studied microbial degradation of low-density polyethylene by *Aspergillus clavatus* strain JASK1 isolated from land field soil and observed that 35% weight loss after 90 days of incubation.

Table 22.9 % loss in OBP film by promising fungal isolate *Aspergillus versicolor* strain GSF-A1 collected from Aravali hot spring.

Organisms	% degradation after 10 days	% degradation after 20 days	% degradation after 30 days
GSF-A1	73	90	100

Above reports demonstrated that a fungal strain isolated from the Aravali hot spring *Aspergillus versicolor* GSF-A1 acts as a potential source of bioplastic degrading microbial scavenger. There is a strong need to develop high technology capable of degrading bioplastic without having any effect on the environment for the sustainable development of the ecosystem.

References

Aamer Ali, S., Ahmed, N., Kanwal, L., Hasan, F., Khan, S., Malik, B., 2015. Degradation of poly(ε-caprolactone) by a thermophilic bacterium Ralstonia sp. strain MRL-TL isolated from hot spring. Int. Biodeterior. Biodegradation 98, 35–42.

Boratyn, et al., 2013. BLAST: a more efficient report with usability improvements. Nucleic Acids Res. 41, W29–W33.

Brock, T.D., 1978. Thermophillic Micro-organisms and Life at High Temperatures. Springer Verlag, New York.

Chandrasekharam, D., Parthasarathy, A., 1978. Geochemical and tectonic studies on the coastal and inland Deccan Trap volcanics and a model for the evolution of Deccan Trap volcanism. N. Jb. Mineral. (Abh.) 132, 214–229.

Das, M.P., Kumar, S., 2015. An approach to low-density polyethylene biodegradation by Bacillus amyloliquefaciens. 3 Biotech 5 (1), 81–86.

Dey, U., Mondal, N.K., Das, K., Dutta, S., 2012. An approach to polymer degradation through microbes. IOSR J. Pharm. 2, 385–388.

Gajendiran, A., Krishnamoorthy, S., Abraham, J., 2016. Microbial degradation of low-density polyethylene (LDPE) by Aspergillus clavatus strain JASK1 isolated from landfill soil. 3 Biotech 6 (1).

Geed, S.R., Kureel, M.K., Shukla, A.K., Singh, R.S., Rai, B.N., 2016. Biodegradation of malathion and evaluation of kinetic parameters using three bacterial species. Resour. Efficient Technol. 2, S3–S11.

Glass, J.E., Swift, G., 1989. Agricultural and Synthetic Polymers, Biodegradatio and Utilization. ACS Symposium Series, vol. 433 American Chemical Society, Washington, DC, pp. 9–64.

Gupta, M.L., Saxena, V.K., Sukhija, B.S., 1976. An analysis of the hot spring activity of the Manikaran area, Himachal Pradesh, India; by geochemical studies and tritium concentration of spring waters. In: Proceed. 2nd UN Symp. Development and Use of Geothermal Resources, San Francisco, 1, pp. 741–744.

Hadad, D., Geresh, S., Sivan, A., 2005. Degradation of polyethylene by the thermophilic bacterium Brevibacillus borstelensis. J. Appl. Microbiol. 98, 1093–1100.

Horikoshi, K., Akiba, T., 1982. Alkalophilic Microorganisms: A New Microbial World. Springer-Verlag KG, Heidelberg, Germany.

Kambura, A.K., Mwirichia, R.K., Kasili, R.W., Karanja, E.N., Makonde, H.M., Boga, H.I., 2016. Diversity of fungi in sediments and water sampled from the hot springs of Lake Magadi and Little Magadi in Kenya. Afr. J. Microbiol. Res. 10 (10), 330–338.

McClenny, N., 2005. Laboratory detection and identification of *Aspergillus species* by microscopic observation and culture. Med. Mycol. 1, 125s–128.

Nishida, H., Tokiwa, Y., 1993. Distribution of poly (β-hydroxybutyrate) and poly (ε-caprolactone) aerobic degrading microorganisms in different environments. J. Environ. Polym. Degr. 1, 227–233. https://doi.org/10.1007/BF01458031.

Ojha, N., Pradhan, N., Singh, S., Barla, A., Shrivastava, A., Khatua, P., Bose, S., 2017. Evaluation of HDPE and LDPE degradation by fungus, implemented by statistical optimization. Sci. Rep. 7, 39515.

Patel, J.B., 2001. 16S rRNA gene sequencing for bacterial pathogen identification in the clinical laboratory. Mol. Diagn. 6 (4), 313–321.

Sambrook, J., Fritsch, E.F., Maniatis, T., 1989. Cold Spring Harbor. Cold Spring Harbor Laboratory Press, New York, p. 1626.

Samson, R.A., Evans, H.C., Latgé, J.P., 2013. Atlas of Entomopathogenic Fungi. Springer Science & Business Media, p. 528.

Tambe, A., Inamdar, M., Indonesiawala, Q., Mutalik, K., Deokar, S., 2016. Synthesis of microbial biodegradable starch-based bioplastic. Asian J. Multidiscip. Stud. 4 (9).

Tamura, K., Stecher, G., Peterson, D., Filisky, A., Kumar, S., 2013. MEGA6: molecular evolutionary genetics analysis version 6.0. Mol. Biol. Evol. 30, 2725–2729.

Thussu, J.L., 2002. Geothermal Energy Resources of India. Geol. Surv. India, Spl. Pub. No. 69, 210 p.

Todd, D.K., 1980. Ground Hydrology, second ed. Wiley, New York.

Vouk, V., 1950. Grundriß zu einer Balneobiologie der Thermen. Birkhanser, Basel. 88 pp.

CHAPTER

23 Specialized microbial metabolites: Their origin, functions, and industrial applications

Annie Jeyachristy Sam[a], Jannathul Firdous[a], and Gokul Shankar Sabesan[b]

[a]Faculty of Medicine, Royal College of Medicine Perak, Universiti Kuala Lumpur, Ipoh, Perak, Malaysia,
[b]Faculty of Medicine, Manipal University College Malaysia, Melaka, Malaysia

23.1 Introduction

Microorganisms produce antibiotics, growth hormones, antitumor agents, and pigments which are known as microbial secondary metabolites. Specialized metabolites, also known as secondary metabolites, are small organic molecules produced by microorganisms, including bacteria, fungi, and other microbes. They are not essential for the growth and development of microorganisms but play essential roles in microbial survival and interactions with their environment. Microbiomes, which are the communities of microorganisms living in a particular habitat, are rich sources of diverse specialized metabolites with various origins, functions, and industrial applications. Secondary metabolites have profound applications in the medical and pharmaceutical industries.

Secondary metabolites from bacteria, such as actinobacteria and fungi are produced by activating cryptic gene clusters, usually inactive under normal conditions. Endophytic microorganisms are another candidate that produces vital secondary metabolites that have industrial applications like other microorganisms. This chapter will focus on the source, functions, and applications of biologically relevant secondary metabolites produced by different microorganisms.

23.2 Microbial metabolites

Microbes are the primary source of nutrients and are essential to human, plant, and animal life. The natural products obtained from the microbes, known as metabolites, have demonstrated their use in healthcare including nutrition and pharmaceuticals, textile, paper, cosmetics, and agriculture (Singh et al., 2017).

Primary metabolites are produced by the microbes and are required for their cell maintenance, survival, development, and growth. Primary metabolites include amino acids, nucleotides, and fermentation end products such as ethanol and organic acids (Pinu et al., 2017).

Secondary metabolites are known as specialized metabolites, that are nonessential to microbes. These metabolites help the organism by providing nutrients, protection against external stress, and help

Entrepreneurship with Microorganisms. https://doi.org/10.1016/B978-0-443-19049-0.00022-0

to interact with other organisms. They do not contribute to the growth, development, and reproduction of microorganisms. The specialized metabolites are formed during the end of the stationary growth phase. The specialized metabolites are predominantly used in healthcare activities as antimicrobial agents, antiparasitic agents, antitumor agents, enzyme inhibitors, immunosuppressive, anesthetics, antiinflammatory agents, anticoagulants, anabolics, hemolytics, hypocholesterolemics, and vasodilators. Secondary metabolites possessing the properties of plant growth stimulants, herbicides, and insecticides have also been reported. Around 53% of the FDA-approved drugs from natural products originate from microbes (Patridge et al., 2016).

Specialized metabolites play a vital role in microbial competition by inhibiting the growth of other microorganisms. Antibiotics produced by bacteria and fungi are examples of specialized metabolites that help microbes compete for limited resources (Hibbing et al., 2010). Specialized metabolites are produced by microbes as a defense mechanism against predators or pathogens (Künzler, 2018). Specialized metabolites also serve as signaling molecules in microbial communities, enabling intercellular communication and coordination of collective behaviors. Quorum sensing molecules produced by bacteria are examples of specialized metabolites involved in communication (Rutherford and Bassler, 2012).

23.3 Source of specialized metabolites in microbiomes

Specialized metabolites are synthesized through secondary metabolic pathways. These pathways are regulated by complex genetic networks, and the expression of specialized metabolite genes is often influenced by environmental factors, such as nutrient availability, temperature, pH, and interactions with other microorganisms. Specialized metabolites are produced by bacteria, fungi, and endophytes. Interaction between different microorganisms also contributes to the production of specialized metabolites.

Bacteria: Many specialized microbial metabolites are produced by bacteria. These metabolites are synthesized by bacteria through complex metabolic pathways encoded in their genomes. Antibiotics like penicillin and streptomycin, and other bioactive compounds like siderophores (iron-chelating molecules) and cyanobacterial toxins.

Fungi: Fungal specialized metabolites possess diverse chemical structures and biological activities. Antibiotics like tetracycline and erythromycin, and mycotoxins, such as aflatoxins and trichothecenes.

Microbial interactions: Microbiomes are a complex network of microorganisms where the interaction of microorganisms leads to the exchange of genetic material. During this interaction, specialized metabolite biosynthetic genes are also transferred between different microorganisms and this horizontal gene transfer contributes to the diversity and origin of specialized metabolites in microbiomes.

23.4 Synthesis of microbial specialized metabolites

Specialized metabolites include antibiotics (de Lima Procópio et al., 2012; Jakubiec-Krzesniak et al., 2018), siderophores (Kramer et al., 2020), quorum sensing molecules (Rodrigues and Černáková, 2020), immunosuppressants (Feng et al., 2022; Singla et al., 2014), and degradative enzymes

(Adlan et al., 2020). The ecological and biogeochemical effects of specialized metabolites have established their prospective uses in the field of medicine and biotechnology. Most of the antibiotics are from actinobacteria, proteobacteria, and firmicutes (Al-Shaibani et al., 2021; Srinivasan et al., 2021). Biosynthetic gene clusters (BGC) are the group of genes that are responsible for the production of secondary metabolites. The two classes of BGCs are the nonribosomal peptide synthetases (NRPS) and polyketide synthases (PKS) that comprise most of the antibiotics and antifungals. The condensation (CD) and the adenylation (AD) domains in NRPS and the ketosynthase (KS) and other enzymatic domains in PKS help to identify the novel NRPS and PKS gene clusters which are used as a proxy for biosynthesis (Payne et al., 2016; Tang et al., 2017). Bacterial pathogens like *Escherichia coli*, *Shigella* spp., and *Salmonella* spp. produced toxins are usually plasmid encoded NRPS or PKS. The presence of these gene clusters in the intestinal tract and its participation in host-microbe interactions serves as a proof that secondary microbial metabolites play a crucial role in the biology of all living things due to their practically ubiquitous distribution of these complicated pathways (Davies, 2013). Also, they are known to manufacture a variety of antibiotics and immunosuppressants with tremendous pharmaceutical potential, in which NRPS and PKS are popular targets for natural product discovery (Tillett et al., 2000). Surfactins, iturins, and fengycins, the NRPS mediated products from *Bacillus* sp. are known for their potent antimicrobial activities (Fazle Rabbee and Baek, 2020). Genomic mining identifies the genetic characteristics of the biological functions and bioactive secondary metabolites, directly or indirectly (Lee et al., 2020).

23.5 Techniques to identify and analyze metabolites

PKS and NRPS gene clusters offer the perquisite of rapid screening for secondary metabolism biosynthetic pathways. They are also an effective tool as a proxy for investigating the metabolic potential of endophytes. The bioinformatic methods to identify the gene clusters include "NP.searcher," "ClustScan," "CLUSEAN," "antiSMASH," "SMURF," "MIDDAS-M," "ClusterFinder," "CASSIS/SMIPS," "C-Hunter," "NRPSpredictor," and "SBSPKS" (Chavali and Rhee, 2018). The bioinformatics tools such as SMURF (Secondary Metabolite Unknown Regions Finder), PRISM (PRediction Informatics for Secondary Metabolomes), and antiSMASH (Antibiotics and Secondary Metabolite Analysis Shell) play a vital role in the identification of specific gene clusters involved in the synthesis of bioactive metabolites (Khaldi et al., 2010; Skinnider et al., 2020; Weber et al., 2015). The metabolites are analyzed using advanced techniques such as chromatography, liquid chromatography (LC) and gas chromatography (GC), and mass spectrometry (MS). Nuclear Magnetic Resonance (NMR) spectroscopy and MS are used to identify and quantify the metabolites but have limitations toward the conditions of the culture and metabolite types (Dona et al., 2016). High performance liquid chromatography tandem mass spectrometry (HPLC-MS/MS) and GC-MS are widely used analytical methods. GC-MS is used only for volatile metabolites and hence more advantageous than LC-MS which shows dynamic coverage, specificity, and simple preparation methods. Recent advanced MS technology such as triple quadruple tandem MS (QQQ-MS/MS) and quadruple time-of-flight MS (QTOF-MS) produce results with accuracy, sensitivity, and specificity. QQQ-MS/MS is used to identify and analyze known compounds and used for targeted studies. QTOF-MS offers accurate results with better resolution in the identification of unknown compounds hence used for untargeted studies (Mohd Kamal et al., 2022).

23.6 Specialized metabolites from marine microbes—Sources and applications

Marine natural products can be categorized either by the phylum that produces them or by their chemical structures such as mycosporine-like amino acids, polysaccharides, carotenoids, polyphenolic compounds, fatty acids, peptides, and alkaloids (Ghareeb et al., 2020). Marine species like sponges have been involved in the formation of drug such as xestospongin C and several manoalides. Antimalarials like salinipostins, cytotoxics such as marinomycins to antibacterials such as abyssomicins were isolated from actinobacteria (Dhakal et al., 2017). Monensin, a chelating agent produced by *Dermacoccus nishinomiyaensis* and dermacozines from *Dermacoccus abyssi* showed cytotoxicity against leukemia (Santos et al., 2019). Commercialized products such as marine-derived mycosporine-like amino acids (MAAs) aim at photoprotection and antiaging effects (Brunt and Burgess, 2018). From *Kappaphycus, Gigartina, Chondrus*, and many more species, there are two types of marine-derived polysaccharides such as sulfated polysaccharides (such as fucoindans, carrageenans, laminarans, galactans, and ulvan) and nonsulfated polysaccharides (e.g., alginates and agars) that have many industrial applications (Ruocco et al., 2016). The DPPH and reducing power assays revealed that marine extracts containing carotenes were a promising source of antioxidants, and they also protected human keratinocytes from UV-induced damage (Jeong et al., 2022). The polyphenolic compounds from marine extracts, *Padina boergesenii* possess antioxidant potential to act as UV shields on UV-induced damage of keratinocytes (Soleimani et al., 2023). Fatty acids isolated from single-cell organisms, such as yeast and mold, and denoted single-cell oils (SCOs), have higher oil content with an increased number of antioxidant molecules comparatively higher than fish oils. In addition, marine collagen is an alternative to bovine collagen for skin repairing properties (Coppola et al., 2020). *Streptomyces* sp., *Pontibacter korlensis*, *Pseudomonas* sp., *Bacillus* sp., and *Vibrio* sp. are bacterial species isolated from maritime sediments or seawater that produce a variety of colored chemicals, including prodigiosin, astaxanthin, pyocyanin, melanin, and beta carotene, respectively. These pigments come from a variety of chemical classes, such as carotenes, an unsaturated subclass of carotenoids, prodiginines, which have a pyrrolyldipyrromethene core structure, tambjamines, which are alkaloid molecules, and violacein compounds, which are indole derivatives produced by the metabolism of tryptophan (Nawaz et al., 2021).

23.7 Specialized metabolites from soil microbes—Sources and applications

23.7.1 Anticancer activity

Streptomycetes were reported to synthesize 75% of the actinobacterial secondary metabolites, and actinomycetes were responsible for about half of the documented microbial natural products. Streptomycin, rifamycin, and gentamycin are antimicrobial drugs obtained from actinomycetes, whereas mitomycin, aclarubicin, doxorubicin, mithramycin, neocarzinostatin, and carzinophilin are anticancer agents. Actinomycetes produce secondary metabolites via various metabolic pathways that are stimulated by polyketide synthases and nonribosomal peptide synthases (Osama et al., 2022).

Maklamicin is a novel polyketide of the spirotetronate class, and a polycyclic compound isolated from endophytic *Micromonospora* sp. GMKU326. Maklamicin demonstrated moderate cancer cell cytotoxicity in addition to its antimicrobial activity (Igarashi et al., 2011).

Nomimicin isolated from the *Actinomadura* sp. exhibited antimicrobial activity and cytotoxicity potential. Moderate cytotoxicity was demonstrated by nomimicin against human cancer cells (Zhang et al., 2021). Lobophorin F, a novel spirotetronate molecule isolated from *Streptomyces* sp. SCSIO 01127, exhibited significant antibacterial and antitumor properties (Rateb et al., 2011).

23.7.2 Antimicrobial activity

In addition, cephamycin, chloramphenicol, tetracycline, kanamycin, spectinomycin, monensin, and mitomycin C were derived from *S. clavuligerus, S. venezuelae, S. aureofaciens, S. kanamyceticus, S. spectablis, S. cinnamonensis,* and *S. lavendulae,* respectively whereas streptomycin, polypeptide avermectin, and fluorometabolites are produced by *S. griseus, S. avermitilis,* and *S. cattleya.* Moreover, *S. cheonanensis* VUK-A strain secretes two bioactive compounds such as 2-methyl butyl propyl phthalate and diethyl phthalate with broad antimicrobial activity (Nicault et al., 2021). Maklamicin, a novel polyketide of the spirotetronate class, is a polycyclic compound isolated from endophytic *Micromonospora* sp. GMKU326. Maklamicin demonstrated potent antimicrobial activity against Gram-positive bacteria, such as *Micrococcus luteus*, *Bacillus subtilis*, *Bacillus cereus*, *Staphylococcus aureus*, and *Enterococcus faecalis* (Igarashi et al., 2011). Antimicrobial activity of nomimicin was observed against *M. luteus*, *Candida albicans*, and *Kluyveromyces fragilis* (Zhang et al., 2021). Chaxamycins A-D were isolated from the *Streptomyces* sp. strain from the hyper-arid soil of the Atacama Desert in Chile which exhibited the highest antibacterial activity against *S. aureus* and *E. coli* (Rateb et al., 2011).

Echinomycin, a cyclic depsipeptide antibiotic produced by *Streptomyces echinatus* possesses antimicrobial activity against bacteria, fungi, viruses, and anticancer potential. The quinoxaline chromophores in the structure of echinomycin intercalate with the base pair of duplex DNA and incorporate its cyclic depsipeptide backbone in the minor groove. The known mechanism of action of echinomycin is DNA damage, inhibition of hypoxia-inducible factor 1, potential FKBP12 binding effect, inhibition of bacterial RNA synthesis, inhibition of HIV-1 Tat transactivation, cell apoptosis, and cell signaling inhibition. Echinomycin showed antitumor activity against colon cancer and ovarian cancer (Chen et al., 2021).

Echinomycin is effective against a variety of Gram-positive bacteria, including vancomycin-resistant Enterococci and methicillin-resistant *S. aureus*, Gram-negative bacteria, such as *Shigella dysenteriae*, viruses including poliomyelitis virus, HIV, and bacteriophage, malarial parasite, *Plasmodium falciparum* and thrombosis. It is also known to possess activity against acute peritoneal infections caused by both methicillin-sensitive and methicillin-resistant *S. aureus* strains and is more effective than vancomycin. The strain *Streptomyces* sp. LS462 isolated from the soil sample collected from Yaoli Virgin Forest of Jiangxi Province, China contains echinomycin. This possesses antimycobacterium tuberculosis H37Rv activity and antifungal effect with a greatly reduced dosage of posaconazole on *C. albicans* SC5314 (Chen et al., 2021).

Streptomyces secretes the secondary metabolites namely mycangimycin, frontalamide A, and frontalamide B where both compounds have antifungal activity against the antagonistic fungus *Ophiostoma minus* (Ye et al., 2017). Actinomycin, fungichromin, thailandin B, and antifungalmycin are other antifungal compounds secreted by a *Streptomyces* strain isolated from rhizospheric soils (Peng et al., 2020). According to a previous research study, *S. halstedii* K122 contains antifungal compounds such as bafilomycin B1 and C1, which inhibit the growth of fungi such as *Penicillium roqueforti, Aspergillus fumigatus, Paecilomyces variotii,* and *Mucor hiemalis* (Frandberg et al., 2000).

Antiviral compounds derived from *Streptomyces,* Virantmycin B from *Streptomyces* sp. AM-2504, showed efficacy against the dengue virus (Kimura et al., 2018). The bioactive chemical xiamycinD, which was isolated from a culture of *Streptomyces* sp. had the most significant inhibitory effect on the proliferation of porcine epidemic diarrhea virus (PEDV) (Kim et al., 2016).

The antiviral activity of *Streptomyces* sp. was found to be both wide and strong against a number of influenza viruses, including the H1N1 and H3N2 subtypes, in addition to the influenza B virus. An antiviral butanolide known as (4S)-4-hydroxy-10-methyl-11-oxo-dodec-2-en-1,4-olide was produced by the *Streptomyces* sp. SMU03 strain of bacteria. This butanolide possesses broad and robust activity against influenza viruses, including H1N1 and H3N2 subtypes, as well as influenza B virus (Alam et al., 2022). In general, the antifungal activity of compounds acts by binding competitively to the mitochondrial complex I (Li et al., 2019b), potential quorum sensing inhibitors (QSIs) (Kang et al., 2016), promote an exchange of $K+$ for $H+$, which changes the ion gradient across membranes, by downregulating the expression of the genes involved in biofilm (Ortega et al., 2019).

23.7.3 Antiparasitic activity

The antiprotozoal action of the secondary metabolites produced by microbes has only been explored in a relatively limited number of studies. More than forty distinct *Photorhabdus* and *Xenorhabdus* bacterial species have been identified for their antiparasitic activity against important human parasites *Acanthamoeba castellanii, Entamoeba histolytica, Trichomonas vaginalis, Leishmania tropica* and *Trypanosoma cruzi* (Gulsen et al., 2022). Mutants of *X. budapestensis, X. cabanillasii, X. hominickii, X. stockiae, and X. szentirmaii* were tested for fabclavines. Fabclavines are peptide/polyketide hybrids that are biochemically attached to a polyamine moiety and because of their broad-spectrum activity; fabclavines might serve as protective agents against saprophytic food competitors or microorganisms. Fabclavines 1a and 1b exhibit various bioactivities against different bacterial, fungal, and protozoal organisms. Since fabclavines are structurally extremely like the zeamines that were discovered in *Serratia plymuthica* and it is possible that they have the same effect on artificial membrane models of bacteria. At low concentrations, a fabclavine derivative from *X. innexi* (Xlt) promotes membrane breakdown in selected mosquito cell lines, which ultimately leads to apoptosis. The production of fabclavine is common in *Xenorhabdus* strains, whereas other *Photorhabdus* species, apart from *Photorhabdus asymbiotica*, did not create fabclavines (Tobias et al., 2017).

The actinomycete *Streptomyces avermitilis* is responsible for the manufacture of the antihelmintic and insecticidal agent avermectin. The primary application of avermectin is the management of fire ants. As a semisynthetic derivative of avermectin, ivermectin finds widespread application in veterinary medicine for the improvement of animal health as well as for the elimination of chocerciasis. The secondary metabolite production of *Streptomyces avermitilis* in the fermentation broth was lacking antibacterial and antifungal activity. According to reports, avermectins are a class of macrocyclic lactones that lack significant antibacterial and antifungal activity (Siddique et al., 2014).

23.7.4 Antimalarial drugs

Coronamycin isolated from *Streptomyces* was able to inhibit plasmodial replication in *P. falciparum* (Ezra et al., 2004). Munumbicin D, munumbicins E-4 and E-5, kakadumycin A were other compounds from *Streptomyces* with antimalarial activity against *P. falciparum* (Castillo et al., 2002, 2006). Gancidin-W isolated from *Streptomyces* sp. was effective against *Plasmodium berghei* (Zin et al., 2017). Within 48h of entering a human erythrocyte, *P. falciparum* will have begun to grow and will have begun to divide into a new parasite. Through a protozoan organelle called the cytosome that is responsible for phagocytosis, this parasite can consume up to 80% of the host's hemoglobin. Cytosome is responsible for transporting the hemoglobin into an acidic vacuole and within this vacuole, a proteolytic enzyme breaks down hemoglobin into smaller peptides, which the parasite can use as a source of nutrients. The breakdown of hemoglobin would result in the formation of a free hazardous substance known as free heme which in a later stage, the detoxication of free heme is altered through the polymerization of free heme into inert hemozoin (Fong and Wright, 2013).

23.7.5 Cholesterol lowering drugs

A powerful hypercholesterolemic medication used to decrease blood cholesterol is lovastatin which exerts its effect by competitively inhibiting the enzyme 3-hydroxy-3-methylglutaryl A reductase involved in cholesterol production. In addition, it decreases LDL levels and triglyceride levels in hypertriglyceridemia patients more than other cholesterol lowering medications. The antiatherosclerotic benefits of lovastatin are favorably correlated with the percentage reduction in LDL cholesterol. HMG-CoA reductase inhibition has advantageous pleiotropic effects since the mevalonate metabolism generates a variety of essential isoprenoids for various cellular processes, from cholesterol synthesis to the regulation of cell growth and differentiation. As a secondary metabolite, lovastatin is produced commercially by a variety of filamentous fungi, including *Penicillium* species, *Monascusruber,* and *Aspergillus terreus* (Seenivasan et al., 2008).

23.8 Specialized metabolites from endophytes—Sources and applications

Medicinal plants are colonized by infrequent and fascinating endophytes with high metabolic potential that possess medicinal properties. Recently, endophytic microorganisms have gained their attention of the researchers due to their rich metabolic potential that can be used to cure diseases. Endophytic bacteria from different tissues of the medicinal plant display different antibiotic resistance profile and antagonistic interactions.

23.8.1 Lipopeptides

Lipopeptides, composed of cyclic or short linear peptides, are produced by endophytes and possess antibiotic activity against several pathogens. The three main classes of lipopeptide antibiotics produced by endophytic bacteria are ecomycins, pseudomycins, and kakadumycins (Gupte et al., 2017). *Bacillus* and *Paenibacillus* species such as *B. amyloliquefaciens* (Jiao et al., 2021) and *B. subtilis* (Stein, 2005) are reported to synthesize a high level of lipopeptides. Polyketide antibiotics such as bacillomycin, fengycin, iturin, lichensyn, mycosubtilin, plipastin, pumilacidin, and surfactin are also produced by *B. subtilus* (Farace et al., 2015; Labiadh et al., 2021; Li et al., 2019a). Most of the industrial antibiotics constitute polyketides which are small peptide antibiotics.

Medicinal plants with antimicrobial activities against the bacterial pathogens *E. faecalis*, *S. aureus*, *Klebsiella pneumoniae*, *E. coli*, *Acinetobacter baumanni*, *Pseudomonas aeroginasa*, and some multidrug-resistant human pathogens in the mangrove forest of China is rich in endophytic actinomycetes (Jiang et al., 2018). Endophytic bacteria, *Bacillus, Pseudomonas, Peaenibacillus, Acidomonas, Streptococcus*, *Ralstonia*, *Micrococcus*, *Staphylococcus*, and *Alcaligenes* of a medicinal plant have been identified in the northern part of India (Sharma and Mallubhotla, 2022). Most of the isolates belonged to the class, Bacillales, Enterobacterales, and Pseudomonadales and showed antimicrobial activity against *B. subtilis* and *K. pneumoniae* (Sharma & Mallubhotla, 2022). The isolates of these classes produce antibiotics and peptides that possess antimicrobial, antiviral, and antitumor activities.

Endophytic *B. thuringiensis* isolated from gymnosperms and angiosperms in Iran also produces a wide variety of antimicrobial compounds (Beiranvand et al., 2017). Malaysian and African Endophytes are rich sources of antimicrobials against *S. aureus, B. cereus*, *E. coli*, and *P. aeruginosa* (Diale et al., 2018; Fikri et al., 2018). Antibiotics produced by *Streptomyces* species mainly belong to kakadumycins, munumbicinsxiamycins, and coronomycins (Castillo et al., 2003; Castillo et al., 2002; Christina et al., 2013; Ding et al., 2010; Ezra et al., 2004). Table 23.1 shows the antibiotics produced from the endophytes and their potential antimicrobial activity.

23.8.2 Amino rich peptides

There are few lipopeptides that are associated with usual and unusual amino acids and are called as amino acid rich peptides.

Ecomycins produced by *Pseudomonas viridiflava* are novel lipopeptides containing unusual amino acids such as homoserine and β-hydroxy aspartic acid and common amino acids, alanine, serine, threonine, and glycine (Miller et al., 1998). Ecomycins are effective against human fungal pathogens. *Pseudomonas syringae* produces pseudomycins, a group of antifungal peptides containing unusual amino acids effective against human and plant pathogenic fungi (Harrison et al., 1991). Bacterial endophytes of *Plectranthus tenuiflorus*, a medicinal plant showed strong inhibitory action against many human pathogens including *E. coli*, *K. pneumoniae*, *Proteus mirabilis*, *S. typhi*, *S. aureus*, *Streptococcus agalactiae*, and *C. albicans* (El-Deeb et al., 2012). The endophytic bacteria of the genera *Staphylococcus, Micrococcus,* and *Sphingomonas* also possess PKS and NRPS genes, indicating its substantial potential for secondary metabolites (Akter et al., 2022).

23.8.3 Cyclic cationic lipopeptides

Polymyxins, antibiotics synthesized by endophytic *Paenibacillus polymyxa* are produced by a nonribosomal peptide synthetase. Polymyxins are effective against most of the *Enterobacteriaceae* family, including *E. coli, Enterobacter* spp., *Klebsiella* spp., *Citrobacter* spp., *Salmonella* spp., and *Shigella* spp. Endophytic *Paenibacillus polymyxa* possess NRPS and PKS gene (Shaheen et al., 2011). Four endophytes from the medicinal plant *Origanum vulgare* L. are known to produce a diverse range of antibiotics such as paeninodin, polymyxins, and paenicidin. The crude extracts of these endophytes were found to be effective inhibitors of ten strains of a *Burkholderia cepacia* complex known to worsen the

Table 23.1 Endophytes derived antibiotics—their sources and antimicrobial activity.

Isolated Strain	Source	Antibiotics	Antimicrobial activity	References
Streptomyces strains NRRL30566	*Grevillea pteridifolia*	Kakadumycins and xiamycins	Effective against Gram-positive organisms *Bacillus anthracis* strains	Castillo et al. (2003) and Christina et al. (2013)
Streptomyces strains GT2002/1503	Mangrove plant, *Bruguiera gymnorrhiza*	Xiamycins	Anti-HIV	Ding et al. (2010)
Streptomyces sp. strain SUK06	*Thottea grandiflora* (Malaysia)	Antibacterial agents	*B. cereus*, *B. subtilus*, *P. shigelloides*, *P. aeruginosa*, and *S. aureus*	Ghadin et al. (2008)
Streptomyces sp. BT01	*Boesenbergia rotunda* (L.)	Flavonoids—7-methoxy-3, 3′,4′,6-tetrahydroxyflavone and 2′,7-dihydroxy-4′,5′-dimethoxyisoflavone, fisetin, naringenin, 3′-hydroxydaidzein and xenognosin B	*B. cereus* and *B. subtilus*	Taechowisan et al. (2014)
Bacillus mojavensis	*Bacopa monnieri*	Lipopeptides consisting of fengycin	*E. coli*, *S. aureus*, *K. pneumoniae*, and *S. typhi*	Jasim et al. (2016)
Streptomyces strain NRRL 30662	Snake vine	Munumbicin D	Gram-positive and Gram-negative bacteria including *B. anthracis, S. pneumoniae, E. faecalis, S aureus,* and multiple drug-resistant strains of *Mycobacterium tuberculosis*	Castillo et al. (2002)

genetic disease cystic fibrosis (Maida et al., 2014). Genomic analysis shows that three biogenetic gene clusters (BGCs), lasso peptide genes, NRP genes, and lanthipeptide genes are present in these strains (Semenzato et al., 2022).

23.8.4 Nanoparticles

Eco-friendly and nontoxic methods employing biological organisms like plants, algae, yeast, bacteria, actinomycetes, and fungi are excellent sources for the synthesis of nanoparticles, also known as green synthesis. Using endophytic microorganisms is one of the widely used methods to synthesize nanoparticles with low metallic ions (Meena et al., 2021). Metal-based nanoparticles with therapeutic uses can be created by using bacterial cell extracts. *Bacillus safensis* strain TEN12 generated spherical AgNPs with a size range of 22–42 nm intracellularly (Ahmed et al., 2020). The endophytic bacterium *B. cereus* isolated from the tropical evergreen tree *Garcinia xanthochymus* produces silver nanoparticles (Sunkar and Nachiyar, 2012). Silver nanoparticles are useful in the treatment of viral infections such as HIV-1, hepatitis B, respiratory syncytial virus, and herpes simplex virus. The utilization of bacterial endophyte-produced nanoparticles in the creation of novel antiviral drugs holds out a lot of potentials. The nanoparticles produced by bacterial endophytes show a high potential for antiviral compounds (Singh et al., 2017).

23.8.5 Extracellular metabolites

Exopolysaccharides (EPS) are known to possess anticancer potential. An endophytic *Bacillus amyloliquefaciens* strain isolated from *Ophiopogon japonicus*, a Chinese medicinal plant, produced EPS that inhibited the growth of human gastric cancer cell lines MC-4 and SGC-7901. EPS-treated cells had abnormal cell morphology and cell death, possibly caused by a mitochondrial dysfunction (Chen et al., 2013). Phenolic compounds have also been reported to be involved in various bioactive properties, including anticancer activity. Two biphenyl producing *Streptomyces* sp. isolated from the root tissue of *Boesenbergia rotunda* (L.) Mansf A. and *Boesenbergia pandurata* showed strong cytotoxicity against three cancer cell lines (HeLa, HepG2, and Huh7) and less toxicity toward normal cells (L929) (Nguyen et al., 2020; Taechowisan et al., 2017).

23.8.6 Pigments

Endophytic bacteria produce pigments that serve as sources of new drugs against antibiotic resistant pathogens. The extract from bacterial endophyte *Burkholderia* sp. WYAT7, of the medicinal plant *Artemisia nilagirica* (Clarke) Pamp., is effective against several Gram-positive and Gram-negative bacteria (Ashitha et al., 2019). The growth of *S. typhi* (MTCC733), *S. aureus* (MTCC1430), *P. aeruginosa* (MTCC2453), *K. pneumoniae* (MTC 432), *E. coli* (MTCC160), *S. paratyphi* (3220), *B. subtilus* (441), and *Acinetobacter baumannii* (12,889) was inhibited by the pigment extracts of the endophytic bacteria (Ashitha et al., 2020). Bacterial pigments are known to have profound applications in pharmaceutical industries. The advantages of employing biological methods over physical and chemical methods in the synthesis of pigments are an added benefit to the industrial applications of pigments. Biological methods reduce the use of high-energy inputs and toxic waste production, and render simple, cost-effective, and eco-friendly productions of pigments.

23.8.7 Antitumor agents

Mangrove-derived streptomycetes are promising producers of compounds with anticancer properties, some of which are new compounds such as streptocarbazoles A and B, streptomyceamide C, neoantimycins A, and B. In addition to that, extracts of these mangrove-derived streptomycetes have demonstrated great anticancer potential (Law et al., 2020).

Endophytic bacteria are reported to produce different classes of bioactive anticancer compounds such as anthracyclines, glycopeptides, aureolic acids, anthraquinones, enediynes, polysachharides, carzinophilin, mitomycins, alnumycin, pterocidin, napthomycin, and alkyl salicylic acids (salaceyins)

(Ek-Ramos et al., 2019). Many studies report the anticancer potential of endophytic actinomycetes bacteria. Brazilian medicinal plant, Lychnophoraericoides, endophytes, and Streptomyces strains possessed strong cytotoxic activity against human cancer cell lines. Around 16 bioactive compounds have been isolated of which 3-hydroxy-4-methoxybenzamide and 2,3-dihydro-2,2-dimethyl-4(1H)-quinazolinone were identified for the first time in their study. Many secondary metabolites produced by endophytic bacteria have been characterized after growing them in vitro (Conti et al., 2016). Characterization of endophytic Streptomyces lacey MS53 showed two new anticancer agents, salaceyins (A and B), inhibited the growth of the human breast cancer line SKBR3 (Kim et al., 2006). Streptomyces sp. strain DSM11575 from root nodules of *Alnus glutinosa* produced the compound alnumycin, which was cytotoxic to K562 human leukemia cells (Bieber et al., 1998).

Streptomyces albidoflavus CMRP4852 and *Verrucosispora* sp. CMR P4860 isolates exhibited anti-melanoma activity without affecting normal cells (Assad et al., 2021). The metabolites produced by the cinnamon plant endophyte, actinomycete strain YBQ59 were effective against human lung cancer cells (Vu et al., 2018). *Streptomyces hygroscopicus* TP-A0451 isolated from Pteridium aquilinum produced pteridines that demonstrated cytotoxic activity to human cancer cell lines NCI-H522, OVCAR-3, SF539, and LOX-IMVI (Igarashi et al., 2006). Napthomycin produced by *Streptomyces* sp. CS isolated from *Maytenus hookeri* was effective against P388 and A549 human tumor cells (Lu and Shen, 2007). Nine new naphthacemycins along with one known naphthacemycin were isolated from the culture of *Streptomyces* sp. N12W1565 which was a potential anti-protein tyrosine phosphatase inhibitor (Huo et al., 2020). *Acinetobacter guillouiae* and *Raoultella ornithinolytica,* the crude extracts of bacterial endophytes isolated from *Crinum macowanii* baker bulbs, reduced the growth of U87MG brain cancer cell line and strongly inhibited lung carcinoma cells, respectively (Sebola et al., 2019).

Maytansine, an anticancer agent effective against breast cancer is produced by *Putterlickia verrucosa* and *Putterlickia retrospinosa*. It was believed that this metabolite was produced by plants but is produced by the endophytes (Kusari et al., 2014; Newman and Cragg, 2015). The roots are the potential source of maytansine. Maytansine produced by *Streptomyces* sp. Is9131 from the medicinal plant *Maytenus hookeri* produces maytansine, which shows potential inhibitory activity to cancer cell lines including leukemia, lung, gastric, and liver cancers (Zhao et al., 2005).

Endophytic fungi secrete bioactive compounds such as paclitaxel, podophyllotoxin, camptothecin, vinblastine, hypericin, and diosgenin. They are isolated from *Thielavia subthermophila, Catharanthus roseus, Sinopodophyllum hexandrum, Dysosma veitchii, Rhizopus oryzae* (94Y-01), *Chaetomella raphigera, Aspergillus fumigatus, Rhizopus oryzae,* and *Seimatoantlerium nepalense* and are known to possess anticancer potential and are used as therapeutic agent for different diseases. Endophytic fungi are effective against human papillary thyroid carcinoma (IHH4), human pancreatic (PANC-1), ovarian (OVCAR-3), hepatic (HepG2), lung (A-549), human lymphoma (U937), human skin carcinoma (A431), breast (MCF-7), and Kaposi's sarcoma (Kousar et al., 2022).

23.9 Specialized metabolites in the cosmetic industry

Microbes are rich in plenty of metabolites and enzymes that find useful applications in several cosmetic and personal care products. The varied properties of these metabolites find applications in a wide range of skincare, hair care, body care, deodorants, and other personal hygiene products (Table 23.2) (Elmarzugi et al., 2016; Gupta et al., 2019). The metabolites derived from several microbes act as

Table 23.2 Microbial metabolites—Origins, properties, and their applications in cosmetic industries.

Compound	Group	Strains of organisms used	Functional attributes	Applications
Cyclodextrins	Cyclic oligosaccharides	*Bacillus subtilis, Brevibacillus brevis* (produce Cyclodextrinase enzyme) *Bacillus agaradhaerensis* (produce Cyclodextrin glucanotransferase)	Sustained release of aroma, odor control in diapers and napkins, reduce volatility in room freshers/perfumes	Enzymatic transformation using bacteria is more useful than chemical synthesis
Mannosylerythritol lipid	Biosurfactant	*Pseudozyma* spp., *Bacillus* spp.	Antiwrinkle	Cosmetics and cleansing products
Rhamnolipids		*Pseudomonas putida*	Foaming and emulsifying	Food and pharmaceutical products
Sophorolipid		*Candida* spp.	Skin hydration	Skin care products
Dextran	Exopolysaccharides	*Streptococcus mutans, Leuconostoc mesenteriodes*	Antiinflammatory property	Skin care formulations (antiwrinkle and antiaging)
Alginates		*Azotobacter vinelandii, Pseudomonas aeruginosa*	Retain water, gelling agent, thickener	Used in skin-smoothing and moisturizing products
Xanthan		*Xanthomonas* sp.	Reduces the trans-epidermal water loss, gelling agent, thickener	Moisture retention skin care products.
Keratinases	Enzymes	*B. subtilis*, dermatophytes (*Microsporum* and *Trichophyton* species)	Acts on keratin	Hair removal products.
Collagenases		*Bacillus cereus, Clostridium histolyticum*	Regeneration of skin	Antiaging and skin care products
Catalases, superoxide dismutase		Many aerobic bacteria	Free radical scavenging, antioxidant function	Antiaging products
Azelaic acid	Natural acid	*Malassezia* spp.	Melanin inhibitor, antiinflammatory	Skin lightening and skin care
Astaxanthin	Pigment	*Agrobacterium* sp.	Improves skin texture	Skin care
Botulinum toxin	Natural toxin	*Clostridium botulinum* (also *Clostridium baratii, Clostridium butyricum*)	Antiwrinkle	Antiaging and skin care

do-good ingredients and deliver functional benefits to these products. The major label claims on many popular cosmetic and personal care products are "naturally derived" and "skin friendly." The microbial derived ingredients fit well in the ideal description of "naturally derived" or "natural." Further, most of these ingredients are skin friendly.

23.10 Conclusion

Many drugs and therapeutic compounds, such as antibiotics, antifungals, anticancer agents, and immunosuppressants, are derived from specialized metabolites produced by microorganisms. Yet, there is an ongoing demand to look-out for new antimicrobial metabolites to meet the challenges of newer variants of existing pathogens and the emergence and reemergence of newer infectious and opportunistic infective agents. Specialized microbial metabolites are also used in agriculture for enhancing crop yield, plant protection, and growth promotion. Biopesticides and biofertilizers derived from microbiome metabolites offer sustainable alternatives to synthetic chemical pesticides and fertilizers.

Microbiome-derived specialized metabolites contribute to the production of flavors, fragrances, and food additives such as the production of fermented foods and beverages due to the metabolic activities of microorganisms and the production of specific metabolites to meet the recent advancements in special culinary needs and functional foods. Microbiome-derived specialized metabolites are also employed to produce biofuels and biochemicals. Certain microorganisms can produce biofuels like ethanol or butanol through their metabolic pathways, offering renewable alternatives to fossil fuels. Microbe derived metabolites also have profound applications in the field of cosmetic and personal care industries.

The contribution of microorganisms to the field of medicine, agriculture, food, and cosmetics are plentiful. The microbes should be continued to be screened and explored for more potential specialized metabolites that would have profound industrial applications offering cost-effective and eco-friendly products.

One of the major sources of potential candidates for the synthesis of secondary metabolites is the endophytic microorganisms. More focused research on this area by subjecting plants to biotic stress to create newer endophytic microorganisms which in turn could yield desired secondary metabolites than plants under normal conditions would be a future prospective.

What has been explored and discovered so far, in this ever-growing area of microbial metabolites research, is just the tip of a humongous iceberg. In the era of advanced molecular biology, there are profound opportunities and ways for exploring novel microbial-based metabolites using gene editing tools, genome mining, manipulation and optimization of specialized pathways using codons and activating the unknown potential of nonactive genome.

References

Adlan, N.A., Sabri, S., Masomian, M., Ali, M.S.M., Rahman, R.N.Z.R.A., 2020. Microbial biodegradation of paraffin wax in Malaysian crude oil mediated by degradative enzymes. Front. Microbiol. 11. https://doi.org/10.3389/FMICB.2020.565608/FULL.

Ahmed, T., Shahid, M., Noman, M., Bilal Khan Niazi, M., Zubair, M., Almatroudi, A., Khurshid, M., Tariq, F., Mumtaz, R., Li, B., 2020. Bioprospecting a native silver-resistant Bacillus safensis strain for green synthesis and subsequent antibacterial and anticancer activities of silver nanoparticles. J. Adv. Res. 24, 475–483. https://doi.org/10.1016/J.JARE.2020.05.011.

Akter, Y., Barua, R., Uddin, N., Muhammad Sanaullah, A.F., Marzan, L.W., 2022. Bioactive potentiality of secondary metabolites from endophytic bacteria against SARS-COV-2: an in-silico approach. PloS One 17 (8). https://doi.org/10.1371/JOURNAL.PONE.0269962.

Alam, K., Mazumder, A., Sikdar, S., Zhao, Y.M., Hao, J., Song, C., Wang, Y., Sarkar, R., Islam, S., Zhang, Y., Li, A., 2022. Streptomyces: the biofactory of secondary metabolites. Front. Microbiol. 13. https://doi.org/10.3389/FMICB.2022.968053/FULL.

Al-Shaibani, M.M., Mohamed, R.M.S.R., Sidik, N.M., El Enshasy, H.A., Al-Gheethi, A., Noman, E., Al-Mekhlafi, N.A., Zin, N.M., 2021. Biodiversity of secondary metabolites compounds isolated from phylum actinobacteria and its therapeutic applications. Molecules 26 (15). https://doi.org/10.3390/MOLECULES26154504.

Ashitha, A., Midhun, S.J., Sunil, M.A., Nithin, T.U., Radhakrishnan, E.K., Mathew, J., 2019. Bacterial endophytes from Artemisia nilagirica (Clarke) Pamp., with antibacterial efficacy against human pathogens. Microb. Pathog. 135. https://doi.org/10.1016/J.MICPATH.2019.103624.

Ashitha, A., Radhakrishnan, E.K., Mathew, J., 2020. Antibacterial potential and apoptosis induction by pigments from the endophyte Burkholderia sp. WYAT7. Curr. Microbiol. 77 (9), 2475–2485. https://doi.org/10.1007/S00284-020-02013-3/METRICS.

Assad, B.M., Savi, D.C., Biscaia, S.M.P., Mayrhofer, B.F., Iantas, J., Mews, M., de Oliveira, J.C., Trindade, E.S., Glienke, C., 2021. Endophytic actinobacteria of Hymenachneamplexicaulis from the Brazilian Pantanal wetland produce compounds with antibacterial and antitumor activities. Microbiol. Res. 248, 126768. https://doi.org/10.1016/J.MICRES.2021.126768.

Beiranvand, M., Amin, M., Hashemi-Shahraki, A., Romani, B., Yaghoubi, S., Sadeghi, P., 2017. Antimicrobial activity of endophytic bacterial populations isolated from medical plants of Iran. Iranian J. Microbiol. 9 (1), 11. /pmc/articles/PMC5533999/.

Bieber, B., Nuske, J., Ritzau, M., Grafe, U., 1998. Alnumycin a new naphthoquinone antibiotic produced by an endophytic Streptomyces sp. J. Antibiot. 51 (3), 381–382. https://doi.org/10.7164/ANTIBIOTICS.51.381.

Brunt, E.G., Burgess, J.G., 2018. The promise of marine molecules as cosmetic active ingredients. Int. J. Cosmet. Sci. 40 (1), 1–15. https://doi.org/10.1111/ICS.12435.

Castillo, U., Harper, J.K., Strobel, G.A., Sears, J., Alesi, K., Ford, E., Lin, J., Hunter, M., Maranta, M., Ge, H., Yaver, D., Jensen, J.B., Porter, H., Robison, R., Millar, D., Hess, W.M., Condron, M., Teplow, D., 2003. Kakadumycins, novel antibiotics from Streptomyces sp NRRL 30566, an endophyte of Grevillea pteridifolia. FEMS Microbiol. Lett. 224 (2), 183–190. https://doi.org/10.1016/S0378-1097(03)00426-9.

Castillo, U.F., Strobel, G.A., Ford, E.J., Hess, W.M., Porter, H., Jensen, J.B., Albert, H., Robison, R., Condron, M.A.M., Teplow, D.B., Stevens, D., Yaver, D., 2002. Munumbicins, wide-spectrum antibiotics produced by Streptomyces NRRL 30562, endophytic on Kennedianigriscans. Microbiology 148 (9), 2675–2685. https://doi.org/10.1099/00221287-148-9-2675.

Castillo, U.F., Strobel, G.A., Mullenberg, K., Condron, M.M., Teplow, D.B., Folgiano, V., Gallo, M., Ferracane, R., Mannina, L., Viel, S., Codde, M., Robison, R., Porter, H., Jensen, J., 2006. Munumbicins E-4 and E-5: novel broad-spectrum antibiotics from Streptomyces NRRL 3052. FEMS Microbiol. Lett. 255 (2), 296–300. https://doi.org/10.1111/J.1574-6968.2005.00080.X.

Chavali, A.K., Rhee, S.Y., 2018. Bioinformatics tools for the identification of gene clusters that biosynthesize specialized metabolites. Brief. Bioinform. 19 (5), 1022. https://doi.org/10.1093/BIB/BBX020.

Chen, C., Chen, X., Ren, B., Guo, H., Abdel-Mageed, W.M., Liu, X., Song, F., Zhang, L., 2021. Characterization of Streptomyces sp. LS462 with high productivity of echinomycin, a potent antituberculosis and synergistic antifungal antibiotic. J. Ind. Microbiol. Biotechnol. 48, 79. https://doi.org/10.1093/jimb/kuab079.

Chen, Y.T., Yuan, Q., Shan, L.T., Lin, M.A., Cheng, D.Q., Li, C.Y., 2013. Antitumor activity of bacterial exopolysaccharides from the endophyte Bacillus amyloliquefaciens sp. isolated from Ophiopogon japonicus. Oncol. Lett. 5 (6), 1787. https://doi.org/10.3892/OL.2013.1284.

Christina, A., Christapher, V., Bhore, S.J., 2013. Endophytic bacteria as a source of novel antibiotics: An overview. Pharmacogn. Rev. 7, 11–16. https://doi.org/10.4103/0973-7847.112833.

Conti, R., Chagas, F.O., Caraballo-Rodriguez, A.M., Melo, W.G., do Nascimento, A.M., Cavalcanti, B.C., de Moraes, M.O., Pessoa, C., Costa-Lotufo, L.V., Krogh, R., Andricopulo, A.D., Lopes, N.P., Pupo, M.T., 2016. Endophytic actinobacteria from the Brazilian medicinal plant Lychnophoraericoides Mart. and the biological potential of their secondary metabolites. Chem. Biodivers. 13 (6), 727–736. https://doi.org/10.1002/CBDV.201500225.

Coppola, D., Oliviero, M., Vitale, G.A., Lauritano, C., D'Ambra, I., Iannace, S., de Pascale, D., 2020. Marine collagen from alternative and sustainable sources: extraction, processing and applications. Mar. Drugs 18 (4). https://doi.org/10.3390/MD18040214.

Davies, J., 2013. Specialized microbial metabolites: functions and origins. J. Antibiotics 66 (7), 361–364. https://doi.org/10.1038/ja.2013.61.

Dhakal, D., Pokhrel, A.R., Shrestha, B., Sohng, J.K., 2017. Marine rare actinobacteria: isolation, characterization, and strategies for harnessing bioactive compounds. Front. Microbiol. 8, 1106. https://doi.org/10.3389/FMICB.2017.01106.

Diale, M.O., Ubomba-Jaswa, E., Serepa-Dlamini, M.H., 2018. The antibacterial activity of bacterial endophytes isolated from Combretum molle. Afr. J. Biotechnol. 17 (8), 255–262. https://doi.org/10.5897/AJB2017.16349.

Ding, L., Münch, J., Goerls, H., Maier, A., Fiebig, H.H., Lin, W.H., Hertweck, C., 2010. Xiamycin, a pentacyclic indolosesquiterpene with selective anti-HIV activity from a bacterial mangrove endophyte. Bioorg. Med. Chem. Lett. 20 (22), 6685–6687. https://doi.org/10.1016/J.BMCL.2010.09.010.

Dona, A.C., Kyriakides, M., Scott, F., Shephard, E.A., Varshavi, D., Veselkov, K., Everett, J.R., 2016. A guide to the identification of metabolites in NMR-based metabonomics/metabolomics experiments. Comput. Struct. Biotechnol. J. 14, 135–153. https://doi.org/10.1016/J.CSBJ.2016.02.005.

Ek-Ramos, M.J., Gomez-Flores, R., Orozco-Flores, A.A., Rodríguez-Padilla, C., González-Ochoa, G., Tamez-Guerra, P., 2019. Bioactive products from plant-endophytic Gram-positive bacteria. Front. Microbiol. 10 (MAR), 463. https://doi.org/10.3389/FMICB.2019.00463/BIBTEX.

El-Deeb, B., Fayez, K., Gherbawy, Y., 2012. Isolation and characterization of endophytic bacteria from Plectranthustenuiflorus medicinal plant in Saudi Arabia desert and their antimicrobial activities. J. Plant Interact. 8 (1), 56–64. https://doi.org/10.1080/17429145.2012.680077.

Elmarzugi, N., Abdulhamid, M., Malek, R.A., Sarmidi, M., El Enshasy, H., 2016. Microbial metabolites. In: Gupta, V., Sharma, G., Touhy, M., Gaur, R. (Eds.), Cosmetic Industries-The Hand Book of Microbial Bioresources, first ed. CABI, Oxfordshire.

Ezra, D., Castillo, U.F., Strobel, G.A., Hess, W.M., Porter, H., Jensen, J.B., Condron, M.A.M., Teplow, D.B., Sears, J., Maranta, M., Hunter, M., Weber, B., Yaver, D., 2004. Coronamycins, peptide antibiotics produced by a verticillate Streptomyces sp. (MSU-2110) endophytic on Monstera sp. 150 (Pt 4), 785–793. https://doi.org/10.1099/MIC.0.26645-0. Microbiology.

Farace, G., Fernandez, O., Jacquens, L., Coutte, F., Krier, F., Jacques, P., Clément, C., Barka, E.A., Jacquard, C., Dorey, S., 2015. Cyclic lipopeptides from Bacillus subtilis activate distinct patterns of defence responses in grapevine. Mol. Plant Pathol. 16 (2), 177–187. https://doi.org/10.1111/MPP.12170/SUPPINFO.

Fazle Rabbee, M., Baek, K.H., 2020. Antimicrobial activities of lipopeptides and polyketides of bacillus velezensis for agricultural applications. Molecules 25 (21). https://doi.org/10.3390/MOLECULES25214973.

Feng, Z., Zhang, X., Wu, J., Wei, C., Feng, T., Zhou, D., Wen, Z., Xu, J., 2022. Immunosuppressive cytochalasins from the mangrove endophytic fungus Phomopsis asparagi DHS-48. Mar. Drugs 20 (8). https://doi.org/10.3390/MD20080526/S1.

Fikri, A.S.I., Rahman, I.A., Nor, N.S.M., Hamzah, A., 2018. Isolation and identification of local bacteria endophyte and screening of its antimicrobial property against pathogenic bacteria and fungi. AIP Conf. Proc. 020072. https://doi.org/10.1063/1.5027987.

Fong, K.Y., Wright, D.W., 2013. Hemozoin and antimalarial drug discovery. Future Med. Chem. 5 (12), 1437. https://doi.org/10.4155/FMC.13.113.

Frandberg, E., Petersson, C., Lundgren, L.N., Schnurer, J., 2000. Streptomyces halstedii K122 produces the antifungal compounds bafilomycin B1 and C1. Can. J. Microbiol. 46 (8), 753–758. https://doi.org/10.1139/CJM-46-8-753.

Ghadin, N., Zin, N.M., Sabaratnam, V., Badya, N., Basri, D.F., Lian, H.H., Sidik, N.M., 2008. Isolation and characterization of a novel endophytic Streptomyces SUK 06 with antimicrobial activity from Malaysian plant. Asian J. Plant Sci. 7 (2), 189–194. https://doi.org/10.3923/AJPS.2008.189.194.

Ghareeb, M.A., Tammam, M.A., El-Demerdash, A., Atanasov, A.G., 2020. Insights about clinically approved and preclinically investigated marine natural products. Curr. Res. Biotechnol. 2, 88–102. https://doi.org/10.1016/J.CRBIOT.2020.09.001.

Gulsen, S.H., Tileklioglu, E., Bode, E., Cimen, H., Ertabaklar, H., Ulug, D., Ertug, S., Wenski, S.L., Touray, M., Hazir, C., Bilecenoglu, D.K., Yildiz, I., Bode, H.B., Hazir, S., 2022. Antiprotozoal activity of different Xenorhabdus and Photorhabdus bacterial secondary metabolites and identification of bioactive compounds using the easyPACId approach. Sci. Rep. 12 (1). https://doi.org/10.1038/S41598-022-13722-Z.

Gupta, P.L., Rajput, M., Oza, T., Trivedi, U., Sanghvi, G., 2019. Eminence of microbial products in cosmetic industry. Nat. Prod. Bioprospect. 9 (4), 267–278. https://doi.org/10.1007/S13659-019-0215-0/TABLES/3.

Gupte, S., Kaur, M., Kaur, M., 2017. Novel approaches to developing new antibiotics. Bacteriol. Mycol. https://doi.org/10.15406/jbmoa.2017.04.00089.

Harrison, L., Teplow, D.B., Rinaldi, M., Strobel, G., 1991. Pseudomycins, a family of novel peptides from Pseudomonas syringae possessing broad-spectrum antifungal activity. J. Gen. Microbiol. 137 (12), 2857–2865. https://doi.org/10.1099/00221287-137-12-2857.

Hibbing, M.E., Fuqua, C., Parsek, M.R., Peterson, S.B., 2010. Bacterial competition: surviving and thriving in the microbial jungle. Nat. Rev. Microbiol. 8 (1), 15–25. https://doi.org/10.1038/nrmicro2259.

Huo, C., Zheng, Z., Xu, Y., Ding, Y., Zheng, H., Mu, Y., Niu, Y., Gao, J., Lu, X., 2020. Naphthacemycins from a Streptomyces sp. as protein-tyrosine phosphatase inhibitors. J. Nat. Prod. 83 (5), 1394–1399. https://doi.org/10.1021/ACS.JNATPROD.9B00417/SUPPL_FILE/NP9B00417_SI_001.PDF.

Igarashi, Y., Miura, S.S., Fujita, T., Furumai, T., 2006. Pterocidin, a cytotoxic compound from the endophytic Streptomyces hygroscopicus. J. Antibiotics 59 (3), 193–195. https://doi.org/10.1038/ja.2006.28.

Igarashi, Y., Ogura, H., Furihata, K., Oku, N., Indananda, C., Thamchaipenet, A., 2011. Maklamicin, an antibacterial polyketide from an endophytic Micromonospora sp. J. Nat. Prod. 74 (4), 670–674. https://doi.org/10.1021/NP100727H.

Jakubiec-Krzesniak, K., Rajnisz-Mateusiak, A., Guspiel, A., Ziemska, J., Solecka, J., 2018. Secondary metabolites of actinomycetes and their antibacterial, antifungal and antiviral properties. Pol. J. Microbiol. 67 (3), 259–272. https://doi.org/10.21307/PJM-2018-048.

Jasim, B., Sreelakshmi, S., Mathew, J., Radhakrishnan, E.K., 2016. Identification of endophytic Bacillus mojavensis with highly specialized broad spectrum antibacterial activity. 3 Biotech 6 (2). https://doi.org/10.1007/S13205-016-0508-5.

Jeong, S.W., Yang, J.E., Choi, Y.J., 2022. Isolation and characterization of a yellow xanthophyll pigment-producing marine bacterium, Erythrobacter sp. SDW2 strain, in coastal seawater. Mar. Drugs 20 (1). https://doi.org/10.3390/MD20010073.

Jiang, Z.K., Tuo, L., Huang, D.L., Osterman, I.A., Tyurin, A.P., Liu, S.W., Lukyanov, D.A., Sergiev, P.V., Dontsova, O.A., Korshun, V.A., Li, F.N., Sun, C.H., 2018. Diversity, novelty, and antimicrobial activity of endophytic actinobacteria from mangrove plants in Beilun Estuary National Nature Reserve of Guangxi, China. Front. Microbiol. 9 (MAY). https://doi.org/10.3389/FMICB.2018.00868/FULL.

Jiao, R., Cai, Y., He, P., Munir, S., Li, X., Wu, Y., Wang, J., Xia, M., He, P., Wang, G., Yang, H., Karunarathna, S.C., Xie, Y., He, Y., 2021. Bacillus amyloliquefaciens YN201732 produces lipopeptides with promising biocontrol activity against fungal pathogen Erysiphe cichoracearum. Front. Cell. Infect. Microbiol. 11. https://doi.org/10.3389/FCIMB.2021.598999.

Kang, J.E., Han, J.W., Jeon, B.J., Kim, B.S., 2016. Efficacies of quorum sensing inhibitors, piericidin A and glucopiericidin A, produced by Streptomyces xanthocidicus KPP01532 for the control of potato soft rot caused by Erwinia carotovora subsp. atroseptica. Microbiol. Res. 184, 32–41. https://doi.org/10.1016/J.MICRES.2015.12.005.

Khaldi, N., Seifuddin, F.T., Turner, G., Haft, D., Nierman, W.C., Wolfe, K.H., Fedorova, N.D., 2010. SMURF: genomic mapping of fungal secondary metabolite clusters. Fungal Genet. Biol. 47 (9), 736–741. https://doi.org/10.1016/J.FGB.2010.06.003.

Kim, S.H., Ha, T.K.Q., Oh, W.K., Shin, J., Oh, D.C., 2016. Antiviral Indolosesquiterpenoid Xiamycins C-E from a halophilic actinomycete. J. Nat. Prod. 79 (1), 51–58. https://doi.org/10.1021/ACS.JNATPROD.5B00634/SUPPL_FILE/NP5B00634_SI_001.PDF.

Kim, N., Shin, J.C., Kim, W., Hwang, B.Y., Kim, B.S., Hong, Y.S., Lee, D., 2006. Cytotoxic 6-alkylsalicylic acids from the endophytic Streptomyces laceyi. J. Antibiot. 59 (12), 797–800. https://doi.org/10.1038/JA.2006.105.

Kimura, T., Suga, T., Kameoka, M., Ueno, M., Inahashi, Y., Matsuo, H., Iwatsuki, M., Shigemura, K., Shiomi, K., Takahashi, Y., Ōmura, S., Nakashima, T., 2018. New tetrahydroquinoline and indoline compounds containing a hydroxy cyclopentenone, virantmycin B and C, produced by Streptomyces sp. AM-2504. J. Antibiotics 72 (3), 169–173. https://doi.org/10.1038/s41429-018-0117-0.

Kousar, R., Naeem, M., Jamaludin, M.I., Arshad, A., Shamsuri, A.N., Ansari, N., Akhtar, S., Hazafa, A., Uddin, J., Khan, A., Al-Harrasi, A., 2022. Exploring the anticancer activities of novel bioactive compounds derived from endophytic fungi: mechanisms of action, current challenges and future perspectives. Am. J. Cancer Res. 12 (7), 2897. /pmc/articles/PMC9360238/.

Kramer, J., Özkaya, Ö., Kümmerli, R., 2020. Bacterial siderophores in community and host interactions. Nat. Rev. Microbiol. 18 (3), 152–163. https://doi.org/10.1038/S41579-019-0284-4.

Künzler., 2018. How fungi defend themselves against microbial competitors and animal predators. PLoS Pathog. 14 (9), e1007184. https://doi.org/10.1371/journal.ppat.1007184.

Kusari, S., Lamshöft, M., Kusari, P., Gottfried, S., Zühlke, S., Louven, K., Hentschel, U., Kayser, O., Spiteller, M., 2014. Endophytes are hidden producers of maytansine in Putterlickia roots. J. Nat. Prod. 77 (12), 2577–2584. https://doi.org/10.1021/NP500219A/SUPPL_FILE/NP500219A_SI_001.PDF.

Labiadh, M., Dhaouadi, S., Ene Chollet, M., Chataigne, G., Tricot, C., Jacques, P., Flahaut, S., Kallel, S., 2021. Antifungal lipopeptides from Bacillus strains isolated from rhizosphere of Citrus trees. Indian. J. Microbiol. https://doi.org/10.1016/j.rhisph.2021.100399.

Law, J.W.F., Law, L.N.S., Letchumanan, V., Tan, L.T.H., Wong, S.H., Chan, K.G., Mutalib, N.S.A., Lee, L.H., 2020. Anticancer drug discovery from microbial sources: the unique mangrove Streptomycetes. Molecules 25 (22). https://doi.org/10.3390/MOLECULES25225365.

Lee, N., Hwang, S., Kim, J., Cho, S., Palsson, B., Cho, B.K., 2020. Mini review: genome mining approaches for the identification of secondary metabolite biosynthetic gene clusters in Streptomyces. Comput. Struct. Biotechnol. J. 18, 1548–1556. https://doi.org/10.1016/J.CSBJ.2020.06.024.

Li, Y., Héloir, M.C., Zhang, X., Geissler, M., Trouvelot, S., Jacquens, L., Henkel, M., Su, X., Fang, X., Wang, Q., Adrian, M., 2019. Surfactin and fengycin contribute to the protection of a Bacillus subtilis strain against grape downy mildew by both direct effect and defence stimulation. Mol. Plant Pathol. 20 (8), 1037. https://doi.org/10.1111/MPP.12809.

Li, Y., Kong, L., Shen, J., Wang, Q., Liu, Q., Yang, W., Deng, Z., You, D., 2019. Characterization of the positive SARP family regulator PieR for improving piericidin A1 production in Streptomyces piomogeues var. Hangzhouwanensis. Synth. Syst. Biotechnol. 4 (1), 16. https://doi.org/10.1016/J.SYNBIO.2018.12.002.

de Lima Procópio, R.E., da Silva, I.R., Martins, M.K., de Azevedo, J.L., de Araújo, J.M., 2012. Antibiotics produced by Streptomyces. Brazil. J. Infect. Dis. 16 (5), 466–471. https://doi.org/10.1016/J.BJID.2012.08.014.

Lu, C., Shen, Y., 2007. A novel ansamycin, naphthomycin K from Streptomyces sp. J. Antibiotics 60 (10), 649–653. https://doi.org/10.1038/ja.2007.84.

Maida, I., Lo Nostro, A., Pesavento, G., Barnabei, M., Calonico, C., Perrin, E., Chiellini, C., Fondi, M., Mengoni, A., Maggini, V., Vannacci, A., Gallo, E., Bilia, A.R., Flamini, G., Gori, L., Firenzuoli, F., Fani, R., 2014. Exploring the anti-Burkholderiacepacia complex activity of essential oils: a preliminary analysis. Evid. Based Complement. Alternat. Med. 2014. https://doi.org/10.1155/2014/573518.

Meena, M., Zehra, A., Swapnil, P., Harish, Marwal, A., Yadav, G., Sonigra, P., 2021. Endophytic Nanotechnology: An Approach to Study Scope and Potential Applications. Front. Chem. 9, 47. https://doi.org/10.3389/FCHEM.2021.613343/BIBTEX.

Miller, C.M., Miller, R.V., Garton-Kenny, D., Redgrave, B., Sears, J., Condron, M.M., Teplow, D.B., Strobel, G.A., 1998. Ecomycins, unique antimycotics from Pseudomonas viridiflava. J. Appl. Microbiol. 84 (6), 937–944. https://doi.org/10.1046/J.1365-2672.1998.00415.X.

Mohd Kamal, K., Mahamad Maifiah, M.H., Abdul Rahim, N., Hashim, Y.Z.H.Y., Abdullah Sani, M.S., Azizan, K.A., 2022. Bacterial metabolomics: sample preparation methods. Biochem. Res. Int. 2022. https://doi.org/10.1155/2022/9186536.

Nawaz, A., Chaudhary, R., Shah, Z., Dufossé, L., Fouillaud, M., Mukhtar, H., Haq, I.U., 2021. An overview on industrial and medical applications of bio-pigments synthesized by marine bacteria. Microorganisms 9 (1), 1–24. https://doi.org/10.3390/MICROORGANISMS9010011.

Newman, D.J., Cragg, G.M., 2015. Endophytic and epiphytic microbes as "sources" of bioactive agents. Front. Chem. 3 (MAY), 34. https://doi.org/10.3389/FCHEM.2015.00034.

Nguyen, S.T., Do, N.M., Tran, D.H.-K., To, N.B., Vo, P.H., Nguyen, M.T.T., Nguyen, N.T., Nguyen, H.X., Truong, K.D., Van Pham, P., 2020. Isopanduratin A isolated from boesenbergiapandurata reduces HepG2 hepatocellular carcinoma cell proliferation in both monolayer and three-dimensional cultures. Adv. Exp. Med. Biol. 1292, 131–143. https://doi.org/10.1007/5584_2020_523.

Nicault, M., Zaiter, A., Dumarcay, S., Chaimbault, P., Gelhaye, E., Leblond, P., Bontemps, C., 2021. Elicitation of antimicrobial active compounds by streptomyces-fungus co-cultures. Microorganisms 9 (1), 1–13. https://doi.org/10.3390/MICROORGANISMS9010178.

Ortega, H.E., Ferreira, L.L.G., Melo, W.G.P., Oliveira, A.L.L., Alvarenga, R.F.R., Lopes, N.P., Bugni, T.S., Andricopulo, A.D., Pupo, M.T., 2019. Antifungal compounds from Streptomyces associated with attine ants also inhibit Leishmania donovani. PLoS Negl. Trop. Dis. 13 (8). https://doi.org/10.1371/JOURNAL.PNTD.0007643.

Osama, N., Bakeer, W., Raslan, M., Soliman, H.A., Abdelmohsen, U.R., Sebak, M., 2022. Anti-cancer and antimicrobial potential of five soil Streptomycetes: a metabolomics-based study. R. Soc. Open Sci. 9 (2). https://doi.org/10.1098/RSOS.211509.

Patridge, E., Gareiss, P., Kinch, M.S., Hoyer, D., 2016. An analysis of FDA-approved drugs: natural products and their derivatives. Drug Discov. Today 21 (2), 204–207. https://doi.org/10.1016/J.DRUDIS.2015.01.009.

Payne, J.A.E., Schoppet, M., Hansen, M.H., Cryle, M.J., 2016. Diversity of nature's assembly lines—recent discoveries in non-ribosomal peptide synthesis. Mol. Biosyst. 13 (1), 9–22. https://doi.org/10.1039/C6MB00675B.

Peng, F., Zhang, M.Y., Hou, S.Y., Chen, J., Wu, Y.Y., Zhang, Y.X., 2020. Insights into Streptomyces spp. isolated from the rhizospheric soil of Panax notoginseng: isolation, antimicrobial activity and biosynthetic potential for polyketides and non-ribosomal peptides. BMC Microbiol. 20 (1), 1–16. https://doi.org/10.1186/S12866-020-01832-5/TABLES/4.

Pinu, F.R., Villas-Boas, S.G., Aggio, R., 2017. Analysis of intracellular metabolites from microorganisms: quenching and extraction protocols. Metabolites 7 (4). https://doi.org/10.3390/METABO7040053.

Rateb, M.E., Houssen, W.E., Arnold, M., Abdelrahman, M.H., Deng, H., Harrison, W.T.A., Okoro, C.K., Asenjo, J.A., Andrews, B.A., Ferguson, G., Bull, A.T., Goodfellow, M., Ebel, R., Jaspars, M., 2011. Chaxamycins A-D, bioactive ansamycins from a hyper-arid desert Streptomyces sp. J. Nat. Prod. 74 (6), 1491–1499. https://doi.org/10.1021/NP200320U.

Rodrigues, C.F., Černáková, L., 2020. Farnesol and tyrosol: secondary metabolites with a crucial quorum-sensing role in candida biofilm development. Genes 11 (4). https://doi.org/10.3390/GENES11040444.

Ruocco, N., Costantini, S., Guariniello, S., Costantini, M., 2016. Polysaccharides from the marine environment with pharmacological, cosmeceutical and nutraceutical potential. Molecules 21 (5). https://doi.org/10.3390/MOLECULES21050551.

Rutherford, S.T., Bassler, B.L., 2012. Bacterial quorum sensing: its role in virulence and possibilities for its control. Cold Spring Harb. Perspect. Med. 2. https://doi.org/10.1101/cshperspect.a012427.

Santos, J.D., Vitorino, I., De La Cruz, M., Díaz, C., Cautain, B., Annang, F., Pérez-Moreno, G., Martinez, I.G., Tormo, J.R., Martín, J.M., Urbatzka, R., Vicente, F.M., Lage, O.M., 2019. Bioactivities and extract dereplication of actinomycetales isolated from marine sponges. Front. Microbiol. 10 (APR). https://doi.org/10.3389/FMICB.2019.00727.

Sebola, T.E., Uche-Okereafor, N.C., Tapfuma, K.I., Mekuto, L., Green, E., Mavumengwana, V., 2019. Evaluating antibacterial and anticancer activity of crude extracts of bacterial endophytes from Crinum macowanii Baker bulbs. Microbiol. Open 8 (12). https://doi.org/10.1002/MBO3.914.

Seenivasan, A., Subhagar, S., Aravindan, R., Viruthagiri, T., 2008. Microbial production and biomedical applications of lovastatin. Indian J. Pharm. Sci. 70 (6), 701. https://doi.org/10.4103/0250-474X.49087.

Semenzato, G., Alonso-Vásquez, T., Del Duca, S., Vassallo, A., Riccardi, C., Zaccaroni, M., Mucci, N., Padula, A., Emiliani, G., Piccionello, A.P., Puglia, A.M., Fani, R., 2022. Genomic analysis of endophytic bacillus-related strains isolated from the medicinal plant Origanum vulgare L. revealed the presence of metabolic pathways involved in the biosynthesis of bioactive compounds. Microorganisms 10 (5). https://doi.org/10.3390/MICROORGANISMS10050919.

Shaheen, M., Li, J., Ross, A.C., Vederas, J.C., Jensen, S.E., 2011. Paenibacilluspolymyxa PKB1 produces variants of polymyxin B-type antibiotics. Chem. Biol. 18 (12), 1640–1648. https://doi.org/10.1016/J.CHEMBIOL.2011.09.017.

Sharma, M., Mallubhotla, S., 2022. Diversity, antimicrobial activity, and antibiotic susceptibility pattern of endophytic bacteria sourced from Cordia dichotoma L. Front. Microbiol. 13. https://doi.org/10.3389/FMICB.2022.879386.

Siddique, S., Syed, Q., Adnan, A., Ashraf Qureshi, F., 2014. Isolation, characterization and selection of avermectin-producing streptomyces avermitilis strains from soil samples. Jundishapur J. Microbiol. 7 (6), 10366. https://doi.org/10.5812/jjm.10366.

Singh, R., Kumar, M., Mittal, A., Mehta, P.K., 2017. Microbial metabolites in nutrition, healthcare and agriculture. Biotech 7, 15. https://doi.org/10.1007/s13205-016-0586-4.

Singh, M., Kumar, A., Singh, R., Pandey, K.D., 2017. Endophytic bacteria: a new source of bioactive compounds. 3 Biotech 7 (5). https://doi.org/10.1007/S13205-017-0942-Z.

Singla, A.K., Gurram, R.K., Chauhan, A., Khatri, N., Vohra, R., Jolly, R.S., Agrewala, J.N., 2014. Caerulomycin A suppresses immunity by inhibiting T cell activity. PloS One 9 (10). https://doi.org/10.1371/JOURNAL.PONE.0107051.

Skinnider, M.A., Johnston, C.W., Gunabalasingam, M., Merwin, N.J., Kieliszek, A.M., MacLellan, R.J., Li, H., Ranieri, M.R.M., Webster, A.L.H., Cao, M.P.T., Pfeifle, A., Spencer, N., To, Q.H., Wallace, D.P., Dejong, C.A., Magarvey, N.A., 2020. Comprehensive prediction of secondary metabolite structure and biological activity from microbial genome sequences. Nat. Commun. 11 (1). https://doi.org/10.1038/S41467-020-19986-1.

Soleimani, S., Yousefzadi, M., Babaei Mahani Nezhad, S., Pozharitskaya, O.N., Shikov, A.N., 2023. Potential of the ethyl acetate fraction of Padina boergesenii as a natural UV filter in sunscreen cream formulation. Life 13 (1). https://doi.org/10.3390/LIFE13010239.

Srinivasan, R., Kannappan, A., Shi, C., Lin, X., 2021. Marine bacterial secondary metabolites: a treasure house for structurally unique and effective antimicrobial compounds. Mar. Drugs 19 (10). https://doi.org/10.3390/MD19100530.

Stein, T., 2005. Bacillus subtilis antibiotics: structures, syntheses and specific functions. Mol. Microbiol. 56 (4), 845–857. https://doi.org/10.1111/J.1365-2958.2005.04587.X.

Sunkar, S., Nachiyar, C.V., 2012. Biogenesis of antibacterial silver nanoparticles using the endophytic bacterium Bacillus cereus isolated from Garcinia xanthochymus. Asian Pac. J. Trop. Biomed. 2 (12), 953. https://doi.org/10.1016/S2221-1691(13)60006-4.

Taechowisan, T., Chaisaeng, S., Phutdhawong, W.S., 2017. Antibacterial, antioxidant and anticancer activities of biphenyls from Streptomyces sp. BO-07: an endophyte in Boesenbergia rotunda (L.) Mansf A. Food Agric. Immunol 28 (6), 1330–1346. https://doi.org/10.1080/09540105.2017.1339669.

Taechowisan, T., Chanaphat, S., Ruensamran, W., Phutdhawong, W.S., 2014. Antibacterial activity of new flavonoids from Streptomyces sp. BT01; an endophyte in Boesenbergia rotunda (L.) Mansf. J. Appl. Pharmaceut. Science 4 (04), 8–013. https://doi.org/10.7324/JAPS.2014.40402.

Tang, G.L., Zhang, Z., Pan, H.X., 2017. New insights into bacterial type II polyketide biosynthesis. F1000Research 6, 172. https://doi.org/10.12688/F1000RESEARCH.10466.1.

Tillett, D., Dittmann, E., Erhard, M., Von Döhren, H., Börner, T., Neilan, B.A., 2000. Structural organization of microcystin biosynthesis in Microcystis aeruginosa PCC7806: an integrated peptide-polyketide synthetase system. Chem. Biol. 7 (10), 753–764. https://doi.org/10.1016/S1074-5521(00)00021-1.

Tobias, N.J., Wolff, H., Djahanschiri, B., Grundmann, F., Kronenwerth, M., Shi, Y.M., Simonyi, S., Grün, P., Shapiro-Ilan, D., Pidot, S.J., Stinear, T.P., Ebersberger, I., Bode, H.B., 2017. Natural product diversity associated with the nematode symbionts Photorhabdus and Xenorhabdus. Nat. Microbiol. 2 (12), 1676–1685. https://doi.org/10.1038/S41564-017-0039-9.

Vu, H.N.T., Nguyen, D.T., Nguyen, H.Q., Chu, H.H., Chu, S.K., Van Chau, M., Phi, Q.T., 2018. Antimicrobial and cytotoxic properties of bioactive metabolites produced by Streptomyces cavourensis YBQ59 isolated from Cinnamomum cassia Prels in Yen Bai Province of Vietnam. Curr. Microbiol. 75 (10), 1247–1255. https://doi.org/10.1007/S00284-018-1517-X/METRICS.

Weber, T., Blin, K., Duddela, S., Krug, D., Kim, H.U., Bruccoleri, R., Lee, S.Y., Fischbach, M.A., Müller, R., Wohlleben, W., Breitling, R., Takano, E., Medema, M.H., 2015. AntiSMASH 3.0-a comprehensive resource for the genome mining of biosynthetic gene clusters. Nucl. Acids Res. 43 (W1), W237–W243. https://doi.org/10.1093/NAR/GKV437.

Ye, L., Zhao, S., Li, Y., Jiang, S., Zhao, Y., Li, J., Yan, K., Wang, X., Xiang, W., Liu, C., 2017. Streptomyces lasiicapitis sp. nov., an actinomycete that produces kanchanamycin, isolated from the head of an ant (Lasiusfuliginosus L.). Int. J. Syst. Evol. Microbiol. 67 (5), 1529–1534. https://doi.org/10.1099/IJSEM.0.001756/CITE/REFWORKS.

Zhang, Z., Zhou, T., Yang, T., Fukaya, K., Harunari, E., Saito, S., Yamada, K., Imada, C., Urabe, D., Igarashi, Y., 2021. Nomimicins B–D, new tetronate-class polyketides from a marine-derived actinomycete of the genus Actinomadura. Beilstein J. Org. Chem. 17, 2194. https://doi.org/10.3762/BJOC.17.141.

Zhao, P.J., Fan, L.M., Li, G.H., Zhu, N., Shen, Y.M., 2005. Antibacterial and antitumor macrolides from Streptomyces sp. Is9131. Arch. Pharm. Res. 28 (11), 1228–1232. https://doi.org/10.1007/BF02978203/METRICS.

Zin, N.M., Baba, M.S., Zainal-Abidin, A.H., Latip, J., Mazlan, N.W., Edrada-Ebel, R.A., 2017. Gancidin W, a potential low-toxicity antimalarial agent isolated from an endophytic Streptomyces SUK10. Drug Des. Devel. Ther. 11, 351. https://doi.org/10.2147/DDDT.S121283.

Index

Note: Page numbers followed by *f* indicate figures and *t* indicate tables.

N

O

P

Q

R

S

Printed in the United States
by Baker & Taylor Publisher Services